YOUR ACCESS TO SUCCESS

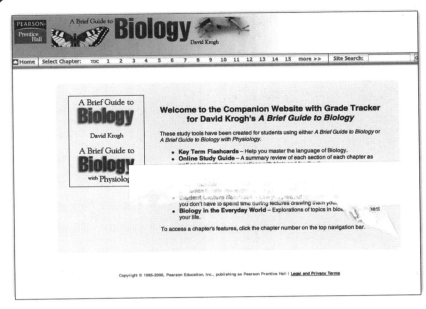

Students with new copies of Krogh, *A Brief Guide to Biology* and *A Brief Guide to Biology with Physiology 1st Ed.*, have full access to the book's Companion Website with Grade Tracker—a 24/7 study tool with quizzes, animated tutorials, and other features designed to help you make the most of your limited study time.

Just follow the easy website registration steps listed below...

Registration Instructions for
www.prenhall.com/krogh

1. Go to *www.prenhall.com/krogh*
2. Click the cover for Krogh, *A Brief Guide to Biology* or *A Brief Guide to Biology with Physiology, 1e.*
3. Click "Register".
4. Using a coin (not a knife) scratch off the metallic coating below to reveal your Access Code.
5. Complete the online registration form, choosing your own personal Login Name and Password.
6. Enter your pre-assigned Access Code exactly as it appears below.
7. Complete the online registration form by entering your School Location information.
8. After your personal Login Name and Password are confirmed by e-mail, go back to *www.prenhall.com/krogh*, enter your new Login Name and Password, and click "Log In".

Your Access Code is:

R

If there is no metallic coating covering the access code above, the code may no longer be valid and you will need to purchase online access using a major credit card to use the website. To do so, go to *www.prenhall.com/krogh*, click the cover for Krogh, *A Brief Guide to Biology, 1e*, and follow the instructions under "Students".

Important: Please read the Subscription and End-User License Agreement located on the "Log In" screen before using the Krogh, *A Brief Guide to Biology, 1e*, Companion Website with Grade Tracker. By using the website, you indicate that you have read, understood, and accepted the terms of the agreement.

Minimum system requirements

PC Operating Systems:

Windows 2000/XP

Pentium II 233 MHz processor. 64 MB RAM In addition to the minimum memory required by your OS.

Internet Explorer(TM) 5.5 or 6, Netscape(TM) 7, Firefox 1.0.

Macintosh Operating Systems:

Macintosh Power PC with OS X (10.2 and 10.3)

In addition to the RAM required by your OS, this application requires 64 MB RAM, with 40MB Free RAM, with Virtual Memory enabled.

Netscape(TM) 7 (OS 9 or OS X), Safari 1.3 (OS X only), Firefox 1.0.

Macromedia Shockwave(TM)

8.50 release 326 plugin

Macromedia Flash Player 6.0.79 & 7.0

Acrobat Reader 6.0.1

4x CD-ROM drive

800 x 600 pixel screen resolution

D1636880

...hone support is available Monday–Friday, 8am to 8pm and Sunday 5pm to 12am, Eastern time. Visit our support site at *http://247.pearsoned.com*. E-mail support is available 24/7.

Krogh
A Brief Guide to Biology
A Brief Guide to Biology with Physiology
Student CD
ISBN 0-13-227578-3
CD License Agreement
© 2007 Pearson Education, Inc.
Pearson Prentice Hall
Pearson Education, Inc.
Upper Saddle River, NJ 07458
All rights reserved.
Pearson Prentice Hall™ is a trademark of Pearson Education, Inc.

YOU SHOULD CAREFULLY READ THE TERMS AND CONDITIONS BEFORE USING THE CD-ROM PACKAGE. USING THIS CD-ROM PACKAGE INDICATES YOUR ACCEPTANCE OF THESE TERMS AND CONDITIONS.

Pearson Education, Inc. provides this program and licenses its use. You assume responsibility for the selection of the program to achieve your intended results, and for the installation, use, and results obtained from the program. This license extends only to use of the program in the United States or countries in which the program is marketed by authorized distributors.

LICENSE GRANT

You hereby accept a nonexclusive, nontransferable, permanent license to install and use the program ON A SINGLE COMPUTER at any given time. You may copy the program solely for backup or archival purposes in support of your use of the program on the single computer. You may not modify, translate, disassemble, decompile, or reverse engineer the program, in whole or in part.

TERM

The License is effective until terminated. Pearson Education, Inc. reserves the right to terminate this License automatically if any provision of the License is violated. You may terminate the License at any time. To terminate this License, you must return the program, including documentation, along with a written warranty stating that all copies in your possession have been returned or destroyed.

LIMITED WARRANTY

THE PROGRAM IS PROVIDED "AS IS" WITHOUT WARRANTY OF ANY KIND, EITHER EXPRESSED OR IMPLIED, INCLUDING, BUT NOT LIMITED TO, THE IMPLIED WARRANTIES OR MERCHANTABILITY AND FITNESS FOR A PARTICULAR PURPOSE. THE ENTIRE RISK AS TO THE QUALITY AND PERFORMANCE OF THE PROGRAM IS WITH YOU. SHOULD THE PROGRAM PROVE DEFECTIVE, YOU (AND NOT PEARSON EDUCATION, INC. OR ANY AUTHORIZED DEALER) ASSUME THE ENTIRE COST OF ALL NECESSARY SERVICING, REPAIR, OR CORRECTION. NO ORAL OR WRITTEN INFORMATION OR ADVICE GIVEN BY PEARSON EDUCATION, INC., ITS DEALERS, DISTRIBUTORS, OR AGENTS SHALL CREATE A WARRANTY OR INCREASE THE SCOPE OF THIS WARRANTY.

SOME STATES DO NOT ALLOW THE EXCLUSION OF IMPLIED WARRANTIES, SO THE ABOVE EXCLUSION MAY NOT APPLY TO YOU. THIS WARRANTY GIVES YOU SPECIFIC LEGAL RIGHTS AND YOU MAY ALSO HAVE OTHER LEGAL RIGHTS THAT VARY FROM STATE TO STATE.

Pearson Education, Inc. does not warrant that the functions contained in the program will meet your requirements or that the operation of the program will be uninterrupted or error-free.

However, Pearson Education, Inc. warrants the CD-ROM(s) on which the program is furnished to be free from defects in material and workmanship under normal use for a period of ninety (90) days from the date of delivery to you as evidenced by a copy of your receipt.

The program should not be relied on as the sole basis to solve a problem whose incorrect solution could result in injury to person or property. If the program is employed in such a manner, it is at the user's own risk and Pearson Education, Inc. explicitly disclaims all liability for such misuse.

LIMITATION OF REMEDIES

Pearson Education, Inc.'s entire liability and your exclusive remedy shall be:
1. the replacement of any CD-ROM not meeting Pearson Education, Inc.'s "LIMITED WARRANTY" and that is returned to Pearson Education, or
2. if Pearson Education is unable to deliver a replacement CD-ROM that is free of defects in materials or workmanship, you may terminate this agreement by returning the program.

IN NO EVENT WILL PEARSON EDUCATION, INC. BE LIABLE TO YOU FOR ANY DAMAGES, INCLUDING ANY LOST PROFITS, LOST SAVINGS, OR OTHER INCIDENTAL OR CONSEQUENTIAL DAMAGES ARISING OUT OF THE USE OR INABILITY TO USE SUCH PROGRAM EVEN IF PEARSON EDUCATION, INC. OR AN AUTHORIZED DISTRIBUTOR HAS BEEN ADVISED OF THE POSSIBILITY OF SUCH DAMAGES, OR FOR ANY CLAIM BY ANY OTHER PARTY.

SOME STATES DO NOT ALLOW FOR THE LIMITATION OR EXCLUSION OF LIABILITY FOR INCIDENTAL OR CONSEQUENTIAL DAMAGES, SO THE ABOVE LIMITATION OR EXCLUSION MAY NOT APPLY TO YOU.

GENERAL

You may not sublicense, assign, or transfer the license of the program. Any attempt to sublicense, assign or transfer any of the rights, duties, or obligations hereunder is void.

This Agreement will be governed by the laws of the State of New York.

Should you have any questions concerning this Agreement, you may contact Pearson Education, Inc. by writing to:

ESM Media Development
Higher Education Division
Pearson Education, Inc.
1 Lake Street
Upper Saddle River, NJ 07458

Should you have any questions concerning technical support, you may write to:

New Media Production
Higher Education Division
Pearson Education, Inc.
1 Lake Street
Upper Saddle River, NJ 07458

or contact:

Pearson Education Product Support Group at (800) 677-6337 Monday through Friday, 8 AM to 8 PM, Sunday, 5 PM to 12 AM, Eastern time, or anytime at http://247.prenhall.com.

YOU ACKNOWLEDGE THAT YOU HAVE READ THIS AGREEMENT, UNDERSTAND IT, AND AGREE TO BE BOUND BY ITS TERMS AND CONDITIONS. YOU FURTHER AGREE THAT IT IS THE COMPLETE AND EXCLUSIVE STATEMENT OF THE AGREEMENT BETWEEN US THAT SUPERSEDES ANY PROPOSAL OR PRIOR AGREEMENT, ORAL OR WRITTEN, AND ANY OTHER COMMUNICATIONS BETWEEN US RELATING TO THE SUBJECT MATTER OF THIS AGREEMENT.

START UP INSTRUCTIONS:

Windows users: If the CD does not auto launch, locate the file "starthere.html" on the CD-ROM and double-click on it.
Macintosh OSX users: Locate the file "startOSX.html" on the CD-ROM and double-click on it.
Macintosh OS9.x users: Locate the file "startOS9.html" on the CD-ROM and double-click on it.

When you run the program, you can click the "Browser Tune-up" button on the splash page, which will help you determine whether you have the browser and plugin you need to view the content.

SYSTEM REQUIREMENTS:

PC Operating Systems: Windows 2000/XP
Pentium II 233 MHz processor. 64 MB RAM In addition to the minimum memory required by your OS.
Internet Explorer(TM) 5.5 or 6, Netscape(TM) 7, Firefox 1.0.

Macintosh Operating Systems: Macintosh Power PC with OS X (10.2 and 10.3) or Macintosh OS 9 (9.2.2)
In addition to the RAM required by your OS, this application requires 64 MB RAM, with 40MB Free RAM, with Virtual Memory enabled.
Netscape(TM) 7 (OS 9 or OS X), Internet Explorer(TM) 5.1 (for OS 9) & 5.2 (for OS X), Safari 1.3 (OS X only), Firefox 1.0.

Macromedia Flash Player 6.0.79 & 7.0
Acrobat Reader 5.0 (for OS 9.2) & 6.0.1
4x CD-ROM drive
800 x 600 pixel screen resolution

ACC Library Services
Austin, Texas

Brief Contents

This textbook is available in two versions:

A Brief Guide to Biology (0-13-185965-X) consists of *Chapters 1–24*, which provide non-majors biology students with a thorough overview of the foundations of modern biological science: cell biology, genetics, evolution, and ecology.

A Brief Guide to Biology with Physiology (0-13-185964-1), an expanded version of the text, contains *Chapters 1–30*: the twenty-four chapters noted above, plus six chapters on human and plant anatomy and physiology.

A BRIEF GUIDE TO
BIOLOGY
WITH PHYSIOLOGY

David Krogh

PEARSON

Prentice
Hall

Upper Saddle River, New Jersey 07458

Library of Congress Cataloging-in-Publication Data

Krogh, David.
  A brief guide to biology with physiology / David Krogh.
    p. cm.
  ISBN 0-13-185964-1
  1. Biology. 2. Physiology. I. Title.
  QH308.2.K767 2007
  570—dc22

                                    2005030027

QH308.2 .K767 2007
A brief guide to biology
with physiology /
NO CD
SAC   - as @ SAC 7/1/19
liquid damage noted 8/27/14 SAC -ps

Executive Editor: Teresa Ryu Chung
Development Editor: Erin Mulligan
Production Editor: Tim Flem/PublishWare
Senior Media Editor: Patrick Shriner
Assistant Editor: Andrew Sobel
Editor in Chief, Science: Daniel Kaveney
Editor in Chief of Development: Carol Trueheart
Executive Managing Editor: Kathleen Schiaparelli
Art Director: John Christiana
Director of Marketing, Science: Patrick Lynch
Senior Managing Editor, Art Production and Management: Patricia Burns
Manager, Production Technologies: Matthew Haas
Managing Editor, Art Management: Abigail Bass
Art Development Editors: Kim Quillin, Jay McElroy
Art Production Editor: Jess Einsig
Illustrations: Precision Graphics
Creative Director: Juan R. Lopez
Director of Creative Services: Paul Belfanti
Page Composition: PublishWare
Manager of Formatting: Allyson Graesser
Manufacturing Manager: Alexis Heydt-Long
Manufacturing Buyer: Alan Fischer
Managing Editor, Science Media: Nicole Jackson
Media Production Editors: Aaron Reid, Tim Flem
Assistant Managing Editor, Science Supplements: Karen Bosch
Editorial Assistants: Gina Kayed; Nancy Bauer
Cover and Interior Designer: Anderson Creative
Director, Image Resource Center: Melinda Reo
Manager, Rights and Permissions: Zina Arabia
Interior Image Specialist: Beth Boyd-Brenzel
Cover Specialist: Karen Sanatar
Image Permission Coordinator: Joanne Dippel
Photo Researcher: Jerry Marshall
Cover Photograph: © Steven Burr Williams/Workbook Stock/Getty Images, Inc.

© 2007 by Pearson Education, Inc.
Pearson Prentice Hall
Pearson Education, Inc.
Upper Saddle River, NJ 07458

All rights reserved. No part of this book may be reproduced, in any form or by any other means, without permission in writing from the publisher.

Pearson Prentice Hall™ is a trademark of Pearson Education, Inc.

Printed in the United States of America

10 9 8 7 6 5 4 3 2

ISBN 0-13-185964-1

Pearson Education Ltd., *London*
Pearson Education Australia Pty., Limited, *Sydney*
Pearson Education Singapore, Pte. Ltd.
Pearson Education North Asia Ltd., *Hong Kong*
Pearson Education Canada, Ltd., *Toronto*
Pearson Educación de Mexico, S.A. de C.V.
Pearson Education—Japan, *Tokyo*
Pearson Education Malaysia, Pte. Ltd.

Preface

David Krogh has been writing about science for 20 years in newspapers, magazines, books, and for educational institutions. He is the author of a best-selling non-majors biology textbook, *Biology: A Guide to the Natural World*, first published in 2000 and now in its third edition. He is also the author of a trade book, *Smoking: The Artificial Passion,* an account of the pharmacological and cultural motivations behind the use of tobacco. Published in 1991, *Smoking* was nominated for the *Los Angeles Times* Book Prize in Science and Technology. David has written on physics and on technology issues, but his primary interest has been in biology. He has written on the possible effect methane may be having on global warming; on early research into the role that growth factors may play in neural regeneration following injury; on the synthesis of naturally occurring neurotoxins and their possible use in heart disease; on the use of imported drugs to treat cancer; and on the relationship between alcohol and mood states in women. He has a particular interest in the history of biology and in the relationship between biological research and modern American culture. He holds bachelor's degrees in both journalism and history from the University of Missouri.

The galloping pace of knowledge acquisition in biology was driven home to me recently by an article in *The Scientist* regarding the Wellcome Trust Sanger Institute in Great Britain. As Stuart Blackman reported, Sanger is using its 70-plus DNA sequencing machines to read 60 million genomic bases per day. When Sanger's own data, from about 100 species, are combined with data submitted to it from the rest of the sequencing world, the result is on-site storage of about 150 terabytes of unique data—an amount of information so large it would take 187,000 CDs to accommodate it all. Needless to say, Sanger faces a tremendous task in just organizing such a volume of information. As Tony Cox, Sanger's head of software services, told Blackman, "We have to integrate things in a sensible fashion; otherwise we get crushed under the weight of it."

Cox was referring only to information management at Sanger, but he could be speaking for everyone in biology today—whether in research or education—when he talks about the need to organize information optimally. We who work on the education side have our own information flow to keep up with, which is to say the information flowing from places like Sanger. Since biology courses and textbooks are relatively fixed entities—courses meet only so many hours and books can only be so long—the challenge becomes one of *winnowing*; of deciding what stays and what doesn't, given the flow of new findings from the research side. Just when reproductive cloning (à la Dolly) has been woven into a course, along comes therapeutic cloning (à la the Korean research teams). And this latter work is directly tied, of course, to stem cells, with all their medical and political ramifications. Long-established paradigms are being challenged in such areas as genetic regulation, the etiology of cancer, and the role of inflammation in disease. Global warming and human evolution are topics that are high on the priority list of most teaching faculty, but just keeping up with developments in these fields can

be a challenge, to say nothing of integrating information about them into a course.

Textbooks can be thought of as tools that assist faculty in teaching and students in learning. But in the contemporary world of biology, a tool that works well in one course may not work so well in another. In the winnowing process taking place today, different faculty are going to make different decisions about which topics in biology are critical, and which are merely valuable or interesting. Most introductory biology courses are likely to cover meiosis; but is it also desirable to cover the egg and sperm formation that are related to it? Most introductory courses are likely to cover microevolution; but is it desirable to introduce the Hardy-Weinberg principle as part of this coverage? In recent years, it seemed to me that a sizable group of faculty was giving me a pretty consistent message about topics such as these, which was *not* to cover them in a textbook, as they were merely making books larger, rather than more useful. The general message from these faculty was that they wanted a book that covered fewer topics and that covered the topics that remained in less detail. Furthermore, these faculty wanted a book that more closely tied its coverage to the everyday lives of students. Prentice Hall and I took this message to heart, and the end-result is the book you see in front of you.

What's Different about *A Brief Guide to Biology*?

So, what is the nature of the book—what's the product like here? Well, carrying out topic reduction meant making real choices. If you do review Hardy-Weinberg in your course, then this may not be the book for you. Likewise, those who cover animal behavior, cladistics, animal development, the human senses, or plant physiology in detail might want to look elsewhere. (I have an excellent book for you, though; it's called *Biology: A Guide to the Natural*

World.) You can get a complete picture of what's covered in the book by looking at its detailed table of contents. Suffice it to say here that, as I considered each chapter, deciding on what should be covered and what shouldn't, I thought of myself as a kind of wood-carver, trying to shape things such that the faculty I had in mind—those who wanted a briefer, more applied book—got exactly what they wanted.

I wasn't making these choices on my own, of course, but instead was guided at every step by faculty who are in the classroom each week. All textbooks receive guidance from faculty reviewers, but *A Brief Guide to Biology* had something else going for it: a panel of three faculty who provided advice solely on topic selection. At the "take off" stage of each chapter—that is, before completion of the first draft that went to reviewers—this group looked over the topics I proposed to include and gave me its advice. Sitting on the panel were Cedric O. Buckley of Jackson State University, David Mirman of Mt. San Antonio College, and Gregory J. Podgorski of Utah State University. I am greatly indebted to them for their advice on what should be put into the book, and what should be left out.

Helping to Produce Biologically Literate Citizens

Topic selection for the book was guided more generally by my sense of the overarching goal most faculty have for their non-majors courses. Overwhelmingly, it seems to me, this goal is to give students the tools they need to be biologically literate citizens. Faculty want students to have a grounding in biology that will allow them to do such things as make better-informed choices at the ballot box; make sense of what they read in the newspaper; and separate truth from fiction in advertising.

Given that faculty have aspirations such as these, it's not surprising that many of them work hard to make their classrooms places where biology and the "real world" come together. Does a "right to die" case in the news have some relevance for the brain or kidney function that's being studied in class? Then that right-to-die case may lead off the discussion on Wednesday. Making linkages such as these, faculty report, is valuable in two ways: It sparks student interest and it demonstrates to students that biology is indeed going to be part of their lives once they leave college.

Science in the News

Knowing the interest that faculty have in bringing the real world into the classroom, those of us who produced *A Brief Guide to Biology* thought: Well, why not bring the real world into the *textbook*? As you will see when you flip through the book's pages, each chapter starts off with an excerpt from one, or sometimes two actual newspaper stories. And in each case, the subject matter of the stories is tied to the central topic of the chapter. Thus the chapter on Mendel, starting on page 166, begins with stories on the real power of recessive alleles (sudden infant death syndrome in an Amish community) and the imagined power of them (the idea that blonde hair will eventually disappear). Our hope is that these chapter beginnings will not only spark student interest and demonstrate the real-world relevance of biology, but perhaps provide some ready-made topics for classroom discussion as well.

A Broader View

The opening newspaper stories represent one kind of pedagogical device in the book, but not the only such device. Flip through any given chapter and you will find, on the left and right sides of many pages, brief text passages headlined *A Broader View*. These are small "perspective" pieces, intended to provide students with broader perspectives on the chapter's central subject. As an example, biotechnology is in large part about money, but if students are to make it through even the rudiments of the science side of biotechnology, then there simply isn't room to say much about the business side. And yet . . . what about the business side? Chapter 15's answer is to have a *Broader View* entry on the monetary value of biotechnology's Cohen-Boyer patent (see page 240). In a similar vein, students can learn something about the connection between natural forms and art in Chapter 20's *Broader View* on diatoms (page 344). And they can learn something about the relation between polyploidy and an item they see in the grocery store—seedless watermelons—in a *Broader View* that appears on page 299 of Chapter 18.

In-Chapter Reviews and Modern Myths

Beyond newspaper stories and *Broader Views*, each chapter also contains a set of mini-reviews of the chapter's core material. At three or four points in each chapter, the text is interrupted by a series of three questions—headlined *So Far*—whose purpose is to allow students to see if they understand what they've just read. These are not questions that test students' abilities to synthesize or to think critically; they are questions designed to let students see if they are getting the gist of a given chapter, section by section.

Sprinkled throughout the book you'll also find a series of boxes called *Modern Myths*, which are just what they sound like: short sidebars reporting on beliefs that manage to be both widely held and mistaken. If you look on page 193, for example, you can see a box on the myth that baldness can be predicted by looking at near-relatives. Meanwhile, on page 405, you will find a box on the myth that lemmings commit suicide en masse by throwing themselves into the ocean.

Verbal Learners and Visual Learners

I've been talking about some of the pedagogical features in *A Brief Guide to Biology*, but in any textbook, the pedagogical load is borne primarily by its main text on the one hand and its illustration program on the other. To the extent that students fall into the camps of "verbal learners" or "visual learners," they will, of course, rely more heavily on one of these elements than the other. With this in mind, I invite faculty to go over a given chapter in two ways—with visual learners and then with verbal learners in mind. I think you'll find that both types of students will be well served by *A Brief Guide to Biology*. In connection with visual learners, look, on page 214, at Chapter 14 as these students would—moving, figure by figure, from the nature of proteins, to the overarching steps of protein production, to the nature of RNA and DNA. A student need not march meticulously through the text to understand what is going on in this chapter; touching on the text while relying on the illustrations will do.

Students who do follow texts closely have responded nicely to my writing in the past and I hope that this will continue to be the case with *A Brief Guide to Biology.* My writing is variously described as "informal" or "storytelling" in style, or "metaphorical," and I have taken all these descriptions as compliments. Learning always involves crossing a bridge, going from what we know to what we don't know. And I believe that informality, metaphor, and a narrative style all help to facilitate that crossing.

When the elements I've been discussing are considered together, I think the result, in *A Brief Guide to Biology,* is a book that will provide a useful alternative for faculty who may have been frustrated by the fit between their needs and the books they've seen so far. Our goal was straightforward: Select the right topics, limit the details, put in more of the real world, add some helpful pedagogical devices, and provide a solid foundation in text and art. It's not for me to say whether we achieved all this, but as I finish up on *A Brief Guide to Biology,* I like our chances.

Acknowledgments

Textbooks are such large undertakings that being the author of one is something like being the person who is the beneficiary of an old-fashioned barn raising: There are so many hands involved that it's useless to pretend you know everyone and the jobs they're doing. All you can do is stand back and be grateful. Given this, there are people who have worked hard on *A Brief Guide to Biology* who I am indebted to, but who I am powerless to thank, because their work didn't directly intersect with mine.

Of those I was in contact with, I have already mentioned our panel of three content reviewers who kept me on track with respect to topic coverage. Joining them in reviewing the text were 112 faculty from across the country who provided advice on everything from factual accuracy to the "toughest topics" for their students, to the approaches that work best for these students. The names of these reviewing faculty can be found on pages viii and ix.

The artwork in the book was produced by Kim Quillin, an able artist and biologist who I have worked with now for more than 10 years. Erin Mulligan did a fine job as the book's developmental editor—the person who looked over all the words and images that were candidates for placement in the book and then provided judgments on whether they worked or not. Chris Thillen copyedited the manuscript, making sure that typos did not creep in and that the English language was being used correctly. And how was it that we were able to obtain timely reviews from 112 faculty? Through the work of assistant editor Andrew Sobel and editorial assistant Gina Kayed, who managed the considerable job of seeing the review process from inception to completion. The design of the book—its cover and colors and layout—sprang from the creative mind of John Christiana. The supplemental print and electronic resources that are so important to books these days were produced by Patrick Shriner and Andrew Sobel. And pulling all these disparate elements together and creating a coherent work out of them was a production editor I am happy to have met with this book, Tim Flem. Finally, executive editor Teresa Chung not only oversaw the book on its largest scale, but conceived of its most innovative pedagogical elements. All in all then, this was a great team who it was a pleasure to work with, and my thanks go out to all of them.

—David Krogh

A Brief Guide to Biology Review Panel

Cedric O. Buckley, *Jackson State University*

David Mirman, *Mt. San Antonio College*

Gregory J. Podgorski, *Utah State University*

Reviewers

We express sincere gratitude to the expert reviewers who worked so carefully with the author in reviewing manuscript to ensure the scientific accuracy of the text and art.

Robert Boyd, *Auburn University*

Mark D. Decker, *University of Minnesota*

Wayne D. Frasch, *Arizona State University*

Loren Knapp, *University of South Carolina*

Andrea Lloyd, *Middlebury College*

Harvey Pough, *Rochester Institute of Technology*

Stephen J. Salek, *Fayetteville State University*

Gray Scrimgeour, *University of Toronto*

Christine Tachibana, *University of Washington*

David Wessner, *Davidson College*

A Brief Guide to Biology Reviewers

Felix Akojie, *West Kentucky Community & Technical College*

Sylvester Allred, *Northern Arizona University*

Mark Alston, *University of Tennessee*

Brenda Alston-Mills, *North Carolina State University*

Susan Aronica, *Canisius College*

Bert Atsma, *Union County College*

Andrew S. Baldwin, *Mesa Community College*

Heidi Banford, *University of West Georgia*

Donald Baud, *University of Memphis*

Paul Beardsley, *Idaho State University*

Carla B. Benejam, *California State University, Monterey Bay*

Dan Bickerton, *Ogeechee Technical Institute*

Mimi Bres, *Prince George's Community College*

Peggy Brickman, *University of Georgia*

George Brooks, *Ohio University*

David Byres, *Florida Community College, Jacksonville*

Thomas Chen, *Santa Monica College*

Genevieve Chung, *Broward Community College*

Clay E. Corbin, *Bloomsburg University*

Dave Cox, *Lincoln Land Community College*

Donald G. DeHay, *Tri-County Technical College*

Elizabeth Desy, *Southwest Minnesota State University*

Donald Dorfman, *Monmouth University*

Claudia Douglass, *Central Michigan University*

Ernest DuBrul, *University of Toledo*

Shannon Dullea, *North Dakota State College of Science*

Patrick J. Enderle, *East Carolina University*

William Epperly, *Robert Morris College*

Marirose Ethington, *Genesee Community College*

Paul Evans, *Metropolitan Community College*

Brandon Foster, *Wake Tech Community College*

Diane Fritz, *Northern Kentucky University*

Michael Fultz, *Morehead State University*

Michelle Geary, *West Valley College*

Alexandros Georgakilas, *East Carolina University*

Cynthia Ghent, *Towson University*

Mita Ghosh, *Florida Community College, Jacksonville*

Tammy Gillespie, *Eastern Arizona College*

Elliot Goldstein, *Arizona State University*

Andrew Goliszek, *North Carolina Agricultural and Technical State University*

John P. Harley, *Eastern Kentucky University*

Wendy Hartman, *Palm Beach Community College*

Catherine J. Hurlbut, *Florida Community College, Jacksonville*

Stan S. Ivey, *Delaware State University*

Robert Iwan, *Inver Hills Community College*

Carl Johansson, *Fresno City College*

Gregory A. Jones, *Santa Fe Community College*

Arnold Karpoff, *University of Louisville*

Jennifer Katcher, *Pima Community College, Downtown*

Dawn G. Keller, *Hawkeye Community College*

Kerry Kilburn, *Old Dominion University*

Marcia Kribs, *Motlow State Community College*

Aaron R. Krochmal, *University of Houston, Downtown*

Ellen S. Lamb, *University of North Carolina, Greensboro*

Kathleen Lavoie, *State University of New York, Plattsburgh*

Brenda G. Leicht, *University of Iowa*

Kristin Lenertz, *Black Hawk College*

Harvey Liftin, *Broward Community College*

Kevin Lyon, *Jones County Junior College*

Mark Manteuffel, *St. Louis Community College*

David A. McClellan, *Brigham Young University*

Thomas H. Milton, *Richard Bland College*

Michael Muller, *University of Illinois at Chicago*

Richard L. Myers, *Missouri State University*

Camellia Okpodu, *Norfolk State University*

Arnas Palaima, *University of Louisville*

Eckle Peabody, *Tulsa Community College*

Brian K. Penney, *Saint Anselm College*

Marcia M. Pierce, *Eastern Kentucky University*

Angela R. Porta, *Kean University*

Erin Rempala-Kim, *San Diego Mesa College*

Jennifer Roberts, *Lewis University*

Lyndell Robinson, *Lincoln Land Community College*

Troy Rohn, *Boise State University*

Tony Rothering, *Lincoln Land Community College*

Allison Shearer, *Grossmont-Cuyamaca Community College District*

Cara Shillington, *Eastern Michigan University*

Mark Smith, *Chaffey College*

Jay Snaric, *St. Louis Community College*

Meredith A. Somerville, *University of North Carolina, Charlotte*

Jacqueline Stevens, *Jackson State University*

Peter Svensson, *West Valley College*

Christine Tachibana, *University of Washington*

Kip Thompson, *Ozarks Technical Community College*

Rani Vajravelu, *University of Central Florida*

Christa Voss, *Tulsa Community College*

David Wassmer, *Bloomsburg University*

Michael Wenzel, *California State University, Sacramento*

David Wessner, *Davidson College*

Allison Wiedemeier, *University of Missouri, Columbia*

Eric Wong, *University of Louisville*

Kenneth Wunch, *Sam Houston State University*

Mark L. Wygoda, *McNeese State University*

Calvin Young, *Fullerton College*

Contributors to Instructor and Student Resources

Carol Britson, *University of Mississippi*

David Byres, *Florida Community College, Jacksonville*

Carol Chihara, *University of San Francisco*

Elizabeth Desy, *Southwest Minnesota State University*

Claudia Douglass, *Central Michigan University*

Anne Galbraith, *University of Wisconsin, La Crosse*

Robert Iwan, *Inver Hills Community College*

Mark Smith, *Chaffey College*

Allison Wiedemeier, *University of Missouri, Columbia*

Reviewers of Instructor and Student Resources

Carla B. Benejam, *California State University, Monterey Bay*

Brenda G. Leicht, *University of Iowa*

John B. McGill, *York Technical College*

Gregory J. Podgorski, *Utah State University*

Carol Welsh, *Long Beach City College*

Student Focus Group Participants

California State University, Fullerton:

   Danielle Bruening

   Leslie Buena

   Andrés Carrillo

   Victor Galvan

   Jessica Ginger

   Sarah Harpst

   Robin Keber

   Ryan Roberts

   Melissa Romero

   Erin Seale

   Nathan Tran

   Tracy Valentovich

   Sean Vogt

Fullerton College:

   Michael Baker

   Mahetzi Hernandez

   Heidi McMorris

   Daniel Minn

   Sam Myers

   David Omut

   Jonathan Pistorino

   James W. Pura

   Samantha Ramirez

   Tiffany Speed

   Tristan Terry

Support Your Biology Lectures with the Best Animated Resources...

PEARSON
Prentice Hall

BLAST!!

BIOLOGY LECTURE ANIMATION & SIMULATION TOOL

69 Topics   Cells   Plants

Chemistry   Genetics   Animals

Energy   Evolution   Ecology

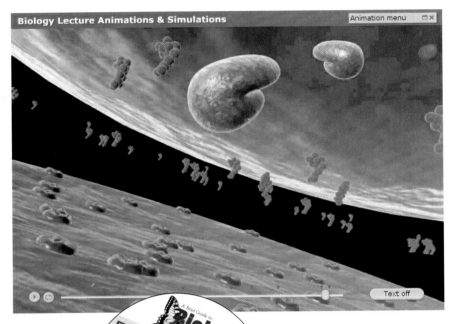

Biology Lecture Animations & Simulations   Animation menu

Text off

Teach biological concepts the way you want with Prentice Hall BLAST!!

- Designed to support individual teaching approaches and class needs

- Interactive variable controls and stunning 3-D graphics engage students and enliven lectures

- Available only with Prentice Hall textbooks

All the Support You Need to Teach Your Course the Way You Want:

- **Krogh Instructor Resource Center on CD/DVD** — 4 CD/1 DVD set plus TestGen® Test Item File CD – a full suite of JPEG and PowerPoint® options, 100+ Instructor Animations, and additional lecture and course support items (0-13-196217-5)

- **Test Item File** — printed version (0-13-196218-3)

- **Instructor Resource Guide** — lecture outlines and activities (0-13-196216-7)

- **Transparency Pack** — 300 transparencies (0-13-173200-5)

—PLUS—

- **Prentice Hall BLAST!!** — 3 CD/1 DVD Biology Lecture Animation and Simulation Tool set – *see above* (0-13-237034-4)

...and the Most Complete JPEG and PowerPoint® Options Available

All found on the Krogh Instructor Resource Center CD/DVD

JPEG Files

- Every **figure** and **table** from the textbook
- All illustrations **optimized** for clear viewing
- All figures offered in **labeled**, **unlabeled**, and **editable-label** versions

PowerPoint® Resources

- All text figure and table JPEG files in an **Image Gallery** file for each chapter
- Two suggested **Lectures** for each chapter
- 100+ **Instructor Animations** pre-loaded into PowerPoint® in PC and Mac versions
- **Classroom Response System** files for each chapter

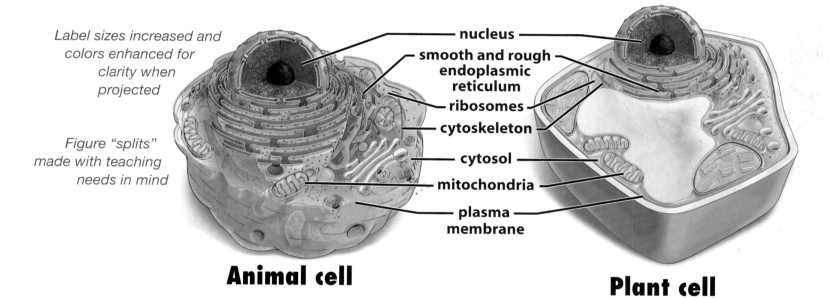

Label sizes increased and colors enhanced for clarity when projected

Figure "splits" made with teaching needs in mind

nucleus
smooth and rough endoplasmic reticulum
ribosomes
cytoskeleton
cytosol
mitochondria
plasma membrane

Animal cell

Plant cell

Student Support Items

- **Companion Website with Grade Tracker** — Online Study Guide, 57 Web Tutorials, key term tool, and printable PDF files of all line drawings appearing in the text: www.prenhall.com/krogh
- **Student CD** — Interactive quizzes and 30 Web Tutorials (packaged with textbook)
- **Audio Study Guide** — 20+ minutes of study tools on MP3 files for each chapter: www.prenhall.com/krogh
- **Student Study Companion** — Printed study guide (0-13-173206-4)
- **Student Lecture Notebook** — All the line drawings from the text in print in full color (0-13-196220-5)
- **SafariX Textbook Online** — eTextbook (0-13-173199-8)

SafariX TEXTBOOKS ONLINE

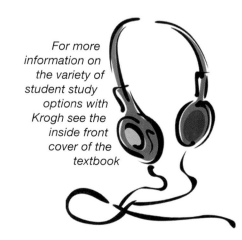

For more information on the variety of student study options with Krogh see the inside front cover of the textbook

Course Management Solutions

Visit http://cms.prenhall.com for details.

OneKey
OneKey is all you need

WebCT

Bb
Blackboard
www.blackboard.com

Science in the News

Modern Myths

Contents

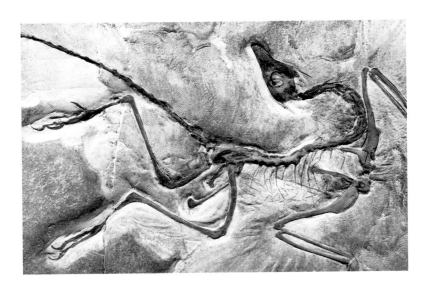

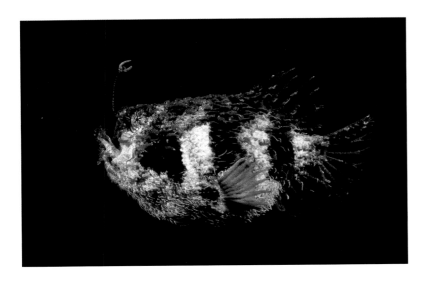

Unit 6 What Makes the Organism Tick? Human Anatomy and Physiology

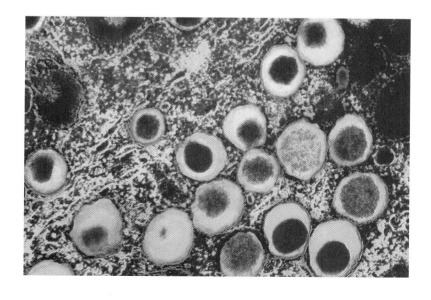

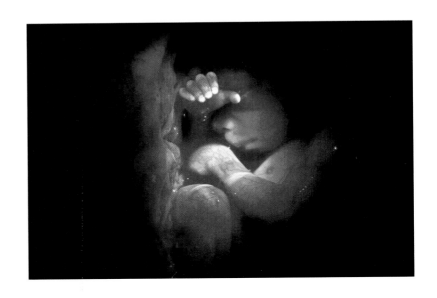

Unit 7 Plant Anatomy and Physiology

Science and the Real World

SCIENCE IN THE NEWS

THE WASHINGTON POST

ALARM SOUNDED ON GLOBAL WARMING

RESEARCHERS SAY DANGERS MUST BE ADDRESSED IMMEDIATELY

By Juliet Eilperin
June 16, 2004

Ten of the nation's top climate researchers warned yesterday that policymakers must act soon to address the dangers associated with global warming, which they described as a looming threat that will hit hardest and soonest at the world's poor and at farmers.

SCIENCE IN THE NEWS

THE WASHINGTON TIMES

GOP DISPUTES GLOBAL-WARMING CAUSE

By Stephen Dinan
July 30, 2003

Senate Republicans are pushing back on the issue of global warming, with the chairman of the Environment and Public Works Committee questioning not only the evidence for warming but also the link between human actions and climate change.

1.1 How Does Science Impact the Everyday World?

Global warming: Should alarms be going off about it, or is it really not much of a threat? To judge by the *Washington Post* newspaper story, above, it is high time the federal government acted on the issue. To judge by the *Washington Times* story, the Earth isn't appreciably warmer now than it has been in human history, and human activity hasn't affected climate much anyway. So, how does a person know what to believe?

This is not just an academic question, as practical decisions can hinge upon it. Who gets your vote for president: the candidate who says we have to act now on global warming, or the candidate who thinks the issue isn't that important? What kind of car do you buy? If global warming is here, then shouldn't you think about buying a car that gets more miles per gallon—and thus put fewer "greenhouse" gases into the air? Where do you buy a house? If the worst of the global-warming scenarios come true, then the answer is: Not in low-lying coastal areas, because, in your lifetime, they could end up underwater.

From these examples, it's easy to see that in some instances, we might decide how to act based on our view of an issue like global warming. But how does a person arrive at a view? If what we seek is *factual information* about global warming—information about whether it is taking place and what might be causing it—then it is not politics, or business, or art that is going to inform us. It is science. Just as we look to journalists to tell us about current events, and to economists to tell us about the economy, so we look to scientists to tell us about the natural world. Indeed, we look to them not only to tell us about the natural world's current state, but to predict its future. (How bad might global warming get?)

If we ask, then, whether science matters, this is just one of the ways in which it does. Scientists are society's eyes and ears on the natural world. They report to society on practical issues related to the natural world. What are the benefits and drawbacks of a high-protein diet? Is a given antidepressant drug safe for children? Does thinning out forests reduce the risk of catastrophic forest fires? In all these instances, society looks to science to provide answers.

But if society brings issues to scientists, things also work the other way around—scientists bring issues to society. We might say that, through their discoveries, scientists present society with *options* from which society then chooses, either in the marketplace or in the political arena. To take an example of a political choice, as little as five years ago, almost no one outside the research world had heard the term *embryonic stem cells*. But today, these cells lie at the center of a heated national debate. Some research breakthroughs in the late 1990s opened up a possibility for treating conditions such as Alzheimer's disease in a new way: Try to generate healthy tissue from cells taken from embryos when they are a tiny ball of about 200 cells. These cells have a remarkable capacity to develop into almost any tissue in the body (hence the name "stem" cells). On the other hand, they are harvested from human embryos that would otherwise be discarded by fertility clinics. So, should society fund research on these cells? In 2001, President Bush decided the answer is a qualified yes: The federal government will fund embryonic stem cell research, he announced, but only in a highly restricted manner. The issue underwent a new round of controversy in 2004, with the death of former President Ronald Reagan from Alzheimer's disease and subsequent calls for President Bush to loosen the restrictions on stem cell research. Of course, stem cells are not the only issue of this sort brought to society by science. Think of genetically modified foods, DNA fingerprinting, and the dizzying possibility of a human clone (see FIGURE 1.1).

Science and its sister discipline, technology, have always been in the business of providing options to society, but the impact of these options has increased greatly in recent years. Typical senior citizens of the 1960s had nothing more than aspirin to keep them fully functional; today's seniors might take half a dozen different drugs a day. Prior to the 1970s, a woman who was having trouble conceiving had essentially two choices: adopt a child or keep taking fertility drugs and hope for the best. Today a range of "assisted reproductive technologies" includes not only fertilization outside the body, but selection of an embryo

FIGURE 1.1   No Shortage of Controversies Scientific innovations can spark societal disagreements. The protesters in the photo are marching to the Statehouse in Montpelier, Vermont, to call for a time-out on genetically modified organisms (GMOs), including genetically engineered food crops.

that has the best chance to develop into a healthy baby. A standard office computer is a tool of astounding versatility compared to anything that existed in the workplace until the 1980s. Then again, we need not go back to the '80s to see how fast technological change is coming. On a typical day, do you e-mail someone, talk on a cell phone, or visit a website? As late as the mid-1990s, the average person would not have done any of these things (see **FIGURE 1.2**).

If we take a step back from all this, the message is that science and technology are now woven more tightly into the fabric of society than at any time in the past. Accordingly, to fully participate in the workforce, to make informed choices at the ballot box, or simply to make routine lifestyle decisions, the average person must now be more scientifically literate than at any time in the past. Why do business, or fine arts, or history majors need to know something about science? Because science is going to be a major influence on their lives.

1.2 What Is Science?

Having looked a little at the impact that science has, it might be helpful to consider the question of what science *is*. The point here is to give you some sense of the underpinnings of science—to review something about the how and why of it before you begin looking at the nature of one of its disciplines, biology.

Science as a Body of Knowledge

Science is in one sense a process—a *way* of learning. In this respect, it is an activity carried out under certain loosely agreed-to rules, which you'll get to shortly. **Science** is also a body of knowledge about the natural world. It is a collection of insights about nature, the evidence for which is an array of facts. The insights of science are commonly referred to as *theories*.

It's unfortunate but true that *theory* means one thing in everyday speech and something almost completely different in scientific communication. In everyday speech, a theory can be little more than a hunch. It is an unproven idea that may or may not have any evidence to support it. In science, meanwhile, a **theory** is a related set of insights, supported by evidence, that explains some aspect of nature. There is, for example, a Big Bang theory of the universe, whose insights explain how the universe came to be and how it developed. It holds that a cataclysmic explosion occurred about 14 billion years ago, and that after this, matter first developed in the form of gases that then coalesced into the stars we can see all around us. Numerous facts support these insights, such as the current size of the universe and its average temperature.

As you might imagine, with any theory this grand, some *pieces* of it are in dispute; some facts don't fit with the theory, and scientists disagree about how to interpret this piece of information or that. On the whole, though, these general insights have withstood the questioning of critics, and together they stand as a scientific theory.

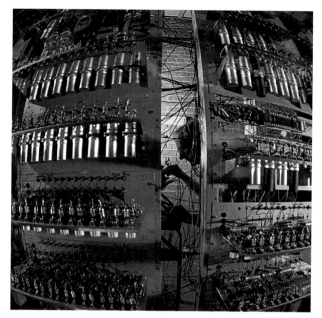

FIGURE 1.2 **Then and Now**

(**a**) A technician enters data into the world's first programmable computer, run initially in 1948. Called "Baby," the computer was more than 6 feet tall and almost 16 feet wide, but had a total memory of only 128 bytes.

(**b**) By comparison, a handheld product called the LifeDrive, released in 2005 by PalmOne, has a total memory of 4 gigabytes—four billion bytes, compared to the 128 bytes of the 1948 computer. The LifeDrive plays music, displays still photos and video, allows web browsing and e-mail communication, and contains a suite of word processing, graphic display, and spreadsheet programs.

A BROADER VIEW

A Misunderstanding of "Theory"

Practical consequences can result from a misunderstanding of scientific terminology. In 2002, the Cobb County, Georgia, School Board voted to place stickers inside district biology textbooks stating that "Evolution is a theory, not a fact, regarding the origin of living things." There is indeed a theory of evolution, but the school board seemed not to understand that it is a theory in the scientific sense: a set of insights well supported by evidence.

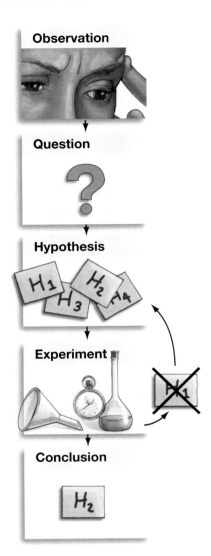

Observation

Question

Hypothesis

H_1 H_2 H_3 H_4

Experiment

H_1

Conclusion

H_2

FIGURE 1.3   **The Scientific Method** The scientific method enables us to answer questions by testing hypotheses.

The Importance of Theories   Far from being a hunch, a scientific theory actually is a much more valued entity than is a scientific fact, for the theory has an *explanatory* power, while a fact is generally an isolated piece of information. That the universe is 14 billion years old is a wonderfully interesting fact, but it explains very little in comparison with the Big Bang theory. Facts are important; theories could not be supported or refuted without them. But science is first and foremost in the theory-building business, not the fact-finding business.

Science as a Process: Arriving at Scientific Insights

So how does a body of facts and theories come about? What is the process of scientific investigation, in other words? When **science** is viewed as a process, it could be defined as a means of coming to understand the natural world through observation and the testing of hypotheses. This process is referred to as the **scientific method**. The starting state for scientific inquiry is always *observation:* A piece of the natural world is observed to work in a certain way. Then follows the *question*, which broadly speaking is one of three types: a "what" question, a "why" question, or a "how" question. Biologists have asked, for example, What are genes made of? Why does the number of land species decrease as we move from the equator to the poles? How does the brain make sense of visual images?

Formulating Hypotheses, Performing Experiments   Following the formulation of the question, various hypotheses are proposed that might answer it. A **hypothesis** is a tentative, testable explanation for an observed phenomenon. In almost any scientific question, several hypotheses are proposed to account for the same observation. Which one is correct? Most frequently in science, the answer is provided by a series of *experiments*, meaning controlled tests of the question at hand (**see** FIGURE 1.3). It may go without saying that scientists don't regard all hypotheses as being equally worthy of undergoing experimental test. By the time scientists arrive at the experimental stage, they usually have an idea of which is the most promising hypothesis among the contenders, and they then proceed to put that hypothesis to the test. Let's see how this worked in an example from history.

The Test of Experiment: Pasteur and Spontaneous Generation   Does life regularly arise from anything *but* life, or can it be generated "spontaneously," through the coming together of basic chemicals? The latter idea had a wide acceptance from the time of the ancient Romans forward, and as late as the nineteenth century it was championed by some of the leading scientists of the day. So how could the issue be decided? The famous French chemist and medical researcher Louis Pasteur formulated a hypothesis to address this question (**see** FIGURE 1.4). He believed that many alleged examples of life arising spontaneously were simply instances of airborne microscopic organisms landing on a suitable substance and then multiplying in such profusion that they could be seen. Life came from life, in other words, not from spontaneous generation. But how could this be demonstrated? In 1860, Pasteur put a meat broth in some glass flasks and then proceeded to sterilize the broth—to kill off all the life in it—by heating the flasks. Meanwhile, he also heated the glass *necks* of the flasks, after which he bent the necks into a "swan" or S-shape. Initially he found that, though the ends of the flasks remained open to the air, inside the flasks there was not a sign of life. Why? The broth remained sterile because microbe-bearing dust particles got trapped in the bend of the flask's neck. If Pasteur broke the neck off before the bend, however, the flask soon had a riot of bacterial life growing within it. In another test, Pasteur tilted the flask so that the broth *touched* the bend in the neck, a change that likewise got the microbes growing.

Elements in Pasteur's Experiments   Now, note what was at work here. Pasteur had a preconceived notion of what the truth was, and he designed experiments to test his hypothesis. Critically, he performed the same set of steps several times in the experiments, keeping all the elements the same each time—

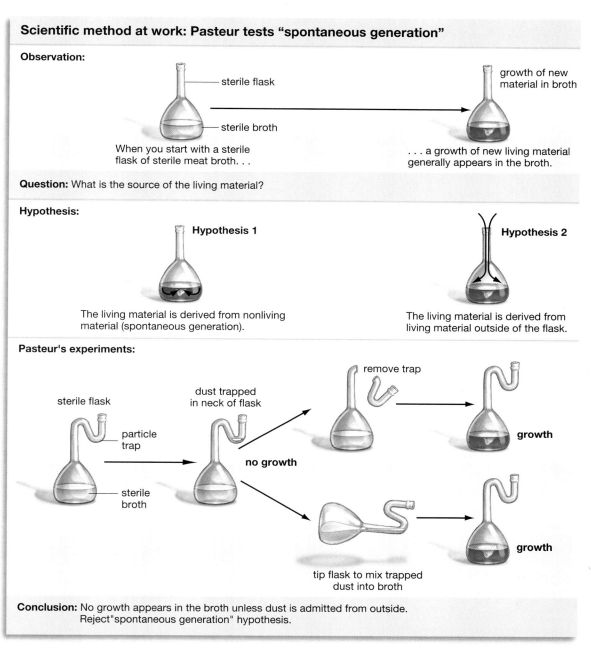

Scientific method at work: Pasteur tests "spontaneous generation"

Observation:

sterile flask

sterile broth

When you start with a sterile flask of sterile meat broth...

growth of new material in broth

... a growth of new living material generally appears in the broth.

Question: What is the source of the living material?

Hypothesis:

Hypothesis 1

The living material is derived from nonliving material (spontaneous generation).

Hypothesis 2

The living material is derived from living material outside of the flask.

Pasteur's experiments:

sterile flask

particle trap

sterile broth

dust trapped in neck of flask

no growth

remove trap

growth

tip flask to mix trapped dust into broth

growth

Conclusion: No growth appears in the broth unless dust is admitted from outside. Reject "spontaneous generation" hypothesis.

FIGURE 1.4 **Pasteur's Experiments** Pasteur's spontaneous generation experiments and the scientific method. Nineteenth-century observation made clear that life would appear in a medium, such as broth, that had been sterilized. But what was the source of this life? One hypothesis was that it arose through "spontaneous generation," meaning it formed from the simple chemicals in the broth. Conversely, Pasteur hypothesized that it originated from airborne microorganisms. He was able to design an experiment that offered evidence for this hypothesis. The device he used was an S-shaped flask, which enabled air to enter the flask freely while trapping all particles (including invisible microorganisms) in a bend in the neck.

except for one. The nutrient broth was the same in each test; it was heated the same amount of time and in the same kind of flask. What *changed* each time was one critical **variable**: an adjustable condition in an experiment. In this case, the variable was either the shape of the flask neck, or the tilt of the flask. Given that all the other elements of the experiments were kept the same, the experiments had rigorous controls: All conditions were held constant over several trials, except for a single variable. A control *condition* can be thought of as an experimental condition that exists prior to the introduction of any variables that are being tested for. Pasteur was testing for what happened with a broken-necked flask and a tilted flask. The control condition, therefore, was the broth-filled flask left sitting straight up with its particle trap intact. Pasteur's finding that no life grew in this condition is interesting, but tells us very little by itself. We learn something only by comparing this finding to the result Pasteur got when he introduced his variables: that life *did* grow when the neck was broken or the flask was tilted.

A BROADER VIEW
What Is a Control Group?

What are the *control groups* often referred to in newspaper accounts of medical studies? In a study that is looking at, for example, the effectiveness of a new drug, the control group usually is composed of people who are *not* being given the drug in question. These people are then compared to a group of people who do get the drug. Only by looking at how both groups fare medically can researchers get a sense of whether an "intervention," such as a new drug, has been effective. The trick in designing any medical study is to match the control group with the intervention group as closely as possible for such factors as age and general health. Only by making sure that both groups are the same in every respect except one—the intervention—will researchers be able to tell if the intervention has been effective.

Note also that the idea of spontaneous generation was not banished with this one set of experiments—nor should it have been. Pasteur's experiments provided one of the *facts* mentioned earlier, in this case the fact that flasks of liquid will remain sterile under certain conditions. The idea that life arises only from life is, however, one of the scientific *theories* noted earlier, meaning that it requires the accumulation of many facts pointing in the same direction.

Other Kinds of Support for Hypotheses Some scientific questions are difficult or impossible to test purely through experiment. For example, there currently is a controversy over whether birds are the direct descendants of dinosaurs. What kind of experiment could be run to test this hypothesis? Certain modern-day evidence is available to us—the DNA of living birds, for example—but examining DNA does not amount to an experiment. Instead it is observation, which is another valid way to test a hypothesis. Evidence from the past can also be observed, of course, which in this case means the observation of dinosaur and bird fossils. Indeed, fossils have been the key evidence in convincing most experts that birds are the descendants of dinosaurs.

When Is a Theory Proven?

At what point does a theory become proven? An irony of the orderly undertaking called science is that there's nothing orderly about this transition. No scientific supreme court exists to make a decision. Scientists aren't polled for their views on such questions, and even if they were, at what point would we say something had been "proven"? When more than 50 percent of the experts in the field assent to it? When there are no dissenters left?

Provisional Assent to Findings: Legitimate Evidence and Hypotheses
This lack of finality in science fits in, however, with one of its central tenets, which is that nothing is ever finally proven. Instead, every finding is given only *provisional assent*, meaning it is believed to be true for now, pending the addition of new evidence. This principle is so deeply embedded in science that scientists rarely have reason to think about it (just as drivers would seldom contemplate why they are driving on the right-hand side of the road). Yet it is profoundly important because it is the thing that most starkly separates science from belief systems, such as those that operate in culture, politics, or religion. Every theory and "fact" in science is subject to modification, based solely on the best evidence available. All laws are subject to challenge, and there are no unquestioned authority figures. This means there is a paradox in science: Its only bedrock is that there is no bedrock; everything scientists "know" is subject to change.

In practice, this is a difficult ideal to live up to. Even when a body of evidence starts to point in a new direction, scientists—like anybody else—may be reluctant to give up old ways of thinking. Recognizing this tendency in human nature, Charles Darwin's friend Thomas Henry Huxley, writing in 1860, gave a beautiful description of the attitude scientists should have when investigating nature:

> Sit down before fact as a little child, be prepared to give up every preconceived notion, follow humbly wherever and whatever abysses nature leads, or you will learn nothing.

This principle of science's openness to revision is one of three important scientific principles having to do with hypotheses and evidence. Here are all three, stated briefly:

- Every assertion regarding the natural world is subject to challenge and revision, based on evidence.

- Results obtained in experiments must be *reproducible*. Different investigators must be able to obtain the same results from the same sets of procedures and materials.

- Any scientific hypothesis or claim must be *falsifiable*, meaning open to negation through means of scientific inquiry. The assertion that "UFOs are visiting the Earth" does not rise to the level of a scientific claim, because there is no way to prove that this is *not* so.

So Far...

1. Science can be defined in two ways. What are they?
2. What is a theory in the scientific sense of the term?
3. Pasteur's finding that life grew in the broken-necked flask was most meaningful when compared to what?

1.3 The Nature of Biology

Let us shift now from an overview of science to a more narrow focus on **biology**, which can be defined as the study of life. But what is life? It may surprise you to learn that there is no standard, short answer to this question. Indeed, the only agreement among scholars seems to be that there is not, and perhaps cannot be, a short answer to this question. The main obstacle to developing one is that any given quality common to all living things is likely to exist in some nonliving things as well. Some living things may "move under their own power," but so does the wind. Living things may grow, but crystals and fire do the same thing. Therefore, biologists generally define life in terms of a group of characteristics possessed by living things. Looked at together, these characteristics are sufficient to separate the living world from the nonliving. We can say that living things:

- Can assimilate and use energy
- Can respond to their environment
- Can maintain a relatively constant internal environment
- Possess an inherited information base, encoded in DNA, that allows them to function
- Can reproduce, through use of the information encoded in DNA
- Are composed of one or more cells
- Evolved from other living things
- Are highly organized compared to inanimate objects

Every one of these qualities exists in all the varieties of Earth's living things. The simplest bacterium needs an energy source no less than any human being does. Our energy source is the food that's familiar to us; the bacterium's might be the remains of vegetation in the soil. The bacterium responds to its environment, and so do we. You would take action if you smelled gas in your house; the bacterium would move away if it encountered something it regarded as noxious. Humans maintain a relatively stable internal environment by, for example, sweating when they get hot. When the bacterium's external environment gets too hot, it has certain genes that will switch on to keep it functioning. Both humans and bacteria use the molecule DNA as a repository of the information necessary to allow them to live. Bacteria and human beings both reproduce—bacteria by simple cell division, human beings through the use of two kinds of reproductive cells (egg and sperm). A bacterium is a single-celled life-form, while humans are a 100-trillion-celled life-form. Bacteria and humans both evolved from complex living things and ultimately share a single common ancestor.

There are some exceptions to these "universals." For example, the overwhelming majority of honeybees and ants are sterile females; they can't reproduce, but no one would doubt that they're alive. Generally, however, it's safe to say that if something is living, it has all these qualities.

So Far... Answers:

1. *As a body of knowledge about the natural world, and as a means of coming to understand the natural world through observation and the testing of hypotheses*
2. *A related set of insights, supported by evidence, that explains some aspect of nature*
3. *The control condition in his experiment: the condition into which no variables being tested for had been introduced*

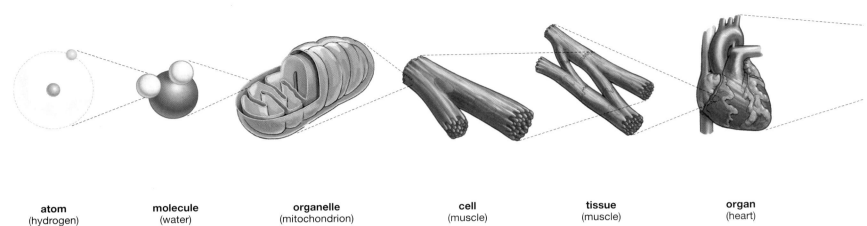

| atom (hydrogen) | molecule (water) | organelle (mitochondrion) | cell (muscle) | tissue (muscle) | organ (heart) |

FIGURE 1.5 **Levels of Organization in Living Things**

Life Is Highly Organized, in a Hierarchical Manner

One item on the list of qualities requires a little more explanation. It is that living things are highly organized compared to inanimate matter. More specifically, they are organized in a "hierarchical" manner: one in which lower levels of organization are integrated to make up higher levels. The main levels in this hierarchy could be compared to the organization of a business. In a corporation, there may be individuals making up an office, several offices making up a department, several departments making up a division, and so forth. In life, there is one set of organized "building blocks" making up another (**see FIGURE 1.5**).

Actually, life is not just "highly" organized. Nothing else comes *close* to it in organizational complexity. The sun is a large thing, but it is an uncomplicated thing compared to even the simplest organism, to say nothing of human beings. Consider that you have about 100 trillion cells in your body and that, with some exceptions, each type of cell has in it a complement of DNA that is made up of chemical building blocks. How many building blocks? Three billion of them. Now, you probably know that most cells divide regularly, one cell becoming two, the two becoming four, and so on. Each time this happens, each of the 3 billion DNA building blocks must be faithfully *copied*, so that both of the cells resulting from cell division will have their own complete copy of DNA. And this copying of the molecule—before anything is actually done with it—is taking place in about 25 million separate cells in your body each second. Complex indeed. Let's see what life's levels of organization are.

Levels of Organization in Living Things The building blocks of matter, called *atoms*, lie at the base of life's organizational structure. (See Chapter 2 for an account of them.) Atoms come together to form *molecules*: entities consisting of a defined number of atoms that exist in a defined spatial relationship to one another. A molecule of water is one atom of oxygen bonded to two atoms of hydrogen, with these atoms arranged in a very precise way. Molecules in turn form what are called *organelles*, which are "tiny organs" in a cell. Each of your cells has, for example, some organelles in it called *mitochondria* that transform the energy from food into an energy form your body can use. Such an organelle is not just a collection of molecules that exist close to one another. It is a highly organized structure, as you can tell just from looking at the rendering of it in Figure 1.5.

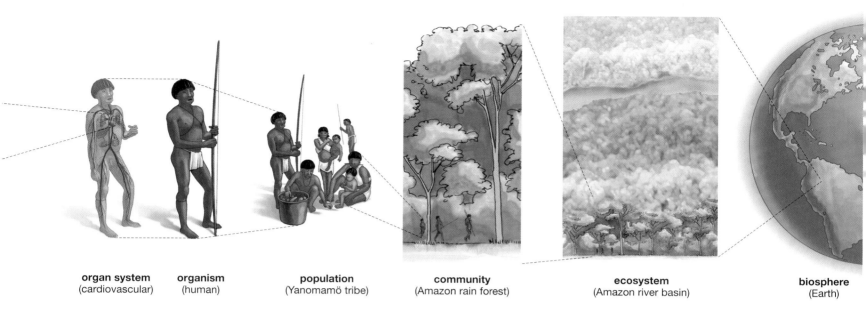

organ system
(cardiovascular)

organism
(human)

population
(Yanomamö tribe)

community
(Amazon rain forest)

ecosystem
(Amazon river basin)

biosphere
(Earth)

At the next step up the organizational chain, we get to entities that are actually *living*, as opposed to entities that are components of life. *Cells* are units that can do all of the things listed earlier: assimilate energy, reproduce, react to their environment, and so forth (**see** FIGURE 1.6). Indeed, most experts would agree that cells are the only place that life exists. You may say: But isn't there a lot of material in between my cells? The answer is yes; it's mostly water with a good number of other

FIGURE 1.6   **Seeing Life at Its Various Levels**
(a) Several kinds of organelles can be seen within a single white blood cell. These organelles include the cell's nucleus (the dark circle) and its energy-transforming mitochondria (the smaller blue circles at the periphery). **(b)** Here many cells can be seen making up a single tissue—in this case a tissue that surrounds an air passageway in a human lung. The nucleus of each cell is stained red; some of the cells have hair-like cilia that can be seen extending into the passageway. **(c)** Many populations can go into making up a community; this one is centered on a coral reef in the South Pacific.

(a)

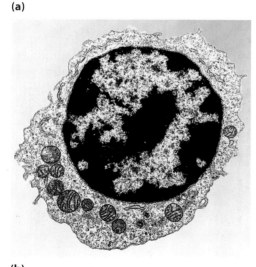

(b)

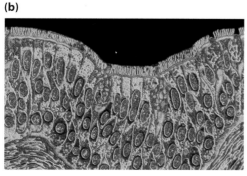

(c)

molecules in the mix. But if all the cells were removed from this watery environment, there would be nothing resembling life left in it.

The next step up is to a *tissue*: a collection of similar cells that serve a common function. Your biceps are collections of a set of similar cells—muscle cells—that are serving the common function of contracting. Each bicep, then, is made up of muscle tissue. Several kinds of tissues can come together to form a functioning unit known as an *organ*. Your heart, for example, is an organ; it is a collection of muscle tissue and connective tissue, among other types. Several organs and related tissues then can be integrated into an *organ system*. Contractions of your heart push blood into a system of blood vessels. The heart, blood, and blood vessels form the cardiovascular system. An assemblage of cells, tissues, organs, and organ systems can then form a multicelled *organism*. (Of course, back down at the cell level, a one-celled bacterium is also an organism; it's just not one with organs and so forth.)

From here on out, life's levels of organization all involve *many* organisms. Members of a single type of living thing (a species), living together in a defined area, make up what is known as a *population*. When you look at *all* the kinds of living things in a given area, you are looking at a *community*. When you consider a community and the *non*living elements with which they interact, such as climate and water, the result is an *ecosystem*. Finally, all the ecosystems of the Earth make up the *biosphere*.

Biology's Chief Unifying Principle

Almost all biologists would agree that the most important thread that runs through biology is **evolution**, meaning the gradual modification of populations of living things over time, with this modification sometimes resulting in the development of new species. Evolution is central to biology, because every living thing has been shaped by it. (There are no known exceptions to this universal.) Given this, the explanatory power of evolution is immense. Why do peacocks have their finery, or frogs their coloration, or trees their height (**see** FIGURE 1.7)? All these things stand as wonders of nature's diversity. But with knowledge of evolution,

A BROADER VIEW

Complex at Any Size

The bacterium *E. coli* is fairly humble as living things go, yet it has about 4,600 genes and its complement of DNA is made up of about 4.6 million of the chemical building blocks called nucleotides. And this level of complexity is required just for life in a single cell—an entity so small that 600 *E. coli* could fit side by side within the period at the end of this sentence.

FIGURE 1.7 Evolution Has Shaped the Living World

(a) A peacock displaying his plumage

(b) A poison dart frog in Colombia

(c) Pine trees in the South Pacific

(a) Golden northern bumblebee

(b) Sandhills hornet

FIGURE 1.8 **Similar Enough to Yield a Benefit** These are two of the many stinging insects that have the black-and-yellow-striped coloration that warns away predators.

they are wonders of diversity that *make sense*. For example, why do so many unrelated stinging insects look alike? Evolutionary principles suggest they *evolved* to look alike because of the general protection this provides from predators. Think of yourself for a moment as a bee predator. Having once gotten stung, would you annoy *any* roundish insect that had a black-and-yellow-striped coloration? You probably learned your lesson about this in connection with one species, but many species of insects are now protected from you simply by virtue of the coloration they share with the others (**see FIGURE 1.8**). Thus, there were reproductive benefits to individuals that, through genetic chance, happened to get a slightly more striped coloration: They left more offspring, because they were bothered less by predators. Over time, entire populations moved in this direction. They evolved, in other words.

The means by which living things can evolve is a topic this book takes up beginning in Chapter 16. Suffice it to say for now that a consideration of evolution is never far from most biological observations. So strong is evolution's explanatory power that, in uncovering something new about, say, a sequence of DNA or the life cycle of a given organism, one of the first things a biologist will ask is: Why would evolution shape things in this way?

On to Chemistry and Life

This book has something in common with the levels of organization you just looked at, in that it too goes from constituent parts to the larger whole. Recall that the smallest constituent part of life is the unit called an atom. A single atom might seem as important to an organism as a thimble full of water is to an ocean wave. Yet it is the interaction of atoms that controls chemistry; and it is chemistry that largely controls life, as you will see. So, we'll begin our tour of the living world by starting very small. Coming up: atoms, chemistry, and life.

A BROADER VIEW

Biology's Moment of Insight

In mystery novels, there often is a key insight that allows a detective to make sense of all the various facts he or she has uncovered. The theory of evolution served as the greatest aha! moment of all for biology. With it, the stripes of zebras and the location of kangaroos and the height of trees made sense: All these features, scientists saw, came about because of evolution's channeling effects. Before the theory of evolution, biology had facts; after the theory, biology had a way of understanding those facts.

So Far...

1. Name three distinguishing characteristics of life. Living things . . .
2. Order the following units of matter and life from smallest to largest: cell, molecule, population, tissue.
3. Biologists find the theory of evolution so useful because it has a great power to _____ .

So Far... Answers:

1. can assimilate and use energy, can respond to their environment, can maintain a relatively constant internal environment, possess an inherited DNA information base, can reproduce, are composed of one or more cells

2. molecule, cell, tissue, population

3. explain the observations that biologists make

Chapter Review

Summary

1.1 How Does Science Impact the Everyday World?

- **The role of science**—Science is playing an increasingly important role in the everyday lives of Americans. Society turns to scientists to answer practical questions having to do with the natural world. In addition, scientific discoveries are transmitted to society in the form of products and processes, which can be thought of as options for carrying out life. Scientific advances regularly confront society with choices that have an ethical dimension to them.   (page 2)

1.2 What Is Science?

- **Science as knowledge and process**—In one of its facets, science is a body of knowledge, a unified collection of insights about nature, the evidence for which is an array of facts.   (page 3)

- Science can also be defined as a way of learning: a process of coming to understand the natural world through observation and the testing of hypotheses.   (page 4)

- **Scientific theories**—A theory is a related set of insights, supported by evidence, that explains some aspect of nature.   (page 3)

- **The scientific method**—Science works through the scientific method, in which an observation leads to the formulation of a question about the natural world. Then comes a hypothesis—a tentative, testable explanation that has not been proven to be true. The hypothesis may be tested through observation or through a series of experiments.   (page 4)

- **Principles of evidence**—In science, every assertion regarding the natural world is subject to challenge and revision; results obtained in ex-periments must be reproducible by other experimenters; and scientific claims must be falsifiable, meaning open to negation by means of scientific inquiry.   (page 6)

Web Tutorial 1.1 Science as a Process: Arriving at Scientific Insights

1.3 The Nature of Biology

- **What is life?** Biology is the study of life. Life is defined by a group of characteristics possessed by living things. These are that living things can assimilate energy, respond to their environment, maintain a relatively constant internal environment, and possess an inherited information base, encoded in DNA. Living things can also reproduce, are composed of one or more cells, are evolved from other living things, and are highly organized compared to inanimate objects.   (page 7)

- **Life's complex organization**—Life is organized in a hierarchical manner, existing in increasing complexity from atoms to molecules and then in sequence to organelles, cells, tissues, organs, organ systems, organisms, populations, communities, ecosystems, and the biosphere.   (page 8)

- **Evolution as biology's unifying theory**—Biology's chief unifying theory is evolution, which can be defined as the gradual modification of populations of living things over time, with this modification sometimes resulting in the development of new species. Evolution provides the means for making sense of the forms and processes seen in living things on Earth today.   (page 10)

Web Tutorial 1.2 Hierarchical Organization of Life

Key Terms

biology p. 7

evolution p. 10

hypothesis p. 4

science p. 3

scientific method p. 4

theory p. 3

variable p. 5

Testing Your Understanding

In Your Own Words *(answers in the back of the book)*

1. What is science? In what ways is science different from a belief system such as religious faith?

2. What is a controlled experiment? Why is it important to keep all variables but one constant in a scientific experiment?

3. How did Louis Pasteur cast doubt on the idea of spontaneous generation?

Thinking about What You've Learned

1. Science doesn't just inform us about critical societal issues such as global warming. It provides us with information about smaller, more down-to-earth questions as well. For example, is it true that there is a moderate level of physical exercise, usually called a "fat-burning zone," that takes weight off faster than a more vigorous level of exercise does? The answer is no; we lose weight by burning calories, and vigorous exercise burns more calories than moderate exercise does. How do we know this? A group of biologists, called exercise physiologists, have studied the subject. Using what you have learned in this chapter, describe how you can separate fact from fiction with respect to claims about how the body works.

2. Can science offer any special insights into moral or aesthetic questions?

3. Is it harder to prove a hypothesis than to disprove it? Imagine you wanted to establish that cheetahs are the fastest land animals, and assume you have the ability to clock any animal moving at its top speed. Now, what would it take to disprove the idea that cheetahs are the fastest land animals? What would it take to prove that cheetahs are the fastest land mammals, meaning no other land mammal can run faster than cheetahs do?

Chemistry and Life

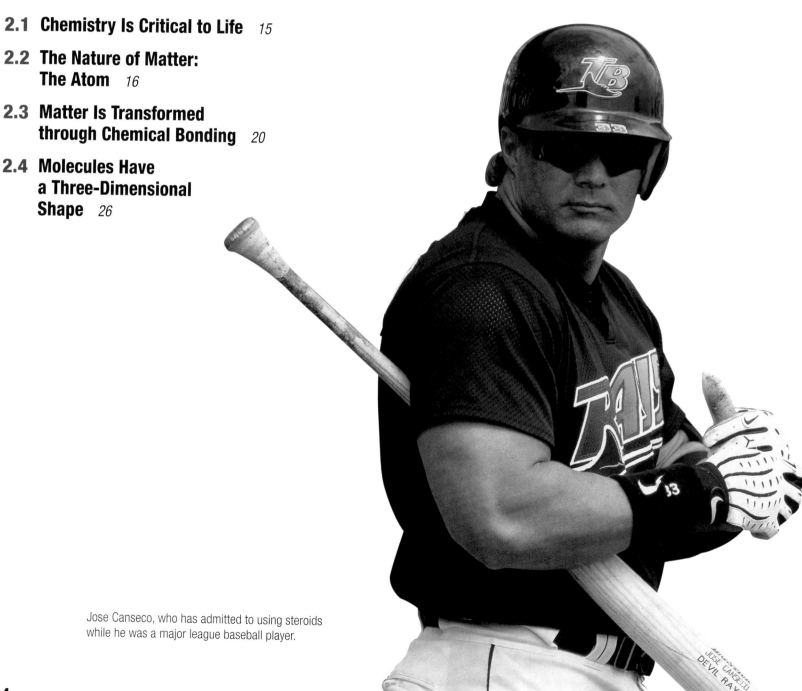

Jose Canseco, who has admitted to using steroids while he was a major league baseball player.

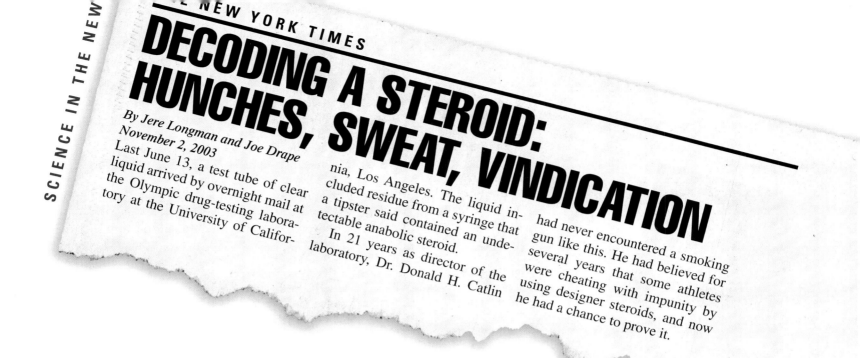

THE NEW YORK TIMES

DECODING A STEROID: HUNCHES, SWEAT, VINDICATION

By Jere Longman and Joe Drape
November 2, 2003

Last June 13, a test tube of clear liquid arrived by overnight mail at the Olympic drug-testing laboratory at the University of Califor- nia, Los Angeles. The liquid in- cluded residue from a syringe that a tipster said contained an unde- tectable anabolic steroid.

In 21 years as director of the laboratory, Dr. Donald H. Catlin had never encountered a smoking gun like this. He had believed for several years that some athletes were cheating with impunity by using designer steroids, and now he had a chance to prove it.

2.1 Chemistry Is Critical to Life

Imagine being an athlete like the one who used the drug, called THG, mentioned in the *New York Times* story, above. You're well aware that you're cheating in your sport. You know you're doing something that could strip you of any awards you receive in it. And to top it all off, you're *injecting* yourself with a drug—one that has never been tested in any rigorous way for harmful effects. What could push you over the line into committing such an act? In all likelihood, the lure of an edge. Athletes take drugs like THG because, in the modern world of sports competition, the difference between winning and losing can be razor thin. In a sport like sprinting, the margin might be a few hundredths of a second. And many athletes believe that drugs like THG might speed them up by that much—or let them throw slightly farther or jump slightly higher. So, which will it be: Train hard for years and be left at home watching the Olympics on television, or train hard for years, inject a drug, and have a chance at stand- ing on the winners' platform?

1

The chemical investigators who analyzed THG found that it was a designer anabolic steroid—"designer" because some chemist deliberately designed it (or at least recognized it) as a drug that could not be detected in existing tests; "anabolic" from the biological term for "building up" and "steroid" from a class of substances whose structure includes a set of four interlocked carbon rings. Testosterone is an anabolic steroid too, but in the cat-and-mouse world of sport doping, it is a banned substance that can be tested for. THG had the remarkable quality of disintegrating in standard tests. Had the anonymous tipster not sent in the syringe containing it, it would probably still be in use today. Instead, its discovery sparked a sports doping scandal whose repercussions are being felt even now.

But what does a drug like THG do? It is made up entirely of three elements—carbon, hydrogen, and oxygen—which are the same elements that make up simple table sugar. Yet put together in the right way, these ingredients add up to a substance that makes muscle grow stronger, at a faster rate, and that can make injuries heal more quickly. Thus do we get a seeming paradox: In a realm of life that we think of as hinging on talent and training and motivation, chemistry makes a difference. But then again, that's just the point. Whenever anything biological is involved, chemistry makes a difference. Indeed, looked at one way, chemistry makes all the difference. The bone that is growing inside a child does not form as a result of conscious activity on the part of bone cells. What's at work is pure chemistry: repulsion and bonding as calcium and proteins interact. When a nervous-system signal arrives from your hand telling you that you've touched a hot stove, the whole thing is handled through a series of chemical reactions. Just as a game of pool is, at one level, a matter of how billiard balls interact with each other, so life is, to a considerable degree, a matter of how chemical elements interact with each other. In the chapter that follows, you'll learn the basics of these chemical interactions.

2.2 The Nature of Matter: The Atom

The central subject of chemistry is matter and its transformations. For our purposes, we can think of matter as being composed of fundamental units that we call atoms. If you look at the right side of FIGURE 2.1, you can see how an atom takes on its building-block role in one specific kind of matter—in this case, a bird's feather. You'll also see that atoms are themselves composed of some constituent parts.

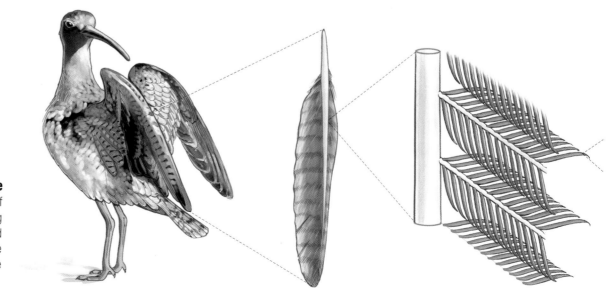

FIGURE 2.1   The Building Blocks of Life
Viewing this idealized feather at different levels of magnification, we eventually arrive at the building block of all matter, the atom. The atom selected here, from among a multitude that make up the feather, is a single hydrogen atom, composed of one proton and one electron.

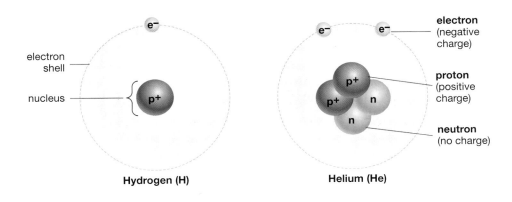

Hydrogen (H)

Helium (He)

electron
shell

nucleus

electron
(negative
charge)

proton
(positive
charge)

neutron
(no charge)

FIGURE 2.2 **Representations of Atoms** One conceptualization of two separate atoms, hydrogen and helium. The model is not drawn to scale; if it were, the electrons would be perhaps a third of a mile away from the nuclei. The model also is simplified, giving the appearance that electrons exist in track-like orbits around an atom's nucleus. In fact, electrons spend time in volumes of space that have several different shapes.

Figure 2.1 shows the simplest kind of atom, the hydrogen atom, which has only two of these constituent parts (a proton and an electron); but if you look at the right side of FIGURE 2.2, you can see a helium atom, which has on display all three kinds of constituent parts that interest us.

Protons, Neutrons, and Electrons

These three parts are protons, neutrons, and electrons, and they exist in a spatial arrangement that is the same in all matter. Protons and neutrons are packed tightly together in a core (the atom's **nucleus**), and electrons move around this core some distance away. The one variation on this theme is the hydrogen atom already mentioned, which has no neutrons, but instead has only a single proton and an electron.

These three "subatomic" particles have mind-bending sizes and proportions. As the chemist P. W. Atkins has pointed out, an atom is so small that 100 million carbon atoms would lie end to end in a line of carbon about this long: _____ (3 centimeters). Things are just as disorienting when we consider the size of the atom as a whole, relative to the nucleus. The whole atom, with electrons at its edge, is 100,000 times bigger than the nucleus. If you were to draw a model of an atom *to scale* and began by sketching a nucleus of, say, half an inch, you'd have to draw some of its electrons more than three-quarters of a mile away.

The components of atoms have another quality that interests us: electrical charge. **Protons** are positively charged subatomic particles, and **electrons** are

A BROADER VIEW

Atom Smashers

There is some irony in the fact that nature's smallest entities—atoms and their constituent parts—can be understood only with the help of science's largest experimental machines. Particle accelerators or "atom smashers" often have a size that can be measured in miles. The Fermilab accelerator in Illinois has an underground magnetic ring that is 3.8 miles in circumference, and it has particle detectors that are three stories high. Subatomic particles are accelerated around the ring and then slammed into each other, or into a fixed target. The scientists who do this work could be compared to people who, in trying to find out what parts a watch has, throw it on the ground and record the way its mechanisms fly out upon impact.

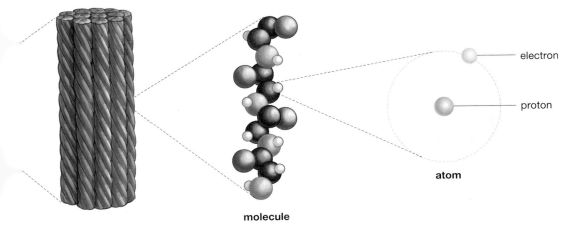

electron

proton

atom

molecule

FIGURE 2.3 **Pure Gold** Gold is an element because it cannot be reduced to any simpler set of substances through chemical means. Each gold bar and nugget is made up of a vast collection of identical atoms—those with 79 protons in their nuclei.

negatively charged subatomic particles. Meanwhile, **neutrons**—as their name implies—are particles that have no charge, but are electrically neutral. Because all these particles do not exist separately, but *combine* to form an atom, as a whole the atom may be electrically neutral as well. The negative charge of the electrons balances out the positive charge of the protons. Why? Because in this state the number of protons an atom has is exactly equal to the number of electrons it has (though we'll see a different, "ionic" state later in this chapter). In contrast, the number of *neutrons* an atom has can vary in relation to the other two particles.

With this picture of atoms in mind, we can begin to answer a fundamental question: What is matter? We certainly have a commonsense answer to this question. Matter is any substance that exists in our everyday experience. For example, the iron that goes into cars is matter. But what is it that makes iron different from, say, gold? The answer is that an iron atom has 26 protons in its nucleus, while a gold atom has 79.

Fundamental Forms of Matter: The Element

Gold is an **element**—a substance that is "pure" because it cannot be reduced to any simpler set of component substances through chemical processes. And the thing that defines each element is the number of protons it has in its nucleus. A solid-gold bar, then, represents a huge collection of identical atoms, each of which has 79 protons in its nucleus (FIGURE 2.3).

Given what you've just read about protons, neutrons, and electrons, you may wonder why gold—or any other element—cannot be reduced to any "simpler set of component substances." Aren't protons and neutrons components of atoms? Yes, but they are not component *substances*, because they cannot exist by themselves as matter. Rather, protons and neutrons must *combine* with each other to make up atoms.

Elements don't have to stay separate from each other, of course. In making gold jewelry, an artist may combine gold with another metal such as silver or copper to form a mixture that is stronger than pure gold. But the gold atoms are still present, all of them retaining their 79-proton nuclei. At root, then, all matter is composed either of pure elements, or of elements that exist in combination with each other. If you look at FIGURE 2.4, you can see the most important elements that go into making up both the Earth's crust and human beings.

FIGURE 2.4 **Constituent Elements** The major chemical elements found in Earth's crust (including the oceans and the atmosphere) and in the human body.

Assigning Numbers to the Elements In the same way that buildings can be defined by a location, and thus have a street number assigned to them, these elements, which are defined by protons in their nuclei, have an *atomic number* assigned to them. We have observed that hydrogen has only one proton in its nucleus, and it turns out that scientists have constructed the atomic numbering system so that it goes from smallest number of protons to largest. Thus, hydrogen has the atomic number 1. The next element, helium, has two protons, so it is assigned the atomic number 2. Continuing on this scale all the way up through the elements found in nature, we would end with uranium, which has an atomic number of 92.

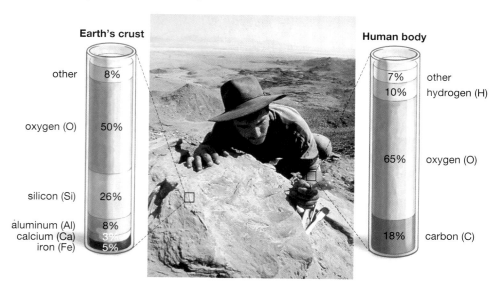

Earth's crust

other	8%
oxygen (O)	50%
silicon (Si)	26%
aluminum (Al)	8%
calcium (Ca)	3%
iron (Fe)	5%

Human body

other	7%
hydrogen (H)	10%
oxygen (O)	65%
carbon (C)	18%

Isotopes All this seems like a nice, tidy way to identify elements—one element, one atomic number, based on number of protons—except that we're leaving out something. Recall that atoms also

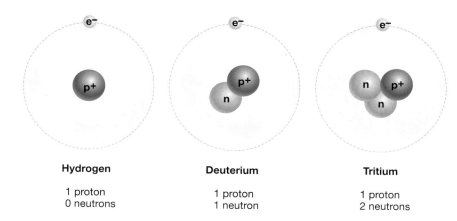

FIGURE 2.5 Same Element, Different Forms Pictured are three isotopes of hydrogen. Like all isotopes, they differ in the number of neutrons they have.

have neutrons in their nuclei and that the number of neutrons can vary independently of the number of protons. What this means is that, because the number of neutrons in an element's nucleus may vary, we can have various *forms* of individual elements, called **isotopes**. Most people have heard of one example of an isotope, whether or not they recognize it as such. The element carbon has six protons, giving it an atomic number of 6. In its most common form, it also has six neutrons. However, a relatively small amount of carbon exists in a form that has *eight* neutrons. Well, the element is still carbon, and in this form the number of its protons and neutrons equals 14, so the *isotope* is carbon-14, which is used in determining the ages of fossils and geologic samples.

Most elements have several isotopes. Hydrogen, for example, which usually has one proton and one electron, also exists in two other forms: deuterium, which has the proton, electron, and one neutron; and tritium, which has one proton, one electron, and two neutrons (**see** FIGURE 2.5). We are familiar mostly with radioactive isotopes, which is to say isotopes that give off radiation in the process of decaying. Carbon-14 is one of these, as are a host of isotopes used in medicine. But there are stable isotopes as well, meaning isotopes that are not radioactive. FIGURE 2.6 provides an example of how isotopes are used in medicine.

The Importance of Electrons

In our account so far of the subatomic trio, we have had much to say about protons and neutrons, but little to say about electrons. This was necessary because we needed to go over the nature of matter,

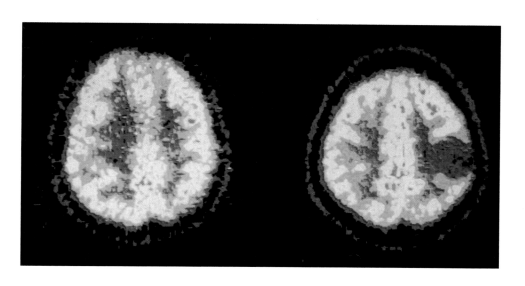

FIGURE 2.6 Using Isotopes to Spot a Brain Tumor Positron emission tomography, or PET scanning, works by means of collisions. A radioactive isotope introduced into the body emits particles called positrons, which collide with electrons that are produced as a result of metabolic activity. The more intense the metabolic activity is, the greater the number of collisions that occur. In this PET scan of a human brain (seen from above), differences in metabolic activity register as different colors—red for the greatest amount of activity, then yellow, green, and finally blue for the least amount of activity. In the image at right, the large blue area on the right represents a region that has been rendered inactive by a tumor.

So Far... Answers:

1. *protons, positively charged, in the nucleus; neutrons, no charge, in the nucleus; electrons, negatively charged, circling around the nucleus*

2. *the number of protons in its nucleus*

3. *the number of neutrons in its nucleus*

but in a sense you can regard what has been set forth to this point as so much stage-setting, because what's most important in biology is the way elements *combine* with other elements. And in this combining, it is the outermost electrons that play a critical role.

So Far...

1. What are the three primary component parts of atoms, what charge does each of them possess, and where is each of them located?

2. What is it that distinguishes one element from another?

3. What is it that distinguishes one isotope of an element from another?

2.3 Matter Is Transformed through Chemical Bonding

The process of chemical combination and rearrangement is called **chemical bonding**, and for us it represents the heart of the story in chemistry. When the outermost electrons of two atoms come into contact, they can reshuffle themselves in such a way that the atoms they are part of become attached to one another. This can take place in two ways: One atom can *share* one or more electrons with another atom, or one atom can *give up* one or more electrons to another atom. Sharing electrons is called covalent bonding, while giving up electrons is called ionic bonding. A third type of bond, which we'll get to shortly, also is important for our purposes: the hydrogen bond.

FIGURE 2.7 Electron Configurations in Some Representative Elements The concentric rings represent energy levels or "shells" of the elements, and the dots on the rings represent electrons. Hydrogen has but a single shell and a single electron within it, while carbon has two shells with a total of six electrons in them. Helium, neon, and argon have filled outer shells and are thus unreactive. Hydrogen, carbon, and sodium do not have filled outer shells and are thus reactive—they readily combine with other elements.

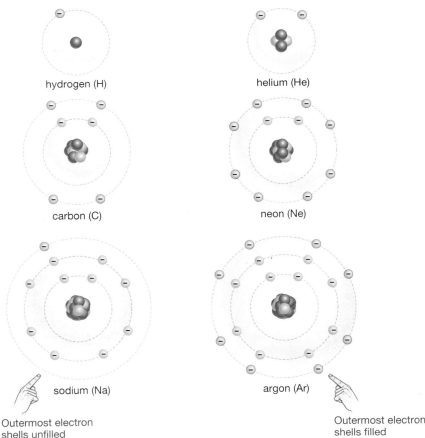

Unstable, very reactive atoms

hydrogen (H)

carbon (C)

sodium (Na)

Outermost electron shells unfilled

Stable, unreactive atoms

helium (He)

neon (Ne)

argon (Ar)

Outermost electron shells filled

Energy Always Seeks Its Lowest State

Atoms that bond with one another do so because they are in a more *stable* state after the bonding than before it. A frequently used phrase is helpful in understanding this kind of stability: Energy always seeks its lowest state. Imagine a boulder perched precariously on a hill. Thanks to gravity, a mere shove might send it rolling toward its lower energy state—at the bottom of the hill. It would not then roll *up* the hill, either spontaneously or with a light shove, because it is now existing in a lower energy state than it did before—one that is clearly more stable than its former precarious perch. When we turn to electrons, the energy is not gravitational, but electrical. Atoms bond with one another to the extent that doing so moves them to a lower, more stable energy state. The critical thing for our purposes is that atoms move to this more stable state by filling what is known as their outer shells.

Seeking a Full Outer Shell: Covalent Bonding

What are these "outer shells"? As it happens, electrons reside in certain well-defined "energy levels" outside the nuclei of atoms. The number of these energy levels varies depending on the element in question. Here we need only note the practical effect of these levels on bonding: *Two* electrons are required to fill the first energy level (or shell) of any given atom, but *eight* usually are required to fill all the levels thereafter. If you look at the electron configurations pictured in **FIGURE 2.7**, you can see that two elements—

Essay 2.1 Getting to Know Chemistry Symbols

In depicting elements and molecules, chemistry employs its own special set of symbols. Here's a primer on a few of them. Our starting point is that each chemical element has its own letter-based symbol, so that hydrogen becomes H, carbon C, and platinum Pt. When we begin to combine these elements into molecules, it is necessary to specify how *many* atoms of each element are part of the molecule. If we have two atoms of oxygen together—which is the way oxygen is usually packaged in our atmosphere—we have the molecule O_2. Three molecules of O_2 is written as $3 O_2$. This kind of notation is known as a *molecular formula*.

To see a molecular formula is to learn a lot about what atoms are in a molecule, but nothing about the way the atoms are *arranged* in relation to one another. To convey this ordering information, chemists and biologists use what are known as *structural formulas*—two-dimensional representations of a given molecule. Methane (CH_4) is a very simple molecule composed, as the molecular formula shows, of one atom of carbon and four of hydrogen. In a structural formula, these constituent parts are conceptualized like this:

$$H - \underset{\underset{\displaystyle H}{|}}{\overset{\overset{\displaystyle H}{|}}{C}} - H$$

methane

Note that there is a single line between each hydrogen atom and the central carbon atom. This has been done because the bond between any of the hydrogen atoms and carbon is a single bond: Each line represents *one* pair of electrons being shared. But, there can also be double bonds and triple bonds. When carbon dioxide forms, there are two oxygen atoms. Each shares two pairs of electrons with a lone carbon atom. Here's how we would notate the double bond in a carbon dioxide (CO_2) molecule:

$$O = C = O$$

carbon dioxide

Though structural (or "skeletal") formulas can tell us a good deal about the ordering of atoms in a molecule, they tell us very little about the three-dimensional arrangements of atoms. For this, we rely on two other kinds of representations, the *ball-and-stick model* and the *space-filling model*. An ammonia molecule (NH_3) is pictured next in both forms, with the molecular and structural formulas added to show the progression. The dots in the structural formula represent an unshared pair of electrons in the nitrogen atom.

NH_3

$$H - \overset{\displaystyle \cdot\cdot}{\underset{\underset{\displaystyle H}{|}}{N}} - H$$

molecular formula	structural formula	ball-and-stick model	space-filling model

hydrogen and helium—have so few electrons in orbit around them that they have nothing *but* a first energy level, while the other elements pictured have two or three energy levels. This means that hydrogen and helium require only two electrons in orbit around their nuclei to have filled outer shells, but that the other elements pictured require eight electrons to have this kind of complete outer electron complement.

How Chemical Bonding Works in One Instance: Water

To see how chemical bonding works in connection with this concept of filled outer shells, take a look at the bonding that occurs with the constituent parts of one of the most common (and important) substances on Earth, water. In so doing, you'll see one of the kinds of bonding mentioned earlier—covalent bonding.

The familiar chemical symbol for water is H_2O. This means that two atoms of hydrogen (H_2) have combined with one atom of oxygen (O) to form water. (See Essay 2.1, "Getting to Know Chemistry Symbols," above, for an explanation of the notation used in chemistry.) Recall that hydrogen has only one electron running around in its single energy level. Also recall, however, that this first level is not completed until it has *two* electrons in it. Next, consider the oxygen atom, which has

A BROADER VIEW

Inert

What does it mean when we say that something is *inert*? The word actually comes from chemistry—specifically from a group of elements, all gases, whose filled outer shells render them incapable of bonding in nature with any other elements. Among these gases are helium and neon. *Inert* is sometimes used in a metaphorical sense, however, to mean something that is incapable of interaction.

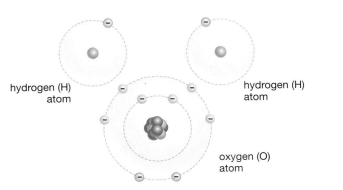

(a) Two hydrogen atoms and one oxygen atom

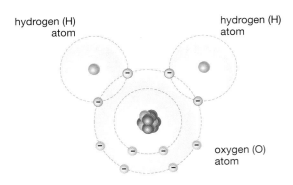

(b) One water molecule

FIGURE 2.8 **Covalent Bonding** A covalent bond is formed when two atoms share one or more pairs of electrons.

(a) The starting state in this reaction is two hydrogen atoms and one oxygen atom. Note that none of these atoms has outer-shell stability—each hydrogen atom would need one more electron to achieve stability, while the oxygen atom would need two more.

(b) In bonding, all the atoms achieve this stability; in this case, two pairs of electrons are shared—one pair between one hydrogen atom and the oxygen atom and the other pair between the second hydrogen and the oxygen.

eight electrons. Looking at FIGURE 2.8, you can see what this means: Two electrons fill oxygen's first energy level, which leaves six left over for its second. But remember that the second shells of most atoms are not completed until they hold eight electrons. Thus oxygen, like hydrogen, would welcome a partner. Only it needs two electrons to fill its outer shell—something that two *atoms* of hydrogen could provide. The outcome is a bonding of two hydrogen atoms with one atom of oxygen. And each of the hydrogen atoms is linked to the oxygen in a **covalent bond**—a chemical bond in which atoms share pairs of electrons. The oxygen atom and first hydrogen atom donate one electron each for the first pair, and these electrons can now be found orbiting the nuclei of both atoms. Then the oxygen and second hydrogen atom each donate one electron for the second pair. The result? Three atoms covalently bonded together and all of them "satisfied" to be in that condition. (Occasionally in nature, covalent bonding will take place in a way that leaves one atom with an unpaired electron, a potentially harmful phenomenon you can read about in Essay 2.2, "Free Radicals," on page 23.)

What Is a Molecule?   When two or more atoms bond covalently, the result is a **molecule**: a compound of a defined number of atoms in a defined spatial relationship. Here, one atom of oxygen has combined with two atoms of hydrogen to create *one water molecule*. (What we commonly think of as water, then, is an enormous, linked collection of these individual water molecules.) A molecule need not be made of two different elements, however. Two hydrogen atoms can covalently bond to form one hydrogen molecule. On the other side of the coin, a molecule could contain many different elements bonded together. Consider sucrose, or regular table sugar, which is $C_{12}H_{22}O_{11}$ (12 carbon atoms bonded to 22 hydrogen atoms and 11 oxygen atoms).

Reactive and Unreactive Elements

The elements considered so far all welcome bonding partners, because all of them have incomplete outer shells. This is not true of all elements, however. There is, for example, the helium atom, which has two electrons. It thus *comes equipped*, we might say, with a filled outer shell. As such, it is extremely stable—it is unreactive with other elements. At the opposite end of the spectrum are elements that are extremely reactive. Look again at the representation of the sodium molecule in

A BROADER VIEW

Counting Up Molecules

It's difficult to conceive of how little matter there is in a single water molecule—one H_2O. To give you some idea, an 8-ounce glass of water has been calculated to contain 7.5×10^{24} water molecules. The place to look in that number is the 10^{24}. If there were 1 million individual water molecules in the glass, it would be written as 10^6—a one with six zeros behind it. The number 10^{24} represents a one with 24 zeros behind it.

Essay 2.2 **Free Radicals**

Why do we get old? What happens to our skin, our hearts, our energy? One factor at work is a set of molecules whose name makes them sound like a group of '60s activists now let out of jail. These are the *free radicals*. As you'll see, these compounds represent a damaging exception to the rules of chemical bonding we've been going over.

Every living thing needs a supply of energy to remain alive, and living things such as ourselves get this energy from food. But our cells need to convert the energy contained in food into a form our bodies can use—a substance called ATP. Not surprisingly, then, we have tiny power plants within our cells that do this energy converting. These structures, called mitochondria, take in food and turn out ATP, just as a power plant takes in coal and turns out electricity. And just as coal cannot be burned in a power plant except in the presence of oxygen, so food cannot be burned in our mitochondria without oxygen.

Oxygen may sustain us, but it also causes trouble.

And therein lies the problem. Oxygen may sustain us, but it also causes trouble. If you have ever seen a rusted hinge, you have seen a piece of metal that has been *oxidized*—oxygen atoms have pulled electrons right off the metal. In the same way, inside our cells, oxygen acts like a molecular magnet, pulling electrons toward it and sharing these electrons with other atoms in the process of chemical bonding. But remember that bonding of this sort means sharing *pairs* of electrons. So what happens if one atom in a bonded pair ends up with an *un*paired electron—a solo flyer, we might say, that is not paired with any other electron? The resulting molecule is a free radical. Oxygen can come together with nitrogen, for example, to form nitric oxide (NO). Recall that oxygen has six outer electrons (and thus needs two for stability), while nitrogen has five outer electrons (and thus needs three). When oxygen and nitrogen hook up, they can share only two electrons with one another before oxygen's outer shell is filled. This, however, leaves nitrogen with an unpaired electron—and a free radical has been created. Though this molecule will be short lived, it begets more of itself: Seeking partners, one free radical leads to more, which in turn lead to more.

Where's the harm in this? Well, take a look at the two mice in **FIGURE E2.2.1**. The mouse in the upper photo is just a normal mouse, while the mouse in the lower photo was genetically engineered to lack a protein that repairs damage to the DNA in its mitochondria. You would expect that mice with damaged DNA would suffer some kind of problem, but the fascinating thing is that, when they reach the equivalent of adolescence, these modified mice begin to show signs of *aging*: hair loss, decreased

energy, curvature of the spine, and osteoporosis. This is evidence that damage to mitochondrial DNA helps bring about the constellation of afflictions we call aging. And, in a natural state, what can damage mitochondrial DNA? Free radicals. They appear to cause breaks in the DNA and to take it out of its normal shape.

Free radicals are a natural product of metabolism in human beings; they are the price we pay for using oxygen to stay alive. However, they can be created in us in greater numbers in accordance with our behavior. Some of the usual suspects seem to be involved here—cigarette smoking, alcohol consumption, and sunlight exposure. Against this production of free radicals, however, nature has also provided its own set of free-radical scavengers, among them beta-carotene and vitamins C and E. (These are so-called antioxidants. Now you can see how they got their name.) Claims are made all the time about the value of buying these substances in pill form. For now, however, experts say that the best bet is to get them through a diet rich in citrus fruits, whole grains, and vegetables of the green leafy, orange, and yellow variety.

FIGURE E2.2.1 **Free Radicals and Aging** The mice in these photos are both about 40–45 weeks old. Yet the mouse in the lower photo shows clear signs of aging—its spine is curved, its weight has dropped, and its hair density has been reduced. All this came about because of damage to the DNA in its mitochondria, the tiny structures that are the sites of energy transformation in cells. Such damage is caused largely by free radicals.

Polar water molecule

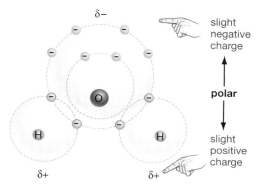

FIGURE 2.9 **Polar Covalent Bonding** In the water molecule, the oxygen atom exerts a greater attraction on the shared electrons than do the hydrogen atoms. Thus the electrons are shifted toward the oxygen atom, giving the oxygen atom a partial negative charge (because electrons are negatively charged) and the hydrogen atoms a partial positive charge. (*Partial* here is indicated by the Greek delta symbol, δ) The molecule as a whole is polar, meaning it has a difference in charge at one end as opposed to the other.

A BROADER VIEW

Negative Ion Generators

What is a negative ion generator? It usually is a device that generates negative oxygen ions by attaching an extra electron to oxygen molecules. One of the uses for these devices is to help clean indoor air. The ions the machines generate impart an electrical charge to pollutant air particles. This makes the particles attach to surfaces, rather than remaining airborne.

Figure 2.7. It has 11 electrons, two in the first shell and eight in the second, which leaves but one electron in the third shell—a very unstable state. Between the extremes of sodium and helium are elements with a range of outer (or *valence*) electrons. Thus, there is a spectrum of stability in the chemical elements, based on the number of outer-shell electrons each element has—from 1 to 8, with 1 being the most reactive and 8 being the least reactive.

Polar and Nonpolar Bonding

Not all covalent bonds are created alike. When two hydrogen atoms bond to each other, the result is a hydrogen molecule (H_2). Now, in the hydrogen molecule, the electrons are shared *equally*. That is, the two electrons the hydrogen atoms are sharing are equally attracted to each hydrogen atom. This is not the case, however, with the water molecule.

Look at the representation of the water molecule in **FIGURE 2.9**. As it turns out, the oxygen atom has a greater power to attract electrons to itself than do the hydrogen atoms. The term for measuring this kind of pull is **electronegativity**. Because the oxygen atom has more electronegativity than do the hydrogen atoms, it tends to pull the shared electrons away from the hydrogen and toward itself. When this happens, the molecule takes on a **polarity** or a difference in electrical charge at one end as opposed to the other. Because electrons are negatively charged, and because they can be found closer to the oxygen nucleus, the oxygen region of the molecule takes on a partial negative charge, while the hydrogen regions take on a partial positive charge. We still have a covalent bond, but it is a specific type: a **polar covalent bond**.

To grasp the importance of this, consider the water molecule, with its positive and negative regions. What's going to happen when it comes into contact with *other* polar molecules? The oppositely charged regions of the molecules will attract and the similarly charged regions will repel. It's like having a bar magnet and trying to bring its positive end into contact with the positive end of another magnet: Left on its own, the second magnet just flips around, so that positive is now linked to negative. In the same way, molecules flip around in relation to their polarity. In sum, some molecules are polar while others are nonpolar, and this difference has significant consequences for chemical bonding.

Ionic Bonding: When Electrons Are Lost or Gained

So we've gone from nonpolar covalent bonding, where electrons are shared equally, to polar covalent bonding, where electrons are pulled to one region of the resulting molecule. What if we carried this just one step further and had instances in which the electronegativity differences between two atoms were so extreme that electrons were pulled *off* of one atom altogether, only to latch onto the atom that was attracting them? This is what happens in our second type of bonding, **ionic bonding**. The classic illustration of this type of bonding involves the sodium atom we looked at earlier and a chlorine atom. Recall that sodium has 11 electrons, meaning that there's a lone electron flying around in its third electron shell. Chlorine, meanwhile, has 17 electrons, so it has 7 electrons in the third shell. Remember that 8 is the magic number for outer-shell stability. Sodium could get to this number by *losing* one electron, while chlorine could get to it by *gaining* one electron. That's just how this encounter goes: Sodium does in fact lose its one electron, chlorine gains it, and both parties become stable in the process (**see** FIGURE 2.10).

What Is an Ion? But this story has a postscript. Having lost an electron (with its negative charge), sodium (Na) then takes on an overall *positive* charge. Having gained an electron, chlorine (Cl) takes on a negative charge. Each is then said to be an **ion**—a charged atom—or, to put it another way, an atom whose number of

FIGURE 2.10 Ionic Bonding

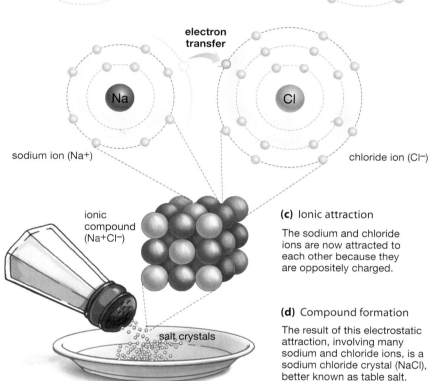

sodium atom (Na)

chlorine atom (Cl)

(a) Initial instability

Sodium has but a single electron in its outer shell, while chlorine has seven, meaning it lacks only a single electron to have a completed outer shell.

electron transfer

(b) Electron transfer

When these two atoms come together, sodium loses its third-shell electron to chlorine, in the process becoming a sodium ion with a net positive charge (because it now has more protons than electrons.) Having gained an electron, the chlorine atom becomes a chloride ion, with a net negative charge (because it has more electrons than protons).

sodium ion (Na+)

chloride ion (Cl⁻)

ionic compound (Na+Cl⁻)

(c) Ionic attraction

The sodium and chloride ions are now attracted to each other because they are oppositely charged.

salt crystals

(d) Compound formation

The result of this electrostatic attraction, involving many sodium and chloride ions, is a sodium chloride crystal (NaCl), better known as table salt.

electrons differs from its number of protons. We denote the ionized forms of these atoms like this: Na^+, Cl^-. If an atom were to gain or lose more electrons than this, we would put a number in front of the charge sign. For example, to show that the magnesium atom has lost two electrons and thus becomes a positively charged magnesium ion, we would write Mg^{2+}.

Note that we now have two ions, Na^+ and Cl^-, with differing charges in proximity to one another. They are thus attracted to one another and have an ionic bond between them. This hardly ever happens with just *two* atoms, of course. Many billions of atoms are bonded together in this way, up, down, and sideways from each other. This whole collection is, likewise, called an ion; or, if two or more elements react together this way, an ionic compound. The particular ionic compound just described actually is very familiar. Sodium and chlorine combine to create sodium chloride, which is better known as table salt.

To take a step back for a second, note that bonding runs a gamut from the nonpolar covalent bonding (where electrons are shared equally) to the slightly charged polar covalent bonding (where they are shared somewhat unequally) to the charged ionic bonding (where electrons are gained or lost altogether).

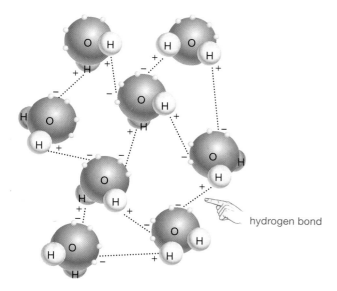

FIGURE 2.11 **Hydrogen Bonding** The hydrogen bond, in this case between water molecules, is indicated by the dotted lines. It exists because of the attraction between hydrogen atoms, with their partial positive charge, and the unshared electrons of the oxygen atom, with their partial negative charge.

—————————

So Far... Answers:

1. *electrons are shared between atoms; atoms gain or lose electrons*

2. *Carbon is in a relatively unstable state. Since it has four electrons in its second energy level, while eight are needed for outer-shell stability, it would need to share four electrons with a bonding partner to become fully stable.*

3. *electrons; charged atom*

So Far...

1. A covalent bond is a bond in which _____, and an ionic bond is a bond in which _____.

2. The element carbon has six electrons. Is it stable in this state? If not, how many more electrons would it need to share with a bonding partner to make it stable?

3. An ion is created when an atom gains or loses one or more _____. An ion is different from an atom in that an ion is a _____.

A Third Form of Bonding: Hydrogen Bonding

Having looked at covalent and ionic bonding, we need to look at one more kind of atomic linkage, hydrogen bonding. Recall that in any water molecule, the stronger electronegativity of the oxygen atom pulls the electrons shared with the hydrogen atoms toward the oxygen nucleus, making the oxygen region of the molecule partially negative and the hydrogen region partially positive. So what happens when you place several water molecules together? The positive hydrogen atom of one molecule is weakly attracted to the negative, *unshared* electrons of its oxygen neighbor. This creates the **hydrogen bond**: a chemical bond that links an already covalently bonded hydrogen atom with an electronegative atom (in this case with oxygen; **see** FIGURE 2.11). Hydrogen bonding is a linkage that, for our purposes, nearly always pairs hydrogen with either oxygen or nitrogen. These relatively weak bonds are important in linking the atoms of a single molecule to one another, but they are just as important in creating bonds *between* molecules, as in the example. The hydrogen bond, indicated by a dotted line, exists in many of the molecules of life—in DNA, proteins, and elsewhere.

2.4 Molecules Have a Three-Dimensional Shape

It is useful to depict molecules as two-dimensional chains and rings and such, but in real life a molecule is as three-dimensional as a sculpture. A fair number of molecular shapes are possible, even in simpler molecules. Atoms may be lined up in a row, or in triangles or pyramid shapes. As an example, look at another representation of the water and methane molecules (**see** FIGURE 2.12). You can see that in

FIGURE 2.12 **Three-Dimensional Representations of Molecules**

(a) In the case of water, there is an angle of 104.5° between hydrogen atoms.

(b) Methane is a molecule with an angle of 109.5° between hydrogen atoms.

(a) Water (H₂O)

(b) Methane (CH₄)

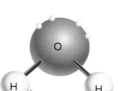

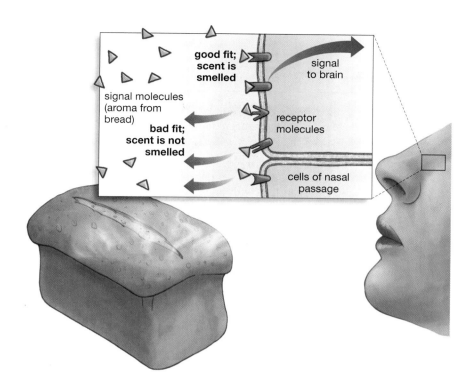

FIGURE 2.13 **The Importance of Shape** Fragrant molecules wafting from bread (the triangles) bind with specific receptors on the surface of the cell, thus acting as signaling molecules that set a cellular process in motion. For this binding to take place, the fragrant molecules and nasal receptor molecules must fit together; this fit is governed by the shape of each molecule.

water there is a very definite spatial configuration: Its hydrogen atoms are splayed out from its oxygen atom at an angle of 104.5°.

Why does molecular shape matter? It is critical in enabling biological molecules to carry out their functions. This is so because molecular shape determines the capacity of molecules to latch onto or "bind" with one another. When, for example, you smell the aroma of fresh-baked bread, fragrant molecules wafting off the bread bind with receptor molecules in your nasal passages, thus sending a message to the brain about the presence of bread. It is the precise shape of the fragrant molecules and nasal receptor molecules that allows them to bind with one another. Look at FIGURE 2.13 to see how this works.

On to Some Detail Regarding Water

One of the molecules we've looked at so far, water, has a tremendous importance in biology. Living things are largely made up of it, and no organism can function over the long run without a supply of it. It is not difficult to see, however, why water should be so prominent in this way. Life got going in water, and living things existed solely in water for billions of years before any of them came onto land. So important to life is this simple, common molecule that we will start the next chapter with a close look at it.

So Far...

1. Hydrogen bonding occurs when a bond is created between an already bonded _____ and an _____ atom.

2. Molecular shape is important in biology because it determines the ability that molecules have to _____.

So Far... Answers:

1. hydrogen atom, electronegative

2. bind with one another

Chapter Review

Summary

2.1 Chemistry Is Critical to Life

- **The role of chemistry**—Chemistry underlies almost all of biology. Most of the processes that take place in living things are chemical reactions. (page 15)

2.2 The Nature of Matter: The Atom

- **The atom**—The fundamental unit of matter is the atom. The three most important constituent parts of an atom are protons, neutrons, and electrons. Protons and neutrons exist in the atom's nucleus, while electrons move around the nucleus. Protons are positively charged, electrons are negatively charged, but neutrons carry no charge. (page 16)

- **Elements**—An element is any substance that cannot be reduced to any simpler set of constituent substances through chemical means. Each element is defined by the number of protons in its nucleus. (page 18)

- **Isotopes**—The number of neutrons in an atom can vary independently of the number of protons. Thus a single element can exist in various forms, called isotopes, depending on the number of neutrons it possesses. (page 18)

2.3 Matter Is Transformed through Chemical Bonding

- **Types of bonds**—Atoms can link to one another in the process of chemical bonding. Among the forms this bonding can take are covalent bonding, in which atoms share one or more electrons; ionic bonding, in which atoms lose and accept electrons from each other; and hydrogen bonding, in which differently charged parts of molecules are attracted to one another. (page 20)

- **The nature of bonding**—Chemical bonding comes about as atoms "seek" their lowest energy state. An atom achieves this state when it has a filled outer electron shell. Hydrogen and helium require two electrons in orbit around their nuclei to have filled outer shells, while most other elements require eight electrons to have filled outer shells. (page 20)

- **Molecules**—A molecule is a compound of a defined number of atoms in a defined spatial relationship. For example, two hydrogen atoms can link with one oxygen atom to form one water molecule. (page 22)

- **Electronegativity**—Atoms of different elements differ in their power to attract electrons. The term for measuring this power is electronegativity. A polar covalent bond exists when shared electrons are not being shared equally among atoms in a molecule, due to electronegativity differences. Through electronegativity, molecules can take on a polarity, meaning a difference in electrical charge at one end compared to the other. (page 24)

- **Ions**—Two atoms will undergo a process of ionization when the electronegativity differences between them are great enough that one atom loses one or more electrons to the other. This process creates ions, meaning electrically charged atoms. The charge differences that result from ionization can produce an attraction between ions—an ionic bond. (page 24)

- **Hydrogen bonds**—Hydrogen bonding links a covalently bonded hydrogen atom with an electronegative atom. In water, a hydrogen atom of one water molecule will form a hydrogen bond with an unshared oxygen electron of a neighboring water molecule. (page 26)

Web Tutorial 2.1 Geometry, Chemistry, and Biology

2.4 Molecules Have a Three-Dimensional Shape

- **Molecular shape**—The three-dimensional shapes of molecules are important in biology because they determine the capacity molecules have to bind with one another. (page 26)

Key Terms

chemical bonding   p. 20

covalent bond   p. 22

electron   p. 17

electronegativity   p. 24

element   p. 18

hydrogen bond   p. 26

ion   p. 24

ionic bonding   p. 24

isotope   p. 19

molecule   p. 22

neutron   p. 18

nucleus   p. 17

polar covalent bond   p. 24

polarity   p. 24

proton   p. 17

Testing Your Understanding

In Your Own Words   *(answers in the back of the book)*

1. As with most elements, carbon comes in several forms, one of which is carbon-14. What are these forms called, and how does one differ from the other?

2. Draw a line and label one end "complete + or − charge" and the other end "no charge" to indicate the charges on the molecules or ions after bonding has occurred. Along the line, indicate where polar covalent bonds, nonpolar covalent bonds, and ionic bonds should be placed.

3. Why are atoms unlikely to react when they have their outer shell filled with electrons?

Thinking about What You've Learned

1. In the Middle Ages, alchemists labored to turn common materials such as iron into precious metals such as gold. If you could journey back in time, how could you convince an alchemist that iron cannot be changed into gold?

2. Why does a balloon filled with helium float? Hydrogen can make balloons float, but it is not used for this purpose today, because it is flammable. Based on chemical principles reviewed in the chapter, can you see why helium is not flammable? (*Hint:* Think about what you are adding to a fire when you blow on it.)

Biological Molecules

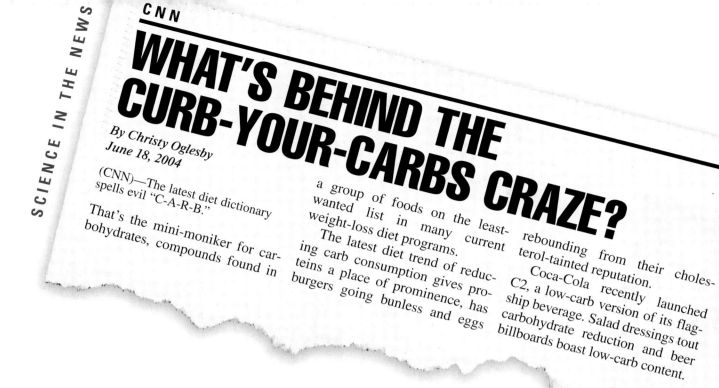

CNN

WHAT'S BEHIND THE CURB-YOUR-CARBS CRAZE?

By Christy Oglesby
June 18, 2004

(CNN)—The latest diet dictionary spells evil "C-A-R-B."

That's the mini-moniker for carbohydrates, compounds found in

a group of foods on the least-wanted list in many current weight-loss diet programs.

The latest diet trend of reducing carb consumption gives proteins a place of prominence, has burgers going bunless and eggs

rebounding from their cholesterol-tainted reputation.

Coca-Cola recently launched C2, a low-carb version of its flagship beverage. Salad dressings tout carbohydrate reduction and beer billboards boast low-carb content.

3.1 Biological Molecules

About the time CNN was running its story on the national impact of the low-carb craze, I was feeling the effects of the craze on a local level. My bakery put up a little handwritten sign saying that, henceforth, it would be selling its bread only two days a week. So many customers had abandoned bread, the bakery said, that it didn't make sense to sell it five days a week anymore. Dietary fads usually have a short shelf life, and the low-carb fad appears to be no exception. Still, it has been a remarkable phenomenon. Millions of people changed their dietary habits, and products ranging from beef to beer were affected in the marketplace. All this was sparked by the idea that people can lose weight more easily by lowering the amount of carbohydrates in their diet while raising the amount of protein and fat.

Given this, here's an interesting question for people of high- or low-carb persuasion to ponder: What's a carb? What is this thing called a carbohydrate? Everyone can give some *examples* of high-carbohydrate foods, but what is it that differentiates a carbohydrate from, say, a protein? Once you start thinking about this, it's easy to see

that carbohydrates and proteins are basic types of substances—or "biological molecules"—that make up the food we eat. But the importance of biological molecules doesn't stop at food. The living world is made up almost entirely of a few types of biological molecules. The leaves on a tree are largely a type of carbohydrate. Your hair is entirely made up of protein. The estrogen that is so important to reproduction is a type of molecule with a familiar name: a steroid. The goal in this chapter is to introduce you to these building blocks of the living world and to a concept that is related to them called pH. Once you've finished, you'll know how a carbohydrate differs from a protein, and what a trans fat is, and what the DNA in your cells is made of. We'll start by reviewing a molecule that is small, but very important.

3.2 Water and Life

Everyone knows that our plants, our pets, and for that matter our own bodies need a steady supply of water. Just broaden that idea out to the living world as a whole and you get the picture regarding water and life. All living things must have a supply of water, usually just to continue living, but certainly to carry out the full range of life's processes. Some bacteria can live for long periods of time without water, essentially by going into a kind of suspended animation. But to grow, to respond to the environment, to reproduce—in other words, to fully function—they and all other living things must have water. This makes sense because life got started in water (in the ancient oceans) and it appears to have existed nowhere *except* water for at least 2.5 billion years after it got started. When large organisms such as plants and animals first made the transition to land, they did so by carrying a watery environment with them—inside themselves. This is as true for us today as it was for the first land dwellers. The human body is about 66 percent water by weight, such that if we have, say, a 128-pound person, about 85 pounds of him or her will be water.

Water Is a Major Player in Many of Life's Processes

Why is water so important to life? Recall last chapter's concept of life being, at one level, a series of chemical reactions. It makes sense that these reactions would best take place in some liquid medium, and that medium turns out to be water. However, water is not just a passive medium in which reactions take place. It *facilitates* many of these reactions thanks to its chemical structure. To understand this, it's important to know something about three terms that start with an *s*: solution, solute, and solvent.

Pour some salt into a container of water, stir the water, and what happens? The salt quickly disappears. It hasn't actually gone anywhere, of course; it has simply mixed with the water. Now, if it has mixed uniformly, so that there are no lumps of salt here or there, you have created a **solution**—a mixture of two or more kinds of molecules, atoms, or ions that is homogenous throughout. The salt is what's being dissolved here, so it is the **solute**. The water is doing the dissolving, so it is the **solvent**. And this gets us to the point about water: It's a terrific solvent, which is to say it has a great ability to dissolve other substances. In fact, over the range of substances, nothing can match it as a solvent. It can dissolve more compounds in greater amounts than can any other liquid.

If you look at FIGURE 3.1, you can get a detailed view of water's solvent power in connection with the salt we've been talking

FIGURE 3.1 **Water's Power as a Solvent**

(a) Attraction   **(b)** Separation   **(c)** Dispersion

water (solvent)
H
O
H

sodium chloride (solute)
Na⁺   Cl⁻

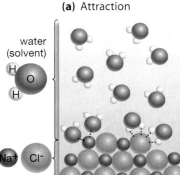

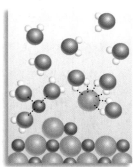

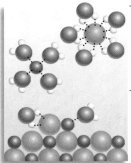

Sodium and chloride ions dissolved in water

Sodium chloride's positively charged sodium ions (Na^+) are attracted to water's negatively charged oxygen atoms, while its negatively charged chloride ions (Cl^-) are attracted to water's positively charged hydrogen atoms.

Pulled from the crystal, and separated from each other by this attraction, sodium and chloride ions become surrounded by water molecules.

This process of separating sodium and chloride ions repeats until both ions are evenly dispersed, making this an aqueous solution.

about. Remember, from last chapter, that water is a "polar" molecule, which is to say a molecule that has differing electric charges at one end as opposed to the other. Attracted by the polar nature of the water molecule, the sodium and chloride ions that make up a salt crystal separate from the crystal—and from each other. Each ion is then surrounded by several water molecules (Figure 3.1b). These units keep the sodium and chloride ions from getting back together. In other words, they keep the ions evenly dispersed throughout the water, which is what makes this a solution (Figure 3.1c). Water works as a solvent here because the ionic compound sodium chloride carries an electrical charge. What *generally* makes water work as a solvent, however, is its ability to form, with other molecules, the hydrogen bonds we talked about at the end of Chapter 2.

Water's Structure Gives It Other Unusual Properties

But solvency power is not the only notable quality of water. Consider the fact that ice floats on water. This is so because the solid form of H_2O is less dense than the liquid form—a strange reversal of nature's normal pattern. Things work this way because, as water cools, individual water molecules slow their motion, which means that the hydrogen bonds between these molecules are not being broken as often. And the more hydrogen bonds that are in place, the *farther apart* water molecules are spaced from each other. Thus, ice is less dense than water. This may seem like some minor, quirky quality, but it actually has the effect of making possible life as we know it. Ice on the surface of water acts to insulate the water beneath it from the freezing surface temperatures and wind above, creating a warmer environment for organisms such as fish (**see** FIGURE 3.2). If ice *sank*, on the other hand, the entire body of water would freeze solid at colder latitudes, creating an environment in which few living things could survive for long.

Water Has a Great Capacity to Absorb and Store Heat
Water serves as an insulator not only when it is frozen, but also when it is liquid or gas. This has to do with a quality called **specific heat**: the amount of energy required to raise the temperature of a substance by 1 degree Celsius. As it turns out, water has a high specific heat. Put a gram container of drinking alcohol (ethyl alcohol) side by side with one of water, heat them both, and it will take almost twice as much energy to raise the temperature of the water 1 degree as it will the alcohol. Having absorbed this much heat, however, water then has the capacity to *release* it when the environment around it is colder than the water itself. The result? Water acts as a great heat buffer for Earth. The oceans absorb tremendous amounts of radiant energy from the sun, only to release this heat when the temperature of the air above the ocean gets colder. Without this buffering, temperature on Earth would be less stable. People who have spent a day and a night in a desert can attest to this effect. The searing heat of the desert day radiates off the desert floor; but at night, with little water vapor in the air to capture this heat, the desert cools off dramatically. In the same way, our *internal* temperature can remain

A BROADER VIEW

Surviving without Water

How long can people live without water? The usual answer to this is "a few days," but as biologist Randall Packer noted in *Scientific American*, it depends on who the person is and what environment he or she is in. An infant in a hot car would not last long, but a healthy adult in a moderate environment probably could last a week or more without drinking water. This outer limit stands in contrast to how long a person can live when water is available but food is not. In the 1980s, a group of Irish Republican Army (IRA) prisoners who went on a hunger strike in Northern Ireland died, on average, about 60 days after they stopped eating.

FIGURE 3.2 **Life under the Ice** Water is unusual in that its solid form (ice) is less dense than its liquid form. This means that ice floats, and this in turn means that life can flourish in cold-weather aquatic environments. Pictured are some shrimp-like krill, feeding on algae that grow on the underside of ice in the waters off Antarctica.

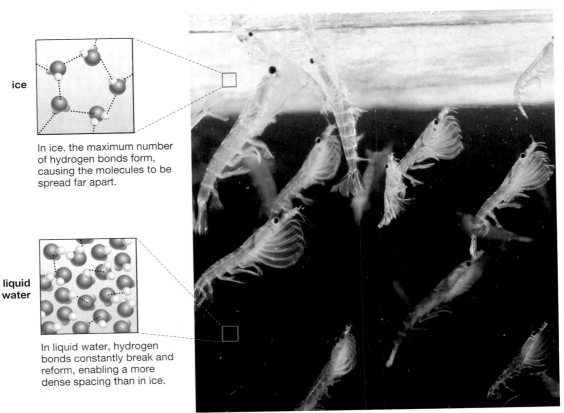

ice

In ice, the maximum number of hydrogen bonds form, causing the molecules to be spread far apart.

liquid water

In liquid water, hydrogen bonds constantly break and reform, enabling a more dense spacing than in ice.

Modern Myths

Eight Glasses of Water a Day

For years, a mantra of nutrition has been that, to achieve optimal health, a person needs to drink at least eight glasses of water a day. But there is no reason to believe this is true, a U.S. Institute of Medicine panel concluded in 2004. A person's best guide for how much water to consume is thirst, the panel said. If people simply drink when thirsty, they are likely to be well "hydrated." So closely does the body regulate its water level that drinking more water than necessary will simply result in urinating more frequently. By the same token, drinking less water than is necessary will cause the body to preserve such water as it has while bringing on thirst. Of course there are times when it is prudent to consume water in advance of being thirsty—for example, if you know you're going to be exercising or simply walking in a hot climate.

The fact that eight glasses of water a day aren't necessary raises the question of where this long-standing advice came from in the first place. So far, the answer is: Who knows? A physiologist who looked into this question in 2002 could find no studies at all that supported the eight-glasses idea. Indeed, such surveys that have been done—on thousands of adults over the years—have strongly suggested that the body does not need such large, fixed amounts of water.

much more stable because the water that makes up so much of us is first able to absorb and then to release great amounts of heat. The sweat that we throw off in exercise has considerable cooling power because each drop of perspiration carries with it a great deal of heat. (Exercise aside, a misconception exists about how much water we need to consume during the course of a normal day. For more on this, see "Modern Myths" above.)

What Water Cannot Do With its great complexity, life requires molecules that *can't* be dissolved by water. Such is the case with molecules that do not carry a charge at one end as opposed to the other—nonpolar molecules, as they're called. Compounds made of hydrogen and carbon (**hydrocarbons**) are nonpolar, and everyday examples of these compounds are all around us in the form of petroleum products. You can see a vivid demonstration of the water-insolubility of hydrocarbons in the oil spill pictured in FIGURE 3.3. Oil doesn't dissolve in water, because the oil carries almost no electrical charge that water can bond with; thus water has no way to separate one oil molecule from another.

Two Important Terms: Hydrophobic and Hydrophilic

The ability of molecules to form bonds with water has a couple of important names attached to it. Compounds that will interact with water—such as the sodium chloride considered earlier—are known as **hydrophilic** ("water-loving"), while compounds that do not interact with water, such as oil, are called **hydrophobic** ("water-fearing"). Both terms are misleading, in that no substance has any emotional relationship with water. *Hydrophobic* is particularly off the mark, because water does not repel hydrophobic molecules; instead, the strong bonding that water molecules do with each other causes them to form circles around concentrations of hydrophobic molecules, as if they had lassoed them.

The importance of hydrophobic molecules can be illustrated in part by the common milk carton. Why is the milk carton important? Because it can keep milk separate from everything else. We living organisms need some kind of "carton" that can separate the world outside of us from ourselves. Likewise, organisms have great use *within* themselves for compartments that can be sealed off to one degree or another. If water broke down every molecule of life it came in contact with, then it would break down all these divisions of living systems. Note, however, that molecules do not have to be completely hydrophilic or hydrophobic. Indeed, as you'll see, a number of important molecules have both hydrophilic and hydrophobic regions.

FIGURE 3.3 **Oil and Water Do Not Mix** When there is an oil spill in the ocean, the oil stays concentrated even as it spreads out, because oil and water do not form chemical bonds with each other. Here trawlers are cleaning up after an oil spill in Great Britain.

3.3 Acids and Bases Are Important to Life

When considering the question of water and solutions, an important concept is that of acids, bases, and the pH scale used to measure their levels.

We've all had experience with acids and bases, whether we've called them by these names or not. Acidic substances tend to be a little more familiar: lemon juice, vinegar, tomatoes. Substances that are strongly acidic have a well-deserved reputation for being dangerous: The word *acid* is often used to mean something that can sear human flesh. It might seem to follow that bases are benign, but ammonia is a base, as are many oven cleaners. The safe zone for living tissue in general lies with substances that are neither strongly acidic nor strongly basic. Science has developed a way of measuring the degree to which something is acidic or basic—the pH scale. So widespread is pH usage that it pops up from time to time in television advertising ("It's pH-balanced!").

Acids Yield Hydrogen Ions in Solution; Bases Accept Them

The "H" in pH stands for hydrogen, while the "p" can be thought of as standing for power. Thus we get "hydrogen power," which describes what lies at the root of pH. An **acid** is any substance that *yields hydrogen ions* when put into a water or "aqueous" solution. A **base** is any substance that *accepts* hydrogen ions in such a solution.

How might this yielding or accepting come about? Recall first that an ion is a charged atom, and that atoms become charged through the gain or loss of one or more electrons. Because electrons carry a negative charge, the loss of an electron leaves an atom with a net *positive* charge. Also recall that the hydrogen atom amounts to one central proton and one electron that circles around it. A hydrogen ion, then, is a lone proton that has become positively charged through loss of its lone electron. The symbol of the hydrogen ion is H^+.

Now, suppose you put an acid—hydrochloric acid (HCl)—into some water. What happens is that HCl *dissociates* or breaks apart into its ionic components, H^+ and Cl^-. The HCl has therefore yielded a hydrogen ion H^+. Now, with a greater concentration of hydrogen ions in it, the water is more acidic than it was.

What about bases? There is a compound called sodium hydroxide (NaOH)—better known as lye—that, when poured into water, dissociates into Na^+ (sodium) and OH^- (hydroxide) ions. The place to look here is the OH^- ions. Negatively charged as they are, they would readily bond with positively charged H^+ ions. In other words, they would *accept* H^+ ions in solution, which is the definition of a base. This accepting of the H^+ ions makes the solution more basic—or, to look at it another way, *less acidic*. Thus, acids and bases are something like a teeter-totter: When one goes up, the other comes down. As you might have guessed, in the right proportions they can balance each other out perfectly. Should H^+ and OH^- ions be poured together into water, for each pair of ions that interacts, the result is:

$$H^+ + OH^- \longrightarrow H_2O$$

Water, which is neutral on the pH scale. The acid and the base have perfectly balanced one another out. The OH^- ion, generally referred to as the **hydroxide ion**, is important because compounds that yield them in quantity are strongly basic and can be used to move solutions from the acidic toward the basic.

Look at FIGURE 3.4 to get an idea of how acidic or basic some common substances are. Following from the notion of what pH amounts to, it's clear that battery acid, for example, is strongly acidic because, when it dissociates in solution, it yields a large number of hydrogen ions. As you move on to lemon juice and then tomatoes, however, you run into weaker acids, which is to say substances that yield fewer H^+ ions in solution. By the time you get to seawater, you've arrived at substances that *accept* hydrogen ions.

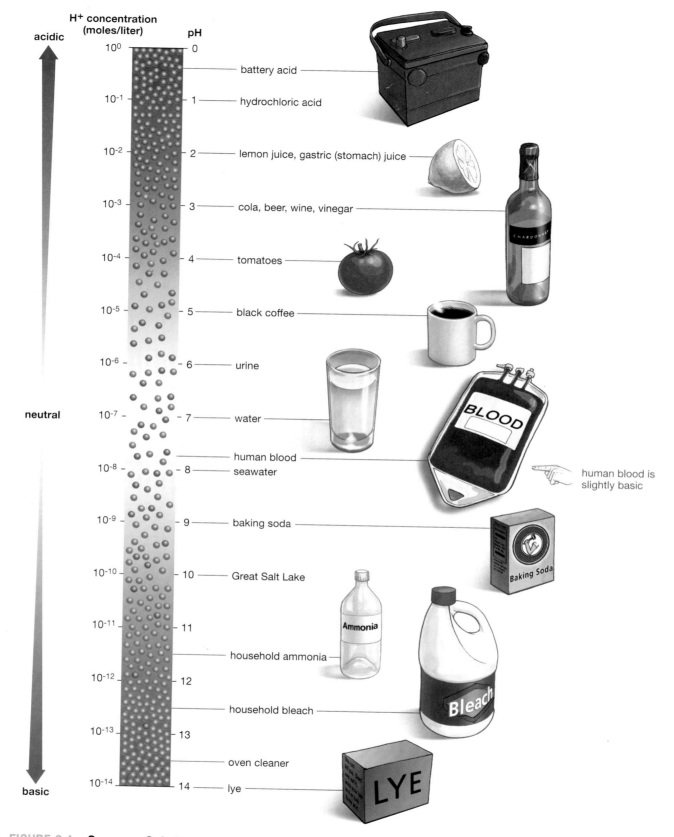

FIGURE 3.4 Common Substances and the pH Scale Chemists use units called moles per liter to measure the concentration of substances in solution. The pH scale, derived from this framework, measures the concentration of hydrogen ions per liter of solution. The most acidic substances on the scale have the greatest concentration of hydrogen ions, while the most basic (or alkaline) substances have the least concentration of hydrogen ions. The scale is logarithmic, so that wine, for example, is 10 times as acidic as tomatoes, and 100 times as acidic as black coffee.

The pH Scale Allows Us to Quantify How Acidic or Basic Compounds Are

The net effect of all this yielding and accepting of hydrogen ions is the concentration of H^+ ions in solution. Through this concentration, the notion of pH can be *quantified*—can have numbers attached to it. What is employed here is the **pH scale**: a scale utilized in measuring the relative acidity of a substance. Look at Figure 3.4 again, this time in connection with the numbers that mark it off. You can see that zero on the scale is the most acidic, while 14 is the most basic. It's important to note that the pH scale is *logarithmic*: A substance with a pH of 9 is 10 times as basic as a substance with a pH of 8 and 100 times as basic as a substance with a pH of 7.

Some Terms Used When Dealing with pH

Here are some notes on pH terminology:

- As a solution becomes more basic, its pH *rises*. Thus the higher the pH, the more basic the solution; and the lower the pH, the more acidic the solution. Oven cleaner is said to have a high pH, while lemon juice has a low pH.

- A solution that is basic is also referred to as an **alkaline** solution.

Why Does pH Matter?

So, why do we care about pH? The brief answer is, because living things are sensitive to its levels in many ways. There is, for example, a class of proteins called enzymes that you'll be reading about later in this chapter. An enzyme is a chemical tool that must retain a very specific shape to function. However, if you put an enzyme in a solution whose pH is too acidic, the enzyme loses its shape. Why? Because the charged nature of the acidic solution starts breaking down the enzyme's chemical bonds. (Remember, lots of positively charged protons are floating around in an acidic solution.) Likewise, cell membranes—which serve the milk carton function noted earlier—can start to break down if pH levels start to go outside normal limits. Membranes and enzymes are so fundamental to life that when they are interfered with, death can result. It is not surprising, then, that many organisms have developed so-called acid-base *buffering systems*, meaning physiological systems that function to keep pH within normal limits. What are these limits? The usual range for living things is about 6–8, with the pH of the human cell being about 7 and that of the blood in our arteries being about 7.4. However, some parts of the human body have special pH requirements. The interior of your stomach, for example, can have a pH as low as 1—an acidic environment that not only helps break down food, but kills most bacteria that ride in on the food. If you look at Essay 3.1, "Acid Rain," on page 38, you can see how pH is important not only to individual organisms, but to life on very large scales.

If you look at Essay 3.1, "Acid Rain," on page 38

A BROADER VIEW

Antacids

The interior of the human stomach is very acidic because cells lining the stomach secrete hydrochloric acid. When the stomach becomes too acidic, however—perhaps because of what we've eaten—the result can be indigestion or "heartburn." To alleviate this condition, we take *antacids*—compounds that raise the pH of the stomach's contents. Some of the active ingredients in these compounds are sodium bicarbonate (baking soda), calcium carbonate (chalk), and aluminum hydroxide. What all these compounds have in common is that they are bases that do their job by neutralizing stomach acids.

So Far...

1. (a) When a liquid can dissolve substances, that liquid is said to be a _____.
   (b) Water has a great ability to serve this function, primarily because it is able to _____.

2. A substance such as motor oil is said to be hydrophobic, meaning that it will not _____.

3. The interior of your stomach can have a pH level that reaches 1; it is thus very _____, a condition that is brought about when there is a high concentration of _____ present.

So Far... Answers:

1. *(a) solvent (b) form chemical bonds, notably hydrogen bonds, with so many substances*

2. *readily dissolve in water*

3. *acidic; hydrogen ions*

Essay 3.1 Acid Rain: When Water Is Trouble

If the concept of pH seems remote from the real world, consider that in parts of New England, more than 70 percent of red spruce forests have been damaged not by natural processes, but by civilization's impact on how acidic the rain is. Likewise, numerous lakes in New England and in New York's Adirondacks region have become so acidic that populations of fish and other wildlife have become reduced or have died off altogether.

What is at work here—a phenomenon called acid rain—shows what can happen when the pH principles we've been reviewing play out on an enormous scale. You've learned something about acids and bases and the importance of *internal* pH to living things. It makes sense, then, that when rainwater itself has a skewed pH, there is going to be widespread trouble. And this is indeed what has happened, not only in the United States but also in Canada, in Eastern Europe and England, in China—in short, worldwide, though industrialized nations are the most affected.

Acid rain comes primarily from two sources: the sulfur dioxide (SO_2) emissions of coal- and oil-burning factories (in particular power plants), and the nitric oxide (NO) and nitrogen dioxide (NO_2) emissions that come mostly from cars, with an assist from factories (**see** FIGURE E3.1.1). When these compounds rise into the air, they come together with a culprit you've seen before: an oxygen radical, specifically the hydroxyl radical (OH). In so doing, they are converted to sulfuric acid and nitric acid. *These* compounds then combine with the rainwater in clouds, and the result is acidic rain.

Acid rain can do its damage directly or indirectly. By simply making a lake more acidic, it begins to interfere with the metabolic processes of the animals living there. It also has the effect, however, of leaching metals from the soil it passes through. Aluminum brought into lakes in this way damages the gills of fish, destroying their ability to absorb oxygen from the water. On land, leached metals can kill plant roots outright or keep them from absorbing nutrients.

When rainwater itself has a skewed pH, there is going to be widespread trouble.

Acid rain will have different effects on different environments, and here again you can see the basics of pH at work. Soils are not neutral in their pH; should acid rain fall on soil that is significantly basic (or "alkaline"), the buffering process you've read about will take place: Basic compounds in the soil will combine with the water and accept hydrogen ions from the rain, thus resulting in water that is less acidic. Likewise, aquatic environments may contain basic ions that can neutralize incoming water.

The extent to which such environments possess basic ions is known as their *acid-neutralizing capacity*. Given what you've read about pH, it might be apparent that this capacity is not limitless but instead can dwindle as additional acidic compounds are added. Indeed, soils can suffer from a long-term depletion of basic compounds, and this has happened in some eastern U.S. forests. A 1990 amendment to the federal Clean Air Act resulted in a reduction in U.S. sulfur emissions, but left some nitrogen emissions unchanged through the middle of the decade. This meant there was an overall decline in acid *rain* in, for example, the Adirondacks in the period 1992–1999.

3.4 Carbon Is a Central Element in Life

You're crossing a bridge of sorts here. To this point, the discussion has largely been about water and hydrogen and various ions. From time to time, mention was made of the element *carbon* (**see** FIGURE 3.5), although only as one substance among many. But carbon is not that at all. Life is based on carbon, or if you will, on carbon compounds immersed in water. You could think of carbon in life in the same way as you do flour in baking. How many times does a baking recipe begin: "Start with two cups of flour"? With the molecules of life, the recipe often begins: "Start with a carbon ring," or "start with a chain of carbon atoms." But just as cinnamon or fruit will be folded into flour, thus imparting special qualities to the basic ingredient, so

FIGURE 3.5 **Pure Carbon, Mostly Carbon** A diamond is pure carbon and the hardest natural material known. Surrounding the diamond is a lump of coal, which has a high carbon content because it is made up mostly of the remains of carbon-rich, ancient vegetation.

Despite this, many Adirondack lakes continued to get more acidic through the end of the 1990s. Why would lakes get more acidic if acid rain has been reduced? Because the acid-neutralizing capacity of the lakes and the soil around them had been so depleted by years of acid rain. More recent measurements have found a welcome decline in the acidity of some lakes in the Northeast, but ecologists remain concerned about the health of both lakes and soils in the region. A return to real health for them would require continued reductions in acid rain over a long period of time, which is far from a sure thing.

(a) How acid rain forms

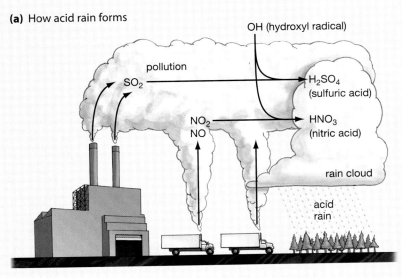

(b) Damage from acid rain

FIGURE E3.1.1   **How Acid Rain Forms**

(a) Sulfur dioxide (SO_2) from coal- and oil-burning power plants rises into the air along with nitric oxide (NO) and nitrogen dioxide NO_2, which come mostly from cars. These compounds combine with hydroxyl radicals (OH) in the atmosphere to produce sulfuric acid H_2SO_4 and nitric acid HNO_3, which combine with atmospheric water, making it acidic.

(b) Trees affected by acid rain in New York State's Harriman State Park.

will other elements join carbon: A little addition of this here, a little of that on the end, and, presto! A complex molecule that serves a very specific purpose.

How does carbon come by its great powers? Linkage. Recall from Chapter 2 that carbon has four outer-shell electrons, but that most elements (except hydrogen and helium) need *eight* outer-shell electrons for maximum stability. This means that each carbon atom achieves maximum stability by linking up with four more electrons, which in turn means that carbon's bonding capacity is great. Moreover, the bonds that carbon creates are covalent—it is *sharing* electrons with other atoms—which means its linkages are stable, compared to other types of bonds. Given the natural forces that can buffet life about, such as ultraviolet radiation and heat, this stability is an important quality in first getting life going and then keeping it going. So important is carbon that it forms one whole area of chemistry—**organic chemistry**, a branch of chemistry devoted to the study of compounds that have carbon as their central element.

In the review that follows, you'll be seeing carbon-based molecules represented in a couple of ways, as shown in FIGURE 3.6. The organic molecule in the figure is glucose, better known as blood sugar. In the drawing on the left, you can see how carbon is literally central to glucose, with the "C" symbols standing for carbon

FIGURE 3.6   **Representations of Carbon-Based Molecules**   One molecule, glucose, is represented here in two ways. In the structural model on the left, note the central ring of carbon atoms with hydrogen atoms and oxygen-hydrogen (OH) groups attached. The space-filling model of glucose, on the right, shows the structure of the molecule in three dimensions.

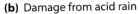

glucose

Table 3.1 Monomers, Polymers

If the monomer is...	The polymer is...
A **monosaccharide** (for example, glucose, fructose) 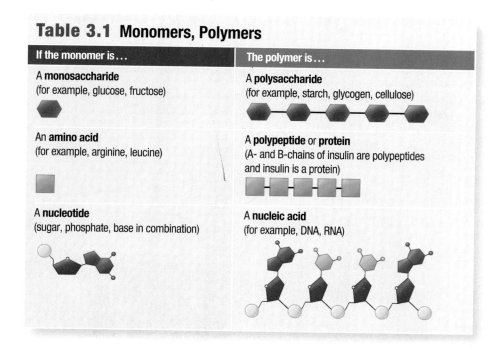	A **polysaccharide** (for example, starch, glycogen, cellulose)
An **amino acid** (for example, arginine, leucine)	A **polypeptide** or **protein** (A- and B-chains of insulin are polypeptides and insulin is a protein)
A **nucleotide** (sugar, phosphate, base in combination)	A **nucleic acid** (for example, DNA, RNA)

atoms that are then linked to the various hydrogen (H) and oxygen (O) atoms. So common are carbon-based "ring" structures such as these that a ring often is drawn with its C's left out, meaning that carbon is *assumed* to exist at each of the bond junctures, unless otherwise noted. You'll soon be seeing this convention in connection with carbohydrates. The model on the right, a so-called space-filling model, gives you a better sense of how glucose is structured in three dimensions, with carbon (in red) largely at the center and the oxygen and hydrogen atoms around the periphery.

So, what are the classes of carbon-based biological molecules? You're about to explore four groupings of these compounds: carbohydrates, lipids, proteins, and nucleic acids. As you go through this section, it will be a great help to keep in mind that *complex* organic molecules often are made from *simpler* molecules. Many of the molecules you'll be reading about have a building-blocks quality to them: Take a simple sugar, or monosaccharide, such as glucose, put it together with another monosaccharide (fructose), and you have a larger *di*saccharide called sucrose (better known as table sugar). Put *many* monosaccharide units together and you have a polysaccharide, such as starch. The starch is an example of a **polymer**—a large molecule made up of many similar or identical subunits. Meanwhile, the glucose is an example of a **monomer**—a small molecule that can be combined with other similar or identical molecules to make a polymer. Look at **Table 3.1** for examples of both.

3.5 Carbohydrates

Happily, for purposes of memory, all the elements in the first type of molecule we'll look at are hinted at in the molecule's name: **Carbohydrates** are organic molecules that always contain carbon, oxygen, and hydrogen and that, in many instances, contain nothing *but* carbon, oxygen, and hydrogen. Furthermore, they usually contain exactly twice as many hydrogen atoms as oxygen atoms. For example, the carbohydrate glucose $C_6H_{12}O_6$ contains 12 atoms of hydrogen and 6 of

FIGURE 3.7 **Carbohydrates in Foods** Breads, cereals, and pasta make up a significant proportion of our diets. These foods are all rich in carbohydrates, one of the four main types of biological molecules.

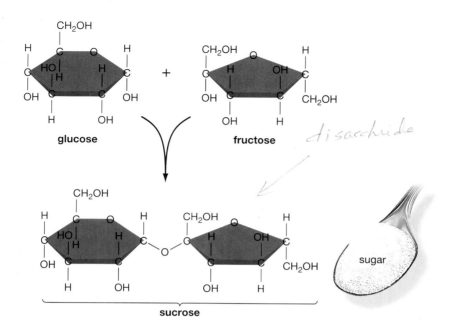

disacchride

FIGURE 3.8 Carbohydrates Follow a Building-Blocks Model In this example, two monosaccharide (or simple sugar) units, glucose and fructose, combine to yield the slightly more complex disaccharide, sucrose, which is common table sugar. An additional product of this reaction is water.

oxygen. Most people think of carbohydrates purely in terms of foods—such as breads and pasta—and food "carbs" certainly have been a subject of great interest in the United States recently. Yet, as you'll see, carbohydrates play more roles in nature than just serving as food (**see** FIGURE 3.7).

The building blocks of carbohydrates are the **monosaccharides** mentioned earlier, which also are known as **simple sugars**. You've already seen several views of one of these molecules—glucose. Glucose has a use in and of itself: Much of the food we eat is broken down into it, at which point it becomes our most important energy source. Glucose can also bond with other monosaccharides, however, to form more complex carbohydrates. If you look at FIGURE 3.8, you can see an example of glucose bonding with a monosaccharide called fructose to create the disaccharide sucrose, better known as table sugar. Monosaccharides also include deoxyribose, which is part of the DNA molecule, while disaccharides also include lactose, which is well known for being in milk. At the risk of pointing out the obvious, note that all these sugars have –*ose* at the end of their name: If it's an –*ose*, it's a sugar (**see** FIGURE 3.9).

FIGURE 3.9 Sugars Come in Many Forms Sucrose or table sugar comes to us from sugarcane or sugar beets, and glucose is found in corn syrup. Fructose comes to us in sweet fruits and in high-fructose corn syrup, which often is used to sweeten soft drinks.

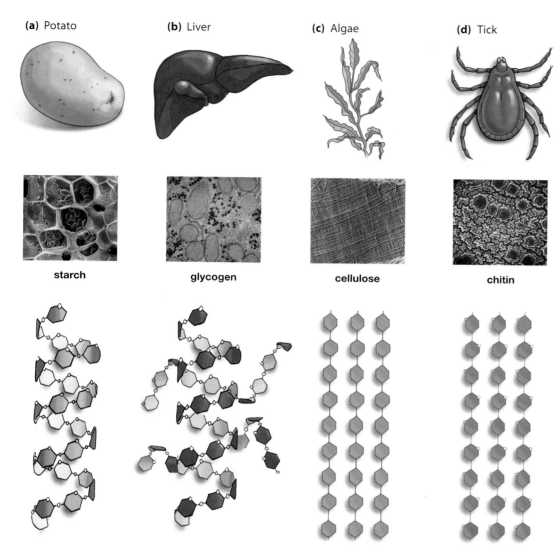

(a) Potato **(b)** Liver **(c)** Algae **(d)** Tick

starch   glycogen   cellulose   chitin

Complex Carbohydrates Are Made of Chains of Simple Carbohydrates

If you add more monosaccharide units to disaccharides, you get to the polymers of carbohydrates, the **polysaccharides**. The *poly* in polysaccharides means "many," while *saccharide* means "sugars," and the combination of these terms is apt. In the polysaccharide molecule cellulose, for example, there may be 10,000 glucose units linked up with one another. The basic unit here is the six-carbon monosaccharide glucose, $C_6H_{12}O_6$, from which chains of glucose units are built up. The complexity of the polysaccharides gives them their alternate name of complex carbohydrates. Four different types of complex carbohydrates interest us: starch, glycogen, cellulose, and chitin (see FIGURE 3.10).

Starch is a complex carbohydrate, found in plants, that exists in the form of foods such as potatoes, rice, carrots, and corn (see Figure 3.10a). In plants, these starches serve as the main form of carbohydrate *storage*, sometimes as seeds (rice and wheat grains), or sometimes as roots (carrots and beets).

Another complex carbohydrate, **glycogen**, serves as the primary form of carbohydrate storage in animals. Thus, glycogen does for animals what starch does for plants. The starches or sugars we eat are broken down, eventually into glucose,

FIGURE 3.10   **Four Examples of Complex Carbohydrates** All complex carbohydrates are composed of chains of glucose, but they differ in the details of their chemical structure.

(a) Starch serves as a form of carbohydrate storage in many plants. Here starch granules can be seen within the cells of a slice of raw potato.

(b) Glycogen serves as a form of carbohydrate storage, as shown in this photo of glycogen globules in the liver.

(c) Cellulose, running as fibers through cell walls, provides structural support for plants. The photo is of sets of cellulose fibers running at right angles to one another in the cell wall of marine algae.

(d) Chitin provides structural support for some animals. The outer "skin" or cuticle of insects is composed mostly of chitin. The photo shows the exoskeleton of a tick.

some of which may be used immediately to fuel our daily activities. Some glucose may not be needed right away, however, in which case it is moved into the muscle cells and liver to be stored as glycogen (Figure 3.10b).

Cellulose is a rigid, complex carbohydrate contained in the cell walls of many organisms. Despite this innocuous-sounding function, cellulose is important because it makes up so much of the natural world. It is easily the most abundant carbohydrate on Earth: Trees, cotton, leaves, and grasses are largely made of it. When cellulose is enmeshed with a hardening compound called *lignin*, the result is a set of cell walls that can hold up giant redwood trees (Figure 3.10c).

Chitin is a complex carbohydrate that forms the external skeleton of the arthropods—all insects, spiders, and "crustaceans," such as crabs. In all these animals, chitin plays a "structural" role similar to that of cellulose in plants: It gives shape and strength to the structure of the organism (Figure 3.10d).

3.6 Lipids

We turn now to the **lipids**, a class of biological molecules whose defining characteristic is that they don't easily dissolve in water. It turns out that lipids are made of the same elements as carbohydrates—carbon, hydrogen, and oxygen. But lipids have much more hydrogen, relative to oxygen, than do the carbohydrates. We're all familiar with some lipids; they exist as fats, as oils, and as cholesterol. Unlike the other biological molecules you'll be studying, a lipid is not a polymer composed of

FIGURE 3.11 **Structural Formula for Stearic Acid**

component-part monomers; no single structural unit is common to all lipids. But the fact that lipids are not easily soluble in water gives many of them a shared function. Remember the earlier discussion about the milk carton and the need life has to create internal containers? Well, thanks to their insolubility, lipids can take on this role. In addition, they have considerable powers to store energy (think of fat cells), to provide insulation (think of bear fat), and some of them form chemically active hormones (for example, estrogen and testosterone).

One Class of Lipids: The Glycerides

The most common kind of lipid, the glyceride, can be thought of as a molecule in two parts. The first part is a "head" composed of a particular kind of alcohol, usually glycerol. The second part is one or more fatty acids, an example of which can be seen in FIGURE 3.11. This is stearic acid, one of the fatty acids found in animal fat. As you can see, it amounts to a long chain of hydrogen and carbon atoms—its "hydrocarbon" portion—that terminates with a COOH chemical group over on the left.

Now look at the other part of this glyceride, the alcohol known as glycerol:

Bringing the —OH group of glycerol together with the COOH part of the fatty-acid chain is what makes a glyceride. But you can see that the glycerol has *three* —OH groups on the right. Thus there is, you might say, docking space on glycerol for three fatty acids, and in the synthesis of many glycerides, this linkage takes place: Three fatty acids link up with glycerol to form a *tri*glyceride, which is an important form of lipid. To get a better idea of the shape of a completed triglyceride, look at the space-filling model in FIGURE 3.12. This particular triglyceride, called tristearin, has three stearic fatty acids stemming like tines on a fork from the glycerol. But this is only one possibility among many, and is exceptional, rather than usual, in that all three fatty acids are the same. Among the dozens of fatty acids that exist, several *different* kinds of fatty acids often will hook up in glycerol's three "slots" to form a triglyceride. Actual fat products, such as butter, are composed of different proportions of various fatty acids.

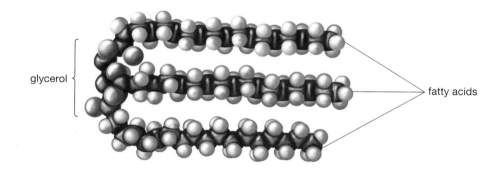

FIGURE 3.12 **The Triglyceride Tristearin** This lipid molecule is composed of three stearic fatty acids, stemming rightward from the glycerol OH "heads." Tristearin is found both in beef fat and in the cocoa butter that helps make up chocolate.

(a) Palmitic acid

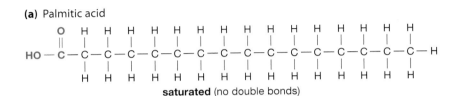

saturated (no double bonds)

(b) Oleic acid

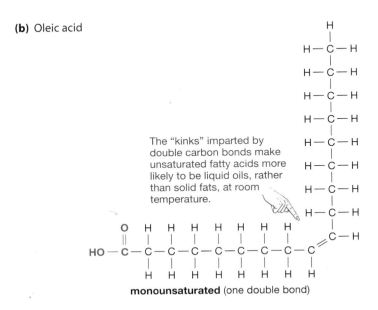

The "kinks" imparted by double carbon bonds make unsaturated fatty acids more likely to be liquid oils, rather than solid fats, at room temperature.

monounsaturated (one double bond)

(c) Linoleic acid

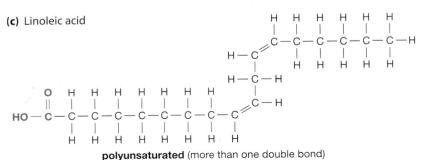

polyunsaturated (more than one double bond)

FIGURE 3.13 **Saturated and Unsaturated Fatty Acids** The degree to which fatty acid hydrocarbon chains are "saturated" with hydrogen atoms has consequences for both the form these lipids take and for human health.

(a) The hydrocarbon "tail" in palmitic acid is formed by an unbroken line of carbons, each with a single bond to the next.

(b) In oleic acid, a double bond exists at one point between two carbon atoms. An additional hydrogen atom could link to each of these carbon atoms instead, which would make this a saturated fatty acid—saturated with hydrogen atoms. As things stand, this is a monounsaturated fatty acid.

(c) The carbons in linoleic acid have double bonds in two locations, making this a polyunsaturated fatty acid.

Triglycerides are the most important of the glycerides, as they constitute about 90 percent of the lipid weight in foods. Thus, the substances we call fats usually are composed mostly of triglycerides. With all this information under your belt, you're ready for a couple of definitions. A **triglyceride** is a lipid molecule formed from three fatty acids bonded to glycerol. A **fatty acid** is a molecule found in many lipids that is composed of a hydrocarbon chain bonded to a COOH group.

Saturated Fats and Trans Fats

If you now look at FIGURE 3.13, you can see three different fatty acids: palmitic, oleic, and linoleic. There is an obvious difference among them, in that the oleic and linoleic hydrocarbon chains are "bent" compared to the straight-line palmitic chain. Upon closer inspection, you can see that the palmitic acid has an unbroken line of single bonds linking its carbon atoms. Meanwhile, the oleic acid has one double bond (represented by the double lines) between one pair of carbons in its chain and the linoleic acid has two double bonds. Furthermore, note that these double bonds exist precisely where the "kinks" appear in these molecules.

What these variations describe are the differences between three *kinds* of fatty acids. First, there is a **saturated fatty acid**— a fatty acid with no double bonds between the carbon atoms of its hydrocarbon chain. Then there is a **monounsaturated fatty acid** (a fatty acid with one double bond between carbon atoms), and a **polyunsaturated fatty acid** (a fatty acid with two or more double bonds between carbon atoms). What a saturated fatty acid is saturated *with* is hydrogen atoms. A given monounsaturated fatty acid could theoretically have its lone carbon-carbon double bond replaced with a *single* bond between the carbons. This change would allow the addition of two more hydrogen atoms at the bond sites. Once that addition occurred, this fatty acid would contain the maximum number of hydrogen atoms possible, meaning it would be a saturated fatty acid.

This may not sound like much of a difference, but it has several important health consequences. Think of the first consequence this way: To the degree that *fatty acids* are saturated, the *fats* they make up will be saturated—and consumption of saturated fats has been linked with heart disease. The linkage here is that saturated fats increase the amount of cholesterol in circulation in the body. When this cholesterol lodges in the arteries of the heart in excessive amounts, heart disease can begin. In contrast to this, the mono- and polyunsaturated fats have little to no effect on circulating cholesterol. The usual source for saturated fats is animal fat—primarily fatty meat and dairy products—but they can also be found in a few plant products, such as coconut oil and cocoa butter. In contrast, unsaturated fats tend to be fats that come to us in the form of **oils**, which are simply fats in liquid form. Monounsaturated fats come in olive, canola, and peanut oil, for example, while polyunsaturated fats can be found in safflower and corn oil.

There is then a twist on this story, which is that food producers have a means of turning naturally occurring oils *into* fats by saturating them—by bubbling hydrogen through the oils in a process called hydrogenation. Why would they do this? Consumers like foods that are creamy, but not oily, and they like foods with a long shelf life, which hydrogenation aids in. One downside of this, as you can see, is that it can result in products with a higher proportion of saturated fats.

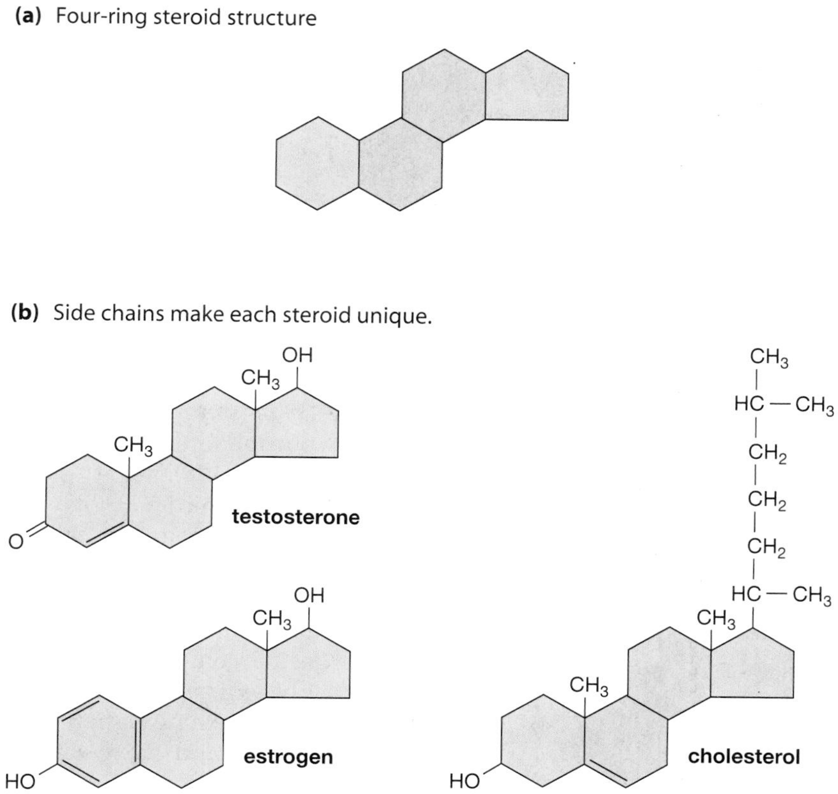

(a) Four-ring steroid structure

(b) Side chains make each steroid unique.

OH
CH₃
CH₃
testosterone
O

OH
CH₃
CH₃
estrogen
HO

CH₃
HC—CH₃
CH₂
CH₂
CH₂
HC—CH₃
CH₃
CH₃
cholesterol
HO

FIGURE 3.14 **Structure of Steroids**

(a) The basic unit of steroids, four interlocked carbon rings.

(b) Types of steroids, each differentiated from the other by the side chains that extend from the four-ring skeleton. Testosterone is a principal "male" hormone, while estrogen is a principal "female" hormone. Both of these steroid hormones actually are found in both men and women, though in differing amounts. Cholesterol is also a prevalent and important steroid in both men and women. Although cholesterol has a bad reputation, it has several important functions—for example, breaking down fats.

Unfortunately, this is only part of the story. In some instances, hydrogenation has the effect not of putting more hydrogen atoms in a fatty acid, but of changing the *orientation* of the hydrogen atoms that already exist. The result is the creation of a particularly unhealthy variety of fat—a *trans fat*. These fats not only raise levels of the "bad" cholesterol molecules that carry cholesterol *to* the heart, but lower levels of the "good" cholesterol molecules that carry cholesterol *away* from the heart. This is a double dose of unhealthy effects that not even saturated fats can manage. Fortunately, the federal Food and Drug Administration is now requiring food producers to list trans fats separately in the ubiquitous "Nutrition Facts" labels on food packages, so that consumers can have a clear idea of what they are eating. We'll review the nature of "good" and "bad" cholesterol later in the chapter.

A Second Class of Lipids Is the Steroids

Steroids are a class of lipid molecules that have, as a central element in their structure, four carbon rings. What separates one steroid from another are the various side chains that can be attached to these rings (**see** FIGURE 3.14). When you see how different steroids are structurally from the triglycerides, you can understand why the monomers-to-polymers framework doesn't apply to lipids.

Among the most well-known steroids are cholesterol and two of the steroid hormones, testosterone and estrogen. For reasons we've just touched on, cholesterol has a bad reputation; but like fats in general, it serves good purposes too. **Cholesterol** is a steroid molecule that forms part of the outer membrane of all animal cells, and that acts as a precursor for many other steroids. One of the steroids formed from cholesterol is the principal "male" hormone, testosterone; another is the principal "female" hormone, estrogen. (Both hormones actually are produced in both sexes, though in differing amounts.) The term *steroids* by itself undoubtedly rings a bell, because the phrase "on steroids" has come to mean artificially bulked up or supercharged. In this common usage, *steroids* refers to manufactured

A BROADER VIEW
Healthy and Unhealthy Fats
One of the great public health issues of the early part of the twenty-first century is the question of what place lipids (fats) have in a healthy diet. A consensus has emerged that a false slogan of "all fats are bad" was sold to the public in the 1980s and 1990s when in fact it is only saturated and trans fats that are unhealthy, while unsaturated fats are neutral or even good for health. The healthiest fats of all are polyunsaturated fats that have a high proportion of "omega-3" fatty acids in them. These fatty acids reduce fat levels in the bloodstream, and they retard the growth of the fatty deposits that clog up heart arteries. Omega-3's are found in greatest abundance in fish such as salmon and albacore tuna.

(a) Natural steroids

(b) Pharmaceutical steroids

FIGURE 3.15 **Steroids, in Uses Natural and Un-natural**

(a) Some steroid hormones, such as estrogen and testosterone, are important in natural processes, such as reproduction.

(b) Other steroid hormones are laboratory-produced versions of natural muscle-building hormones. In 2005, Rafael Palmeiro, pictured, was given a 10-day suspension from baseball after testing positive for this kind of "synthetic" steroid.

drugs that are close chemical cousins of the muscle-building "male" steroid hormones (**see** FIGURE 3.15).

A Third Class of Lipids Is the Phospholipids

The final class of lipids, phospholipids, has something of the same makeup as triglycerides, in that a phospholipid has a glycerol head with fatty acids attached to it. But where triglycerides have three fatty acids stemming from the glycerol, phospholipids have only two (**see** FIGURE 3.16). Linking up with glycerol's third —OH group is a **phosphate group**, meaning a phosphorus atom surrounded by four oxygen atoms. Thus, we can define a **phospholipid** as a charged lipid molecule composed of two fatty acids, glycerol, and a phosphate group.

The combination of a phosphate group and fatty-acid tails is extremely important because it gives phospholipids a dual nature: Being hydrocarbons, the fatty-acid tails are hydrophobic; but the phosphate head is hydro*philic* because it has a *charged* portion. In Figure 3.16, you can see the effect of this structure in solution. Imagine a phospholipid as a marker buoy in deep water. No matter how you push the hydrocarbon tail around, it's going to end up waving free *out* of the water, while the head is going to be submerged in the water, because it bonds with it. You will learn more about these molecules later, when you get to cells; for now, just note that the material on the periphery of cells—the outer membrane of a cell—is largely made of phospholipids. Living things need the kind of partitions described earlier, and these partitions are composed to a significant extent of phospholipids.

So Far...

1. Life on Earth is said to be based on the element _____, which is central to so many molecules of life because it has a great capacity to _____.

2. Carbohydrates are always composed solely of the elements _____, _____, and _____ and exist in forms ranging from the small monomers known as _____ to larger polymers, known as _____.

3. A fatty acid that has only one double bond between carbon atoms in its hydrocarbon chain is said to be a _____. When consumed as fats, are these lipids thought to be harmful to health? If not, which related lipids are?

So Far... Answers:

1. carbon; form stable bonds with other elements

2. carbon, oxygen, hydrogen; monosaccharides, polysaccharides

3. monounsaturated fatty acid; no; saturated fats

3.7 Proteins

Living things must accomplish a great number of tasks just to get through a day, and the diverse biological molecules you've been looking at allow this to happen. You've seen carbohydrates do some things, and you've seen lipids do some more.

FIGURE 3.16 A Dual-Natured Molecule

(a) Phospholipids are composed of two long fatty-acid "tails" attached to a "head" containing a phosphate group (which carries a negative charge) and another variable group (which often carries a charge).

(b) Because the head is polarized, it can bond with water and thus will remain submerged in it; the tails, on the other hand, have no such bonding capability.

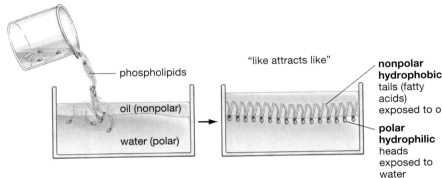

(a) Phospholipid structure

variable group   phosphate group

polar head   **nonpolar tails**

(b) Phospholipid orientation

phospholipids

oil (nonpolar)

water (polar)

"like attracts like"

nonpolar hydrophobic tails (fatty acids) exposed to oil

polar hydrophilic heads exposed to water

But in the range of tasks that molecules accomplish, proteins reign supreme. Witness the fact that almost every chemical reaction that takes place in living things is hastened—or, in practical terms, *enabled*—by a particular kind of protein called an enzyme. These molecules function in nature like some vast group of tools, each one taking on a specific chemical task. Accordingly, an animal cell might contain up to 4,000 different types of enzymes.

We might marvel at proteins solely because of what enzymes can do, but the amazing thing is that enzymes are only *one class* of proteins. Proteins also form the scaffolding, or structure, of a good deal of tissue; they're active in transporting molecules from one site to another; they allow muscles to contract and cells to move; some hormones are made from them. If you factor out water, they account for about half the weight of the average cell. In short, it's hard to overestimate the importance of these molecules. **Table 3.2** lists some of the different kinds of proteins.

Table 3.2 Types of Proteins

Type	Role	Examples
Enzymes	Quicken chemical reactions	Sucrase: Positions sucrose (table sugar) in such a way that it can be broken down into component parts of glucose and fructose
Hormones	Chemical messengers	Growth hormone: Stimulates growth of bones
Transport	Move other molecules	Hemoglobin: Transports oxygen through blood
Contractile	Movement	Myosin and actin: Allow muscles to contract
Protective	Healing; defense against invader	Fibrinogen: Stops bleeding Antibodies: Attack bacterial invaders
Structural	Mechanical support	Keratin: Hair Collagen: Cartilage
Storage	Stores nutrients	Ovalbumin: Egg white, used as nutrient for embryos
Toxins	Defense, predation	Bacterial diphtheria toxin
Communication	Cell signaling	Glycoprotein: Receptors on cell surface

Proteins Are Made from Chains of Amino Acids

Proteins are prime examples of the building-block type of molecule described earlier. The monomers in this case are called amino acids. String a minimum number of them together in a chain—some say 10, some say 30—and you have a **polypeptide**, defined as a series of amino acids linked in linear fashion. When the polypeptide chain *folds up* in a specific three-dimensional manner, you have a **protein**, defined as a large, folded chain of amino acids. As a practical matter, proteins are likely to be made of hundreds of amino acids strung together and folded up. It's not unusual for two or more polypeptide chains to be part of a single protein.

All of this is laid out for you in FIGURE 3.17. If you look at Figure 3.17a, you can see the chemical structure of two amino acids, glycine (gly) and isoleucine (ile). If you then look at Figure 3.17b, you can see that these two amino acids are the first two that occur in one of the two polypeptide chains that make up the unusually small hormonal protein we call insulin. Figure 3.17c then shows a three-dimensional representation of insulin—the folded-up form this protein assumes before taking on its function of moving blood sugar into cells.

A Group of Only 20 Amino Acids Is the Basis for All Proteins in Living Things

Although only two amino acid examples are shown here, it is a group of 20 amino acids that is the basis for all the proteins that occur in living organisms. The millions of proteins that exist can be made from a mere 20 amino acids, because these

FIGURE 3.17   The Structure of Proteins

glycine **(gly)**   isoleucine **(ile)**

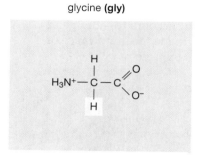   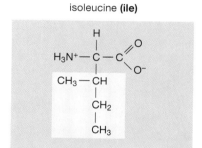

(a) Amino acids

The building blocks of proteins are amino acids such as glycine and isoleucine, which differ only in their side-chain composition (light colored squares).

(b) Polypeptide chain

These amino acids are strung together to form polypeptide chains. Pictured is one of the two polypeptide chains that make up the unusually small protein insulin.

(c) Protein

Polypeptide chains function as proteins only when folded into their proper three-dimensional shape, as shown here for insulin. Note the position of the glycine and isoleucine amino acids in one of the insulin polypeptide chains (colored light green).

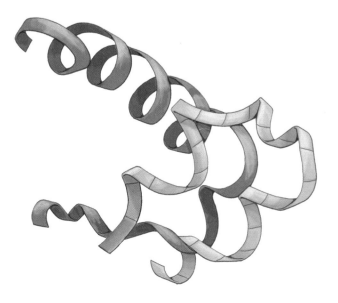

amino acids can be strung together in different *order*. Substitute an alanine amino acid here for a glutamine there, and you've got a different protein. In this respect, amino acids commonly are compared to letters of the alphabet. In English, substituting one letter can take us from *bat* to *hat*. In the natural world, 20 amino acids can be put together in different order to create a multitude of proteins, each with a different function.

Shape Is Critical to the Functioning of All Proteins

Now, recall that a protein is a chain of amino acids that has become *folded up* in a specific way. As the amino acids are being strung together in sequence, all the kinds of chemical forces discussed in Chapter 2 begin to work on the chain; as a result, it begins to twist and turn and fold into a unique three-dimensional shape. And it turns out that in the functioning of proteins, this shape, or *protein conformation*, is utterly crucial. Here's an example of why. If you look at the top illustration in FIGURE 3.18, on the left you can see a computerized model of a protein molecule found on the surface of the influenza virus. If you look on the right, you will see a model of an immune-system protein, called an antibody, that fights foreign invaders such as the influenza virus. Now, how does the antibody interact with the virus? It *binds to it* chemically as shown in the lower illustration, and this process is made possible by the exact fit of antibody to invader. Look at the shape of the two molecules to see how closely they conform to one another in shape. The binding of an antibody to an invader sounds important enough; but recall that proteins are the chemical enablers called enzymes, and they're transport molecules, and so forth. In *all* these functions, shape is critical. The American architect Frank Lloyd Wright had a famous dictum about his designs: "Form follows function." With proteins, we can turn this around and say that function follows form.

Proteins Can Come Undone

As noted, proteins fold up into a precise conformation in order to function. However, proteins can *lose* their shape, and thus lose their functionality. You saw an example of this earlier in connection with pH and enzymes. In the wrong pH environment, an enzyme can unfold, losing its structure and thus losing its ability to speed up a chemical process. Alcohol works as a disinfectant on skin because it *denatures* or alters the shape of bacterial proteins.

Lipoproteins and Glycoproteins

Some molecules in living things are hybrids, or combinations, of the various types of molecules you've been looking at. **Lipoproteins**, as their name implies, are molecules that are a combination of lipids and proteins. Active in transporting fats throughout the body, lipoproteins are transport molecules that amount to a capsule of protein surrounding a globule of fat.

Two kinds of lipoproteins, touched on earlier, have managed to enter public consciousness despite the handicap of having long names. These are low-density lipoproteins (LDLs) and high-density lipoproteins (HDLs), also known respectively as "bad" and "good" cholesterol. LDLs acquired a reputation as villains because they are capsules of protein that carry cholesterol *to* outlying tissues including the coronary arteries of the heart, where this cholesterol may come to reside. HDLs, meanwhile, came to be regarded as the cavalry, because they are capsules that carry cholesterol *away* from outlying cells, to the liver. A high proportion of HDLs in relation to total cholesterol is predictive of keeping a healthy heart.

Glycoproteins are molecules that are combinations of proteins and *carbohydrates*. Where do we find these molecules? One place is the surface of cells, which usually are peppered with a profusion of antenna-like structures, called receptors, that allow a cell to receive signals from outside itself. A typical receptor is likely to be composed of both protein and carbohydrate, the protein forming the "stem" of the receptor and

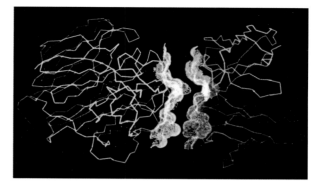

(a)

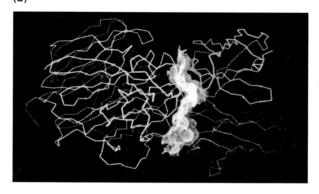

(b)

FIGURE 3.18 **Hand-in-Glove Fit** Molecular shape is a critical element in the functioning of most proteins.

(a) The computer-model on the left is of a molecule found on the surface of the influenza virus. On the right is a model of a portion of human protein, called an antibody, that attacks invaders such as viruses.

(b) The antibody helps disable the virus by binding with it, as shown. The antibody is able to carry out this binding because it has a shape that is complementary to that of the virus molecule.

the carbohydrate forming side-chains that extend from the stem and serve as the actual binding sites for signaling molecules that may pass by. Some hormones are glycoproteins, along with many other proteins released from cells.

3.8 Nucleotides and Nucleic Acids

Nucleotides are the last major type of biological molecule we'll study. They are important molecules in and of themselves and serve as building blocks for a couple of very important molecules.

As molecules in their own right, some nucleotides, called adenosine phosphates, function as chemical energy carriers. When you study energy within cells, you'll find many references to a molecule called adenosine triphosphate, or ATP. As money is to shopping, so ATP is to getting things done in living organisms. In this text, the most detailed discussion of nucleotides, however, comes in connection with their role as monomers in building two very large and important polymers—DNA and RNA.

DNA Provides Information for the Structure of Proteins

You've learned that proteins perform a large number of biological functions, and that one class of proteins, the enzymes, may be represented with up to 4,000 types in a single animal cell. If you had a factory that turned out well over four thousand different kinds of tools, you would obviously need some direction on how each of these tools was to be manufactured: This part of the tool goes here first, and then that goes there, and so on. There is a molecule that in essence provides this kind of information for the construction of proteins. It is **DNA**, or **deoxyribonucleic acid**: the primary information-bearing molecule of life, composed of two linked chains of nucleotides. How many nucleotides? In human beings, about 3 billion of them are strung together to form our main molecule of DNA, and two copies of this molecule exist in each of our cells.

The information contained in the DNA molecule is much like the information contained in a cookbook, only what the DNA "recipes" are calling for are precisely ordered chains of amino acids, the building blocks of proteins. "Start with an alanine, then add a cysteine, then a tyrosine . . ." and so on for hundreds of steps, the DNA-encoded instructions will say, after which—through a series of steps—a protein becomes synthesized and then gets busy on some task. A player in this series of steps is another nucleic acid, **ribonucleic acid** or **RNA**, a molecule composed of nucleotides that is active in the synthesis of proteins. RNA's functions include ferrying the DNA-encoded instructions to the sites in the cell where proteins are put together. It also helps make up a protein-synthesizing workbench you'll look at later, called a ribosome.

The Structural Unit of DNA Is the Nucleotide

Look at FIGURE 3.19a and you'll see the structure of the building blocks that make up DNA. Each nucleic acid **nucleotide** is a molecule in three parts: a *phosphate* group, a *sugar* (deoxyribose), and a nitrogen-containing *base*. One nucleotide then attaches to another, forming a chain. Two of these chains then link together—as if a ladder, split down the middle, were coming together—forming the most famous molecule in all of biology, the DNA double helix.

On to Cells

Before getting to the story of this elegant DNA molecule, you need to learn a little bit about the territory in which it does its work. What you'll be looking at is a profusion of jostling, roiling, ceaselessly working chemical factories that make up all living things. These factories are called cells. Look at **Table 3.3** on page 52, and you'll find a summary of all the types of molecules that were reviewed in this chapter.

A BROADER VIEW

Measuring Our DNA

The DNA molecule is so thin that it can only be seen with an electron microscope. But it is an exceedingly long molecule. If the human complement of DNA were uncoiled, it would stretch out to about six feet. It does not exist as a single long molecule, however, but instead comes "packaged" in 23 separate units, called chromosomes.

(a) Nucleotides are the building blocks of DNA.

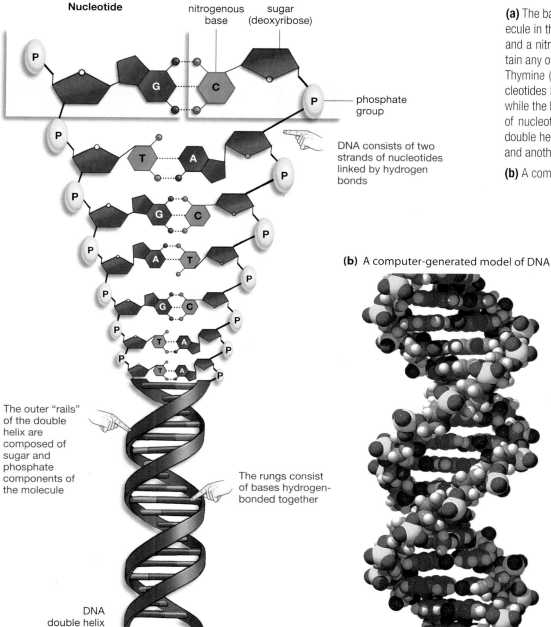

Nucleotide

nitrogenous base

sugar (deoxyribose)

P

G

C

P — phosphate group

P

P

T

A

DNA consists of two strands of nucleotides linked by hydrogen bonds

P

P

G

C

P

P

A

T

P

P

G

C

P

P

T

A

P

T

A

P

The outer "rails" of the double helix are composed of sugar and phosphate components of the molecule

The rungs consist of bases hydrogen-bonded together

DNA double helix

FIGURE 3.19 **Nucleotides Are the Building Blocks of DNA**

(a) The basic unit of the DNA molecule is the nucleotide, a molecule in three parts: a sugar, (deoxyribose), a phosphate group, and a nitrogen-containing base. A given nucleotide might contain any of four bases: Adenine (A), Guanine (G), Cytosine (C), or Thymine (T). The sugar and phosphate components of the nucleotides link up to form the outer "rails" of the DNA molecule, while the bases point toward the molecule's interior. Two chains of nucleotides are linked, via hydrogen bonds, to form DNA's double helix. Note that the hydrogen bond is between one base and another.

(b) A computer-generated space-filling model of DNA.

(b) A computer-generated model of DNA

So Far...

1. A protein is composed of one or more sequences of _____, which link to form one or more _____, which then fold up into their _____.

2. The proteins called enzymes might number in the _____ in animal cells, but all of these proteins are put together from the same starting set of _____ amino acids.

3. The DNA in your body contains the instructions for making _____.

So Far... Answers:

1. amino acids; polypeptide chains; functional shape

2. thousands; 20

3. proteins

Table 3.3 Summary Table of Biological Molecules

Type of Molecule	Subgroups	Examples and Roles
Carbohydrates	Monosaccharides	Glucose: Energy source
	Disaccharides	Sucrose: Energy source
	Polysaccharides	Glycogen: Storage form of glucose Starch: Carbohydrate storage in plants; used by animals in nutrition Cellulose: Plant cell walls, structure; fiber in animal digestion Chitin: External skeleton of arthropods
Lipids	Triglycerides 3 Fatty acids and glycerol	Fats, Oils (butter, corn oil): Food, energy, storage, insulation
	Fatty acids Components of Triglycerides	Stearic Acid: Food, energy sources
	Steroids Four-ring structure	Cholesterol: Fat digestion, hormone precursor, cell membrane component
	Phospholipids Polar head, nonpolar tails	Cell membrane structure
Proteins	Enzymes Chemically active	Sucrase: Breaks down sugar
	Structural	Keratin: Hair
	Lipoproteins Protein-lipid molecule	HDLs, LDLs: Transport of lipids
	Glycoproteins Protein-sugar molecule	Cell surface receptors
Nucleotides	Adenosine phosphates	Adenosine triphosphate (ATP): Energy transfer
	Nucleic acids Sugar, phosphate group, base	DNA, RNAs: Contain information for and facilitate synthesis of proteins

Chapter Review

Summary

3.1 Biological Molecules

- **The importance of biological molecules**—Living things are largely composed of a few types of substances, known as biological molecules. (page 31)

3.2 Water and Life

- **The necessity of water**—All living things must have a supply of water to fully function. (page 32)
- **Water's special qualities**—Water has several qualities that have strongly affected life on Earth. First, water is a powerful solvent. Second, because its solid form (ice) is less dense than its liquid form, bodies of water in colder climates do not freeze solid in winter, which allows life to flourish under the ice. Third, water has a great capacity to absorb and retain heat. Because of this, the oceans act as heat buffers for the Earth, stabilizing Earth's temperature. Water also acts within us to stabilize our internal temperature, by being released from us as perspiration when we get too hot. (page 33)

- **Hydrophilic and hydrophobic**—Some compounds, such as petroleum, do not interact with water and thus are said to be hydrophobic. Water cannot dissolve hydrophobic compounds, which is why oil and water don't mix. Compounds that do interact with water are polar or carry an electric charge and are called hydrophilic compounds. (page 34)

Web Tutorial 3.1 Chemistry and Water

3.3 Acids and Bases Are Important to Life

- **Acids and bases**—An acid is any substance that yields hydrogen ions when put in solution. A base is any substance that accepts hydrogen ions in solution. A base added to an acidic solution makes that solution less acidic, while an acid added to a basic solution makes that solution less basic. (page 35)
- **Measuring pH**—The concentration of hydrogen ions that a given solution has determines how basic or acidic that solution is, as measured on the pH scale. This scale runs from 0 to 14, with 0 being most acidic, 14 being most basic, and 7 being neutral. Living things function best in

a near-neutral pH, though some systems in living things have different pH requirements.   (page 37)

3.4 Carbon Is a Central Element in Life

- **The importance of carbon**—Biological molecules are built on a carbon framework. Carbon plays this central role because, thanks to its four outer-shell electrons, it is able to form stable, covalent bonds with a wide variety of other elements.   (page 38)

3.5 Carbohydrates

- **The nature of carbohydrates**—Carbohydrates are formed from the building blocks or monomers of simple sugars, such as glucose. These can be linked to form the larger carbohydrate polymers such as starch, glycogen, cellulose, and chitin.   (page 40)

3.6 Lipids

- **The nature of lipids**—Lipids are a diverse group of molecules whose main common feature is their relative insolubility in water. Among the most important lipids are the triglycerides, composed of a glyceride and three fatty acids. Most of the fats we consume are triglycerides. Another important variety of lipids is the steroids, which include cholesterol, and such hormones as testosterone and estrogen.   (page 42)

- **Saturated and unsaturated**—A saturated fatty acid is a fatty acid with no double bonds between the carbon atoms of its hydrocarbon chain. Mono- and polyunsaturated fatty acids have, respectively, one or two or more double bonds between the carbon atoms of their hydrocarbon chains.   (page 44)

3.7 Proteins

- **The nature of proteins**—Proteins are a diverse group of biological molecules composed of the monomers called amino acids, which link together to form polypeptide chains. These chains then fold up into completed proteins. Important groups of proteins include enzymes, which hasten chemical reactions, and structural proteins, which make up such structures as hair. The activities of proteins are determined by their final, folded shapes.   (page 46)

Web Tutorial 3.2 Proteins

3.8 Nucleotides and Nucleic Acids

- **The nature of nucleotides and nucleic acids**—Nucleic acids are polymers composed of units called nucleotides. The nucleic acid DNA is composed of nucleotides that contain a phosphate group, a sugar (deoxyribose), and one of four nitrogen-containing bases. DNA is a repository of genetic information. The sequence of its bases encodes the information for the production of the huge array of proteins produced by living things.   (page 50)

Web Tutorial 3.3 Nucleic Acids

Key Terms

acid p. 35	glycoprotein p. 49	nucleotide p. 50	protein p. 48
alkaline p. 37	hydrocarbon p. 34	oil p. 44	ribonucleic acid (RNA) p. 50
base p. 35	hydrophilic p. 34	organic chemistry p. 39	saturated fatty acid p. 44
carbohydrate p. 40	hydrophobic p. 34	pH scale p. 37	simple sugar p. 41
cellulose p. 42	hydroxide ion p. 35	phosphate group p. 46	solute p. 32
chitin p. 42	lipid p. 42	phospholipid p. 46	solution p. 32
cholesterol p. 45	lipoprotein p. 49	polymer p. 40	solvent p. 32
DNA (deoxyribonucleic acid) p. 50	monomer p. 40	polypeptide p. 48	specific heat p. 33
fatty acid p. 44	monosaccharide p. 41	polysaccharides p. 42	starch p. 42
glycogen p. 42	monounsaturated fatty acid p. 44	polyunsaturated fatty acid p. 44	steroids p. 45
			triglyceride p. 44

Testing Your Understanding

In Your Own Words (answers in the back of the book)

1. Describe two ways that water serves as a heat buffer.

2. Both low-density lipoproteins (LDLs) and high-density lipoproteins (HDLs) are involved with carrying fats through the bloodstream. If your LDL count is unusually high, should you be concerned? What if your HDL count is high? Why are they different?

3. List as many functions of proteins as possible.

Thinking about What You've Learned

1. Nutritional experts are constantly stressing the idea that complex carbohydrates, such as whole-wheat bread, are nutritionally preferable to simple carbohydrates, such as white bread, in part because complex carbohydrates don't result in the quick "spikes" in blood sugar that come with simple carbohydrates. Based on what you've learned about carbohydrate structure, why should simple carbohydrates result in blood-sugar spikes, while complex carbohydrates do not?

2. To ensure a healthy heart, is it reasonable to simply avoid all fats as much as possible?

3. Many species of beans produce a poisonous group of glycoproteins called lectins, which cause red blood cells to clump together (agglutinate) and cease to function. (Fortunately, cooking destroys most of them.) Using what you've learned about their structure and function, suggest how glycoproteins might cause agglutination. What benefit do you think plants might get just by causing red blood cells to clump together?

Exploring Life's Home: The Cell

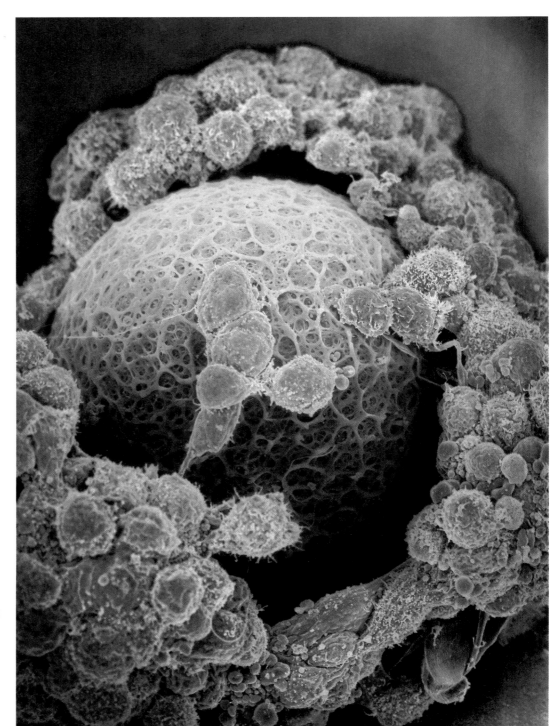

The largest cell in the human body, the egg produced by females, surrounded by a layer of accessory cells.

SCIENCE IN THE NEWS

THE STAR-LEDGER

BRAIN NERVE CELLS TRIGGER ADDICTS' RELAPSE—RUTGERS PROF LINKS ENVIRONMENTAL STIMULI TO MEMORIES THAT REVIVE CRAVING FOR DRUGS

By Angela Stewart
August 14, 2003

Relapse among recovering addicts now can be linked to specific nerve cells in the brain that respond to environmental stimuli associated with prior drug use, a team of Rutgers University researchers has concluded in a new study.

SCIENCE IN THE NEWS

THE CAPITAL TIMES

UW STUDY: STEM CELLS GROW INTO HEART CELLS

By Aaron Nathans The Capital Times
June 26, 2003

Researchers at the University of Wisconsin say they've found evidence that human embryonic stem cells in a lab can grow into the three major kinds of muscle cells found in the heart.

4.1 Cells Are the Working Units of Life

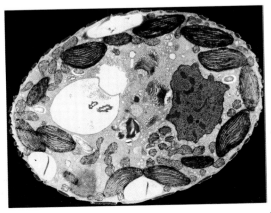

There's a fair amount of territory covered in the two newspaper stories above—the nature of drug addiction and the growing of new heart tissue. But note what was at work in both cases: the activities of cells. This actually is to be expected, however. So central are cells to the living world that it would be a challenge to name any major function of life that is *not* carried out by them. There are hormones that travel through the bloodstream—outside of any cells—but hormones are invariably produced by cells, and they have their effects on cells. The tissue that makes up your bones lies outside of cells, but it is cells that secrete layer after layer of this calcified material. The muscles that allow you to walk and talk are essentially collections of long, thin cells that are lined up together like so many pencils in a box.

Given activities as diverse as these, cells clearly are able to specialize to a great degree—muscle cells are specialized for contraction, nerve cells for signal transmission, and so forth. Running like a thread through this diversity, however, is the unity of cellular life. Every form of life either is a cell, or is composed of cells. The one possible exception to this is viruses, but even they must use the machinery of cells in order to reproduce. There is unity, too, in the way cells come about: Every cell comes *from* a cell. Human beings are incapable of producing cells from scratch in the laboratory, and so far as we can tell, nature has fashioned cells from simple molecules only once—back when life on Earth got started. The fact that all cells come from cells means that each cell in your body is a link in a cellular chain that stretches back more than 3.5 billion years.

Given all this, if we ask how important cells are to the living world, we can see that the answer is: really important. The purpose of this chapter is to give you a basic understanding of cells as the fundamental unit of life (**see** FIGURE 4.1).

4.2 All Cells Are Either Prokaryotic or Eukaryotic

What is the nature of the cell? In answering this question, we first have to make a distinction between two fundamentally different types of cells—prokaryotic and eukaryotic cells. Every cell that exists is one or the other, and this simple either-or quality extends to the organisms that fall into these groups. All prokaryotic cells are either bacteria or another single-celled form of life known as archaea. Setting bacteria and archaea aside, *all other cells* are eukaryotic. This means all the cells in plants, in animals, in fungi, in yourself—all the cells in every living thing except in the bacteria and the archaea. The name *eukaryote* comes from the Greek *eu*, meaning "true," and *karyon,* meaning "nucleus," while *prokaryote* means "before nucleus." These terms describe the most critical distinction between the two cell types. **Eukaryotic cells** are cells whose primary complement of DNA is enclosed within a nucleus. **Prokaryotic cells** are cells whose DNA is not enclosed within a nucleus. To complete this circle, a **nucleus** is a membrane-lined compartment that encloses the primary complement of DNA in eukaryotic cells.

FIGURE 4.1   Nature's Fundamental Unit

(a) Human red and white blood cells inside a blood vessel. The large, dark ovals that can be seen in the background are flat cells that form the interior lining of the blood vessel.

(b) Cells of the fungus brewer's yeast, which is used to make wine, beer, and bread.

(c) Cells of a spinach leaf, with the leaf seen in cross section. Two layers of outer or "epidermal" cells can be seen at top and bottom; many of the cells in between perform photosynthesis.

(a)

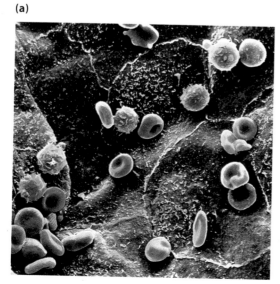

(b)

(c)

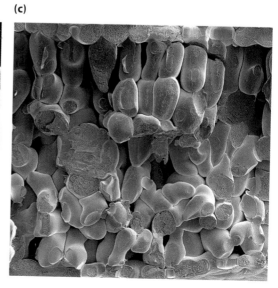

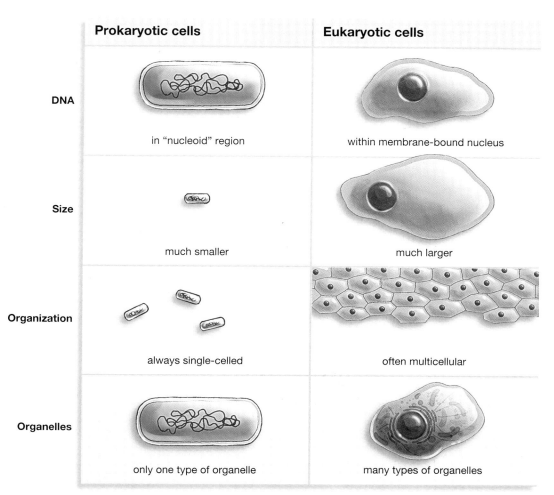

	Prokaryotic cells	Eukaryotic cells
DNA	in "nucleoid" region	within membrane-bound nucleus
Size	much smaller	much larger
Organization	always single-celled	often multicellular
Organelles	only one type of organelle	many types of organelles

FIGURE 4.2 Prokaryotic and Eukaryotic Cells Compared A prokaryotic cell is a self-contained organism, since the prokaryotes—bacteria and archaea—are all single-celled. Eukaryotic organisms, which may be single- or multicelled, include plants, animals, fungi, and another group called protists.

While having a nucleus is the most important difference between eukaryotic and prokaryotic cells, it is not the only difference. It may seem sensible to think of two single-celled creatures—one a prokaryote, the other a eukaryote—as being very similar, but the distance between them as life-forms is immense. Human beings and chimpanzees are nearly identical in comparison. Eukaryotic cells tend to be much larger than their prokaryotic counterparts; indeed, thousands of bacteria could easily fit into an average eukaryotic cell. Eukaryotes are quite often multicelled organisms, while prokaryotes are always single-celled (**see** FIGURE 4.2).

Perhaps the most notable distinction between prokaryotic and eukaryotic cells, though, is that eukaryotic cells are *compartmentalized* to a far greater degree than is the case with prokaryotes. The nucleus in eukaryotic cells turns out to be only one variety of **organelle**: a highly organized structure, internal to a cell, that serves some specialized function. Eukaryotic cells contain several different kinds of these "tiny organs." There are organelles called mitochondria, for example, that transform energy from food, and organelles called lysosomes that recycle the raw materials of the cell. In prokaryotic cells, meanwhile, there is only a single type of organelle—a kind of workbench for producing proteins we'll look at later.

If we look at eukaryotes that are familiar to us, such as trees, and mushrooms, and horses, it's easy to see that there is a fantastic diversity in the *forms* of eukaryotes, compared to the strictly single-celled prokaryotes. It does not follow from this, however, that prokaryotes are uniform, nor that they are unsuccessful. On the contrary, prokaryotes differ greatly from one another in, say, the way they obtain nutrients, and they are extremely successful, if success is defined as living in a lot of places in huge numbers. As Lynn Margulis and Karlene Schwartz have observed, there are more bacteria living in your mouth right now than the number of

A BROADER VIEW

The Diversity of Bacteria

Who is more diverse, animals or bacteria? It's true that animals range from snails to whales while all bacteria are single-celled. But all animals must get their nutrition from outside themselves. Some bacteria do this as well, but others make their own food, through photosynthesis. All animals must have oxygen to exist; some bacteria need it too, but others can take it or leave it, while still others are poisoned by it. Animals have diverse forms, but bacteria have diverse "metabolisms" or means of carrying out life-supporting chemical reactions.

Components of eukaryotic cells

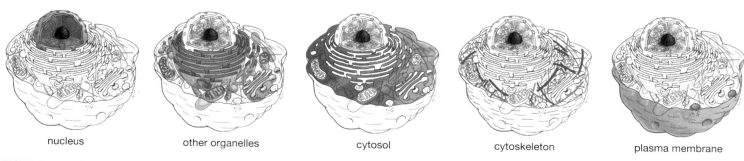

nucleus   other organelles   cytosol   cytoskeleton   plasma membrane

FIGURE 4.3   **The Eukaryotic Cell** This cutaway view of a eukaryotic cell displays the elements that nearly all such cells possess: a nucleus, other membrane-bound organelles, a jelly-like cytosol, a cytoskeleton, and an outer plasma membrane.

people who have ever existed. But for the range of biological structures and processes we want to look at, eukaryotic cells are more diverse, and hence are the initial focus of our study.

The Eukaryotic Cell

What are the constituent parts of eukaryotic cells? Here are five larger structures that in turn have smaller structures within them. The first two have been mentioned already (**see** FIGURE 4.3).

- The nucleus

- Other organelles (which lie outside the nucleus)

- The **cytosol**, a protein-rich, jelly-like fluid in which the cell's organelles are immersed

- The *cytoskeleton*, a kind of internal scaffolding consisting of three sorts of protein fibers

- The **plasma membrane**, the outer lining of the cell

In any discussion of a eukaryotic cell, you are likely to hear the term **cytoplasm**, which simply means the *region* of the cell inside the plasma membrane but outside the nucleus. The cytoplasm is different from the cyto*sol*. If you removed all the structures in the cytoplasm—meaning the organelles and the cytoskeleton—what would be left is the cytosol, which is mostly water. This does not mean that the cytosol is simply a passive medium for the other structures. But it is not an organized structure in the way the organelles are. Almost all the organelles you'll see are encased in their own membranes, just as the whole cell is encased in its plasma membrane. What are membranes? Until you get the formal definition next chapter, think of a membrane as the flexible, chemically active outer lining of a cell or of its compartments. In the balance of this chapter, we'll explore all of the constituent parts listed above except for the plasma membrane, which is so special it gets its own chapter.

We'll now begin looking at a specific type of eukaryotic cell that will occupy us for much of the rest of this chapter. This is the animal cell, which is to say the type of cell that makes up our own bodies. (We'll come back and take a look at the plant cell toward the end of the chapter.) Insofar as we can characterize the typical animal cell (**see** FIGURE 4.4), it is roughly spherical, surrounded by, and linked to, cells of similar type, immersed in water; and about 25 micrometers (μm) in diameter, meaning that about 30 of them could fit side by side within the period at the end of this sentence. (To find out what a micrometer is, and to learn more about the size of different inhabitants of the microworld, see Essay 4.1, "The Size of Cells," on page 60.)

A BROADER VIEW

First Voyager to the Micro-world

Who was the first person to behold a living cell? The Dutchman Anton van Leeuwenhoek, who in the 1670s began to report on what he had seen under the microscopes he built. No one before him thought that anything smaller than a worm could exist in the human body. Hence, the microbes he discovered, in the body and elsewhere, were a revelation to everyone, including him. Here he is, reporting on the scrapings of his own mouth: "I saw, with as great a wonderment as ever before, an inconceivably great number of little animalcules, and in so unbelievably small a quantity of the foresaid stuff, that those who didn't see it with their own eyes could scarce credit it."

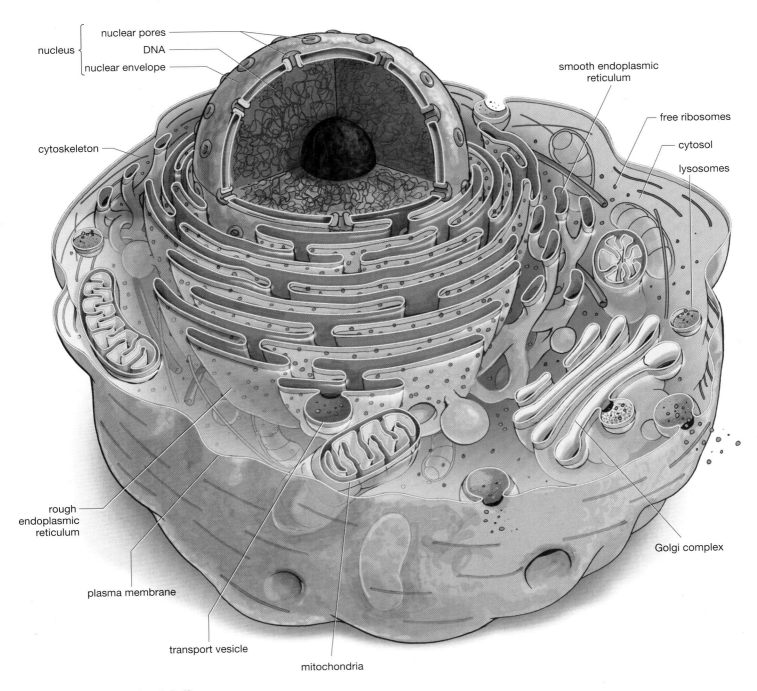

nucleus
nuclear pores
DNA
nuclear envelope
cytoskeleton
smooth endoplasmic reticulum
free ribosomes
cytosol
lysosomes
rough endoplasmic reticulum
plasma membrane
transport vesicle
mitochondria
Golgi complex

FIGURE 4.4 **The Animal Cell**

4.3 A Tour of the Animal Cell: The Protein Pathway

One way to think of a cell is as a living factory. In line with this, we'll begin learning about the animal cell by tracing the way it manufactures a product—a protein—for export outside itself. Just as a new employee might tour an assembly line as a means of learning about factory equipment, so you are going to follow the path of protein production as a means of learning about cell equipment.

Essay 4.1 The Size of Cells

For several chapters, you've been studying atoms and ions and molecules and such—things small enough to be invisible to the unaided eye. For the most part, you've had to imagine what these things look like, simply because most of them are so small that we either have no pictures of them at all (as with electrons) or few clear pictures (as with atoms). If you flip through the pages of this chapter and those to come, however, you begin to see a fair number of actual photographs. They don't have the same quality as summer vacation snapshots, but they are recognizable as pictures, or more properly as micrographs (pictures taken with the aid of a microscope). Micrographs enable us to see surprising things; for example, hundreds of bacteria on the tip of a pin (FIGURE E4.1.1). So, with the cells that are introduced in this chapter, there has obviously been a bump up in size into a world that is more easily visible with the help of various kinds of technology.

How Small Are They?

In taking stock of the microworld, two units of measure are particularly valuable. The smaller of them is the nanometer, which is a billionth of a meter and is abbreviated as nm. The larger is the micrometer, which is a millionth of a meter, abbreviated μm. (The unfamiliar-looking first letter there is the Greek symbol for a small *m*. Scientists are not trying to be purposely obscure in using it; another unit of the metric system, the millimeter, lays claim to the mm abbreviation.) A meter equals about 39.6 inches, or just over a yard, which gives you some starting sense of physical reality in understanding the rest of these sizes.

Now look at FIGURE E4.1.2 to see what size various objects are. Atoms are down at the bottom of the scale, at about a tenth of a nanometer. Something less than a tenfold increase from there gets us to 1 nm, the size of the protein building blocks, called amino acids, that you looked at last chapter. Another tenfold-plus increase to 10 nm and you've reached the upper limit on proteins.

Most of the cells in an elephant are no bigger than those in an ant.

You have to go better than 10 times larger than this, to 100 nm, before you arrive at the size of something that is actually *living*, as opposed to something that is a component part of life. Because, as you've seen, life means cells, this means the smallest cells in existence, which are bacteria measuring perhaps 200–300 nm. This extreme on the small side of cells has a counterpart on the large side with the single cells we call chicken eggs, and with certain nerve cells that can stretch out to a meter in length. In general, however, we just cross into the micrometer range with the smaller bacterial cells: about 1–10 μm. The cell size for most plants and animals falls in a range that is a little less than 10 times larger than this, about 10–30 μm. At 10 μm, perhaps a billion cells could fit into the tip of your finger. And how about your whole body? One estimate is 100 trillion.

(a) Bacteria on a pin, magnified x 85

(b) Magnified x 425

(c) Magnified x 2100

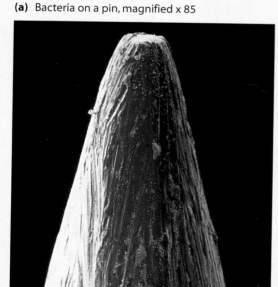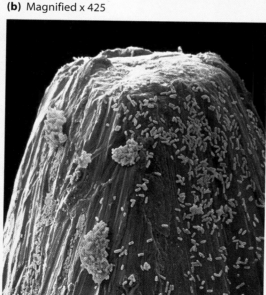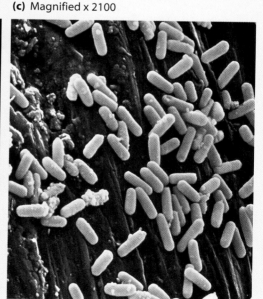

FIGURE E4.1.1 **Hidden Life** Microscope enlargements of the tip of this pin show an abundance of life—in this case bacteria—thriving on an object that we normally think of as being devoid of living organisms.

So, Why So Small?

Having learned how small cells are, you might then well ask: *Why* are they so small? As noted, cells are small chemical factories, and just like any factory, they are constantly shipping things in and out. A primary size-limiting factor for cells is having enough surface area to export and import all that they need.

This limitation comes about because of a fundamental mathematical principle: As the surface area of an object increases, its *volume* increases even more. Say you have a cube, 1 inch long on each of its sides. Its surface area is 6 square inches (length × width × number of sides), while its volume is one cubic inch (length × width × height). Now say you increase the side dimension to 8 inches. The surface area goes from 6 to 384 square inches, but the *volume* goes from 1 to 512 cubic inches. Where, at a 1-inch dimension, there were *six* square inches of surface area for every cubic inch of volume, now there are only *three-quarters* of an inch of surface for every cubic inch of volume. Beyond a certain volume, then, a cell simply would not have enough surface area to import and export all the materials it needs. This effectively sets an upper limit on how big cells can be and helps explain why most of the cells in an elephant are no bigger than those in an ant, though the elephant does have more cells than the ant. **FIGURE E4.1.3** shows you some size comparisons in the microworld.

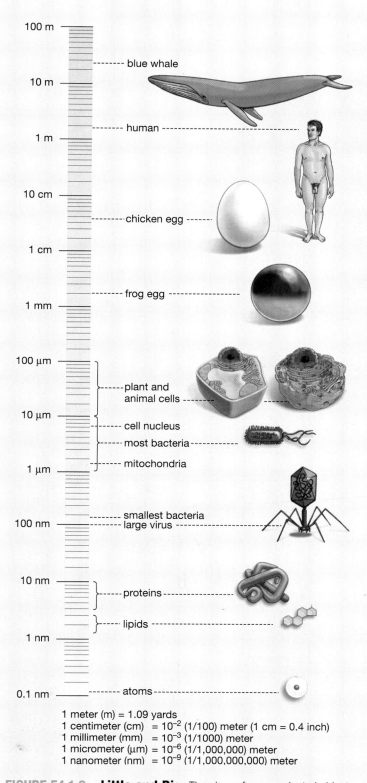

1 meter (m) = 1.09 yards
1 centimeter (cm) = 10^{-2} (1/100) meter (1 cm = 0.4 inch)
1 millimeter (mm) = 10^{-3} (1/1000) meter
1 micrometer (µm) = 10^{-6} (1/1,000,000) meter
1 nanometer (nm) = 10^{-9} (1/1,000,000,000) meter

FIGURE E4.1.2   Little and Big The sizes of some selected objects in the natural world.

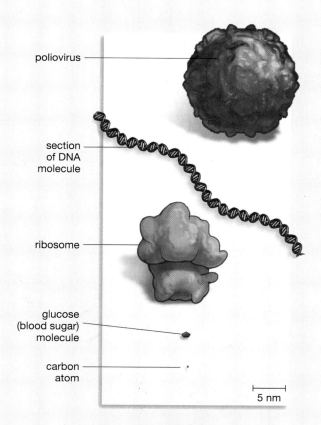

FIGURE E4.1.3   Small Is a Relative Thing The sizes and shapes of five natural-world entities, each magnified a million times.

FIGURE 4.5 **Path of Protein Production in Cells**

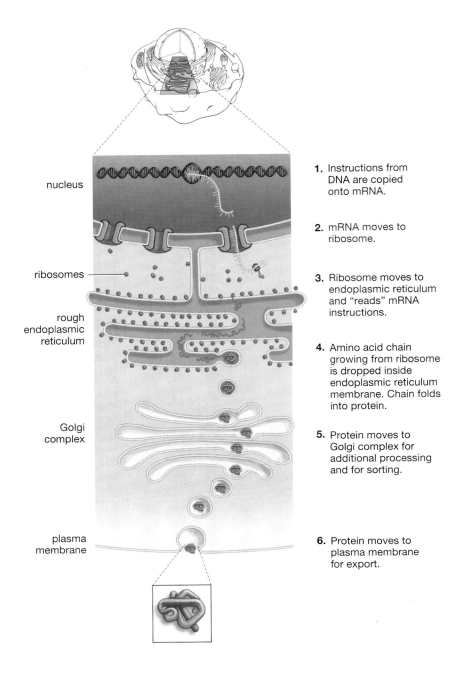

nucleus

1. Instructions from DNA are copied onto mRNA.

2. mRNA moves to ribosome.

ribosomes

3. Ribosome moves to endoplasmic reticulum and "reads" mRNA instructions.

rough endoplasmic reticulum

4. Amino acid chain growing from ribosome is dropped inside endoplasmic reticulum membrane. Chain folds into protein.

Golgi complex

5. Protein moves to Golgi complex for additional processing and for sorting.

plasma membrane

6. Protein moves to plasma membrane for export.

FIGURE 4.5 shows the path you'll be taking, from nucleus to the outer edge of the cell. Don't be bothered by the unfamiliar terms in Figure 4.5, because they'll all be explained in the text. The important thing is that, before we begin our tour, you have some sense of the path that protein production takes.

Beginning in the Control Center: The Nucleus

As noted in Chapter 3, proteins are critical working molecules in living things, and DNA contains the information for producing these proteins. Our entire complement of DNA is like a cookbook that contains individual recipes, which we call genes. Each gene's chemical building blocks in effect say, "Now give me some of this, now some of this, then some of this," the final result being the specifications for a protein. (See Chapter 3, p. 50.) In the eukaryotic cell, as you've seen, DNA is largely confined within a nucleus bound by a membrane. If you look at FIGURE 4.6, you can see the nature of this membrane. It is the **nuclear envelope**: the double membrane that lines the nucleus in eukaryotic cells.

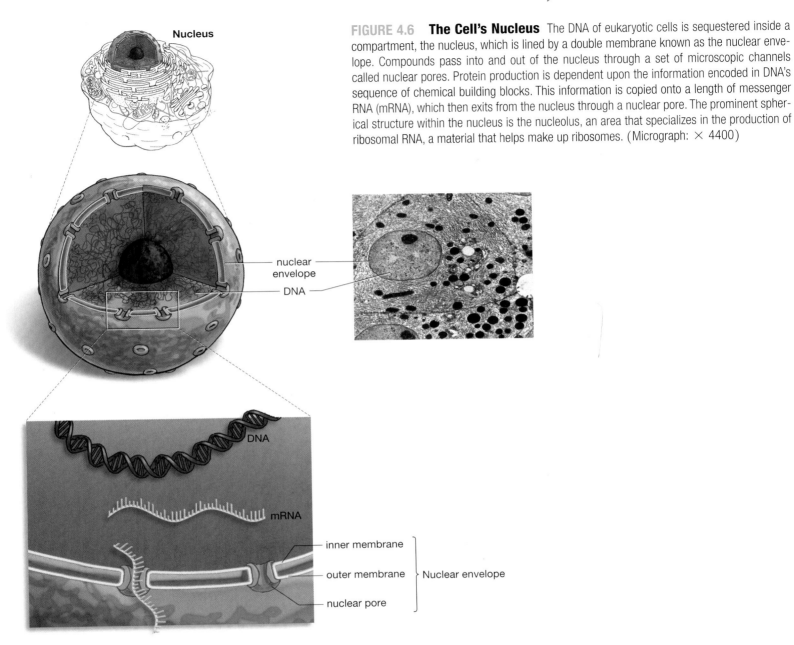

FIGURE 4.6 **The Cell's Nucleus** The DNA of eukaryotic cells is sequestered inside a compartment, the nucleus, which is lined by a double membrane known as the nuclear envelope. Compounds pass into and out of the nucleus through a set of microscopic channels called nuclear pores. Protein production is dependent upon the information encoded in DNA's sequence of chemical building blocks. This information is copied onto a length of messenger RNA (mRNA), which then exits from the nucleus through a nuclear pore. The prominent spherical structure within the nucleus is the nucleolus, an area that specializes in the production of ribosomal RNA, a material that helps make up ribosomes. (Micrograph: × 4400)

There comes a point in the life of most cells when they divide, one cell becoming two. Because (with a few exceptions) all cells must possess the set of instructions that are contained in DNA, it follows that when a cell divides, its original complement of DNA must *duplicate*, so that both cells that result from the cell division can have their own DNA. The nucleus, then, is not just the site where DNA exists; it is the site where new DNA is put together, or "synthesized," for this duplication.

Messenger RNA

At the end of Chapter 3, you saw that the process of protein synthesis requires that DNA's instructions first get copied onto another long-chain molecule, RNA. This step is akin to having a cassette tape (of DNA) and then making a copy of it (onto RNA). Our RNA "tape" then moves out to the cell's cytoplasm to continue the process of protein synthesis. As it turns out, RNA comes in several forms. The one that the DNA instructions are copied onto is called *messenger RNA* (mRNA). Given

FIGURE 4.7 **Proteins Taking Shape: Ribosomes and the Rough Endoplasmic Reticulum** The steps of the lower figure begin with a messenger RNA "tape," which has migrated from the nucleus of a cell. (**1**) The tape binds with a ribosome that is free-standing in the cytoplasm. The ribosome then starts to "read" the length of the mRNA, which bears instructions for producing a protein. The result is an amino acid (or polypeptide) chain that begins to grow from one part of the ribosome. The ribosome soon halts this activity, however, and moves toward the rough endoplasmic reticulum (rough ER). (**2**) The ribosome docks on the outside face of the rough ER and resumes reading the mRNA tape. When it is completed, the resulting amino acid chain drops into the cisternal space of the rough ER. (**3**) The amino acid chain folds up, thus becoming a protein. (**4**) The protein undergoes processing (in this example, by having a side-chain added). (**5**) The protein is then encased in a membrane vesicle, which will bud off from the rough ER and be transported to the Golgi complex. (Micrograph: × 90,500)

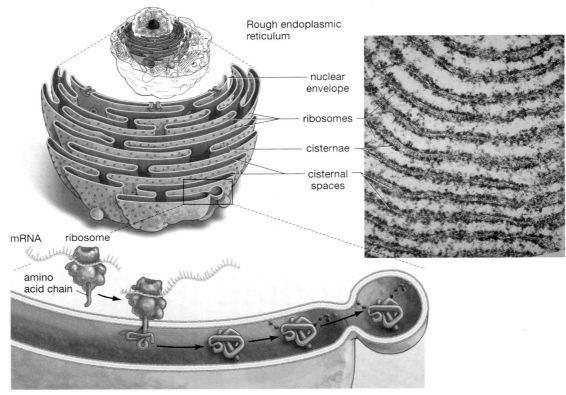

Rough endoplasmic reticulum

nuclear envelope

ribosomes

cisternae

cisternal spaces

mRNA   ribosome

amino acid chain

1. mRNA docks on ribosome. Amino acid chain production begins.

2. Ribosome docks on ER. Amino acid chain moves into cisternal space as it is completed.

3. Amino acid chain folds up, making a protein.

4. Side chains added to protein.

5. Vesicle formed to house protein while in transport.

that mRNA goes to the cytoplasm, it must, of course, have some way of getting out of the nucleus; its exit points turn out to be thousands of channels that stud the surface of the nuclear envelope—the *nuclear pores*. Materials can go the other way through the nuclear pores as well: Proteins and other materials pass from the cytoplasm into the nucleus by way of them.

Ribosomes

So where do our messenger RNA tapes go once they've left the nucleus? Their destinations are a set of small structures called ribosomes. For now, we can define the **ribosome** as an organelle that serves as the site of protein synthesis in the cell. In line with this, ribosomes are commonly described as the "workbenches" of protein synthesis, which is a fine metaphor. But following the notion of mRNA as a cassette tape, you might look at a ribosome as a kind of *playback head* on a cassette deck. What does such a head do in an actual deck? As a tape is run through it, it reads signals that have been laid down on it (as magnetic bits) and turns these into sound. Likewise, each ribosome acts as a site that an mRNA tape runs through, only the information on this tape results in the production of an *object* that grows from the ribosome: a chain of amino acids that folds up into the molecule we call a protein (**see FIGURE 4.7**).

You may remember that the kind of protein we are tracking will eventually be exported out of the cell altogether. The synthesis of this kind of protein actually stops when only a very short sequence of the amino acid chain has grown from the ribosome. Why? Amino acid chains destined to become "export" proteins need to be processed within other structures in the cell before they can become fully functional proteins. The first step in this process is for the ribosome, and its associated amino acid chain, to migrate a short distance in the cell and then attach to another cell structure.

The Rough Endoplasmic Reticulum

If you look again at Figure 4.4, you can see that, though the nucleus cuts a roughly spherical figure out of the cell, there is, in essence, a folded-up continuation of the nuclear envelope on one side. This mass of membrane has a name that is a mouthful: the **rough endoplasmic reticulum**. This structure can be defined as a network of membranes that aids in the processing of proteins in eukaryotic cells. The rough endoplasmic reticulum is rough because it is studded with ribosomes; it is endoplasmic because it lies within (endo) the cyto*plasm;* and it is a reticulum because it is a network, which is what *reticulum* means in Latin. Understandably, it is generally referred to as the rough ER or the RER.

Our ribosome, bound up with its mRNA and amino acid chain, will migrate to the rough ER and dock on its outside face, thus joining many other ribosomes that have done the same thing. Remember, the amino acid (or polypeptide) chain that is being output from the ribosome is, in essence, an unfinished protein that needs to go through more processing before it can be exported. The first step in this processing leads only to the other side of the ER wall the ribosome is embedded in. As the ribosome goes on with its work, the amino acid chain it is producing drops into chambers inside the rough ER.

If you look at Figure 4.7, you can see that the whole of the rough ER takes the shape of a set of flattened sacs (called *cisternae*). The membrane that the rough ER is composed of forms the periphery of these sacs. As amino acid chains drop into the sacs, they first fold up into their protein shapes, as you saw in Chapter 3. Beyond this, most proteins that are exported from cells have sugar side-chains added to them here.

Elegant Transportation: Transport Vesicles

The proteins that have been processed within the rough ER need to move out of it and to the next station of the assembly line before being exported. But how do proteins move from one location to another within the cell? Recall that the rough ER and the nuclear membrane amount to one long, convoluted membrane. And, as just noted, all the organelles in the cell except ribosomes are "membrane-enclosed," meaning they have membranes surrounding them in the way that icing surrounds a cake. Each of these membranes has its own chemical structure, but collectively the membranes of all the organelles in a cell have an amazing ability to work together: A piece of one membrane can *bud off*, as the term goes, carrying inside it some of our proteins-in-process. Moving through the cytosol, this tiny sphere of membrane can then *fuse* with another membrane-bound organelle, releasing its protein cargo in the process. The membrane spheres that move within this network, carrying proteins and other molecules, are called **transport vesicles**. The network itself is known as the **endomembrane system**: an interactive group of membrane-enclosed organelles and transport vesicles within eukaryotic cells.

This system gives cells a remarkable capability. One minute a piece of membrane may be an integral part of, say, the rough ER; the next it is separating off as a spheroid and moving through the cytosol, carrying proteins within. It is this system that makes it possible for our proteins-in-process to move out of the rough ER. Note, though, that many different *kinds* of proteins are being processed at any one time in the rough ER cisternae. It is as if the cellular factory has a lot of different assembly lines working at once. Most of the proteins under construction are, however, initially bound for the same place—the Golgi complex.

Downstream from the Rough ER: The Golgi Complex

What happens to a transport vesicle, bearing proteins, after it has budded off from the rough ER? It moves through the cytosol to fuse with the membrane of another organelle, the **Golgi complex**: a network of membranes that processes and distributes

A BROADER VIEW

What Are Membranes Really Like?

Christian de Duve, who won the Nobel Prize for his work on the component parts of cells, has suggested that common soap bubbles give us an idea of the nature of membranes. Soap bubbles are like membranes in that they are self-sealing—when you cut a soap bubble in half, you don't get two half-bubbles; you get two smaller whole bubbles. Likewise, two soap bubbles can come together and fuse to create a single larger bubble (a trick you can carry out with bubble-blowing liquid). Soap bubbles and membranes differ, however, in terms of flexibility. Though we think of soap bubbles as extremely flexible, they actually are rigid compared to cellular membranes.

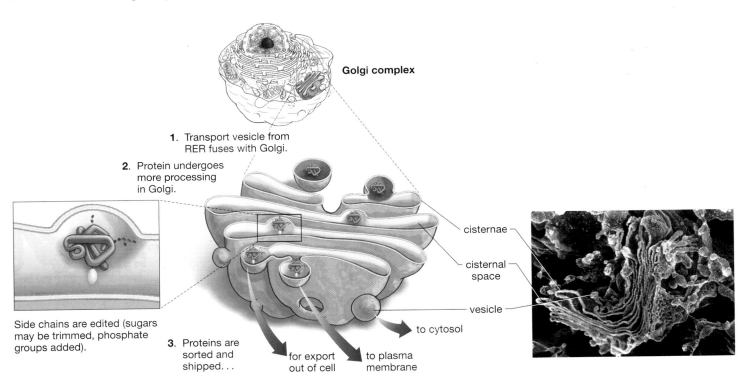

Golgi complex

1. Transport vesicle from RER fuses with Golgi.

2. Protein undergoes more processing in Golgi.

Side chains are edited (sugars may be trimmed, phosphate groups added).

3. Proteins are sorted and shipped. . .

for export out of cell

to plasma membrane

to cytosol

cisternae

cisternal space

vesicle

FIGURE 4.8 **Processing and Routing: The Golgi Complex** Transport vesicles from the rough endoplasmic reticulum (RER) move to the Golgi complex, where they unload their protein contents by fusing with the Golgi membrane. The protein is then passed, within other vesicles, through the layers of Golgi cisternae, where editing of the protein may occur. At the part of the Golgi furthest from the rough ER, the proteins are sorted, packed in vesicles, and shipped to sites mostly in cell membranes or outside the cell altogether. The vesicles in the micrograph are the pink and purple spheres.

proteins that come to it from the rough endoplasmic reticulum. Some side-chains of sugar may be trimmed from proteins here, or phosphate groups may be added. But the Golgi complex does something else as well. Recall that some proteins in this production line are bound for export outside the cell, while others will end up being used within various membranes in the cell. It follows that proteins have to be *sorted and shipped* appropriately, and the Golgi does just this, acting as a kind of distribution center. Chemical "tags" that are part of the proteins often allow for this routing.

The Golgi is similar to the ER in that it amounts to a series of cisternae, or connected membranous sacs with internal spaces (see FIGURE 4.8). Proteins arrive at the Golgi housed in transport vesicles that fuse with the Golgi "face" nearest the RER, at which point the vesicles release their protein cargo into the Golgi cisternal space for processing. Once processed, proteins of the sort we are following eventually bud off from the outside face of the Golgi, now housed in their final transport vesicles.

From the Golgi to the Surface

For export proteins, the journey that began with the copying of DNA information onto messenger RNA is almost over. Once a vesicle buds off from the Golgi, all that remains is for it to make its way through the cytosol to the plasma membrane at the outer reaches of the cell. There, the vesicle fuses with the plasma membrane of the cell and the protein is ejected into the extracellular world. This last step, called exocytosis, is a process you'll be looking at next chapter. With it, one finished product of the cellular factory has rolled out the door.

So Far...

1. What forms of life exist outside of cells?

2. Which of the following metaphors best describes a cell's DNA: workbench, cookbook, shipping center.

3. Ribosomes are the sites in cells at which _____ are put together.

So Far... Answers

1. *Arguably no forms at all; the only possible exception is viruses, and even they cannot reproduce outside of cells.*

2. *Cookbook; a cell's DNA contains "recipes" that specify different proteins.*

3. *proteins*

4.4 Other Cell Structures

A functioning cell engages in more activities than the protein synthesis and shipment just reviewed, for the simple reason that cells do a lot more than produce proteins.

The Smooth Endoplasmic Reticulum

If you look back to Figure 4.4, you can see that there actually are *two* kinds of endoplasmic reticuli. The part of the ER membrane, farther out from the nucleus, that has no ribosomes is called the smooth endoplasmic reticulum or smooth ER. It's "smooth" because it is not peppered with ribosomes, and this very quality means it is not a site of protein synthesis. Instead, the **smooth endoplasmic reticulum** is a network of membranes that is the site of the synthesis of various lipids, and a site at which potentially harmful substances are detoxified within the cell. The tasks the smooth ER undertakes, however, will vary in accordance with cell type. The lipids we normally think of as "fats" are synthesized and stored in the smooth ER of liver and fat cells, while the "steroid" lipids reviewed last chapter—testosterone and estrogen—are synthesized in the smooth ER of the ovaries and testes. The detoxification of potentially harmful substances, such as alcohol, takes place largely in the smooth ER of liver cells.

Tiny Acid Vats: Lysosomes and Cellular Recycling

Any factory must be able to get rid of some old materials, while recycling others. A factory also needs new materials, brought in from the outside, that probably will have to undergo some processing before being used. A single organelle in the animal cell aids in doing all these things. It is the **lysosome**: an organelle found in animal cells that digests worn-out cellular materials and foreign materials that enter the cell. Several hundred of these membrane-bound organelles may exist in any given cell. You could think of them as sealed-off acid vats that take in large molecules, break them down, and then return the resulting smaller molecules to the cytosol. What they cannot return, they retain inside themselves or expel outside the cell. They carry out this work not only on molecules entering the cell from the outside (say, invading bacteria) but on materials that exist inside the cell—on worn-out organelle parts, for example (see FIGURE 4.9).

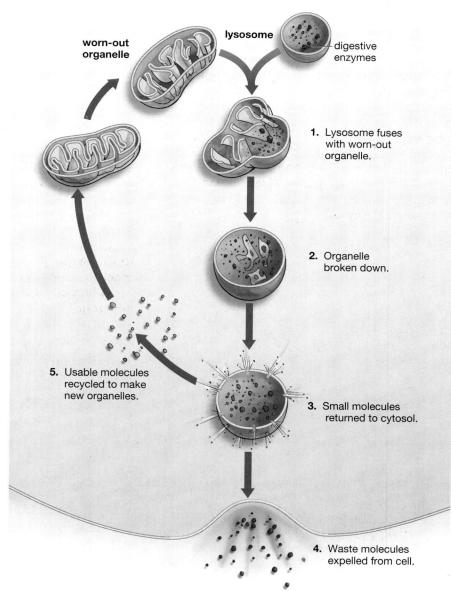

worn-out
organelle

lysosome

digestive enzymes

1. Lysosome fuses with worn-out organelle.

2. Organelle broken down.

3. Small molecules returned to cytosol.

4. Waste molecules expelled from cell.

5. Usable molecules recycled to make new organelles.

FIGURE 4.9   Cellular Recycling: Lysosomes Lysosomes are membrane-enclosed organelles that contain potent enzymes capable of digesting large molecules, such as worn-out organelles. The useful parts of such organelles will be returned to the cytosol and used elsewhere—a form of cellular recycling. If a lysosome cannot digest a given material, it may expel it outside the cell, though, in multicelled organisms, lysosomes generally will hold on to such materials, so as not to harm structures outside the cell. Lysosomes also digest small particles, such as food, that come from outside the cell, along with invading organisms, such as bacteria.

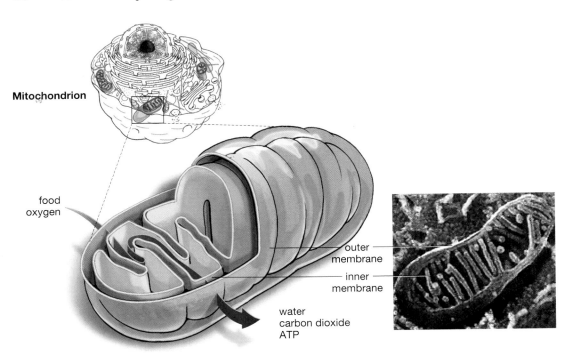

Mitochondrion

food
oxygen

outer
membrane

inner
membrane

water
carbon dioxide
ATP

A given lysosome may be filled with as many as 40 different enzymes that can break larger molecules into their component parts—an enzymatic array that allows each lysosome to break down most of what comes its way. A lysosome gets ahold of its prey through the endomembrane system. A lysosome will fuse with the membrane surrounding a worn-out organelle part. Proceeding to engulf it, the lysosome then goes to work breaking the organelle down. The small molecules that result then pass freely out of the lysosome and into the cytosol for reuse elsewhere. Thus, there is recycling at the cellular level.

FIGURE 4.10 Energy Transformers: Mitochondria Just as a power plant converts the energy contained in coal into useful electrical energy, mitochondria convert the energy contained in food into a useful chemical form of energy for the cell—the molecule ATP. Cells can contain anywhere from a single mitochondrion to several thousand.

Extracting Energy from Food: Mitochondria

Just as there is no such thing as a free lunch, there is no such thing as free lysosome activity, or ribosomal action, or protein export. There is a price to be paid for all these things, and it is called energy expenditure. The fuel for this energy is contained in the food that cells ingest. But the energy in this food has to be converted into a molecular *form* that the cell can easily use, just as the energy in, say, coal needs to be converted by a power plant into a form that home appliances can easily use—electricity. The place to look for most of this conversion is inside the **mitochondria**, which can be defined as organelles that are the primary sites of energy conversion in eukaryotic cells. While cells that don't use much energy might have one or two mitochondria within them, an energy-ravenous liver cell might have a thousand. Most of the heat in our bodies is generated within mitochondria, and almost all the food we eat is ultimately consumed in them. FIGURE 4.10 shows a single mitochondrion.

Few details about mitochondria are included here, because much of Chapter 7 is devoted to them. Suffice it to say that to carry out their work, mitochondria need not only food, but oxygen. (Ever wonder why you need to breathe?) The most important *product* of mitochondrial activity, meanwhile, is the energy molecule that cells use for their activities, adenosine triphosphate, or ATP. Another thing of note about mitochondria is that they are the descendants of resident aliens—bacteria that invaded eukaryotic cells long ago, eventually taking up residence in them. You can read more about this in Essay 4.2, "The Stranger within," on page 71.

So Far...

So Far... Answers:

1. has no ribosomes attached to it and hence is not a site of protein production

2. lysosomes

3. energy; food

1. The smooth endoplasmic reticulum is smooth because it _____.

2. The tiny acid-filled organelles involved in cellular recycling are called _____.

3. Mitochondria are organelles in which the _____ contained in _____ is put into a chemical form the body can use.

FIGURE 4.11 Structure and Movement: The Cell's Cytoskeleton
Three types of fibers form the inner scaffolding or cytoskeleton of the eukaryotic cell: microfilaments, intermediate filaments, and microtubules.

(a) Microfilaments

(b) Intermediate filaments

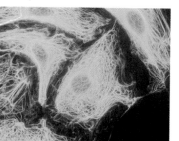

(c) Microtubules

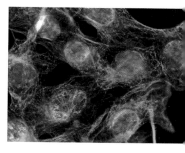

7 nm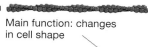
Main function: changes in cell shape

10 nm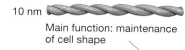
Main function: maintenance of cell shape

25 nm
Main functions: maintenance of cell shape, movement of organelles, cell mobility (cilia and flagella)

4.5 The Cytoskeleton: Internal Scaffolding

When you see a set of cells in action, the word that comes to mind is *hyperactive*. Cells are not passive entities. Jostling, narrowing, expanding, moving about, capturing objects and bringing them in, expelling other objects: This is life as a bubbling cauldron of activity. It was once thought that cells did all these things with pretty much the equipment you've looked at so far, meaning that the Golgi complex, the ribosomes, and so forth were thought to be floating in an undifferentiated soup that was the cytosol. But improved work with microscopes showed that there is, in fact, a dense forest within the cytoplasm. It is the **cytoskeleton**: a network of protein filaments that functions in cell structure, cell movement, and the transport of materials within the cell. This network is found in its full form only in eukaryotic cells, but in the last few years scientists have confirmed that bacteria have at least one type of cytoskeletal fiber within them.

The cytoskeleton consists of three types of fibers. Ordered by size, going from smallest to largest in diameter, they are microfilaments, intermediate filaments, and microtubules (**see** FIGURE 4.11). *Microfilaments* are cytoskeleton fibers that play a support or "structural" role in almost all eukaryotic cells. Microfilaments can also help cells move or capture prey, essentially by growing very rapidly at one end—in the direction of the movement or extension—while breaking down rapidly at the other end. You can see a vivid example of microfilament-aided cell extension in FIGURE 4.12. Some bacteria have been shown to have a microfilament-like set of cytoskeleton fibers that helps give them their shape. *Intermediate filaments* are filaments of the cytoskeleton intermediate in diameter between microfilaments and microtubules. These in-between-sized proteins are the most permanent of the cytoskeletal elements, perhaps coming closest to our everyday notion of

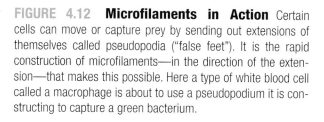

FIGURE 4.12 Microfilaments in Action Certain cells can move or capture prey by sending out extensions of themselves called pseudopodia ("false feet"). It is the rapid construction of microfilaments—in the direction of the extension—that makes this possible. Here a type of white blood cell called a macrophage is about to use a pseudopodium it is constructing to capture a green bacterium.

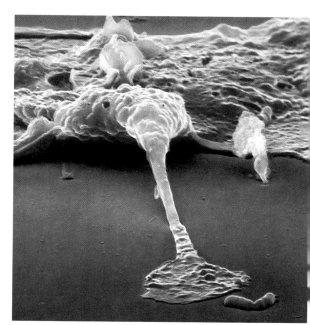

(a) Transport monorails

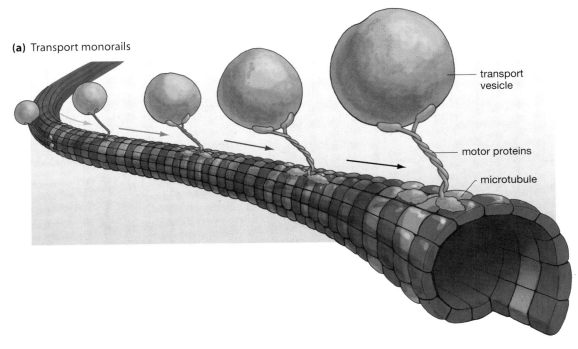

transport vesicle

motor proteins

microtubule

(b) Cilia

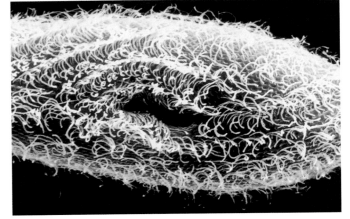

(c) Flagellum

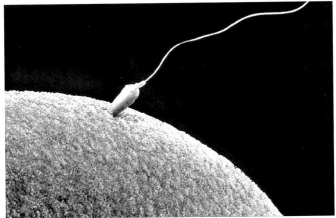

FIGURE 4.13 Several Functions for Microtubules

(a) They are the "rails" on which vesicles move through the cell, carried along by "motor proteins."

(b) They exist outside the cell in the form of cilia, which are profuse collections of hair-like projections that beat rapidly, forming currents that can propel a cell or move material around it.

(c) They are also found outside cells in the form of flagella. The flagellum on this sperm cell is enabling it to seek entry into an egg.

what a skeleton is like. *Microtubules,* the largest of the cytoskeletal elements, take the form of tubes. Microtubules play a structural role in the cell, but they also serve as what we could think of as monorails. Recall that protein-laden vesicles move from one organelle to another in the cell. These spheres move along the "rails" of microtubules, while sitting atop the "engine" of one of the so-called motor proteins (**see** FIGURE 4.13a).

Cell Extensions Made of Microtubules: Cilia and Flagella

Microtubules also form the underlying structure for two kinds of cell extensions, cilia and flagella. **Cilia** are microtubular extensions of cells that take the form of a large number of active, hair-like growths stemming from them. The function of cilia is simple: Move back and forth very rapidly, perhaps 10 to 40 times per second. The effect of this movement can be either to propel a cell or to move material *around* a cell (**see** FIGURE 4.13b). Cilia are extremely common among single-celled organisms and in some of the cells of simple animals (sponges, jellyfish). Our own lungs are lined with cilia, whose job it is to sweep the lungs

Essay 4.2 The Stranger within: Endosymbiosis

In modern eukaryotic cells, mitochondria are just one type of organelle among many. But these tiny "power plants" are, in fact, the descendants of free-standing *cells*—bacterial cells—that invaded ancient eukaryotic cells, only to take up residence within them over time. How do we know? The evidence is, you might say, written all over mitochondria. They have their own ribosomes and their own DNA, both of a bacterial type. And they reproduce under a bacteria-like division, partly under their own genetic control. All this makes them look like once-independent cells that now find themselves living inside other cells. The question this raises, of course, is: Why the merger? Mutual benefit seems to lie at the root of it. The bacterial ancestors of mitochondria could use or "metabolize" oxygen, and they were living in a period, about 1.5 billion years ago, in which oxygen was making up an ever-increasing proportion of Earth's atmosphere. The eukaryotic cells they invaded, however, were fairly intolerant of oxygen. Thus was born a working relationship: The "host" eukaryotes provided food to the bacterial invaders, and the bacteria allowed the hosts to live in an oxygenated world. When two organisms of different species live in close association, symbiosis has occurred. When symbiosis takes place inside a cell, it is called endosymbiosis. In mitochondria, which function in almost all eukaryotes, we can see the lasting effects of an ancient endosymbiosis. Plant and algae cells then have an additional set of bacterial descendants within them—the organelles called chloroplasts that carry out photosynthesis.

clean of whatever foreign matter has been inhaled. These cilia, like most others, all beat at once in the same direction, acting like rowers in a crew.

Cilia grow from eukaryotic cells in great profusion, but it is a different story with **flagella**—the relatively long, tail-like extensions of some cells that function in cell movement. It is sometimes the case that several flagella will sprout from a given cell, but often there is but a single flagellum. Only one kind of animal cell is flagellated, and it scarcely needs an introduction: A sperm is a single cell that whips its flagellum in a corkscrew motion in order to get to an unfertilized egg.

In Summary: Structures in the Animal Cell

In your tour of the cell so far, you've pictured a cell as a factory, one that synthesizes proteins in a "production line" that starts with DNA in the nucleus and then goes to the ribosomes (via mRNA), to the rough ER, to the Golgi complex, and finally to the protein's destination (the plasma membrane, export, and so on). You've also seen that cells have other structures such as lysosomes for digestion and recycling, mitochondria for energy transformation, the smooth endoplasmic reticulum for detoxification and lipid synthesis, and the cytoskeleton for structure and movement. If you look at FIGURE 4.14, you can see, in metaphorical form, a "map" of these component parts within the cell. **Table 4.1** (on page 75, following the discussion of plant cells) lists cellular elements found in plants and animals, as well as some elements found only in plant cells.

FIGURE 4.14 **The Cell as a Factory** This comparison may help you to remember some roles of the different parts of the cell.

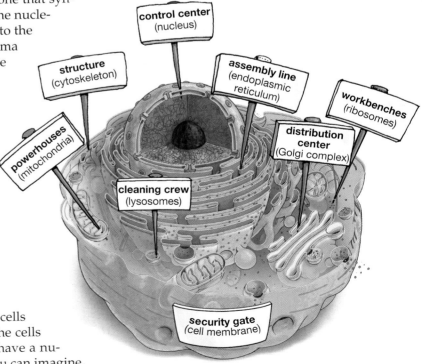

control center (nucleus)

structure (cytoskeleton)

assembly line (endoplasmic reticulum)

workbenches (ribosomes)

powerhouses (mitochondria)

distribution center (Golgi complex)

cleaning crew (lysosomes)

security gate (cell membrane)

4.6 Another Kind of Cell: The Plant Cell

The animal cell you just looked at has lots in common with the cells that you'd see in any eukaryotic living thing. If you looked at the cells that make up plants or fungi, for example, you'd see that they have a nucleus, and ribosomes, and mitochondria, and so forth. But, as you can imagine,

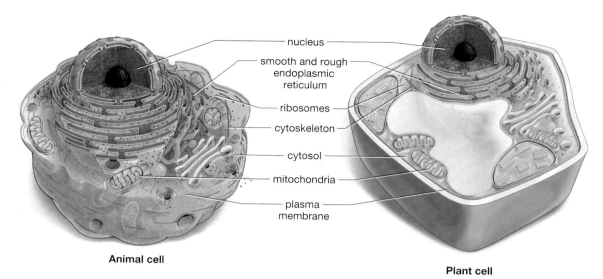

Animal cell

- nucleus
- smooth and rough endoplasmic reticulum
- ribosomes
- cytoskeleton
- cytosol
- mitochondria
- plasma membrane

Plant cell

FIGURE 4.15 **Structures Common to Both Animal and Plant Cells**

the cells of a fungus have to be *somewhat* different from those of an animal, since fungi and animals are such different types of organisms. To give you some idea of some of the kinds of differences found in the cells of life's various kingdoms, we'll look briefly now at the plant cell, which differs from an animal cell in several basic ways.

A quick look at FIGURE 4.15 will confirm for you the first part of this story—how structurally similar plant and animal cells are. As you see, a plant cell has a nucleus, the smooth and rough ERs, a cytoskeleton. Indeed, there is only one structure present in the animal cells you've looked at that plant cells don't have: the lysosome. What jumps out at you when you look at plant cells is not what they lack compared to animal cells, but what they *have* that animal cells do not. As you can see in FIGURE 4.16, these additions are

- A thick cell wall
- Structures called chloroplasts
- A large structure called a central vacuole

The Central Vacuole

The central vacuole pictured in Figure 4.16 is so prominent it appears to be a kind of organelle continent surrounded by a mere moat of cytosol. And indeed, in a mature plant cell, one or two central vacuoles may comprise 90 percent of cell volume. Although animal cells can have vacuoles, the imposing central vacuole in plant cells is different. For a start, it is composed mostly of water, which demonstrates just how watery plant cells are. A typical animal cell may be 70 to 85 percent water, but for plant cells the water proportion is likely to be 90 to 98 percent. The central vacuole helps to store nutrients in plant cells, degrade waste products and balance the cell's pH. Despite the vacuole's size, however, it is the other two structures noted above that interest us most, as they tell us the most about differences among the cells of various types of living things. Let's have a look at the cell wall and at chloroplasts.

The Cell Wall

Plant cells have an outer protective lining, called a **cell wall**, that makes their plasma membrane—just inside the cell wall—look rather thin and frail by comparison. This is because it *is* thin and frail by comparison; the plasma membrane of a plant cell may be 0.01 μm thick, while the combined units of a cell wall may be 700 times this width—7 μm or more. Cell walls are nearly always present in plant cells. And they exist in many organisms that are neither plant nor animal—bacteria, fungi, and a group of living things called protists—though the cell walls of these life-forms differ in chemical composition from those of plants. Thus cell walls are the rule, rather than the exception, in nature. Animals are the one major group of living things whose cells never have cell walls.

What do cell walls do for plant cells? They provide them with structural strength, put a limit on their absorption of water (as you'll see next chapter), and generally protect plants from harmful outside influences. So, if they're so useful,

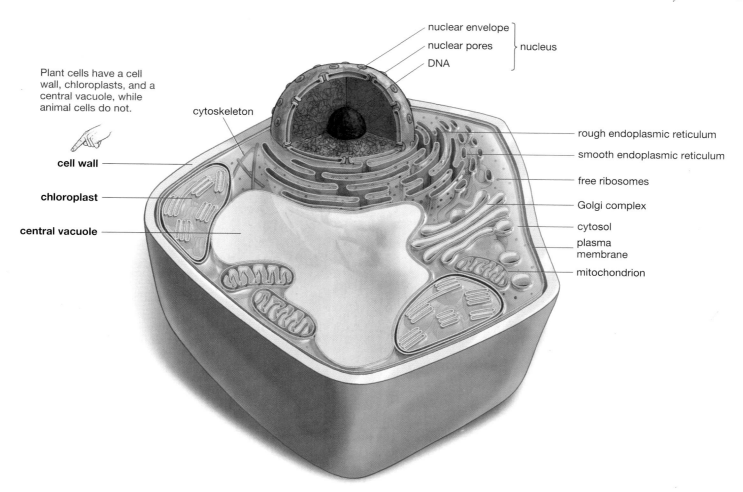

Plant cells have a cell wall, chloroplasts, and a central vacuole, while animal cells do not.

FIGURE 4.16   The Plant Cell The cell wall, central vacuole, and chloroplasts do not exist in animal cells, but all other components are common to both plant and animal cells.

why don't *animal* cells have them? Cell walls make for a rather rigid, inflexible organism—like plants, which are stationary. Animals, meanwhile, are mobile and thus must remain flexible.

Cell walls in plants can come in several forms, but all such forms will be composed chiefly of a molecule you were introduced to last chapter: cellulose, a complex sugar or "polysaccharide" that is embedded within cell walls in the way reinforcing bars run through concrete. In some cell walls, cellulose is joined by a compound called lignin, which imparts considerable structural strength. You can see a vivid demonstration of this in wood, which is largely made of cell walls.

Chloroplasts

There is a diverse group of organelles, known as plastids, that are found only in plants and algae. Some plastids gather and store nutrients for plant cells; others are pigment-containing organelles that give us, for example, the red color of the tomato skin. The best known of the plastids, however, are the chloroplasts.

Human beings are indebted, in a sense, to chloroplast-containing organisms, because most of the food we eat and the oxygen we breathe comes from them. This is so because organisms with chloroplasts carry out photosynthesis, the process by which organisms produce their own food (a sugar), using nothing but the energy contained in sunlight and the starting materials of carbon dioxide,

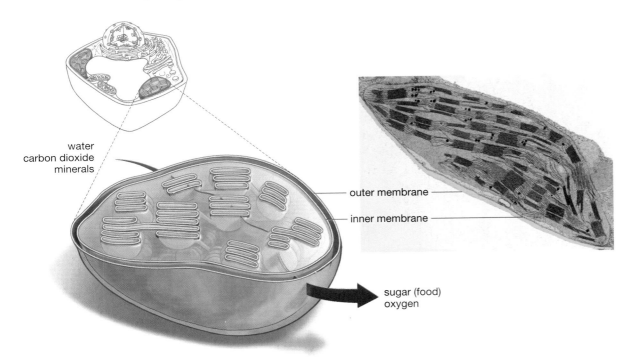

water
carbon dioxide
minerals

outer membrane

inner membrane

sugar (food)
oxygen

FIGURE 4.17 **Food Source for the World** Chloro-plasts, the tiny organelles that exist in plant and algae cells, are sites of photosynthesis—the process that provides food for most of the living world. Using the starting materials of water, carbon dioxide, and a few minerals, these organisms capture energy from sunlight to produce their own food. A double mem-brane lines the chloroplasts. (Micrograph: × 13,000)

water, and a few minerals. As it turns out, the **chloroplast** is the organelle that is the site of photosynthesis in plant and algae cells (**see** FIGURE 4.17). This may not seem like such a big deal until you try to name a food you eat that is not a plant or does not itself eat plants. A by-product of photosynthesis is oxygen, which we can-not exist without. If we take a step back from this, we can see that, in thinking about how life-forms differ from each other, one major way is that some organ-isms produce both oxygen and their own food, while others do not. And these dif-ferences in capabilities are made possible by differences in cell equipment.

On to the Periphery

Having looked at what is inside the cell, you've arrived at the cell's periphery, the plasma membrane, which is where you'll be staying for awhile. It may at first seem strange to devote a whole chapter to an outer boundary. How much attention would you pay, after all, to a factory's wall as opposed to its contents? The answer is: A lot, if that wall could facilitate communication with the outer world, continually renew itself, and let some things in while keeping others out. Such is the case with the plasma membrane, a slender lining that manages to make one of the most fundamental distinctions on Earth: Inside, life goes on; outside, it does not.

So Far...

So Far... Answers:

1. *False; the cytoskeleton exists in the cytoplasm, which is the area that lies outside the nucleus.*

2. *animals*

3. *chloroplasts*

1. True or false: The cytoskeleton is a set of protein fibers that exist inside a cell's nucleus.

2. Among the major forms of life, only the cells of _____ never have cell walls.

3. Photosynthesis in plant and algae cells takes place in the organelles called _____.

Table 4.1 Structures in Plant and Animal Cells

Name	Location	Function
Cytoskeleton	Cytoplasm	Maintains cell shape, facilitates cell movement and movement of materials within cell
Cytosol	Cytoplasm	Protein-rich fluid in which organelles and cytoskeleton are immersed
Golgi complex	Cytoplasm	Processing, sorting of proteins
Lysosomes (in animal cells only)	Cytoplasm	Digestion of imported materials and cell's own used materials
Mitochondria	Cytoplasm	Transform energy from food
Nucleus	Inside nuclear envelope	Site of most of the cell's DNA
Ribosomes	Rough ER, free-standing in cytoplasm	Sites of protein synthesis
Rough endoplasmic reticulum	Cytoplasm	Protein processing
Smooth endoplasmic reticulum	Cytoplasm	Lipid synthesis, storage; detoxification of harmful substances
Vesicles	Cytoplasm	Transport of proteins and other cellular materials
Cell walls (in plant cells only)	Outside plasma membrane	Limit water uptake, maintain cell membrane shape, protect from outside influences
Central vacuole (in plant cells only)	Cytoplasm	Cell metabolism, pH balance, digestion, water maintenance
Chloroplasts (in plant cells only)	Cytoplasm	Photosynthesis

Chapter Review

Summary

4.1 Cells Are the Working Units of Life

- **The importance of cells to life**—With the possible exception of viruses, every form of life on Earth either is a cell or is composed of cells. Cells come into existence only through the activity of other cells. (page 55)

4.2 All Cells Are Either Prokaryotic or Eukaryotic

- **Prokaryotic and eukaryotic**—All cells can be classified as prokaryotic or eukaryotic. Prokaryotic cells are either bacteria or another single-celled life-form called archaea. Setting bacteria and archaea aside, all other cells are eukaryotic. Eukaryotic cells have most of their DNA contained in a membrane-lined compartment, called the cell nucleus, whereas prokaryotic cells do not have a nucleus. (page 56)

- **Components of the eukaryotic cell**—There are five principal components to the eukaryotic cell: the nucleus, other organelles, the cytosol,

the cytoskeleton, and the plasma membrane. Organelles are "tiny organs" within the cell that carry out specialized functions, such as energy transfer and materials recycling. The cytosol is the fluid outside the nucleus in which these organelles are immersed. The cytoskeleton is a network of protein filaments that function in cell structure, cell movement, and the transport of materials. The plasma membrane is the outer lining of the cell. (page 58)

Web Tutorial 4.1 A Comparison of Prokaryotic and Eukaryotic Cells

4.3 A Tour of the Animal Cell: The Protein Pathway

- **The production of proteins**—Information for the construction of proteins is contained in the DNA located in the cell nucleus. This information is copied onto an informational "tape" of messenger RNA (mRNA) that departs the cell nucleus through its nuclear pores and goes to the sites of protein synthesis, structures called ribosomes, which lie in the cytoplasm. Many ribosomes that receive mRNA tapes process only a short stretch of them before migrating to, and then embedding in, one of a series of sacs in a membrane network called the rough endoplasmic reticulum (RER). The amino acid chains produced by the ribosomal "reading" of the mRNA tapes are dropped from ribosomes into the internal spaces of the RER; there, they fold up, thus becoming proteins, and undergo editing.   (page 59)

- **The endomembrane system**—Materials move from one structure to another in the cell via the endomembrane system, in which a piece of membrane, with proteins or other materials inside, can bud off from one organelle, move through the cell, and then fuse with another membrane-lined structure. Membrane-lined structures that carry cellular materials are called transport vesicles.   (page 65)

- **The Golgi complex**—Once protein processing is finished in the rough endoplasmic reticulum, proteins undergoing processing move, via transport vesicles, to the Golgi complex, where they are processed further and marked for shipment to appropriate cellular locations. (page 65)

Web Tutorial 4.2 A Protein Pathway

4.4 Other Cell Structures

- **The smooth ER, lysosomes, and mitochondria**—The smooth endoplasmic reticulum is a network of membranes that functions to synthesize lipids and to detoxify potentially harmful substances. Lysosomes are organelles that break down worn-out cellular structures or foreign material that comes into the cell. Once this digestion is completed, the lysosomes return the molecular components of these materials to the cytoplasm for further use. Mitochondria are organelles that function to extract energy from food and to transform this energy into a chemical form the cell can use.   (page 67)

4.5 The Cytoskeleton: Internal Scaffolding

- **The cytoskeleton**—Cells have within them a web of protein strands, called a cytoskeleton, that provide the cell with structure, facilitate the movement of materials inside the cell, and facilitate cell movement. There are three principal types of cytoskeleton elements. Ordered by size, going from smallest to largest in diameter, they are microfilaments, intermediate filaments, and microtubules. Microfilaments help the cell move and capture prey. Intermediate filaments provide support and structure to the cell. Microtubules play a structural role in cells and serve as transport "rails."   (page 69)

4.6 Another Kind of Cell: The Plant Cell

- **Life-forms and their cells**—Eukaryotic cells are likely to have many features in common, no matter what kind of organism they exist in. But cells that make up different life-forms do differ in significant ways. Plant cells have a cell wall; and they have organelles called plastids, one variety of which is the chloroplast. The cell wall gives the plant structural strength and helps regulate the intake and retention of water, while chloroplasts are the sites of photosynthesis. (page 71)

Key Terms

cell wall   p. 72

cilia   p. 70

chloroplast   p. 74

cytoplasm   p. 58

cytoskeleton   p. 69

cytosol   p. 58

endomembrane system   p. 65

eukaryotic cell   p. 56

flagella   p. 71

Golgi complex   p. 65

lysosome   p. 67

mitochondria   p. 68

nuclear envelope   p. 62

nucleus   p. 56

organelle   p. 57

plasma membrane   p. 58

prokaryotic cell   p. 56

ribosome   p. 64

rough endoplasmic
   reticulum   p. 65

smooth endoplasmic
   reticulum   p. 67

transport vesicle   p. 65

Testing Your Understanding

In Your Own Words   *(answers in the back of the book)*

1. Suppose your entire body were just one gigantic cell with one central nucleus and lots of organelles outside it to perform the various functions. What problems can you envision with this system?

2. Which is the correct ranking of these small things, from smallest to largest? (Use typical sizes): animal cells, atoms, bacteria, proteins, amino acids.

3. Insulin, the hormone that controls sugar levels in the body, is a protein that is exported from special cells in the pancreas. Trace the path from a length of DNA in the nucleus that carries the code for insulin to the release of the hormone in the bloodstream.

Thinking about What You've Learned

1. The earliest organisms in the fossil record are all single-celled. Why do you think life had to "start small" like this?

2. Why do membranes figure so prominently in eukaryotic cells? What essential function do they serve?

On Life's Edge: The Plasma Membrane

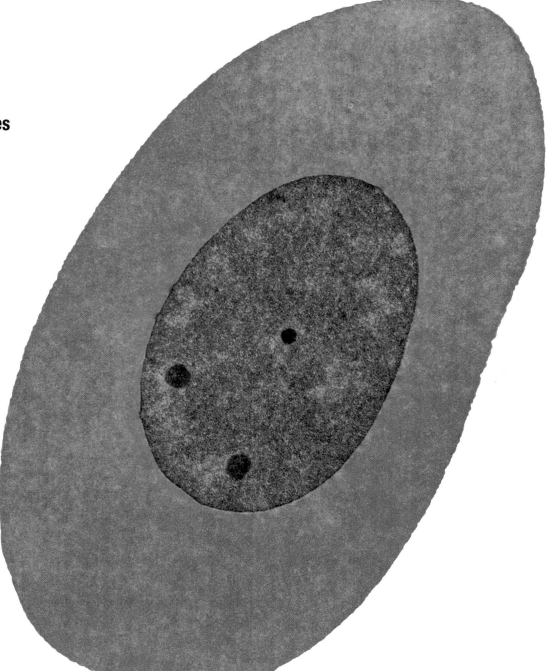

An immature human red-blood cell.

AKRON BEACON JOURNAL

EX-NFL STAR ON MARK IN CYSTIC FIBROSIS FIGHT;

BOOMER ESIASON MAKES ROUNDS TO CHAMPION CAUSE FOR SON, GUNNAR

By Jewell Cardwell
December 10, 2003

It's been a few years since Boomer Esiason retired from the National Football League. Ironically, the 42-year-old—selected four times to the Pro Bowl and named the league's most valuable player in 1988—is still a quarterback of sorts. Only this time the battleground is far different and the stakes much higher.

5.1 The Nature of the Plasma Membrane

During his 14 years with the Cincinnati Bengals, New York Jets, and Arizona Cardinals, quarterback Boomer Esiason made it once to the Super Bowl and four times to the Pro Bowl. His toughest challenge, however, has come not on the field but at home, where he and his wife have struggled to care for their son Gunnar, who was born with the disease cystic fibrosis.

Like other children who have cystic fibrosis, Gunnar is plagued by a mucus that accumulates in his lungs. The lungs of healthy children are lined by mucus too, but in a layer that is thin and wet enough to be regularly swept away by the hair-like cilia that extend into lung passages like so many tiny brooms. In cystic fibrosis patients, the mucous layer is thicker and drier and becomes a site for repeated bacterial infections. In time, these infections can result in the lung passages being destroyed. The average life expectancy for a cystic fibrosis patient is 35 years.

The mother of a child with cystic fibrosis provides gentle "percussive" therapy to dislodge mucus from the child's lungs.

The difference between a healthy child and Gunnar Esiason comes down to this: Because of a faulty gene, the substance chloride cannot be transported in sufficient quantity from the inside of Gunnar's cells to the outside of them. Healthy individuals have a protein that acts as a channel for chloride—a kind of passageway that spans the cell's outer membrane. In contrast, the cells of cystic fibrosis patients have either defective chloride channels or none at all. The result is a lack of chloride outside their cells, which gets the mucus buildup going.

As noted last chapter, life goes on only inside cells. The fact that the cell's outer, or "plasma," membrane is out on life's edge may prompt the thought that it's *merely* at the edge, as if the real action is taking place deep within the cell. But consider, as in cystic fibrosis, the effect of having a *defective* plasma membrane. The focus in this chapter is to learn more about this important cell lining.

Component Parts of the Plasma Membrane

So, what is the nature of the plasma membrane? First, it's worth noting that it's very much like the membranes described in Chapter 4. Much of what follows about the plasma membrane could also be said of the various membranes that are internal to the cell—those that line lysosomes or that make up the Golgi apparatus, for example. There are some important differences between the plasma and other membranes, however, because the plasma membrane is constantly interacting with the world outside the cell, whereas internal membranes are not.

When we start to consider the qualities of the plasma membrane, we find that, even in the microworld of cells, it is an extremely *thin* entity. If you stacked 10,000 of them on top of one another, their combined width would be about that of a sheet of paper. Next, the plasma membrane has a fluid and somewhat fatty make-up. If we looked for a counterpart to it in our everyday world, a soap bubble might come close, meaning this membrane is very flexible. Yet it is stable enough to stay together despite being constantly re-formed, due to the endless movement of materials in and out of it. If you imagine half the surface of your skin remaking itself every 30 minutes or so, you begin to get the picture about how dynamic this membrane is. Let's look now at the most important component parts of it as they exist in animal cells. All of these components are illustrated in FIGURE 5.1. They are:

- The phospholipid bilayer. Just as bread is made mostly from flour, the plasma membrane is made mostly from two layers of phospholipid molecules.

FIGURE 5.1 **The Plasma Membrane** The cell's outer lining or plasma membrane is composed of four main structural elements, as shown in the figure.

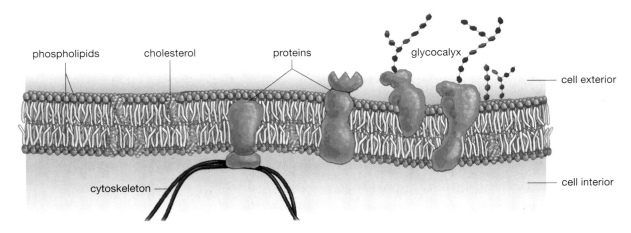

phospholipids   cholesterol   proteins   glycocalyx

cell exterior

cell interior

cytoskeleton

Phospholipid bilayer: a double layer of phospholipid molecules whose hydrophilic "heads" face outward, and whose hydrophobic "tails" point inward, toward each other.

Cholesterol molecules that act as a patching substance and that help the cell maintain an optimal level of fluidity.

Proteins, which can either be bound to the membrane's interior (as with the protein on the right) or not bound to it (as with the protein on the left).

Glycocalyx: sugar chains that attach to proteins and phospholipids, serving as protein binding sites and as cell lubrication and adhesion molecules.

- Cholesterol. Like mortar between bricks, cholesterol acts as a "patching" material for the membrane; it also keeps the membrane at an optimal level of fluidity.

- Proteins. These molecules span the membrane and lie on either side of it, serving as support structures, signaling antennas, identification markers, and molecular passageways.

- Glycocalyx. This is the collective term for a set of branched carbohydrate chains that layer the outside of the membrane.

Now let's look at each of these components in greater detail.

First Component: The Phospholipid Bilayer

You may recall the discussion in Chapter 3 of phospholipids—molecules that have two long fatty-acid chains linked to a phosphate-bearing group (**see** FIGURE 5.2). You may also recall that, because fatty-acid chains are lipids, they are hydrophobic or "water-fearing," meaning they will not bond with water. Conversely, phosphate groups are charged, hydro*philic* molecules, meaning they are "water-loving" and will *readily* bond with water. When such phosphate and lipid components are put together, the resulting phospholipid is a molecule with a dual nature. Its phosphate "head" will seek water, but its fatty-acid "tail" will avoid it. Drop a group of phospholipids in water, and they will arrange themselves into the form you see in Figure 5.2: two *layers* of phospholipids sandwiched together, the tails of each layer pointing inward (thus avoiding water) and the heads pointing outward (thus bonding with the watery environment that lies both inside and outside the cell). With this, we can define a **phospholipid bilayer** as a chief component of the plasma membrane, composed of two layers of phospholipids, arranged with their fatty-acid chains pointing toward each other.

The fact that the phospholipid bilayer is structured like this has an important consequence for the plasma membrane: It determines which substances will pass through the membrane with relative ease and which substances will not. Since the membrane includes two layers of hydrophobic tails, this means that the only substances that can pass through with ease are *other* hydrophobic substances or a few very small molecules. Why should this be? It stems from a rule of thumb about solvents: Like dissolves like. Fats dissolve in fats, charged molecules dissolve in charged molecules. Thus, various steroid hormones (which, remember, are lipids) gain fairly easy entry into the cell, as do fatty acids, because these substances will dissolve in the lipid bilayer. Conversely, hydro*philic* substances—ions, amino acids, all kinds of polar molecules—will not dissolve in the membrane and thus cannot make it past the fatty-acid chains of the phospholipids without help. What kind of help? Stay tuned.

Second Component: Cholesterol

Molecules of the lipid material cholesterol nestle between phospholipid molecules throughout the plasma membrane, performing two functions. First, they act as a kind of patching substance on the bilayer, keeping some small molecules from getting through. Second, they help keep the membrane at an optimum level of fluidity, ensuring that it doesn't become too rigid in cold temperatures or too fluid in hot ones.

Third Component: Proteins

Embedded within, or lying on, the phospholipid bilayer is a third major group of membrane components, membrane proteins. It is hard to

A BROADER VIEW

Bad Molecules

Is cholesterol a "bad" molecule? We hear about it all the time as just that, in connection with heart disease. Yet cholesterol is a necessary component of all our cells. As such, it stands as a case in point regarding a more general concept: that the component parts of living things almost always serve some life-*sustaining* function. It would be very rare for nature to produce a molecule whose chief function is to harm its owner. In this same vein, "cancer genes" were not developed during the course of evolution to cause cancer. Instead, they are genes whose normal function has been subverted (usually by mutation) such that they help set this disease in motion.

FIGURE 5.2   Dual-Natured Lining

(a) The essential building block of the cell's plasma membrane is the phospholipid molecule, which has two elements: a hydrophilic "head," containing a charged phosphate group that bonds with water; and hydrophobic "tails," composed of fatty acids that do not bond with water.

(b) Two layers of phospholipids sandwich together to form the plasma membrane. The phospholipids' hydrophobic tails form the interior of the membrane, while their hydrophilic heads jut out toward the watery environments that exist on both sides of the membrane. Though the bilayer forms a barrier to all but the smallest hydrophilic molecules, hydrophobic molecules can pass through it fairly easily.

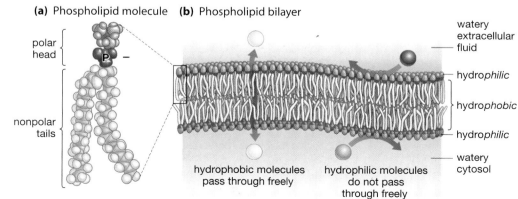

(a) Phospholipid molecule   **(b)** Phospholipid bilayer

polar head

nonpolar tails

watery extracellular fluid

hydro*philic*

hydro*phobic*

hydro*philic*

watery cytosol

hydrophobic molecules pass through freely

hydrophilic molecules do not pass through freely

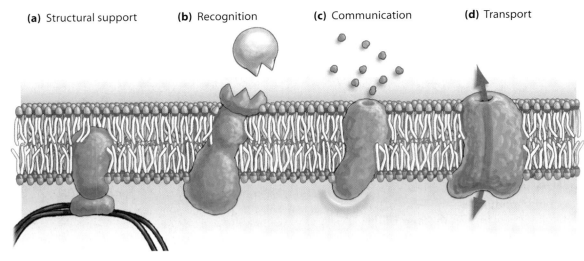

(a) Structural support

(b) Recognition

(c) Communication

(d) Transport

Membrane proteins can provide structural support, often when attached to parts of the cell's scaffolding or "cytoskeleton."

Binding sites on some proteins can serve to identify the cell to other cells, such as those of the immune system.

Receptor proteins, protruding from the plasma membrane, can be the point of contact for signals sent to the cell via traveling molecules, such as hormones.

Proteins can serve as channels through which materials can pass in and out of the cell.

FIGURE 5.3 **Roles of Membrane Proteins** Plasma membrane proteins serve four principal functions, as shown in the figure.

overstate the importance of proteins to the plasma membrane. By itself, the lipid bilayer has a handful of capabilities. With the addition of proteins, it is turned into a structure with a huge array of capabilities. Here is a selected set of the roles that membrane proteins play (**see** FIGURE 5.3).

Structural Support   Proteins that lie on the interior or "cytoplasmic" side of the membrane often are attached to elements of the cell's scaffolding or "cytoskeleton." Thus anchored, they help to stabilize various parts of the cell and play a part in giving animal cells their characteristic shape. (Why not plant cells? Because their shape is determined by the cell wall.)

Recognition   Much like military sentries, some cells need to know the answer to a critical question: Who goes there, friend or foe? The sentries are immune-system cells, moving through the body and interacting with proteins that extend from cell surfaces. These proteins have "binding sites" on them whose shape can convey a message to immune cells: "Self here, pass on by." Conversely, a cell with a *foreign* set of binding sites on its proteins can convey the message: "Invader here, begin an immune-system attack." Have you ever wondered what it means for people to have, say, type A blood? It means that they have type A recognition proteins extending from the plasma membranes of their red blood cells. These proteins have binding sites on them that allow the body's immune-system cells to recognize "self" and pass on by.

Communication   Cells communicate with one another in various ways. Signals can be sent from one cell to a neighboring cell, and communication over longer distances is possible through the chemical messengers known as hormones. These signals are likely to be channeled through a **receptor protein**—a plasma membrane protein that binds with a specific signaling molecule.

Receptor proteins—usually just called receptors—have binding sites on them whose shape is so specific that they generally will bind only to a single type of signaling molecule. To take one example of this, the hormone insulin is a signaling molecule. The message it conveys to cells is: "Glucose levels are rising in the bloodstream; please construct some protein channels in your plasma membranes so that you can take some of this glucose in." Now, how does insulin transmit this message? It binds with an insulin receptor protein that spans a given cell's plasma membrane. This binding causes a chemical change on the portion of the protein that lies *inside* the cell. The result is a cascade of cellular reactions that puts in motion the construction of the protein channels. Note that, through this means, a signaling molecule need not come into the cell to have an effect; it merely needs to bind with a receptor protein that spans the plasma membrane.

Transport   You have seen that most materials cannot simply pass through the cell's plasma membrane. Yet cells need many of these very materials. How do they get in? Different kinds of proteins take on a transport task—they act as channels through which materials pass both into cells and out of them. We'll review all the forms of protein-aided transport shortly. For now, just note the existence of **transport proteins**: proteins that facilitate the movement of molecules or ions from one side of the plasma membrane to the other.

Fourth Component: The Glycocalyx

If you look at Figure 5.1 and focus on the extracellular face of the plasma membrane, you can see, protruding from proteins and membrane lipids alike, a number of short, branched extensions. These are simple carbohydrate chains, meaning sugar chains. It turns out that these chains serve as the actual binding sites for many signaling molecules, including insulin. Sugar chains also serve to lubricate cells, and allow them to stick to other cells by acting as an adhesion layer. Collectively, all these chains form the **glycocalyx**: an outer layer of the plasma membrane, composed of branched carbohydrate chains that attach to membrane proteins and phospholipid molecules.

A General View of the Plasma Membrane

With this review of component parts under your belt, you're ready for a definition of the **plasma membrane**. It is a membrane, forming the outer boundary of many cells, composed of a phospholipid bilayer that is interspersed with proteins and cholesterol and coated on its exterior face with carbohydrate chains. Taking a step back, the general message here is that the plasma membrane is a loose, lipid structure peppered with proteins and coated with sugars. A better image, however, might be of a *sea* of lipids that has proteins floating on it, because as it turns out, the plasma membrane is fluid enough that most of its elements are able to move sideways or "laterally" through it fairly freely. Indeed, membrane proteins frequently are compared to icebergs drifting through an ocean.

So Far...

1. Name the four chief components of an animal cell's plasma membrane.

2. Vitamins A, D, E, and K are all fat-soluble vitamins. Given the chemical makeup of the plasma membrane, are these vitamins likely to pass through it easily?

3. Substances that cannot easily diffuse through the membrane often gain access through
_____.

5.2 Diffusion, Gradients, and Osmosis

It may be obvious that one of the main roles of the plasma membrane is to let in whatever's needed and to keep out whatever's not. Having learned something about the membrane's makeup, you will now see how it carries out this task of passage and blockage.

A BROADER VIEW

Marvelous Membrane

Can you think of any human-made surface that has a fraction of the capabilities of the plasma membrane? It manages to concentrate and protect all the cell's necessary components (proteins, DNA, etc.) while at the same time allowing, or even facilitating, the movement of a huge number of materials into and out of the cell. It's as though it is both a tough security guard and an accommodating host.

So Far... Answers:

1. *phospholipid bilayer, cholesterol, proteins, glycocalyx*

2. *Yes; given that like dissolves like, and that the plasma membrane has a lipid interior, fat-soluble vitamins can dissolve in the membrane and thus pass into the cell.*

3. *transport proteins*

(a) Dye is dropped in. **(b)** Diffusion begins. **(c)** Dye is evenly distributed.

water
molecules

dye
molecules

FIGURE 5.4 From Concentrated to Dispersed
Diffusion is the movement of molecules or ions from areas of
their greater concentration to areas of their lesser concentration. In this sequence of photos and diagrams, a few drops of
red dye, added to a beaker of water, are at first heavily concentrated in one area but then begin to diffuse, eventually becoming evenly distributed throughout the solution.

It is necessary to begin this story, however, not with the cell, but with the more
general notion of how substances go from places where they are more concentrated to places where they are less concentrated. Anyone who has put some food dye
in water has a sense of how this works: A few drops of, say, red dye will tumble
into the liquid, start dispersing, and eventually the result is a solution that is uniformly more reddish than plain water (**see** FIGURE 5.4). The question that will be
addressed here is why this takes place.

Random Movement and Even Distribution

All molecules or ions are constantly in motion, and this motion is random. (The
degree to which molecules are in motion defines their temperature; absolute zero is
the point at which all molecular motion has ceased.) The laws of thermodynamics
dictate that, because the motion of molecules is random, they will naturally move
from any initial *ordered* state—in our example, their concentrated state when first
dropped in the container—to their most disordered state, meaning evenly distributed throughout a given volume.

What is at work here is **diffusion**: the movement of molecules or ions from a
region of their higher concentration to a region of their lower concentration. This
notion carries with it the concept of a gradient. A **concentration gradient** is the
difference between the highest and lowest concentration of a solute within a
given medium. In our example, the solute is the dye and the medium is water. As
with bicycles coasting down a grade, the natural tendency for any solute is to
move *down* its concentration gradient, from higher concentration to lower. Our
dye did just that. Bikes and solutes can move up a grade or gradient; but there is
a price to be paid, and that price is the expenditure of energy. (On its own, would
red dye ever spontaneously come *back* to its concentrated state in the container?
Not without some work being performed, which is to say not without some energy being expended.)

Diffusion through Membranes

So far, you have looked at molecules diffusing in an undivided container. But the subject here is divisions—those provided by the plasma membrane. Let us consider, therefore, what happens to a solution in a container that is divided by a membrane. If that membrane is *permeable* to both water and the solute—that is, if both water and the solute can freely pass through it—and the solute lies only on one side of the membrane, then the predictable happens: The solute moves down its concentration gradient, diffusing right through the membrane, eventually becoming evenly distributed on both sides of it.

Now let's imagine a *semi*permeable membrane, one that water can freely move through but that solutes cannot. **FIGURE 5.5** shows you what happens if more solute is put on the right side of the membrane than the left. Water flows through the membrane both ways, but *more* water flows into the right chamber, which has a greater concentration of solutes in it. The result is that the solution on the right side rises to a higher level than the one on the left.

This seems strange on first viewing, as if gravity were taking a vacation on the right side of the container. But what has been demonstrated here is **osmosis**: the net movement of water across a semipermeable membrane from an area of lower solute concentration to an area of higher solute concentration. Why should this occur? In the case of a solute like salt, water molecules will surround and bond with the sodium and chloride ions that salt separates into in solution. Because these solutes are not free to pass through the membrane, the water molecules bound to them will likewise remain confined to the right side. This means that more "free" water will exist on the left side—water that is free to act on the natural tendency of all molecules to remain in motion. The result is a net movement of water into the right side (see Figure 5.5b). Looked at one way, this is just another example of diffusion—of a substance moving from an area of its higher concentration to an area of its lower concentration. Once the solute binds with the water molecules on the right side of the container, there is a greater concentration of free water molecules on the left than on the right. By moving to the right, the molecules are simply moving down their concentration gradient.

The Plasma Membrane as a Semipermeable Membrane
So what do diffusion and the movement of water have to do with the cell's outer lining? The cell's phospholipid bilayer is itself a semipermeable membrane. It is somewhat permeable to water and lipid substances but not permeable to larger charged substances. Thus, osmosis can take place across the plasma membrane—indeed, it does all the time. It is the primary means by which plants get water, and it is a player in all sorts of routine metabolic processes in animals. In humans, the fluid portion of blood is at one point driven out of the blood vessels called capillaries by the force of blood pressure. How does this fluid get back in? Osmosis. The proteins that remain in the capillaries are like the salt in the container: They bring about an osmosis that pulls the fluid back into the blood vessels.

Or think about a more extreme example. Why are we always told that someone who is stranded at sea should never drink salt water? Because doing so would provide an enormous concentration of sodium chloride ions in the watery fluid that lies *outside* the cells. This is

FIGURE 5.5   Osmosis in Action

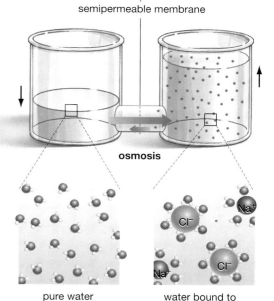

(a) An aqueous solution divided by a semipermeable membrane has a solute—in this case, salt—poured into its right chamber.

salt
solute
solvent

semipermeable membrane

(b) As a result, though water continues to flow in both directions through the membrane, there is a net movement of water toward the side with the greater concentration of solutes in it.

osmosis

(c) Why does this occur? Water molecules that are bonded to the sodium (Na⁺) and chloride (Cl⁻) ions that make up salt are not free to pass through the membrane to the left chamber of the container.

pure water   water bound to salt ions

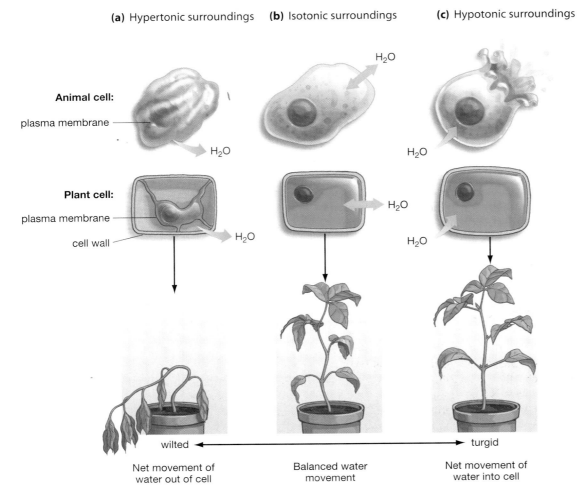

(a) Hypertonic surroundings   **(b)** Isotonic surroundings   **(c)** Hypotonic surroundings

Animal cell:

plasma membrane

H_2O

H_2O

H_2O

H_2O

Plant cell:

plasma membrane

cell wall

H_2O

H_2O

H_2O

wilted   turgid

Net movement of   Balanced water   Net movement of
water out of cell   movement   water into cell

FIGURE 5.6 Osmosis in Cells

(a) When solutes (such as salt) exist in greater concentration outside the cell than inside, water moves out of the cell by osmosis and the cell shrinks. Here, the fluid surrounding the cell is hypertonic to the cell's cytoplasm.

(b) When solute concentrations inside and outside the cell are balanced, there is a balance of water movement into and out of the cell. Here, the fluid surrounding the cell is isotonic to the cell's cytoplasm.

(c) When solutes exist in greater concentration inside the cell than outside, water moves into the cell by osmosis. This influx may cause animal cells to burst, but plant cells are reinforced with cell walls and thus remain turgid—generally a healthy state for them. Here, the fluid surrounding the cell is hypotonic to the cell's cytoplasm.

A BROADER VIEW

Osmosis as a Metaphor

Osmosis is one of those scientific concepts that is used as a metaphor by writers, probably because it seems to imply an effortless transfer. Thus one picks up a skill "by osmosis," just as the container in our example had water flowing into it by osmosis. Use of the metaphor is understandable in that, in actual osmosis, water will flow spontaneously from one side of a membrane to another. But remember that a condition has to be met for this to take place: A solute must be put into the area on one side of the membrane. How did that solute get there? By means of some work being performed.

no different in principle than dumping more solutes into the right side of the container in Figure 5.5. The result would be water flowing out of cells, dehydration of the cells, and in extreme cases, death. (What actually kills people in this situation is a shrinkage of brain cells.)

This is an example of an "osmotic imbalance," meaning a situation in which osmotic pressure is either drying out cells or flooding them. In the case of flooding, plant cells have a great advantage over animal cells—their cell walls. Remember from Chapter 4 that one function of the cell wall is to regulate water uptake. Well, now we can see why this is so valuable. Animal cells, which do not have a cell wall, can expand until they burst when water comes in (**see FIGURE 5.6**). Plant cells, conversely, will expand only until their membranes push up against the cell wall with some force, setting up a pressure, or turgor, that keeps more water from coming in. Such tight quarters actually are an optimal condition for plants. A nice, crisp celery stick is one that has achieved this kind of *turgid* state, while a droopy stick that has lost this quality has cells that are *flaccid*; flowers in the latter condition are *wilted*.

Osmosis and Cell Environments Is a given cell likely to lose water to its surroundings, gain water from them, or have a balanced flow back and forth with them? Any of these things are possible, depending on what the solute concentration is outside the cell as opposed to inside. Three terms are helpful in describing the various conditions that can exist. A fluid that has a higher concentration of solutes than another is said to be a **hypertonic solution**. If a cell's surroundings are hypertonic to the cell's cytoplasm, water will flow out of the cell. Two solutions that have equal concentrations of solutes are said to be **isotonic**. If one of these solutions is the cell's cytoplasm and the other the fluid surrounding the cell, fluid flow will be balanced between cell and surroundings. Finally, a fluid that has a lower concentration of solutes than another is a **hypotonic solution**. If a cell's surroundings are hypotonic to the cell's cytoplasm, water will flow into the cell. Figure 5.6 gives examples of what can happen to cells in all three types of environments.

So Far... Answers:

1. *down; higher concentration; lower concentration*
2. *osmosis; out of*
3. *hypotonic*

So Far...

1. The term *diffusion* describes the tendency of solutes to diffuse _____ their concentration gradients, going from a region of _____ to a region of _____.

2. Think of salt in food as a solute put into your bloodstream and coming to reside in the watery environment outside your cells. The resulting higher concentration of this solute outside the cells brings about an _____ that will result in a net movement of water _____ the cells.

3. In this example, the fluid inside the cells is _____ relative to the fluid outside the cells.

5.3 Moving Smaller Substances In and Out

This excursion into the land of diffusion and its special variant, osmosis, has prepared us to start looking at the ways materials actually move into and out of the cell, across the plasma membrane. The big picture here is that some molecules are able to cross with no more assistance than is provided by a concentration gradient and diffusion. Other molecules require these factors and the protein channels we talked about, and still others require channels and the expenditure of energy to get across. Different terms are applied to these different kinds of transport. **Active transport** is any movement of molecules or ions across a cell membrane that requires the expenditure of energy. **Passive transport** is any movement of molecules or ions across a cell membrane that does not require the expenditure of energy (**see** FIGURE 5.7). Our review of transport begins with two varieties of passive transport.

Passive Transport

Simple Diffusion   Gases such as oxygen and carbon dioxide are such small molecules that they need only move down their concentration gradients to move into or out of the cell. Having been delivered by blood capillaries to an area just outside the cell, oxygen exists there in greater concentration than it does inside the cell. Moving down its concentration gradient, it diffuses through the plasma membrane and emerges on the other side. This is an example of **simple diffusion**, meaning diffusion through a cell membrane that does not require a special protein channel. Molecules that are larger than oxygen, but that are fat soluble, such as the steroid hormones mentioned before, also move into and out of the cell in this manner. Carbon dioxide, which is formed *in* the cell as a result of cellular respiration, has a net movement *out* of the cell through simple diffusion.

Facilitated Diffusion: Help from Proteins   Water can traverse the lipid bilayer through simple diffusion during osmosis, moving through *despite* its hydrophilic nature (perhaps because it's such a small molecule). Yet for some cells that have a high traffic in water—such as

FIGURE 5.7   **Transport through the Plasma Membrane**

Passive transport		Active transport
simple diffusion	facilitated diffusion	

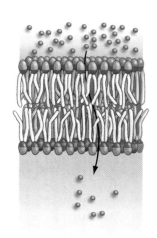
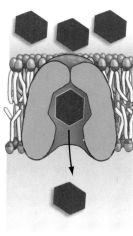

Materials move down their concentration gradient through the phospholipid bilayer.

The passage of materials is aided both by a concentration gradient and by a transport protein.

Molecules again move through a transport protein, but now energy must be expended to move them against their concentration gradient.

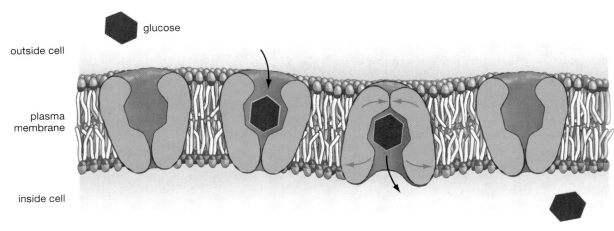

outside cell

glucose

plasma membrane

inside cell

1. The transport protein has a binding site for glucose that is open to the outside of the cell.

2. Glucose binds to the binding site.

3. This binding causes the protein to change shape, exposing glucose to the inside of the cell.

4. Glucose passes into the cell and the protein returns to its original shape.

FIGURE 5.8 **Facilitated Diffusion** Owing to its size and hydrophilic nature, a molecule such as glucose cannot move into a cell through simple diffusion, even when it has a concentration gradient working in its favor. Its movement must be facilitated by a transport protein, such as the one pictured in the figure. The movement of glucose is passive transport, however, in that no expenditure of cellular energy is required to bring it about.

our own kidney cells—simple diffusion is not enough to meet all water needs. Protein channels begin to be required for passage in and out of the cell. With this, the method of passage is no longer *simple* diffusion; it is **facilitated diffusion**, meaning passage of materials through the plasma membrane that is aided by a concentration gradient and a transport protein. With the various materials that traverse the membrane through this process, you can begin to see the protein specificity referred to earlier. Each transport protein acts as a conduit for only one substance, or at most a small family of related substances. Transport through these proteins begins with the kind of binding explained earlier. A circulating molecule of glucose, for example, latches onto the binding site of a glucose transport protein, as you can see in FIGURE 5.8. This binding causes the protein to change its shape, thus allowing the glucose molecule to pass through. In addition to glucose, other hydrophilic molecules, such as amino acids, move through the plasma membrane in this way.

Note that this facilitated diffusion does not require the expenditure of energy, because it has a concentration gradient working in its favor. Glucose exists in greater concentration outside most cells than inside them. It is moving *down* its concentration gradient, then, by passing into the cell through a protein channel. So facilitated diffusion has something in common with simple diffusion. *Both* processes are driven by concentration gradients, meaning that neither requires metabolic energy. Thus, both are specific examples of passive transport.

Though the glucose in our example moved only one way in facilitated diffusion (into the cell), in general transport proteins are a channel for movement *either* way through the membrane. All that's required is a concentration gradient and a binding of the material with the transport protein.

Active Transport

If passive transport—either in simple or facilitated diffusion—were the only means of membrane passage available, cells would be totally dependent on concentration gradients. If, for example, molecules of a given amino acid come to exist in the same concentration on both sides of the plasma membrane, these amino acid travelers will continue to be transported, but they will be flowing out of the cell at the same rate they are flowing in. For cells, the problem is that some solutes are *needed* in greater concentration, say, inside the cell as opposed to outside.

How do cells deal with such a situation? The cell's solution is pumps: It expends energy and pumps molecules against their concentration gradients. This is active transport at work. Many kinds of pumps are in operation in a cell, each of them being specific to one or perhaps two substances. The energy source for this transport often is ATP (adenosine triphosphate), though the cell also harnesses the power of oppositely charged ions to move substances across the membrane. Active transport often operates with the same kind of shape-changing of channel proteins you saw in connection with glucose (in Figure 5.8), but with the important addition that an energetic chemical group will come from ATP to power the operation. You'll see an example of this in Chapter 6.

5.4 Getting the Big Stuff In and Out

Pumps and channels, diffusion and osmosis—these mechanisms move substances in and out of the cell, but as it turns out they only move relatively *small* substances. Remember an earlier discussion about the immune-system cells that check to see if a given cell is friend or foe? Well, if an immune sentry finds a bacterial foe, it may have to ingest this *whole cell*. As you might guess, a cell cannot do this by employing little channels or pumps. You learned a bit in Chapter 4 about what does happen in these cases; now you'll be looking at it in somewhat greater detail. The methods of transport here are endocytosis, which brings materials into the cell; and exocytosis, which sends them out. What these mechanisms have in common is their use of vesicles, the substance-carrying pieces of membrane that alternately bud off from membranes or fuse with them.

Movement Out: Exocytosis

Exocytosis is defined as the movement of materials out of the cell through a fusion of a vesicle with the plasma membrane. As FIGURE 5.9 shows, exocytosis involves a transport vesicle making its way to the plasma membrane and fusing with it, whereupon the vesicle's contents are released into the extracellular fluid. You observed in Chapter 4 that cells use exocytosis when they are exporting proteins. For single-celled creatures, waste products may be released into the extracellular fluid through this process.

FIGURE 5.9 Movement Out of the Cell

(a) In exocytosis, a transport vesicle—perhaps loaded with proteins or waste products—moves to the plasma membrane and fuses with it. This section of the membrane then opens, and the contents of the former vesicle are released to the extracellular fluid.

(b) Micrograph of material being expelled from the cell through exocytosis.

(a) Exocytosis

(b) Micrograph of exocytosis

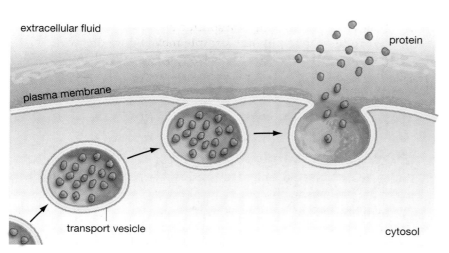

extracellular fluid

plasma membrane

protein

transport vesicle

cytosol

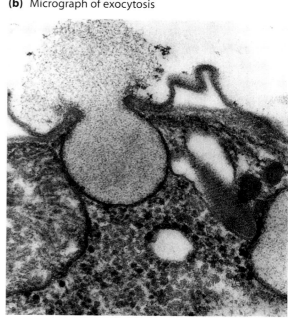

(a) Pinocytosis

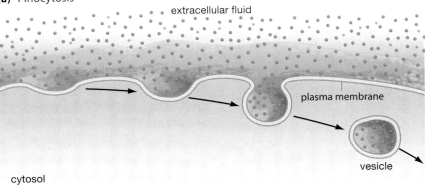

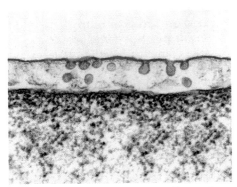

In pinocytosis, the plasma membrane folds inward to create a kind of harbor. The harbor then encloses completely, pinches off as a vesicle, and moves into the cell's cytoplasm, carrying with it whatever material was enclosed.

Micrograph of pinocytosis in a capillary cell.

(b) Receptor-mediated endocytosis

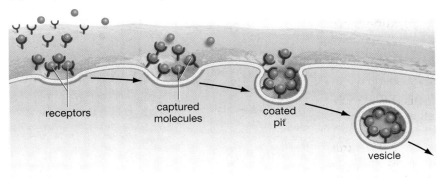

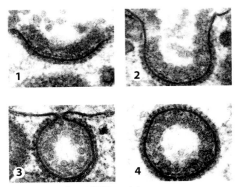

In receptor-mediated endocytosis, many receptors bind to molecules. Then, while holding on to the molecules, the receptors migrate laterally through the cell membrane, arriving at a depression called a coated pit. The coated pit pinches off, delivering its receptor-held molecules into the cytoplasm.

Formation of an RME vesicle.

(c) Phagocytosis

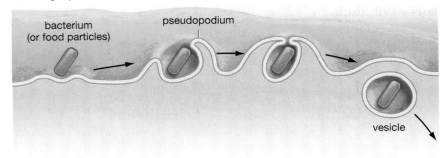

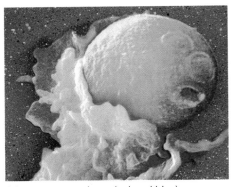

In phagocytosis, food particles—or perhaps whole organisms—are taken in by means of "false feet" or pseudopodia that surround the material. Pseudopodia then fuse together, forming a vesicle that moves into the cell's interior with its catch enclosed.

A human macrophage (colored blue) uses phagocytosis to ingest an invading yeast cell.

FIGURE 5.10 Three Ways to Get Relatively Large Materials into the Cell

Movement In: Endocytosis

Endocytosis is the movement of relatively large materials into the cell by infolding of the plasma membrane. It can take any of three forms: pinocytosis, receptor-mediated endocytosis, and phagocytosis.

Pinocytosis   Pinocytosis means "cell drinking," and if you look at FIGURE 5.10, you can see why this term is fairly accurate. The cell folds inward, creating a kind of harbor on its exterior. Whatever material happens to be enclosed in the harbor when it pinches off to become a vesicle is brought into the cell. What is brought in, of course, is mostly water, with some solutes in it. **Pinocytosis** can be defined as a form of endocytosis that brings into the cell a small volume of extracellular fluid and the materials suspended in it. Remember how it was noted earlier that the plasma membrane is constantly being "remade," due to the movement of materials in and out of it? With pinocytosis you can get some sense of this. Through this form of endocytosis, whole sections of the plasma membrane are continually budding off and moving to the interior of the cell. These sections have to be replaced, of course, with membrane brought up from the cell's interior. The result is a membrane in a constant state of flux.

Receptor-Mediated Endocytosis (RME)

The second form of endocytosis, **receptor-mediated endocytosis** (RME) gets its name from a protein on receptors, whose role is to bind to specific molecules and then hold onto them. Groups of receptors bearing these molecules then make a lateral migration through the cell membrane and congregate in a place that is literally a pit; it is a depression in the cell, referred to as a *coated pit*. Eventually, the pit deepens and pinches off, creating the familiar vesicle moving into the cell. RME is very important in getting nutrients and other substances into cells. It is the way cholesterol gets into your cells, for example.

Phagocytosis

The third form of endocytosis is a means of bringing even larger materials into the cell. It is phagocytosis (literally, "cell eating"). This is the mechanism mentioned earlier, by which a human immune-system cell might ingest a whole bacterium (see Figure 5.10c). This is also the way that many one-celled creatures eat, so it's not surprising that the materials brought into the cell in this process might be 10 times the size of materials brought in with pinocytosis or RME. Not all the materials brought into the cell by phagocytosis are whole cells; parts of cells and large nutrients are fair game as well.

As you can see in Figure 5.10, phagocytosis begins when the cell sends out extensions of its plasma membrane called pseudopodia ("false feet"). These surround the food and fuse their ends together. What was once outside is now inside, encased in a vesicle and moving toward the cell's interior. **Phagocytosis** can be defined as a process of bringing relatively large materials into a cell by means of wrapping extensions of the plasma membrane around the materials and fusing the extensions together.

On to Energy

Your tour of the cell and its plasma membrane is now complete. What's coming up next is something called bioenergetics. In our story of biology so far, there have been continuing references to proteins that are called enzymes and to a molecule called ATP and to energy expenditures. The fact that it's been necessary to mention these things before actually reviewing them gives you some idea of how important they are to biology. Now, however, we'll get to them as part of a tour of bioenergetics, a field that deals with forces so basic they affect every form of life.

So Far...

1. A hydrophilic molecule such as glucose cannot simply diffuse through the plasma membrane. Instead, it passes through in a process of _____, which is always aided by a concentration gradient and a _____.

2. All the mechanisms of moving relatively large materials in and out of the cell have the use of _____ in common.

3. Certain immune-system cells are able to ingest whole bacterial cells through the process of _____.

A BROADER VIEW

Human RME and Phagocytosis

Receptor-mediated endocytosis (RME) may bring needed substances into human cells, but it can also bring in something else: harmful viruses and bacteria. The polio and flu viruses are among those that enter cells through RME. Phagocytosis is seen fairly frequently in one-celled organisms, but in human beings, there are only three types of cells that carry it out, all of them immune-system cells. One of these cells is the macrophage, which not only acts as a warrior of the body, engulfing invading microbes, but as a janitor, cleaning up dead human cells and cellular debris.

So Far... Answers:

1. facilitated diffusion; transport protein

2. membrane-lined vesicles

3. phagocytosis

Chapter Review

Summary

5.1 The Nature of the Plasma Membrane

- **Qualities of the membrane**—The plasma membrane is a thin, fluid, lipid entity that manages to be very flexible and yet stable enough to stay together despite being continually remade, due to the constant movement of materials in and out of it. In animal cells, it has four principal components: (1) a phospholipid bilayer, (2) molecules of cholesterol interspersed within the bilayer, (3) proteins that are embedded in or that lie on the bilayer, and (4) short carbohydrate chains on the cell surface, collectively called the glycocalyx.   (page 79)

- **Components of the membrane**—Phospholipids are molecules composed of two fatty-acid chains linked to a charged phosphate group. They are arranged as a bilayer that composes much of the plasma membrane. Cholesterol molecules interspersed between phospholipid molecules act as a patching material and keep the membrane at an optimal level of fluidity. Proteins in or on the membrane serve in structural support, cell identification, cell communication (as receptors), and transport (as channels for the movement of compounds into and out of the cell). Glycocalyx chains function in cell adhesion and as binding sites on proteins.   (page 80)

5.2 Diffusion, Gradients, and Osmosis

- **Diffusion and concentration**—Diffusion is the movement of molecules or ions from a region of their higher concentration to a region of their lower concentration. A concentration gradient defines the difference between the highest and lowest concentrations of a solute within a given medium. Through diffusion, compounds naturally move from higher to lower concentrations—meaning down their concentration gradients. Energy must be expended, however, to move compounds against their concentration gradients.   (page 83)

- **Membranes and osmosis**—A semipermeable membrane is one that allows some compounds to pass through freely while blocking the passage of other compounds. Osmosis is the net movement of water across a semipermeable membrane from an area of lower solute concentration to an area of higher solute concentration. Because the plasma membrane is a semipermeable membrane, osmosis operates in connection with it. Osmosis, a form of diffusion, is a major force in living things; it is responsible for much of the movement of fluids into and out of cells. Cells will gain or lose water relative to their surroundings in accordance with what the solute concentration is inside the cell as opposed to outside.   (page 85)

Web Tutorial 5.1 Diffusion and Osmosis

5.3 Moving Smaller Substances In and Out

- **Forms of transport**—Some compounds are able to cross the plasma membrane strictly through concentration gradients and diffusion; others require these factors and special protein channels; still others require protein channels and the expenditure of cellular energy. Active transport is any movement of molecules or ions across a cell membrane that requires the expenditure of energy. Passive transport is any movement of molecules or ions across a cell membrane that does not require the expenditure of energy.   (page 87)

- **Passive transport**—Two forms of passive transport are simple diffusion and facilitated diffusion. Simple diffusion is diffusion that does not require a protein channel. Facilitated diffusion is diffusion that requires a protein channel.   (page 87)

- **Active transport**—Cells cannot rely solely on passive transport to move substances across the plasma membrane. It may be that a given cell needs to have, on one side of its membrane, a greater concentration of a given substance than exists on the other side. To deal with such needs, cells employ chemical pumps to move compounds across the plasma membrane against their concentration gradients. These pumps operate in active transport.   (page 88)

Web Tutorial 5.2 Passive and Active Transport

5.4 Getting the Big Stuff In and Out

- **Endocytosis and exocytosis**—Larger materials are brought into the cell through endocytosis and moved out through exocytosis. Both mechanisms employ vesicles, the membrane-lined enclosures that alternately bud off from membranes or fuse with them.   (page 89)

- **Forms of endocytosis**—There are three principal forms of endocytosis: (1) pinocytosis, in which the plasma membrane folds inward, creating an enclosure that pinches off to become a vesicle that moves into the cell, carrying with it extracellular fluid and the materials suspended in it; (2) receptor-mediated endocytosis (RME), in which cell-surface receptors bind with materials to be brought into the cell and then migrate laterally through the cell membrane, congregating in a "coated pit," where vesicle-budding will bring them into the cell; and (3) phagocytosis, in which certain cells engulf whole cells, fragments of them, or other large organic materials, and bring these materials into the cell inside vesicles that bud off from the plasma membrane.   (page 90)

Web Tutorial 5.3 Endocytosis and Exocytosis

Key Terms

active transport   p. 87

concentration gradient   p. 84

diffusion   p. 84

endocytosis   p. 90

exocytosis   p. 89

facilitated diffusion   p. 88

glycocalyx   p. 83

hypertonic solution   p. 86

hypotonic solution   p. 87

isotonic solution   p. 86

osmosis   p. 85

passive transport   p. 87

phagocytosis   p. 91

phospholipid bilayer   p. 81

pinocytosis   p. 91

plasma membrane   p. 83

receptor-mediated endocytosis
(RME)   p. 91

receptor protein   p. 82

simple diffusion   p. 87

transport protein   p. 83

Testing Your Understanding

In Your Own Words *(answers in the back of the book)*

1. Explain why the phospholipid "heads" of the plasma membrane are always pointed toward the cytosol and extracellular fluid, whereas the "tails" are always oriented toward the middle of the membrane.

2. Describe the process of receptor-mediated endocytosis.

3. As you saw in this chapter, all cellular membranes are similar in structure. However, what differences are likely to exist between the plasma membrane and membranes in the interior of the cell?

Thinking about What You've Learned

1. All life-forms on Earth exist either as single cells, or as collections of cells, that have a plasma membrane at their outer edge (though some cells have cell walls outside their plasma membrane). Why do you think the plasma membrane exists in all living things? Why aren't there living things that don't have a membrane at their outer edge?

2. Because materials move from high concentration to low concentration by simple diffusion, and they move the same way by facilitated diffusion, why do cells make proteins to carry out facilitated diffusion? Isn't this just a waste of energy?

3. Place a number of marbles on one end of a cafeteria tray. Shake the tray gently a few times, keeping it level. What happens to the marbles? Start over and shake faster the next time. What happens now? Explain how this is like chemical diffusion.

Powering Life: An Introduction to Energy

THE CINCINNATI POST

ARCTIC PEOPLE EVOLVED TO PRODUCE MORE HEAT

By Associated Press
June 18, 2004

People native to the far north evolved to produce more heat in their cells, a new study says. The researchers suggest this change is a climate-driven effect.

The change occurs in the mitochondria, the parts of human cells that burn fuel to produce heat and energy, according to the team of researchers led by Eduardo Ruiz-Pesini of the University of California at Irvine. The scientists analyzed mitochondria from 1,125 people ranging from Africa to Europe and Arctic Siberia. They found that mutations in mitochondrial DNA, increasing production of heat, though reducing energy production, rise in people living closer to the pole, compared to tropical residents.

6.1 Energy Is Central to Life

How much warmer do the indigenous people of the Arctic tend to be? When Arctic men are at rest, their bodies have a metabolic rate that ranges from 7 to 19 percent higher than what would be expected, and the rate for Arctic women is almost this elevated.

It might not be too surprising that, if your ancestors are from a cold climate, you tend to produce more body heat. But note, in the newspaper story above, where this heat gets produced: in the tiny oval structures within cells called mitochondria. These are the organelles that extract energy from food. So, why should heat get generated there? Equally puzzling, there turns out to be a downside to increased heat production in the mitochondria. Arctic natives tend to have higher rates of "energy deficiency" diseases, such as having sperm that aren't very active (which is a cause of infertility). It seems that there is a trade-off between the production of heat and the availability of energy.

The story of energy use among people of the far north actually draws back a curtain for us on a whole realm of the living world, which is the connection between energy and life. Why does heat get produced in such abundance in the mitochondria? Because the mitochondria are the primary places in the body where energy goes from one form to another, and heat is always generated when this happens. (Ever touch a car engine after the energy in gasoline has been used to make the engine's pistons go up and down?) And why should there be a trade-off between the Arctic people's production of heat and the amount of energy available for use in their cells? Because, just as you can have a middleman in a business transaction, so you can have a middleman—heat—in an energy transaction; and the more this middleman takes, the less there is left for you. Even so, how important is this matter of energy availability? Well, think of it this way: The energy transfer that goes on in our mitochondria is so critical that it prompts an act that seems indistinguishable from life itself—breathing. Why do we need to breathe? So that we can acquire the oxygen that allows our mitochondria to keep transferring energy to us.

Energy is such a given in our lives that we're likely to think of it only at times when it's in short supply—when we're running up a hill or lifting a box, for example. But in fact energy is needed for every single activity we undertake. We read a word on a page, we use some energy; our heart beats once, we use some energy. Indeed, no living thing can exist without a supply of energy. A bacterium cannot move itself a millionth of a meter without energy; a maple tree cannot produce a thimbleful of sap without energy.

And where does this energy come from? For most living things, there is one ultimate source: the sun. Human beings can't use the sun's energy directly, of course, but solar energy can be *converted* to forms that we can use. A peach is a vehicle that first locks up solar energy and then transfers this energy to us. The peach *tree* can perform photosynthesis, and in this activity, solar energy is converted to chemical energy, which exists within the peach. We eat the peach, and now this energy is in us, locked up in the form of glucose. At this point this energy is almost ready for our use, whether in reading a book or making our heart beat.

The tricky thing is to understand how energy could be locked up chemically in something as seemingly unenergetic as a peach. Well, setting aside the water in it, a peach is mostly a carbohydrate; and, as you may recall from Chapter 3, carbohydrates are molecules made of carbon, oxygen, and hydrogen atoms that are linked together in very specific ways. You could think of the sun's rays as having moved the electrons in these atoms up an energy hill, at which point they were locked into place, at their higher level, through the process of chemical bonding. At this point, a given electron is like a rock perched at the top of a hill. And just as a rock can fall down the hill, releasing energy as it goes, so the electrons in the peach can fall down an energy hill, releasing energy as they go.

Now, can our bodies simply use the electrons in food to directly power all the processes that go on within us? The answer is no. Even when a peach has been broken down into glucose, the energy contained in the glucose molecules needs to be put into a different form—one that our bodies can use. This is akin to what happens in a power plant. The coal that a power plant uses contains energy too, once again locked up in chemical bonds. But the energy in coal must be converted into a form that can be used in our homes—electricity. In the same way, the chemical energy contained in food must be put into a chemical form we can use in our *bodies*, and this chemical form is a substance called ATP. This molecule acts as a kind of energy dispenser for most of the activities in the body. (We need something done in the liver, it's ATP that gives up the energy to do it; we need a muscle to contract, it's ATP that's supplying the energy again.)

If we take a step back from this, we can see that there is a hugely important story going on here—one that could properly be called cosmic, given the involvement of the sun. Here's how energy flowed in our example. It went:

$$\text{sun} \longrightarrow \text{peach} \longrightarrow \text{glucose} \longrightarrow \text{ATP} \longrightarrow \text{muscle activity}$$

This chapter and the next two in the book are aimed at describing every link in this chain. In this chapter, we'll consider some basics of energy transfer, what the nature of ATP is, and how it is put to use in reactions that are sped up through substances called enzymes. Next chapter we'll consider how energy is extracted from food and transferred to ATP. And in Chapter 8, we'll look at how the sun's energy comes to reside in food in the first place, which is another way of saying we'll look at photosynthesis.

Some Basic Energy Concepts

You may have noticed that, so far, most of this talk about energy has involved the notion of energy *transfer*. We may also talk about energy being used, but that's not the same thing as energy being used *up*, because that cannot happen. A fundamental insight about energy is something called the **first law of thermodynamics**, which states that energy is never created or destroyed, but instead can only be transformed. The solar energy of the sun is transformed into the chemical energy in the peach; the chemical energy in coal is transformed into the electrical energy that comes through our wall sockets, and so forth.

Now, every time one of these energy "transactions" takes place, there is a kind of middleman who is taking a cut of the available energy, and that middleman is a player we've already been introduced to: *heat*. The reason we say that energy is never destroyed is that heat itself is a form of energy. It is, however, referred to as the *lowest* form of energy because it is the most easily dispersed, random form of energy that exists. A lump of coal is a very ordered object, containing carbon atoms bound to each other in a precise spatial relationship. Once we bring a flame to the coal, however, the ordered, concentrated chemical energy in the coal becomes the disordered, dispersed energy of heat. And this gets us to the critical thing. Energy transformations will run spontaneously only in this way—from greater order to lesser order. Once the energy in the coal has dispersed to heat, there is no chance that it will spontaneously convert to anything *but* heat. We expect to see a lump of coal burn to ashes, and to feel the heat that results. But we do not expect to see the ashes re-form into a lump of coal, nor to feel the hot air concentrate itself again. The scientific principle that speaks to this is called the **second law of thermodynamics**: Energy transfer always results in a greater amount of disorder in the universe.

If you look at **FIGURE 6.1**, you can see how this law plays out in the workings of a steam engine, with the disorder (or "entropy") of energy increasing at every step, while the total amount of energy remains constant. The importance of the second law, however, is not limited to something as abstract as power in a steam engine. The effects of this law are all around us. The trick comes in understanding that things we take for granted actually are the result of it. Why does a substance like salt dissolve in water? Why does a bicycle roll downhill? Why will our sun burn itself out but never re-form? Because of the second law's dictum that energy always tends toward its lowest state, with the lowest state of them all being heat.

On a more practical level, the loss of energy to heat is considerable, as you can see in **FIGURE 6.2**. If we look at an automobile engine, we find that it is capable of converting only about 25 percent of energy in the chemical bonds of gasoline into movement of the car. Such an example allows us to see what was at work with the indigenous Arctic people mentioned at the start of the chapter. Evolution has shaped their metabolism such that, in their conversion of the energy in food, they produce more heat relative to chemical energy than do, say, African people. The upside of this is that they stay warmer in a cold climate; the downside is that they have less energy available for other uses.

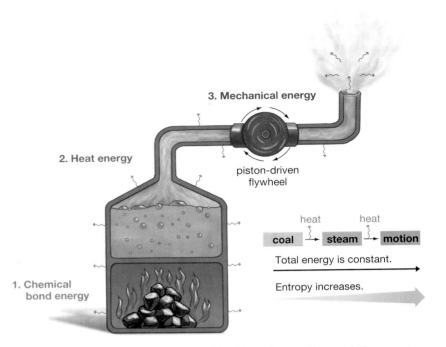

FIGURE 6.1 The Transformations of Energy In a steam engine, energy locked up in the chemical bonds of coal is transformed into heat energy and mechanical energy. There is no loss of energy in this process, but energy is transformed from a more-ordered, concentrated form (the chemical bonds of coal) to a less-ordered, more dispersed form (heat). Thus, the amount of disorder—or entropy—increases in the transaction.

FIGURE 6.2 Energy Efficiency The efficiency of several energy systems, as measured by the proportion of energy they receive relative to what they then make available to perform work. In measuring the efficiency of the car engine, the question is, how much of the energy contained in the chemical bonds of gasoline is converted by the car into the kinetic energy of wheel movement? In each system, most of the energy not available for work is lost to heat.

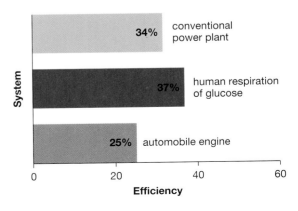

FIGURE 6.3 **Energy Stored, Energy Released** It takes energy to build up a more complex molecule (in this case glycogen) from simpler molecules (in this case individual glucose units). Energy is stored in the chemical bonds of the more complex molecule, however, with the result that, when this molecule is broken down, energy is released.

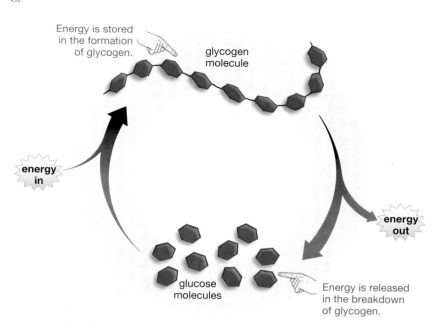

Energy is stored in the formation of glycogen.

glycogen molecule

energy in

energy out

glucose molecules

Energy is released in the breakdown of glycogen.

A BROADER VIEW

Perpetual Motion Machines

You may have heard that there can be no such thing as a perpetual motion machine. Now you can understand why. Any machine transfers energy. But every time energy is transferred, there is some loss of the original energy to heat. So any original quantity of energy that a machine has must diminish over time. Eventually there will not be enough energy left to run the machine.

Energy Storage and Release

Though every energy transaction results in some loss of energy to heat, a starting quantity of energy can be large enough to make it through several transactions, during which time this energy can be stored first in one form and then in another before being used. Thus, the sun's energy we looked at earlier was first stored in a peach, then stored in us as glucose, then transferred to ATP, which dispensed it to be *used* in us for making our heart beat, allowing our kidneys to function, and so forth. A similar pattern of energy storage and use could be seen in all living things. But here's the thing: It takes some energy to store away energy. Think of it as akin to what could happen with a boulder, a person, and a hill. It would take some energy (from the person's arms and legs) to push a boulder up a hill, but once this rock is perched at the top of the hill, it is storing energy (potential energy) that can be released if the boulder is given a light push down the hill. Likewise, the glucose that quickly enters our bloodstream when we eat a peach can be stored away in our muscles and liver as a more complex molecule called glycogen. But, as with pushing the boulder, it is a trip *up* the energy hill to turn a set of small, individual glucose molecules into one larger glycogen molecule. Why? Because a large glycogen molecule is a more complex, ordered chemical object than the individual glucose molecules it is made of. The flip side of this is that the breakdown of a larger molecule such as glycogen into its component parts does not require energy; in fact, some energy will be released in such a reaction (**see** FIGURE 6.3).

In Chapter 8, you'll see how solar energy is used to store away energy in plants through photosynthesis. But right now, we're interested in the uphill process by which the energy in food gets stored, for a brief time, in the energy-dispensing molecule we talked about, ATP. Learning about this process will show you how ATP is able to play its central role in powering our activities.

So Far...

So Far... Answers:

1. chemical; solar

2. created; destroyed; transformed

3. heat

1. The energy contained in the food we eat comes in the form of _____ energy that existed originally as _____ energy.

2. According to the first law of thermodynamics, energy can never be _____ or _____, but it can be _____.

3. According to the second law of thermodynamics, every contraction of a muscle or firing of a car cylinder results in some loss of energy to _____.

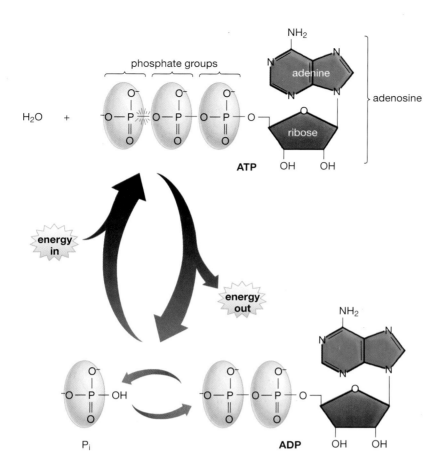

FIGURE 6.4 Energy Release from Breakdown of ATP ATP (adenosine triphosphate) is life's most important energy transfer molecule. It stores energy in the form of chemical bonds between its phosphate groups. When the bond between the second and outermost phosphate group is broken, the outermost phosphate separates from ATP and energy is released. This separation transforms ATP into adenosine diphosphate or ADP, which then goes on to pick up another phosphate group, becoming ATP again.

6.2 The Energy Molecule: ATP

ATP is short for adenosine triphosphate. If you look at the upper drawing in FIGURE 6.4, you can see where this formidable name comes from: ATP has a sugar (called ribose) and a nitrogen-containing molecule (called adenine) that together make up a compound called adenosine. Phosphate groups are then linked to adenosine—three phosphate groups, which gives us adenosine triphosphate. Now, the place to look for energy storage in ATP is in the phosphate groups, particularly the third one, out on the left. Notice that all the phosphate groups are negatively charged. Remember that like charges repel each other, and in this case the three phosphate groups are doing just that. Now recall that it is the energy in food that is transferred to ATP, thus making it energetic. With all this in mind, you're ready to understand *how* the energy from food energizes ATP: It powers the process by which ATP's third phosphate group is added to it. This is a trip up the energy hill, because the phosphate groups are repelling each other. Thus it takes energy to overcome that repulsion and get that third phosphate group onto ATP. But once this is done, the phosphate group is like a rock at the top of the hill: It can *release* energy—by splitting off from the rest of the ATP molecule.

If you now look at the lower drawing in Figure 6.4, you can see what happens once the third phosphate group is released. ATP has become a molecule containing only *two* phosphate groups, adenosine *di*phosphate or ADP. It is now free to have a third phosphate group added to it again. This is just what happens: The energy contained in food once again drives the uphill process by which a phosphate group is added onto ADP, and the resulting ATP is capable of providing energy for yet another reaction. This conversion from ATP to ADP and back takes place constantly in cells.

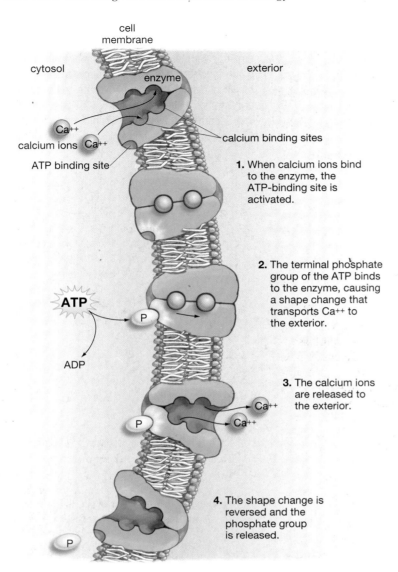

cell membrane

cytosol

exterior

enzyme

Ca++

calcium ions Ca++

ATP binding site

calcium binding sites

1. When calcium ions bind to the enzyme, the ATP-binding site is activated.

ATP

P

ADP

2. The terminal phosphate group of the ATP binds to the enzyme, causing a shape change that transports Ca++ to the exterior.

3. The calcium ions are released to the exterior.

P

Ca++

Ca++

4. The shape change is reversed and the phosphate group is released.

P

FIGURE 6.5 How ATP Functions By transferring a phosphate group to the enzyme shown in this figure, ATP causes the enzyme to change shape in a manner that transports calcium ions across the cell membrane.

But what is it that the phosphate group in ATP is actually doing to power our activities? As a starting point, consider an action, shown in FIGURE 6.5, that ATP molecules carry out within muscle cells. Under the right conditions ATP will bind to a type of protein, called an enzyme, that spans the membrane of this cell. The ATP molecule is then split, with its outermost phosphate group breaking off from the rest of the molecule. The subsequent attachment of this phosphate group to the protein causes the protein to change its shape. This shape change happens to drive the transport of calcium ions across the protein. For our purposes, however, what's important is that something that would not have happened on its own took place because of the energy provided by the ATP molecule. Summing up, **ATP** can be defined as a molecule that functions as the most important energy transfer molecule in living things. Note the term energy *transfer* molecule. While it's true that ATP stores energy, this is very brief storage. For long-term energy storage in the body, we look to the glycogen mentioned earlier, stored in muscles and the liver, and to fat, which is stored in fat cells.

Next chapter, we will go over the steps by which the body channels the energy from food to drive the attachment of the phosphate group onto ADP. But before we get to that subject, we need to review a group of molecules that living things use to control energy.

6.3 Efficient Energy Use in Living Things: Enzymes

Lactose—better known as milk sugar—is composed of two simple sugars, glucose and galactose. Lactose will split into its glucose and galactose parts, and this splitting is a downhill reaction, which is to say it doesn't require any input of energy from a molecule such as ATP. If we actually took a small amount of lactose, however, and put it in some plain water, it would certainly be hours—it might be days—before *any* of the lactose molecules would break down into glucose and galactose. How can this be? Most of us drink milk and it doesn't seem to pile up in us.

Accelerating Reactions

The secret is that when lactose is metabolized in living things, it isn't being split up in plain water. Something else is present that speeds up this process immensely, accelerating it perhaps a billionfold. That something is an **enzyme**, a type of protein that accelerates a chemical reaction. The enzyme that works on lactose is called lactase, but it is merely one of thousands of enzymes known to exist. Each of these compounds carries out some chemical process, but not all are involved in splitting molecules. Some enzymes combine molecules, some rearrange them.

Given their numbers, it may be apparent that enzymes are involved in a lot of different activities; but this doesn't begin to give them the recognition they deserve. Enzymes facilitate nearly every chemical process that takes place in living things. No organism could survive without them. Technically, these compounds are only *accelerating* chemical reactions that would happen anyway, as with lactase splitting lactose. But in practical terms, they are *enabling* these reactions, because no living thing could wait days or months for the milk sugar it ingests to be broken down, or for hormones to be put together, or for bleeding to stop.

To get an idea of the importance of enzymes, consider the millions of Americans who are known as lactose intolerant because, after childhood, their bodies reduced their production of lactase. When these people consume milk, the lactose in it *does* simply stay in their intestines. It is eventually digested not by them, but by the bacteria that live in the human digestive tract (something that causes bloat and gas).

Troublesome as this affliction may be, it is minor compared to other conditions caused by enzyme dysfunction. All newborn babies in the United States are screened for a list of inherited diseases, three of which are Tay-Sachs disease, phenylketonuria, and congenital adrenal hyperplasia. Each one of these diseases is caused by an enzyme deficiency. In the case of Tay-Sachs disease, the lack of an enzyme leads to a buildup of fat deposits within nerve cells, and the result usually is death by age 4.

Specific Tasks and Metabolic Pathways

From the sheer number of enzymes that exist, you can probably get a sense of how specifically they are matched to given tasks. Although some enzymes can work on *groups* of similar substances, a specific enzyme usually facilitates a specific reaction: It works on one or perhaps two molecules and no others. Thus, lactase breaks down lactose—and *only* lactose—and the products of its activity are always glucose and galactose. The substance that is being worked on by an enzyme is known as its **substrate**. Thus, lactose is the substrate for lactase. The *–ase* ending you see in *lactase* is common to most enzymes.

Certain activities in living things, such as blood clotting, are much more complex than the breakdown of lactose. They are multistep processes, *each step of which* requires its own enzyme. Most large-scale activities in living things work this way: leaf growth, digestion, hormonal balance. The result is a **metabolic pathway**: a set of enzymatically controlled steps that results in the completion of a product or process in an organism. In such a sequence, each enzyme does a particular job and then leaves the succeeding task to the next enzyme, with the product of one reaction becoming the substrate for the next (**see** FIGURE 6.6). The sum of all the chemical reactions that a cell or larger organism carries out is known as its **metabolism**. To put things simply, enzymes are active in all facets of the metabolism of all living things.

A BROADER VIEW

ATP Is Like Money

It is often said that ATP functions like money. Now you can see why. Imagine working at a typical job, in which you're called upon to do dozens of different things. In the end what you get for all your efforts is money. You can then take this cash and pay the rent, buy some CDs, or go bowling. Now think of animals and their breakdown of food. They have to obtain food in the first place and then digest what they obtain. But what they get in the end from all this work is ATP. It is the final outcome of the energy-acquisition process. And, like money, ATP then can be used to do any number of things, such as contract a muscle or read a sentence in a book.

A BROADER VIEW

What Is Our Metabolism?

We generally think of our "metabolism" in terms of the rate our bodies burn up food—the equivalent of how many miles per gallon our car gets as opposed to someone else's. But when we begin to think of metabolism as the sum of all the chemical reactions in the body—each one facilitated by an enzyme—we begin to get a better sense of what metabolism is, as opposed to merely having in mind a featureless process that burns calories at one rate or another.

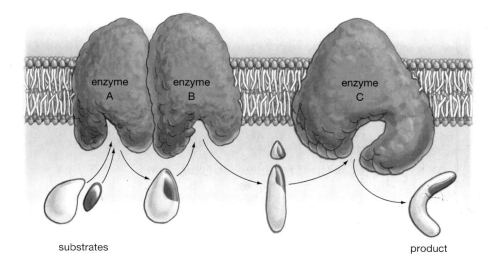

substrates   product

FIGURE 6.6 Metabolic Pathway: Sequence of Enzyme Action Most processes in living organisms are carried out through a metabolic pathway—a sequential set of enzymatically controlled reactions in which the product of one reaction serves as the substrate for the next. Enzymes perform specific tasks on specific substrates. In this example, enzyme A combines two substrates, enzyme B removes part of its substrate, and enzyme C changes the shape of its substrate.

(a) Without enzyme

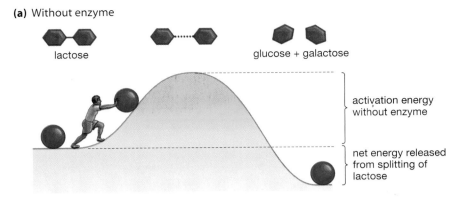

(b) With enzyme

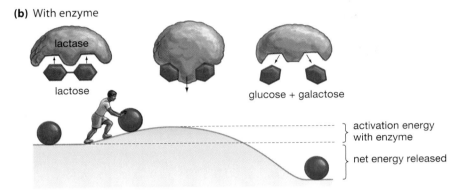

FIGURE 6.7 **Enzymes Accelerate Chemical Reactions** How is lactose split into glucose and galactose? Without an enzyme, the amount of energy necessary to activate this reaction is high. In the presence of the enzyme lactase, however, a low activation energy is sufficient to get the process started. The energy released from the splitting of lactose is the same in both cases.

Glucose molecule binding to active site of hexokinase enzyme.

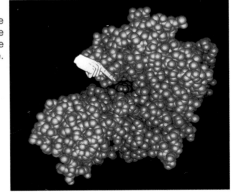

6.4 Lowering the Activation Barrier through Enzymes

All this information about enzymes is well and good, you may be saying at this point, but what does it have to do with energy? The answer is that in carrying out their tasks, enzymes are in the business of lowering the amount of energy needed to get chemical reactions going. And this means that the reactions can get going faster.

An analogy may be helpful here. Consider the now-familiar rock perched at the top of a hill. Situated in this way, it has a good deal of potential energy. If it rolled down the hill, it would release this potential energy. But here is the critical question: What would be required to get this rock going? It is perched at the top of the hill because it is *stable* at the top of the hill. To get it going would require *additional* energy in the form of, say, a push.

Now consider the lactose molecule. It is "perched" like the rock, in a sense, because it lies "uphill" from the glucose and galactose it can break into. But just as the rock is stable because of its position in the ground, so the lactose is stable because it has a strong set of chemical bonds holding its atoms in a fixed position. The question thus becomes: Is there anything that can lower the amount of energy required to break these bonds? The answer is yes—an enzyme. If you look at FIGURE 6.7, you can see this in schematic terms. Imagine that you had to push a boulder up the hill in (a) to get it to roll down the other side. Now imagine that you only had to push a boulder up the hill in (b) to get the same result. In which situation would you be able to get your desired result *faster*? What is at work here is a lowering of **activation energy**: the energy required to initiate a chemical reaction.

How Do Enzymes Work?

How do enzymes carry out this task of lowering activation energy? As you'll see shortly, they bind to their substrates, and in so doing make these substances more vulnerable to chemical alteration. The amazing thing is that they do this without being permanently altered themselves. Enzymes are **catalysts**—substances that retain their original chemical composition while bringing about a change in a substrate. At the end of its cleaving of the lactose molecule, lactase has exactly the same chemical structure as it did before. It is thus free to pick up another lactose molecule and split *it*. All this takes place in a flash: The fastest-working enzymes can carry out 100,000 chemical transformations per second.

Enzymes generally take the form of globular or ball-like proteins whose shape includes a kind of pocket into which the substrate fits. If you look at FIGURE 6.8, you can see a space-filling model of one enzyme, called hexokinase. You can also see, buried there in the middle, its substrate—glucose. As with all proteins, enzymes are made up of amino acids, but only a few of the hundreds of amino acids in an enzyme are typically involved in actually binding with the substrate. Five or six would

FIGURE 6.8 **Shape Is Important in Enzymes** Substrates generally fit into small grooves or pockets in enzymes. The location at which the enzyme binds the substrate is known as the enzyme's active site. Pictured is a computer-generated model of a glucose molecule (in red) binding to the active site of an enzyme called hexokinase (in blue).

be common. These amino acids help form the substrate pocket, known as the **active site**: the portion of an enzyme that binds with and transforms a substrate.

In some cases, the participants at the active site include, along with amino acids, one or more accessory molecules. One variety of these molecules is the **coenzymes**: molecules other than amino acids that facilitate the work of enzymes by binding with them. If you've ever wondered what vitamins do, participation in enzyme binding is a big part of it. Once we have ingested them, many vitamins are transformed into coenzymes that sit in the active site of an enzyme and provide an added chemical attraction or repulsion that allows the enzyme to do its job.

An Enzyme in Action: Chymotrypsin
The details of how enzymes carry out their work are complex, and they vary according to enzyme. Let's consider in a general way, however, the activity of a much-studied enzyme called chymotrypsin.

Chymotrypsin is delivered from the human pancreas to the small intestine, where it works with water to break down proteins we have ingested. Its function is to clip protein chains in between their building-block amino acids. It does this by breaking the single bond that binds one amino acid to the next (**see** FIGURE 6.9). How does it work? After having bound to part of a protein chain, chymotrypsin

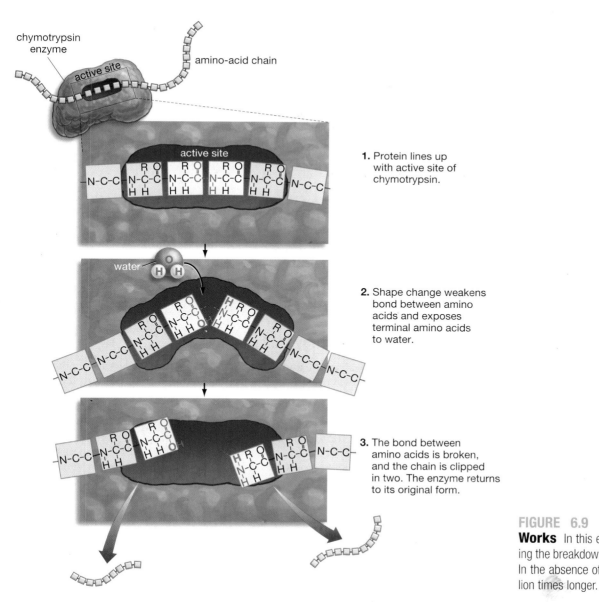

1. Protein lines up with active site of chymotrypsin.

2. Shape change weakens bond between amino acids and exposes terminal amino acids to water.

3. The bond between amino acids is broken, and the chain is clipped in two. The enzyme returns to its original form.

FIGURE 6.9 How the Enzyme Chymotrypsin Works In this example, the enzyme chymotrypsin is facilitating the breakdown of a protein by changing the protein's shape. In the absence of chymotrypsin, this process would take a billion times longer.

then interacts with it to create a transition-state molecule. In effect, it distorts the shape of the protein—and holds it in this new shape briefly—in such a way that the protein becomes vulnerable to bonding with ionized water molecules. This allows carbon and nitrogen atoms to latch onto new partners, and the protein chain is clipped. Chymotrypsin then returns to its original form and proceeds to a new reaction.

On to ATP Production

Our story of energy concluded in this chapter with a look at enzymes and energy. Next chapter, however, we will pick up the story of energy with a more detailed look at a process we touched on earlier—the transfer of energy from food to ATP. The question we will tackle is: What are the steps by which our bodies extract energy from food and then channel this energy into pushing phosphate groups onto ADP? Coming up: the body's energy-harvesting operation.

So Far...

1. The energy in food powers the process by which a phosphate group is added to _____, thus transforming it into the body's main energy transfer molecule, _____.

2. Enzymes are _____ that _____ chemical reactions by lowering their _____.

3. Metabolism is the sum of all the _____ in a cell or larger organism.

So Far... Answers:
1. *ADP; ATP*
2. *proteins; accelerate; activation energy*
3. *chemical reactions*

Chapter Review

Summary

6.1 Energy Is Central to Life

- **Energy and life**—All living things require a source of energy. The sun is the ultimate source of energy for most living things. The sun's energy is captured on Earth by photosynthesizing organisms (such as plants), whose structures (such as fruits) then have this energy locked up in them in the form of chemical energy, which can be passed along to other life-forms, such as ourselves.   (page 95)

- Foods that contain chemical energy may first need to be broken down into a set of simpler molecules (often glucose). Then the chemical energy contained in these simpler molecules will be transferred to another molecule—ATP, which yields the energy that powers most of our activities.   (page 96)

- **The first and second laws of thermodynamics**—A fundamental insight about energy is the first law of thermodynamics: Energy can be neither created nor destroyed, but instead can only be transformed. A second fundamental insight is the second law of thermodynamics: Energy transfer always results in a greater amount of disorder in the universe. All energy transfers result in some loss of energy to the most disordered, dispersed form of energy, heat.   (page 97)

- **Energy storage and release**—It takes energy to combine a simpler set of molecules into a more complex, larger molecule. Energy is released rather than used, however, when a more complex molecule is broken down into its component parts.   (page 98)

Web Tutorial 6.1 Energy and Biology

6.2 The Energy Molecule: ATP

- **The structure and operation of ATP**—Adenosine triphosphate (ATP) is an energy transfer molecule composed of three phosphate groups linked to the sugar ribose and a nitrogen-containing molecule called adenine. In animals, the energy contained in the chemical bonds of food drives the attachment of the third phosphate group onto ATP. This group is then "spent"—by means of splitting off from the rest of the ATP molecule—in reactions that require energy. With this loss of a phosphate group, ATP has become adenosine disphosphate or ADP, which goes on to have a third phosphate group added to it, thus becoming ATP once again.   (page 99)

6.3 Efficient Energy Use in Living Things: Enzymes

- **The importance of enzymes**—An enzyme is a type of protein that increases the rate at which a chemical reaction takes place in an organism. Nearly every chemical process that takes place in living things is facilitated by an enzyme. For example, the enzyme lactase facilitates the splitting of the sugar lactose into its component sugars, glucose and galactose. Enzymes work by lowering activation energy, meaning the energy required to initiate a chemical reaction.   (page 100)

- **Metabolic pathways**—Many activities in living things are controlled by metabolic pathways, in which a series of interrelated reactions is undertaken in sequence, each one of them facilitated by an enzyme. The substance that an enzyme helps transform through chemical reaction is called its substrate. Lactose is the substrate of the enzyme lactase. The sum of all the chemical reactions that a cell or larger living thing carries out is its metabolism.   (page 101)

Web Tutorial 6.2 Enzymes

6.4 Lowering the Activation Barrier through Enzymes

- **The nature of enzymes**—Enzymes work by lowering activation energy, meaning the energy required to initiate a chemical reaction. Enzymes are catalysts: They bring about a change in their substrates without being chemically altered themselves. Enzymes generally take the form of globular or ball-like proteins whose shape includes a pocket into which the enzyme's substrate fits. This pocket is the active site—that portion of an enzyme that binds with and transforms a substrate into a product.   (page 102)

Key Terms

activation energy p. 102	**catalyst** p. 102	**first law of thermodynamics** p. 97	**second law of thermodynamics** p. 97
active site p. 103	**coenzyme** p. 103	**metabolic pathway** p. 101	**substrate** p. 101
adenosine triphosphate (ATP) p. 100	**enzyme** p. 100	**metabolism** p. 101	

Testing Your Understanding

In Your Own Words   *(answers in the back of the book)*

1. Describe the general flow of energy between the sun and an activity, such as the beating of the heart, in the human body.

2. Where does the energy come from that is stored and released by ATP? How does ATP function?

3. Describe what is meant by a "metabolic pathway."

Thinking about What You've Learned

1. In light of the laws of thermodynamics, explain why you get so much hotter when you run than when you are sitting still.

2. Could even a science-fiction writer plausibly invent life-forms that do not require energy? Put another way, is it possible to imagine life without energy, or are the two things so tightly linked in our experience that we can't even conceive of them being separable?

3. In light of the laws of thermodynamics, is there really such a thing as "renewable" energy?

Supplying the Body with Energy

THE WASHINGTON POST

EIGHT SAID TO SUCCUMB ON MT. EVEREST
COPTER SAVES AMERICAN AND TAIWANESE

By David Fogarty
May 14, 1996

Eight climbers, including two Americans, were feared dead in ferocious weather on Mount Everest. Two other climbers were flown to safety today in the high- est helicopter rescue on the world's tallest peak.

7.1 Energy's Vital Role

The newspaper story above was just one of many that chronicled a tragedy that took place on Mt. Everest in Nepal in May of 1996. Disaster struck several groups of mountain climbers who had been attempting to scale Everest, the world's high- est peak. A vicious storm had taken the members of two Everest expeditions by surprise as they were moving down from the mountain's summit, toward the safety of lower altitudes. Within 36 hours, five climbers had lost their lives and another was so badly frostbitten that one of his hands had to be amputated.

One of the expeditions was led by a highly respected guide from New Zealand, Rob Hall. Courageously remaining near the summit in an effort to bring down one of his clients (who had collapsed), Hall ended up spending a night in a howl- ing storm on Everest without a tent. When the sun rose the next day, the factors working against him included cold and wind, which he might have encountered in lots of in- hospitable places on Earth. Also bedeviling him, however, was something that human beings rarely encounter: a lack of oxygen. At his altitude—sitting in the snow at more than 28,000 feet—the oxygen he took in with each breath

was little more than a third of the amount he would have inhaled with each breath at sea level. Technology might have come to his aid, as he had two oxygen bottles with him, but his intake valve for them had become clogged with ice.

Down at Everest's lower altitudes, other climbers—some of them Hall's friends—had to endure the agony of talking to him by two-way radio while not being able to reach him physically. All of them knew that a lack of oxygen was draining not only his muscles but his mind. Here is Hall in one of his radio transmissions asking about another guide on the ill-fated expedition: "Harold was with me last night, but he doesn't seem to be with me now. Was Harold with me? Can you tell me that?" In the end, the mountain without mercy claimed Rob Hall. He never got up from the place where he spent his night on Everest.

The tragedy of the 1996 Everest expeditions drives home a point and raises a paradox: No one needs to be reminded that we need to breathe in order to live, and most people are aware that oxygen is the most important thing that comes in with each breath. That said, of the next 100 people you meet, how many could tell you what oxygen is *doing* to sustain life? Put another way, why do we need to breathe?

The short answer is that breathing and oxygen are in the energy transfer business—the same business food is in, as you saw last chapter. Oxygen is part of a system that allows us to extract, from food, energy that is then used to put together the "energy transfer" molecule, ATP. As you'll recall from Chapter 6, the body uses ATP (adenosine triphosphate) to power activities that range from muscle contraction to thinking to cell repair. Living things need large amounts of ATP to live, and organisms such as ourselves use oxygen to produce most of our ATP. If we don't get enough oxygen to keep making ATP, our bodies and minds start failing.

Oxygen is not required for all the "harvesting" we do of the energy contained in food. When you lift a box, the short burst of energy that is required comes largely from a different sort of energy harvesting, which you'll learn more about shortly. But even here, the essential *product* of energy harvesting is ATP. Keep that in mind and you won't get lost in the byways of energy transfer. How does the body produce enough ATP to allow us to read a book or climb a mountain? In this chapter, you'll find out.

7.2 Electrons Fall Down the Energy Hill to Drive the Uphill Production of ATP

Given the importance of ATP to this story, let us begin with a brief review of how this molecule works. Recall from Chapter 6 that each ATP molecule has three phosphate groups attached to it, and that ATP powers a given reaction by losing the outermost of these phosphate groups. In this process ATP is transformed into a molecule with *two* phosphate groups, adenosine diphosphate or ADP. To return to its more "energized" ATP state, it must have a third phosphate group attached again. This is, however, a trip *up* the energy hill. Think of it as akin to pushing a jack-in-the-box back into the box. It takes energy to do this, and the same is true for pushing a third phosphate group on ADP (see FIGURE 7.1). Where does this energy come from? For animals such as ourselves, it comes from food; energy that is extracted from food powers the phosphate group up the energy hill, and literally onto ADP.

In tracking the extraction of energy from food, we will use as an example one particular food molecule, glucose, to see how energy is harvested from it. Though the details here are complex, the essential story is simple. *Electrons* derived from glucose, which is high in energy, will be running downhill; they will be channeled off, a few at a time, and their downhill drop will power the *uphill* push needed to attach a phosphate group onto ADP. The glucose-derived electrons will be transferred to several intermediate molecules in their downhill journey, but the *final* molecule that will receive them at the bottom of the energy hill is oxygen. We need to breathe because we need oxygen to serve as this final electron acceptor.

A BROADER VIEW

Energy and Death

Physical calamities such as strokes and heart attacks are largely calamities of cells being deprived of energy. Here is surgeon Sherwin Nuland writing in his book *How We Die* about one form of stroke: "It takes a great deal of energy to keep the brain's engine functioning efficiently. Almost all of that energy is derived by the tissue's ability to break down glucose . . . The brain does not have the capacity to keep any glucose in reserve; it depends on a constant immediate supply being brought to it by the coursing arterial blood. Obviously the same is true of oxygen. It takes only a few minutes for [a blood-deprived] brain to run out of both before it suffocates."

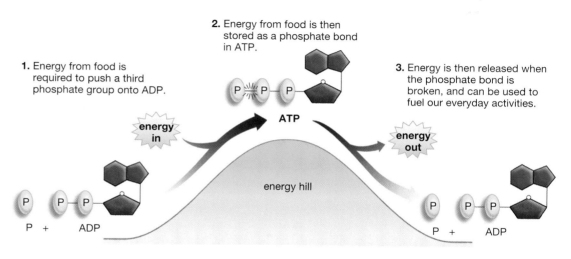

1. Energy from food is required to push a third phosphate group onto ADP.

2. Energy from food is then stored as a phosphate bond in ATP.

3. Energy is then released when the phosphate bond is broken, and can be used to fuel our everyday activities.

FIGURE 7.1 Storing and Releasing Energy Adenosine triphosphate (ATP) is the most important energy-releasing molecule in our bodies. The energy it contains is used to power everything from muscle contraction to thinking.

The Great Energy Conveyors: Redox Reactions

The basis for electron transfers down the energy hill is straightforward: Some substances attract electrons more strongly than do others. A substance that loses one or more electrons to another is said to have undergone **oxidation**. We hear the related word "oxidized" all the time—when paint on an outdoor surface has become dulled, for example, or when metal rusts (see **FIGURE 7.2**). Meanwhile, the substance that *gains* electrons in this reaction is said to have undergone **reduction**. (This seems about as logical as saying a country lost a war by winning, but there is logic in this choice of words. Because electrons carry a negative charge, any substance that gains electrons has had a reduction in its *positive charge*.)

In cells, oxidation and reduction never occur independently. If one substance is oxidized, another must be reduced. It's like a teeter-totter; if one side goes down, the other must come up. The combined operation is known as a reduction-oxidation reaction, or simply a **redox reaction**: the process by which electrons are transferred from one molecule to another. Critically, the substance being oxidized in a redox reaction has its electrons traveling energetically *downhill*.

Many Molecules Can Oxidize Other Molecules

The term *oxidation* might give you the idea that oxygen must be involved in any redox reaction, but this is not the case. Any compound that serves to pull electrons from another is a so-called oxidizing agent. In living things, a large number of molecules are involved in energy transfer, and each has a certain tendency to gain or lose electrons relative to the others.

This is how electrons can be passed down the energy hill: The starting "energetic" molecule of glucose is oxidized by another molecule, which in turn is oxidized by the *next* molecule down the hill. The whole thing might be thought of as a kind of downhill electron bucket brigade. Molecules that serve to transfer electrons from one molecule to another in ATP formation are known as **electron carriers**. The thing that makes their role a little complicated is that many of the electrons they accept are bound up originally in hydrogen atoms. You may remember from Chapter 3 that a hydrogen atom amounts to one proton and one electron. In transferring a hydrogen atom, then, a molecule is transferring a single electron (bound to a proton), which means a redox reaction has taken place.

Redox through Intermediates: NAD The most important electron carrier in energy transfer is a molecule known as **nicotinamide adenine dinucleotide**, or **NAD**, which can be thought of as a city cab. It can exist in two states: loaded with

FIGURE 7.2 An Effect of Oxidation Rust has formed on a hook atop a corroded steel sheet.

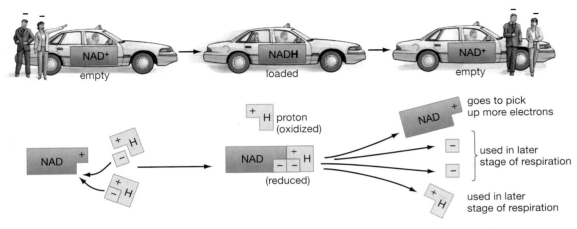

1. NAD⁺ is contained within a cell, along with two hydrogen atoms that are part of the food that is supplying energy for the body.

2. NAD⁺ is reduced to NAD by accepting an electron from a hydrogen atom. It also picks up another hydrogen atom to become NADH.

3. NADH carries the electrons to a later stage of respiration and then drops them off, becoming oxidized to its original form, NAD⁺.

FIGURE 7.3 **The Electron Carrier NAD⁺** In its un-loaded form (NAD⁺) and its loaded form (NADH), this molecule is a critical player in energy transfer, picking up energetic electrons from food and transferring them to later stages of respiration.

passengers or empty. And like a cab, it can switch very easily between those two states. The passengers that NAD picks up and drops off are electrons (**see FIGURE 7.3**).

The "empty" state that NAD comes in is ionic: NAD⁺ Remembering back to the discussion about ions, you can see that NAD⁺ is positively charged, meaning that it has fewer electrons than protons. In a redox reaction, what NAD⁺ does is pick up, in effect, one hydrogen atom (an electron and a proton) and one solo electron (from a second hydrogen atom). The isolated electron that NAD⁺ picks up turns it from positively charged to neutral (NAD⁺ → NAD); the whole hydrogen atom takes it from NAD to NADH. Keeping an eye on redox reactions here, NAD⁺ has become NADH by oxidizing a substance—by accepting electrons from it.

So much for half of NAD's role: picking up passengers. *Now*, as NADH, it is loaded with these passengers. It can proceed down the energy hill to donate them to molecules that have a greater potential to *accept* electrons than it does. Having dropped its passengers off with such a molecule, it returns to being the empty NAD⁺ and is ready for another pickup. Through this process, NAD transfers energy from one molecule to another. Electron transfer through intermediate molecules such as NAD provides the energy for most of the ATP produced. Thus, when looking at the diagrams in this chapter that outline respiration, you'll see:

$$NAD^+ \quad NADH$$

or:

$$NADH \quad NAD^+$$

What these drawings indicate is the electron carrier shifting between its empty state (NAD⁺) and its loaded state (NADH), or vice versa.

So Far...

So Far... Answers:

1. *ATP*
2. *oxidized; electrons*
3. *electron carriers; NAD (nicotinamide adenine dinucleotide)*

1. You need energy to think, to keep your heart beating, to play a sport, and to study this book. This energy is directly supplied by _____, which is produced in the process of energy harvesting.

2. Energy transfer in living things works through redox reactions, in which one substance is _____ by another substance, thereby losing _____ to it.

3. In ATP production, molecules that serve to transfer electrons from one molecule to another are called _____, the most important of which is _____.

Essay 7.1 Glycolysis, Beer, and Muscle "Burn"

For many bacteria—and even sometimes for people—energy harvesting ends with the set of steps called glycolysis, rather than proceeding on through the Krebs cycle and the oxygen-using electron transport chain. When this happens, organisms need a way to recycle their energy transfer molecules in a way that doesn't depend on oxygen. The NADH they produce in glycolysis needs to lose its added electrons and become NAD^+ again, so it can be reused in glycolysis. When energy harvesting proceeds through all three of its major steps, oxygen accepts NADH's added electrons; but when it ends at glycolysis, cells must have an alternative means of dealing with them. The solution to this dilemma is a kind of electron dumping, the *products* of which turn out to be some of the most familiar substances in the world.

Fermentation in Yeast

Yeast, which is a single-celled fungus, provides a good example of how this works. Yeast cells can live by glycolysis alone when oxygen is not present, or they can go through the rest of the stages of aerobic respiration when oxygen is present. Say, then, that yeasts are working away on sugar, going through glycolysis, but doing so in an oxygenless environment. Recall that the final "substrate" product of glycolysis is two molecules of a substance called pyruvic acid. In yeasts, the pyruvic acid they end up with is converted to a molecule called acetaldehyde, and *it* takes on the electrons from NADH, meaning the recycling problem is now solved (since NADH becomes NAD^+ in this process). Having taken on NADH's electrons, though, acetaldehyde now is converted to something very familiar: ethanol, better known as drinking alcohol.

For yeasts, alcohol is simply a by-product of the fermentation they carry out.

Human beings *put* yeasts in environments in which this will happen, of course, because this is how we make wine and beer (**see FIGURE E7.1.1**). Imagine yeast cells inside a dark, airless wine cask, working away on burgundy grape juice, harvesting energy from the grape-juice sugars through glycolysis, but in so doing turning out pyruvic acid, which is turned into alcohol. This continues until the alcohol content of the wine reaches a level (about 14 percent) at which the yeast cells can no longer survive in it. This sequence of events is known as **alcoholic fermentation**: the process by which yeasts produce alcohol as a by-product of glycolysis they perform in an oxygenless environment.

For yeasts, alcohol is simply a waste by-product of the glycolysis they utilize in *anaerobic* energy conversion, meaning conversion in an oxygenless environment. But the gifts that yeasts provide don't stop with alcohol. A second product of their conversion of pyruvic acid is carbon dioxide. When we add yeast to dough, the fermentation that results produces the CO_2 that causes bread to rise and become "light," because of the air holes now in it. (And what happens to the alcohol when we bake bread? It evaporates.)

Fermentation in Animals

Alcoholic fermentation is the solution that fungi (and occasionally plants) employ to sustain glycolysis in the absence of oxygen. But animals take a different tack, because in them the product of glycolysis, pyruvic acid, accepts the electrons from NADH. In this process, however, pyruvic acid is turned into another substance whose name should be familiar, lactic acid. Thus, the kind of fermentation that animals (and certain bacteria) carry out is called **lactate fermentation**. Have you ever experienced the muscle "burn" that comes with, for example, climbing several flights of stairs? If so, you have experienced a buildup of lactic acid in your muscles. But why this should happen? Why should glycolysis ever end in lactate fermentation for us, when, unlike yeast, we always have access to oxygen? The problem turns out to be not a matter of oxygen access, but a matter of oxygen *delivery*. During a quick burst of energy use, oxygen cannot be delivered into our muscle cells fast enough to accommodate the big increase in our energy expenditure. And when a muscle's capacity for aerobic energy transfer has reached its limit, it turns to glycolysis and lactate fermentation to supply more ATP.

FIGURE E7.1.1 Yeast Inside, Performing Fermentation
This man is inspecting a gauge on a huge metal tank of fermenting wine. The alcohol in the wine is a by-product of the glycolysis carried out by yeast in an oxygenless environment.

7.3 The Three Stages of Cellular Respiration

Having seen who the energy-harvesting players are, you're ready to see how energy harvesting works. The big picture here can be written out as a molecular formula that shows what goes into the harvesting reactions and what comes out of them. In simplified form, this formula can be written as:

$$C_6H_{12}O_6 + 6\,O_2 + 36\,ADP + 36\,Pi \rightarrow 6\,CO_2 + 6\,H_2O + 36\,ATP$$

The starting molecule on the left ($C_6H_{12}O_6$) is glucose, which is food, storing chemical energy. You can also see, next to it, the oxygen ($6\,O_2$) that is needed as the final electron acceptor. Then there is ADP, the two-phosphate molecule; and inorganic phosphate molecules (the 36 Pi), which are pushed on ADP to make it ATP. The *products* of this reaction, to the right of the arrow, are carbon dioxide ($6\,CO_2$) and water ($6\,H_2O$) as by-products, and energy in the form of ATP.

Though many separate steps are involved in transforming our starting substances into ATP, the energy harvesting involved can be divided into three main phases. These are known as *glycolysis*, the *Krebs cycle,* and the *electron transport chain.* In overview, energy harvesting through these phases goes like this: We get some ATP yield in glycolysis and the Krebs cycle, and glycolysis is useful in providing small amounts of ATP in quick bursts. But for most organisms, the main function of glycolysis and the Krebs cycle is the transfer of electrons to the electron carriers such as NAD^+. Bearing their electron cargo, these carriers then move to the third phase of energy harvesting—the electron transport chain (ETC). Here the electron carriers themselves are oxidized (lose electrons), and the resulting movement of these electrons through ETC provides the main harvest of ATP. How big a harvest? Well, if you look at the last number in our formula above, you can see: From a single molecule of glucose, all three steps of energy harvesting net a maximum of about 36 molecules of ATP. And of these, 32 are obtained in the ETC.

If you look at FIGURE 7.4, you can see two representations of the three stages of energy extraction. Taken as a whole, all three stages of energy harvesting are known as **cellular respiration**. Another term for the kind of cellular respiration we're looking at is *aerobic* respiration, given oxygen's role in serving as the final electron acceptor.

7.4 First Stage of Respiration: Glycolysis

Glycolysis means "sugar splitting," which is appropriate because, in this first stage of respiration, our starting molecule is the simple sugar glucose, and it will be split. Briefly, here is what happens during glycolysis. First, this one molecule of glucose has to be prepared, in a sense, for energy release. ATP actually has to be *used*, rather than synthesized, in two of the first steps of glycolysis, so that the relatively stable glucose can be put into the form of a less-stable sugar. Then this sugar is split in half. The two resulting molecules have three carbons each (whereas the starting glucose has six). Once this split is accomplished, the steps of glycolysis take place in *duplicate*: It is like taking one long piece of cloth, dividing it in two, and then proceeding to do the same things to both of the resulting pieces. If you look at FIGURE 7.5, you can see the individual steps of glycolysis. (If your class is studying glycolysis in detail, you will need to look carefully at Figure 7.5, as it is the only place in the text where the individual steps of glycolysis are reviewed.)

Now, what are the results of these steps? Glycolysis accomplishes three valuable things in energy harvesting: It nets two ATP molecules, it yields two energized molecules of NADH (with their important electrons), and it results in two

A BROADER VIEW

Lactic Acid and Athletic Performance

The lactic acid buildup that occurs in the muscles of athletes during strenuous exertion was, for decades, assumed to be an important cause of muscle fatigue—so much so that this buildup was thought to put limits on athletic performance. Evidence gathered since 2001 has cast doubt on this idea, however. Indeed, it appears that lactic acid helps to keep fatigued muscles functioning.

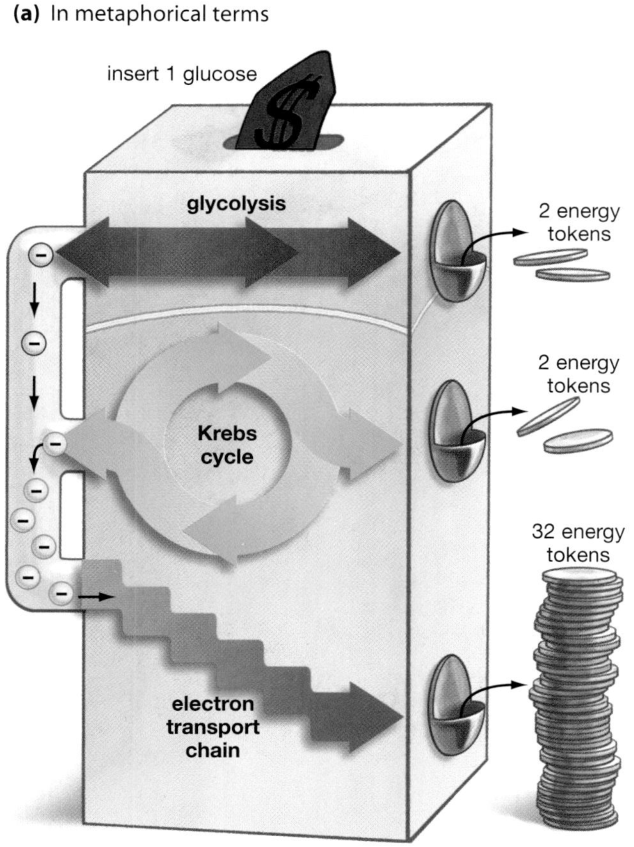

(b) In schematic terms

reactants

products

(a) In metaphorical terms

Just as the video games in some arcades can use only tokens (rather than money) to make them function, so our bodies can use only ATP (rather than food) as a direct source of energy. The energy contained in food—glucose in the example—is transferred to ATP in three major steps: glycolysis, the Krebs cycle, and the electron transport chain. Though glycolysis and the Krebs cycle contribute only small amounts of ATP directly, they also contribute electrons (on the left of the token machine) that help bring about the large yield of ATP in the electron transport chain. Our energy-transfer mechanisms are not quite as efficient as the arcade machine makes them appear. At each stage of the conversion process, some of the original energy contained in the glucose is lost to heat.

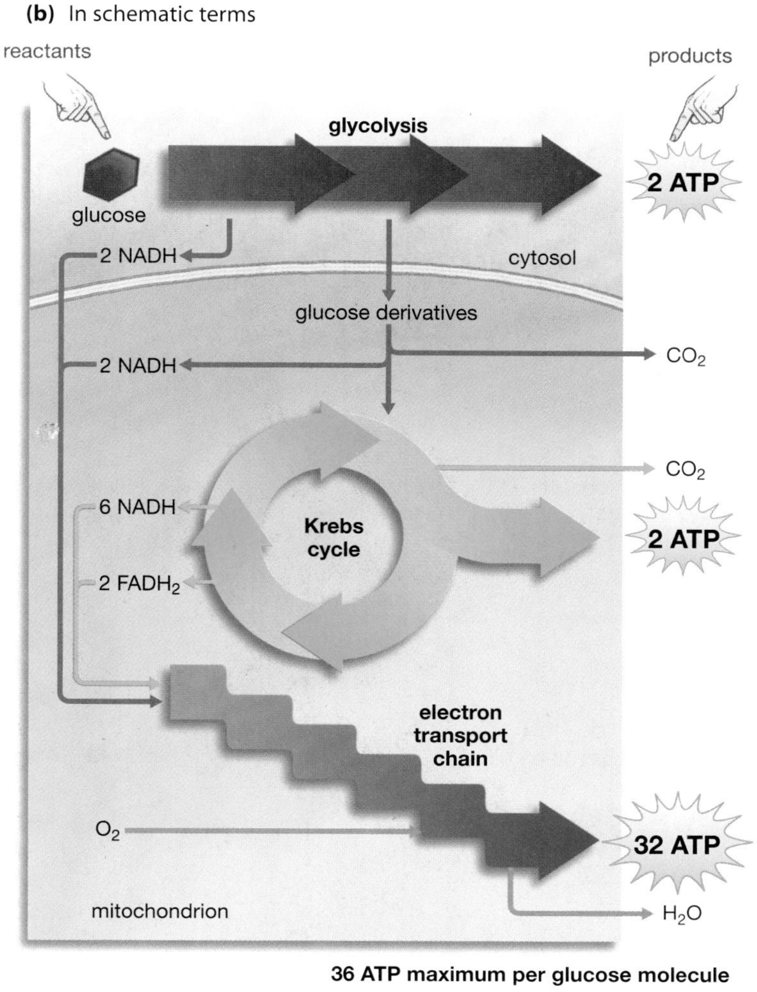

As with the arcade machine, the starting point in this example is a single molecule of glucose, which again yields ATP in three major sets of steps: glycolysis, the Krebs cycle, and the electron transport chain (ETC). These steps can yield a maximum of about 36 molecules of ATP: 2 in glycolysis, 2 in the Krebs cycle, and 32 in the ETC. As noted, however, glycolysis and the Krebs cycle also yield electrons that move to the ETC, aiding in its ATP production. These electrons get to the ETC via the electron carriers NADH and $FADH_2$, shown on the left. Oxygen is consumed in energy harvesting, while water and carbon dioxide are produced in it. Glycolysis takes place in the cytosol of the cell, but the Krebs cycle and the ETC take place in cellular organelles, called mitochondria, that lie within the cytosol.

FIGURE 7.4 Overview of Energy Harvesting

molecules of pyruvic acid. These pyruvic acid molecules are the derivatives of the original glucose molecule. Now *they* are the molecules that will be oxidized—that will lose electrons—in the next stage of energy harvesting.

Before continuing on to this next stage, you may wish to read more about what might be called the consequences of glycolysis. Even oxygen-using organisms such as ourselves rely on glycolysis when short bursts of energy are needed. And for some organisms, glycolysis is the end of the line in energy harvesting. You can read about this in Essay 7.1, "Glycolysis, Beer, and Muscle 'Burn,'" on page 111.

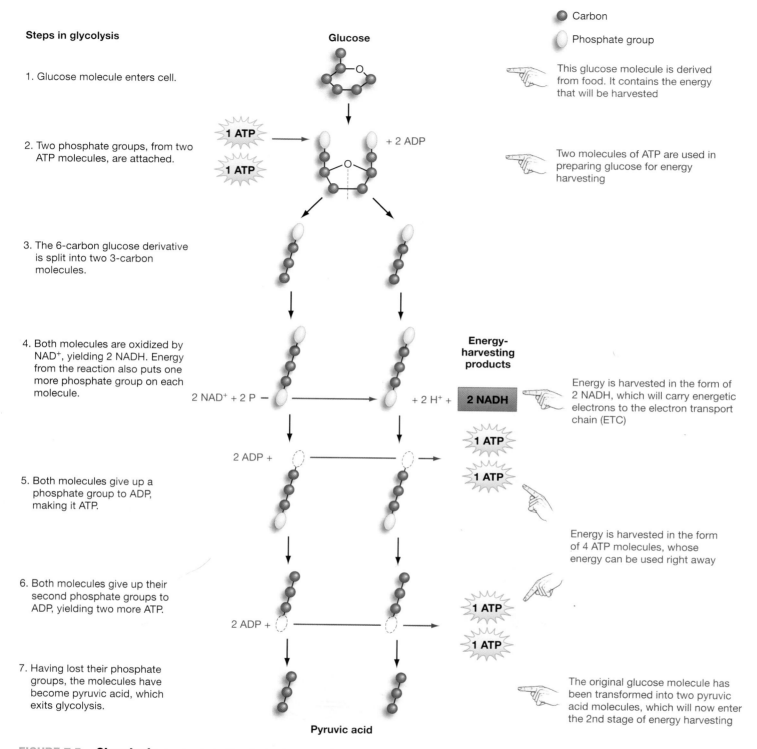

Steps in glycolysis

1. Glucose molecule enters cell.

2. Two phosphate groups, from two ATP molecules, are attached.

3. The 6-carbon glucose derivative is split into two 3-carbon molecules.

4. Both molecules are oxidized by NAD^+, yielding 2 NADH. Energy from the reaction also puts one more phosphate group on each molecule.

5. Both molecules give up a phosphate group to ADP, making it ATP.

6. Both molecules give up their second phosphate groups to ADP, yielding two more ATP.

7. Having lost their phosphate groups, the molecules have become pyruvic acid, which exits glycolysis.

Carbon

Phosphate group

Glucose

1 ATP

1 ATP

+ 2 ADP

This glucose molecule is derived from food. It contains the energy that will be harvested

Two molecules of ATP are used in preparing glucose for energy harvesting

Energy-harvesting products

$2 NAD^+ + 2 P$ → $+ 2 H^+ +$ **2 NADH**

Energy is harvested in the form of 2 NADH, which will carry energetic electrons to the electron transport chain (ETC)

2 ADP +

1 ATP

1 ATP

Energy is harvested in the form of 4 ATP molecules, whose energy can be used right away

2 ADP +

1 ATP

1 ATP

The original glucose molecule has been transformed into two pyruvic acid molecules, which will now enter the 2nd stage of energy harvesting

Pyruvic acid

FIGURE 7.5   Glycolysis   In glycolysis, the single glucose molecule is transformed in a series of steps into two molecules of a substance called pyruvic acid. These two molecules then move on to the next stage of cellular respiration (the Krebs cycle). Glycolysis also produces two molecules of electron-carrying NADH, which move to the electron transport chain. Although two molecules of ATP are used in the early stages of glycolysis, four are produced in the later stages, for a net production of two ATP.

7.5 Second Stage of Respiration: The Krebs Cycle

Since glycolysis yielded only two of the 36 molecules of ATP that eventually are harvested, the glucose derivative that resulted from glycolysis, pyruvic acid, still clearly has a lot of energy left in it. Some of this energy will be harvested in the second stage of cellular respiration, the **Krebs cycle**. This stage is named for the German and English biochemist Hans Krebs, who in the 1930s used the flight muscles of pigeons to find out how aerobic respiration works. Because the first product of the Krebs cycle is citric acid, the cycle is sometimes referred to as the **citric acid cycle**. The ATP yield in this cycle is once again a paltry two molecules. But the Krebs cycle serves to set the stage for the big harvest of ATP in the electron transport chain by supplying electrons that will be carried off by electron carriers to the ETC, as you will see.

Site of Action Moves from the Cytosol to the Mitochondria

Glycolysis yielded its two molecules of pyruvic acid in the cytosol, but the site of energy harvesting quickly shifts, with the pyruvic acid, to the new site of energy transfer, the mitochondria. These organelles are where all the action takes place from here on out, so now would be a good time to look at FIGURE 7.6 and examine their structure. You can see that mitochondria have both an inner and an outer membrane. The Krebs cycle reactions take place to the interior of the inner membrane—in an area known as the inner compartment—while the reactions of the ETC take place *within* the inner membrane.

Between Glycolysis and the Krebs Cycle, an Intermediate Step

There actually is a transition step in respiration between glycolysis and the Krebs cycle. In it, the three-carbon pyruvic acid molecule combines with a substance called coenzyme A, thus forming acetyl coenzyme A (acetyl CoA). One outcome of this reaction is that acetyl CoA enters the Krebs cycle as the derivative of the original glucose molecule. There are, however, two other products of this reaction. One is a carbon dioxide molecule that, like others you'll be seeing, eventually diffuses out of the cell and into the bloodstream. (This is a major source of the

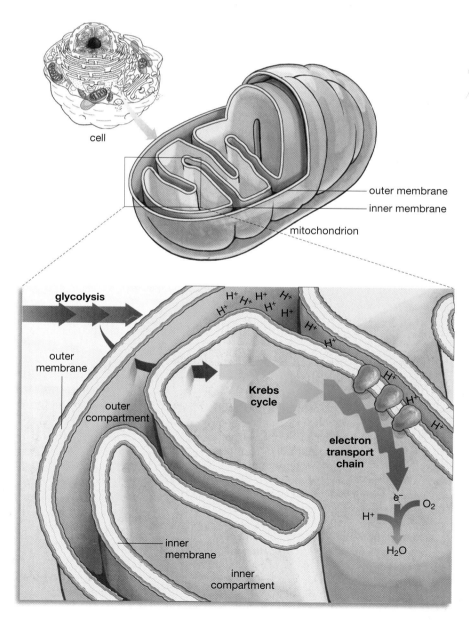

FIGURE 7.6 Energy Transfer in the Mitochondria
Mitochondria are organelles, or "tiny organs," that exist within cells. They are the location for the second and third sets of steps in cellular respiration, the Krebs cycle and the electron transport chain. Following a transitional step (see Figure 7.7), the products of glycolysis—the downstream products of the original glucose molecule—pass into the inner compartment of a mitochondrion, where the Krebs cycle takes place. Electrons derived from the Krebs cycle then migrate, via electron carriers, from the Krebs cycle site into the highly folded inner membrane of the mitochondrion, where the bulk of ATP is produced in the electron transport chain.

FIGURE 7.7 **Transition between Glycolysis and the Krebs Cycle** The pyruvic acid product of glycolysis does not enter directly into the Krebs cycle. Rather, it must first be transformed into acetyl coenzyme A. The consequences of this reaction are the production of CO_2, which diffuses into the bloodstream, and the production of an NADH molecule, which continues onto the electron transport chain. Because one molecule of glucose produces two molecules of pyruvic acid, two molecules of NADH are produced in this step.

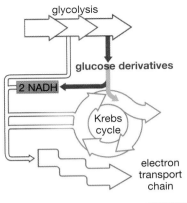

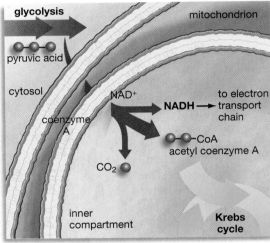

A BROADER VIEW

Energy for Athletics

Where does a football player get the surge of energy to carry out a typical play? Mostly from small stores of ATP and a molecule called phosphocreatine, which acts as a reservoir of the phosphate groups that are used to produce ATP. On longer plays, glycolysis may supply some energy, but the play is likely to be over before the oxygen-using (aerobic) energy-harvesting system can begin to make much of a contribution. Conversely, where does a marathon runner get the energy to run a race? Overwhelmingly from the whole, three-part energy-harvesting process of glycolysis, the Krebs cycle, and the electron transport chain. As a rule of thumb, the longer an activity goes on, the more the energy for it will come from aerobic respiration.

CO_2 that we exhale when we breathe.) The other product is one more molecule of NADH, which will eventually make its way to the final step in aerobic respiration, the ETC. This reaction takes place twice for each glucose molecule (**see** FIGURE 7.7).

Into the Krebs Cycle: Why Is It a Cycle?

Acetyl CoA now enters the Krebs cycle. The reason this is a *cycle* becomes clear when you look at FIGURE 7.8. You can see that, in the first step of the cycle, acetyl CoA combines with a substance called oxaloacetic acid to produce citric acid. This oxaloacetic acid is, however, produced in the *last* step in the cycle (over at about 11 o'clock in the figure.) Thus, a substance that's necessary for this chain of events to take place is itself a product of the chain.

Stroll around the circle now to see how the Krebs cycle functions in a little more detail. In essence, what's happening is that, as the entering acetyl CoA molecule is transformed into these various molecules, it is being oxidized by electron carriers, with the resulting electrons moving on to the ETC. ATP is also derived, and CO_2 is a by-product. The only major player that's unfamiliar here is the FAD–$FADH_2$ that can be seen taking part in a redox reaction at about 8 o'clock. It is simply another electron carrier, similar to $NAD^+/NADH$.

In counting up the Krebs cycle's yield of both ATP and electron carriers (NADHs and so on), recall that *two* turns around this cycle result from the original glucose molecule (because it provided two molecules of the starting acetyl CoA). This means a total yield of 6 NADH, 2 $FADH_2$, and 2 ATP from Krebs for each glucose

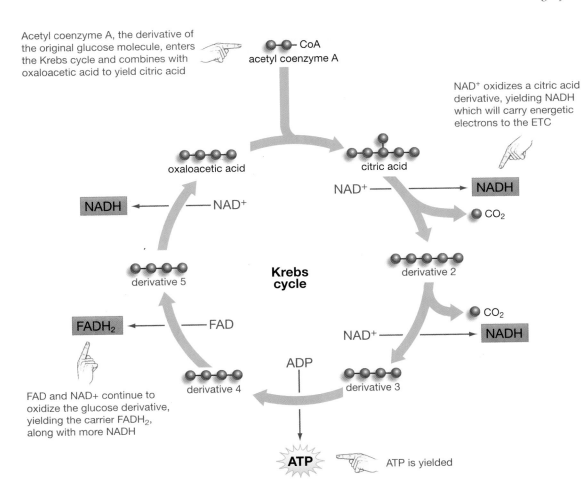

Acetyl coenzyme A, the derivative of the original glucose molecule, enters the Krebs cycle and combines with oxaloacetic acid to yield citric acid

NAD^+ oxidizes a citric acid derivative, yielding NADH which will carry energetic electrons to the ETC

FAD and NAD+ continue to oxidize the glucose derivative, yielding the carrier $FADH_2$, along with more NADH

ATP is yielded

FIGURE 7.8 The Krebs Cycle The Krebs cycle is the major source of electrons that are transported to the electron transport chain. Note, at the top of the circle, how the acetyl coenzyme A produced after glycolysis combines with the oxaloacetic acid (over at about 11 o'clock) to produce the citric acid (over at about 1 o'clock). Now, note how this citric acid is then oxidized by NAD^+ to produce NADH. This NADH is loaded with energetic electrons that will be taken to the electron transport chain. Continuing around the figure, note that this same kind of oxidation happens twice more with NAD^+ and then once with FAD. Finally, at about 6 o'clock, note that a molecule of ATP is produced. Because *two* acetyl CoA molecules enter per molecule of glucose—remember that the original glucose derivative was split in two—the yield per glucose is 6 NADH, 2 $FADH_2$, and 2 ATP.

molecule. Figure 7.8 reviews the Krebs cycle step by step. (If your class is studying the Krebs cycle in detail, you should look carefully at Figure 7.8, as it is the only place in the text where the individual steps of the cycle are reviewed.)

So Far...

1. The three primary stages of cellular respiration are _____, _____, and _____, with most of the yield of ATP coming in _____.

2. In addition to ATP, another product of glycolysis is _____, which are transferred by NADH to the _____.

3. For purposes of cellular respiration, the most important product of the Krebs cycle is _____, which are transferred by _____ and _____ to the _____.

7.6 Third Stage of Respiration: The Electron Transport Chain

Having moved through the Krebs cycle, you've reached the bonanza stage of ATP production in cellular respiration, the **electron transport chain**, the third stage of aerobic energy harvesting. Glycolysis took place in the cytosol, and the Krebs

So Far... Answers:

1. *glycolysis, the Krebs cycle, the electron transport chain (ETC); the ETC*

2. *energetic electrons; ETC*

3. *energetic electrons; NADH and FADH$_2$; ETC*

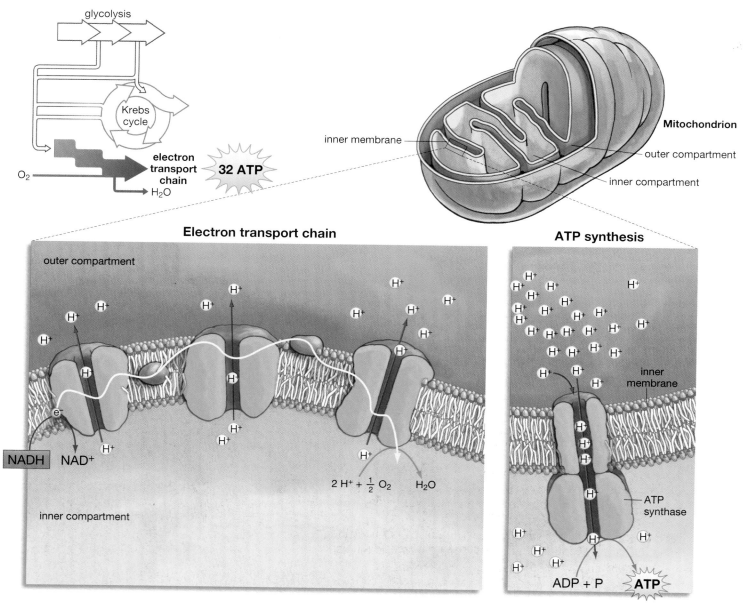

FIGURE 7.9 **The Electron Transport Chain (ETC)**
The movement of electrons through the ETC powers the process that provides the bulk of the ATP yield in respiration. The electrons carried by NADH and $FADH_2$ are released into the ETC and transported along its chain of molecules. The movement of electrons along the chain releases enough energy to power the pumping of hydrogen ions (H^+) across the membrane into the outer compartment of the mitochondrion. It is the subsequent energetic "fall" of the H^+ ions back into the inner compartment that drives the synthesis of ATP molecules by the enzyme ATP synthase.

cycle took place in the inner compartment of the mitochondria. Now, however, the action shifts to the mitochondrial inner membrane (**see FIGURE 7.9**). The "links" in the electron transport chain are a series of molecules. You've been looking for some time now at how NADH and $FADH_2$ carry off the electrons derived from glycolysis and the Krebs cycle. The ETC molecules are the destination of these electron carriers and their cargo.

Upon reaching the mitochondrial inner membrane, the electron carriers donate electrons and hydrogen ions (H^+ ions) to the ETC. Once again, this is a trip down the energy hill, only this time *NADH* is the molecule whose electrons exist at a higher energy level. Thus, when NADH runs into the appropriate enzyme in the ETC, NADH is oxidized, donating its electrons and H^+ ion to the ETC. This process then is simply repeated down the whole ETC, each carrier donating electrons to the next molecule in line. Looking at Figure 7.9, you can see the carriers in the ETC: There are three large enzyme complexes (the big green objects) with two smaller mobile molecules that link them (the small green objects). When NADH arrives at the inner membrane, it bumps into the ETC's first carrier, on the left, and donates electrons to it; this carrier then donates these electrons to the next carrier, and so on down the line to the last electron acceptor, which is oxygen in the inner compartment. (Note the "$\frac{1}{2}O_2$," which will be explained shortly.)

So, Where's the ATP?

At this point you may be tapping your foot, waiting for this promised harvest of ATP to appear. So far, all that's taken place is a transfer of electrons in the ETC. Well, note what happens in the first ETC enzyme complex. The movement of electrons through it releases enough energy to power the movement of hydrogen ions (H^+ ions) through the complex, pushing them from the inner compartment into the outer compartment. (The fall of the electrons causes the enzyme complex to change its shape in a way that facilitates the passage of H^+ ions through it and into the outer compartment.) By the time the electrons have completed their movement through the ETC, this pumping of H^+ ions will occur with the two other large enzyme complexes. Critically, these H^+ ions are being pumped *against* their concentration and electrical gradients (reviewed in Chapter 5). Put another way, the ions are being pumped up the energy hill with energy supplied by the *downhill* fall of electrons through the ETC.

Here's where the ATP is produced at last. The H^+ ions that have been pumped into the outer compartment now move *back down* their concentration and energy gradients, back into the inner compartment, through a special enzyme called ATP synthase. This remarkable enzyme is driven by the H^+ ions flowing through it, which cause part of the enzyme to rotate (as fast as 100 revolutions per second). We could think of it as being something like a waterwheel, except that its movement is driven not by water, but by the movement of H^+ ions that are passing through it, on their way back into the inner compartment. It is this energetic spinning that puts a third phosphate group (Pi) onto ADP, thus making ATP. Lest it pass by too quickly, *this* is the essence of aerobic respiration. The downhill drop of electrons derived from food has powered the uphill synthesis of the molecule that fuels the vast majority of our activities.

Counting up ATP production, recall that we got two net ATP from glycolysis, and another two from the Krebs cycle. Thirty-*four* ATP actually are produced in the ETC, but two ATP are used in moving the NADH produced in glycolysis into the mitochondria. So the net ATP produced in the ETC is 32. When added to the four we got from glycolysis and the Krebs cycle this means a maximum of 36 ATP are netted in all the steps of cellular respiration. Clearly, the ETC accounts for the lion's share of this production, but remember that the earlier steps of respiration yielded the NADH's and FADH$_2$'s, with their energetic electron cargo, that made the big ETC harvest possible.

Finally, Oxygen Is Reduced, Producing Water

The story of the breakdown of glucose—the saga of *one molecule's* transformation— is nearly complete. The only unfinished business lies at the end of the ETC. As noted earlier, oxygen is the final acceptor of the working electrons. In the mitochondrial inner compartment, a single oxygen atom $\left(\frac{1}{2}O_2\right)$ accepts two electrons from the ETC and two H^+ ions. The result is H_2O: water. With its production, the long chain of events in cellular respiration is complete.

Other Foods, Other Respiratory Pathways

By looking at cellular respiration as it applies to glucose, which was used as the starting molecule, you've been able to go through the whole respiratory chain. But there are many kinds of nutrients besides glucose—for example, fats, proteins, and other sugars. All these can provide energy by being oxidized within the chain of reactions you've just gone over. Organisms are able to handle this variability because nutrients and their derivatives can be channeled through pathways in different directions in the respiratory chain in accordance with cell needs. If you look at FIGURE 7.10, you can see a summary of the way foods are broken down through various respiratory pathways.

A BROADER VIEW

The ETC and Free Radicals

When oxygen accepts both the electrons and the hydrogen ions used in the ETC, the result usually is water, or H_2O—one oxygen atom combined with two hydrogen atoms. The problem is that oxygen can combine with atoms to create damaging "free radicals," such as the hydroxyl radical, which is an oxygen atom combined with a *single* hydrogen atom. This is a relatively rare occurrence, but it happens often enough that kilograms of free radicals are produced in a typical human body each year. The body has ways of neutralizing these free radicals, but as it cannot get rid of all of them, the result is damage to our mitochondria—and the aging of our bodies.

FIGURE 7.10 **Many Respiratory Pathways** Glucose is not the only starting material for cellular respiration. Other carbohydrates, proteins, and fats can also be used as fuel for cellular respiration. These foods enter the process at different stages.

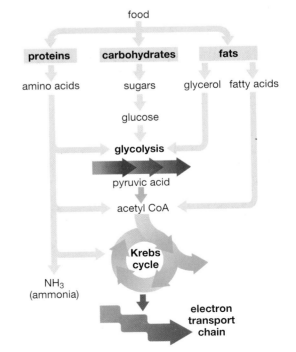

On to Photosynthesis

This long walk through cellular respiration has illustrated how living things harvest energy from food. Recall, however, that almost all living things have one source to thank for this food: the sun's energy, which is trapped by plants in the chemical bonds of carbohydrates. This capture of energy takes place through a process that has a beautiful symmetry with the cellular respiration you just looked at. Coming up: the process of photosynthesis.

So Far...

1. The electrons donated to the molecules in the electron transport chain (ETC) started out in the body as electrons in _____ and then were transferred to _____, which in turn donated them to the ETC.

2. It is the energetic movement of _____ through the enzyme complex ATP synthase that supplies the energy to attach a third phosphate group onto _____, making it _____.

3. At the end of cellular respiration, the electrons that moved through the ETC combine with hydrogen ions and oxygen to produce _____.

_____

So Far... Answers:

1. *food; electron carriers (NAD^+ and FAD)*
2. *hydrogen ions; ADP; ATP*
3. *water*

Chapter Review

Summary

7.1 Energy's Vital Role

- **ATP's importance**—The molecule adenosine triphosphate (ATP) supplies the energy for most of the activities of living things. For ATP to be produced, a third phosphate group must be added to adenosine diphosphate (ADP), a process that requires energy.   (page 107)

7.2 Electrons Fall Down the Energy Hill to Drive Uphill the Production of ATP

- **The route to ATP**—In animals, the energetic fall of electrons derived from food powers the process by which the third phosphate group is attached to ADP, making it ATP.   (page 108)

- **Redox reactions**—Electron transfer in the production of ATP works through redox reactions, meaning reactions in which one substance loses electrons to another substance. The substance that loses electrons in a redox reaction is said to have been oxidized, while the substance that gains electrons is said to have been reduced. In energy transfer in living things, starting food molecules are oxidized, and the electrons they lose power the production of ATP.   (page 109)

- **Electron carriers**—Electrons are carried between one part of the energy-harvesting process and another by electron carriers, the most important of which is nicotinamide adenine dinucleotide or NAD. In its "empty" state, this molecule exists as NAD^+. Through a redox reaction, it picks up one hydrogen atom and another single electron from food, thus becoming NADH, a form it will retain until it drops off its energetic electrons (and a proton) in a later stage of the energy-harvesting process.   (page 109)

7.3 The Three Stages of Cellular Respiration

- **Three stages of respiration**—In most organisms, the harvesting of energy from food takes place in three principal stages: glycolysis, the Krebs cycle, and the electron transport chain (ETC). Taken as a whole, this oxygen-dependent three-stage harvesting of energy is known as cellular respiration.   (page 112)

- **Glycolysis and energy**—Some organisms rely solely on glycolysis for energy harvesting. For most organisms, glycolysis is a primary process of energy extraction only in certain situations—when quick bursts of energy are required, for example—but in them it is a necessary first stage to the Krebs cycle and the ETC.   (page 112)

- **Contributions of the three steps** — Glycolysis takes place in the cell's cytosol, while the Krebs cycle and the ETC take place in cellular organelles, called mitochondria, that lie within the cytosol. Glycolysis yields two net molecules of ATP per molecule of glucose, as does the Krebs cycle. The net yield in the electron transport chain is a maximum of about 32 ATP molecules per molecule of glucose. Glycolysis and the Krebs cycle are critical, however, in that they yield electrons that are carried to the ETC (via electron carriers such as NAD^+) for the final high-yield stage of energy harvesting. (page 113)

7.4 First Stage of Respiration: Glycolysis

- **The products of glycolysis**—When glycolysis begins with a single molecule of glucose, the ultimate products are two molecules of NADH (which move to the ETC, bearing their energetic electrons) and two molecules of ATP (which are ready to be used). Glycolysis also produces two molecules of pyruvic acid—the derivatives of the original glucose molecule—which move on to the Krebs cycle. (page 112)

7.5 Second Stage of Respiration: The Krebs Cycle

- **A transitional step**—There is a transitional step in respiration between glycolysis and the Krebs cycle. In it, each pyruvic acid molecule that was produced in glycolysis combines with coenzyme A, thus forming acetyl coenzyme A (acetyl CoA), which enters the Krebs cycle. There are also two other products of this reaction. One is a molecule of carbon dioxide, which diffuses to the bloodstream; the other is one more molecule of NADH, which moves to the ETC. (page 115)

- **The Krebs cycle**—In the Krebs (or citric acid) cycle, the derivatives of the original glucose molecule are oxidized, with the result that more energetic electrons are transported by the electron carriers NADH and $FADH_2$ to the ETC. The net energy yield of the Krebs cycle is six molecules of NADH, two molecules of $FADH_2$, and two molecules of ATP per molecule of glucose. (page 115)

7.6 Third Stage of Respiration: The Electron Transport Chain

- **Electrons through the ETC**—The ETC is a series of molecules that are located within the mitochondrial inner membrane. Upon reaching the ETC, the electron carriers NADH and $FADH_2$ are oxidized by molecules in the chain. Each carrier in the chain is then reduced by accepting electrons from the carrier that came before it. The last electron acceptor in the ETC is oxygen.   (page 117)

- **Making ATP**—The movement of electrons through the ETC releases enough energy to power the movement of hydrogen ions (H^+ ions) through the three ETC protein complexes, moving them from the mitochondrion's inner compartment to its outer compartment. The movement of these ions down their concentration and charge gradients, back into the inner compartment through an enzyme called ATP synthase, drives the synthesis of ATP from ADP and phosphate. In the inner compartment, oxygen accepts the electrons from the ETC and hydrogen ions, thus forming water.   (page 119)

Web Tutorial 7.1 Harvesting Energy

Key Terms

alcoholic fermentation p. 111	**electron transport chain (ETC)** p. 117	**lactate fermentation** p. 111	**redox reaction** p. 109
cellular respiration p. 112	**glycolysis** p. 112	**nicotinamide adenine dinucleotide (NAD)** p. 109	**reduction** p. 109
citric acid cycle p. 115	**Krebs cycle** p. 115	**oxidation** p. 109	
electron carrier p. 109			

Testing Your Understanding

In Your Own Words (answers in the back of the book)

1. Living things need ATP to power most of the processes that go on within them. In essence, how does cellular respiration yield this ATP?

2. Where do you expect people would have more mitochondria: their skin cells or their muscle cells? Why?

3. Most of the ATP we use is produced in the ETC. So when we have to run across the street to avoid traffic, does the ATP we are using come primarily from the ETC?

Thinking about What You've Learned

1. In Madeleine L'Engle's children's novel *A Wrinkle in Time*, the mitochondria in one of the characters start to die. Describe what would happen to people who lost their mitochondria, and explain why it would happen.

2. High-altitude mountain climbers can experience both weakness in their muscles and impairment in their ability to think. What is going on in them that causes both things?

3. The food we eat is broken down in our digestive tract, but where are most of the calories in this food actually "burned"?

The Green World's Gift: Photosynthesis

HOUSTON CHRONICLE

CHESAPEAKE BAY GRASSES DYING OFF; POLLUTION BLOTTING OUT SUNLIGHT FOR IMPORTANT PLANT LIFE

By David A. Fahrenthold
May 30, 2004

Almost a third of the underwater grasses in the Chesapeake Bay died during 2003, unable to survive as pollution blotted out their sunlight, according to a recent report.

It was the biggest one-year decline for the grasses since the first survey in 1984, according to the report issued by the Chesapeake Bay Program, a partnership of bay states and the U.S. Environmental Protection Agency.

8.1 Photosynthesis and Energy

For 10 points, how are the words to the old spiritual "Dry Bones" something like the flow of energy in the living world? Well, "Dry Bones" goes like this:

The head bone's connected to the neck bone

The neck bone's connected to the shoulder bone

The shoulder bone's connected to the back bone . . .

And the flow of energy? As you can see in the newspaper story above, polluted water kept sunlight—and the solar energy it delivers—from getting to the underwater grasses of

Chesapeake Bay in 2003. This meant the grasses simply did not grow in large portions of the bay. This then had an effect on *animal* life in the bay, given that fish and waterfowl eat the grass while other animals, such as the bay's blue crabs, use it for habitat. As William Baker, president of the Chesapeake Bay Foundation, told the newspaper, "It's not an overstatement to say, 'No grasses, no seafood.'"

This basic story is repeated again and again in the living world: The sun provides energy; plants use this energy to grow; and animals eat the plants. Now, it's not hard to understand how animals eat plants. But most people are a little hazy on the earlier step—how is it that plants grow with the help of sunlight? The short answer is that plants and other members of the green world have a special talent: They can carry out the process called photosynthesis. But what is photosynthesis? The word is a mouthful, and in keeping with this, the process is complicated, involving a long series of chemical reactions. But the *essence* of photosynthesis can be easily understood through a simple act of imagination: Think what life would be like if *we* possessed this green-world talent.

The bone and muscle within us, and the energy to make them work, come from the foods we eat—carbohydrates and fats and proteins. Imagine, though, being able to flourish simply by spending time in the sun and having access to three things: water, small amounts of minerals, and carbon dioxide. In this condition we would not *eat* carbohydrates in the usual forms (bread, sugar, potatoes). Rather, we would transform the simple gas carbon dioxide *into* carbohydrates using nothing more than the water, minerals, and sunlight. Carbohydrates produced in this way are as useful as any food; they can be stored away, used to provide ATP for energy needs, or transformed to make proteins. This, then, is what plants do: They *make their own food*. They use the energy provided by the sun to take a simple gas, carbon dioxide, join it to an energy-poor sugar, and then *energize* that sugar, thus transforming it into food. In this activity, they are joined by a variety of other photosynthesizing organisms (**see** FIGURE 8.1).

From Plants, a Great Bounty for Animals

The photosynthesis that plants carry out provides great benefits not only for the plants themselves, but for animals as well. As noted, almost all animal life is sustained by the food that plants provide. Beyond this, however, a by-product of photosynthesis—a kind of castoff from the work of plants—is oxygen. As part of photosynthesis, plants break water molecules apart. In doing so, they *use* electrons from H_2O, but they *leave behind* oxygen molecules (O_2). This is where the oxygen in our atmosphere comes from; it is the oxygen that we breathe in every minute of every day, or suffer the mortal consequences.

Up and Down the Energy Hill Again

With this, you can step back and view the largest picture of them all with respect to energy flow in living things. It is a great pathway in which energy comes from the sun and then, in photosynthesis, is stored in plants in the complex molecules

(a) Tulips (plants)

(b) Giant kelp (algae)

(c) Cyanobacteria (bacteria)

FIGURE 8.1 **Three Types of Photosyn-thesizers** Photosynthesis is carried out not only by familiar plants, such as these tulips **(a)**, but by algae, such as this giant kelp **(b)**, and by some bacteria, such as these cyanobacteria **(c)**. (magnified × 1025)

we call carbohydrates. Thus stored—in such forms as wheat grains or grass leaves—these carbohydrates can be broken down and used, either by the plants themselves or by the organisms that eat them. The breakdown of this food ends with the "cellular respiration," covered last chapter, that provides ATP.

Looked at one way, photosynthesis and cellular respiration are trips up and down the energy hill that became so familiar in Chapter 7. In respiration, the trip was down the hill, meaning from more stored energy (in food) to less, as the food was being broken down to power the production ATP. Photosynthesis is a trip up the energy hill. Here electrons are being removed from water, boosted to a more energetic state by the power of sunlight, and then brought together with a low-energy sugar and carbon dioxide, resulting in an *energy-rich* sugar (a carbohydrate). The story of how this carbohydrate is produced is the subject of this chapter. To encapsulate this story in a definition, **photosynthesis** is the process by which certain groups of organisms capture energy from sunlight and convert this solar energy into chemical energy that is initially stored in a carbohydrate.

We could think of this process in terms of a kind of recipe—a formula that shows us what "ingredients" go into photosynthesis and what products come out. The simplified formula for photosynthesis looks like this:

sunlight energy

$$6\,CO_2 + 6\,H_2O \rightarrow C_6H_{12}O_6 + 6\,O_2$$

Over on the left, you can see carbon dioxide (CO_2) and water (H_2O) as starting ingredients (or "reactants"), and then on the right are glucose ($C_6H_{12}O_6$) and oxygen (O_2) as products. Intervening in the middle, and driving the whole process, is the energy contained in the rays of the sun.

One fascinating thing about this process is the symmetry it has with the cellular respiration we looked at last chapter. Note that the products of photosynthesis are glucose and oxygen. Now, what were the *starting ingredients* of cellular respiration? Glucose and oxygen—the glucose being the food from which electrons were derived to power ATP production, and the oxygen serving as the final "acceptor" of those electrons at the end of respiration. Flipping this around, what were the final products of respiration? Carbon dioxide and water—the *starting ingredients* of photosynthesis. To put this simply, Nature has fashioned things such that what she produces in photosynthesis, she uses in cellular respiration, while what she produces in respiration, she uses in photosynthesis.

A BROADER VIEW

The Importance of Photosynthesis
The naturalist Edward O. Wilson has said that only three plants—rice, wheat, and corn—stand between humanity and starvation. In the United States, we take abundant food as a given. But we should remember that, directly or indirectly, it is plants that feed us all—and that plants can do this only because of photosynthesis.

FIGURE 8.2 The Electromagnetic Spectrum
Sunlight is composed of rays of many different wavelengths, but we can see only those in the visible light range (represented by the color spectrum). When the light of the sun shines on a green leaf, it is primarily the sun's green rays that are scattered, while frequencies in other ranges are absorbed; frequencies in the red and blue ranges drive photosynthesis most strongly.

8.2 The Components of Photosynthesis

We can think of photosynthesis as starting with the absorption of sunlight by leaves. *Absorption* here means just what it sounds like: Light is taken in by the leaves. Actually, leaves capture only a *portion* of the light that falls on them. The sunlight that makes it to the Earth's surface is composed of a spectrum of energetic rays (measured by their "wavelengths") that range from very short, ultraviolet rays, through visible light rays, to the longer and less-energetic infrared rays. (In FIGURE 8.2 you can see that these wavelengths are, in turn, part of a larger range of electromagnetic radiation.) Photosynthesis is driven by part of the *visible light* spectrum—mainly by blue and red light of certain wavelengths. Tune a laser to emit blue light of a certain wavelength, shine this beam on a plant, and the plant will absorb a great deal of this light. On the other

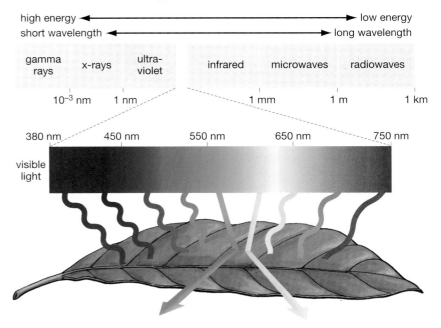

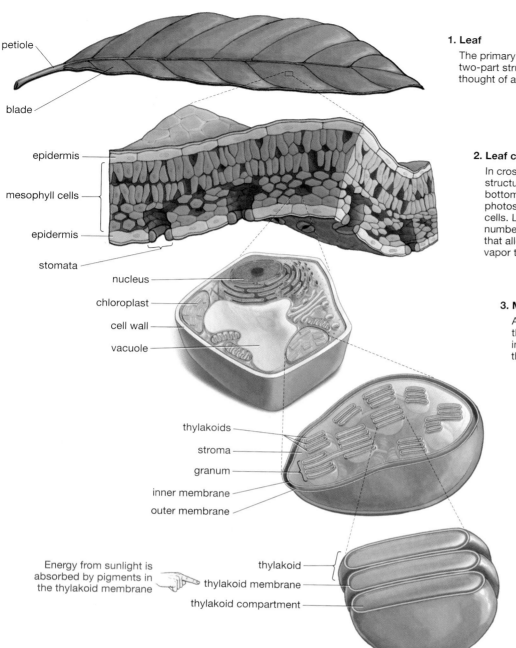

1. Leaf

The primary site of photosynthesis in plants, leaves have a two-part structure: a petiole (or stalk) and a blade (normally thought of as the leaf).

2. Leaf cross-section

In cross section, leaves have a sandwich-like structure, with epidermal layers at top and bottom and mesophyll cells in between. Most photosynthesis is performed within mesophyll cells. Leaf epidermis is pocked with a large number of microscopic openings, called stomata, that allow carbon dioxide to pass in and water vapor to pass out.

3. Mesophyll cell

A single mesophyll cell within a leaf contains all the component parts of plant cells in general, including the organelles—called chloroplasts—that are the actual sites of photosynthesis.

4. Chloroplast

Each chloroplast has an outer membrane at its periphery; then an inner membrane; then a liquid material, called the stroma, that has immersed within it a network of membranes, the thylakoids. These thylakoids sometimes stack on one another to create. . .

5. A Granum

Electrons used in photosynthesis will come from water contained in the thylakoid compartment, and all the steps of photosynthesis will take place either within the thylakoid membrane, or in the stroma that surrounds the thylakoids.

Energy from sunlight is absorbed by pigments in the thylakoid membrane

FIGURE 8.3   Site of Photosynthesis

hand, tune the laser to emit *green* light, and the plant will scatter most of this light. This is why plants are green: They strongly scatter the green portion of the visible light spectrum.

Where in the Plant Does Photosynthesis Occur?

So, where is this light absorption occurring within plants? To answer this question, you need to know something about the playing fields of photosynthesis in most plants, their leaves.

If you look at the idealized leaf in FIGURE 8.3, you can see that it can be likened to a kind of cellular sandwich, with one layer of outer (or epidermal) cells at the top, another layer at the bottom, and layers of mesophyll cells in between. Take time now to read through Figure 8.3's drawings and captions; as you do, you'll go from the leaf's layers of cells, then into the cells themselves, and then into the tiny structures *within* the cells—the chloroplasts—that are the sites of photosynthesis.

Here is some detail on the structures seen in Figure 8.3. The **stomata** seen in the figure are microscopic pores that let carbon dioxide pass into leaves and water vapor pass out of them. There is no shortage of stomata for this work; a given square centimeter of plant leaf may contain from a thousand to a hundred thousand of them.

Going down a couple of levels, **chloroplasts** are the organelles within plant and algae cells that are the sites of photosynthesis. Anywhere from one to several hundred of these oblong structures might exist within a given leaf cell, each capable of carrying out photosynthesis on its own. Moving into the interior of the chloroplast, there is a network of chloroplast membranes, active in photosynthesis, called **thylakoids**. Thylakoids are immersed in the liquid material of the chloroplast, the **stroma**. All the steps of photosynthesis take place either within the thylakoid membrane or in the stroma.

Embedded in the thylakoid membranes are the molecules that actually absorb the sunlight—pigments. Anything that strongly absorbs certain visible wavelengths of sunlight is called a pigment. Thylakoid membranes contain a pigment, called **chlorophyll *a***, that is the primary pigment active in plant photosynthesis. Chlorophyll *a* is then aided by several substances known as *accessory pigments*, which do just what their name implies: They aid chlorophyll *a* in absorbing energetic rays from the sun, after which they pass the absorbed energy along.

There Are Two Essential Stages in Photosynthesis

Keeping these structures and concepts in mind, you can now begin tracing the steps of photosynthesis. It is helpful to divide photosynthesis into two main stages. In the first stage (the *photo* of photosynthesis), the power of sunlight will do two things—strip water of electrons and then boost these electrons to a higher energy level. Thus boosted, the electrons are passed along through a series of the electron *carriers* you were introduced to in Chapter 7. All of this takes place in the thylakoid membranes. This first stage of photosynthesis ends when the original, energized electrons get attached to a mobile electron carrier, called $NADP^+$, that transports them to the second set of reactions. In this second stage (the *synthesis* of photosynthesis), the electrons come together with carbon dioxide and a low-energy sugar. The attachment of the electrons and CO_2 to this sugar produces a high-energy sugar—meaning food. This second stage takes place in the stroma of the chloroplast.

The steps in the first stage of photosynthesis obviously depend directly on sunlight. Thus, these steps are sometimes called the **light reactions**. Meanwhile, the second set of steps are referred to as the *Calvin cycle*, in honor of Melvin Calvin, the scientist who first revealed their details. (These steps are sometimes referred to as the *light-independent reactions*.) You will be following both sets of steps as they occur in time—from the collection of sunlight to the synthesis of carbohydrates.

The Working Units of the Light Reactions

The light reactions take place within a set of working units. Each of these units is a **photosystem**—an organized complex of molecules within a thylakoid membrane that, in photosynthesis, collects solar energy and transforms it into chemical energy. If you look at FIGURE 8.4, you can see the components that make up a photosystem. First, there is a group of a few hundred pigment molecules that serve to absorb the sunlight. Most of these molecules serve only as "antennae" that absorb energy from the sun and pass it on. At the center of the antennae system, however, there is a **reaction center**—a pair of special chlorophyll *a* molecules and associated compounds that first receive the solar energy from photosystem pigments and then transform this solar energy into chemical energy. As you can see in Figure 8.4, the reaction center includes a molecule that could be thought of as the first recipient of the absorbed solar energy, the "primary electron acceptor."

The Transfer of Energy

You could conceptualize photosynthesis as beginning when sunlight is absorbed by any of the hundreds of antennae pigment molecules in a photosystem unit. The absorbed energy is then passed on to the pair of chlorophyll *a* molecules in a reaction

FIGURE 8.4 **The Working Units of Photosynthesis** Photosystems are multipart units that bring together electrons derived from water and energy derived from the sun. Hundreds of antennae pigments absorb sunlight and transfer solar energy to the photosystem's reaction center. This energy gives a boost to an electron within the reaction center in two ways: It physically moves the electron to the center's primary electron acceptor, and it moves the electron up the energy hill to a more energetic state.

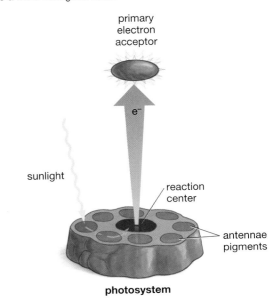

primary
electron
acceptor

e⁻

sunlight

reaction
center

antennae
pigments

photosystem

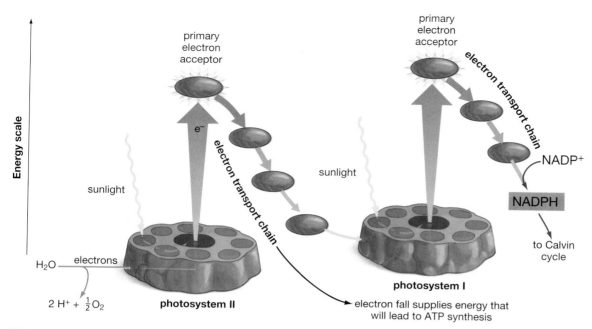

FIGURE 8.5 **Collecting Solar Energy, Boosting Electrons** Solar energy gathered by photosystems II and I is used to energize electrons that exist in the photosystems' reaction centers. Energetically, the electrons are boosted up, and then move down, two energy hills. Physically, they first move to primary electron acceptors and then down electron transport chains until they are at last taken up by NADP⁺ to form NADPH. The NADPH molecules then transfer the electrons to the Calvin cycle, where they are used to make sugars. When an electron is boosted in photosystem II, the resulting energy imbalance drives the process by which water molecules are split. Electrons derived from the water near photosystem II then move to its reaction center, where they are the next to be boosted by the sun's energy. The energetic fall of electrons down the electron transport chain between photosystems II and I provides the energy for the synthesis of ATP, which powers the Calvin cycle.

center. As a result, electrons from this pair are "moved" in a couple of ways, one physical and one metaphorical. Physically, these electrons are transferred to the primary electron acceptor. You also could think of the initial "distance traveled" by the electrons, however, as a movement up the energy hill—solar energy boosts the electrons to a higher energetic state prior to their transfer to the primary electron acceptor. What follows then is a further physical transfer of these electrons through two electron transport *chains*, as you can see in FIGURE 8.5. Note, however, that in terms of energy, this is a trip up and *down* the energy hill. With energy supplied by the sun, the electrons are pumped up a couple of formidable energy gradients (in photosystems II and I) only to come partway back down them, as they "seek" their lowest energy state. Critically, they are releasing energy as they fall—just as a boulder releases energy by rolling downhill.

So Far...

1. Photosynthesis is a process by which the energy of the _____ strips electrons from _____ and boosts their _____, ultimately resulting in the production of _____.

2. In plants, photosynthesis is carried out in tiny structures called _____ that are most abundant in certain cells in leaves.

3. Photosynthesis in plants is primarily driven by light in the _____ and _____ parts of the _____ spectrum.

So Far... Answers:

1. *sun; water; energy; a carbohydrate*
2. *chloroplasts*
3. *blue and red; visible light*

8.3 Stage 1: The Steps of the Light Reactions

With these components and processes in mind, let's trace the steps of the light reactions (see Figure 8.5). The first step is that solar energy, collected by photosystem II's antennae molecules, arrives at the reaction center. This energy then gives a boost to an electron in the reaction center in the two ways noted: The electron *physically* moves to another part of the reaction center complex, the primary electron acceptor. But it is also pumped up the energy hill. With this activity, the reaction center chlorophyll has *lost* an electron. That loss leaves an energy "hole" in this chlorophyll, making it a molecule that can "oxidize" or pull electrons from another molecule. With the energy provided by this pull, a special enzyme in the reaction center splits water molecules that lie within the thylakoid compartment. These water molecules are now being oxidized, which means *they* are losing electrons. The electrons travel to the reaction center, where they will be the next electrons in line for an energy boost.

A Chain of Reactions and Another Boost from the Sun

By following Figure 8.5 in a general way, you can see what happens next to electrons that are boosted in photosystem II. After arriving at the primary electron acceptor, they fall back down the energy hill as they are transferred through a series of electron transport molecules, each one oxidizing its predecessor. At the bottom of this hill they arrive at photosystem I, which includes a slightly different kind of reaction center. This center is also receiving solar energy, and uses it to boost electrons to a higher energy state. From this energetic state, electrons are transferred down the second energy hill—until they are received by the electron carrier NADP$^+$ which is very much like the electron carrier NAD$^+$ reviewed in Chapter 7. In accepting electrons, NADP$^+$ becomes NADPH, an electron carrier that ferries the electrons into the next stage of photosynthesis. This second stage is the Calvin cycle, which will yield the high-energy sugar that is the essential product of photosynthesis.

The Physical Movement of Electrons in the Light Reactions

In this movement up and down the energy hill, the electrons have moved physically through the chloroplast. As FIGURE 8.6 shows, they started out in the water of the thylakoid compartment, moved into and then through the thylakoid membrane—handed off at each step by the various electron transport molecules—and then finally ended up in the stroma, attached to NADPH.

8.4 What Makes the Light Reactions So Important?

The steps just reviewed are the essence of the light reactions. To make sure you don't miss the forest for the trees, let's be clear about two momentous things that have taken place within these steps.

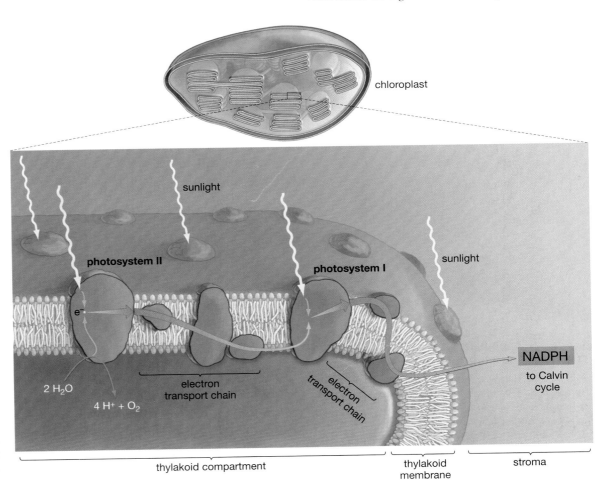

FIGURE 8.6 **The Light Reactions** The light reactions take place in the thylakoid membranes within the chloroplasts. Electrons are donated by water molecules located in the thylakoid compartments. Powered by the sun's energy, these electrons are passed along the electron transport chain embedded in the thylakoid membrane and end up stored in NADPH in the stroma. An additional product of the splitting of water molecules is individual oxygen atoms, which quickly combine into the O_2 form. This is the atmospheric oxygen that we breathe in.

The Splitting of Water: Electrons and Oxygen

The first notable event takes place near the start: the splitting of water. Critically, this reaction provides the traveling electrons whose path you've just followed. But it also provides something else: the oxygen that today accounts for 21 percent of the Earth's atmosphere. In the water splitting that goes on in photosynthesis, hydrogen atoms are removed from H_2O, while the oxygen is left behind. When two liberated oxygen atoms come together in the thylakoid compartment, the result is O_2—the form of oxygen that exists in our atmosphere. *Here* is where we get the substance that is the breath of life itself for most species, all of it contributed by living things. Given its importance, it is no small irony that oxygen is, in another sense, a kind of green-world refuse—leftovers from the process by which plants strip water of the electrons they need to carry out photosynthesis.

The Transformation of Solar Energy to Chemical Energy

The second major feat that takes place during light reactions is the transformation of solar energy to chemical energy. When the energy of sunlight moves to a chlorophyll molecule in a reaction center, it boosts an electron there from what is known as a ground state to an excited state. Thus far, this has been described as a movement up a metaphorical energy hill. In physical terms, however, this electron is moving farther out from the nucleus of an atom. To appreciate what's special about photosynthesis, consider what the *common* fates are for such excited electrons. One fate is for the electrons to drop back down to the original (and more stable) state, in the process releasing as *heat* the energy they have absorbed. This is the process by which black objects get hot on sunny days. Another possible fate is for falling electrons to release part of their energy as light in the process known as fluorescence.

In photosynthesis, however, the energized electrons are *transferred to a different molecule*—the initial electron acceptors of photosystems II and I. They don't fall back to their ground state, releasing relatively useless heat or light in the process. They are passed on. This is the bridge to life as we know it. Without this step, the energy of the sun merely makes the Earth warmer. With it, the green world grows profusely; it captures some of the sun's energy and with it builds trunks and grains and leaves. By one reckoning, photosynthesis produces up to 155 billion tons of material each year. This is an amazing amount—about 25 tons for every person on Earth. Electrons are taken from water, boosted to a higher energy state by the sun, and transferred to other molecules. The result is the food that sustains nearly all living things.

Production of ATP

It's also worth noting that a primary function of the fall of electrons between photosystems II and I is a release of energy that is used for the production of ATP. Put together through a process similar to the one you looked at last chapter, this ATP will be used to power the reactions that are coming up in the second stage of photosynthesis. Thus, the end result of the light reactions is that the sun's power has provided energy that is stored in *two* forms: ATP, and the energetic electrons in NADPH. Looking at the big picture, you can see that the *photo* of photosynthesis is an energy-capturing operation.

8.5 Stage 2: The Steps of the Calvin Cycle

In the second stage of photosynthesis, the energy captured in the light reactions will be *used*. It will power a process by which carbon dioxide, taken from the atmosphere, will be joined to a low-energy sugar, with the resulting product energized, thus creating a high-energy sugar—which is to say food. To see how this happens, let's begin by taking stock of where the light reactions left off. The short answer is, in the stroma. That's where $NADP^+$ has become NADPH by accepting the energized electrons pouring out from photosystem I. The stroma also contains the ATP that was made during the light reactions.

A BROADER VIEW

The Effects of Photosynthesis

At one time, Earth's atmosphere contained almost no oxygen; but now it is about 21 percent oxygen, nearly all of it having come from living things performing photosynthesis. Meanwhile, the food that photosynthesis has provided has likewise had momentous consequences. It's extremely unlikely that organisms as big and complex as animals could have evolved without photosynthesis, for the simple reason that there would have been so little to eat on Earth.

Energized Sugar Comes from a Cycle of Reactions

Even with the right ingredients, it takes more than one step to go from carbon dioxide and a sugar to a sugar that is energetic enough to serve as food. It takes a set of chemical reactions that make up a cycle. This is the **Calvin cycle** or the C_3 cycle, the set of reactions in photosynthesis in which energetic electrons are brought together with carbon dioxide and a sugar to produce an energetic carbohydrate. In summary, these reactions break down into four sets of steps, which are detailed in FIGURE 8.7.

The first steps of the cycle can be thought of as a process of **fixation**—of a gas being incorporated into an organic molecule. Specifically, carbon dioxide, which comes in through the leaves of the plant, is being fixed into the starting sugar, which is called RuBP. (From the time plants are embryos, they have small amounts of this sugar in them as a legacy from their parent plants.) This is a terrifically important step, for it is the way life builds itself up with materials that lie *outside* of life. Note that carbon is being taken in from the atmosphere and made part of a living thing—a plant.

The next reactions in the cycle are the *energizing* steps of the process (which you can see in section 2 of the cycle). Here is where our low-energy sugar receives the energetic products of the light reactions. The RuBP derivatives first interact with ATP and then receive energetic electrons from NADPH—the electrons that came originally from water and that were boosted up the energy hills by sunlight. With this, a relatively low-energy sugar has been energized—it has been moved up the energy hill—from which position it can now serve as food. The energized sugar that is produced, called G3P, is the essential product of photosynthesis. One molecule of it is netted with each turn of the Calvin cycle, as you can see in section 3. The remainder of the cycle can be thought of as a preparation of molecules for another trip around the cycle. Over in section 4, several reactions are carried out that result in the formation of more RuBP, which then enters the cycle again.

FIGURE 8.7   The Calvin Cycle CO_2, ATP, and electrons from NADPH are the input into the Calvin cycle, while a sugar (G3P) is its output.

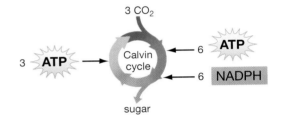

1. **Carbon fixation.** An enzyme called rubisco brings together three molecules of CO_2 with three molecules of the five-carbon sugar RuBP; the three resulting six-carbon molecules are immediately split into six three-carbon molecules named 3-PGA (3-phosphoglyceric acid).

2. **Energizing the sugar.** In two separate reactions, six ATP molecules react with six 3-PGA, in each case transferring a phosphate onto the 3-PGA. The six 3-PGA derivatives oxidize (gain electrons from) six NADPH molecules; in so doing, they are transformed into the energy-rich sugar G3P (glyceraldehyde 3-phosphate).

3. **Exit of product.** One molecule of G3P exits as the output of the Calvin cycle. This molecule, the product of photosynthesis, can be used for energy or transformed into materials that make up the plant.

4. **Regeneration of RuBP.** In several reactions, five molecules of G3P are transformed into three molecules of RuBP, which enter the cycle.

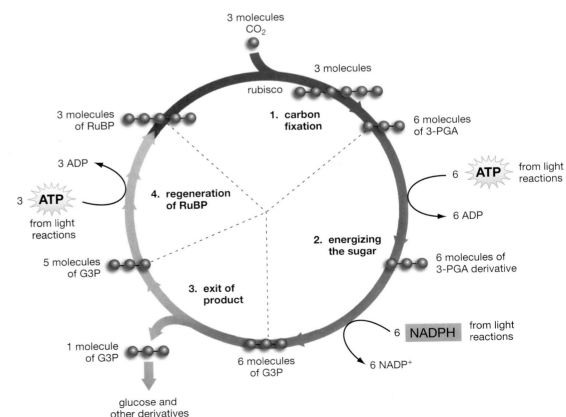

FIGURE 8.8 **Nature's Bounty, Through Photosynthesis** Living things produce a tremendous amount of material or "biomass" through the process of photosynthesis. All of the greenery visible in this Malaysian rainforest came about as a result of photosynthesis.

A BROADER VIEW

The Most Abundant Protein

Remember how, in Chapter 6, you learned about the importance of the proteins called enzymes that facilitate chemical reactions? Well, an enzyme called rubisco, which facilitates the incorporation of carbon dioxide into the Calvin cycle's starting sugar, is really important. Indeed, it's been called the most important protein on Earth because of its role in bringing carbon into the living world. At a minimum, rubisco is the most *abundant* protein on Earth. It constitutes about 30 percent of the protein in a typical plant leaf, and plants account for a great deal of the dry weight or "biomass" of the natural world.

The Ultimate Product of Photosynthesis

The importance of photosynthesis becomes clear only with an understanding of what G3P can give rise to once it exits the Calvin cycle. This sugar has something in common with steel coming out of a factory, in that it can be turned into many things. Put two molecules of G3P together and you get the more familiar six-carbon sugar, glucose. In turn, many molecules of glucose can come together to form the large storage molecules known as starches, familiar to us in such forms as potatoes or wheat grains. Beyond this, sugar is used to make proteins, which are then used as structural components of the plant or as enzymes. Given all this, if we ask what the *ultimate* product of photosynthesis is, the answer turns out to be: *the whole plant* (see FIGURE 8.8). If you look at FIGURE 8.9, you can see a summary of the process of photosynthesis.

FIGURE 8.9 **Summary of Photosynthesis in the Chloroplasts of Plant Cells** In the light reactions, solar energy is converted to chemical energy in the thylakoids, and the chemical energy is stored temporarily in the form of ATP and NADPH. Water is required for this reaction, and oxygen is a by-product. The stored chemical energy is in turn used in the Calvin cycle, taking place in the stroma, in which a high-energy sugar is made from carbon dioxide and the sugar RuBP. The sugar can be used for food, or it may become part of the plant's structure.

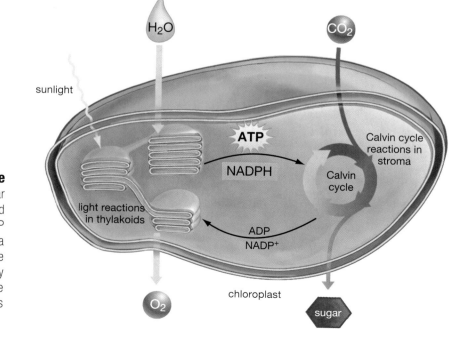

Closing Thoughts on Photosynthesis and Energy

Readers who have stayed the course on this account of photosynthesis now know at least one thing very well about it: how complicated it is, with its many components and long metabolic pathways. It's also probably clear by this point why scientists continue to study photosynthesis in such detail. It is the foundation of plant growth, and upon plant growth hinges nothing less than the survival of most animals—including human beings. Without an understanding of this linkage, it's easy to see plants as a set of mute fixtures whose main contribution to human life is aesthetic. But with this knowledge, you can begin to see the central position that plants occupy in the interconnected web of life.

Over the last three chapters, you have learned about the endless back-and-forth of oxygen, carbon dioxide, and energy in the living world. *Cycle* has been a recurring word in this long discussion, because the only one-way trip you've encountered has been the relentless "spillage" of energy from the sun down into heat. Looked at in a cynical way, Earth and its inhabitants constitute a kind of leaky holding tank for energy that comes from the sun. Looked at another way, however, the living world has been able, through photosynthesis, to take the sun's energy and build a remarkable edifice with it. Think of the forms and sheer mass of living material on Earth. One of the most amazing things about this structure is that we humans get to be both a part of it and witnesses to it.

So Far...

1. In the Calvin cycle, _____ from the atmosphere is incorporated into a low-energy _____ called RuBP.

2. The light reactions provide two products that are used in the "energizing" steps of the Calvin cycle. These products are _____ and the _____ contained in NADPH.

3. The essential product of the Calvin cycle is a _____ named G3P.

So Far... Answers:

1. carbon dioxide; sugar

2. ATP; electrons

3. sugar

Chapter Review

Summary

8.1 Photosynthesis and Energy

- **The importance of photosynthesis**—Photosynthesis has made possible life as we know it on Earth, because the organic material produced in photosynthesis (a sugar) is the source of food for most of Earth's living things. Photosynthesis also is responsible for the atmospheric oxygen used by many living things in cellular respiration. (page 123)

8.2 The Components of Photosynthesis

- **Light and leaves**—In plants and algae, photosynthesis takes place in the organelles called chloroplasts, which can exist in great abundance in the mesophyll cells of leaves. The energy for photosynthesis comes mostly from various blue and red wavelengths of visible sunlight that are absorbed by pigments in the chloroplasts. Plant leaves contain microscopic pores called stomata that can open and close, letting carbon dioxide in and water vapor out.   (page 125)

- **Two stages of photosynthesis**—There are two primary stages to photosynthesis. In the first stage, called the light reactions, electrons derived from water are energetically boosted by the power of sunlight. These electrons physically move in this process; they are passed along through a series of electron carriers, ending up as part of the electron carrier NADPH, which carries them to the second stage of photosynthesis. In this second stage, the Calvin cycle, the electrons are brought together with carbon dioxide and a sugar to produce a high-energy sugar.   (page 127)

8.3 Stage 1: The Steps of the Light Reactions

- **Energy from the sun, electrons from water**—In its first stage, photosynthesis works through a pair of molecular complexes, photosystems II and I, which receive solar energy and use it to transfer electrons derived from water. These electrons are boosted up the "energy hills" of photosystems II and I, only to come back down them, in the process yielding both ATP (in the fall between photosystems II and I) and high-energy electrons that are ultimately attached to the electron carrier NADPH. (page 128)

8.4 What Makes the Light Reactions So Important?

- **Oxygen and chemical energy**—Two actions of great consequence take place in the light reactions. The first is that water is split, yielding both the electrons that move through the light reactions and the oxygen that organisms such as ourselves breathe in. The second is that the electrons that are derived from the water—and then given an energy boost by the sun's rays—are transferred to a different molecule: the initial electron acceptor in photosystem II. This is the bridge between solar energy and the chemical energy locked up in food. (page 129)

8.5 Stage 2: The Steps of the Calvin Cycle

- **The synthesis stage**—In the second stage of photosynthesis, carbon dioxide from the atmosphere is brought together with a low-energy sugar called RuBP. The resulting compound is energized with the addition of ATP phosphate groups and electrons supplied by the first stage of photosynthesis. The result is the high-energy sugar G3P, which is the product of photosynthesis. All these steps are carried out within a process known as the Calvin cycle, or C_3 cycle. (page 130)

Web Tutorial 8.1 Photosynthesis

Key Terms

Calvin cycle   p. 131

chlorophyll *a*   p. 127

chloroplast   p. 127

fixation   p. 131

light reactions   p. 127

photosynthesis   p. 125

photosystem   p. 127

reaction center   p. 127

stomata   p. 127

stroma   p. 127

thylakoids   p. 127

Testing Your Understanding

In Your Own Words *(answers in the back of the book)*

1. Sunlight reaching Earth is composed of a spectrum of energetic rays. Describe the rays that make up this spectrum. Which portion of the spectrum drives photosynthesis?

2. What are the component parts of a photosystem?

3. Since plants capture energy through photosynthesis, do they need to carry out cellular respiration as well? Why or why not?

Thinking about What You've Learned

1. One of the most dramatic environmental changes today is the increase in carbon dioxide levels in the atmosphere. We usually hear about it in connection with global warming, but that's not the only issue. What effect would increased carbon dioxide have on plants? If more trees are planted, would this stand to have any effect on atmospheric CO_2 levels?

2. In general, the deeper parts of the ocean, out beyond the continental shelves, are desert-like with respect to the amount of life they contain. Why should this be?

3. Keeping in mind that, to a first approximation, all the plants, animals, and fungi in the world depend on photosynthesis in order to live, what would the world look like without photosynthesis?

Genetics, Mitosis, and Cytokinesis

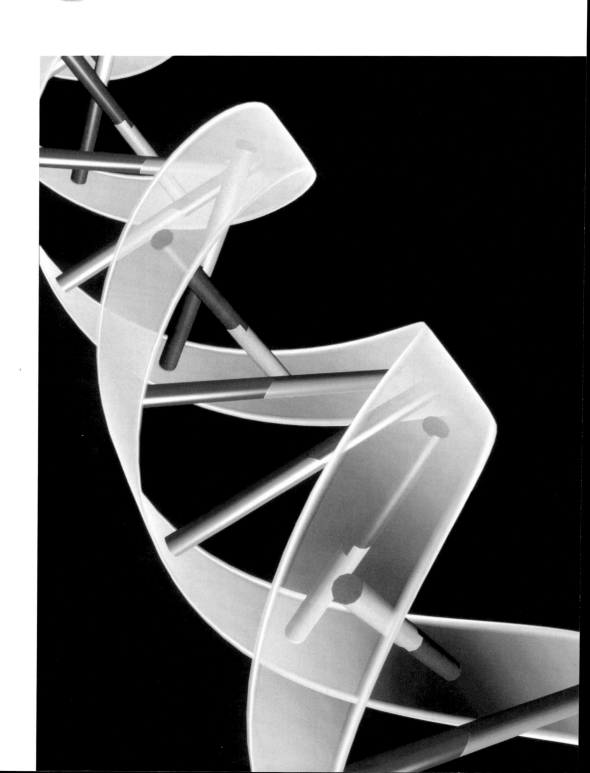

SCIENCE IN THE NEWS

THE MARYLAND GAZETTE

GENETIC MUTATION PRODUCES TODDLER WITH TWICE THE MUSCLE

By The Associated Press
June 30, 2004

Somewhere in Germany is a baby Superman, born in Berlin with bulging arm and leg muscles. Not yet 5, he can hold seven-pound weights with arms extended, something many adults cannot do. He has muscles twice the size of other kids his age and half their body fat. DNA testing showed why: The boy has a genetic mutation that boosts muscle growth.

SCIENCE IN THE NEWS

SAN FRANCISCO CHRONICLE

RESEARCHERS DISCOVER 'JEKYLL AND HYDE' CANCER GENE
AMOUNT OF A SPECIFIC PROTEIN DETERMINES WHETHER A TUMOR IS CREATED OR SUPPRESSED

By Keay Davidson
September 13, 2004

Scientists in France and Marin County say they have discovered a third fundamental type of cancer-influencing gene, which acts as a sneaky double agent in the never-ending biological Cold War between cancer-triggering and cancer-suppressing agents.

9.1 An Introduction to Genetics

What kind of power is in this thing called genetics? A four-year-old can hold seven-pound weights straight out from his body because of a genetic mutation? And a single gene can act both to suppress cancer and to help cause it? In genetics, we're clearly dealing with something that has great power over life.

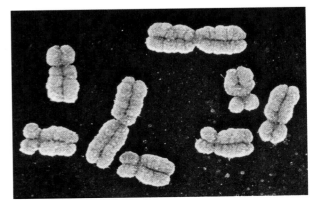

Human chromosomes, in a color-enhanced photo.

FIGURE 9.1 **The Power of a Protein** The bulging muscles of this Belgian blue bull have come about because its breed has a genetic mutation that results in the production of a dysfunctional form of the protein myostatin. This protein functions in cattle (and in humans) to retard muscle growth. The lack of functional myostatin also accounts for the extraordinary lack of fat in the Belgian blues, whose calves are so large they must be delivered through C-section.

FIGURE 9.2 **Information-Bearing DNA Molecule**

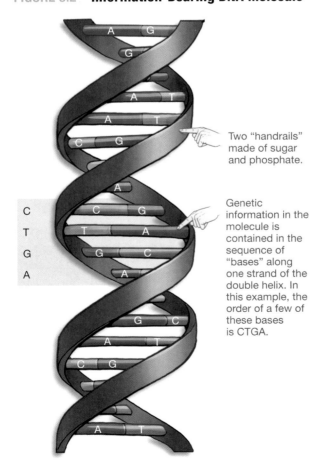

Two "handrails" made of sugar and phosphate.

Genetic information in the molecule is contained in the sequence of "bases" along one strand of the double helix. In this example, the order of a few of these bases is CTGA.

To see what kind of power, consider the "baby Superman" noted in the first newspaper story on the previous page. If we ask, how is it that this boy's body is *maintained*, such that he has such rippling muscles and low body fat, the answer is: through the power of the mutated gene he has (which allows his muscles to develop with no brakes on their growth). If we then ask, how did the boy come to possess this gene, the answer is: he *inherited* it. Doctors in Germany found that his mother has one copy of the mutated gene, and it appears likely that his father likewise has a copy. Thus, the genetic capacity for his unusual muscles was passed on, from one generation to the next, through transmission of the unit of inheritance known as a gene.

Now if you think about it, the capabilities of passing on and maintaining traits go a long way toward describing what it means to be a living thing, as opposed to an inanimate object. Living things can *transmit* their traits—from parent to offspring or from cell to cell—and they can *maintain* their internal states, taking care of such business as muscle growth and digestion and vision. Since they do this largely through genetics, it's not too much to say that genetics is central to what it means to be alive. How is it that any trait is passed on; indeed, how is it that *life* is passed on? Through genetic transmission. And how is it that life's unbelievable complexity is managed? Through genetics—through the daily operation of tens of thousands of genes.

The question this raises is: What are genes doing that makes them so crucial? The most important thing they are doing is serving in the role of . . . cookbook recipes. And the primary products of these recipes turn out to be *proteins*.

Proteins? Most of us think of them as something that we need a certain amount of in our diet. But as readers of Chapters 3 and 6 know, proteins function in living things like some vast group of building materials and tools. There are structural proteins, such as those that make up hair and cartilage; there are signaling proteins that bring messages to cells; and there are hormonal proteins and transport proteins. And then there are the enzyme proteins, which speed up—or in practical terms, *enable*—almost all the chemical reactions in living things. To get a visceral sense of the power of proteins, recall that baby Superman has a mutated gene. As such, he does not produce a protein, called myostatin, that inhibits muscle growth. Now, if you look at FIGURE 9.1, you can see what happens when a lack of myostatin unleashes muscle growth.

So, how is it that myostatin and other proteins are produced with the help of genes? Well, as most people know, genes exist along the length of a long, thin molecule called DNA (which stands for deoxyribonucleic acid). If you look at FIGURE 9.2, you can see that, in its basic structure, DNA looks something like an open, spiral staircase, the handrails of which are a repeating series of sugar and phosphate molecules (colored red in the figure). But then look at the steps of the staircase—for example, the ones highlighted in the blue box that are labeled C, T, G, and A. These are the so-called bases of DNA: the chemical units adenine, thymine, guanine, and cytosine (which are almost always referred to as A, T, G, and C). Here is where the protein recipe information is contained. If we were to walk along the handrails of the DNA in your own cells, we would eventually come to a sequence of bases that starts out like this: A G A T T C A C T G. This is, in fact, the start of the recipe for myostatin. Meanwhile, elsewhere in your DNA, there is a sequence that starts out: G G A C T C T G T C. This is the starting sequence for one of the forms of the protein keratin, which is what your hair is made of. Now, if you look at the keratin and myostatin sequences, you can see that they are different. And that's just the point: One series of A's, T's, C's, and G's contains the information for the production of one protein, but a *different* sequence of A's, T's, G's, and C's specifies a different protein. These separate sequences of bases are separate genes.

But how can a gene "specify" a given protein? Well, just as a gene is made of building blocks (the A, T, G, and C bases), so proteins are made of building blocks—the chemical units known as amino acids that were reviewed in Chapter 3. When a set of amino acids is strung together in the proper sequence, they make up a specific protein. The base sequence of a gene, then, is like a recipe that says, "Give me this amino acid, now this one, now this one . . ." and so on for hundreds of amino acids until a protein is created that gets busy on some task.

If you look at FIGURE 9.3, you can see where all this happens. The DNA you have been reading about is contained in the nucleus of the cell. The first step is that a stretch of it unwinds there, and its message—the order of a string of A's, T's, G's, and C's—is copied onto a molecule called messenger RNA (mRNA). This length of mRNA, which could be thought of as an information tape being dubbed off a "master" DNA tape, then exits from the cell nucleus. Its destination is a molecular workbench in the cell's cytoplasm, a structure called a ribosome. It is here that both message (the mRNA tape) and raw materials (amino acids) come together to make the product (a protein). The mRNA tape is "read" within the ribosome, and as this happens a growing chain of amino acids is linked together in the ribosome in the order called for by the mRNA. When the chain is finished, a protein has come into existence.

Genetics as Information Management

The story you have been reading about is that of **genetics**, meaning the study of physical inheritance among living things. In reflecting on this story, the first thing to note is that it concerns the storage, transfer, and use of *information*. To grasp this fact, consider the myostatin protein we looked at earlier, which acts as a signaling molecule on muscle cells. In line with what we've noted, there is a gene that prompts the production of myostatin. This gene's message is copied onto mRNA, which then migrates to a ribosome in the cytoplasm. Myostatin is put together there, amino acid by amino acid and, once synthesized, it gets busy carrying signals to muscle cells. Now note the role of the myostatin *gene* in this story. It didn't directly affect the muscle cell at all; it didn't even migrate to the ribosome to take part in myostatin synthesis. Its role really was like that of a cookbook recipe: It simply contained information that could be read by the cellular machinery in putting together this protein.

When we consider not just one gene, but rather the entire *collection* of genes in a living thing, it is easy to see that we are dealing with a vast *library* of information. The complete

FIGURE 9.3 **The Path of Protein Synthesis**

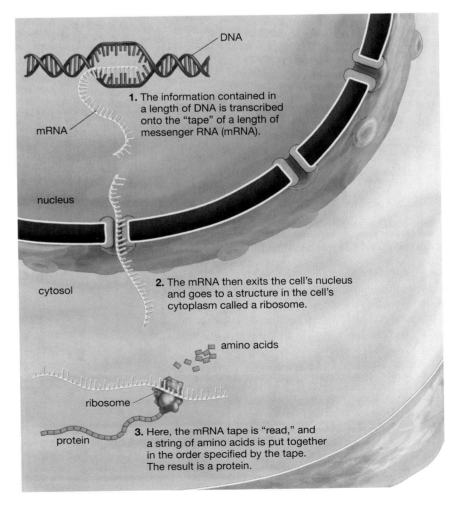

1. The information contained in a length of DNA is transcribed onto the "tape" of a length of messenger RNA (mRNA).

DNA

mRNA

nucleus

cytosol

2. The mRNA then exits the cell's nucleus and goes to a structure in the cell's cytoplasm called a ribosome.

amino acids

ribosome

protein

3. Here, the mRNA tape is "read," and a string of amino acids is put together in the order specified by the tape. The result is a protein.

A BROADER VIEW

The Importance of Genetics

The baby Superman story in the text made clear how a body can be made super-strong through the activity of genes. Anytime scientists see unusual genetic twists such as this, however, they don't just see oddities; they see medical possibilities. Imagine if, through genetic engineering, muscular dystrophy patients could be like baby Superman and have less muscle-retarding protein delivered to their muscles. The result would not be super-strong individuals, but instead patients who have had their strength returned to them. Biotech firms are trying to manage this feat now.

A BROADER VIEW

Genetics as Information Management

Interested in computers or in such media as DVDs and MP3 players? Then take note: Like a computer, a gene stores its information digitally. It is a sequence of 0's and 1's that codes for, say, the sound of a guitar in an MP3 file. Likewise, it is a sequence of A's, T's, G's, and C's that codes for a given protein in a living thing. This is why computer scientists are as highly valued in biotech as they are in high-tech: Genetic information can be sorted and searched for in the same way that computer information can be.

So Far... Answers:

1. genes

2. proteins

3. bases; A, T, C, G; amino acids

collection of an organism's genetic information is known as its **genome**. The sizes of genomes can vary, but to give you some idea, the human genome is estimated to include 20,000–25,000 genes. We generally think that information comes to us after birth and from outside the body, but in fact we are born with a huge volume of information that has been amassed and edited over 3.8 billion years of evolution.

The amazing thing about such collections is that there is not just one copy of them in living things; there may be trillions. Most cells within an organism contain a complete copy of that organism's genome. And since human beings, as an example, have trillions of cells, this means that each of us has trillions of copies of our genome within us. A given type of cell puts to use (or "expresses") only some parts of this genome, while another type of cell utilizes other parts; but this is merely a way of saying that different genes are active in different cells.

Cells duplicate, of course; one cell divides to become two, those two divide to become four, and so on. Given this, if each cell in the body is to have its own copy of the genome, then each time a cell duplicates, the genome that lies within that cell must duplicate as well, prior to the actual cell division. This is just what happens, in a process we'll be reviewing shortly. First, however, let's bring this introduction to genetics to a close by seeing how genetics as a whole will be covered in the rest of this chapter and in the chapters coming up.

The Path of Study in Genetics

Genetics is so important to biology that eight chapters in this book are devoted to the subject. Here's how the lineup of these chapters will go. First, beginning in this chapter, you will study DNA mostly as it comes packaged in units called chromosomes. Then you will learn how genetics functions in whole organisms by way of Gregor Mendel's famous pea plants. Only later will you return to DNA proper—to its structure and its protein-coding function. Finally, you will see how basic knowledge about genetics is being applied in the brave new world of biotechnology. Let's begin our tour of these component parts of genetics by seeing how life goes from one cell to many.

So Far...

1. Traits are passed on between generations through the transmission of _____ from parents to offspring.

2. Genes contain information for the production of _____.

3. The information-bearing building blocks of genes are the chemical units called _____, whose abbreviations are _____; the building blocks of proteins are called _____.

9.2 An Introduction to Cell Division

How does a baby grow, or a plant develop, or a wound heal? Always through cell division. As noted in Chapter 4, with the possible exception of viruses, life exists only inside cells. Further, we know that cells come only from other cells. And the *way* cells come from cells is by dividing. To understand the continuity in life, then, we need to have some understanding of how cells divide.

The sheer numbers involved in this process are mind-boggling. In your body, as many as 25 million cell divisions are completed each second. It is likely that every one of these divisions serves some function in maintaining your body; yet there are instances in which cells divide excessively. How does the disease we call cancer take hold? Always through the *unrestrained* division of cells. As you can see in Essay 9.1, "The Cell Cycle Runs Amok: Cancer" (page 148), cancer is intimately related to cell division.

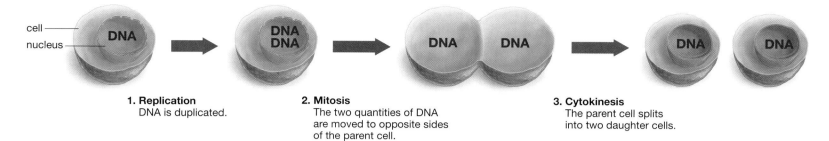

1. Replication
DNA is duplicated.

2. Mitosis
The two quantities of DNA are moved to opposite sides of the parent cell.

3. Cytokinesis
The parent cell splits into two daughter cells.

FIGURE 9.4 **Overview of Cell Division**

Not all cells divide throughout their existence. Most human brain cells are formed in the first three months of embryonic existence and then live for decades, with relatively few of them ever dividing again. Much the same is true of the leaf cells in plants; they divide only when the leaf is very small, then grow for a time, and then function at this mature size for as long as the leaf lives. At the other extreme, cells located in human bone marrow never *stop* dividing as they produce red blood cells (each one of which only lives for about two months). The output here is staggering: about 180 million new red blood cells are produced inside us each minute.

"Cell division" may be a somewhat misleading term if it is taken to mean a simple separation of cellular material, akin to a candy bar being split between two friends. Indeed, a splitting does occur in cell division, but certain parts of the cell must *duplicate* before this happens. Then the duplicated material is parceled out with fine precision, half of it going to one "daughter" cell and half to the other.

What's being duplicated and divided is DNA. Because, as we've seen, a cell's full complement of DNA contains such critical information, it would not do for a cell to be left with 50 or 75 percent of this information; rather, it needs the whole thing—no more and no less.

With this in mind, here is the big picture on cell division. First, there is a duplication of DNA; then there is the movement of two precisely matched quantities of it to opposite sides of the "parent" cell; finally there is the splitting of the parent cell into two daughter cells. The duplication of DNA is known as *replication*, the apportioning of it into two identical quantities is known as *mitosis*, and the splitting of the cellular material is known as *cytokinesis* (**see** FIGURE 9.4). The goal for the remainder of this chapter is to learn a little bit about replication and a good deal more about mitosis and cytokinesis.

The Replication of DNA

So how does the first part of cell division work—the DNA replication? If you look at FIGURE 9.5, you can see a simplified representation of the DNA molecule shown in Figure 9.2. Notice that the strands of the double helix have started to unwind, which is an initial step in DNA replication (Figure 9.5, step 1). Each of the two resulting single strands then serves as a template or pattern upon which a new strand is created (Figure 9.5, step 2). Earlier, DNA structure was compared to a spiral staircase. Now think of that staircase as splitting right down the middle of its steps (where the A and T or the G and C bases meet). Then a new *half*-staircase is created that bonds with one of the old DNA strands; a second half-staircase is likewise created that bonds with the *second* original strand. (These new strands are the blue ones in Figure 9.5, step 2.) How does this happen? Free-floating DNA bases will bond, one base at a time, with bases on the original strand. This process continues down the line until two strands have been created, each composed of one "old" half-strand and one "new" half-strand. This replication process, which doubles the cell's DNA content, takes place in advance of the cell dividing.

FIGURE 9.5 **DNA Replication**

(1) A DNA molecule unwinds.

(2) Each of the single strands of the original molecule (in red) serves as a template or pattern for the creation of a second DNA strand (in blue). Bases on the red strand pair, one base at a time, with free-floating bases until an entire second strand is created. With this, the quantity of DNA has effectively doubled.

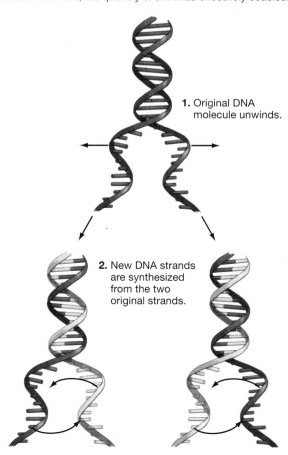

1. Original DNA molecule unwinds.

2. New DNA strands are synthesized from the two original strands.

In Chapter 13, you will metaphorically walk along the rungs of the double helix to grasp the details of this doubling of DNA strands. Right now, however, you'll be considering DNA on a larger scale, as it comes *packaged*.

A BROADER VIEW
Frequency of Division
Note the great diversity of the cells in your own body. Almost all the muscle cells in your heart have existed since before you were born. Thus, the heart cells you had working for you at 8 are the cells you will have working for you at 80. But of all the red blood cells coursing through your blood vessels right now, not one will be alive in five months; they all will have been replaced by new cells.

9.3 DNA Is Packaged in Chromosomes

You have thus far conceptualized each cell's DNA as spiraling out in *one* long double helix; but it does not, in fact, take that form. Rather, the DNA in each cell comes divided up and packaged into individual units of DNA called **chromosomes**. Different organisms have different numbers of chromosomes; human cells have 46, for example, while onion cells have 16. The chromosomes of eukaryotic organisms, such as ourselves, amount to DNA "packages" in several senses. They consist not only of DNA itself, but of a protein around which DNA is wrapped. The result is **chromatin**: a molecular complex, composed of DNA and associated proteins, that makes up the chromosomes of eukaryotic organisms. Thus, a number of chromosomes, composed of chromatin, exist in a cell's nucleus, detached but close to one another, and the *collection* of chromosomes makes up nearly the entire complement of a cell's DNA—its genome.

If you look at FIGURE 9.6a, you can see how to think about the double helix being packaged into chromosomes. FIGURE 9.6b shows you how to relate the "staircase-splitting" DNA replication just described to what this means at the chromosomal level. When we say that DNA is replicating, this is another way of saying that the chromosomes that DNA helps make up are *duplicating*. Once that has happened, the result is chromosomes in "duplicated state," which means individual chromosomes, each of which is made up of two sister chromatids. Each chromatid is one of the newly replicated DNA double helices, plus its associated proteins, as you can see in Figure 9.6b. More formally, a **chromatid** is one of the two identical strands of chromatin that make up a chromosome in its duplicated state.

Matched Pairs of Chromosomes

Individual chromosomes are detached from one another, but that does not mean they are completely different from one another. In eukaryotes such as ourselves, in fact, chromosomes tend to come in *pairs* that are close, but not exact matches. The 46 chromosomes we have come to us as 23 chromosomes from *each parent*. Critically, with one exception that you'll soon get to, these are 23 matched pairs of chromosomes, each chromosome from the mother matching with one from the father.

Defining a "Matched Pair" What is a "matched pair" of chromosomes? We have a chromosome 1 that we inherit from our mother, and it contains a set of genes very similar to those that lie on the chromosome 1 we inherit from our father. The same is true for the maternal and paternal copies of chromosome 2, chromosome 3, and so on. Each of us, then, has 23 pairs of **homologous chromosomes**, *homologous* here meaning "the same in size and function."

However, homologous chromosomes are not *exactly* alike. Any two of them will contain genes for the same kinds of protein products. If a given paternal chromosome has a gene that codes for hair color, it is a safe bet that the matching maternal chromosome will have a gene that codes for hair color. But there can be variations on these genes. A gene on the paternal chromosome may help code for *red* hair, while the gene on the matching maternal chromosome may help code for blonde hair. Nevertheless, both chromosomes have genes that code for hair color, as well as for thousands of other traits, and as such, are said to be homologous.

X and Y Chromosomes The one exception to the matched-pairs rule in human chromosomes is the so-called sex chromosomes of males. Human females have the 23 pairs of homologous chromosomes mentioned before; 22 of these are *autosomes*, or nonsex chromosomes, and one is a homologous pair of X chromosomes that, when present, means a growing embryo will be a female. Males, on

(a) DNA is packaged in units called chromosomes

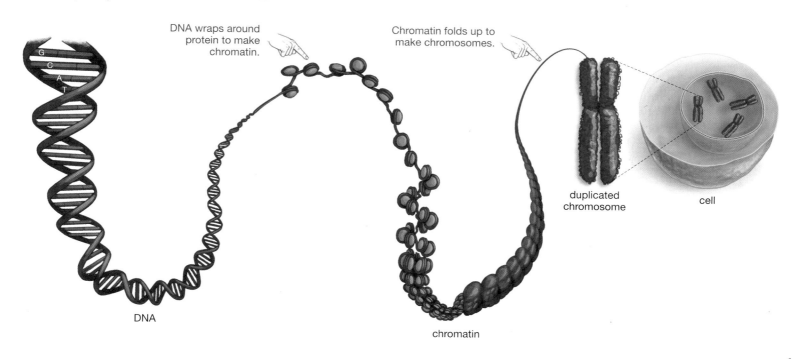

DNA wraps around protein to make chromatin.

Chromatin folds up to make chromosomes.

DNA

chromatin

duplicated chromosome

cell

(b) DNA replication at two levels

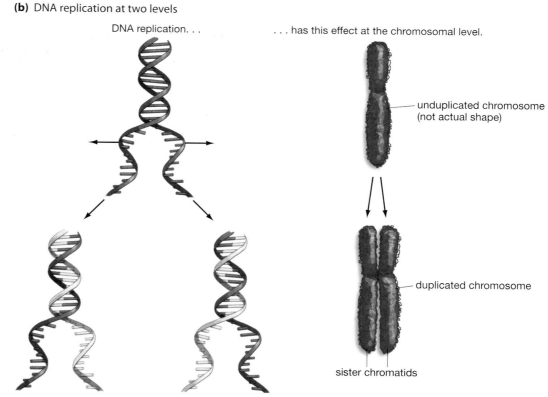

DNA replication. . .

. . . has this effect at the chromosomal level.

unduplicated chromosome (not actual shape)

duplicated chromosome

sister chromatids

FIGURE 9.6 **Chromosomes and DNA Replication**

(a) The DNA molecule is bound up with proteins, with the resulting combination known as chromatin. This chromatin then makes up structures called chromosomes.

(b) When DNA replicates, the result is two copies of the original DNA molecule. At the chromosomal level, the result of this replication is a single chromosome in duplicated state. The chromosome is composed of two sister chromatids—the two copies of the original DNA molecule. An unduplicated chromosome doesn't actually have the well-defined shape of the chromosome in the figure. It's shown this way merely for purposes of comparison with the duplicated chromosome.

FIGURE 9.7   A Karyotype Displays a Full Set of Chromosomes One member of each chromosome pair comes from the individual's father and the other member from the mother. Each paired set of chromosomes is said to be "homologous," meaning the same in size and function. (The two chromosomes over the number 1 are a homologous pair, the two over number 2, and so forth.) Notice that this karyotype is from a human male; there are 22 pairs of homologous chromosomes and then one X and one Y chromosome (which are not homologous). A female would have two X chromosomes. All the chromosomes are in the duplicated state.

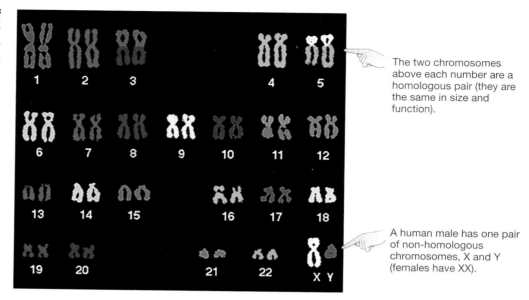

The two chromosomes above each number are a homologous pair (they are the same in size and function).

A human male has one pair of non-homologous chromosomes, X and Y (females have XX).

the other hand, have the 22 pairs of homologous autosomes and one X chromosome; but then they also have one Y chromosome that, when present, means a growing embryo will be a male.

This whole scheme is laid out for you in FIGURE 9.7. There you can see a **karyotype**, or pictorial arrangement of a full set of human chromosomes. There are 46 chromosomes in all, 22 of them matched pairs. In this case, one pair, at lower right, has one X chromosome and one Y chromosome, meaning this is a male.

The chromosomes in Figure 9.7 are in the duplicated state noted earlier, each of them being composed of two sister chromatids. For the sake of clarity, it would be nice to see a picture of some *un*duplicated chromosomes, but we have no pictures that are the counterpart to Figure 9.7. This is so because chromosomes take on an easily discernable shape only after they duplicate, prior to cell division. As cell division approaches, the DNA-protein complex changes its form; it tightens up, condensing mightily to produce well-defined chromosomes (FIGURE 9.8b). After cell division is complete, the cell's chromosomes return to their relatively formless state, shown in FIGURE 9.8a.

FIGURE 9.8   DNA Can Be Arranged in Two Ways

(a) Replication occurs when the chromosomes DNA comes packaged in are in a relatively formless state. In this state, the chromatin that makes up the chromosomes has yet to condense into its compact arrangement.

(b) Mitosis occurs after chromatin condenses into the easily discernable shapes of duplicated chromosomes. The micrograph is of color-enhanced duplicated human chromosomes.

Why this change in form? This condensing before cell division has the same effect as you taking all your scattered belongings and packing them into boxes just before moving from one apartment to another. Remember that DNA is packing up to leave as well, in this case for life in a successor cell. Were it not to tighten into its duplicated chromosomal form, its elongated fibers would get tangled up in the move.

(a) DNA in uncondensed form

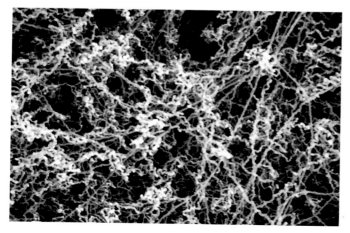

(b) DNA condensed into duplicated chromosomes

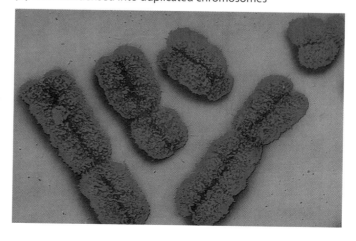

Chromosome Duplication as a Part of Cell Division

With the chromosomes duplicated, you are now ready to look at the process by which they will split up. Before following this path, though, recognize that this separation occurs within the larger process that we are following: the division of the cell as a whole. As you saw in Chapter 5, there are many more things inside a cell than its complement of chromosomes. There are mitochondria, lysosomes, ribosomes, and so forth, and the fluid in which they are immersed. Together, these things lie in the cell's cytoplasm—the area *outside* the nucleus, as opposed to chromosomes, which lie *inside* the nucleus. With cell division, about half of the cytoplasmic material goes to one daughter cell and half to the other. It is helpful, then, to conceptualize cell division as having two separable components. **Mitosis** is the separation of a cell's duplicated chromosomes prior to cytokinesis. **Cytokinesis** is the physical separation of one cell into two daughter cells. These two processes are part of the big picture of cell division sketched earlier, which now can be stated in a slightly different way. First, chromosomes duplicate; then chromosomes separate and move to opposite sides of the parent cell (mitosis); then the parent cell splits into two daughter cells (cytokinesis).

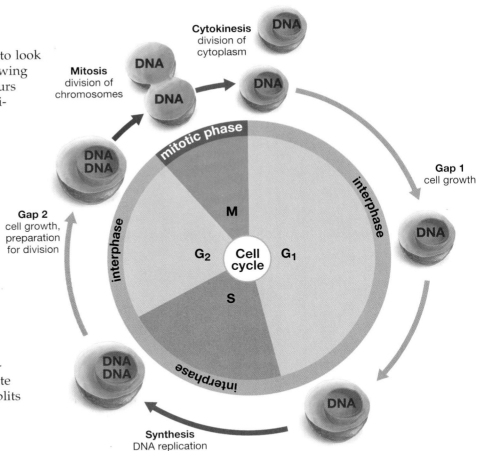

FIGURE 9.9 **The Cell Cycle** The three main stages of cell division—DNA replication, mitosis, and cytokinesis—can be seen here in the context of a complete cell cycle. This cycle traditionally is divided into two main phases, interphase and mitotic (or "M") phase, which are in turn divided into other phases. Note that DNA replication—and hence chromosomal duplication—takes place within the S phase of interphase. The average cell spends most of its time in G_1 phase, growing and carrying out its normal cellular functions. Some types of cells, however, cease going through the cell cycle at all, while others progress through it repeatedly and rapidly.

So Far...

1. Chromosomes whose DNA has replicated are said to be in _____ state, in which a single chromosome is composed of two _____.

2. Human beings have _____ pairs of chromosomes. With one exception (in males) each of these pairs is said to be _____ because they are the same in _____. One member of each pair comes from _____, while the other member comes from _____.

3. The lone exception to the pairs rule in human chromosomes comes with the male _____ and _____ chromosomes.

The Cell Cycle If you look at FIGURE 9.9, you can see how all these processes are linked over time. The figure illustrates the **cell cycle**: the repeating pattern of growth, genetic duplication, and division seen in most cells. There are two main phases in this cycle. First there is **interphase**: that portion of the cell cycle in which the cell simultaneously carries out its work and, in preparation for division, duplicates its chromosomes. Second, there is **mitotic phase** (or M phase): that portion of the cell cycle that includes both mitosis and cytokinesis.

Interphase and M phase are in turn subdivided into smaller phases. In interphase, first there is G_1, standing for "gap-one." What goes on here are normal cell operations and cell growth. The cell then enters the S or "synthesis" phase, which is the synthesis of DNA, resulting in the duplication of the chromosomes. When this ends, interphase's gap-two or G_2 period begins, during which there is more cytoplasmic growth and a preparation for cell division.

The length of the cell cycle varies greatly from one type of cell to another. In a typical animal cell, the total cell-cycle length averages about 24 hours. Within this cycle, mitotic phase takes up only about 30 minutes. Within interphase, an animal cell spends roughly 12 hours in G_1, 6 hours in S, and 6 hours in G_2.

So Far... Answers:

1. *duplicated; sister chromatids*
2. *23; homologous; size and function; the mother; the father*
3. *X; Y*

9.4 Mitosis and Cytokinesis

It is the second overarching phase of the cell cycle, mitotic phase, that is the focus for the rest of this chapter. Now the concern is not with how a cell manages its general functions, but with how it *divides*—how it carries out both mitosis and cytokinesis. Mitosis turns out to have four phases, which we'll look at in order, after which we'll review cytokinesis. We will first consider mitosis and cytokinesis as they occur in animal cells.

The Phases of Mitosis

Prophase The beginning of mitosis marks the end of the cell's interphase (**see FIGURE 9.10**). The starting state is that the cell's DNA, which replicated during the S phase, has just begun to pack itself into well-defined chromosomes. When we can finally *see* such chromosomes with a microscope, we can say that mitosis has begun, with prophase. Now the packing job continues; before it is done, the DNA will have

FIGURE 9.10 **Mitosis and Cytokinesis** The micrographs at the bottom of the figure are pictures of mitosis and cytokinesis in a whitefish embryo.

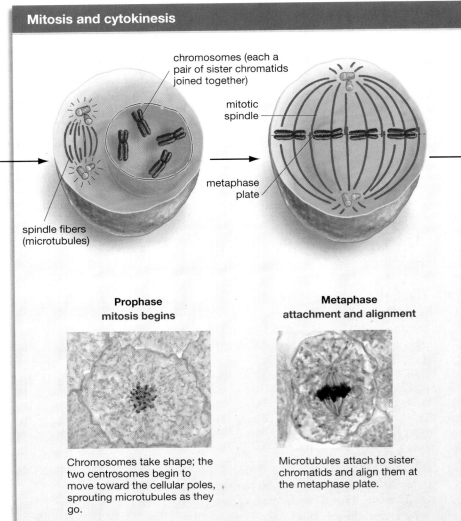

Mitosis and cytokinesis

chromosomes (each a pair of sister chromatids joined together)

mitotic spindle

metaphase plate

spindle fibers (microtubules)

pair of centrosomes

nucleus

replicated, uncondensed DNA

End of interphase

DNA has already duplicated back in S phase. Centrosome has doubled.

Prophase
mitosis begins

Chromosomes take shape; the two centrosomes begin to move toward the cellular poles, sprouting microtubules as they go.

Metaphase
attachment and alignment

Microtubules attach to sister chromatids and align them at the metaphase plate.

coiled and condensed into 1 eight-thousandth of its S-phase length. When this is over in human beings, there are 46 well-defined chromosomes in the nucleus in duplicated state (meaning they are composed of 92 chromatids). Meanwhile, the nuclear envelope—the double membrane surrounding the nucleus—begins to break up.

While this is going on, big changes are taking place outside the nucleus. Recall from Chapter 4 the structures called microtubules: protein fibers that are part of the cell's cytoskeleton or internal scaffolding network. In some of their roles, microtubules can be compared to tent poles that can shorten or lengthen as need be. Throughout mitosis, microtubules stretch the cell as a whole, and physically move the cell's chromosomes around.

In an interphase cell there exists, just outside the nucleus, a **centrosome**: a cellular structure that acts as an organizing center for the assembly of microtubules. This centrosome duplicates, so that now there are two microtubule organizing centers. Now, in prophase, these two centrosomes start to move apart. Where are they going? To the poles. From here on out, it is convenient to think of mitosis and cytokinesis in terms of a global metaphor: The centrosomes migrate to the cellular poles, while the chromosomes first align, and then separate, along a cellular equator

A BROADER VIEW

What We Inherit

Isn't it interesting to ponder the fact that, in each of your cells, there is a chromosome 1 that came from your father and a counterpart chromosome 1 that came from your mother; then a chromosome 2, 3, and so forth, on down the line? Of course, the original paternal chromosome 1 existed in the sperm that gave rise to you, after which it has been copied over and over, with each new cell division (with the same thing happening on the maternal side). Nevertheless, if we ask what it is that we carry within us from both parents, this is it: the chromosomes that they passed on to us.

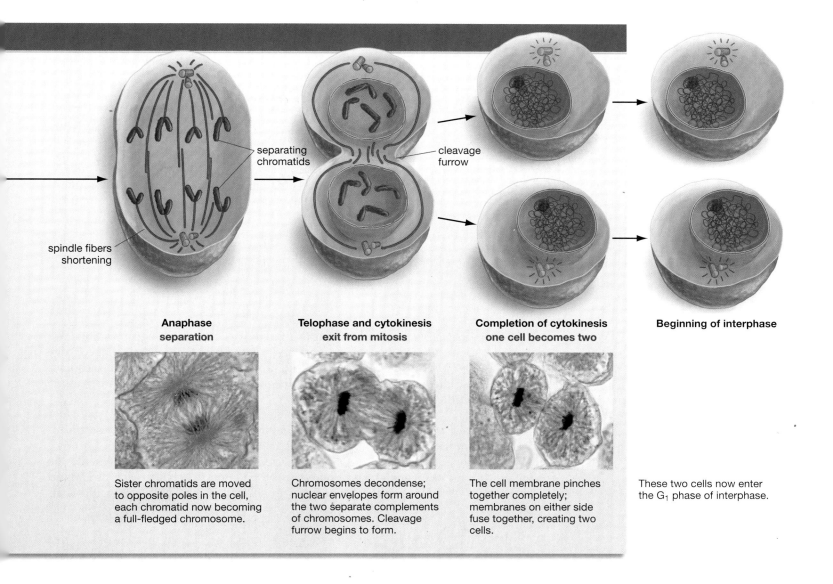

separating chromatids

cleavage furrow

spindle fibers shortening

Anaphase separation

Telophase and cytokinesis exit from mitosis

Completion of cytokinesis one cell becomes two

Beginning of interphase

Sister chromatids are moved to opposite poles in the cell, each chromatid now becoming a full-fledged chromosome.

Chromosomes decondense; nuclear envelopes form around the two separate complements of chromosomes. Cleavage furrow begins to form.

The cell membrane pinches together completely; membranes on either side fuse together, creating two cells.

These two cells now enter the G$_1$ phase of interphase.

Essay 9.1 The Cell Cycle Runs Amok: Cancer

The cell cycle reviewed in this chapter is, in one sense, a common natural process: Cells grow; they duplicate their chromosomes; these chromosomes separate; one cell divides into two. This goes on like clockwork, millions of times a second in each one of us. But then one day we learn that an aunt or a grandfather or a friend is experiencing an *unrestrained* division of cells. Things aren't explained to us in this way, of course. We are simply told that someone we know has cancer.

At root, all cancers are failures of the cell cycle. Put another way, all cancers represent a failure of cells to limit their multiplication in the cell cycle. What is liver cancer, for example? It is a damaging multiplication of liver cells. First one, then two, then four, then eight liver cells move repeatedly through the cell cycle, and as their numbers increase, they destroy the liver's working tissues. Given that cancer manifests in this way, it's not surprising that a large portion of modern cancer research is *cell-cycle* research. The logic here is simple: To the extent that uncontrolled cell division can be stopped, cancer can be stopped.

A number of ideas are being tested for how cancers get going in the first place, but a common thread that runs through these ideas is that, for cells to be brought to a cancerous state, two things are required: Their accelerators must get stuck and their brakes must fail. The control mechanisms that *induce* cell division must become hyperactive, and the mechanisms that *suppress* cell division must fail to perform. You may have heard a couple of terms used to describe the genetic components of this process. There are normal genes that induce cell division, but that when mutated can cause cancer; these are the stuck-accelerator genes, called oncogenes. Then there are genes that normally suppress cell division, but that can cause cancer by acting like failed brakes. These are tumor suppressor genes. Note that *both* kinds of genes must malfunction for cancer to

get going; indeed, it usually takes a long succession of genetic failures to induce cancer. This is why cancer is most often a disease of the middle-aged and elderly: It can take decades for the required series of mutations to fall into line in a single cell, such that it becomes cancerous.

For cells to be brought to a cancerous state, two things are required: Their accelerators must get stuck and their brakes must fail.

How do oncogenes interact with tumor suppressor genes? In the normal case, cells will not begin division until prompted to do so by a signal from outside themselves. A protein (called a growth factor) will bind to a cell, setting off a cascade of chemical reactions inside it that triggers division. One of the links in this chemical cascade is a protein called Ras that could be thought of as an old-time railway switch. When Ras is chemically pointed one way (toward "on"), the cell moves through the cell cycle. When it is pointed the other (toward "off"), the cell stays in G_1. The gene that codes for Ras can become mutated, however, and when this happens, the Ras protein changes shape and points in the "on" direction all the time—no matter what signals it is getting from the outside. Thus, *ras* is an example of an oncogene. It normally prompts the cell to divide intermittently; when mutated, it prompts the cell to divide continuously.

A cell with a mutated *ras* gene is not doomed to become cancerous, however. What can save it is a good set of brakes, in the form of tumor suppressor genes. The most important of

called the *metaphase plate*. As the centrosomes move apart, they begin sprouting microtubules in all directions. One variety of them forms a football-shaped cage around the nuclear material, while a second variety attaches to the chromosomes themselves. Taken together, the microtubules active in cell division are known as the **mitotic spindle**.

Metaphase   By the time metaphase begins, the nuclear envelope has disappeared completely, and the microtubules that were growing toward the chromosomes now *attach* to them. Through a lively back-and-forth movement, the microtubules align the chromosomes at the equator. With this, each chromatid now faces the pole *opposite* that of its sister chromatid, and each chromatid is attached to its respective pole by perhaps 30 microtubules.

Anaphase   At last the genetic material divides. As you may have guessed, this is a parting of sisters. The sister chromatids are pulled apart, each now becoming a full-fledged chromosome. All 46 chromatid pairs divide at the same time, and

these genes—one known as *p53*—is so vital to human health that it is sometimes referred to as the "guardian of the genome." In the presence of certain kinds of mutations, *p53* protein levels rise in the cell, and these levels start turning selected genes on and off. The result is the cell's first line of defense against cancer—the cell cycle is shut down until the mutation has been repaired. This shutdown doesn't happen at just any point in the cycle, however. As cancer researchers Leland Hartwell and Ted Weinert discovered in the 1980s, cells have specific *checkpoints* in their cycle. Just as NASA mission control will stop at a defined point in a countdown to see if "all systems are go" for a launch, so a cell has specific points at which it makes sure that all its systems are healthy enough for cell division to continue. The first of its checkpoints comes in G_1, as it is about to enter S phase (during which it doubles its DNA). The second point comes in G_2, as it is about to enter into mitosis and cytokinesis. Thus, if a dividing skin cell, for example, has acquired some mutations in G_1, it will not enter S phase until its DNA repair enzymes have fixed the problem.

But what happens if the DNA damage spotted in G_1 can't be fixed? Then, prompted by *p53*, the cell goes to level-two of its emergency responses: It commits suicide. Through an orderly process called *apoptosis*, the cell shuts down its activities, breaks up, and dies. If you have ever been sunburned, you have probably seen the effects of apoptosis up close. The ultraviolet light in the sun damages the DNA in skin cells. When this damage cannot be fixed with repair enzymes, these cells undergo apoptosis; their remains are the peeling skin that comes with a bad sunburn.

In sum, when genes such as *ras* become mutated, they can lead to an out-of-control cell cycle, but this process can be halted in several ways by tumor suppressor genes such as *p53*. With this, you can probably see what comes next. What will stave off cancer if a cell has *both* a mutated *ras* gene and a mutated *p53* gene? Perhaps nothing, because now the accelerator is stuck and the brakes have failed. As noted, there is generally

more to cancer than two mutated genes, but when both *ras* and *p53* malfunction, a cell is well on its way to cancer.

As it turns out, *ras* is probably the most important human oncogene among the hundred or so that have been identified. And *p53* certainly is the most important tumor suppressor gene among the two dozen that have been identified. A mutated *ras* gene is found in about 30 percent of all human cancers, while a mutated *p53* is found in half of all human cancers. As you can imagine, there is intense research interest in both these genes. From 1989 to 2000, more than 17,000 scientific publications were written on *p53* alone. The war on cancer may not have been won, but it certainly is being waged.

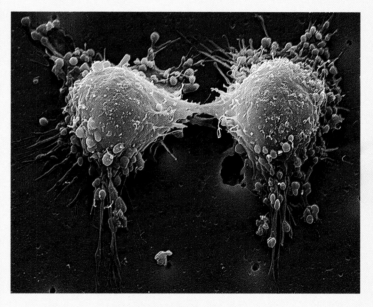

FIGURE E9.1.1 **Harmful Division** Pictured are two prostate cancer cells in the final stages of cell division.

each member of a chromatid pair moves toward its respective pole, pulled by a shortening of the microtubules to which it is attached.

Telophase Telophase represents a return to things as they were before mitosis started. The newly independent chromosomes, having arrived at their respective poles, now unwind and lose their clearly defined shape. New nuclear membranes are forming. When this work is complete, there are two finished daughter nuclei lying in one elongating cell. Even as this is going on, though, something else is taking place that will result in this one cell becoming two.

Cytokinesis

Cytokinesis actually began back in anaphase and is well under way by the time of telophase. It works through the tightening of a cellular waistband that is composed of two sets of protein filaments working together. These filaments—the same type that allow your muscles to contract—form a ring that narrows along

FIGURE 9.11 **Cytokinesis in Animals** Cytokinesis in animal cells begins with an indentation of the cell surface, a cleavage furrow, shown here in a dividing frog egg. (×85)

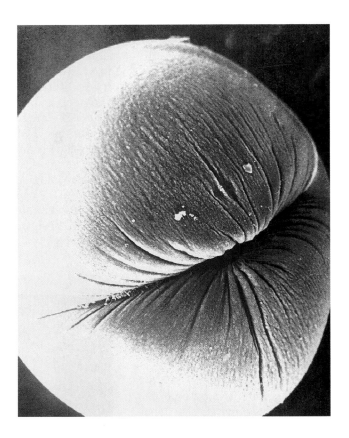

the cellular equator (**see** FIGURE 9.11). An indentation of the cell's surface (called a cleavage furrow), results from the ring's contraction; consequently, the fibers in the mitotic spindle are pushed closer and closer together, eventually forming one thick pole that is destined to break. The dividing cell now assumes an hourglass shape; as the contractile ring continues to pinch in, one cell becomes two by means of something you looked at in Chapter 4: membrane fusion. The membranes on each half of the hourglass circle toward each other and then fuse. With this, the two cells become separate. Mitosis and cytokinesis are over, and the two daughter cells slip back into the relative quiet of interphase.

9.5 Variations in Cell Division

Not all cells go through mitosis and cytokinesis in the same way. To get an idea of one kind of difference among cells, consider plant cells, which carry out mitosis as animal cells do, but then perform cytokinesis differently. In the splitting of cytokinesis, plant cells must deal with something that animal cells don't have: the cell wall (see page 72 in Chapter 4). The way animal cells carry out cytokinesis—pinching the plasma membrane inward via a fiber ring—wouldn't work with plant cells, because their thick cell wall lies *outside* the plasma membrane.

The plant cell's solution to cytokinesis is to grow a new cell wall and plasma membrane that run roughly down the middle of the parent cell. If you look at FIGURE 9.12, you can see how this works. A series of membrane-lined spheres (called vesicles) begins to accumulate near the metaphase plate and fuse together. Eventually they will form a flat *cell plate* that runs from one side of the parent cell to the other. The membrane portion of the cell plate then fuses with the original plasma membrane of the parent cell, and the result is two new adjacent plasma membranes that split the parent cell in two. Meanwhile, the material enclosed *inside* the cell-plate membrane is the foundation for the new cell-wall segments of the daughter cells.

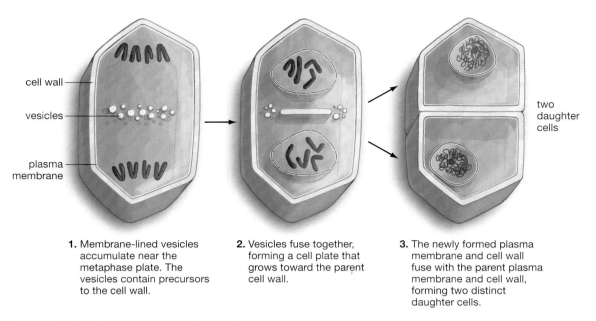

1. **Membrane-lined vesicles accumulate near the metaphase plate. The vesicles contain precursors to the cell wall.**

2. **Vesicles fuse together, forming a cell plate that grows toward the parent cell wall.**

3. **The newly formed plasma membrane and cell wall fuse with the parent plasma membrane and cell wall, forming two distinct daughter cells.**

cell wall

vesicles

plasma membrane

two daughter cells

FIGURE 9.12   Cytokinesis in Plants The cell walls of plants are too rigid to form cleavage furrows, as animal cells do when dividing. Therefore, plant cells use a different strategy in cytokinesis. They build a new plasma membrane and cell wall down the middle of the parent cell to separate the two daughter cells from the inside out.

On to Meiosis

The cell division reviewed in this chapter concerns "somatic" cells—the kind that form bone, muscle, nerve, and many other sorts of tissue. Given a distribution this wide, you might well ask: What kind of cells are *not* somatic? The answer is only one kind—the kind that forms the basis for each succeeding generation of sexually reproducing living things.

So Far...

1. The separation of a cell's duplicated chromosomes that takes place prior to cell division is called _____, while the separation of one cell into two new daughter cells is called _____.

2. The life of a typical cell can be divided into two main phases. During the first of these, called _____, a cell carries out normal operations and replicates its DNA. In the second of these, called _____, the cell divides, which is to say it carries out both _____ and _____.

3. The division of genetic material in mitosis involves a separation of _____, which migrate to opposite _____ in the parental cell.

So Far... Answers:

1. *mitosis; cytokinesis*

2. *interphase; mitotic phase; mitosis; cytokinesis*

3. *sister chromatids; poles*

Chapter Review

Summary

9.1 An Introduction to Genetics

- **The importance of genetics**—Genetics governs two key capabilities in living things: how they pass on their traits—from parent to offspring or from cell to cell—and how they maintain their internal environments. (page 138)

- **What is a gene?**—DNA is a molecule that contains or "encodes" information in the chemical form of substances called "bases" that are laid out along the DNA double helix. These bases come in four varieties: adenine (A), thymine (T), guanine (G), and cytosine (C). One series of bases contains information for the production of one protein, while a different series of bases specifies a different protein. Each group of bases that specifies a given protein is known as a gene. (page 138)

- **The protein production path**—Protein synthesis begins with the information in a sequence of DNA bases being copied onto a molecule called messenger RNA (mRNA). This molecule moves out of the cell's nucleus to a structure in its cytoplasm called a ribosome. There, the mRNA "tape" is brought together with the building blocks of proteins, amino acids. As the ribosome "reads" the mRNA tape, it strings together a sequence of amino acids called for by the tape. The result is a protein. (page 139)

- **Genetics as information management**—Genes function as repositories of genetic information, but it is proteins that carry out most of the tasks that maintain living things. The complete collection of an organism's genetic information is its genome. Most cell types in an organism contain a complete copy of that organism's genome. Before cells divide, their genome must first be copied and the resulting copies apportioned evenly into what will become two daughter cells. (page 139)

9.2 An Introduction to Cell Division

- **Overview of cell division**—Cell division includes the duplication of DNA (replication); the apportioning of the copied DNA into two quantities in a parent cell (mitosis); and the physical splitting of this parent cell into two daughter cells (cytokinesis). In DNA replication, the two strands of the double helix unwind, after which each single strand serves as a template for construction of a second, complementary strand of DNA. The result is a doubling of the original quantity of DNA. (page 140)

- **Frequency of division**—Cells exhibit great diversity in how frequently they divide. Some cells never divide after coming into being, while others never stop dividing. (page 141)

9.3 DNA Is Packaged in Chromosomes

- **The nature of chromosomes**—Chromosomes are composed of DNA and its associated proteins—a combined chemical complex called chromatin. Chromosomes exist in an unduplicated state until such time as DNA replicates, prior to cell division. DNA replication results in chromosomes that are in duplicated state, meaning one chromosome composed of two identical sister chromatids. (page 142)

- **Homologous chromosomes**—Chromosomes in human beings (and many other species) come in matched pairs, with one member of each pair inherited from the mother, and the other member of each pair inherited from the father. Such homologous chromosomes have closely matched sets of genes on them, though many of these genes are not identical. A given paternal chromosome may have genes that code, for example, for different hair or skin color than the counterpart genes on the homologous maternal chromosome. Human beings have 46 chromosomes—22 matched pairs and either a matched pair of X chromosomes (in females) or an X and a Y chromosome (in males). (page 142)

- **The cell cycle**—Cell division fits into the larger framework of the cell cycle, meaning a repeating pattern of growth, genetic replication, and cell division. The cell cycle has two main phases. The first is interphase, in which the cell carries out its work, grows, and duplicates its chromosomes in preparation for division. The second is mitotic phase, in which the duplicated chromosomes separate and the cell splits in two. (page 145)

Web Tutorial 9.1 The Cell Cycle

9.4 Mitosis and Cytokinesis

- **Mitotic stages**—There are four stages in mitosis: prophase, metaphase, anaphase, and telophase. The essence of the process is that duplicated chromosomes line up along an equatorial plane of the parent cell, called the metaphase plate, with the sister chromatids that make up each duplicated chromosome lying on opposite sides of the plate. The sister chromatids are then pulled apart, by fibers called microtubules, to opposite poles of the parent cell. Once cell division is complete, sister chromatids that once formed a single chromosome will reside in separate daughter cells, with each sister chromatid now functioning as a full-fledged chromosome. (page 146)

- **The nature of cytokinesis**—Cytokinesis in animal cells works through a ring of protein filaments that tightens at the middle of a dividing cell. Membranes on the portions of the cell being pinched together then fuse, resulting in two daughter cells. (page 149)

Web Tutorial 9.2 Mitosis

9.5 Variations in Cell Division

- **Cytokinesis in plants**—The cell wall that plants possess means they must synthesize new cell walls and plasma membranes that divide the parent cell in two; a fusion of these structures with the existing cell walls and plasma membranes in the parent cell splits this cell in two, yielding two daughter cells. (page 150)

Key Terms

cell cycle   p. 145

centrosome   p. 147

chromatid   p. 142

chromatin   p. 142

chromosome   p. 142

cytokinesis   p. 145

genetics   p. 139

genome   p. 140

homologous chromosomes
   p. 142

interphase   p. 145

karyotype   p. 144

mitosis   p. 145

mitotic phase   p. 145

mitotic spindle   p. 148

Testing Your Understanding

In Your Own Words *(answers in the back of the book)*

1. In what way are the 23 pairs of human chromosomes "matched" chromosomes?

2. How do mitosis and cytokinesis differ?

3. Name the four phases of mitosis and describe the major events occurring in each phase.

4. The drawing below shows a cell in a phase of mitosis. What phase is shown? How many chromatids are present in this stage? What is the total number of chromosomes each daughter cell will have?

Thinking about What You've Learned

1. Why is it accurate to think of each human being as the owner of a library of ancient information?

2. We generally think of information as coming to us in the form of words. Yet a baby's smile also conveys information, as do the A's, T's, G's, and C's of the DNA molecule. In what other forms does information exist?

3. Given the conceptualization of genetics as information management, what would you predict about the relative size of the human genome as opposed to the genome of a bacterium?

Preparing for Sexual Reproduction: Meiosis

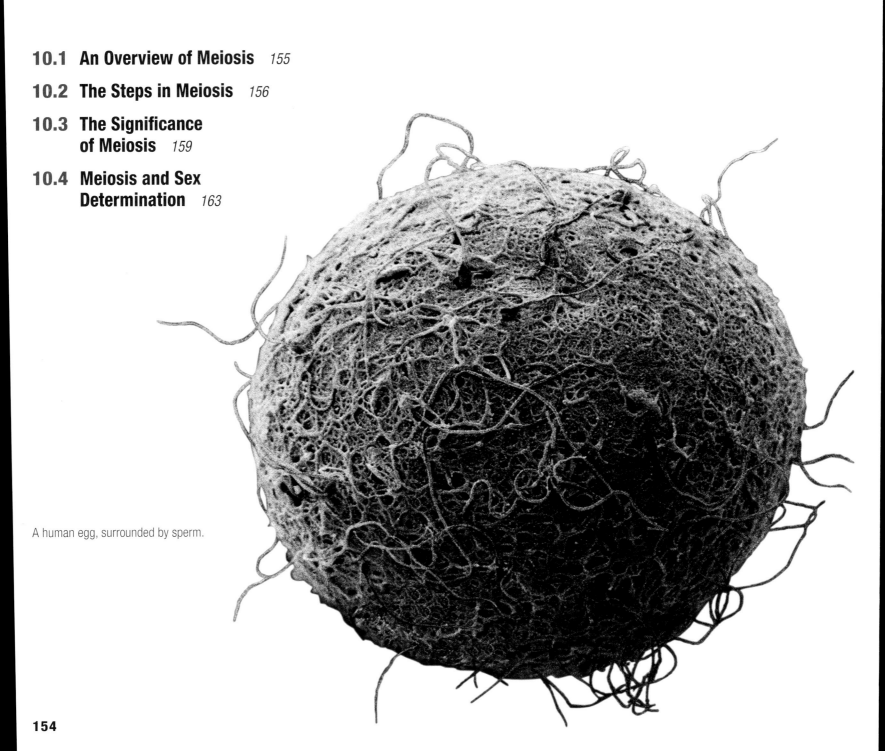

A human egg, surrounded by sperm.

THE MIAMI HERALD

WOMEN'S SUPPLIES OF EGGS MAY NOT BE FIXED NUMBER

By Rosie Mestel, Los Angeles
Times Service
March 11, 2004

The long-held biological dogma that females are born with all the eggs they will ever have is wrong, according to a study being published today in the journal *Nature*. Instead—at least in mice—eggs are renewed throughout life, probably from a store of stem cells in the ovary.

10.1 An Overview of Meiosis

Talk about stopping the presses. In the spring of 2004, presses that were printing biology textbooks may literally have been stopped when their authors heard the news you can see in the story above. For decades, every biology textbook confidently told its readers that all the eggs a female mammal makes are formed within her prior to her birth—while she herself is an embryo, in other words. This initial store of eggs then slowly dwindles over time, the textbooks said, and this is the primary reason that women can't get pregnant past early middle age: Their supply of eggs runs out. By contrast, men remain fertile throughout life because their sperm are produced each day by "stem cells," which is to say cells that never lose their ability to produce more sperm.

Then came the news that, at least in mice, female mammals *do* possess stem cells that are capable of giving rise to eggs and that these stem cells remain active throughout life. Now, imagine the implications if such cells could be found or coaxed into development in *human* females: Fertility might be prolonged for older women or enhanced for women of any age. This news was big, and it was good.

Buried within it, however, was something that would be easy to miss. Note the implication that eggs and sperm develop *from cells*. Indeed, eggs and sperm are cells themselves. In Chapter 9, you went over the process by which one cell becomes two, and you might think that this kind

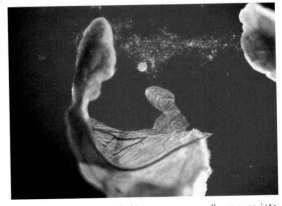

A human egg, surrounded by accessory cells, moves into a woman's Fallopian tube during the process of ovulation.

of cell division would be at work in producing eggs and sperm, but that's not the case. Eggs and sperm can't develop the way regular cells do, because eggs and sperm are destined to *come together*—sperm fertilizing egg—to create offspring. Why should this affect the way they develop? Well, a little story from regular life can show you the problem that would ensue if sperm and egg were produced through normal cell division.

A real-life couple—let's call them Jack Fennington and Jill Kent—combine their last names when they get married and thus become Jack and Jill Fennington-Kent. Now, what would happen if the Fennington-Kent's children were to continue in this tradition? Their daughter Susie might marry, say, Ralph Reeson-Dodd, which would make her Susie Fennington-Kent-Reeson-Dodd. If *her* daughter, Alicia, were to keep this up, she might be fated to become Alicia Fennington-Kent-Reeson-Dodd-Garcia-Lee-Minderbinder-Green, and so on.

Now consider our cells. Remember that regular human cells have 23 pairs of chromosomes or 46 chromosomes in all. If an egg and sperm each brought 46 chromosomes to their *union*—to their merger in conception—the result would be an embryo with 92 chromosomes. The *next* generation down would have 184 chromosomes, the next after that 368, and so on. This would be as functional as having Fennington-Kent-Reeson-Dodd-Garcia-Lee-Minderbinder-Green as a last name.

Thus, eggs and sperm have to be produced in a special way. What way is this? In sexual reproduction, chromosome union is preceded by chromosome *reduction*. The reduction occurs in the cells that give rise to sperm and egg. When these cells divide, the result is sperm or egg cells that have only *half* the usual number of chromosomes. Human sperm or eggs, in other words, have only 23 chromosomes in them. Each 23-chromosome sperm can then unite with a 23-chromosome egg to produce a 46-chromosome fertilized egg that develops into a new human being. In each generation, then, there is first a halving of chromosome number (when egg and sperm cells are produced), followed by a coming together of these two halves (when sperm and egg unite). In this chapter, you'll learn about this process and its relation to the fantastic variety we see in the living world.

Some Helpful Terms

The kind of cell division that results in the halving of chromosome number is called *meiosis*. This stands in contrast to mitosis, which you looked at in Chapter 9. The cells that reproduce through mitosis are known as **somatic cells**. And which cells are these? In animals, all the cells in the organism, *except* for the reproductive cells—the eggs and sperm—which are known as **gametes**.

Egg and sperm are said to be in the haploid state, the term *haploid* meaning "single number." When egg and sperm unite, however, it marks a return to the *diploid*, or "double number" state of cellular existence. By definition, **haploid** cells possess a single set of chromosomes while **diploid** cells possess two sets of chromosomes. In human beings, haploid cells have 23 chromosomes, while diploid cells have 46 chromosomes. **Meiosis** can be defined as a process in which a single diploid cell divides to produce haploid reproductive cells. Diploid cells are also sometimes referred to as **2n** cells (the "2" here standing for a doubled number of chromosomes), while haploid cells are said to be **1n**. In what follows, we will be looking at meiosis as it occurs in human beings.

10.2 The Steps in Meiosis

How does the chromosomal halving take place? Let's go over the process of meiosis and see. FIGURE 10.1 shows meiosis, in a stripped-down form, as compared to the mitosis described in Chapter 9. Two essential differences between the two processes can be seen in the figure. First, you may remember that the formula for mitosis was: duplicate once, divide once. Duplicate the chromosomes once, then divide the original cell once. With meiosis, on the other hand, the formula is: duplicate once, divide *twice*. Meiosis includes one chromosome duplication followed

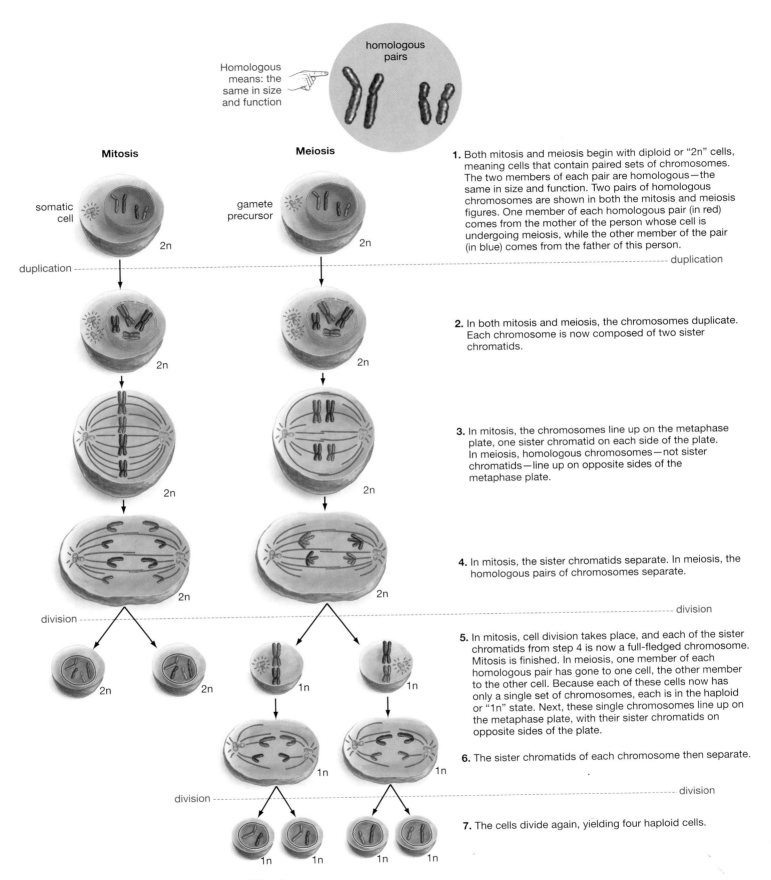

homologous pairs

Homologous means: the same in size and function

Mitosis

Meiosis

somatic cell

gamete precursor

2n

2n

duplication

duplication

2n

2n

2n

2n

2n

2n

division

division

2n

2n

1n

1n

1n

1n

1n

1n

division

division

1n 1n

1n 1n

1. Both mitosis and meiosis begin with diploid or "2n" cells, meaning cells that contain paired sets of chromosomes. The two members of each pair are homologous—the same in size and function. Two pairs of homologous chromosomes are shown in both the mitosis and meiosis figures. One member of each homologous pair (in red) comes from the mother of the person whose cell is undergoing meiosis, while the other member of the pair (in blue) comes from the father of this person.

2. In both mitosis and meiosis, the chromosomes duplicate. Each chromosome is now composed of two sister chromatids.

3. In mitosis, the chromosomes line up on the metaphase plate, one sister chromatid on each side of the plate. In meiosis, homologous chromosomes—not sister chromatids—line up on opposite sides of the metaphase plate.

4. In mitosis, the sister chromatids separate. In meiosis, the homologous pairs of chromosomes separate.

5. In mitosis, cell division takes place, and each of the sister chromatids from step 4 is now a full-fledged chromosome. Mitosis is finished. In meiosis, one member of each homologous pair has gone to one cell, the other member to the other cell. Because each of these cells now has only a single set of chromosomes, each is in the haploid or "1n" state. Next, these single chromosomes line up on the metaphase plate, with their sister chromatids on opposite sides of the plate.

6. The sister chromatids of each chromosome then separate.

7. The cells divide again, yielding four haploid cells.

FIGURE 10.1 **Meiosis Compared to Mitosis**

by two cellular divisions, which means the process produces four cells (instead of the two that come about in mitosis). The other big difference you can see is that the first of the meiotic divisions is a separation of homologous chromo*somes*, not the separation of chrom*atids* that you saw in mitosis. The subsequent meiotic division is, however, just like mitotic division: It separates the chromatids that make up the homologous chromosomes.

But what is this term *homologous*? You may remember from Chapter 9 that homologous chromosomes are chromosomes that are the same in size and function. These chromosomes do similar things, which is to say they have genes on them that code for similar proteins. We have a chromosome 1 that we inherit from our mother, and it is homologous to the chromosome 1 we inherit from our father. Likewise we inherit a chromosome 2 from our mother that is homologous with chromosome 2 from our father, and so forth for 23 pairs. The lone exception to this pairs rule is that the X and Y chromosomes males have are not homologous with one another.

The steps of meiosis are separated into two multistep stages, called meiosis I and meiosis II. The big picture regarding these stages is that, in meiosis I, homologous chromosomes are positioned close together, on opposite sides of the cellular equator called the metaphase plate. Then these chromosomes move apart into different daughter cells. In meiosis II, the chrom*atids* of these now-separated chromosomes separate into different daughter cells. Let's walk through the steps of meiosis now, as illustrated in FIGURE 10.2 on page 160.

Meiosis I

Prophase I Meiosis I begins with the same appearance of 46 identifiable chromosomes that you observed in mitosis. The first big difference between meiosis and mitosis appears early on, however. It is that, in meiosis, *homologous chromosomes pair up*. In mitosis, 46 individual chromosomes became visible, but did not pair up in any way. In meiosis, however, each pair of homologous chromosomes links up. Maternal chromosome 5, for example, intertwines with *paternal* chromosome 5, maternal chromosome 6 with paternal 6, and so on. (See part (b) in Figure 10.2.)

A critical bit of part-swapping then takes place between the non-sister chromatids of the paired chromosomes. This process is called *crossing over* (or *recombination*), and you'll be looking at it in detail later. Once this crossing over has finished, the homologous chromosomes begin to unwind from one another, though they remain overlapped.

Metaphase I Still paired up, the homologous chromosomes, attached to microtubules, are moved to the metaphase plate; in this step of meiosis, the maternal member of a given pair lies on one side of the plate and the paternal member on the other. A critical point about this, however, is that the alignment adopted by any one pair of chromosomes bears no relation to the alignment adopted by any other pair. It may be, for example, that paternal chromosome 5 will line up on what we might call side A of the metaphase plate; if so, then maternal chromosome 5 ends up on side B. Shift to chromosome 6, however, and things could just as easily be reversed: The *maternal* chromosome might end up on side A, and the paternal on side B. It is thus a throw of the dice as to which side of the plate a given chromosome lines up on. As you will see, this randomness is critical in the shuffling of genetic material.

Anaphase I The paired, homologous chromosomes now begin to move away from each other, toward their respective poles, pulled through the disassembly of the microtubule spindles they are attached to. Though homologous chromosomes have now separated, each chromosome is still made up of a joined pair of sister chromatids.

A BROADER VIEW

Two Phases of Meiosis

Look at what a complicated process we sexual beings go through in producing offspring, compared to the humble bacterium, which simply doubles its single chromosome and then divides into two cells. One challenge for science has been to figure out why sexual reproduction got going in the first place. Compare two manufacturing operations that turn out the same product, except that one makes the product with a single factory (think: bacteria) while the other requires two factories (think: males and females). Which is going to manufacture the product at a lower cost? The single-factory operation. So how did sexual reproduction manage to break into the business of life, given its higher costs? Several ideas have been put forward, but none has won universal acceptance.

Telophase I With chromosome movement toward the poles completed, the original cell now undergoes cytokinesis, dividing into two completely separate daughter cells. With this, there are two haploid cells, whereas in the beginning there was one diploid cell. (Why are the cells now haploid? Meiosis I moved one set of chromosomes to one cell and the second set to another cell. Therefore, each cell now has only one set of chromosomes.)

Meiosis II

What happens now is another division—or set of divisions, because there are now two cells. Each of these cells now has 23 duplicated chromosomes in it. These chromosomes now line up at a new metaphase plate, with sister chromatids on opposite sides of each plate. What happens next is exactly the same thing that happened in mitosis. The *sister chromatids* now separate, moving toward opposite poles; with this, they assume the role of full-fledged chromosomes. Once this separation is completed, cytokinesis occurs. Where once there were two cells, there are now four. The difference between this process and mitosis is that each of these cells has 23 chromosomes in it instead of 46.

So Far...

1. In meiosis, the separation of genetic material that comes with the first cellular division is a separation of _____.

2. The number of cells that meiosis starts with is _____, while the number of cells that result from the process is _____.

3. The egg and sperm that are produced in humans by meiosis contain _____ chromosomes each, while human somatic cells each contain _____ chromosomes.

10.3 The Significance of Meiosis

You have now examined the mechanics of meiosis. But what are the effects of this process? First of all, the "Fennington-Kent" problem has been solved. There will be no 92-chromosome zygotes, because egg and sperm will each bring only 23 chromosomes to their meeting, thus yielding a combined 46 chromosomes in human somatic cells. Just as notable, however, are two kinds of diversity that meiosis brings about. The first of these these is genetic diversity in offspring; the second is a large-scale diversity in the natural world.

Genetic Diversity through Crossing Over

Why are siblings visibly different from each other? Indeed, why don't children look just like their parents? The visible diversity we can see in children is based on a *genetic* diversity that comes about through two processes that occur during meiosis. The first of these processes is crossing over.

Recall that, back when meiosis first gets started, homologous chromosome pairs intertwine for a time and then engage in some part-swapping. This is crossing over (or recombination). Look at part (b), "Crossing over," in Figure 10.2 to see how this important process works. There is first a physical breaking of non-sister chromatids and then a "reunion" of these chromosome sections onto new chromatid partners. With this, what had been a "maternal" chromosome now has a portion of the paternal chromosome within it, and vice versa. Formally, then, **crossing over** can be defined as a process, occurring during meiosis, in which homologous chromosomes exchange reciprocal portions of themselves.

So Far... Answers:

1. *homologous chromosomes*

2. *1; 4*

3. *23; 46*

The linkage between crossing over and genetic diversity probably is apparent. The ability of reciprocal lengths of DNA to be exchanged between chromosomes provides a means by which the genetic deck can be reshuffled prior to the formation of eggs and sperm. Maternal and paternal chromosomes don't get passed on intact from one generation to the next; they exchange parts with each other. This

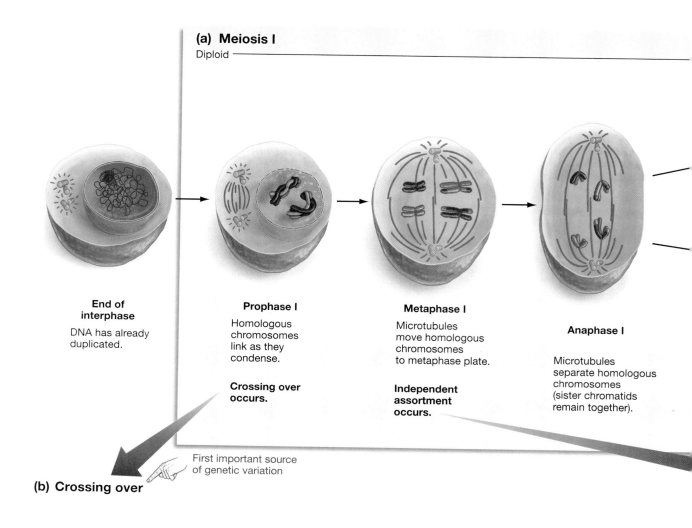

(a) Meiosis I

Diploid

End of interphase

DNA has already duplicated.

Prophase I

Homologous chromosomes link as they condense.

Crossing over occurs.

Metaphase I

Microtubules move homologous chromosomes to metaphase plate.

Independent assortment occurs.

Anaphase I

Microtubules separate homologous chromosomes (sister chromatids remain together).

First important source of genetic variation

(b) Crossing over

Exchange of parts of non-sister chromatids.

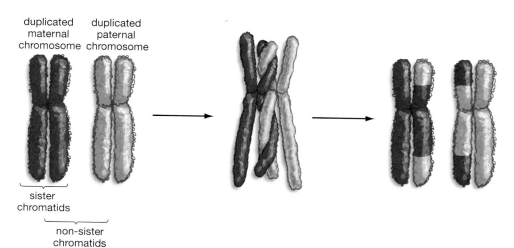

duplicated maternal chromosome

duplicated paternal chromosome

sister chromatids

non-sister chromatids

FIGURE 10.2   Meiosis

(a) The Steps of Meiosis

(b) Crossing Over

(c) Independent Assortment

reshuffling has practical consequences because of what you saw in Chapter 9 about homologous chromosomes: While the genetic information on any two will be similar, it will not be identical. A gene on one chromosome may code for red hair color, but the gene on its homologous chromosome may code for blonde hair color. In crossing over, such genetic variants are being swapped between chromosomes.

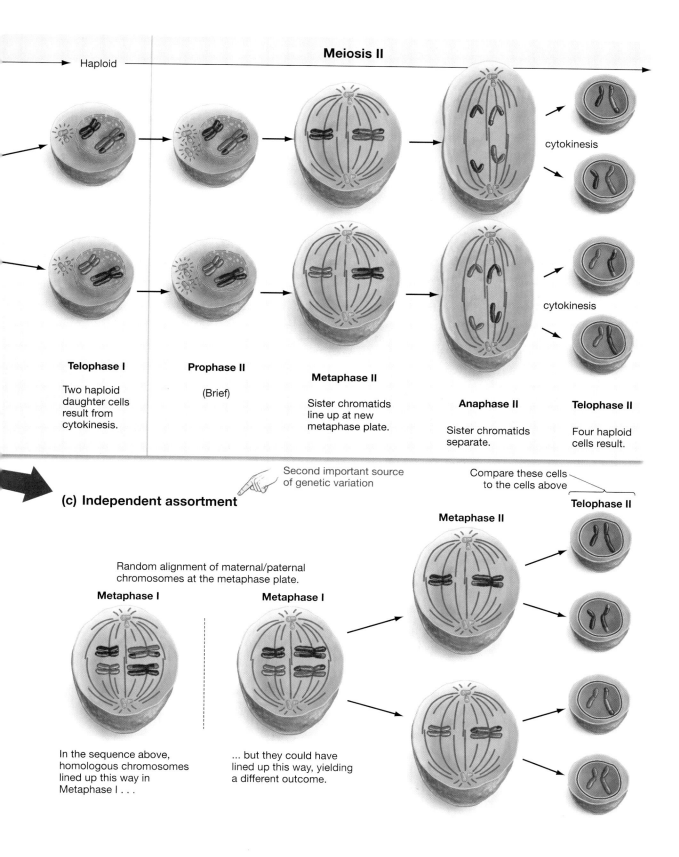

Meiosis II

Haploid

Telophase I

Two haploid daughter cells result from cytokinesis.

Prophase II

(Brief)

Metaphase II

Sister chromatids line up at new metaphase plate.

cytokinesis

cytokinesis

Anaphase II

Sister chromatids separate.

Telophase II

Four haploid cells result.

Second important source of genetic variation

(c) Independent assortment

Compare these cells to the cells above

Telophase II

Metaphase II

Random alignment of maternal/paternal chromosomes at the metaphase plate.

Metaphase I

Metaphase I

In the sequence above, homologous chromosomes lined up this way in Metaphase I . . .

. . . but they could have lined up this way, yielding a different outcome.

Genetic Diversity through Independent Assortment

Another way that meiosis ensures genetic diversity is a crucial event that takes place right after crossing over. It is the random alignment or *independent assortment* of chromosomes at the metaphase plate. See part (c), "Independent assortment," in Figure 10.2. Recall that the alignment of any one pair of homologous chromosomes along the plate bears no relation to the alignment of any other homologous pair. If you looked at, say, chromosome 5, the paternal member of the pair might line up on "side A" of the metaphase plate, meaning maternal chromosome 5 will be on side B. Shift to chromosome 6, however, and things could just as easily be reversed. This random alignment of chromosomes means that maternal and paternal chromosomes will randomly end up in separate gametes. Why? Because a chromosome that lines up on *side* A of the plate will end up in *cell* A, while a chromosome that lines up on side B will end up in cell B. **Independent assortment** can thus be defined as the random distribution of homologous chromosome pairs during meiosis.

If we wonder how traits from a mother and father can get "mixed up" in their children, independent assortment joins crossing over in providing an answer. Indeed, for complex creatures such as ourselves, independent assortment assures that, with the exception of identical twins, offspring produced through sexual reproduction will not just be diverse. Each offspring will be *unique*—no other offspring will be exactly like it in genetic terms. The sheer number of ways in which chromosomes can line up in independent assortment makes this a mathematical certainty. Think about it: we have 2 locations for any chromosome (side A or side B of the metaphase plate), and we have 23 chromosomes. This means there are 2^{23} different ways that chromosomes could line up in meiosis, which is to say *8 million* different ways. The egg from your mother that produced you may have been, say, variation 1,527,000 in this range, while the egg that produced your brother or sister may have been variation 4,573,000.

When we add independent assortment to crossing over, it's easy to see why separate children from the same parents can come out looking so different—from each other as well as from the parents themselves (**see** FIGURE 10.3). Overall, meiosis generates genetic *diversity*, while its counterpart, mitosis, does not. Mitosis makes genetically exact copies of cells; it *retains* the qualities that cells have from one generation to the next. Meiosis mixes genetic elements each time it produces reproductive cells and thus brings about genetic variation in succeeding generations.

A BROADER VIEW

Generating Diversity

Notice the balance that evolution has struck between cellular "sameness" and diversity in the different functions of life. Outside of reproduction, sameness is the rule, in mitosis. In our own bodies, mitosis produces millions of daughter cells each second that are nearly identical to their parent cells. When it comes time to create reproductive cells, however, diversity is the order of the day, through meiosis.

FIGURE 10.3 **Ensuring Variety** The shuffling of genetic material that occurs during meiosis is the primary reason that children look different from their parents, and from each other.

From Genetic Diversity, a Visible Diversity in the Living World

This genetic variation has an importance way beyond that of making siblings look different from each other. It turns out that this diversity is responsible, in significant part, for the fantastically diverse natural world around us. Think of it: two-foot ferns, 100-yard redwood trees, bats, bees, whales, mushrooms. How did the natural world come to be such a remarkably diverse place? The short answer is that evolution is spurred on by the differences in offspring that meiosis brings about.

To see why, just think about meiosis producing, in one generation after the next, *this* tree that grows a little taller than others around it, or *this* fish that dives a little deeper. How does meiosis bring about such individuals? Through the same means it brings about, say, a boy who has darker hair than any of his siblings. In the parental generation, crossing over and independent assortment do their genetic shuffling, and the eggs and sperm that result carry within them millions of variants on the parents' genetic makeup. One of these variants happens to produce a boy with darker hair than that of his siblings. In the same way, meiotic shuffling can produce one tree that is slightly taller than all the others around it, or one fish that can dive deeper than others in its population.

Now, if you think about it, some traits can result in an organism producing more *offspring* than the others in its area. To use trees as an example, a taller tree catches more sunlight, performs more photosynthesis as a result, and is thus so well nourished that it may give rise to more trees than others around it. As a consequence, in the next generation of these trees, there are relatively more trees that are a bit taller. This population of trees has thus evolved in the direction of being taller. The trait of greater height was *selected*, in a sense, to be passed on with more frequency than the trait of lesser height. But how did it happen that nature had a range of heights to select *from*? Because meiosis produced a range of heights in this generation of trees. Meiosis assures variations in offspring, and variations are the basis for life evolving down different pathways. How did the natural world come to be such a diverse place? In part through the very small-scale process you've been looking at, meiosis.

10.4 Meiosis and Sex Determination

Let us think now about what meiosis means in relation to determining the sex of offspring. How do humans, in particular, come to be male or female? In humans there is one exception to the rule that chromosomes come in homologous pairs. Human females do indeed have 23 pairs of matched chromosomes, including 22 pairs of "autosomes" and one matched pair of **sex chromosomes**, meaning the chromosomes that determine what sex an individual will be. The sex chromosomes in females are called X chromosomes, and each female possesses two of them. In males, conversely, there are 22 autosomes, one X sex chromosome, and then one Y sex chromosome. It is this Y chromosome that confers the male sex. (See **FIGURE 10.4** for the differences in the X and Y sizes.)

In meiosis I in a female, the female's two X chromosomes, being homologous, line up together at the metaphase plate. Then these chromosomes separate, each

A BROADER VIEW

Becoming a Male

So what is it about the Y chromosome that channels a fertilized egg into becoming a male? The critical element is a gene the Y has, called *SRY*. It switches on other genes that turn the embryonic gonad, which is genderless at that point, into male testes. The testes in turn secrete hormones (including testosterone) that bring about all the other male characteristics. *SRY* is short for sex-determining region Y.

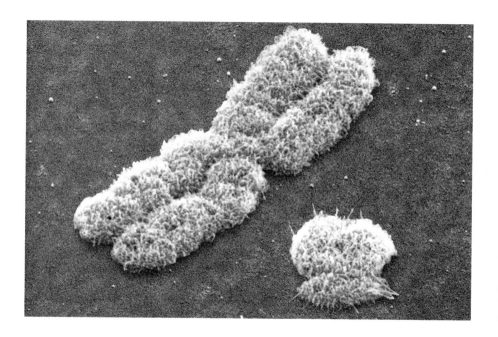

FIGURE 10.4 The X and the Y The human X chromosome can be seen on the left, while the human Y chromosome is on the right. The difference in size between them is in part a reflection of the difference in the number of genes each of them contains. The Y chromosome only has 78, while the X has about 1,500. (Magnified × 10,000)

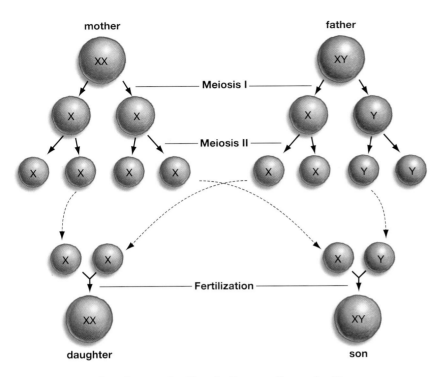

mother

father

Meiosis I

Meiosis II

Fertilization

daughter

son

1. Early in meiosis I, the mother's two X chromosomes line up at the metaphase plate (XX). Meanwhile, in the father's meiosis, his X and Y chromosomes line up (XY).

2. The X-X and X-Y pairs then separate into different cells.

3. The chromatids that made up the duplicated chromosomes separate, yielding individual eggs and sperm.

4. Should an X-bearing sperm from the male reach the egg first, the child will be a girl; should the male's Y-bearing sperm reach the egg first, the child will be a boy.

FIGURE 10.5   **Sex Determination in Human Reproduction**

of them going to different cells (**see** FIGURE 10.5). In males, the non-homologous X and Y chromosomes line up as if they were homologues. Then one resulting cell gets an X chromosome while the other gets the Y chromosome.

Sex determination is then simple. Each of the eggs produced by a female bears a single X chromosome. If this egg is fertilized by a *sperm* bearing an X chromosome, the resulting child will be a female. If, however, the egg is fertilized by a sperm bearing a Y chromosome, the child will be a male.

Because females don't have Y chromosomes to pass on, it follows that the Y chromosome that any male has had to come from his father. Meanwhile, the single X chromosome that any male has had to come from his mother. Females, conversely, carry one X chromosome from their mother and one from their father.

On to Patterns of Inheritance

The tour of cell division over the last couple of chapters has yielded not only a look at how one cell becomes two, but at how chromosomes operate within this process. Given the paired nature of chromosomes and their precisely ordered activity during meiosis, it stands to reason that there would be some predictability or pattern to the passing on of traits. This is the case, as it turns out. And the person who first recognized this pattern was a monk who worked in nineteenth-century Europe in an obscurity he did not deserve.

So Far...

So Far... Answers:

1. *homologous chromosomes; reciprocal portions of themselves*
2. *genetic diversity*
3. *False; half contain a Y chromosome, half contain an X chromosome.*

1. Crossing over is a process occurring during meiosis in which _____ exchange _____.

2. Crossing over and independent assortment assure _____ among offspring.

3. True or false: all the sperm produced by human males contain a Y chromosome.

Chapter Review

Summary

10.1 An Overview of Meiosis

- **Diploid and haploid**—The cells that take part in sexual reproduction are special in that they contain only a single set of chromosomes, as opposed to the paired set of chromosomes that most regular or "somatic" cells possess. Cells that have a paired set of chromosomes are diploid, while cells that have a single set of chromosomes are haploid. Meiosis is the process by which a single diploid cell divides to produce four haploid cells.   (page 155)

- **Gametes**—The haploid cells produced through meiosis are called gametes. Female gametes are eggs; male gametes are sperm. They are the reproductive cells of humans and many other organisms.   (page 156)

10.2 The Steps in Meiosis

- **Two phases of meiosis**—In meiosis, there is one round of chromosome duplication, followed by two rounds of cell division. There are two primary stages to meiosis, meiosis I and meiosis II. In meiosis I, chromosome duplication is followed by a pairing of homologous chromosomes with one another, during which time they exchange reciprocal sections of themselves. Homologous chromosome pairs then line up at the metaphase plate—one member of each pair on one side of the plate, the other member on the other side. These homologous pairs then separate in the first round of cell division, into different daughter cells. In meiosis II, the chromatids of the duplicated chromosomes separate into different daughter cells.   (page 156)

Web Tutorial 10.1 Meiosis

10.3 The Significance of Meiosis

- **Generating diversity**—Meiosis generates genetic diversity by ensuring that the gametes it gives rise to will differ from one another. In this, it is unlike regular cell division, or mitosis, which produces daughter cells that are exact genetic copies of parent cells.   (page 159)

- **Diversity in two ways**—Meiosis generates genetic diversity in two ways. First, homologous chromosomes pair with each other and, in the process called crossing over or recombination, exchange reciprocal segments with one another. Second, following crossing over, there is a random alignment or independent assortment of maternal and paternal chromosomes on either side of the metaphase plate. This chance alignment determines which daughter cell each chromosome will end up in.   (page 159)

- **Diversity in the living world**—The genetic diversity brought about by meiosis is responsible, to a significant extent, for the great diversity of life-forms seen in the living world today. Evolution is spurred on by differences among offspring, and meiosis and sexual reproduction ensure such differences.   (page 162)

10.4 Meiosis and Sex Determination

- **Chromosomes in females and males**—Human females have 23 matched pairs of chromosomes—22 autosomes and two X chromosomes. Human males have 22 autosomes, one X chromosome, and one Y chromosome. Each egg that a female produces has a single X chromosome in it. Each sperm that a male produces has either an X or a Y chromosome within it. If a sperm with a Y chromosome fertilizes an egg, the offspring will be male. If a sperm with an X chromosome fertilizes the egg, the offspring will be female.   (page 163)

Web Tutorial 10.2 Meiosis and Sex Outcome

Key Terms

1n   p. 156
2n   p. 156
crossing over   p. 159

diploid   p. 156
gamete   p. 156

haploid   p. 156
independent assortment   p. 162

meiosis   p. 156
sex chromosome   p. 163

Testing Your Understanding

In Your Own Words   (answers in the back of the book)

1. In what two ways does meiosis ensure genetic diversity in offspring?

2. Define and distinguish between somatic cells and gametes.

3. Describe the role that chromosomes play in sex determination in human beings.

Thinking about What You've Learned

1. Mammals have now been cloned by scientists (think of Dolly the sheep). Imagine a science-fiction world in which half the world's human births were achieved through cloning. How would our world be different?

2. Bacteria reproduce through a simple duplication of their single chromosome and then a splitting of the original parent bacterial cell into two daughter cells. Forms of asexual reproduction such as this become less and less common, however, with movement up the complexity scale among life-forms. Why should it be that complexity and sexual reproduction seem to go hand in hand?

3. All the bananas we eat come from trees that are produced through "cuttings"—a stem from an existing tree is planted in the ground, resulting in a new tree. Thus, each tree is a clone of another. Why would growers find it advantageous to produce an enormous series of identical clones, rather than using trees that reproduce sexually?

Mendel and His Discoveries

Amish children riding a buggy in eastern Pennsylvania.

THE ADVOCATE

BABIES' DEATHS IN AMISH AREA TIED TO SIDS GENE

By Susan Fitzgerald
July 20, 2004

PHILADELPHIA—The deaths of Amish babies in central Pennsylvania were a mystery.

For two generations, 21 infants from nine families had died unexpectedly, inexplicably.

Researchers announced Monday they had found the reason: a gene that causes a form of sudden infant death syndrome that may help explain some SIDS cases in the general population.

ST. PAUL PIONEER PRESS

FROM BOTTLE BLONDE TO HONEST BRUNETTE

By Laura Billings
October 6, 2002

London's *Daily Star* called it a "blonde-shell study": a new World Health Organization report claiming that blondes were now an endangered species, eventual victims of a gene so rare and recessive that in 200 years' time, there won't be a single one left on the face of the planet.

11.1 Mendel and the Black Box

Genes have not only a real power over our lives, but an imagined power as well. If you look at the *Advocate* newspaper story, above, you can see the real power of genes: Over the course of two generations, 21 children from nine central-Pennsylvania Amish families died from a disease, akin to sudden infant death syndrome (SIDS), that had a faulty gene as its primary cause. Moreover, the gene in question was said by experts

FIGURE 11.1 **Austrian Monk and Naturalist Gregor Mendel**

to be a "recessive" one that, in its normal form, helps regulate how children develop before they are born. When it comes to the imagined power of genes, look at the *Pioneer Press* story for what people around the world were willing to believe: that a gene for blonde hair is "so recessive" that natural blondes will eventually disappear from the planet. Though this story turned out to be based on a hoax, note what it shares with the all-too-real story out of Pennsylvania: Both are about not just genes, but genes that are recessive. But what are recessive genes? To judge from the story about blondes, you'd think that recessive genes can die out, like members of an endangered species, when confronted with dominant genes. But do things actually work this way? Could blondes in fact cease to exist? On the other hand, if recessive genes have so little power, how did they cause the SIDS-like disease in the Amish children?

As it happens, the answers to these questions are based on concepts that were first discovered by someone who had no knowledge of genes as that term is understood today. Indeed, Gregor Johann Mendel (FIGURE 11.1) did his research long before any of the elements of genetics we've talked about had been discovered. DNA, chromosomes, mitosis, meiosis: None of these things were known to scientists of the mid-nineteenth century, when Mendel did his work. And Mendel's scientific instruments were tweezers and an artist's paintbrush, which he used on a species of common pea, *Pisum sativum*. Yet this unassuming monk, the son of eastern European peasant farmers, generally is accorded the title of the father of genetics. What contribution earned him this honor?

Mendel's achievement was to comprehend what was going on inside what might be called the "black box" of genetics without ever being able to look inside that box himself. Science is filled with so-called black-box problems, in which researchers know what goes *into* a given process and what comes *out*. It is what is going on in between—in the black box—that is a mystery. In the case of genetics, what lies inside the black box is DNA and chromosomes and meiosis, and so forth—all the component parts of genetics, in other words, and the way they work together. Due to the timing of his birth, Mendel had no knowledge of any of these things. Until Mendel, however, nobody had looked carefully at even the starting and ending points of this black-box problem: at what went in and what came out. Mendel's original pea plants represented the "input" side to the black box of genetics, while the offspring he got from breeding these plants represented the output. By looking carefully at generations of parents and offspring—at both sides of the box, in a sense—Mendel was able to infer something about what had to be going on within. As you will see, his main inferences were correct: (1) that the basic units of genetics are material elements; (2) that these elements come in pairs; (3) that these elements (today called genes) can retain their character through many generations; and (4) that gene pairs *separate* during the formation of gametes.

These insights may sound familiar, because all of them were approached from another direction in Chapters 9 and 10. Here are the lessons you went over then:

1. Genes are material elements—lengths of DNA.

2. In human beings, genes come in pairs, residing in pairs of homologous chromosomes.

3. Chromosomes make copies of themselves, thus giving the genes that lie along them the ability to be passed on intact through generations.

4. In meiosis, homologous chromosomes line up next to each other and then *separate*, with each member of a pair ending up in a different egg or sperm cell.

By observing generations of pea plants and applying mathematics to his observations, Mendel inferred that something like this had to be happening in reproduction. Because Mendel was the first to perceive a set of principles that govern inheritance, we date our knowledge of genetics from him. In this chapter, we'll go over the basics of what Mendel found. Once we're done, you'll understand a great deal more about genetics—about what *dominant* and *recessive* mean, about the nature of gene pairs, and about how traits are passed on by living things down through generations.

11.2 The Experimental Subjects: *Pisum sativum*

Beginning work in a monastery in what is now the Czech Republic, Mendel managed to pick, in *Pisum sativum*, a nearly perfect species on which to carry out his experiments. If you look at FIGURE 11.2, you can see something of the life cycle of this garden pea. Note that what we think of as peas in a pod are seeds in this plant's ovary. Each of these seeds begins as an unfertilized egg, just as a human baby begins as a maternal egg that is unfertilized. Sperm-bearing pollen, landing on the plant's stigma, then set in motion the fertilization of the eggs.

Importantly, *Pisum* plants can *self*-pollinate. The anthers of a given flower release pollen grains that land on that flower's stigma. Each seed that develops from the resulting fertilizations can then be planted in the ground and give rise to a new generation of plant. The way to think of the seeds in a pod is as multiple offspring from separate fertilizations—one pollen grain fertilizes this seed, another pollen grain fertilizes another. This is why, as you'll see, seeds that are in the same pod can have different characteristics.

A BROADER VIEW

Mendel and His Insights

The physicist Richard Feynman once said that being a scientist is like being someone who is allowed to watch a complicated game as it's being played, and who is then asked to state what the rules of the game are. Gregor Mendel worked in just this way; he watched generations of pea plants pass along their traits and then offered a set of insights about the rules by which this process operated.

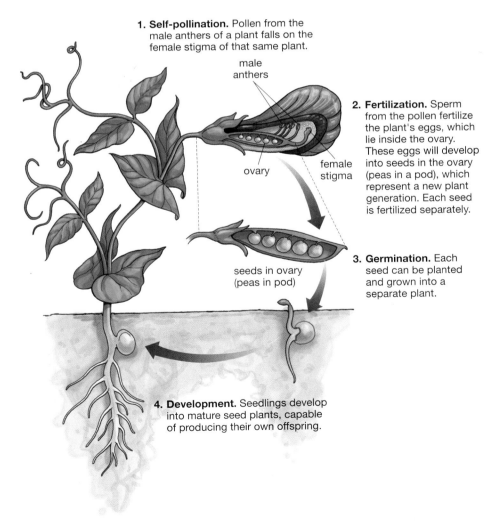

1. **Self-pollination.** Pollen from the male anthers of a plant falls on the female stigma of that same plant.

male anthers

ovary

female stigma

2. **Fertilization.** Sperm from the pollen fertilize the plant's eggs, which lie inside the ovary. These eggs will develop into seeds in the ovary (peas in a pod), which represent a new plant generation. Each seed is fertilized separately.

seeds in ovary (peas in pod)

3. **Germination.** Each seed can be planted and grown into a separate plant.

4. **Development.** Seedlings develop into mature seed plants, capable of producing their own offspring.

FIGURE 11.2 Life Cycle of the Pea Plant

How to cross-pollinate pea plants

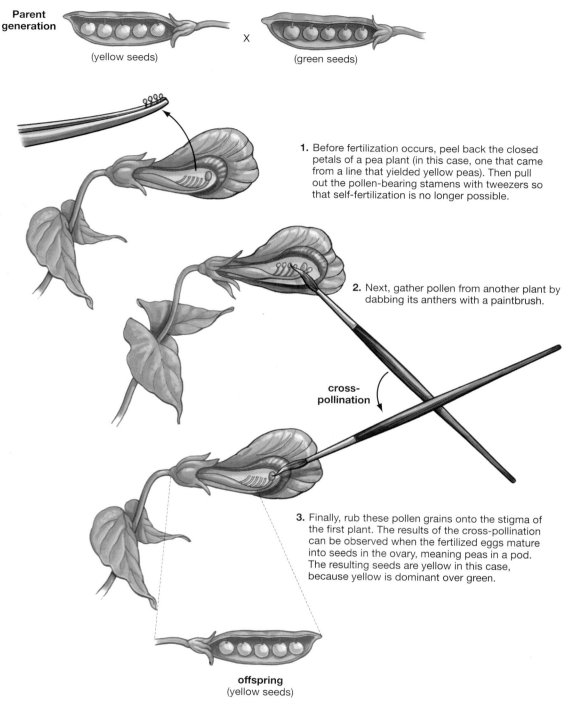

Parent generation

(yellow seeds) X (green seeds)

1. Before fertilization occurs, peel back the closed petals of a pea plant (in this case, one that came from a line that yielded yellow peas). Then pull out the pollen-bearing stamens with tweezers so that self-fertilization is no longer possible.

2. Next, gather pollen from another plant by dabbing its anthers with a paintbrush.

cross-pollination

3. Finally, rub these pollen grains onto the stigma of the first plant. The results of the cross-pollination can be observed when the fertilized eggs mature into seeds in the ovary, meaning peas in a pod. The resulting seeds are yellow in this case, because yellow is dominant over green.

offspring
(yellow seeds)

FIGURE 11.3   How to Cross-Pollinate Pea Plants

Though the pea plants can self-pollinate, Mendel could also **cross-pollinate** the plants at will—he could have one plant pollinate another—by going to work with his tweezers and paintbrushes, as shown in FIGURE 11.3.

In directing pollination, Mendel could control for certain attributes in his pea plants—qualities now referred to as *characters*. If you look at **Table 11.1**, over on the right, you can see that he was looking at seven characters in all—such things as stem length, seed color, and seed shape. Note that each of these characters comes in two varieties. There are yellow or green seeds, for example, and purple or white flowers. Such character variations are known as *traits*. Mendel referred to each of these variations as being either a "dominant" or a "recessive" trait, as the table shows. For now, you can think of a recessive trait as one that tends to remain hidden in certain generations of pea plants. A dominant trait, meanwhile, can be thought of as a trait that tends to appear in those same generations. You'll get a formal definition of dominant and recessive later.

Phenotype and Genotype

Now let's put these traits in another context. Taken together, traits represent what are called **phenotypes**. Broadly speaking, a phenotype is a physiological feature, bodily characteristic, or behavior of an organism. In the context of Mendel, phenotype means the pea plant's visible physical characteristics. Purple flowers are one phenotype, white flowers another; yellow seeds are one phenotype, green seeds another. Meanwhile, any phenotype is in significant part determined by an organism's underlying **genotype**, meaning its genetic makeup.

11.3 Starting the Experiments: Yellow and Green Peas

Mendel began by making sure that his starting plants "bred true" for the phenotypes under study. That is, he assured himself that, for example, all of his purple-flowered plants would produce nothing but generations of purple offspring if self-pollinated.

Parental, F_1, and F_2 Generations

The *parental generation* in such experiments is often referred to in shorthand as the *P* generation. Meanwhile, the offspring of the parental generation are known as the *first filial generation*, the word "filial" indicating a son or a daughter. The shorthand for first filial is F_1. The F_x form can be used for any succeeding generation. You will be seeing a lot of the second filial, or F_2 generation, for example.

So how did Mendel proceed with his plants? Let's start with just one example. Mendel took, for his P generation, some plants from a line that bred true for yellow seeds and other plants from a line that bred true for green seeds. Then he took pollen from one variety of the plants and fertilized the other variety, as seen in Figure 11.3. The plants fertilized in this way then produced their own seeds—every one of which was yellow. (Remember that although these seeds were developing within the pods of the parental generation, they represented the new generation—the F_1 generation.) Getting all yellow seeds was an interesting result

Table 11.1 Pea-Plant Characters Studied by Mendel

Characters studied	Dominant trait	Recessive trait
Seed shape	smooth	wrinkled
Seed color	yellow	green
Pod shape	inflated	wrinkled
Pod color	green	yellow
Flower color	purple	white
Flower position	on stem	at tip
Stem length	tall	dwarf

Some of these peas have a smooth texture, while others are wrinkled.

FIGURE 11.4 Variation within a Pea Pod Since each garden pea is fertilized separately, individual peas within a pod can have different character traits.

A BROADER VIEW

Genes and Alleles

From today's perspective it may seem strange to regard as an "insight" the statement that there are units of genetics (genes) that are material elements. You might say, what else could heredity be based on? But in the mid-nineteenth century, things weren't so clear. Only in the late eighteenth century did microscopes banish the notion that eggs and sperm contained fully formed, tiny individuals (who in turn contained even smaller individuals, the whole thing functioning like a series of Russian dolls). By Mendel's time, the few scientists who thought about the issue held to vague notions such as a "living formative force" that existed in men and women.

in itself. Because each seed in a pea-plant pod is fertilized separately, it *can* be the case that a given pod will contain both yellow and green seeds. (Other pea characters can be variable within a pod as well, as you can see in FIGURE 11.4.) Yet all the seeds in this generation were yellow, indicating that yellow was a dominant trait and green a recessive one.

Having viewed these results, Mendel then planted his F_1 seeds and let plants grow from them. These plants were then allowed to *self*-pollinate—sperm from a given plant fertilized the eggs from that same plant, so that instead of "crossing" one plant with another, Mendel was crossing a plant with itself. What he got from this self-pollination, in the F_2 generation, was 6,022 yellow seeds and 2,001 green seeds.

The Power of Counting

In counting the seeds, Mendel was taking a giant step forward. Why? Experimenters before Mendel had gotten the kind of results he had. What they did not do, however, was undertake a careful *counting* of their results and then analyze the results in terms of proportions.

Looking at the specific results, two things stand out. First, green seeds disappeared in F_1, but came back in F_2. Recall that there was not a green seed to be found in the F_1 generation pods, but in F_2, there are 2,001 of them. Second, green seeds came back in F_2 as *one-fourth* of the seeds as a whole. The F_2 generation isn't divided 50/50, half yellow and half green; instead, there are roughly 3 yellow seeds for every 1 green seed—or to put it another way, a 3:1 ratio of yellow to green seeds. As it turned out, Mendel got the same result with each of the seven characters he was studying, as you can see in **Table 11.2**. Note that the 3:1 ratio is a proportion of dominant to recessive traits—more yellow seeds than green, more purple flowers than white. It was *solely* the dominant traits that appeared in the F_1 generation, but in the F_2 generation the dominants are simply appearing in greater proportion than the recessives.

Interpreting the F_1 and F_2 Results

What did Mendel learn from these results?

No "Blending" in Inheritance

For one thing, Mendel saw that inheritance for his peas was not a matter of the "blending" of their characteristics. For example, the flower colors purple and white were retained as just that. No intermediate phenotypes—say, pink flowers—resulted from the cross in any generation. This finding ran contrary to the notion, popular in Mendel's time, that two given traits would blend into a homogenous third entity, as coffee and milk will blend together.

Dominant and Recessive Elements Come in Pairs

Beyond this finding, it was apparent that plants could retain the *potential* for recessive phenotypes, even though those phenotypes might not appear in a given generation. Mendel's F_1 plants had no green seeds, but in F_2 the green seeds were back. It was reasonable to assume, therefore, that the yellow-seed F_1 plants retained a green-seed *element*, which got expressed only in the F_2 generation. Finally, because his phenotypes came in pairs, it was reasonable for Mendel to hypothesize that the elements likewise came in pairs.

If you think Mendel's *pairs of elements* sound suspiciously like the "pairs of genes" on homologous chromosomes we've seen before, you are right. However, it's time to start referring to these "matched pairs of genes" in scientific terminology. It is more accurate to think of matched pairs of genes as alternative forms of a single gene. The proper name for an alternative form of a gene is an **allele**. Thus the pea plant had a single gene for seed color that came in two alleles—one of which coded for yellow seeds, the other of which coded for green seeds. These alleles resided on separate homologous chromosomes.

Table 11.2 Ratios of Dominant to recessive in Mendel's Plants

Dominant trait	Recessive trait	Ratio of dominant to recessive in F_2 generation
Smooth seed	Wrinkled seed	2.96:1 (5,474 smooth, 1,850 wrinkled)
Yellow seed	Green seed	3.01:1 (6,022 yellow, 2,001 green)
Inflated pod	Wrinkled pod	2.95:1 (882 inflated, 299 wrinkled)
Green pod	Yellow pod	2.82:1 (428 green, 152 yellow)
Purple flower	White flower	3.14:1 (705 purple, 224 white)
Flower on stem	Flower at tip	3.14:1 (651 along stem, 207 at tip)
Tall stem	Dwarf stem	2.84:1 (787 tall plants, 277 dwarfs)
	Average ratio, all traits:	**3:1**

So Far...

1. A physical characteristic, such as the height of a person or the color of one of Mendel's peas, represents a _____ in these organisms, which is in large part based on their genetic makeup or _____.

2. In Mendel's experiments, the first generation of plants that he got, from crossing parental generation plants, exhibited only _____ traits, while the next generation of plants, which resulted from self-pollination of F_1 plants, exhibited both _____ and _____ traits, in a 3:1 ratio.

3. Genes come in alternative forms known as _____, which reside on separate _____ chromosomes.

11.4 Mendel's Generations Illustrated

Let's make these points clearer by reviewing all three generations of experiments schematically.

The F_1 Generation

It will be helpful to introduce a convention here. Let us refer to dominant and recessive alleles by uppercase and lowercase letters, with uppercase used for dominant types and lowercase used for recessive types. Thus a "pure" yellow-seeded plant, having *two* yellow alleles, would be symbolized as YY. Meanwhile, pure green-seeded plants are yy, while mixed seeds are Yy. Let us say, just as an example, that female gametes are being supplied by a plant that is YY, while male gametes are coming from a plant that is yy. In meiosis, as you've seen, homologous chromosomes separate. This happens in the pea plants, meaning that each of these plants will contribute *one member* of its gene pair to its respective gametes—the YY female contributing a Y gamete, and the yy male contributing a y gamete. When these gametes fuse in the moment of fertilization, the result is a Yy hybrid in the F_1

So Far... Answers:

1. *phenotype; genotype*

2. *dominant; dominant; recessive*

3. *alleles; homologous*

generation. If you look at FIGURE 11.5, you will be introduced to a time-honored way to represent such outcomes, the Punnett square, and see Mendel's F_1 results symbolized as well.

Note that, because Y is dominant, all the seeds in the F_1 offspring pods will have a yellow *phenotype*, even though every one of these seeds contains a mixed *genotype*. (Every one is Yy.) It takes only one Y allele for a seed to be yellow; for a seed to be green, then, it must have two y alleles.

The F₂ Generation

Next come the F_1 crosses that yielded the F_2 generation. The starting point here is the F_1 seeds, which are all of the mixed Yy type, as you can see in FIGURE 11.6. Meiosis then occurs in the gamete precursors, which results in a separation of these alleles: Half the gametes now contain Y alleles and the other half y. You can see from the figure why this cross can now give us back the green-seed phenotype, on average in a 1:3 proportion with yellow seeds. FIGURE 11.7 illustrates how the three genotypes discussed here yield only the two phenotypes of yellow or green.

The Law of Segregation

For inheritance to work this way, Mendel saw something we noted earlier: That, though plant cells may contain *two* copies (alleles) of a gene relating to a given character, these copies must *separate* in gamete formation. How else could two Yy parents ever give rise to yy (or YY) progeny, unless the Yy elements could first separate (in the parents) and then recombine in different ways (in the offspring)? Thus did Mendel derive his insight, sometimes called Mendel's First Law or the **Law of Segregation**: Differing traits in organisms result from two genetic elements (alleles) that separate in gamete formation, such that each gamete gets only one of the two alleles. As noted, the physical basis for this law, which Mendel knew nothing of, is the separation of homologous chromosomes during meiosis.

Homozygous and Heterozygous Conditions

It's time to add a couple more terms to the concepts you have been considering. There is scientific terminology for the genotypically "pure" and "mixed" organisms noted earlier. An organism that has two identical alleles of a gene for a given character is said to be **homozygous** for that character (as with YY or yy). An organism that has differing alleles for a character is said to be **heterozygous** for that character (as with Yy). You often see homozygous used in combination with the terms dominant and recessive. For example, a yy plant is a *homozygous recessive*, while a YY is a *homozygous dominant* (**see** FIGURE 11.8). Knowledge of these terms puts you in a position to understand formal definitions of dominant and recessive. **Dominant** is a term used to designate an allele that is expressed in the heterozygous condition. In a heterozygous pea plant (Yy), the yellow allele (Y) is expressed, meaning

---

FIGURE 11.5 **Visualization of Crosses**

(a) Mendel's P-Generation Crosses

(b) Examples of Punnett Squares

(a) P-generation crosses

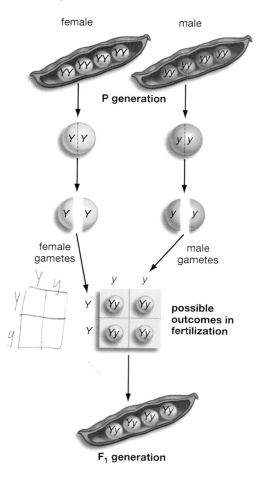

1. Female gametes are being provided by a plant that has the dominant, yellow alleles (YY); male gametes are being provided by a plant that has the recessive, green alleles (yy).

2. The cells of the pea plants that give rise to gametes start to go through meiosis.

3. The two alleles for pea color, which lie on separate homologous chromosomes, separate in meiosis, yielding gametes that each bear a single allele for seed color. In the female, each gamete bears a Y allele; in the male, each bears a y allele.

4. The Punnett square shows the possible combinations that can result when the male and female gametes come together in the moment of fertilization. (If you have trouble reading the Punnett square, see Figure 11.5b). The single possible outcome in this fertilization is a mixed genotype, Yy.

5. Because Y (yellow) is dominant over y (green), the result is that all the offspring in the F_1 generation are yellow, because they all contain a Y allele.

(b) How to read a Punnett square

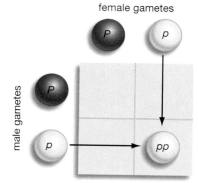

1. A p gamete from the male combines with a p gamete from the female to produce an offspring of pp genotype (and white color).

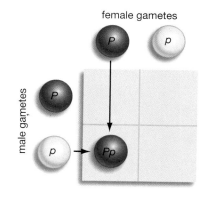

2. A p gamete from the male combines with a P gamete from the female to produce an offspring of Pp genotype (and purple color).

F₁ generation

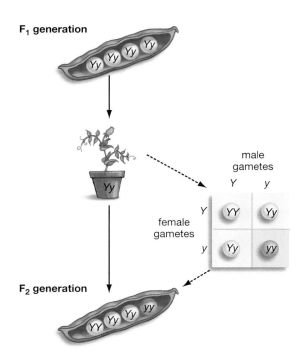

F₂ generation

From the F₁ to the F₂ generation

The starting point is the F₁ generation, a set of seeds that all have the *Yy* genotype. These seeds are planted and the plants go through meiosis, yielding the gametes shown around the Punnett square. When these gametes come together in self-fertilization, the possibilities include *YY* and *yy* combinations, as well as the *Yy* combination seen in the F₁ generation. The existence of *yy* individuals is the reason green seeds reappear in the F₂ generation. Because *Y* is dominant, the green phenotype could not appear in seeds that had even a single *Y* allele.

FIGURE 11.6 From the F₁ to the F₂ Generation

it is dominant over the green allele (*y*). **Recessive** is a term used to designate an allele that is not expressed in the heterozygous condition. The green allele (*y*) is recessive because it is not expressed when it exists heterozygously with the yellow allele (*Yy*).

Now, having acquired knowledge of what *recessive* means, you're in a position to understand how a recessive allele could have brought about the deaths, mentioned earlier, of the children in the Pennsylvania Amish community. All these children inherited *two* copies of the same faulty allele—one from their father and one from their mother. In the terms we've been discussing, each child was homozygous for this allele. Meanwhile, any child who had even a single "good" allele—any child who was at least heterozygous for it—was not susceptible to this SIDS-like death. Recessive alleles have their effects when they are paired in an individual.

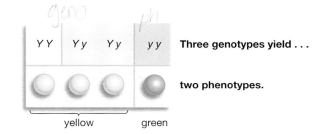

Three genotypes yield . . .

two phenotypes.

yellow   green

FIGURE 11.7 Three Genotypes Yield Two Phenotypes The two alleles for seed color (*Y* = yellow and *y* = green) can result in three genotypes (*YY, Yy, yy*), but these can yield only two phenotypes (yellow and green).

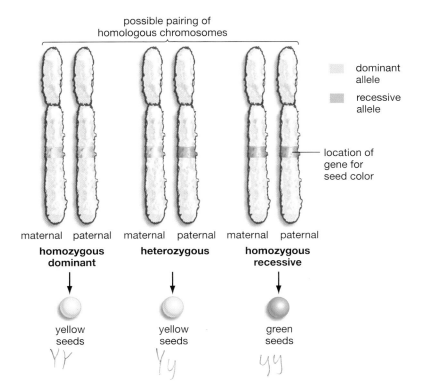

FIGURE 11.8 Chromosomes and Phenotypes The figure shows how alleles on chromosomes yielded the pea-color phenotypes that Mendel observed.

Modern Myths

Blonde Hair Will Disappear

As the newspaper story at the start of the chapter makes clear, people seem ready to believe that, because alleles for blonde hair are recessive to those for dark hair, blonde hair is fated to disappear from the human race over time. Contrast this belief, however, with two things that Mendel's research revealed about genes: They are material elements that *persist* from one generation to the next; and a recessive allele for a given gene will have effects when paired with another recessive allele in an individual. Thus, the recessive alleles for blonde hair don't get "wiped out" when they are paired with dominant alleles; they persist across generations. Second, these alleles retain their potential to have an effect, but recessive must be paired with recessive in an individual for this to happen. So, to the extent that men and women with blonde-haired alleles continue to come together to have children, blonde hair will continue to exist in human beings.

You'll be reading more about alleles and human health in Chapter 12. Meanwhile, if you look at the "Modern Myths" box, above, you can see what the concept of recessive alleles means in connection with the issue of blonde hair.

11.5 Crosses Involving Two Characters

Thus far, the subject has been how pea plants come to differ in *one* of their characters, seed color. When people breed organisms for a single difference such as this, looking to see how the offspring will come out, the procedure is known as a *monohybrid cross*. Mendel, however, went on to ask: What happens if you breed organisms for *two* characters? What happens, in other words, if you undertake what is known as a *dihybrid cross*? The question Mendel was seeking to answer in carrying out his dihybrid crosses was: Do given characters travel together in heredity, or do they travel separately? If you get one character passed on to a plant, do you necessarily get the other?

Crosses for Seed Color *and* Seed Shape

One dihybrid cross that Mendel performed involved, for its first character, the yellow and green seeds you've become so familiar with. In addition, this cross involved a second character of these seeds: their shape, which can be smooth or wrinkled (as you saw in Figure 11.4). It's clear that yellow color is dominant to green and, as Table 11.1 shows, smooth seed shape is dominant to wrinkled. We have been denoting the alleles for seed color as Y for yellow and y for green. In the same fashion, let us now denote seed-shape alleles as S for smooth and s for wrinkled.

You can see, in FIGURE 11.9a, what the phenotypic outcomes were for these crosses from the P to F_1 generations. In the F_1 generation, all the seeds were smooth and yellow. When the F_1s were self-pollinated, however, the F_2 phenotypes that resulted came out in the formidable numbers you can see in the figure (315 smooth yellow seeds; 108 smooth green seeds; 101 wrinkled yellow seeds; and 32 wrinkled green seeds). These numbers work out to a 9:3:3:1 ratio, meaning 9 parts smooth yellow; 3 parts smooth green; 3 parts wrinkled yellow; and 1 part wrinkled green.

A Hidden, Underlying Ratio For Mendel, these figures represented another opportunity to perceive something about how the black box of genetics operated. He realized early on that this ratio might be hiding an underlying reality for each of the *single* characters he was studying. Think for a second, as Mendel did, of only one of these characters, the color of the seeds. Here, the result is:

315 (smooth) yellow seeds	108 (smooth) green seeds
101 (wrinkled) yellow seeds	32 (wrinkled) green seeds
416 yellow seeds	140 green seeds

A BROADER VIEW

Alleles and First Cousins

Why are there laws in some states banning the marriage of first cousins? Because many genetically based illnesses are caused by recessive alleles, which is to say alleles that must exist in two copies in the same individual—one inherited from the father and one from the mother. Closely related individuals (i.e., first cousins) are more likely to have a copy of the same allele than are individuals chosen at random from the population. At root, then, prohibitions against first cousins marrying are attempts to keep harmful recessive alleles apart.

(a) Results of Mendel's dihybrid cross

(b) Why Mendel got these results

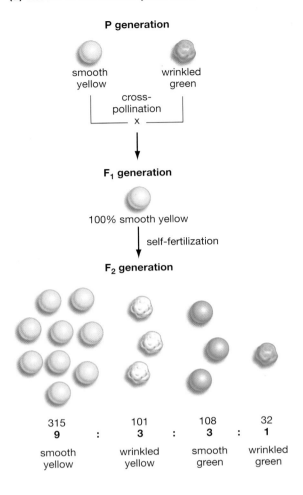

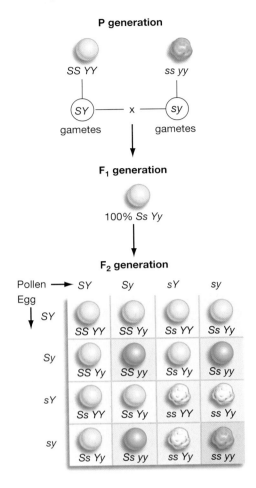

In one of his dihybrid crosses, Mendel cross-bred plants that had smooth yellow seeds with those that had green wrinkled seeds. The result was a generation of plants that all had smooth yellow seeds. When these plants self-fertilized, the result was an F_2 generation that had the phenotypes shown in a 9:3:3:1 ratio.

The Punnett square demonstrates why Mendel got the 9:3:3:1 phenotypic ratio in his dihybrid cross. Nine combinations yield smooth yellow seeds, 3 yield smooth green seeds, 3 yield wrinkled yellow seeds, while only 1 results in a wrinkled green seed.

FIGURE 11.9 **Phenotype Ratios in a Dihybrid Cross**

This equals a yellow:green ratio of about 3:1. In other words, the familiar 3:1 ratio in the F_2 generation still holds. It held as well when Mendel looked only at seed *shape*. This suggested something to Mendel that turned out to be another of his major insights: that characters—in this case, seed shape and seed color—are transmitted *independently* of one another. When these two characters were being crossed at the same time, the results for each were the same as when they were crossed as single characters in the earlier experiments. Thus, one character's transmission did not appear to affect the other's.

Understanding the 9:3:3:1 Ratio Still, what's the basis of the formidable 9:3:3:1 ratio that Mendel got in his dihybrid cross, with its mixture of wrinkled yellows, smooth greens, and so forth? This is simply a more complex example of the kind of outcomes you saw earlier with the aid of the Punnett square. If you look at the square in **FIGURE 11.9b** and start adding up internal squares *by phenotype*, Mendel's 9:3:3:1 ratio starts making sense. You can see, for example, where the 1 in the ratio comes from: There is only 1 part wrinkled green seeds, because it is only the *sy* (pollen) and *sy* (egg) combination that yields this phenotype. Likewise, you will get three parts wrinkled yellow seeds by adding up the three possible male and female gamete combinations that could bring this about.

The Law of Independent Assortment

This outcome could result, however, only if Mendel's fundamental insight about dihybrid crosses was correct: that characters were being transmitted independently of one another. If one character had affected another's transmission, these ratios would have been very different. This insight of Mendel's is now known as Mendel's Second Law, or the **Law of Independent Assortment**. It states that during gamete formation, gene pairs assort independently of one another.

Independent Assortment and Chromosomes   Now recall that the underlying physical basis for this law was set forth in Chapter 10: In meiosis, pairs of homologous chromosomes *assort independently* from one another at the metaphase plate. It may be, for example, that paternal chromosome 5 will line up on side A of the metaphase plate; if so, then maternal chromosome 5 ends up on side B. For chromosome 6, however, things could just as easily be reversed: The *maternal* chromosome might end up on side A, and the paternal on side B (see Figure 10.2, page 161.) The genes for Mendel's seed color exist on the plant's chromosome 1, while the genes for seed shape exist on its chromosome 7. Because these are separate chromosomes, they assort independently at the metaphase plate, meaning they are passed on independently to future generations.

So Far...

1. Mendel's law of segregation states that differing traits in organisms result from two differing _____ that _____ in gamete formation, such that each gamete gets only one of the two.

2. Homozygous refers to an organism that has two _____ alleles for a character, while heterozygous refers to an organism that has two _____ alleles for the character.

3. An allele that is expressed in the heterozygous condition is said to be _____, while an allele that is not expressed in the heterozygous condition is said to be _____.

11.6 Reception of Mendel's Ideas

When Mendel finished his experiments, he delivered two lectures on them to a local scientific society in 1865. It would be nice to report that the society members' jaws dropped open once Mendel let them in on how inheritance worked in the living world, but no such thing happened. His findings were published, in a scientific journal that was distributed in Germany, Austria, the United States, and England. And Mendel even took it upon himself to get 40 reprints of his paper, sending some out to various scientists in an effort to spark some exchange on his findings. All to no avail. His work sank nearly without a trace, finally to be rediscovered in 1900, 16 years after his death. Today, the consensus within the scientific community is that nobody cared about Mendel's findings in his own time simply because nobody grasped their significance.

This poor early reception notwithstanding, Mendel's insights have stood the test of time. For certain kinds of phenotypes, his rules of inheritance are extremely reliable. To this day, biologists will begin an observation by noting that "Mendelian rules tell us that . . ." or "Mendelian inheritance operates here." Even allowing that he had intellectual predecessors, the fact is that Mendel did not just *add* to the discipline of genetics; he founded it.

A BROADER VIEW

Unrecognized

To what extent does achievement go unrecognized in any field? In Mendel, we have a spectacular example of unrecognized achievement in science, but we know of examples of brilliance overlooked in other fields. Jazz cornetist Buddy Bolden probably was never recorded and poet Emily Dickinson was never published during her lifetime.

So Far... Answers:

1. alleles; separate

2. identical; differing

3. dominant; recessive

11.7 Multiple Alleles and Polygenic Inheritance

Multiple Alleles

Despite the importance of Mendel's findings, there are lots of instances in which the inheritance patterns he discovered do not operate in nature. In fact, the kind of predictable, either-or Mendelian inheritance we've been reviewing is the exception rather than the rule in the living world. Note that every one of Mendel's peas was either yellow or green—there were no other colors. And this was so because, in the peas he was using, there were only two alleles for seed color (Y and y). But in the living world, there generally are *more* than two alleles for a given trait, as you'll see in this example from the world of human health.

When people speak of blood "types," what they mean is variations on types of "glycoproteins"—proteins with carbohydrate chains attached—that cover the surface of red blood cells. These surface molecules come in many different varieties, but the two most important of them are designated A and B (**see** FIGURE 11.10). Blood types are completely under genetic control, with a single gene that lies on chromosome 9 determining what blood type a person will have—A, B, AB, or O. There are two alleles of this gene in each individual, because there are two copies of chromosome 9 in each individual—one inherited from the mother, the other inherited from the father. So, a single person could have two alleles that code for type A molecules, or a person could have both alleles coding for type B molecules. In the first case a person would have type A blood, and in the second type B blood. However, a person could have one allele that codes for the type A molecule *and* one that codes for type B, in which case that person would have type AB blood. Finally, a person could have two alleles that are inactive—that code for neither type of surface protein—and would thus have type O blood.

Now, note what's at work here. First, unlike the case with pea color, blood type doesn't have one allele that is dominant and one that is recessive. People who are heterozygous for the A and B alleles get *both* the A and B proteins on the surface of their blood cells (and are thus type AB). Alleles that have this kind of independent effect are said to be *codominant*. For our purposes, however, the more important thing to note is that there are *three* alleles for this character (A, B, and O), not the two that existed in Mendel's peas. Since alleles come in pairs in an individual, no one person can have more than two of these alleles. (A given person may be, say, A and B; but he or she cannot be A, B, and O.) But in a *population* of humans, the full range of all three alleles can be found. And therein lies the lesson. In a population, alleles can come in many variants. When three or more alleles of the same gene exist in a population, they are known as **multiple alleles**. And multiple alleles are the rule, rather than the exception, in nature. As you can imagine, a greater diversity of alleles leads to a greater phenotypic diversity in nature.

Polygenic Inheritance

Now, let's examine a second way in which the living world tends to be more complex than Mendel's laws imply. Recall that each character in Mendel's peas was governed by a single gene. There was one gene for flower color, for example. Though it came in variant alleles (yielding purple or white flowers), this one gene completely determined the flower-color character. Likewise, there was one gene for seed color, one gene for stem height, and so forth. But in nature, characters tend to be influenced by *many* genes acting together, rather than just one. The human characters of height, weight, eye color, and skin color are each controlled by several genes. The color of a wheat grain, the length of an ear of corn, and the amount of milk a cow gives are all controlled by many genes acting together. The term for such genetic influence is **polygenic inheritance**, meaning the inheritance of a genetic character that is determined by the interaction of multiple genes, with each gene having a small additive effect on the character.

Blood type   **Surface glycoproteins on red blood cells**

A

B

AB

O

no surface glycoproteins

FIGURE 11.10   Human Blood Types   The familiar A, B, O human blood-typing system refers to glycoproteins on the surface of red blood cells. People whose blood is "type A" have type A glycoproteins on their blood cells. It is also possible to have only B molecules (and be type B); to have both A and B molecules (and be type AB); or to have none of these molecules (and be type O).

A BROADER VIEW

Multiple Alleles

How many allelic variants might a gene have? To give one example, a gene called human *Pax6* that is involved in the development of the eyes has 302 known variants.

(a) Continuous variation in human height

(b) The bell curve

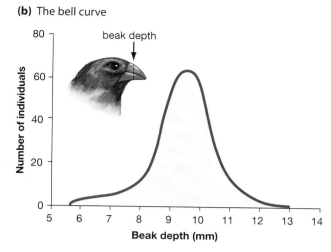

FIGURE 11.11 **Continuous Variation and the Bell Curve**

(a) Mendel's pea seeds may have been green or yellow, but traits such as human height do not have this either-or quality. Human heights exist in a range, with no fixed increments between heights of individuals. Such "continuous variation" results from polygenic inheritance, in which each of several genes contributes a small additive effect to a character. University of Connecticut students in the picture have been arranged by height to show how continuous variation works in this one human trait. Note that the group as a whole takes on the shape of a bell. The students' heights are distributed in a pattern that creates a bell curve.

(b) A bell curve can also be seen in this graph, which plots the average beak depth of a population of Darwin's finches from the Galápagos Islands. Note how most of the finches had beak depths close to the average depth of the population as a whole.

When we have many genes contributing some small increment to a trait, the result is what you see in FIGURE 11.11. Human beings don't come in two heights, or three or four; they display what is known as a "continuous variation" in height, each person being just barely taller or shorter than the next. Continuous variation also holds true for human skin color. Human beings don't really have "black" or "white" skin. Instead they have skin that comes in a *range* of hues, in which one color shades imperceptibly into another. Likewise, trees of a given species come in a range of heights, and the beaks of birds come in a range of lengths. Polygenic inheritance works with multiple alleles to create this fine-grained diversity.

Now, note something else about Figure 11.11a. Most of the students fall in the middle range of heights in the group. Put another way, there are fewer tall or short students than there are students of medium height. This height distribution means that the group as a whole takes on a kind of shape—a bell shape. This is, in fact, the famous **bell curve**, meaning a distribution of values that is symmetrical, and largest around the average. Most biological traits manifest in this way. Look, for example, at the beak depths in a group of Darwin's finches in Figure 11.11b. Graphs like this just confirm what we know intuitively—that traits in living things cluster around what is average, rather than what is extreme.

In polygenic inheritance, with its many genes and alleles, gene interactions are so complex that predictions about phenotype are a matter of *probability*, not certainty. Human height is largely under genetic control, and there are means of trying to predict the height of children based on the height of their parents. (The starting point generally is to add together the height of the parents and divide the result by two.) All that such predictions can do, however, is specify that if parents are of a given height, then there is a certain probability that their children will fall into a given range of heights. Think how different this is from the situation with Mendel's peas. There, you could confidently predict that a given cross would yield, say, all yellow peas.

11.8 Genes and Environment

Scientists work awfully hard at trying to determine the probabilities for certain polygenic outcomes—those related to disease. In carrying out this work, they are essentially doing what Mendel did: observing traits in a parental generation and then observing traits in the parents' offspring. Such work is the basis for the warnings we sometimes hear about how much a person's risk for a particular cancer goes up if that person has a relative who has contracted the disease. With polygenic illnesses such as cancer, however, it is difficult to separate the genetic factors from what are called "environmental" factors. What's an environmental factor? Any external influence that is favorable or unfavorable for the development of a trait in an organism. Smoking, for example, is an environmental factor that influences development of lung cancer. Indeed, smoking is responsible for about 90 percent of all lung-cancer cases. Even so, only a minority of smokers ever contract lung cancer. Why do some smokers get this disease while others are spared? A large part of the answer is genetic susceptibility: the genetic makeup of some individuals makes them more susceptible to the environmental influence of smoking. Thus, we can see genes and environment working together to create the outcome of lung cancer. And this is almost always the case in nature. For polygenic traits—which is to say, most traits—it is rare for genes to be the sole determining factor.

A clear example of how genes and environment work together can be seen in **FIGURE 11.12**, which shows pictures of the most popular species of hydrangea flower, *Hydrangea macrophylla*. Any gardener could start with a blue hydrangea like the one in the picture on the left, take a "cutting" from it, do a little preparatory work, plant the cutting in the ground, and get a whole new hydrangea plant as a result. For our purposes, the important point is that this second plant is a clone of the first. It is not produced by mixing eggs and sperm; it is an exact genetic replica of the first plant. It can, however, be a plant with *pink* flowers instead of blue if some garden lime is mixed into its surrounding soil. It's possible, therefore,

FIGURE 11.12 Flower Color Is Affected by Environment Both the blue hydrangeas in the background and the pink hydrangeas in the foreground are from the same species (*Hydrangea macrophylla*). Yet they have different coloration because of different chemical conditions in their soil. (The soil the blue plants are in is more acidic than the soil the pink plants are in.) This is but one example of how genes and environment work together to produce the traits we see in living things. The plants are blooming along a wall in Salem, Oregon.

to have two hydrangea plants with exactly the same genotype, but with very different phenotypes—pink or blue flowers. The message here is one of genetic limitation. Genotype specifies only so much about phenotype. Protein products that bring about a blue plant in one environment will bring about a pink plant in another. Across the natural world, things generally work this way. Genes do not come with an unvarying ability to bring about an effect. Rather, genes and environment work together to create the phenotypes we see in living things.

On to the Chromosome

This completes the survey of Mendel's work and the variations on his ideas of inheritance. As you have seen, Mendel's work lay dormant for years in the nineteenth century. What sent scientists scurrying back to his findings were experiments on those tiny entities you have looked at in detail before: chromosomes, the subject of the next chapter.

A BROADER VIEW

Genes and Environment Work Together

You might think that a trait such as human height would be completely under genetic control, but consider the fact that U.S. adult men and women are about an inch taller now than they were in 1960. Since genes change very little over a few generations, why should U.S. height have increased? Some hypotheses have to do with better nutrition and health care. Throughout history, such environmental influences have played a major role in determining height. Worldwide, the tallest people in a society have tended to be those with higher incomes and better educations.

So Far...

1. The usual case in nature is that a given character in an organism is influenced by _____ genes. To put this another way, the character is produced through _____ inheritance.

2. True or false: This form of inheritance tends to produce fixed increments of difference between organisms.

3. The hydrangea example in the book shows that the same genes can produce different phenotypes depending on _____ influences that act on an organism.

So Far... Answers:

1. *many; polygenic*

2. *False; polygenic inheritance tends to produce continuous variation among organisms.*

3. *environmental*

Chapter Review

Summary

11.1 Mendel and the Black Box

- **Mendel and his insights**—Gregor Mendel was the first person to comprehend some of the most basic principles of genetics. He reached these understandings in the mid-nineteenth century, in what is now the Czech Republic, by means of breeding generations of pea plants, observing, first, the "input" (the parental generation) and then the "output" (their offspring). (page 167)

11.2 The Experimental Subjects: *Pisum sativum*

- **Characters and traits**—Mendel looked at seven characters in his plants—such attributes as seed color and texture. In his plants, each of these characters came in two varieties or traits, one of them dominant, the other recessive. His experiments involved breeding pea plants, starting with plants that had a given set of traits and then observing which of those traits showed up in succeeding generations. (page 170)

- **Phenotype and genotype**—A phenotype is any physiological feature, bodily characteristic, or behavior of an organism. In Mendel's plants, purple flowers were one phenotype and white flowers were another.

Phenotypes in any organism are in significant part determined by that organism's genotype, meaning its genetic makeup. Mendel realized that the phenotypes in his plants were being controlled by what we would today call their genotypes. (page 171)

11.3 Starting the Experiments: Yellow and Green Peas

- **Genes and alleles**—One of Mendel's central insights was that the basic units of genetics are material elements that, in his pea plants, came in pairs. These elements, today called genes, come in alternative forms called alleles. One member of an allele pair resides on one chromosome, while the other allele resides on a second chromosome that is homologous to the first. (page 172)

- **Persistence of genes**—Another of Mendel's insights was that genes retain their character through many generations, rather than being "blended" together. Genes that coded for green pea color, for example, were retained in their existing form over many generations. (page 172)

Web Tutorial 11.1 Mendel's Experiments

11.4 Mendel's Generations Illustrated

- **Law of segregation**—A third insight of Mendel's was that alleles separate prior to the formation of gametes. Though Mendel did not know it, the physical basis for this is that the alleles he observed resided on homologous chromosomes, which always separate in meiosis.   (page 174)

- **Homozygous and heterozygous**—An organism that has two identical alleles of a gene for a given character is said to be homozygous for that character. An organism that has differing alleles for a character is said to be heterozygous for that character. *Dominant* means "expressed in the heterozygous condition" (as with the yellow color of peas, expressed in the heterozygous *Yy* condition). *Recessive* means "not expressed in the heterozygous condition" (as with the green color of peas in the *Yy* condition).   (page 174)

Web Tutorial 11.2 Independent Assortment

11.5 Crosses Involving Two Characters

- **Independent assortment**—Mendel observed that the genes for the different characters he studied were passed on independently of one another. This was so because the genes for these characters resided on separate, non-homologous chromosomes. The physical basis for what he found is the independent assortment of chromosomes during meiosis.   (page 176)

11.6 Reception of Mendel's Ideas

- **Unrecognized**—Gregor Mendel published his work, but the significance of it was never recognized in his lifetime. It was not rediscovered until 16 years after his death, in 1900.   (page 178)

11.7 Multiple Alleles and Polygenic Inheritance

- **Multiple alleles**—Human beings and many other species can have no more than two alleles for a given gene, each allele residing on a separate, homologous chromosome. Many allelic variants of a gene can, however, exist in a population.   (page 179)

- **Polygenic inheritance**—Most traits in living things are governed not by one gene but by many, with these genes often having several allelic variants. The term for such genetic influence is polygenic inheritance: the inheritance of a genetic character that is determined by the interaction of multiple genes, with each gene having a small additive effect on the character.   (page 179)

- **Continuous variation**—Polygenic inheritance produces continuous variation in phenotypes, meaning there are no fixed increments of difference between individuals. Human skin, for example, comes in a range of colors in which one color shades imperceptibly into the next.   (page 180)

- **The bell curve**—The traits produced in polygenic inheritance tend to manifest in bell-curve distributions, in which most individuals display near-average trait values, rather than extreme trait values. Gene interactions and gene–environment interactions are so complex in polygenic inheritance that predictions about phenotypes are a matter of probability, not certainty.   (page 180)

11.8 Genes and Environment

- **Genes and environment work together**—The effects of genes can vary greatly in accordance with the environment in which the genes are expressed. An organism's genotype and environment interact to produce that organism's phenotype.   (page 181)

Key Terms

allele   p. 172

bell curve   p. 180

cross-pollinate   p. 170

dominant   p. 174

genotype   p. 171

heterozygous   p. 174

homozygous   p. 174

Law of Independent Assortment   p. 178

Law of Segregation   p. 174

multiple alleles   p. 179

phenotype   p. 171

polygenic inheritance   p. 179

recessive   p. 175

Testing Your Understanding

In Your Own Words *(answers in the back of the book)*

1. What are the differences between:
   a. phenotype and genotype?
   b. dominant and recessive?

2. What two laws are credited to Mendel and his research?

3. In a case of Mendelian genetics where one gene affects one trait, how many different genotypes can bring about a dominant phenotype (for example, yellow pea color)?

Thinking about What You've Learned

1. Stands of aspen trees often are a series of genetically identical individuals, with each succeeding tree growing from the severed shoot of another tree. Using what you've learned of genetics in this chapter, would you expect one aspen tree in a stand to differ greatly from another in its phenotype? Would you expect each to look exactly like the next in terms of phenotype?

2. The ABO blood-typing system mentioned in the text has great practical significance because the human immune system will attack blood-cell surface proteins that it recognizes as "foreign"—that is, proteins that a person does not have on his or her own blood cells. Given this, what is the "universal donor" blood type? What is the "universal recipient" type?

3. If, as the text states, the effects of genes can be expected to vary in accordance with the environment in which these genes work, would you expect this phenomenon to be applicable to human genes, such as those that help produce such traits as height, weight, or introversion and extroversion?

Genetics Problems *(answers in the back of the book)*

One of Mendel's many monohybrid (single character) crosses was between a true-breeding (homozygous) parent bearing purple flowers and a true-breeding (homozygous) parent bearing white flowers. All the offspring had purple flowers. When these offspring were allowed to self-fertilize, Mendel observed that their offspring occurred in the ratio of approximately 3 purple-flowered:1 white-flowered plants. The allele determining purple flowers is symbolized P, and the allele determining white flowers is symbolized p. Use this information to answer the following questions.

1. What are the genotypes of the parents?
   a. $PP; pp$
   b. $PP; Pp$
   c. $Pp; Pp$
   d. $pp; pp$

2. Which generation has only purple-flowered plants?
   a. P
   b. F_1
   c. F_2
   d. F_3

3. Which generation tells you which of the traits is dominant?
   a. P
   b. F_1
   c. F_2
   d. F_3

4. How many phenotypes are present in the F_2 generation?
   a. 1
   b. 2
   c. 3

5. How many different genotypes are present in the F_2 generation?
   a. 1
   b. 2
   c. 3

6. Mendel allowed the white-flowered F_2 plants to self-fertilize. Approximately what proportion of these F_2 plants produced only white-flowered offspring?
   a. none
   b. 1/4
   c. 1/3
   d. 3/4
   e. all

7. Mendel allowed the purple-flowered F_2 plants to self-fertilize. What proportion of these F_2 plants produced only purple-flowered offspring?

    a. none

    b. 1/4

    c. 1/3

    d. 3/4

    e. all

8. If a heterozygous purple-flowered pea plant is crossed with a homozygous white-flowered plant, what proportion of offspring will be white-flowered?

    a. none

    b. 1/4

    c. 1/2

    d. 3/4

    e. all

9. Given what you know about the alleles that determine purple and white color in pea flowers, why is it possible to have a purple-flowered heterozygote, but not a white-flowered heterozygote?

10. In considering flower color in pea plants, we are dealing with _____ gene(s) and _____ allele(s).

    a. 1; 1

    b. 1; 2

    c. 1; 4

    d. 2; 1

    e. 2; 2

11. If two individuals with AB blood type have children together, what proportion of their children are expected to have AB blood type?

    a. none

    b. 0.25

    c. 0.50

    d. 0.75

    e. all

Genetic Inheritance and Health

Ronalda Pierce in December, 2003.

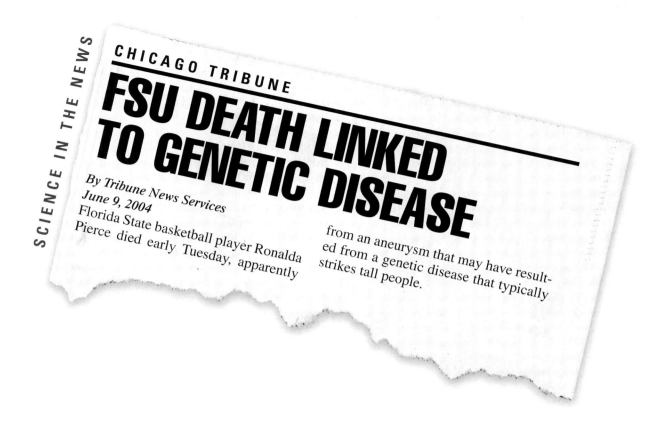

SCIENCE IN THE NEWS

CHICAGO TRIBUNE

FSU DEATH LINKED TO GENETIC DISEASE

By Tribune News Services
June 9, 2004

Florida State basketball player Ronalda Pierce died early Tuesday, apparently from an aneurysm that may have resulted from a genetic disease that typically strikes tall people.

12.1 Tall, Fit, and Vulnerable

What killed Ronalda Pierce? She was only 19 and in some ways was in superb physical condition. After all, she had been a standout freshman basketball player at Florida State in the year leading up to her death. Given her fitness and her age, the news of her passing was surprising, to say the least. Yet this news also was eerily reminiscent of a similar tragedy that took place almost two decades earlier. In 1984, a woman named Flo Hyman starred on the U.S. volleyball team that won the Olympic silver medal in Los Angeles. Two years later, while playing in a match against Japan, Hyman collapsed on the volleyball court and died.

Like Ronalda Pierce, Flo Hyman was tall—both she and Pierce stood six-foot-five. But the very thing that gave these women their height appears to have brought about their deaths. Both suffered from a condition called Marfan's syndrome that weakens the so-called connective tissue in the body, which could be thought of as the glue that holds other kinds of tissue together. And the connective tissue for both women failed in the same place: the aorta, which is a massive artery leading from the heart.

So, how did Marfan's syndrome weaken these womens' arteries to the point that they gave way? There are fibers that run through connective tissue in the way that steel bars run through reinforced concrete. One component part of these fibers is a protein called fibrilin. Like all proteins, fibrilin is coded for by a gene. But Marfan victims have a mutated form of this gene, and *its* product is fibrilin that either is misshapen or is absent altogether. The end result is connective tissue that can rupture without warning. Marfan's syndrome is not a particularly common affliction, but it is common enough that modern researchers have tried to identify historical figures who suffered from it. One of these possible Marfan victims was a tall, gaunt President of the United States, Abraham Lincoln.

Marfan's turns out to be a "genetic disorder," which is to say a type of illness that is inheritable. Though such illnesses are rare, they provide a kind of window onto how inheritance works in human beings. As such, this whole chapter is devoted to explaining these disorders. What are the odds of passing on a "dominant" disorder such as Marfan's, as opposed to a "recessive" disorder, such as sickle-cell anemia? How can such afflictions be tracked across generations in a family? And what happens when a person inherits not a faulty gene, as with Marfan's, but the wrong number of chromosomes? In this chapter, you'll find out. Note, however, that almost all the disorders we'll review are transmitted through "Mendelian" inheritance, meaning they are all disorders in which a single gene or chromosome makes the difference between sickness and health. In contrast, many widespread diseases, such as heart disease, are polygenic—they are based on many genes working in concert. We'll start our review with a look at genetic conditions that manifest one way in men, but another way in women.

12.2 X-Linked Inheritance

Minor cuts are simply an irritation to most of us, but for people suffering from the condition known as hemophilia, such cuts can be life threatening. Hemophilia is a failure of blood to clot properly. A group of proteins interact to make blood clot, but about 80 percent of hemophiliacs lack a functioning version of just one of these proteins (called Factor VIII). Genes contain the information for the production of proteins, of course, so at root hemophilia is a genetic disease—a faulty gene is causing the faulty Factor VIII protein. This same process is at work in a couple of other well-known afflictions: the disease called Duchenne muscular dystrophy, and a far less serious condition, red-green color blindness. These disorders have more in common than being genetically based, however. All of them are known as *X-linked* disorders, because the genes that cause them lie on the X chromosome. As it turns out, X-linked disorders claim more male victims than female. Why should this be so? A look at color blindness can provide the answer.

The Genetics of Color Vision

Despite its name, "color blindness" almost never means a *total* inability to perceive color. Rather, it most often means an inability to distinguish among certain shades of red and green (FIGURE 12.1). The reason we can perceive any color—red and green included—is that we have cells toward the back of our eyes that contain substances, called pigments, that absorb different colors of light. These pigments are molecules composed of two main parts, one of which is a protein. Now, recalling that genes code for proteins, think about how color vision is genetically based: We have genes that are coding for the protein portions of our light-absorbing pigments. And as it turns out, the genes that code for our red and green pigment proteins lie very close to one another on a specific chromosome—the X chromosome. A red-green color-blind person, then, is one in whom these X-located genes fail to code for the proper proteins.

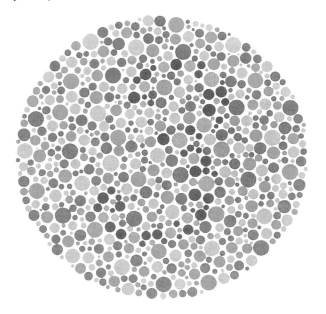

FIGURE 12.1 **Typical Test for Red-Green Color Blindness** If you do not see a number inside the large circle, you may have this recessive trait.

Alleles and Recessive Disorders

In the normal case, as you've seen, a given gene can come in two variant forms or *alleles* in each person. One allele for a gene might reside on, say, the chromosome 1 we inherit from our mother; if so, the second allele for this gene will reside on the chromosome 1 we inherit from our *father*. Thus there is a kind of backup system in human genetics: Genes come in pairs of alleles that lie on separate, homologous chromosomes. If one allele for a given trait is defective, there will usually exist a second, functional allele for that same trait on a homologous chromosome.

This genetic structure usually works fine when a condition is a so-called **recessive disorder**, meaning a genetic disorder that will not exist in the presence of a functional allele. In such a condition, a person does not have to have *two* functional alleles; a single "good" allele will do. And red-green color blindness is a recessive disorder: A person who has even one functional allele for red pigment and one functional allele for green pigment will not be color blind. So why do men suffer more from this condition than women? Because men have no backup; they have only one allele for red pigment and one allele for green pigment because they have only one *X chromosome* (and one Y). Meanwhile, women have two X chromosomes. If the pigment alleles are defective on one of their chromosomes, the alleles on their other X chromosome will still provide them with full color vision.

Given this, think about the interesting way that color blindness is passed along. Say there is a mother who is not color blind herself, but who is heterozygous for the trait. That is, she has functional color alleles on one of her X chromosomes but dysfunctional alleles on her other X chromosome. Should her son happen to inherit this second X chromosome, he will be color blind, because the X chromosome he got from his mother is his *only* X chromosome (**FIGURE 12.2**). Meanwhile, a daughter who inherited this chromosome would likely be protected by her second X chromosome—the one she got from her father. Not surprisingly, then, more males are color blind than females. About 8 percent of the male population has some degree of color blindness, while for females the figure is about half a percent. Women enjoy a similar protection in connection with other X-linked disorders, such as hemophilia and muscular dystrophy.

A BROADER VIEW

Male Vulnerability

Why don't males enjoy special *benefits* conferred by their Y chromosome? After all, women don't even have a Y chromosome. The reason the Y does so little for health is that the Y does so little in general. There are only 78 genes on the Y (compared to about 1,500 on the X), and almost all these Y genes function within the narrow realm of bringing about maleness, rather than assisting in general processes such as vision and blood clotting.

12.3 Autosomal Genetic Disorders

The X and Y chromosomes are, of course, only two of the chromosomes in the human collection, and any chromosome can have a malfunctioning gene on it. The chromosomes other than the X and Y chromosomes are called *autosomes*. A

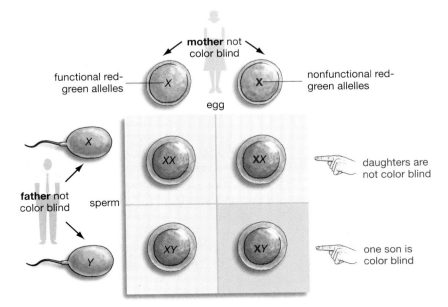

FIGURE 12.2 **The Value of Having Two X Chromosomes** The mother in the figure is not herself color blind but has one dysfunctional set of red-green alleles (represented by the red X), which she passes on to one son and one daughter. The daughter is not color blind because she has inherited, from her father, a second X chromosome—one that has functional red-green alleles. Meanwhile, the son is color blind because his only X chromosome is the flawed one he inherited from his mother. (The other son, shown in the square's lower left cell, is not color blind because he has inherited his mother's functional alleles.)

Table 12.1 Selected Examples of Human Genetic Disorders

Type	Name of condition	Effects
X-linked recessive disorders	Hemophilia	Faulty blood clotting
	Duchenne muscular dystrophy	Wasting of muscles
	Red-green color blindness	Inability to distinguish shades of red from green
Autosomal recessive disorders	Albinism	No pigmentation in skin
	Sickle-cell anemia	Decreased oxygen to brain and muscles
	Cystic fibrosis	Impaired lung function, lung infections
	Phenylketonuria	Mental retardation
	Tay-Sachs disease	Nervous system degeneration in infants
	Werner syndrome	Premature aging
Autosomal dominant disorders	Marfan's syndrome	Ruptured blood vessels
	Huntington disease	Brain tissue degeneration
	Polydactyly	Extra fingers or toes
Aberrations in chromosome number	Down syndrome	Mental retardation, shortened life span
	Turner syndrome	Sterility, short stature
	Klinefelter syndrome	Dysfunctional testicles, feminized features

recessive dysfunction related to an autosome is thus known as an autosomal recessive disorder. (For a list of all the disorders you'll be looking at and more, **see Table 12.1.**)

Recessive Disorders

A well-known example of an autosomal recessive disorder is sickle-cell anemia, which affects populations derived from several areas on the globe, including Africa. In the United States it is, of course, most widely known as a disease affecting African Americans. The "sickle" in the name comes from the curved shape that is taken on by the red blood cells of its victims. Red blood cells carry oxygen to all parts of the body; in their normal shape, they look a little like doughnuts with incomplete holes. When they take on a sickle shape, however, red cells clog up capillaries, resulting in decreased oxygen supplies to brain and muscle (**see** FIGURE 12.3). In the United States, the average life expectancy for men with the condition is 42 years; for women, it is 48 years.

The question is, what causes a red blood cell to take on this lethal, sickled shape? There is a protein, called hemoglobin, that carries the oxygen within red blood cells. The vast majority of people in the world have one form of hemoglobin, called hemoglobin A; but sickle-cell anemia sufferers have another form of this protein, hemoglobin S, which coalesces into crystals that distort the cell.

Now, how is sickle-cell anemia passed from parents to offspring? Because it is a recessive condition, any child who gets it must inherit *two* defective alleles—one allele from the mother, one from the father, each of them coding for the hemoglobin S protein. This means, of course, that both parents must themselves have at least one hemoglobin S allele; both parents must be at least heterozygous for the trait. In this situation, the laws governing inheritance of this disorder are simply those of Mendel's monohybrid cross: A child of these two parents has a 25 percent

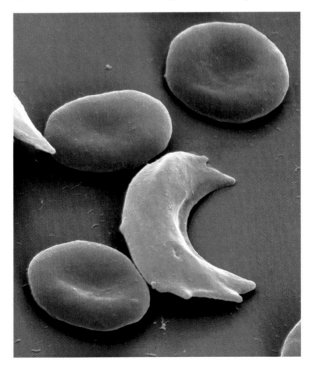

FIGURE 12.3 Healthy Cells, Sickled Cell Three normal red blood cells flank a single sickled cell. (×7400)

chance of inheriting the disorder. You can look at the Punnett square in **FIGURE 12.4a** to see how this works out.

It's important to note the status of the parents in this example. They are heterozygous for a recessive debilitation, meaning they do not suffer from the condition themselves because they are protected by their single "good" allele. Each of them is, however, a **carrier** for the condition: a person who does not suffer from a recessive genetic debilitation, but who carries an allele for it that can be passed along to offspring. In this respect, they are just like the mother in the color-blindness example.

Dominant Disorders

Though sickle-cell anemia is an autosomal disorder and red-green color blindness an X-linked disorder, both are recessive disorders. That is, a person with even a single properly functioning allele will not suffer from them. However, there are also **dominant disorders**: genetic conditions in which a single faulty allele can cause damage, even when a second, functional allele exists. This is the kind of disorder that killed Ronalda Pierce and Flo Hyman. If you look at the Punnett square in **Figure 12.4b**, you can see how the inheritance pattern of an autosomal dominant disease, such as Marfan's syndrome, differs from that of an autosomal recessive illness. Because a parent need only pass on a single Marfan's allele for a son or daughter to suffer from the condition, the chances of any given offspring getting the disease from a single affected parent are one out of two. Perhaps the best-known dominant disorder is one called Huntington disease that results in mental impairment, uncontrollable spastic movements, and death. You may wonder how people afflicted with such a terrible disease could live long enough to pass it along to their children. The answer is that Huntington victims usually don't begin to show symptoms of the disease until well into adulthood—after they have had children.

(a) Sickle-cell anemia: transmission of a recessive disorder.

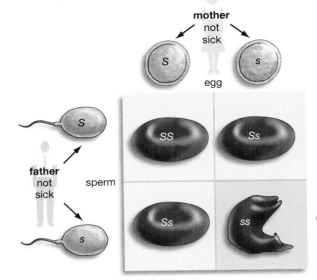

Sickle-cell anemia is a recessive autosomal disorder; both the mother and father must carry at least one allele for the trait in order for a son or a daughter to be a sickle-cell victim. When both parents have one sickle-cell allele, there is a 25 percent chance that any given offspring will inherit the condition.

25% probability of inheriting the disorder

(b) Marfan's syndrome: transmission of a dominant disorder.

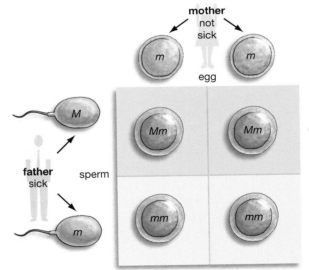

50% probability of inheriting the disorder

In Marfan's syndrome, if only a single parent has a Marfan's allele there is a 50 percent chance that a son or daughter will inherit the condition.

FIGURE 12.4 Transmission of Recessive and Dominant Disorders

So Far...

1. A recessive genetic disorder is one that will not exist in the presence of a _____, while a dominant disorder is one in which a _____ can cause damage, even in the presence of a _____.

2. True or false: A woman who has one X chromosome containing dysfunctional alleles for red-green color vision is not color blind.

3. To inherit an autosomal recessive disorder, both parents of an offspring must be at least _____ for the condition, meaning each parent must have at least _____.

So Far... Answers:

1. *functional allele; single, faulty allele; functional allele*

2. *True; her second X chromosome contains functional alleles, which are enough to keep her from being color blind.*

3. *heterozygous; one faulty allele*

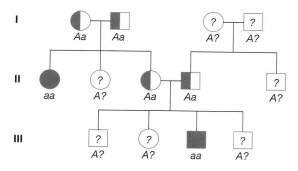

female   male

○ □   normal

◑ ◧   carrier

● ■   albino

FIGURE 12.5   A Hypothetical Pedigree for Albinism through Three Generations

12.4 Tracking Traits with Pedigrees

Confronted with a medical condition, such as Marfan's, that is running through a family, scientists find it helpful to construct a medical **pedigree**, defined as a familial history. Normally set forth as diagrams, medical pedigrees do more than give a family history of a disease. They can be used to ascertain whether a condition is dominant or recessive, and X-linked or autosomal. This can help establish probabilities for *future* inheritance of the condition—something that can be very helpful for couples thinking of having a child.

If you look at **FIGURE 12.5**, you can see a simple pedigree for albinism: a lack of skin pigmentation, which is known to be an autosomal recessive condition. In the figure, you can see some of the standard symbols used in pedigrees. A circle is used for a female and a square for a male. Parents are indicated by a horizontal line connecting a male and a female, while a vertical line between the parents leads to a lower row that denotes the parents' offspring. This second row—a horizontal line of siblings—has an order to it: oldest child on the left, youngest on the right. A circle or square that is filled in indicates a family member who has the condition (in this case albinism), while a symbol that is half filled in indicates a person known to be heterozygous for a recessive condition (a carrier). One strength of a pedigree is that it can sometimes tell researchers which persons are heterozygous carriers of a recessive condition. Remember that a carrier does not display *symptoms* of the condition. So how can a pedigree reveal this?

Take a look at Figure 12.5. The only thing that would be apparent from actually seeing any of the people in the pedigree is that two of them—a female in generation II and a male in generation III—have the condition of albinism. But a little knowledge of Mendelian genetics also allows some other deductions. If the condition had been dominant, then it would have manifested itself in at least one of the parents on the left in generation I. Because this was not the case, it is fair to deduce that this is a recessive condition. The fact that it is recessive, however, means that *both* parents in generation I had to be carriers for the allele, which is why their symbols can be half shaded in. Things are less clear with the parents on the right in generation I. One of their sons did not manifest the condition himself, but went on to have a son who did. From this, we know that the son in generation II had to have been heterozygous for the condition. But from which parent did this son get his albinism allele? Either, or both, of his parents in generation I could have been heterozygous for the condition and yet between them, have passed along only a single albinism allele.

This mixture of certainty and uncertainty leads to the genotype labeling you see in the figure, with *A* representing the dominant "normal" allele and *a* representing the recessive albinism allele. We know, for example, that both parents on the left in generation I had to be *Aa*, but all we can say for sure about the parents on the right in the same generation is that at least one of them was a carrier.

A BROADER VIEW

Tracking Color Blindness

Males who are red-green color blind: Ask your mother if she is color blind. Chances are she is not. Then inform her of her genetic makeup and yours: She possesses faulty alleles on one of her X chromosomes for color vision, which she passed on to you. Her other red-green alleles, on her second X chromosome, are fine, which is why she is only a carrier for color blindness, rather than being color blind herself. In the roll of the genetic dice, you didn't happen to inherit her "good" alleles, because you didn't get her "good" X chromosome.

Modern Myths

Baldness Is Strictly Predictable

Even among some scientists, an idea persists that male "pattern baldness"—losing hair from the top and crown of the head—is a trait that is passed on through a single gene in simple, Mendelian inheritance. In fact, researchers in the field are agreed that baldness is polygenic: Several genes are involved in determining whether a man (or occasionally a woman) will go bald. Where did the single-gene idea come from? In the 1980s, two German researchers looked into this question and found that it is based on one poorly controlled study done by a scientist named Dorothy Osborn in 1916. Despite the weakness of Osborn's study, Wolfgang Kuster and Rudolf Happle wrote, "Her conclusions were adopted, cited, and re-cited" at least up through the mid-1980s by dermatologists and geneticists. Even today, it's easy to find her thesis expressed as fact on some academic web pages.

Osborn's idea was that there is a single gene with a baldness allele, *B*, that is dominant in men but recessive in women, such that if a man is *BB* or *Bb*, he will be bald, while a woman who is *Bb* would not be bald. At least in the popular imagination, this idea then morphed into the belief that if a young man wants to know whether he is going to go bald, he should look at his mother's brother. (This is a strange notion even if you buy into the single-gene hypothesis. Under it, you wouldn't be certain about your future by looking at your maternal uncle; even if he were bald, you could still receive two *b* alleles and have a full head of hair.) The persistence, over nine decades, of the notion that baldness is a single-gene trait is particularly odd given that baldness manifests several clear markers of a polygenic trait. Consider that men don't fall into one of two categories: either "bald" or "full-haired." Rather, there is a *range* of hair loss in which any given man loses just slightly more or less hair than any other. Such "continuous variation" is a hallmark of a polygenic trait.

12.5 Aberrations in Chromosomal Sets: Polyploidy

All of the maladies you've looked at so far have resulted from dysfunctional genes that exist among a standard complement of chromosomes. In humans, this complement is 22 pairs of autosomes and either an XX combination (in females) or an XY combination (in males). Meanwhile, Mendel's peas had 7 pairs of chromosomes, and the common fruit fly has 4 pairs. Whatever the *number* of chromosomes, note the similarity in all these species: Their chromosomes come in pairs; or to put it another way, they all have two *sets* of chromosomes, meaning the organisms that posses them are diploid.

In some situations, however, organisms don't end up with the two sets of chromosomes that are standard for their species. This is called **polyploidy**: a condition in which one or more entire sets of chromosomes have been added to the genome of a diploid organism.

You may wonder how a human being could end up with more than two sets of chromosomes. After all, in meiosis, sperm and egg end up with just one set of chromosomes in them. When they later fuse, in the moment of conception, the result is then two sets of chromosomes in the fertilized egg. But what if *two* sperm manage to fuse with one egg? The result is a fertilized egg with three sets of chromosomes, and the outcome is polyploidy. Normally an egg lets in only one sperm, and it blocks the others hovering about with a kind of chemical gate-slamming mechanism. But occasionally more than one sperm manages to make it in. Human polyploidy can also get started before fertilization, when meiosis malfunctions and gives a single egg or sperm two sets of chromosomes, meaning there will be three sets in total when egg and sperm fuse.

Polyploidy actually is tolerated well in some organisms. Indeed, it has come about countless times in plant species—cotton, soybeans, peanuts, and bananas are all polyploid plants. In human beings, however, polyploidy is a disaster. Perhaps only 1 percent of human embryos with the condition will survive to birth,

A BROADER VIEW

Polyploidy and Pasta

Had any pasta lately? If so, you almost certainly were enjoying the outcome of polyploidy in plants, since the durum wheat that pasta is made from is one of the many polyploid species in the plant world. (Durum began with 14 chromosomes but then underwent a polyploidy event that left it with 28.) Far from being harmful, polyploidy can be so useful in plants that growers often introduce it in them, so as to derive more robust species.

and none of the babies lives long. Concerns about chromosome number in living persons, then, center not on addition or deletion of whole sets of chromosomes, but rather on the gain or loss of *individual* chromosomes.

12.6 Incorrect Chromosome Number: Aneuploidy

The condition called **aneuploidy** is one in which an organism has either more or fewer chromosomes than normally exist in its species' full set. Put another way, aneuploidy is a condition in which an organism has one chromosome too many, or one too few. Among human genetic malfunctions, aneuploidy is unusual in that it occurs quite commonly and yet goes largely unrecognized. This is so because it most often occurs not in fully formed human beings, but in embryos, in which case it can go unnoticed. A would-be mother may know only that she is having a hard time getting pregnant. What she may not know is that she actually *has* been pregnant—perhaps several times—but that aneuploidy has doomed the embryo in each case. Doomed it in what way? The most common outcome of aneuploidy is a pregnancy that ends in miscarriage. And many of these miscarriages happen so early in a pregnancy that a woman never realizes she was pregnant; the microscopic embryo may simply be part of material that is discharged, perhaps with the woman's next menstrual period. A small proportion of embryos do manage to survive aneuploidy, but even in these instances there are consequences. Down syndrome is one of the outcomes of aneuploidy.

Aneuploidy's Main Cause: Nondisjunction

What brings about the harmful condition of aneuploidy? The cause usually is a phenomenon known as **nondisjunction**, which simply means a failure of homologous chromosomes or sister chromatids to separate during meiosis. If you look at FIGURE 12.6, you can see how this works. Note that it can occur either in meiosis I or in meiosis II. In meiosis I, two homologous chromosomes, on the opposite side

FIGURE 12.6 A Mistake in Meiosis Brings about an Abnormal Chromosome Count Nondisjunction can occur either in meiosis I or meiosis II, when either homologous chromosomes or chromatids fail to separate properly. When it occurs in meiosis I, 100 percent of the resulting gametes will be abnormal; when it takes place in meiosis II, only 50 percent will be abnormal.

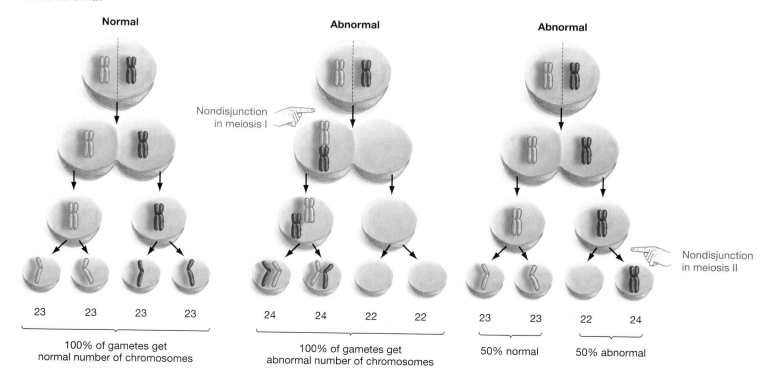

of the metaphase plate, can be pulled to the *same* side of a dividing cell, producing daughter cells with imbalanced numbers of chromosomes. Conversely, nondisjunction can take place in meiosis II, by means of sister chromatids going to the same daughter cell after failing to "disjoin" (hence the cumbersome term *nondisjunction*). Such actions then produce an egg or sperm that has 24 or 22 chromosomes instead of the standard haploid number of 23; the fertilized egg that results from the union of this egg and sperm would then have either 47 or 45 chromosomes, instead of the standard 46. Since every cell in a human being is, in a sense, a copy of this original fertilized egg, this means that every cell in the body would have this incorrect number of chromosomes.

But why does this incorrect number matter? Because it creates a genetic imbalance—a condition that perhaps can be made clear by an analogy. Imagine that while baking chocolate chip cookies, you decided to increase the amount of flour called for in the recipe. This change might be of little consequence if you increased all the other ingredients in the same proportion. If, however, you increased nothing but the flour, the final product would be hard and flavorless because the ingredients in it would be unbalanced. An added chromosome has this same effect: It unbalances the proportions of *biological* ingredients (the proteins) that help produce and maintain a living thing.

The Consequences of Aneuploidy

As noted, embryos with the wrong number of chromosomes are not likely to survive. Setting aside aneuploidies that affect the X and Y chromosomes, *all* aneuploidies result in miscarriage except for those that give an embryo an additional chromosome 13, 18, or 21. It is the gain of an additional chromosome 21 that causes the most well-known outcome of aneuploidy in human beings. **Down syndrome** is, in some 95 percent of cases, a condition in which a person has three copies of chromosome 21, rather than the standard two. Ninety percent of these cases come about because of nondisjunction in egg formation, with the remaining 10 percent caused by nondisjunction in sperm formation. Seen in about 0.1 percent of all live births, Down syndrome results in an array of effects: smallish, oval heads, IQs that are well below normal, infertility in males, short stature and reduced life span in both sexes.

It is well known that when women pass the age of about 35, their risk of giving birth to a Down syndrome child increases dramatically. It seems less well known that even at maternal age 40, the odds of conceiving such a child are less than 1 in 100 (FIGURE 12.7). Scientists are not certain why the mother's age should figure so prominently in Down syndrome, though several hypotheses have been put forward to account for this. Most scientists agree that one piece of the puzzle has to

(a) Maternal age and Down syndrome risk

Mother's age	Chances of giving birth to a child with Down syndrome
20	1 in 1925
25	1 in 1205
30	1 in 885
35	1 in 365
40	1 in 110
45	1 in 32

(b) Karyotype of a person with Down syndrome

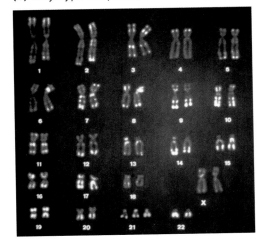

FIGURE 12.7 Down Syndrome: Increasing Risk with Age (a) The risk of giving birth to a Down syndrome child increases dramatically past maternal age 35. **(b)** A karyotype is a visual display of an entire set of chromosomes. Pictured is a karyotype of a person with Down syndrome. Note the three chromosomes, stained blue-green, above the number 21. The karyotype also shows that this person is a female, as indicated by the two X chromosomes.

Essay 12.1 PGD: Screening for a Healthy Child

"Is the baby OK?" In a life full of questions, it's possible that none carries more weight than this one. In most cases the answer will be reassuring, but in some it will be devastating. Small wonder, then, that scientists constantly are trying to perfect ways to know in advance whether a newborn will be healthy. You have seen that a large number of debilitating human conditions have genetic causes. Further, you know that nearly every kind of cell in a person's body contains a complete copy of that person's genome, or set of genes. Therefore, to check on someone's genetic well-being, all that is necessary is to have access to a small collection of that person's cells. And such cells exist, of course, in embryos as well as in newborn babies and adults. Thus, it's not hard to see what the procedure would be for prenatal genetic testing: Gather embryonic cells and examine their DNA and chromosomes.

This kind of testing has been going on, in ever-more refined forms, since the late 1960s. From that time until quite recently, however, most prospective parents who got bad news from a genetic test had two choices: abort the fetus, or bring a child with an incurable condition into the world. Now, an ever-growing number of reproductive clinics are providing a third choice. Parents who are willing to undertake so-called in vitro fertilization—fertilization of the mother's eggs in a laboratory—can have each one of the early-stage embryos that results from this process tested for genetic trouble. Of this initial group of embryos, only those found to be genetically healthy are candidates to be inserted back into the mother. The hope is that at least one of them will implant in the mother's uterus and result in a child.

This process, called preimplantation genetic diagnosis or PGD, can be used by couples who have no known genetic risk factors, but who are simply having a hard time conceiving a child. As the main text notes, a large proportion of human embryos have genetic defects (such as aneuploidy). Moreover, some couples may be particularly prone to producing such embryos, which generally are washed away in miscarriages. PGD allows couples to screen their embryos for genetic problems before implantation in the uterus, thus greatly increasing their chances of bringing a child to term.

PGD can also be used, however, in connection with couples who are *likely* to produce a child with genetic defects. In an extreme example, imagine a prospective father who knows he is fated to fall victim to an illness such as Huntington disease, which inevitably leads to dementia and death, but usually doesn't start showing up in people until they reach the age of 35 or so. Since Huntington is a dominant disorder, any person who carries the Huntington allele has a one-in-two chance of passing the disease along to a child. With PGD, only those embryos found *not* to have the harmful allele would be selected for implantation.

Right now, some clinics allow parents to use PGD to choose the sex of their child.

Of course, more conventional means of genetic screening, such as amniocentesis, can provide information about conditions such as Huntington, but there is a difference. In amniocentesis, cells are obtained from an embryo that has been developing in a mother's uterus for a minimum of 14 weeks. By the time the results of amniocentesis are in, more than four months of a nine-month pregnancy may have elapsed. Most prospective parents would make a distinction between aborting a four-month-old fetus on the one hand or not implanting a four-day-old embryo on the other. Of course, not everyone feels this way. For some people, all embryos represent human life; therefore, producing an embryo, and then not using it, amounts to taking a human life. And in PGD, most embryos have two fates: those that are not implanted are eventually destroyed (though a small number of "embryo adoptions" are now taking place across the country).

Beyond this issue, PGD also raises another ethical question: What limits should be placed on choosing from among embryos? Right now, some clinics allow parents to use PGD to choose the *sex* of their child. If this seems disturbing, consider that, with our expanding knowledge of genetics, it may be possible in the future

do with the way women's eggs proceed through meiosis. A given egg (actually an "oocyte") starts meiosis before a woman is even born but then does not *complete* this meiosis for years or even decades. Instead, it remains paused in the cell cycle. It completes meiosis I only upon being ovulated—upon moving out of the ovaries and down the uterine tube, which usually happens to only one egg every 28 days. Such an egg completes meiosis II only if it is fertilized by sperm while proceeding through the uterine tube. Thus, an egg that started meiosis I while a woman was still an embryo may not complete this meiosis until the woman is, say, in her thirties. In the intervening decades, this egg's cellular machinery has aged, perhaps to the point that it can no longer separate its chromosomes or chromatids properly.

to select for a tall child or a blonde-haired child. Prospective parents who would never abort a fetus to get such an outcome might not object to choosing from among embryos to get it, particularly since more embryos must be produced than will be used. Parents may think: Since a selection process is going to take place anyway (for healthy genes), why not select for things like hair color as well?

Such questions become less abstract with a little knowledge of how PGD works. Many of its steps are those that occur with any in vitro fertilization (IVF). The starting point for IVF is to administer to the prospective mother a group of hormones that stimulate formation of a large group of eggs in her ovaries. (These eggs actually are "oocytes," which develop into eggs through the combined processes of meiosis, ovulation, and fertilization.) Anywhere from a week to two weeks later, a physician uses a needle attached to an ultrasound guide to remove the mature eggs from the ovaries. Though numbers vary, it would not be uncommon for 10 or 12 eggs to be recovered in this process. The husband then provides sperm to fertilize the eggs, with this fertilization taking place in a laboratory.

Three days after fertilization, each egg has divided into an eight-cell embryo. If you look at **FIGURE E12.1.1a**, you can see what comes next: Using a tiny hollow tube called a pipette, a physician gently suctions one of the eight cells away from the others. (Embryonic cells are so undifferentiated at this point that the loss of any one of them does not interfere with development of the others into a child.) The cell that is removed provides the DNA for testing. If you look at **FIGURE E12.1.1b**, you can see the results of one of the tests. Recall that children born with Down syndrome generally have a third copy of chromosome 21. It is possible to *test* for a third chromosome 21 by preparing DNA "bases" that will latch onto the unique set of DNA bases that exist on chromosome 21. Moreover, these "homing" DNA bases are fluorescent—they light up. The result is what you see in the figure: The prepared bases have latched onto three chromosome 21's in these embryonic cells; they have revealed that, had this embryo been implanted in the mother, the result would have been a child with Down syndrome.

By use of this test and others, embryos produced through IVF are ranked according to their health. Parents then choose from among them, selecting perhaps three or four that will be transferred back into the mother by a physician.

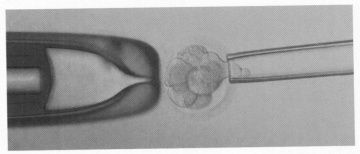
(a)

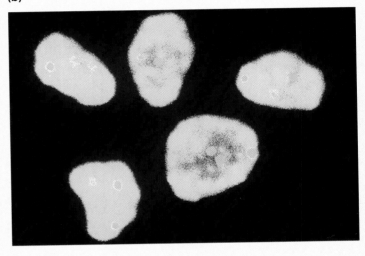
(b)

FIGURE E12.1.1

(a) Removing a Cell for Testing A human embryo at the eight-cell stage (center), being manipulated to have one of its cells removed for preimplantation genetic diagnosis. A pipette at left holds the embryo while a smaller pipette at right gently suctions off one cell from it. The cell's genetic material is then checked for abnormalities. If this embryo is determined to be normal, it may be inserted back into the mother's uterus, where it can develop into a baby.

(b) Down Syndrome Diagnosed Genetic testing of this embryo has revealed that each of its cells has three copies of chromosome 21, indicated by the pink areas in the nuclei of the cells. Such an embryo would develop into a child with Down syndrome. The pink areas represent fluorescently labeled DNA that has bound or "hybridized" with regions of DNA specific to chromosome 21.

Abnormal Numbers of Sex Chromosomes

Aneuploidy can affect not only autosomes but sex chromosomes as well, meaning the two X chromosomes that females have and the X and Y chromosomes that males have. Embryos often survive sex chromosome aneuploidies, but the result, once again, is usually debilitations. An example is Turner syndrome, which produces people who are phenotypically female, but who have only one X chromosome. Such females, then, have only 45 chromosomes, rather than the usual 46. Their state is sometimes referred to as XO, the "O" signifying the missing X chromosome. The absence of a second sex chromosome causes a range of afflictions.

Females with Turner syndrome have ovaries that don't develop properly (which causes sterility), they are generally short, and they often have brown spots (called nevi) on their skin.

Turner syndrome results from a loss of a chromosome, but all manner of sex chromosome *additions* can also take place. There are, for example, XXY men, who while phenotypically male in most respects, tend to have a number of feminine features: some breast development, a more feminine figure, and lack of facial hair. When coupled with other characteristics (such as tall stature and dysfunctional testicles), the result is a condition called Klinefelter syndrome.

Beginning in the late 1960s, it became possible to test developing fetuses for genetic abnormalities. In recent years, however, reproductive technology has moved beyond testing. Prospective parents can now choose, from a group of embryos they have created, embryos that do not have any recognizable genetic defects. You can read more about this in Essay 12.1, "PGD: Screening for a Healthy Child," on page 196.

Aneuploidy and Cancer

We've thus far looked at varying conditions that can result when aneuploidy takes place in meiosis, which is to say the cell divisions that produce eggs or sperm. But it's possible for aneuploidy to take place in *mitosis*, meaning regular cell division in non-sex or "somatic" cells. As one cell divides into two, a given chromatid can fail to migrate to its proper "pole," and one resulting daughter cell ends up with one chromosome too many, while the other ends up with one chromosome too few.

Now, note that this kind of aneuploid event need not take place in an embryo—it might take place in a child, a teenager, or an adult. If so, the only cells affected would be the "line" of cells that stem from the original two cells that underwent aneuploidy. Put another way, not every cell in the body would be affected, as is the case with aneuploidy that comes about in meiosis; only the daughter cells of the original aneuploid cells would have the wrong number of chromosomes. You might think: Well, this is bound to be less serious than aneuploidy resulting from meiosis; and in general, this is the case. Most aneuploid cells resulting from mitosis will die, but a few can survive. Indeed, aneuploidy is often seen in a destructive kind of "super cell" found in the body, the cancer cell.

Researchers have known about aneuploidy in cancer cells for decades, but for many years it has been assumed that these aneuploidies are the *result* of a cancer in progress, rather than the *cause* of the cancer. In recent years, however, this idea has been turned on its head. Some prominent cancer researchers believe that aneuploidy may be a cause of cancer. To be sure, this idea differs from the leading hypothesis about cancer: that it results from a series of mutations to individual genes on chromosomes. (For an account of this view, see Essay 9.1, "The Cell Cycle Runs Amok: Cancer," on page 148.) The problem with this idea is that, despite decades of research, no one has yet identified a series of mutations that will predictably cause any of the most serious forms of cancer. The hypothesis about aneuploidy presents an alternative, but only time will tell whether it's correct. Don't be surprised, however, if in the near future you start seeing newspaper stories about aneuploidy and cancer. We're certain that the cellular mistake of aneuploidy has devastating consequences every day. The question is whether cancer is one of them.

On to DNA

For several chapters, you've been looking at genetics from the viewpoint of the genetic packages called chromosomes. Important as chromosomes are, it's probably been apparent that the real genetic control lies with the units that help make them up: the DNA sequences we call genes. As you've seen, it is genes that make pea-plant flowers purple or white—and that give a person Marfan's syndrome or not. It's time to explore just what these genes are and how they operate.

So Far...

1. A medical pedigree is used to identify _____ disorders that run in _____.

2. The most common cause of aneuploidy is a failure of homologous chromosomes or sister chromatids to _____ in meiosis, which leads to one daughter cell having _____, while the other daughter cell has _____.

3. Which of the following can be a result of aneuploidy: (a) miscarriage (b) Down syndrome (c) Turner syndrome?

So Far... Answers:

1. *genetic; families*
2. *separate; one chromosome too many; one too few*
3. *a, b, and c*

Chapter Review

Summary

12.1 Tall, Fit, and Vulnerable

- **Mendelian genetic disorders**—A number of human illnesses are genetic disorders that are caused by a single dysfunctional gene or chromosome. These disorders provide insights into the way heredity works in human beings. (page 187)

12.2 X-Linked Inheritance

- **Male vulnerability**—Certain human conditions, such as red-green color blindness and hemophilia, are called X-linked conditions because they stem from dysfunctional genes located on the X chromosome. Men are more likely than women to suffer from these conditions because men have only a single X chromosome. A woman with, for example, a dysfunctional blood-clotting allele on one of her X chromosomes usually will be protected from hemophilia by a functional allele that lies on her second X chromosome. (page 188)

- **Recessive conditions**—Hemophilia and red-green color blindness are examples of recessive genetic conditions, meaning conditions that will not exist in the presence of functional alleles. (page 189)

- **Carriers**—Given the nature of recessive genetic conditions, persons who do not suffer from such conditions themselves can still possess an allele for them, which they can pass on to their offspring. Such persons, referred to as carriers of the condition, are heterozygous for it—the alleles they have for the trait differ, one of them being functional, the other being nonfunctional. (page 190)

Web Tutorial 12.1 X-Linked Recessive Traits

12.3 Autosomal Genetic Disorders

- **Recessive disorders**—Sickle-cell anemia is an example of a recessive autosomal disorder. It is autosomal because the genetic defect that brings it about involves neither the X nor Y chromosome. It is recessive because persons must be homozygous for the sickle-cell allele in order to suffer from the condition—they must have two alleles that code for the same sickle-cell hemoglobin protein. (page 190)

- **Dominant disorders**—Some genetic disorders are referred to as dominant disorders, meaning those in which a single allele can bring about the condition, regardless of whether a person also has a normal allele. Marfan's syndrome is an example of a dominant disorder. (page 191)

12.4 Tracking Traits with Pedigrees

- **Medical pedigrees**—In tracking inherited diseases, scientists often find it helpful to construct medical pedigrees, meaning familial histories that normally take the form of diagrams. Pedigrees allow experts to make deductions about the genetic makeup of several generations of family members. (page 192)

12.5 Aberrations in Chromosomal Sets: Polyploidy

- **Polyploidy**—Human beings and many other species have diploid or paired sets of chromosomes. In human beings, this means 46 chromosomes in all: 22 pairs of autosomes and either an XX chromosome pair (for females) or an XY pair (for males). The state of having more than two sets of chromosomes is called polyploidy. Many plants are polyploid, but the condition is inevitably fatal for human beings. (page 193)

12.6 Incorrect Chromosome Number: Aneuploidy

- **The nature of aneuploidy**—Aneuploidy is a condition in which an organism has either more or fewer chromosomes than normally exist in its species' full set. Aneuploidy is responsible for a large proportion of the miscarriages that occur in human pregnancies. A small proportion of embryos survive aneuploidy, but the children that result from these embryos are born with such conditions as Down syndrome. The cause of aneuploidy usually is nondisjunction, in which homologous chromosomes or sister chromatids fail to separate correctly in meiosis, leading to eggs or sperm that have one too many or one too few chromosomes. (page 194)

- **Aneuploidy and cancer**—Aneuploidy can come about in simple cell division or mitosis as well as in meiosis. A growing body of researchers believe that mitotic aneuploidy is at least involved in the promotion of some kinds of human cancers, with some of these researchers speculating that aneuploidy can be the triggering event in the formation of these cancers. (page 198)

Web Tutorial 12.2 Mistakes in Meiosis

Key Terms

aneuploidy p. 194

carrier p. 191

dominant disorder p. 191

Down syndrome p. 195

nondisjunction p. 194

pedigree p. 192

polyploidy p. 193

recessive disorder p. 189

Testing Your Understanding

In Your Own Words *(answers in the back of the book)*

1. How can the serious genetic defect known as aneuploidy be both widespread and relatively unrecognized?

2. In sex-linked diseases, which gender is more frequently affected—male or female? Why?

3. The text notes the possibility that some forms of human cancer may result from aneuploidy. What is the fundamental difference between an aneuploidy that might lead to cancer and one that results in Down syndrome?

Thinking about What You've Learned

1. The text notes that the technique known as preimplantation genetic diagnosis (PGD) allows prospective parents to screen embryos not only for serious genetic defects such as Down syndrome, but for traits that may be regarded as more or less desirable, such as male or female gender. What limits, if any, do you think should be placed on parents' ability to choose from among embryos for traits such as this?

2. One dominant genetic condition, Huntington disease, sends all of its victims into dementia, physical immobility, and ultimately death, but usually does not start doing so until the person is at least 35. Imagine yourself as the adolescent child of a parent who has started to manifest Huntington symptoms. Your chances of inheriting this incurable condition are one out of two, but you yourself have thus far shown no symptoms of it. The question is: Do you get yourself tested to see if you inherited the harmful Huntington allele? Do you find out for sure whether you will, or will not, have this illness, or do you live with uncertainty? Assuming you've arrived at an answer, consider it now in the context of the conclusion that most people in this situation come to. Ninety-five percent of such people decline to be tested.

3. A female friend of yours is red-green color blind. She tells you she inherited the condition from her father, who is also red-green color blind. Do you agree with her analysis of how she came to be color blind?

Genetics Problems *(answers in the back of the book)*

1. A man who is a carrier for sickle-cell anemia, a recessive genetic disease, has a child with a normal, noncarrier woman. What proportion of their children are expected to be afflicted with sickle-cell anemia?

2. Hemophilia is an X-linked recessive disease that prevents blood clotting. If a woman who is a carrier for hemophilia has children with a normal man, what proportion of their sons are expected to have hemophilia?

3. Neurofibromatosis is a genetic disorder associated with an allele of one autosomal gene. Individuals with neurofibromatosis have uneven skin pigmentation and skin tumors. A man with neurofibromatosis has children with a normal woman who does not carry the allele for this disorder. The couple has 5 children, 3 of whom have neurofibromatosis. The most likely explanation for this outcome is that neurofibromatosis is a _____ trait and the man is _____.

   a. recessive; homozygous recessive

   b. recessive; heterozygous

   c. dominant; homozygous dominant

   d. dominant; heterozygous

   e. a and b are both possible explanations

Consider the following pedigree for a human autosomal trait. (Note that in pedigrees, generations are indicated by Roman numerals, and individuals within a generation are numbered from left to right with Arabic numerals.)

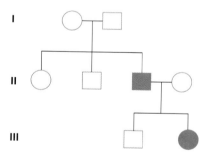

4. Is the allele that determines this trait dominant or recessive? Explain your reasoning.

5. Using symbols *A* and *a* for the dominant and recessive alleles, respectively, what is the genotype of the affected male in the second generation (individual II-3)?

6. What is the genotype of this individual's mate?

7. If female III-2 has children with a heterozygous man, what proportion of their children are expected to have this trait?

8. Consider the following pedigree, which posits a fantasy gene for ear shape, symbolized by *P* for the regular allele or *p* for a pointy-shaped allele if this is an autosomal trait; or by X^+ for the regular allele or X^p for the pointy allele if this is an X-linked trait.

   This pedigree indicates the trait is most likely inherited as an:

   a. autosomal dominant
   b. autosomal recessive
   c. X-linked dominant
   d. X-linked recessive

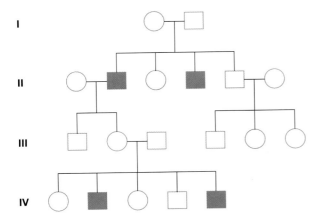

9. What is the genotype of individual I-1?

10. If a woman who is a carrier for this trait had children with individual III-1, what proportion of their children are expected to show this trait? What proportion of their sons are expected to show this trait?

11. Construct a pedigree to answer the following question: Red-green color blindness is an X-linked recessive trait. A woman who has a color-blind mother and a father with normal color vision has a son with a man with normal vision. What is the chance that the son is color blind?

12. Hyperphosphatemia, a disease that causes a form of rickets (abnormal bone growth and development), is inherited as an X-linked dominant trait. A woman who is heterozygous for the disease allele has children with a normal man. What proportion of their sons will have hyperphosphatemia? What proportion of their daughters will have hyperphosphatemia? If hyperphosphatemia were an X-linked recessive trait, how would your answer differ?

DNA Structure and Replication

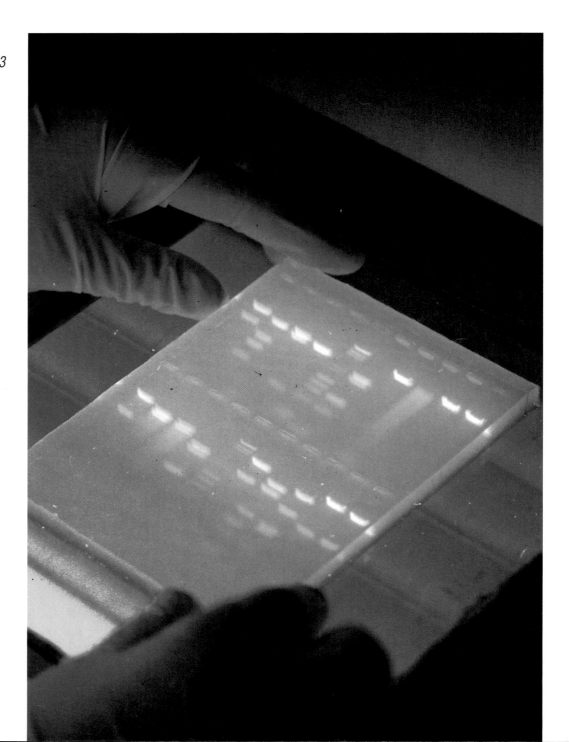

LONG BEACH PRESS-TELEGRAM

CRICK MODEL CRACKED CODE OF LIFE
OBITUARY: 1950S WORK WITH JAMES WATSON LED TO DNA 'DOUBLE HELIX'

By Michelle Morgante,
Associated Press
July 30, 2004

Nobel Prize-winning scientist Francis Crick, who co-discovered the spiral, "double-helix" structure of DNA in 1953 and opened the way for everything from gene-spliced crops and medicines to DNA fingerprinting and the genetic detection of diseases, has died. He was 88.

13.1 What Do Genes Do, and What Are They Made of?

Obituaries are not places to be stingy with praise, but even so, could the obituary above be correct about Francis Crick? Did Crick and his colleague, James Watson, crack the "code of life"? And did their discovery set the stage for gene-spliced crops, DNA fingerprinting, and more? The answer is yes. To the extent that there is a code of life, Watson and Crick cracked it. And their achievement was momentous in the sense that it *opened doors*. It allowed science to go down paths that previously could not have been imagined. Researchers who had spent

their careers puzzling over how life could be passed on from one generation to the next had an answer following Watson and Crick's discovery. Likewise, scientists had long wondered how even the humblest of living things could turn out a fabulous array of the molecular tools called proteins. With Watson and Crick's discovery, a key piece of the puzzle fell into place. And insights such as this led directly to biology as we know it today—to DNA fingerprinting, genetically engineered medicines, genetically engineered foods, the Human Genome Project, animal cloning, and more.

Francis Crick, left, and James Watson, photographed in 1953, shortly after their discovery of the structure of DNA.

To get a feel for what Watson and Crick did, it's helpful to consider the state of biological knowledge prior to the time they made their discovery. By 1920, it had been demonstrated beyond any doubt that genetic information resided on chromosomes. Within a decade or so, by looking at some abnormally large fly chromosomes, scientists could observe in great detail such chromosomal processes as crossing over. By viewing the banding patterns on these chromosomes, they could even identify the rough location of some genes.

Yet what was a gene? What was the physical nature of this unit of heredity, and how did it work? No one knew. Because genes are much smaller than the chromosomes on which they reside, there was no hope of simply viewing one under a microscope. In observing chromosomes, scientists were only "looking" at genes in the way that any of us is "looking" at lunar rocks by glancing up at the moon. Through the 1920s and 1930s, then, genes could be described only in vague, functional terms, as in: A gene is an entity that lies along a chromosome and brings about a phenotypic trait in an organism (for example, the green or yellow peas of Mendel's experiments).

Things were to clear up somewhat in the ensuing years. The seminal achievement in genetics in the late 1930s and early 1940s was a more concrete description of what genes do: They bring about the production of proteins. The central achievement in genetics from the mid-1940s to the early 1950s was strong evidence indicating what genes are composed of: deoxyribonucleic acid, or DNA for short.

DNA Structure and the Rise of Molecular Biology

By the early 1950s, a key question became *how* DNA carried out its genetic function. To find out, scientists had to piece together its exact chemical structure. This investigation turned out to be a watershed event in biology. Just as opening a mechanical watch and observing how its parts fit together would allow a person to understand how the watch works, so deciphering the structure of DNA allowed biologists to understand how genetics works at its most fundamental level.

Apart from what this investigation uncovered, the inquiry itself stood as a symbol of a new era in biology. When Gregor Mendel did his experiments with peas, he was looking at whole organisms. When his successors in the early twentieth century were working on genetics, they were looking at whole chromosomes, which they knew to be composed of several types of molecules. In the early 1950s, however, the search turned to a single molecule. How was DNA structured, and how did this structure allow it to carry out its various functions? Biological research of this sort has grown ever more important in the decades since the 1950s. It is today known as **molecular biology**: the investigation of life at the level of its individual molecules.

A BROADER VIEW

A Fundamental Discovery

What's the linkage between Watson and Crick's discovery and today's real-world biology? Well, DNA fingerprinting works by obtaining, for example, blood from a crime scene and comparing the DNA in this blood with DNA taken from a suspect. What police look for in these instances are DNA matches—sequences of A's, T's, G's, and C's that are the same in both the crime-scene DNA and the suspect's DNA. But how is it that we came to have a concept of DNA sequences? Because Watson and Crick's model implied that bases could be laid out this way in the double helix.

13.2 Watson and Crick: The Double Helix

James Watson and Francis Crick may not be instantly recognized scientific names in the way that, say, Albert Einstein or Louis Pasteur are, but there's a certain public awareness that these two researchers did something important in connection with DNA (FIGURE 13.1). What they did was present to the world, in 1953, the structure of DNA: Atom-by-atom, bond-by-bond, this is how DNA fits together, they said in unveiling DNA's now-famous configuration, the double helix.

The two were a seemingly unlikely pair to make an epochal scientific discovery. Watson, an American, was a 23-year-old who was scarcely more than a year out of graduate school, and Crick, an Englishman, was a 35-year-old just then working on his doctorate when the two met in the fall of 1951 at Cambridge University in England. Ending up together by coincidence at the same laboratory, they realized in short order their mutual interest in the structure of DNA and, to the neglect of projects they were supposed to be working on, they began several

FIGURE 13.1 **Young and Famous** James Watson (on the left) and Francis Crick, with a model of the DNA double helix, shortly after they published their paper on the molecule's structure.

rounds of model building and brainstorming that resulted in their breakthrough.

Watson and Crick were greatly aided in their investigation by the work of others. Though the DNA molecule was too small to be seen by even the most powerful microscopes of the time, something about its structure could be inferred from a technique called *X-ray diffraction*. In this process, a purified form of a molecule is bombarded with X rays. The way these rays scatter upon impact then reveals something about the structure of the molecule.

If you look at FIGURE 13.2, you can see the results of some X-ray diffraction. As you can imagine, it takes a highly trained observer to be able to deduce anything about the structure of a molecule from such an image. Fortunately for Watson and Crick, such a person was working just up the road from them. She was Rosalind Franklin, a researcher at King's College in London and one of the handful of individuals then skilled in performing X-ray diffraction on DNA. She and her colleague, Maurice Wilkins, were themselves working on the structure of DNA at the time, as were other researchers in America. Thus did Watson, at least, regard the search for DNA structure to be a race between several teams, a fact that concentrated his efforts wonderfully. In 1962 Watson, Crick, and Wilkins were awarded the Nobel Prize in Medicine or Physiology for their work on DNA. Rosalind Franklin

(a) Rosalind Franklin

(b) X-ray diffraction image of DNA

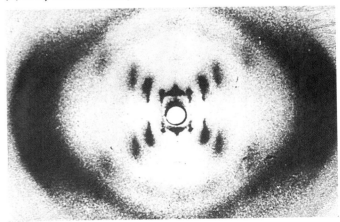

FIGURE 13.2

(a) DNA Investigator Rosalind Franklin, whose work in X-ray diffraction was important in revealing the structure of the DNA molecule.

(b) Imaging DNA One of Franklin's X-ray diffraction images of DNA. The "cross" formed of dark spots indicated the molecule had a helical structure.

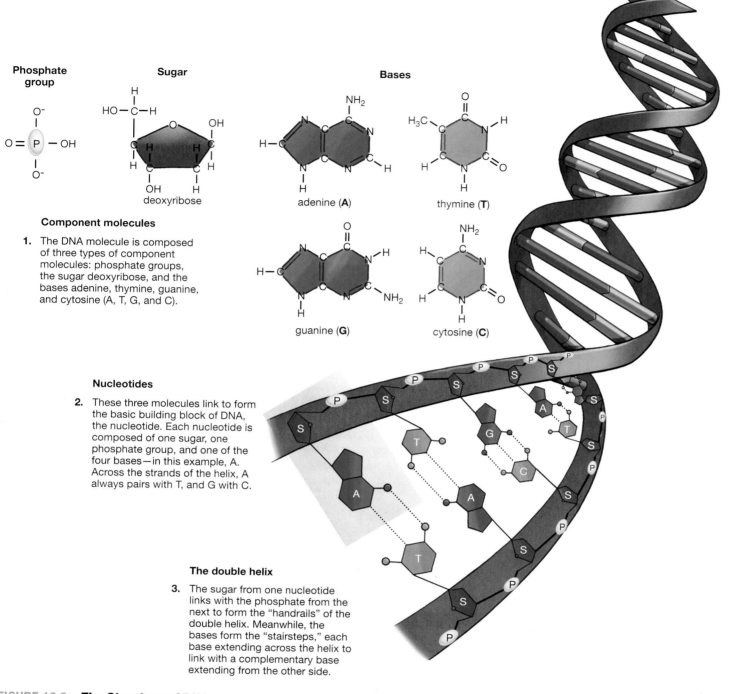

Phosphate group

Sugar

deoxyribose

Bases

adenine (**A**)

thymine (**T**)

guanine (**G**)

cytosine (**C**)

Component molecules

1. The DNA molecule is composed of three types of component molecules: phosphate groups, the sugar deoxyribose, and the bases adenine, thymine, guanine, and cytosine (A, T, G, and C).

Nucleotides

2. These three molecules link to form the basic building block of DNA, the nucleotide. Each nucleotide is composed of one sugar, one phosphate group, and one of the four bases—in this example, A. Across the strands of the helix, A always pairs with T, and G with C.

The double helix

3. The sugar from one nucleotide links with the phosphate from the next to form the "handrails" of the double helix. Meanwhile, the bases form the "stairsteps," each base extending across the helix to link with a complementary base extending from the other side.

FIGURE 13.3 The Structure of DNA

died of cancer in 1958 at the age of 37. Nobel Prizes are not awarded posthumously; it's unknown what would have happened had she lived.

Let's start now to look at DNA's structure, which will put you in a position to appreciate the achievement of Watson, Crick, and their fellow researchers.

13.3 The Components of DNA and Their Arrangement

If you look at **FIGURE 13.3**, you can see the component parts of DNA. First, there is a phosphate group, and second, a sugar called *deoxyribose*. Third, there are four possible DNA "bases": adenine, guanine, thymine, and cytosine—A, G, T, and C, for short. When you link together a phosphate group, a deoxyribose molecule, and one of the four bases, you get the basic building block of DNA—the *nucleotide*, one of which is boxed in the larger DNA molecule in the figure.

When you look at this larger figure as a whole, you can see that the double helix looks something like a spiral staircase. By focusing on its exterior "handrails," you can see how sugar and phosphate fit together to form a kind of chain, with its links going: sugar-phosphate-sugar-phosphate (symbolized in the figure by S and P). The "steps" lying between these handrails are composed of DNA's bases. As the figure shows, each of the bases amounts to a *half*-stair-step, if you will. Each extends inward from one handrail of the double helix and is then joined to a base extending inward from the *other* handrail of the DNA molecule. (The bases are linked via hydrogen bonds, symbolized by the dotted lines in the figure.)

The figure shows some examples of one base being paired with another. At the bottom, for example, an A base on the left is paired with a T base on the right. Go up two sets of nucleotides, and a G on the left is being paired with a C on the right. This turns out to be one of the fundamental rules about DNA structure. If you viewed a billion DNA *base pairs*, as they're called, you would find the same thing: A always pairing with T, and G always pairing with C, across the helix. (Any two bases that can pair together in this way are said to be *complementary*; thus A is complementary to T, and G is complementary to C.)

The Structure of DNA Gives Away the Secret of Replication

It was DNA's very structure that suggested the answer to one of the great questions of genetics: How is genetic information passed on? To put this another way, how does a cell make a copy of its own DNA? As you've seen, a full complement of our DNA is contained in nearly every cell in our body. Yet cells divide, and the daughter cells contain exactly the same DNA complement as the parent cell. This means that genetic information is passed on by means of DNA being copied, with one copy of this molecule ending up in each daughter cell. Indeed, such copying extends to the formation of egg and sperm cells, meaning this is the way genetic information is passed on from one *generation* to the next. But how could this work?

The structure Watson and Crick discovered suggested a way. We've observed that A must always pair with T, and G with C, across the two strands of the double helix. This rule, Watson and Crick saw, meant that each single strand of DNA could serve as a *template* for the synthesis of a new single strand (see FIGURE 13.4). Each A on an old strand would specify the place for a T on the new, each G on the old a place for C on the new, and so forth. All that was required was for the two old strands to separate—splitting the stair steps right down the middle—and for new strands to be synthesized that were complementary to the old. As it turned out, this was exactly how things worked, as you'll see shortly.

The Structure of DNA Gives Away the Secret of Protein Production

The second thing the structure suggested was a partial answer to another great question of genetics: How the molecule of heredity could be versatile enough to specify the dazzling array of proteins that all living things

A BROADER VIEW

Not a "Dumb" Molecule

Given DNA's ability to retain a vast amount of information, it's an irony that, for years, most scientists thought DNA was a "dumb" molecule, incapable of encoding anything. A hypothesis popular in the 1940s held that DNA's bases were simply repeated over and over in the same order (A, T, G, C). But, just as the key quality of an alphabet is that its letters can be variably arranged to spell different words (cat, act), so the key quality of DNA is that its bases can be variably arranged to specify different proteins.

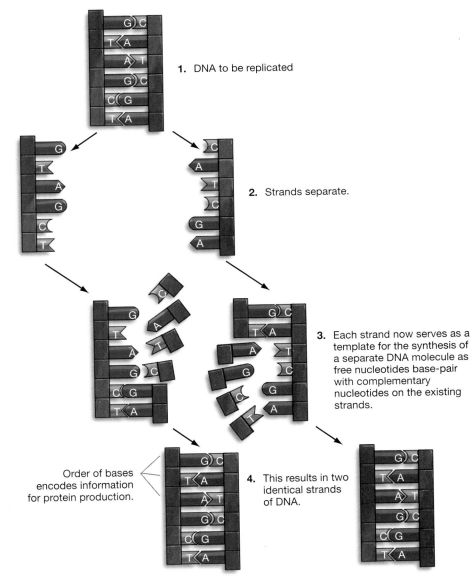

FIGURE 13.4 **DNA Replication** The result of DNA replication is two identical molecules of DNA, whereas the process began with one.

1. DNA to be replicated

2. Strands separate.

3. Each strand now serves as a template for the synthesis of a separate DNA molecule as free nucleotides base-pair with complementary nucleotides on the existing strands.

Order of bases encodes information for protein production.

4. This results in two identical strands of DNA.

So Far... Answers:

1. structure of DNA

2. A; C base

3. bases; protein

produce. The handrails of DNA's double helix are monotonous—phosphate, sugar, phosphate, sugar. But the bases can be laid out along these handrails in an extremely varied manner. A has to pair with T and G with C *across* the helix; but *along* the handrails, the bases can come in any order. Look at the top drawing in Figure 13.4 and see the order this hypothetical group of bases comes in; starting from the bottom right, A-G-C-T-A-C. Strung together by the thousands, these bases can specify a particular protein, just as in Morse code a series of short and long clicks can specify a particular word.

So Far...

1. James Watson and Francis Crick discovered the _____.

2. The key to DNA replication is that each _____ base on an original strand of DNA serves as a template for the addition of a T base on the new strand, while each _____ on the original strand serves as a template for the addition of a G base.

3. A particular sequence of DNA _____ can specify a particular _____.

The Building Blocks of DNA Replication

You have already looked briefly at the basic process by which DNA is copied or "replicated," but we now need to review some of the details of this process. The basic steps here are straightforward. The essential building block of DNA is the unit called the **nucleotide**, which can be seen in the box in Figure 13.3. Every nucleotide contains one sugar and one phosphate group. What differentiates one nucleotide from another is the base that is attached to it; in this case the base is A, but as you can see, others have T, C, or G.

Figure 13.4 shows the process of DNA replication in overview. Note the first thing that happens: The joined strands of the double helix unwind, separating from one another. The nucleotides on each of the single strands are then paired with free-floating nucleotides that line up in new, complementary strands. Because *both* strands of the original double helix are being paired with new strands, the end product is two double helices, where before there was one.

You can see in Figure 13.4 that the addition of nucleotides moves in differing directions on the two strands. This is so because of something you can see in Figure 13.3: The two strands of the double helix have opposite orientations. Notice that the sugars of the right-hand strand, for example, are upside down relative to the sugars of the left-hand strand. (If the strands were people, we would think of them as lying head to *foot* relative to each other.) Because new nucleotides can be added to only one end of a DNA strand—what's known as the three-prime end—the nucleotides are added in different directions.

DNA Replication: Something Old, Something New It's worth drawing a little finer point on one aspect of this process: Each resulting double helix is a combination of the old and the new. Each has one "parental" strand of DNA and one newly synthesized complementary strand. This combination is conceptually important because it is how life builds upon itself. You can see this illustrated in FIGURE 13.5.

Eventually, the two double helices are fully formed. A little later they part company, during mitosis, with each newly formed double helix moving into a separate cell and thereafter serving as an independently functioning segment of DNA. (At the chromosomal level, each newly replicated helix is, when joined by appropriate proteins, one of the sister chromatids you've been seeing so much of. Sister chromatids eventually separate, of course, thus becoming independent chromosomes.)

However simple this process may sound in overview, its details are enormously complicated. As you might imagine, such a process could not proceed without enzymes to catalyze it. To name just two groups of them, there are enzymes called

FIGURE 13.5 **How Life Builds on Itself** Each newly synthesized DNA molecule is a combination of the old and the new. An existing DNA molecule unwinds, and each of the resulting single strands (the old) serves as a template for a complementary strand that will be formed through base pairing (the new).

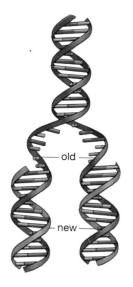

helicases that unwind the double helix, separating its two strands to make the bases on them available for base pairing. There is also another group of enzymes, collectively known as **DNA polymerases**, that move along each strand of the double helix, joining together nucleotides as they are added—one by one—to form the new, complementary strands of DNA.

Editing Out Mistakes   The base pairing that goes on in replication happens hundreds, then thousands, then millions of times in a given stretch of DNA: Free-standing A's are aligned with complementary T's, while C's are aligned with G's. Given the numbers of base pairs involved, the amazing thing is how few mismatches there are by the time the process is completed. The error rate in DNA replication—the rate at which the *wrong* bases have been brought together—might be only one in every billion bases by the end of replication. Yet *during* replication, such a mistake might be made once in every 100,000 bases. Obviously, to start with one number of mistakes but to end up with far fewer, the cell's genetic machinery has to be capable of correcting its errors.

This happens partly through the services of the versatile DNA polymerases, which are able to perform a kind of DNA editing: They remove a mismatched nucleotide and replace it with a proper one. Interestingly enough, this happens through a kind of backspacing. Normally the DNA polymerases are moving along a DNA chain, linking together recently arrived nucleotides. When an error is detected, however, they stop, move "backward," remove the incorrect nucleotide, put in the correct one, and then move forward again.

A BROADER VIEW

The Value of Template Copying

Make a photocopy of, say, a pencil sketch; then make a copy of the copy, and keep going with this process for 15 generations. What have you got? A blurry mess that has lost most of the qualities of the original. Now, take a given gene in a cell and let it get copied over and over during the process of cell division. Keep this going for say, 10,000 generations, and what have you got? Exactly the same gene, with all the qualities of the original. It's the template process that makes this high-fidelity copying possible. The complementary strands that result are newly synthesized—they all start out "young" in a sense—and each of these strands contains exactly the same information as its parental strand.

13.4 Mutation: Another Name for a Permanent Change in a DNA Sequence

When you start thinking about alterations in DNA structure, you have arrived at an important concept: that of a **mutation**, which can be defined as a permanent alteration of a DNA base sequence. Such alterations come about because the cell's various DNA error-correcting mechanisms are not foolproof; they do not correct all the mistakes that occur. For the balance of this chapter, we'll focus on these genetic mistakes, which have a powerful effect both on human health and on evolution.

DNA's makeup can be altered in many ways. A slight change in the chemical form of a base might, for example, cause a G to link up across the helix with a T (instead of with its normal partner, C), as you can see in FIGURE 13.6. Then, in a subsequent round of DNA replication, the cell might "repair" this error in such a way that a permanent mistake is introduced: An A-T pair now exists, whereas the original sequence had a G-C pair. Permanent mistakes like these are called **point mutations**, meaning a mutation of a single base pair in the genome.

In what way are mutations "permanent"? Think of how DNA replication works. Before any cell can divide, it must first make a copy of its complement of DNA. Should this DNA contain an uncorrected mistake, it too will be copied again and again with each succeeding cell division. Most mutations have no noticeable effect on an organism. But the concept of mutation is quite rightly fearful to us, because in relatively rare instances, mutations can have disastrous effects.

FIGURE 13.6   **One Route to a Mutation**

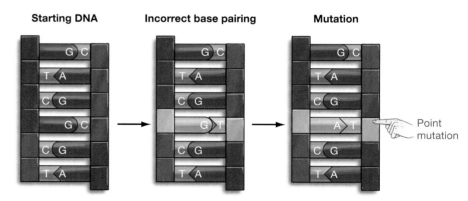

Starting DNA   **Incorrect base pairing**   **Mutation**

Point mutation

1. In replicating a cell's DNA, mistakes are sometimes made, such that one base can be paired with another base that is not complementary to it (G with T in this case).

2. The next time a cell replicates its DNA, the replication repair mechanism may "fix" this error in such a way that a permanent alteration in the DNA sequence results. The original G will be replaced, instead of the wrongly added T. The result is an A-T base pair, whereas the cell started with a G-C base pair.

Examples of Mutations: Cancer and Huntington

A cancerous growth is a line of cells that has undergone some special kinds of mutations—ones that cause the affected cells to proliferate wildly. As an example, the skin cancer known as melanoma causes skin cells called melanocytes to start dividing very rapidly. For our purposes, the question is: How does this process get going? In 2002, British scientists looked at numerous lines of melanoma cells that had been taken from cancer patients. They found that 59 percent of these cells contained a mutation in a gene called *BRAF*. This gene, like all others, amounts to a sequence of DNA nucleotides—a sequence of C's, T's, A's, and G's. When the British scientists looked at how the *BRAF* gene had actually mutated, they found that in 92 percent of cases, there had been a single change: An A had been substituted for a T at *BRAF*'s 1,796th nucleotide. What they found, in other words, was a point mutation, nearly two thousand bases within the sequence of nucleotides that makes up the *BRAF* gene. And the effect of this mutation? It produces a protein that is a slightly altered version of the normal BRAF protein. This protein is altered enough, however, that it keeps melanocytes moving through the cell-division cycle, causing them to multiply wildly.

Cell proliferation is not the only kind of trouble mutations can bring about. You may have heard of an affliction, called Huntington disease, that causes spastic movements, severe dementia, and death among its victims. In all people there is a gene, called *IT15*, that has within it a number of repeats of a particular "triplet" of bases, CAG. People with 34 or fewer CAG repeats in their *IT15* gene are fine; they will suffer no illness at all from the gene. People with 35–39 CAG repeats, however, may develop Huntington, though this is not a certainty. Meanwhile, people who have more than 40 repeats definitely will develop the disease, and the more repeats they have, the younger they will be when its symptoms first appear. (Those who develop the disease as juveniles generally have more than 55 repeats.) Unlike the case with cancer, however, the Huntington mutation does not cause cells to multiply wildly. Instead, the mutated *IT15* gene creates a faulty version of protein called huntingtin, which cannot be broken down by nerve cells and ends up building up inside them, eventually killing them. Because Huntington is caused by a repeating group of three nucleotides, it is referred to as a "trinucleotide repeat" disease. At least eight other diseases of the nervous system fall into this category.

Heritable and Non-Heritable Mutations

Though Huntington disease and melanoma are both caused by mutations, there is an important distinction to be made between them. Most mutations come about in the body's **somatic cells**, which is to say cells that do not become eggs or sperm. This is the case with the mutations that bring about melanoma. Conversely, some mutations arise in **germ-line cells**, meaning the cells that do become eggs or sperm. This is the case with the Huntington mutation. The important point here is that germ-line cell mutations are *heritable*—they can be passed on from one generation to the next, in the ways reviewed in Chapter 12. In contrast, though a line of melanoma cells may be quite harmful, it is separate from the line of cells that gives rise to eggs or sperm. For this reason, melanoma cannot be passed on from one generation to the next.

What Causes Mutations?

A critical question, of course, is what causes DNA to mutate? One answer is so-called environmental insults. The chemicals in cigarette smoke are powerful *mutagens*, meaning substances that can mutate DNA. So is the ultraviolet light that comes from the sun. Not all mutations are caused by environmental influences, however. Mutations happen simply as random, spontaneous events. The collision of water molecules with DNA can remove nucleotide bases, or alter them in such a way that their base-pairing properties are changed. The very process of eating

A BROADER VIEW

The Fear of Mutation

How is it that the word *mutant*, so loved in science fiction, can strike such panic in us? Perhaps it's because of the human fear not just of being different, but of being permanently different in the way implied by a mutation.

A BROADER VIEW

Heritable Mutations

Huntington disease serves as a case in point for our knowledge of genes and the proteins they code for. We know what the *faulty* version of the huntingtin protein does: It kills nerve cells. But what does the normal version of this protein do in the body? We have no idea.

and breathing produces so-called free radicals that can damage DNA. And the DNA replication machinery itself may introduce errors, irrespective of any outside influences.

The Value of Mistakes: Evolutionary Adaptation

While it's true that, for individuals, mutations can have a negative effect, for the living world as a whole mutations have been vitally important because of a role they play in evolution. It turns out that germ-line mutations are the primary means by which completely new genetic information can be added to a species' genome, in the form of new alleles (meaning variant forms of a gene). Organisms can combine *existing* alleles, in myriad ways. Think of meiosis, with its chromosomal part-swapping (crossing over) and reshuffling of chromosomes (independent assortment). Valuable as these processes are, no amount of genetic recombination could have produced, for example, the eyes that some living things possess. To go from no eyes to eyes, there had to have been some mutations along the line—some accidental reorderings of DNA sequences such that entirely new proteins were produced. Such adaptations are vital to living things, given their struggle to get along in environments that are constantly changing. (Temperatures shift; streams dry up.) If environments change, species need to change too, in order to survive. And the primary way major changes can come about is through mutations. You'll be looking at this topic again in the evolution unit of this book. For now, however, isn't it interesting to ponder the fact that the living world adapts partly through its mistakes?

On to How Genetic Information Is Put to Use

Since antiquity, human beings have speculated about how one generation of living things can give rise to another—about *how* it is that life goes on. Questions posed by the Greeks, by scientists in the Renaissance, by Gregor Mendel, and by others were finally answered in the 1950s and 1960s, when scientists came to understand DNA replication at the molecular level. The something-old, something-new quality to this replication—parental DNA strands serving as templates for new strands—was the detailed answer to the ancient question of how the qualities of living things can be passed down through generations.

Splendid as this function is, it is only one of the two great tasks carried out by our genetic machinery. You have just reviewed the process by which genetic information is replicated; now we'll look at the process by which this information is used. How can a stretch of DNA bring about the production of a protein? You'll see in the chapter coming up.

So Far...

1. In DNA replication, both of the strands of the double helix serve as templates for _____.

2. The _____ are a group of enzymes that add nucleotides to a replicating DNA chain.

3. A mutation is a _____ of a DNA base sequence.

Chapter Review

Summary

13.1 What Do Genes Do, and What Are They Made of?

- **A fundamental discovery**—James Watson and Francis Crick discovered the chemical structure of DNA in 1953. This event ushered in a new era in biology because it allowed researchers to understand some of the most fundamental processes in genetics. (page 203)

- **Molecular biology**—In trying to decipher the structure of DNA, Watson and Crick were performing work in molecular biology. This is the investigation of life at the level of its individual molecules. Molecular biology has grown greatly in importance since the 1950s. (page 204)

13.2 Watson and Crick: The Double Helix

- **History of the double-helix investigation**—Watson and Crick met in the early 1950s at Cambridge University in England and set about to decipher the structure of DNA. Their research was aided by the work of others, including Rosalind Franklin, who was using X-ray diffraction to learn about DNA's structure. (page 204)

13.3 The Components of DNA and Their Arrangement

- **DNA's structure**—The DNA molecule is composed of building blocks called nucleotides, each of which consists of one sugar (deoxyribose), one phosphate group, and one of four bases: adenine, guanine, thymine, or cytosine (A, G, T, or C). The sugar and phosphate groups are linked together in a chain that forms the "handrails" of the DNA double helix. Bases then extend inward from the handrails, with base pairs joined to each other in the middle by hydrogen bonds. In this base pairing, A always pairs with T across the helix, while G always pairs with C. (page 206)

- **DNA replication**—DNA is copied by means of each strand of DNA serving as a template for the synthesis of a new, complementary strand. The DNA double helix first divides down the middle. Each A on an original strand then specifies a place for a T in a new strand, each G specifies a place for a C on the new strand, and so forth. (page 208)

- **How life builds on itself**—Each double helix produced in replication is a combination of one parental strand of DNA and one newly syn-

thesized complementary strand. This is how life builds upon itself. A group of enzymes known as DNA polymerases are central to DNA replication; they move along the double helix, bonding together new nucleotides in complementary DNA strands. (page 208)

- **Coding for proteins**—DNA can encode the information for the huge number of proteins utilized by living things because the sequence of bases along DNA's handrails can be laid out in an extremely varied manner. A collection of bases in one order encodes the information for one protein, while a different sequence of bases encodes the information for a different protein. (page 207)

Web Tutorial 13.1 DNA

13.4 Mutation: Another Name for a Permanent Change in a DNA Sequence

- **What is a mutation?**—The error rate in DNA replication is very low, partly because repair enzymes are able to correct mistakes. When such mistakes are made and then not corrected, the result is a mutation: a permanent alteration in a cell's DNA base sequence. (page 209)

- **Mutation and cancer**—Most mutations have no effect on an organism, but when they do have an effect, it is generally negative. Cancers result from a line of cells that have undergone types of mutations that cause them to proliferate wildly. (page 210)

- **Heritable mutations**—Some mutations come about in the body's germ-line cells, meaning cells that become eggs or sperm. Such mutations are heritable: They can be passed on from one generation to another. The gene for Huntington disease, which is expressed in nerve cells, is a heritable, mutated form of a normal gene. (page 210)

- **Mutation and evolution**—Mutations have played a key role in evolution, in that they are the primary means by which new proteins come into existence. (page 211)

Web Tutorial 13.2 Mutations

Key Terms

DNA polymerase   p. 209

germ-line cell   p. 210

molecular biology   p. 204

mutation   p. 209

nucleotide   p. 208

point mutation   p. 209

somatic cell   p. 210

Testing Your Understanding

In Your Own Words *(answers in the back of the book)*

1. How does the phrase "something old, something new" describe the method of DNA replication employed by the cell?

2. What did Watson and Crick's discovery reveal that was so important?

3. What is a mutation? Are all mutations harmful?

Thinking about What You've Learned

1. Given the following DNA sequence, determine the complementary strand that would be added in replication:

   ATTGCATGATAGCC

2. Why does a germ-line mutation carry greater potential significance than a somatic mutation?

3. Would you expect cancer to arise more often in types of cells that divide frequently (such as skin cells) or in types of cells that divide rarely or not at all (such as nerve cells)? Explain your reasoning.

How Proteins Are Made

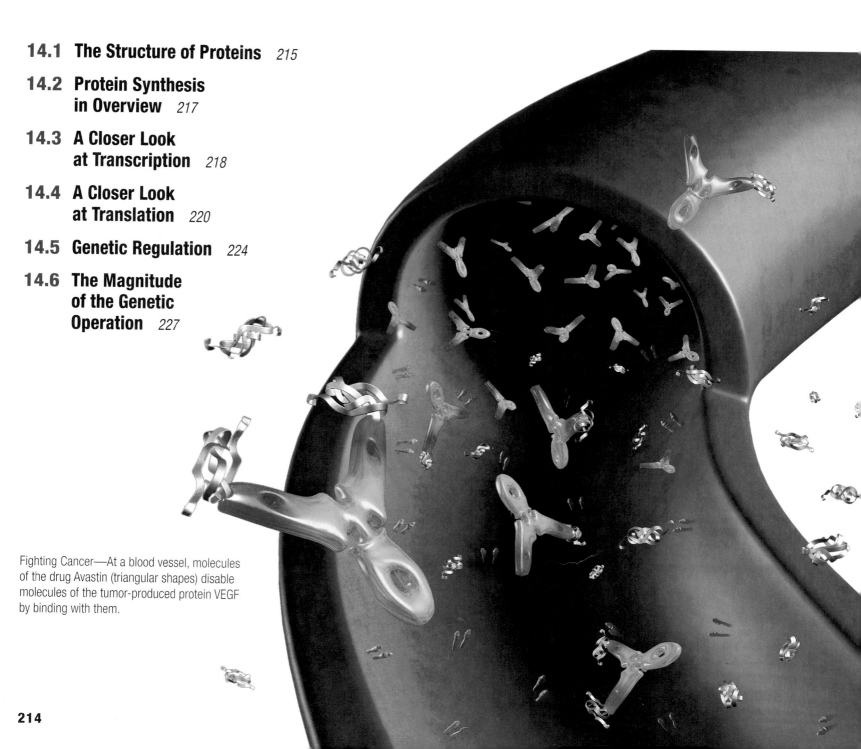

Fighting Cancer—At a blood vessel, molecules of the drug Avastin (triangular shapes) disable molecules of the tumor-produced protein VEGF by binding with them.

CONTRA COSTA TIMES

GENENTECH WINS CANCER DRUG OK
AVASTIN APPROVAL BY FDA DRIVES UP STOCK, IS SEEN AS BIG PROFIT DRIVER

By Judy Silber
February 27, 2004

Ushering in what could prove to be a new era in the fight against cancer, the U.S. Food and Drug Administration on Thursday approved Genentech Inc.'s drug Avastin for colorectal cancer.

Doctors enthusiastically welcomed the news, hailing the drug as a significant advance not only for its success in treating cancer, but because it is the first in a new category of drugs known as anti-angiogenesis treatments. "It's just a huge day for patients with colon cancer," said Robert Mass, a director of clinical oncology at South San Francisco–based Genentech.

14.1 The Structure of Proteins

As the newspaper story above makes clear, it was indeed a red-letter day when the FDA approved the drug Avastin for use in colon cancer. This is so because, in the war on cancer, Avastin represents a new kind of weapon. Most cancer drugs work directly on the cells in a cancer tumor. But Avastin works on the tumor's lifeline: its blood supply. Cancerous tumors can keep growing because they are able to surround themselves with networks of blood vessels. They accomplish this by inducing nearby blood vessels to sprout new vessels in their direction. The tumors convey the message, "more vessels needed over here," by sending a chemical signal to the existing vessels. But Avastin has a counterstrategy for this: It blocks the signal. The result is a tumor that is deprived of oxygen and nutrients—a tumor that is suffocated and starved. While Avastin is not a cure for cancer, it is a fundamentally new way of attacking it.

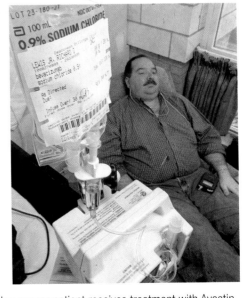

A colon cancer patient receives treatment with Avastin.

For our purposes, the thing to note is the nature of the players in this contest, both on the "dark side" (the cancerous cells) and the heroic side (the cancer treatment). How do cancerous cells send their signal to blood vessels? They export a protein (called VEGF) that binds with the blood vessels and gets the process of new vessel formation going. Now, what is the nature of Avastin? It too is a protein—a type of protein, called an antibody, that can selectively disable other proteins. That's what Avastin does: It binds to VEGF, and in so doing disables VEGF as a signal carrier.

This kind of protein-to-protein interaction may seem exotic, but in fact it is quite common in the body. Wherever you find complex signaling going on, proteins are likely to be involved. But then again, proteins are players in so many facets of our functioning that the challenge would be to name an area in which they are *not* involved. Hastening chemical reactions, acting as cellular passageways, transporting substances through the bloodstream—proteins do all this and more. Last chapter we reviewed the means by which genetic information is stored in DNA. But the vast bulk of this information is *used* to create the molecules called proteins. DNA may be a kind of playbook, but proteins are the actual players out in the field. The question is, how do we go from one to the other? How do we get from the information contained in DNA to an actual working protein? In this chapter, you'll find out.

Because you'll be dealing extensively with proteins here, now may be a good time to review some basic information about them. As you saw in Chapter 3, proteins fit

FIGURE 14.1　**The Structure of Proteins**

glycine **(gly)**

isoleucine **(ile)**

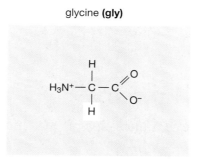
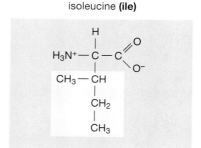

(a) Amino acids

The building blocks of proteins are amino acids such as glycine and isoleucine, which differ only in their side-chain composition (light-colored squares).

H₃N⁺— gly | ile | val | glu | gln | cys | cys | ala | ser | val | cys | ser | leu | tyr | gln | leu | glu | asn | tyr | cys | asn —C

(b) Polypeptide chain

These amino acids are strung together to form polypeptide chains. Pictured is one of the two polypeptide chains that make up the unusually small protein insulin.

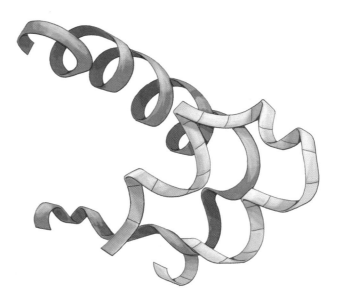

(c) Protein

Polypeptide chains function as proteins only when folded into their proper three-dimensional shape, as shown here for insulin. Note the position of the glycine and isoleucine amino acids in one of the insulin polypeptide chains (colored light green).

into the "building blocks" model of biological molecules. The blocks in this case are amino acids. String a number of these together and you have a polypeptide chain, which then folds up in a specific three-dimensional manner, resulting in a protein. Proteins are likely to be made of hundreds of amino acids strung together, often in several linked polypeptide chains.

Synthesizing Many Proteins from 20 Amino Acids

Though there are hundreds of thousands of different proteins, each one of them is put together from a starting set of a mere 20 amino acids. If you wonder how such diversity can proceed from such simplicity, think of the English language, which has thousands of words, but only 26 letters in its alphabet. It is the *order* in which letters occur that determines whether, for example, "cat" or "act" is spelled out; just so, it is the order of amino acids that determines what protein is synthesized.

If you look at FIGURE 14.1a, you can see the chemical structure of two free-standing amino acids, glycine (gly) and isoleucine (ile). If you look at FIGURE 14.1b, you can see that these two amino acids are the first two that occur in one of the two polypeptide chains that make up the unusually small protein we call insulin. FIGURE 14.1c then shows a three-dimensional representation of insulin—the folded-up form this protein assumes before taking on its function of helping to move blood sugar into cells. A list of all 20 primary amino acids and their three-letter abbreviations can be found in **Table 14.1** (next page).

For our purposes, the question is: How do the chains of amino acids that make up a protein come into being? How do gly, ile, val, and so forth come to be strung together in a specific order to create this protein? You know from what you've studied so far that genes contain the information for producing proteins and that genes amount to a series of chemical "bases"—the A's, T's, C's, and G's that lie along the DNA strands. How, then, do we get from *this* series of DNA bases to *that* series of amino acids—to a protein? Let's find out.

14.2 Protein Synthesis in Overview

In overview, the process of protein synthesis can be described fairly simply. In eukaryotic organisms such as ourselves, the DNA just referred to is contained in the nucleus of the cell. The first step is that a stretch of it unwinds there, and its message—the order of a string of A's, T's, C's, and G's—is copied onto a molecule called messenger RNA (mRNA). This length of mRNA, which could be thought of as an information tape being copied off a "master" DNA tape, then exits from the cell nucleus (**see** FIGURE 14.2).

The destination of this mRNA is a molecular workbench in the cell's cytoplasm, a structure called a ribosome. More formally, a **ribosome** is an organelle, located in the cell's cytoplasm, that is the site of protein synthesis. At the ribosome, both the informational tape (the mRNA chain) and the raw materials (amino acids) come together to make the product (a protein). As the mRNA tape is "read" within the ribosome, something grows from it: a chain of amino acids that have been linked together in the ribosome in the order specified by the mRNA. When the chain is finished and folded up, a protein has come into existence. And how do amino acids get to the ribosomes? They are brought there by a second type of RNA, transfer RNA (tRNA).

As may be apparent from this account, protein synthesis divides neatly into two sets of steps. The first set is

FIGURE 14.2 **The Two Major Stages of Protein Synthesis**

Transcription

1. In transcription, a section of DNA unwinds and nucleotides on it form base pairs with nucleotides of messenger RNA, creating an mRNA "tape."

2. This segment of mRNA then leaves the cell nucleus, headed for a ribosome in the cell's cytoplasm, where translation takes place.

Translation

3. Joining the mRNA tape at the ribosome are amino acids, brought there by transfer RNA molecules. The length of messenger RNA is then "read" within the ribosome. The result? A chain of amino acids is linked together in the order specified by the mRNA tape.

4. When the chain is finished and folded up, a protein has come into existence.

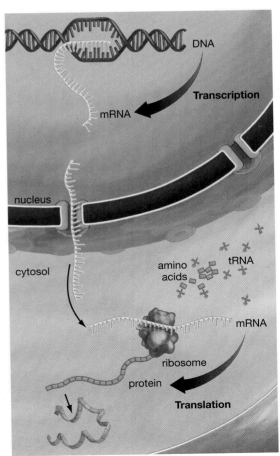

Table 14.1 Amino Acids

Amino Acid	Abbreviation
Alanine	ala
Arginine	arg
Asparagine	asn
Aspartic acid	asp
Cysteine	cys
Glutamine	gin
Glutamic acid	glu
Glycine	gly
Histidine	his
Isoleucine	ile
Leucine	leu
Lysine	lys
Methionine	met
Phenylalanine	phe
Proline	pro
Serine	ser
Threonine	thr
Tryptophan	trp
Tyrosine	tyr
Valine	val

So Far... Answers:

1. *20; amino acids*
2. *messenger RNA (mRNA); ribosome*
3. *ribosome; transfer RNA (tRNA)*

called **transcription**: the process by which the genetic information encoded in DNA is copied onto messenger RNA. The second set is called **translation**: the process by which information encoded in messenger RNA is used to assemble a protein at a ribosome. Let's look in more detail now at both of these processes, starting with transcription.

So Far...

1. Though there are hundreds of thousands of proteins active in living things, all of them are put together from a starting set of _____ of the building blocks known as _____.

2. The information for building proteins encoded in DNA is passed on first to _____, which then migrates to an organelle called a _____, where the protein is put together.

3. Amino acids are brought to the _____ by another form of RNA, called _____.

14.3 A Closer Look at Transcription

From what's been reviewed so far, you can see that a key player in transcription (and translation) is RNA, whose full name is ribonucleic acid. If you look at FIGURE 14.3, you can see just how similar RNA is to DNA. For one thing, RNA has the sugar and phosphate "handrail" components you saw in DNA last chapter. Then, in both molecules, this two-part structure is joined to a third element, a base.

Recall that any given DNA building block or "nucleotide" has one of four bases: adenine, guanine, cytosine, or thymine (A, G, C, or T). RNA utilizes the first three of these, but then substitutes uracil (U) for the thymine (T) found in DNA.

Passing on the Message: Base Pairing Again

Given the chemical similarity between DNA and RNA, it's not hard to see how DNA's genetic message can be passed on to messenger RNA: Base pairing is at work again. Recall from Chapter 13 how DNA is replicated. The double helix is unwound, after which bases along the now-single DNA strands are paired up with complementary DNA bases. Every T on a single DNA strand is paired with a free-floating A, and every C is linked with a G, thus yielding a complementary DNA *chain*. Because of RNA's similarity to DNA, the bases RNA has can also form base pairs with DNA. The twist to this RNA-DNA base pairing is that each A on a DNA strand links up with a *U* on the RNA strand, instead of the T that would be A's partner in DNA-to-DNA base pairing. You can see how base pairing plays out

FIGURE 14.3 RNA and DNA Compared
The "handrails" of both RNA and DNA are composed of linked sugar and phosphate molecules (thus the S-P-S-P labeling). The "stair steps" stemming from the handrails are formed by bases, as with the G and C bases extending inward at the top of the DNA strand. While DNA is double-stranded, RNA generally is single-stranded and utilizes the base uracil (U) instead of the thymine (T) that DNA utilizes.

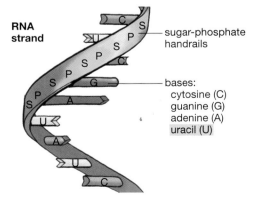

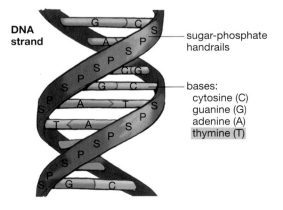

1. RNA polymerase unwinds a region of the DNA double helix.

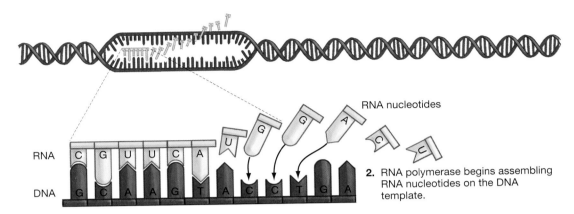

RNA nucleotides

RNA

DNA

2. RNA polymerase begins assembling RNA nucleotides on the DNA template.

RNA

3. The completed portion of the RNA chain separates from the DNA. Meanwhile, RNA polymerase unwinds more of the untranscribed region of the DNA.

RNA

4. The RNA chain is released from the DNA, and the DNA is rewound into its original form. Transcription is completed.

FIGURE 14.4 **Transcription Works through Base Pairing** Thanks to their chemical similarity, DNA and RNA can engage in base pairing, and this base pairing is how RNA chains are synthesized. The enzyme complex RNA polymerase undertakes two tasks in transcription: It unwinds the DNA sequence to be transcribed, and it brings together RNA nucleotides with their complementary DNA nucleotides, thus producing an RNA chain.

in transcription by looking at **FIGURE 14.4**. It will come as no surprise to you that an enzyme is critically involved in this process. **RNA polymerase**, as this complex of enzymes is known, actually undertakes two critical tasks: It unwinds the DNA strand and strings together the chain of RNA nucleotides that is complementary to it.

In organisms such as ourselves (the eukaryotes), the length of RNA that results from this process is not actually the finished messenger RNA we've been talking about. Instead, it is an RNA chain that must undergo some modification or "editing" before it qualifies as mRNA. You can read more about this editing in section 14.5, on genetic regulation. For now, we'll set it aside and follow the fate of a completed mRNA chain. Before leaving the subject of mRNA, however, here's a more formal definition of it: **Messenger RNA (mRNA)** is a type of RNA that encodes, and carries to ribosomes, information for the synthesis of proteins.

The Triplet Code

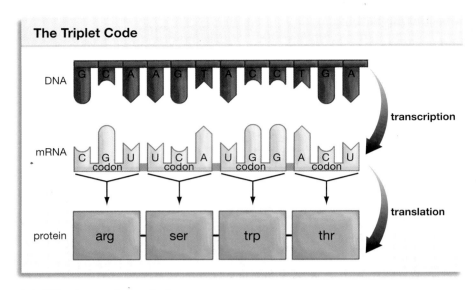

FIGURE 14.5 Triplet Code Each triplet of DNA bases codes for a triplet of mRNA bases (a codon), but it takes a complete codon to code for a single amino acid.

A Triplet Code

Thus far, we have been looking at what might be called the flow of genetic information (from DNA to mRNA). A separate issue in protein synthesis, however, has to do with the linkage between DNA and RNA on the one hand and amino acids on the other. The question here is: How many DNA bases does it take to code for an amino acid? As FIGURE 14.5 shows, the answer is three. Each three bases in a DNA sequence pairs with three mRNA bases, but each group of three mRNA bases then codes for a *single* amino acid. Each coding triplet of mRNA bases is known, appropriately enough, as a **codon**.

A question that follows is: If you have a given three bases, *which* amino acid do they specify? In Morse code, the sounds . — . (short-long-short) code for the letter *R*. But if we know that a given mRNA codon has the base sequence UCC, what *amino acid* does this code for? Today, we know that the answer is serine (ser), but at one time this was not clear. Indeed, it took years for scientists to figure out all the linkages between codons and amino acids. When this work was completed, the result was the **genetic code**, which can be defined as the inventory of linkages between nucleotide triplets and the amino acids they code for. If you look at Essay 14.1, "Cracking the Genetic Code" on page 228, you can learn more about the importance of this inventory.

14.4 A Closer Look at Translation

With a length of RNA having first been transcribed from a length of DNA, and having then moved to a ribosome as mRNA, many of the players are in place for translation—the second stage of protein production. What's also needed, however, are the building blocks of proteins, amino acids. As noted, they are brought to ribosomes by a second form of RNA, called transfer RNA (or tRNA).

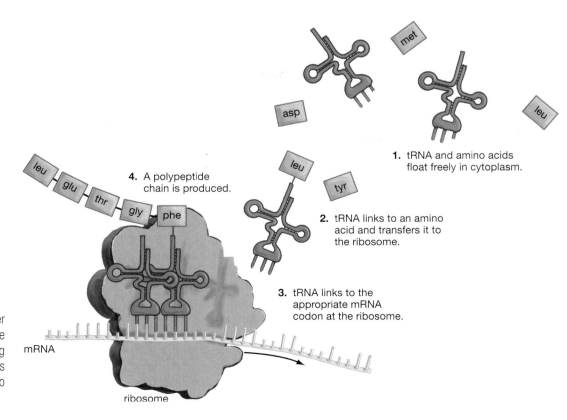

4. A polypeptide chain is produced.

1. tRNA and amino acids float freely in cytoplasm.

2. tRNA links to an amino acid and transfers it to the ribosome.

3. tRNA links to the appropriate mRNA codon at the ribosome.

FIGURE 14.6 Bridging Molecule Transfer RNA (tRNA) molecules link up with amino acids on the one hand and mRNA codons on the other, thus forming a chemical bridge between the two kinds of molecules in protein synthesis. They also transfer amino acids to ribosomes, as shown in the steps of the figure.

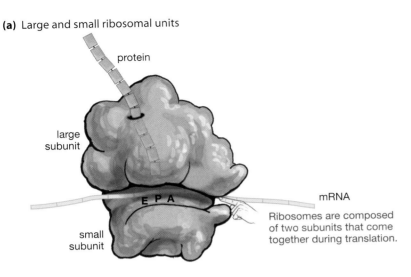

(a) Large and small ribosomal units

protein

large
subunit

E P A

small
subunit

mRNA

Ribosomes are composed
of two subunits that come
together during translation.

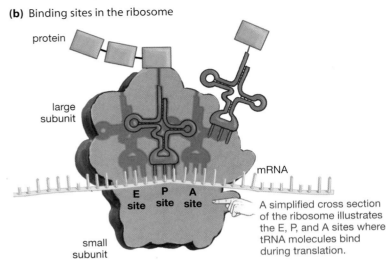

(b) Binding sites in the ribosome

protein

large
subunit

mRNA

E   P   A
site site site

small
subunit

A simplified cross section
of the ribosome illustrates
the E, P, and A sites where
tRNA molecules bind
during translation.

FIGURE 14.7 **The Structure of Ribosomes**

The Nature of tRNA

If you look at FIGURE 14.6, you can see why tRNA is aptly placed within the translation phase of protein synthesis. When we think of a translator, we think of someone who can communicate in *two* languages. Transfer RNA effectively does this. One end of each tRNA molecule links with a specific *amino* acid, which it finds floating free in the cytoplasm. Then, transferring this amino acid to the ribosome, the other end of this tRNA molecule bonds with a *nucleic* acid—a triplet of bases on the mRNA that is moving through the ribosome. Thus, tRNA is bonding with two very different kinds of molecules—amino acids on the one hand, nucleic acids on the other—thereby serving as a translator between them. **Transfer RNA (tRNA)** can be defined as a form of RNA that, in protein synthesis, bonds with amino acids, transfers them to ribosomes, and then bonds with messenger RNA.

The Structure of Ribosomes

You've now been introduced to almost all the players that will be active in protein synthesis at the ribosome. But what is the structure of the ribosome itself? If you look at FIGURE 14.7, you can begin to get an idea. Note that ribosomes are composed of two "subunits"—one larger than the other—both made of a mixture of proteins and yet another type of RNA, **ribosomal RNA (rRNA)**. The two subunits may float apart from one another in the cytoplasm until prompted to come together by the process of translation. You can see that when the subunits have been joined, there exist three binding sites (which it is convenient to think of as slots), one of them an "E" site, the next a "P" site, and the third an "A" site. You'll be looking at their roles shortly. **Table 14.2** sets forth the three types of RNA reviewed so far that are active in protein synthesis.

The Steps of Translation

With all the players introduced, let's see how translation works. To keep things simple, we'll follow the process as it occurs in prokaryotes. Our starting point is an mRNA chain that is ready to begin binding with a ribosome. Meanwhile, nearby tRNA molecules have linked to their appropriate amino acids.

Table 14.2 Types of RNA

Type of RNA	Functions in	Function
Messenger RNA (mRNA)	Nucleus, migrates to ribosomes in cytoplasm	Carries DNA sequence information to ribosomes
Transfer RNA (tRNA)	Cytoplasm	Provides linkage between mRNA and amino acids; transfers amino acids to ribosomes
Ribosomal RNA (rRNA)	Cytoplasm	Structural component of ribosomes

The steps of translation

1. A messenger RNA transcript binds to the small subunit of a ribosome as the first transfer RNA is arriving. The mRNA codon AUG is the "start" sequence for most polypeptide chains. The tRNA, with its methionine (met) amino acid attached, then binds this AUG codon.

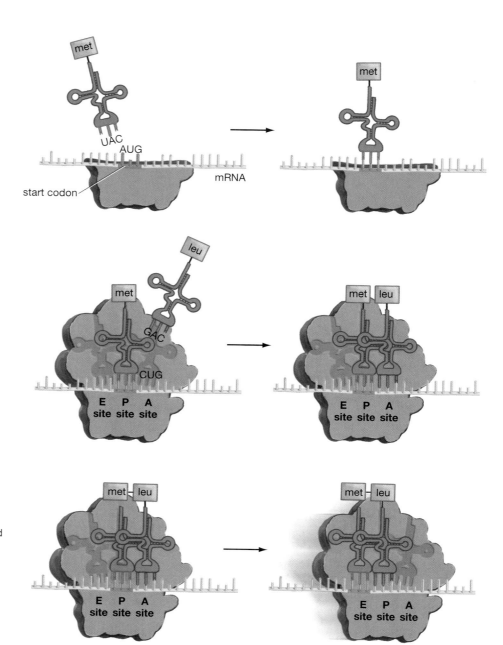

2. The large ribosomal subunit joins the ribosome, as a second tRNA arrives, bearing a leucine (leu) amino acid. The second tRNA binds to the mRNA chain, within the ribosome's A site.

3. A bond is formed between the newly arrived leu amino acid and the met amino acid, thus forming a polypeptide chain. The ribosome now effectively shifts one codon to the right, relocating the original P-site tRNA to the E site, the A-site tRNA to the P site, and moving a new mRNA codon into the A site.

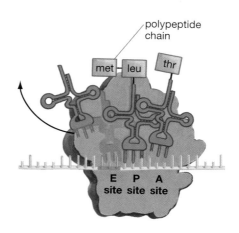

4. The E-site tRNA leaves the ribosome, even as a new tRNA bonds with the A-site mRNA codon, and the process of elongation continues.

FIGURE 14.8 The Steps of Translation

mRNA Binds to Ribosome, First tRNA Arrives

In this first step, the mRNA chain arrives at the ribosome and binds to the ribosome's small subunit (**see** FIGURE 14.8). The mRNA codon AUG is the usual "start" codon for a polypeptide chain. Next, a tRNA molecule arrives that has a complementary base sequence—UAC—that allows it to bind this AUG codon. This tRNA is bearing its amino acid, which is methionine (met). Following this, the large ribosomal subunit becomes part of the ribosome, providing the ribosome's A, P, and E binding sites.

Polypeptide Chain Is Elongated

Now amino acids will start to be joined together in a chain. As you can see, this elongation process begins with a second incoming tRNA molecule binding to an mRNA codon in the A site. Because it happens to be a CUG codon, a tRNA with a complementary GAC sequence binds to it. This tRNA comes bearing the amino acid leucine (leu). Next, the met amino acid attached to the tRNA in the P site bonds with the leu amino acid attached to the tRNA in the A site. In this process, the bond is broken between met and its original tRNA.

Once this occurs, a kind of molecular musical chairs ensues: The ribosome effectively shifts one codon to the right. With this, the tRNA that had been in the P site is relocated to the E site. What does the *E* stand for? Exit. The tRNA in this site bears no amino acid now, since its amino acid has joined the polypeptide chain; soon this tRNA will be ejected from the ribosome altogether. Meanwhile, the tRNA that had been in the A site moves to the P site, and a new mRNA codon shifts into the now-vacated A site. You can see what comes next: A tRNA bonds with this new codon in the A site. Soon, the growing polypeptide chain will bond with *this* tRNA's amino acid, and the process will continue.

Termination of the Growing Chain

There are three separate codons that don't code for any amino acid, but that instead act as stop signals for polypeptide synthesis. Any time one of these "termination" codons moves into the ribosome's A site, it doesn't bind with an incoming tRNA, but instead brings about a severing of the linkage between the P-site tRNA and the polypeptide chain. Indeed, the whole translation apparatus comes apart at this point, with the polypeptide chain being released to fold up and be processed as a protein. Translation has been completed.

Speed of the Process; Movement through Several Ribosomes

How fast does this process go? Very fast. An *E. coli* bacterium can string together up to 40 amino acids per second, meaning that an average-sized protein of about 400 amino acids could be put together in 10 seconds. Note, however, that mRNA chains often are read not by one ribosome, but by many. As you can see in FIGURE 14.9, several ribosomes—perhaps scores of them—might move over a given mRNA sequence, with the result that identical polypeptide chains grow out of each ribosome. This greatly increases the number of proteins that can be put together in a given period of time.

(a)

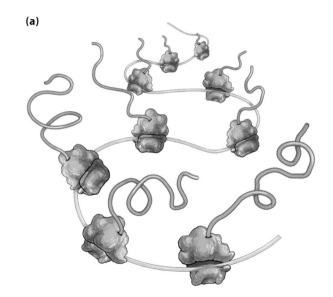

(b)

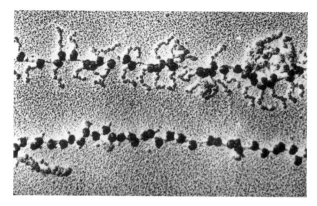

FIGURE 14.9 **Mass Production**

(a) An mRNA chain can be translated by many ribosomes at once, resulting in the production of many copies of the same protein.

(b) A micrograph of this process in operation. The figure shows two mRNA strands with ribosomes spaced along their length. In the upper strand, translation is under way and polypeptides can be seen emerging from the ribosomes.

So Far...

1. DNA passes on its information to messenger RNA by means of _____.

2. Each _____ DNA bases code for _____ RNA bases, which code for _____ amino acid(s).

3. A transfer RNA molecule attaches to an _____ on one end and a _____ on the other.

So Far... Answers:

1. *complementary base pairing*

2. *three; three; one*

3. *amino acid; messenger RNA codon (triplet)*

14.5 Genetic Regulation

From what you've read so far, you might have gotten the impression that transcription never ends for genes—that all genes are constantly being transcribed, thus bringing about a constant production of the proteins they code for. But a genome that worked like this would be like a restaurant that ceaselessly turned out every dish on its menu, no matter what its customers were ordering. Life is complex, which means that organisms need to finely tune their production of proteins. Think of insulin, a protein that helps put blood sugar into storage in our cells. Until we get the blood-sugar surge caused by, say, swigging a soft drink, we don't need much insulin in our bloodstream. Given this, what sense would it make for the insulin gene to simply stay "on" all the time? None at all, which is why the activity of the insulin gene is *regulated*, as is the case with all genes. Because genetic regulation is so complex, you'll only be introduced to a couple of its aspects here. Before you can learn about regulation, however, you need to learn a little more about the entity that gets regulated, the genome.

A Genome Contains Lots More than Genes

At one time, scientists thought that genes in the genome sat right beside each other like pearls on a string. Beginning in the 1970s, however, scientists began to realize that the human genome contains large segments of DNA that do not code for proteins. Indeed, over time it became apparent that, in all eukaryotic organisms, it is the *coding* sections of the genome that are few and far between. Completion of the Human Genome Project in 2004 revealed that less than 1.2 percent of the human genome codes for proteins. To gain some perspective on this number, consider that although the DNA in any of our cells would stretch to about 6 feet in length if uncoiled, less than 1 inch of this length is protein-coding sequence.

Surprisingly, perhaps, some of this non-coding DNA is interspersed *within* coding gene sequences. If the text of a novel read like the bases in one of our genes, the text would go something like this: "It was the best of times, it was the pzknlku likh uiop nklkj nndhfoz worst of times." Such non-coding sequences are passed on to the initial RNA chain transcribed from a gene. However, these are the sequences, referred to earlier, that are edited out of the initial chain, thus yielding messenger RNA. If you look at FIGURE 14.10, you can see how this editing takes place. The sequences that are removed from the original RNA chain are called introns (for *in*tervening sequences), while the sequences that are left after this editing takes place are called exons (because most of them are *ex*pressed as proteins). An average gene might have half a dozen introns, which may comprise more than 90 percent of the gene's sequence.

Introns represent only a part of the non-coding DNA in the genome, however. Vast stretches of non-coding sequence also lie *in between* genes. So, if what genes "do" is code for proteins, what is all this non-coding DNA doing in the genome? At one time, scientists thought it wasn't doing anything at all. Hence, they gave it a name—junk DNA—that may accurately describe portions of it. Yet some non-coding DNA clearly is regulatory, and recent evidence suggests that most of it may be regulatory in one way or another. In what follows, we'll look at two kinds of genetic regulation that are based on non-coding DNA sequences.

A BROADER VIEW

The Genome as a Program

Computer enthusiasts: The human genome is sometimes compared to a set of cookbook recipes, but a more apt metaphor would be a computer program with a huge number of subroutines.

FIGURE 14.10 Editing Out Non-coding Sequences In eukaryotes, lengths of DNA that are transcribed contain sequences that do not code for amino acids, as well as sequences that do. Both kinds of sequences are copied onto an initial RNA chain. Once this is done, however, the non-coding sequences (called introns) are removed by enzymes through the editing process shown, while the coding sequences (called exons) are spliced back together. The result is a messenger RNA molecule.

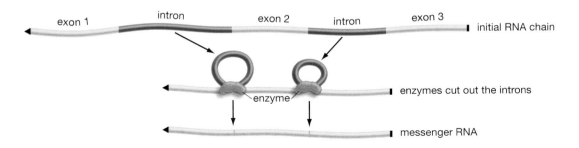

exon 1 · intron · exon 2 · intron · exon 3 · initial RNA chain

enzyme · enzymes cut out the introns

messenger RNA

Proteins Regulate DNA

If you look at FIGURE 14.11, you can see a much-simplified version of some real-life genetic regulation. When organisms develop as embryos, there are genes in them that control the development of their mid-body or "thoracic" structures, such as the vertebrae that make up their backbones. One of these genes, called *Hoxc8*, is nearly identical in creatures as diverse as chickens, mice, and snakes. Yet a chicken has 7 vertebrae, while a mouse has 13. So the same developmental gene is at work in these two animals, yet it is yielding very different outcomes. If we ask what accounts for this, the answer is that the mouse *Hoxc8* is being *transcribed* more than the chicken *Hoxc8*. More transcription means the mouse has more of the protein that *Hoxc8* codes for; this in turn means a broader distribution of this protein in the mouse embryo, and the result is more vertebrae in the mouse.

But why would *Hoxc8* get transcribed more in a mouse than a chicken? You may recall that for any gene to be transcribed, an enzyme complex called RNA polymerase has to go down the line on the gene's sequence, pairing up its DNA

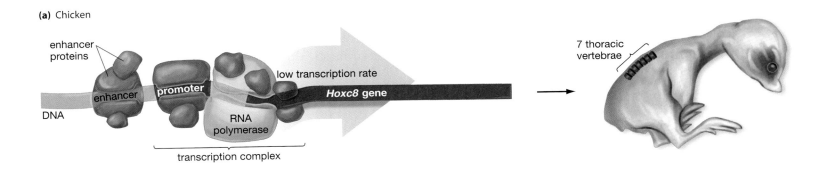

(a) Chicken

enhancer proteins
enhancer
promoter
DNA
RNA polymerase
transcription complex
low transcription rate
Hoxc8 gene
7 thoracic vertebrae

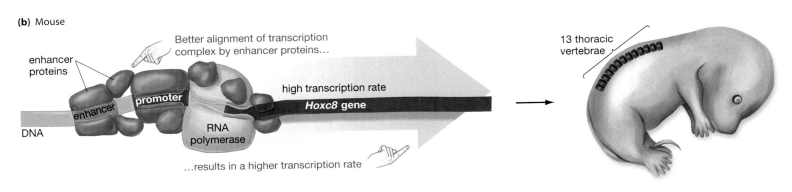

(b) Mouse

enhancer proteins
Better alignment of transcription complex by enhancer proteins...
enhancer
promoter
DNA
RNA polymerase
high transcription rate
Hoxc8 gene
...results in a higher transcription rate
13 thoracic vertebrae

FIGURE 14.11   **Genetic Regulation in Action**   The *Hoxc8* gene regulates chest-level or "thoracic" development in the embryos of a wide range of vertebrate animals, including the chicken and the mouse. The mouse ends up with more vertebrae in its backbone than the chicken does, however, because the mouse *Hoxc8* gene is transcribed with greater frequency.

(a) For transcription of any gene to begin, RNA polymerase must be aligned at a sequence of DNA called the promoter, which lies just upstream from a given gene—in this case the *Hoxc8* gene. A multi-part protein complex, seen here surrounding RNA polymerase, is active in getting a base level of transcription going. Another DNA sequence, called an enhancer, can facilitate transcription by acting as a binding site for other proteins that can interact with the promoter-site protein complex.

(b) The enhancer in the mouse has a slightly different DNA sequence than that of the enhancer in the chicken—a sequence that binds a different set of proteins. These proteins interact with the promoter sequence proteins in such a way as to better align RNA polymerase at the promoter. The result is increased transcription of the *Hoxc8* gene in the mouse, a broader distribution of the Hoxc8 protein in the mouse embryo, and more vertebrae in the mouse.

A BROADER VIEW

"Junk" and Regulatory DNA

If any DNA really is "junk," think what a strange entity it is. This is DNA that has been replicated over and over, down millions of years of evolution, and yet does nothing for an organism. It's like a lazy relative who gets to stay on in a household simply because getting rid of him would be more trouble than keeping him.

A BROADER VIEW

microRNAs and Medicine

Don't be surprised if you pick up the newspaper soon and find a story that mentions a "micro" RNA of the sort noted in the text. These small RNAs are one of the hottest topics in biology right now, not so much because of their possible role in large-scale gene regulation but because they are giving scientists a fantastic research tool while holding out a promise of helping to fight diseases such as cancer. These RNAs can be used to diminish or even halt the production of specific proteins—something that would be very useful in fighting a disease like cancer, which depends on the work of specific proteins.

bases with complementary RNA bases. But this is not an all-or-nothing proposition. RNA polymerase is *more or less* likely to get busy transcribing a gene depending on how it is chemically positioned on a DNA sequence that lies just "upstream" from the gene. This is the "promoter" sequence that you can see in the figure. For transcription to take place in any gene, a multi-part protein complex must align RNA polymerase at the promoter. The question then becomes: What factors stand to facilitate or impede this alignment? If you look just upstream from the promoter in the figure, you can see a facilitating factor: a sequence that is called an enhancer. This sequence is a segment of DNA, but it is not a gene; its sole role is to serve as a binding site for proteins that help get RNA polymerase positioned at the promoter. Now, remember that the *Hoxc8* gene is nearly identical in lots of animals. But this is not true of the *Hoxc8* enhancer; *its* sequence differs markedly from chickens, to mice, to snakes. Accordingly, the mouse's enhancer binds with a different set of aligning proteins than does the chicken's. And these mouse proteins more strongly induce transcription than do the chicken's. The result is more *Hoxc8* transcription in the mouse, and ultimately more vertebrae in it.

This peek into the regulation of one gene shows you several elements that are common in gene regulation. First, note that there are sequences of DNA that do not code for protein but that are instead purely regulatory—the promoter and the enhancer, in this case. Second, note that transcription gets going by means of *proteins* binding with these DNA sequences. Now, where did these proteins come from? They were produced through transcription and translation, just like any other protein. Thus, DNA codes for proteins that then *feed back* on DNA itself, controlling its transcription. At root, then, the whole system is self-regulating.

RNA in Genetic Regulation

The message "DNA makes RNA makes protein" has come through loud and clear in this chapter, but this message actually needs a little modification. DNA does indeed code for RNA, but not all RNA codes for protein. Think of, for example, the transfer RNA you looked at earlier that ferries amino acids to ribosomes. Where does it come from? It is coded for by DNA. A segment of DNA unwinds and then forms base pairs with RNA nucleotides to produce an RNA sequence. Only this is a *transfer* RNA sequence that doesn't code for any protein; it simply migrates out of the nucleus to become a transfer RNA molecule. Ribosomal RNA is produced in this same way. Now, these two forms of RNA happen to be structural—they help make up the structures of ribosomes and tRNA molecules—but transcription can also produce non-coding RNA segments that are *regulatory*.

One class of regulatory RNAs, called microRNAs, was shown in 2004 to help regulate the production of the insulin we talked about earlier. Researchers induced mice to produce more of a specific microRNA and watched as the animals' production of insulin dropped off. Conversely, when production of this micro-RNA was reduced, insulin production rose in the mice. All microRNAs identified to date have this effect: They reduce the production of specific proteins. And the means by which they do this is well known. They interfere with *messenger* RNAs, actually targeting them for destruction. In this case, a microRNA targets insulin messenger RNAs, and the result is diminished insulin production. For our purposes, the main message is that there is a small section of DNA that does not code for protein, but that does code for an RNA sequence that regulates the production of a protein.

The question that follows from this is: How important is RNA *generally* in genetic regulation? The short answer is that we don't know yet; but as a lawyer might say, there is circumstantial evidence that the importance may be great indeed. In 2004, researchers announced a new, lowered estimate for the number of genes we humans have: between 20,000 and 25,000. If you look at FIGURE 14.12, you can see four other organisms whose genomes have been sequenced. Note that we humans—with our vision, speech and 100 trillion cells—have, at most, 6,000 more genes than the roundworm *C. elegans*, which is eyeless, speechless, and

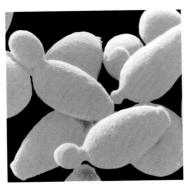

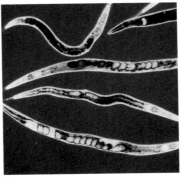

Saccharomyces cerevisiae (baker's yeast)	*Drosophila melanogaster* (fruit fly)	*Caenorhabditis elegans* (roundworm)	*Arabidopsis thaliana* (mustard plant)
Estimated number of genes: **6,034**	**13,061**	**19,099**	**25,000**

FIGURE 14.12 **Not as Much Difference as We Thought** At one time, scientists assumed that the human genome contained about 100,000 genes. Genome sequencing revealed, however, that the human genome probably contains about 20,000–25,000 genes. This is, at most, about 6,000 more genes than the tiny roundworm *C. elegans* has, and about 12,000 more than the *Drosophila* fruit fly has.

made up of 959 cells. Indeed, we probably have *fewer* genes than a mustard plant. It's pretty clear that there is little relationship between the number of genes an organism has and the complexity of that organism. Fascinatingly, what does seem to go hand in hand is complexity and the proportion of *non*-coding DNA an organism has. And here, humans do shine. Remember how less than an inch of our 6 feet of DNA codes for proteins? Well, what is the rest of that DNA doing? Some of it contains regulatory sequences such as the enhancers and promoters we looked at earlier. And some of it codes for regulatory RNA of the sort we just went over. But how much of the genome codes for such regulatory RNA? By one count, mammalian genomes transcribe thousands of RNAs that do not go on to be translated into proteins. Some researchers feel we are just beginning to understand the regulatory value of these RNAs. Indeed, these scientists are suggesting that it is not so much genes, but a finely tuned RNA-based regulation of genes that accounts for a great deal of the difference between us and, say, a roundworm. This is a cutting-edge idea, and scientific opinion is mixed on it. Only time will tell whether RNA-based regulation does indeed play a large role in allowing us to be the complex organisms we are.

14.6 The Magnitude of the Genetic Operation

Our 20,000–25,000 genes are contained in a genome that is about 3.2 billion base pairs long. To make this large number a little less abstract, consider the following exercise. If we took the base sequence of the human genome and simply arrayed the single-letter symbols for the bases on a printed page, going like this:

AATCCGTTTGGAGAAACGGCCCTATTG
GCAGCAAGGCTCTCGGGTCGTCAACG
CGTATTAAACATATTTCAAGGCTCTA . . .

it would take about 1,000 telephone books, each of them 1,000 pages long, merely to record it all. We're talking, then, about an unbroken series of these base symbols that goes on for a million pages. That is one measure of the size of the human genome. Simply maintaining this genome—making sure that it is passed on from one cell and generation to another—requires that all of these 3.2 billion base pairs be faithfully copied each time a cell divides. (They have to be copied twice, actually, because

Essay 14.1 Cracking the Genetic Code

Like scholars trying to extract meaning from an ancient written language, biologists in the early 1960s were trying to extract meaning from an ancient molecular language. They knew that triplets of mRNA bases were coding for amino acids, but what was the linkage? If you had a given mRNA triplet, *which* amino acid did it code for? By 1968, scientists had figured out the whole puzzle. If you look at **FIGURE E14.1.1**, you can see a summation of their work—the genetic code in its entirety.

What was the importance of deciphering this code? For one thing, it handed scientists a powerful tool. If they knew one "end" of a genetic chain, they were instantly given insight into the other. Knowing something about a DNA sequence meant knowing something about a protein sequence—and vice versa. For example, in the 1980s, researchers wanted to produce a genetically engineered version of a protein, called erythropoietin, that is manufactured in the kidneys and that stimulates the production of red blood cells. (Kidney disease

first base	second base U	second base C	second base A	second base G	third base
U	UUU } phe	UCU } ser	UAU } tyr	UGU } cys	U
	UUC	UCC	UAC	UGC	C
	UUA } leu	UCA	UAA **Stop**	UGA **Stop**	A
	UUG	UCG	UAG **Stop**	UGG trp	G
C	CUU } leu	CCU } pro	CAU } his	CGU } arg	U
	CUC	CCC	CAC	CGC	C
	CUA	CCA	CAA } gln	CGA	A
	CUG	CCG	CAG	CGG	G
A	AUU } ile	ACU } thr	AAU } asn	AGU } ser	U
	AUC	ACC	AAC	AGC	C
	AUA	ACA	AAA } lys	AGA } arg	A
	AUG met (start)	ACG	AAG	AGG	G
G	GUU } val	GCU } ala	GAU } asp	GGU } gly	U
	GUC	GCC	GAC	GGC	C
	GUA	GCA	GAA } glu	GGA	A
	GUG	GCG	GAG	GGG	G

FIGURE E14.1.1 The Genetic Code Dictionary If we know what a given mRNA codon is, how can we find out what amino acid it codes for? This dictionary of the genetic code provides a way. Say you want to confirm that the codon CGU codes for the amino acid arginine (arg). Looking that up here, C is the first base (go to the C row along the "first base" line), G is the second base (go to the G column under the "second base" line), and U is the third (go to the codon parallel with the U in the "third base" line).

we have two copies of each chromosome.) When we start thinking about *using* this genome—about producing proteins from it—all the layers of genetic regulation sketched out earlier come into play. Genes help bring about proteins; some of these proteins feed back on other genes and control their transcription; RNA sequences help out with regulation, and so forth.

The obvious message here is that the human genetic operation is unimaginably complex. But it turns out that humans are not special in this regard. If we were to walk through the genetic operation of even a humble bacterium, you would see that its complexity is only slightly less stunning than ours. What's important to recognize is that this complexity is necessary because it *enables life*. Living things must obtain energy, they must grow, they must respond constantly to their environment, they must reproduce, and they must coordinate all these activities. Only an incredibly sophisticated system of information storage and use could allow a self-contained entity to do all these things. And genetics is exactly such a system. With each twitch of a muscle or swig of a soft drink, information repositories called genes are opened or closed within us, setting in motion the delivery of the tiny working entities called proteins—or stopping delivery of them, if that's what is called for. Not surprisingly, we are nowhere close to having a complete understanding of how any genetic system works. But, as one scientist has said, we are at "the end of the beginning" of our understanding—we understand the process in outline; now it's on to the details.

patients often suffer from anemia—a lack of red blood cells—because their kidneys can't produce enough erythropoietin.) The researchers started with a known entity, which was a portion of the amino acid sequence of erythropoietin. With this in hand, the genetic code allowed them to "work backwards" to learn the DNA sequence that gives rise to this protein. If they saw a tryptophan (trp) amino acid at one point in the erythropoietin amino acid chain, that meant that a UGG triplet had to exist at a corresponding point in erythropoietin's messenger RNA chain. And each mRNA triplet has, of course, a counterpart DNA triplet. Thus the researchers could work backwards—from protein to mRNA to DNA—to piece together portions of the sequence of the erythropoietin *gene*. Doing this allowed them to find this gene within the human genome, after which they cloned it. Today, erythropoietin is manufactured not just in the human body, but in biotech facilities.

> ## We humans share in an informational linkage that stretches back billions of years and runs through all the contemporary living world.

Apart from providing the critical linkage between amino acid and DNA sequences, the cracking of the genetic code also had the effect of revealing how unified life is at the genetic level. Once the code was in hand, biologists examined various species and found that with only a few exceptions, the code is universal in all living things. This means that the base triplet CAC codes for the amino acid histidine, whether this coding is going on in a bacterium or in a human being.

This insight turned out to be important in two ways. First, it is evidence that all life on Earth is derived from a single ancestor. How else can we explain all of life's diverse organisms sharing this very specific code? Such a complex molecular linkage—this triplet equals that amino acid—is extremely unlikely to have evolved *more* than once. What seems likely is that it came to be employed in an ancient common ancestor, after which it was passed on to all of this organism's descendants—every creature that has subsequently lived on Earth. Note, then, that this code has been passed on from one generation to the next, and from one evolving species to the next, for billions of years. We humans share in an informational linkage that stretches back billions of years and runs through all the contemporary living world.

Apart from this, the universality of the genetic code has a very practical consequence. It means that genes from one organism can function in another. This has both good and bad consequences for human beings. First the bad: Viruses that cause diseases ranging from colds to AIDS can "hijack" the human cellular machinery for their own benefit, precisely because their genes function in human cells. (The human cellular machinery will put together proteins whether these proteins are called for by human or viral DNA.) Now the good news: Using "biotechnology" processes you'll learn about next chapter, human beings can today use viruses and bacteria to manufacture all kinds of products, including medicines such as human insulin and human growth hormone. In these cases, human genes are being put to work inside microorganisms. This transferability of genes from one species to another has, as its basis, the universality of the genetic code.

Biotechnology Is Next

If each gene in a genome is something like a book in a library, think what the effect would be if organisms could, in effect, check books out of each other's libraries. Actually, you don't have to imagine what this would be like; it's happening right now with great frequency. Soybeans in the fields of Indiana and Iowa have the genes of bacteria spliced into them. Likewise, human genes have been spliced into bacteria so that these one-celled creatures can turn out human medicines. And we are now able to read the books in our own genetic library with an efficiency that lets us catch criminals on the basis of flakes of dandruff they leave at crime scenes. This may sound fanciful, but it's not. It's biotechnology, the subject of Chapter 15.

So Far...

1. True or false: Most of the DNA in the human genome codes for proteins.
2. In eukaryotic organisms such as ourselves, RNA chains that code for proteins must undergo _____ on their way to becoming _____.
3. A modern hypothesis holds that, rather than being "junk" DNA, most DNA that does not code for proteins serves a _____ function.

So Far... Answers:

1. *False; less than 1.2 percent of the human genome codes for proteins.*
2. *editing; messenger RNA*
3. *regulatory*

Chapter Review

Summary

14.1 The Structure of Proteins

- **The nature of proteins**—Proteins are composed of building blocks called amino acids. A string of amino acids is called a polypeptide chain. Once such a chain has folded into its working three-dimensional shape, it is a protein. Though there are hundreds of thousands of different proteins, all of them are put together from a starting set of 20 amino acids. It is the order in which the amino acids are linked in a polypeptide chain that determines which protein will be produced. (page 216)

14.2 Protein Synthesis in Overview

- **Transcription and translation**—There are two principal stages in protein synthesis. The first is transcription, in which the information encoded in DNA is copied onto a length of messenger RNA (mRNA), which in eukaryotes moves from the cell nucleus to structures in the cytoplasm called ribosomes. The second is translation, in which amino acids are linked together at the ribosomes in the order specified by the mRNA sequence. (page 217)

14.3 A Closer Look at Transcription

- **DNA-RNA base pairing**—The information in DNA is transferred to messenger RNA through complementary base pairing. Each C nucleotide in a segment of DNA being transcribed results in a G nucleotide being added to a segment of RNA, and so forth. The enzyme RNA polymerase unwinds the DNA sequence to be transcribed while stringing together a chain of RNA nucleotides that is complementary to it. In eukaryotes, this initial RNA chain must undergo some editing before it is referred to as messenger RNA. (page 218)

- **Three bases, one amino acid**—Each three coding bases of DNA pair with three RNA bases, but each group of three mRNA bases then codes for a single amino acid. Each triplet of mRNA bases that codes for an amino acid is called a codon. The inventory of linkages between base triplets and the amino acids they code for is called the genetic code. (page 220)

Web Tutorial 14.1 Transcription

14.4 A Closer Look at Translation

- **Transfer RNA**—Transfer RNA serves as a bridging molecule in protein synthesis, thanks to its ability to bond with both amino acids and nucleic acids (in the form of mRNA). A given tRNA molecule bonds with a specific amino acid in the cell's cytoplasm and then transfers that amino acid to a ribosome in which an mRNA chain is being "read." There, another portion of the tRNA molecule bonds with the appropriate codon in the messenger RNA chain. (page 220)

- **Steps of translation**—Translation works by means of a succession of tRNA molecules arriving at a ribosome, bound to their appropriate amino acids, and then bonding to their appropriate codon in the mRNA chain. As this process takes place, the succession of amino acids is linked together into a polypeptide chain. (page 221)

Web Tutorial 14.2 Translation

14.5 Genetic Regulation

- **The need for regulation**—Protein production is carefully controlled or "regulated" in living things. Genes do not simply stay "on," but instead are transcribed in accordance with the needs of an organism. (page 224)

- **The nature of the genome**—Only a small portion of the DNA of eukaryotes codes for protein; the rest of the eukaryotic genome is made up of non-coding DNA. Some non-coding genome sections occur within gene sequences. These non-coding sequences, called introns, get copied onto the initial RNA chain that is transcribed from a gene sequence. They are subsequently edited out of the RNA chain, however. The sequences that remain once this editing is completed are called exons. The chain that exists once editing is completed is messenger RNA. Extensive non-coding sections of the genome also occur between gene sequences. (page 224)

- **"Junk" and regulatory DNA**—All the non-coding portions of the genome were once thought to serve no function and hence were named "junk DNA." At least some of this DNA is regulatory, however, and recent research indicates that most of it may be regulatory in one way or another. (page 224)

- **Proteins regulate DNA**—One form of genetic regulation involves proteins that influence transcription. All gene transcription requires that RNA polymerase be properly aligned at a sequence of bases, called a promoter, that lies just "upstream" from a gene sequence. Various combinations of proteins can facilitate or impede this alignment. To the extent that the alignment is facilitated, transcription will increase. The proteins that operate in this system are themselves coded for by DNA, meaning the whole system is self-regulating.   (page 225)

- **RNA in genetic regulation**—Several varieties of RNA are transcribed from DNA, but of these, only messenger RNA codes for protein. Noncoding forms of RNA include various regulatory RNAs. The regulatory RNAs found to date have all had the effect of reducing the production of particular proteins by targeting their mRNAs for destruction. It is thus far unclear how big a role RNAs play in genetic regulation. Some scientists believe, however, that their role may be so extensive that it accounts for much of the complexity seen in organisms such as ourselves.   (page 226)

14.6 The Magnitude of the Genetic Operation

- **Complexity and life**—Life as we know it is made possible by the sophisticated system of information storage and use called genetics. Living things require such a system because, simply to stay alive and reproduce, they must carry out so many varied and interrelated activities.   (page 227)

Key Terms

codon   p. 220

genetic code   p. 220

messenger RNA (mRNA)
   p. 219

ribosomal RNA (rRNA)   p. 221

ribosome   p. 217

RNA polymerase   p. 219

transcription   p. 218

transfer RNA (tRNA)   p. 221

translation   p. 218

Testing Your Understanding

In Your Own Words (answers in the back of the book)

1. In the process of protein synthesis reviewed at the start of the chapter, what is the difference between transcription and translation?

2. During transcription, DNA neither unwinds itself nor pairs its own bases with complementary RNA bases. How are these tasks accomplished?

3. What are the three types of RNA discussed in this chapter? What do their abbreviations stand for? What are their functions?

Thinking about What You've Learned

1. Whether the subject is living things, businesses, or music, do complexity and regulation always go hand in hand?

2. Given the following sequence of mRNA, what would the resulting amino acid sequence be?
   AUGAAACGGGGACCAAUGGAUAACUAA

3. In a bacterial gene sequence, all the DNA bases within a gene code for protein, and, relative to eukaryotes, there is very little regulatory DNA in a bacterial genome. Does this imply that evolution has brought about inefficiency in eukaryotes?

Biotechnology

These two kittens, Tabouli (on the left) and Baba Ganoush, were cloned in 2004 from a one-year-old female Bengal cat.

THE CHRISTIAN SCIENCE MONITOR

CLONING KITTY
A CALIFORNIA COMPANY IS SELLING CLONING TECHNOLOGY TO PET OWNERS. BUT OPPONENTS SAY THIS TECHNOLOGY HOLDS FALSE PROMISES

By Michael B. Farrell
November 24, 2004

Three years after scientists in a Texas laboratory successfully cloned the first house cat, a company in California is now selling that same technology to pet owners who want a carbon copy of their cat or dog.

For $50,000 Genetics Savings & Clone can take a cat's DNA and create an exact genetic duplicate.

THE ORLANDO SENTINEL

DANDRUFF, DNA SOLVE 1993 ROBBERY

By the Associated Press
November 23, 2004

LONDON—A British criminal received the longest prison sentence of his career Monday after being caught because of the dandruff he left at the scene of a robbery.

Using DNA profiling, investigators identified Andrew Pearson as a suspect by examining 25 flakes of dandruff found in a stocking he wore as a mask during the June 1993 robbery.

15.1 What Is Biotechnology?

No, the newspaper headlines above are not from supermarket tabloids. Today, a criminal can be convicted of a crime on the basis of 25 flakes of dandruff he left in a mask 11 years earlier. And, yes, if you love your cat so much that you would like to have a genetic replica of it, you can now have exactly that, assuming a $50,000 price tag isn't a deterrent.

We may be left scratching our heads in amazement that such things are possible, but thanks to the emerging

A scientist holds a frame containing silk-like fibers that are a hybrid of synthetic and natural spider proteins, produced with the help of bacteria.

field known as biotechnology, we should start getting used to amazement. In one instance after another, biotechnology is upsetting our notions of what is possible and what is impossible in the world. We may assume that all animals must have both a biological mother and father, but cloned kittens arguably have neither. We may assume that humans and sheep are fundamentally separate entities, but scientists have been adding human stem cells to sheep fetuses, thus producing sheep who have mostly human liver tissues. And the humble bacterium now routinely has human genes inserted into it, which allows it to turn out human proteins for medical use. If you have eaten anything made from soybeans recently, you probably have eaten food from a plant that was bioengineered. And if you're worried about the survival of the panda bear, you may take some comfort in knowing that the Chinese government is investigating ways to clone it as a means of staving off its extinction.

Taken together, achievements such as these amount to a revolution; but like all revolutions, this one seems to be inspiring both hope and fear in about equal measure. For many people, the production of a sheep that is partly human doesn't mean we're living in a world that is better controlled; it means we're living in a world that is out of control. Forget about the "promise" of biotechnology, this view goes; just put the genie back in the bottle. On the other hand, who wouldn't want more food for a hungry world, or more medicines for people who are sick?

In thinking about such questions, it's helpful to view biotechnology not as a monolithic entity, but rather as a collection of separate *capabilities*. There is the capability to clone, the capability to splice genes into organisms, and so forth. The purpose of this chapter is to provide you with a basic understanding of some of these capabilities. Once you see what's at work in them, you'll be in a better position to make judgments about the degree to which biotechnology holds out promise or peril.

Defining Biotechnology

The term *biotechnology* is familiar to most of us, but what does it really mean? **Biotechnology** can be defined as the use of technology to control biological processes as a means of meeting societal needs. One biological process that can be controlled is the way a cell copies its DNA every time it divides. In the 1980s, a California scientist realized there was a way to use technology to exploit this natural copying, such that a tiny DNA sample could rapidly be copied over and over, thus yielding a relatively large DNA sample. One result of this "PCR" process is that police can now recover, from the tiniest specks of blood—or flakes of dandruff, even— enough DNA to get a "match" in a criminal case. Like other biotech processes we'll be looking at, PCR is a marriage of both biology and technology. By the time you finish this chapter, you'll see why *biotechnology* is such an appropriate term.

Because biotechnology is so varied, any account of it has to be limited to a few examples chosen from among thousands that exist. In this chapter, we will consider four aspects of biotechnology:

- Transgenic biotechnology—the splicing of DNA from one species into another

- Reproductive cloning—the production of mammals through cloning

- Forensic biotechnology—the use of biotechnology to establish identities (of criminals, of crime victims, and so forth)

- Stem cells—the use of special cells to produce new human tissues

Let's start now by reviewing an area of biotechnology in which genes are moved from one organism to another.

15.2 Transgenic Biotechnology

Human beings grow to their full height under the influence of human growth hormone (HGH), a protein that normally is secreted by the human pituitary gland. HGH's role in promoting growth is, of course, most important during childhood and adolescence. A faulty pituitary gland can greatly reduce the amount of HGH young people have in their system, leaving them abnormally short. For years, the only way to get HGH was to laboriously extract it from the pituitaries of dead human beings, a practice that not only yielded too little HGH to go around but also turned out to be unsafe.

Enter biotechnology, which in the mid-1980s produced synthetic HGH in the following way. Using collections of human cells, the gene for HGH was snipped out of the human genome and inserted into the *E. coli* bacterium. Each bacterium that took on the HGH gene began transcribing and then translating this gene, which is to say turning out a small quantity of HGH. These bacteria were then grown in vats by the billions. The result? Collectible quantities of HGH, clinically indistinguishable from that produced in human pituitary glands, manufactured by a biotech firm, and shipped to pharmacies worldwide.

Note that in this process, a gene was taken from one species (a human being) and spliced into another species (a bacterium). With this, the bacterium became a **transgenic organism**: an organism whose genome has stably incorporated one or more genes from another species. This same phenomenon is repeated time and time again in biotechnology. If you look at FIGURE 15.1, you can see a fish that is transgenic, in that it has incorporated a gene from another fish into its genome. Likewise, there are transgenic goats that have incorporated human genes and transgenic plants that have incorporated firefly and bacterial genes. But how is this possible? How do you cut DNA out of one genome and paste it into another? To get an answer, we'll take a walking tour of some basic biotech processes. As a starting point imagine that, as with human growth hormone, the goal is to produce quantities of a hypothetical human protein that will be used as a medicine.

A Biotech Tool: Restriction Enzymes

In the early 1970s it became possible for scientists to cut genomes at particular places, with the discovery of **restriction enzymes**. These are enzymes, occurring naturally in bacteria, that are used in biotechnology to cut DNA into desired fragments. (In nature, bacteria use them to cut up the DNA of invading viruses.) In isolating restriction enzymes, scientists found that many of them had a wonderful property: They didn't just cut DNA randomly; they cut it at very specific places. We can see how this works by looking at an actual enzyme called *Bam*HI. The two strands of DNA's double helix are complementary, as you know, and they run in opposite directions. Thus, the sequence GGATCC would look like FIGURE 15.2 if we were viewing both strands of the helix.

FIGURE 15.1 Bigger, Sooner through Biotechnology All the Atlantic salmon in the picture are about 14 months old, but the salmon on the left grew to 3 kilograms in this time frame, while those on the right won't reach this "market weight" for another year. The difference is that the larger salmon is transgenic; a gene from another fish (an ocean pout) was spliced into its genome, thus changing the effect of growth hormone on its size and making it grow bigger at a younger age. At maturity, however, this fish will weigh no more than the natural variety.

A BROADER VIEW

Silk from Goats

Genes can be moved across a startling array of species. Spider genes have been spliced into the genomes of goats, thus allowing the goats to secrete large quantities of spider-silk proteins in their milk. What's the point? Spider silk is lightweight and very strong; one hope is to use it as replacement tissue in humans—for example, in human tendons.

FIGURE 15.2 The Work of Restriction Enzymes

1. A portion of a DNA strand has the highlighted recognition sequence GGATCC.

2. A restriction enzyme moves along the DNA strand until it reaches the recognition sequence and makes a cut between adjacent G nucleotides.

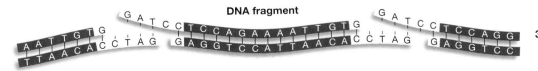

3. A second restriction enzyme makes another cut in the strand at the same recognition sequence, resulting in a DNA fragment.

Now, the *Bam*HI restriction enzyme will move along the double helix, leaving the DNA alone *until* it comes to this series of six bases, known as its *recognition sequence*, and here it will make identical cuts on both strands of the DNA molecule, always between adjacent G nucleotides. When another *Bam*HI molecule encounters another GGATCC sequence, it too makes cuts, which effectively is like making a second cut in a piece of rope, giving us a rope *fragment*.

*Bam*HI's recognition sequence may be GGATCC, but another restriction enzyme will have a *different* recognition sequence, and will make its cuts between a different pair of bases. Indeed, nearly 1,000 different restriction enzymes have been identified so far, which cut in hundreds of different places. The fact that they make cuts at so many specific locations has given scientists a terrific ability to cut up DNA in myriad ways.

Note that with *Bam*HI, each of the resulting DNA fragments has one strand that protrudes. Restriction enzymes that make this kind of cut are particularly valuable, for they produce "sticky ends" of DNA, so named because they have the potential to *stick to* other complementary DNA sequences. In the fragment on the lower left in Figure 15.2, for example, the protruding sequence CTAG could now easily form a base pair with *any* piece of DNA whose sequence is the complementary GATC. So useful are restriction enzymes that they can now be ordered from biochemical suppliers, much as a person might order a set of socket wrenches from a hardware store.

Another Tool of Biotech: Plasmids

From the overview of manufacturing human growth hormone, recall that the human gene for HGH was inserted into *E. coli* bacteria, which then started turning out quantities of this hormone. The question is: How did this human gene get into a bacterium? Several methods of transfer are available, but for now let's focus on a specific kind of DNA delivery vehicle. As it turns out, bacteria have small DNA-bearing units that lie *outside* their single chromosome. These are the **plasmids**, extrachromosomal rings of bacterial DNA that can be as little as 1,000 base pairs in length (**see** FIGURE 15.3). Plasmids can replicate independently of the bacterial chromosome; but just as important for biotech's purposes, they can *move into* bacterial cells.

How do plasmids do this? Through a process called transformation, bacteria are capable of taking up DNA from their surroundings, after which this DNA will function—that is, code for proteins—inside the bacterial cells. Some bacterial cells are naturally adept at transformation, while others, such as *E. coli*, can be induced to perform it by means of chemical treatment. Critically, plasmid DNA can be taken in via transformation and continue to function, as plasmid DNA, inside the bacteria.

Using Biotech's Tools: Getting Human Genes into Plasmids

At this point, you know about a couple of tools in the biotech tool kit: restriction enzymes and the transformation process involving plasmids. Let's now see how they work together. As you can see in FIGURE 15.4, the process starts with a gene of interest in the human genome—in our example, a gene that codes for a protein that is used in a medical treatment. The first step is to use restriction enzymes on this human DNA. Knowing, say, the starting and ending sequence of the gene of interest, a restriction enzyme is selected that allows part of the genome to be cut into a manageable fragment that includes this sequence of interest, preferably in a sticky-ended form.

Here's where the beauty of restriction enzymes really comes into play. If the same restriction enzyme is now used on the DNA of *plasmids*, the result is *complementary* sticky ends of plasmid and human DNA. In other words, through sticky-ended base pairing, these segments of human and plasmid DNA fit together like puzzle pieces.

When the DNA fragments are mixed with the "cut" plasmids, that's just what happens: Human and plasmid DNA form base pairs, and the human DNA is incorporated into the plasmid circle. With this, a segment of DNA that was once part of the human genome has now been *re-combined* with a different stretch of DNA

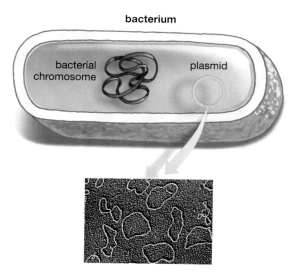

FIGURE 15.3 **Transfer Agent** Plasmids are small rings of bacterial DNA that are not a part of the bacterial chromosome. They can exist outside bacterial cells and then be taken up by these cells. This artificially colored micrograph shows a type of plasmid, from *E. coli* bacteria, that is commonly used in genetic engineering.

bacterium

bacterial chromosome

plasmid

(the plasmid sequence). This process gives us the term **recombinant DNA**, defined as two or more segments of DNA that have been combined by humans into a sequence that does not exist in nature.

Getting the Plasmids Back inside Cells, Turning out Protein

To this point, what has been produced is a collection of independent plasmids that have a human gene within them. Remember, though, the goal is to turn out quantities of the protein the human gene is coding for. To do this requires a vast quantity of plasmids working away, which means working away *back inside* bacterial cells, and it's here that transformation comes into play. If the plasmids are put into a medium containing specially treated bacterial cells, a few of these cells will take up plasmids through transformation (at which point these cells have become transgenic, as noted earlier). Once this happens, the plasmids start *replicating* along with the bacterial cells themselves. As one bacterial cell becomes two, two become four, and so on, the plasmids are replicating away as well. And as the cell count reaches into the billions, collectible quantities of the protein begin to be turned out via instructions from the human gene inserted into the plasmid DNA.

In reviewing this process, we've focused on plasmids as the agent that first takes on and then transfers DNA; but there are other so-called cloning vectors as well, the most important of them being viruses that infect bacteria. They too can have genes spliced into them, and then transfer this DNA into working cells.

Real-World Transgenic Biotechnology

The example we just went over concerned a hypothetical medicine, but lots of actual medicines are being produced through biotechnology today. By one count, more than 175 biotechnology drugs and vaccines have been approved by the U.S. Food and Drug Administration. We've already touched on one biotech medicine—the human growth hormone that is produced inside the *E. coli* bacterium. As it turns out, *E. coli* is something of a workhorse in medical biotech. Human insulin and several kinds of cancer-fighting compounds are also produced through it.

E. coli and its fellow bacteria are not the only "bio-factories" used in medical biotechnology, however. Through various gene-splicing techniques, yeast, hamster cells, and even large mammals are utilized as well. Transgenic goats produce several kinds of human proteins in their milk, including one called AT3 that prevents the formation of blood clots. The reason expensive goats, rather than inexpensive bacteria, are used for this purpose is that goats produce large *quantities* of milk, which means they also produce prodigious amounts of the human protein that comes with the milk. A single goat can turn out over two pounds of human protein per year.

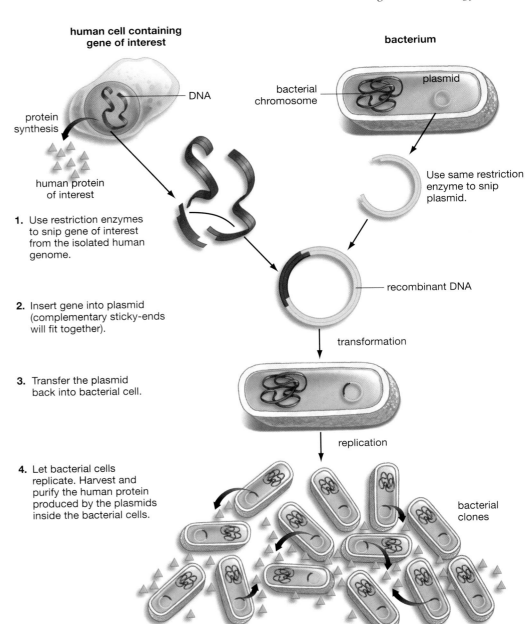

1. Use restriction enzymes to snip gene of interest from the isolated human genome.

2. Insert gene into plasmid (complementary sticky-ends will fit together).

3. Transfer the plasmid back into bacterial cell.

4. Let bacterial cells replicate. Harvest and purify the human protein produced by the plasmids inside the bacterial cells.

FIGURE 15.4 How to Use Bacteria to Produce a Needed Human Protein

A BROADER VIEW

A Transgenic First

What was the first biotech product? Genetically engineered insulin, made in *E. coli* bacteria. It was approved for sale in 1982.

FIGURE 15.5 **A More Nutritious Rice** Grains of the genetically engineered "golden rice" stand out next to grains of ordinary rice. The rice has a golden color because it is able to produce its own beta carotene, which the human body converts into Vitamin A.

Genetically Modified Food Crops Transgenic organisms do more than produce medicines. You may have heard of genetically modified or "GM" food crops. It won't surprise you to hear that all of these crops are transgenic. If you look at FIGURE 15.5, you can see a transgenic rice that has been dubbed "golden rice" because of its ability to produce its own beta-carotene, which the human body coverts to vitamin A. Conventional rice does not contain beta-carotene, and this fact has health consequences. The World Health Organization estimates that half a million children go blind each year from vitamin A deficiency, and that 1 to 2 million children die from it. No one claims that golden rice can completely solve this problem, but the hope is that it can help alleviate it. Golden rice is transgenic in more ways than one: Genes from both a bacterium and a daffodil have been spliced into it, thus enabling it to produce the beta-carotene.

Field tests are just beginning on golden rice, but several other transgenic food crops are in full commercial production right now. Indeed, during 2004 in the United States, 85 percent of all soybeans, 76 percent of all cotton, and 45 percent of all corn crops were genetically modified in one way or another. And China has produced several varieties of insect-resistant transgenic rice, though none has yet been approved for commercial use. In general, the purpose of transgenic crops is to get greater crop yields with less use of chemical pesticides and herbicides. No biotech product has sparked as much international controversy as GM crops, however, and as a result their future is uncertain. We'll review some of the issues surrounding GM crops in the chapter's concluding section.

So Far...

1. A transgenic organism is one whose genome has stably incorporated one or more _____ from _____.

2. DNA is cut into useful fragments through use of the molecular scissors known as _____.

3. The plasmids noted in the text had _____ DNA spliced into them as a first step in using them to turn out a human _____.

So Far... Answers:

1. *genes; another species*
2. *restriction enzymes*
3. *human; protein*

15.3 Reproductive Cloning

You may have noticed that the word *cloning* popped up in the discussion of transgenic biotechnology. Thanks to movies and recent real-life events, this word has taken on some sinister implications, as if biological cloning were inherently a Frankenstein-like procedure. But it need be no more threatening than the process of making copies of a single gene. To **clone** simply means "to make an exact genetic copy of." An individual clone is one of these exact genetic copies. Thus, in the biotech world, the gene for human growth hormone is cloned by means of being snipped out of the human genome and then copied within bacterial cells. A collection of such cells can be thought of as a single clone, because all of them are genetically identical. (Each bacterial cell reproduces by simply making a copy of its DNA and then dividing.)

Cloning can involve not just bacteria, however, but larger organisms as well. Human beings actually have been making clones for centuries, though by low-tech rather than high-tech methods. A "cutting" taken from a plant and put into soil can sometimes grow into a whole new plant. When this happens, there is no

FIGURE 15.6 **Revolutionary Sheep** Dolly, the first mammal ever cloned from an adult mammal, is shown here as a young sheep with her surrogate mother. Note that Dolly is white-faced, like the sheep she was cloned from, while her surrogate mother is black-faced.

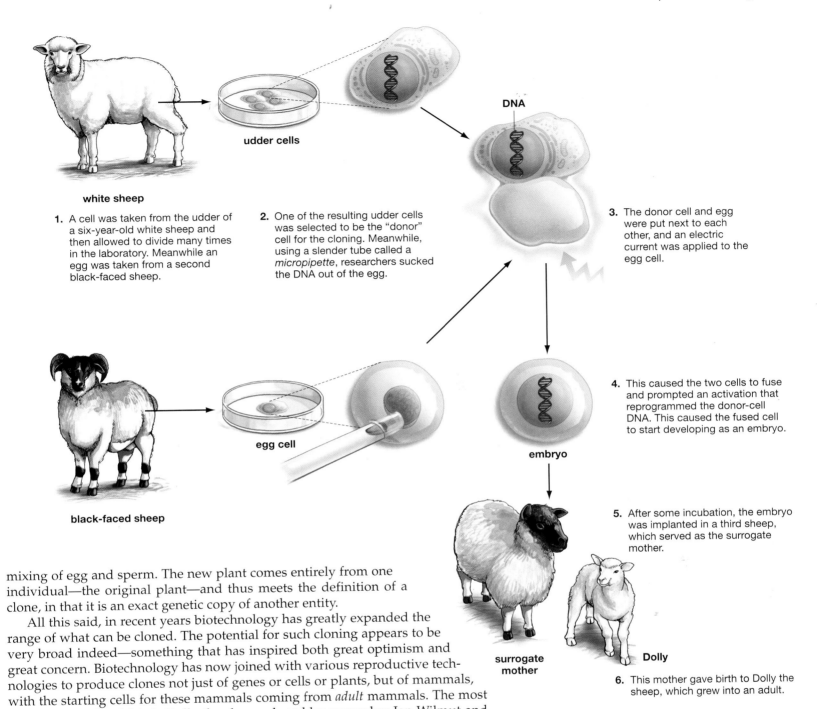

1. A cell was taken from the udder of a six-year-old white sheep and then allowed to divide many times in the laboratory. Meanwhile an egg was taken from a second black-faced sheep.

2. One of the resulting udder cells was selected to be the "donor" cell for the cloning. Meanwhile, using a slender tube called a *micropipette*, researchers sucked the DNA out of the egg.

3. The donor cell and egg were put next to each other, and an electric current was applied to the egg cell.

4. This caused the two cells to fuse and prompted an activation that reprogrammed the donor-cell DNA. This caused the fused cell to start developing as an embryo.

5. After some incubation, the embryo was implanted in a third sheep, which served as the surrogate mother.

6. This mother gave birth to Dolly the sheep, which grew into an adult.

white sheep

black-faced sheep

egg cell

DNA

udder cells

embryo

surrogate mother

Dolly

FIGURE 15.7 **How Dolly Was Cloned**

mixing of egg and sperm. The new plant comes entirely from one individual—the original plant—and thus meets the definition of a clone, in that it is an exact genetic copy of another entity.

All this said, in recent years biotechnology has greatly expanded the range of what can be cloned. The potential for such cloning appears to be very broad indeed—something that has inspired both great optimism and great concern. Biotechnology has now joined with various reproductive technologies to produce clones not just of genes or cells or plants, but of mammals, with the starting cells for these mammals coming from *adult* mammals. The most famous example of this is Dolly the sheep, cloned by researcher Ian Wilmut and his colleagues in Scotland in 1997 (see FIGURE 15.6). This is **reproductive cloning**: cloning intended to produce genetically identical adult animals. Powerful in its own right, it gains added potential when combined with the basic biotech processes you've been reviewing.

Reproductive Cloning: How Dolly Was Cloned

Dolly the sheep, who died in 2003, was a clone as that term is defined: She was, to a first approximation, an exact genetic replica—in this case, of another sheep. Here's how she was produced (see FIGURE 15.7). A cell was taken from the udder of a six-year-old adult sheep and then grown in culture in the laboratory, meaning that the original cell divided into many "daughter" cells. While this was going on, researchers took an *egg* from a second sheep and removed its nucleus, meaning they removed all its nuclear DNA. Then they placed the udder cell (which had

A BROADER VIEW

Biotech Means Business

It pays to be first with a big idea. In the 1970s, Stanley Cohen of Stanford was working with plasmids, while Herb Boyer of the University of California, San Francisco was working with restriction enzymes. The two met at a conference in Hawaii in 1972 and, over hot pastrami and corned beef sandwiches, decided to bring their research interests together. The result was the invention of the gene-transfer process reviewed in the text. In 1980, Stanford and UC were issued the "Cohen-Boyer" patent, under which both universities, along with Cohen and Boyer, were paid royalties by every biotech company that wanted to employ the Cohen-Boyer technique. By the time the patent expired in 1997, the universities' share of the royalties was more than $200 million.

A BROADER VIEW

Why Clone Human Beings?

What are some arguments *for* human cloning? One set of parents whose child died in infancy wanted to clone this child by using DNA from its cells (which they had preserved). Their wish was to have a second child who was as much like their deceased child as possible. A separate argument involves situations in which both members of a couple are infertile. Cloning would allow such couples to have a child who is the biological offspring of one of them.

DNA) next to the egg cell (which no longer had any DNA) and applied a small electric current to the egg. This had two effects: It caused the two cells to fuse into one, and it mimicked the stimulation normally provided when a *sperm* cell fuses with an egg. With this, the udder cell DNA began to be reprogrammed. As a result of this reprogramming, the fused cell started to develop as an embryo. (Though the egg cell had its DNA removed, it still contained all kinds of egg-cell proteins whose normal function is to trigger development of an embryo. It was these factors still in the egg that reprogrammed the donor-cell DNA.) After the embryo had developed to a certain point, the researchers implanted it in a third sheep, which served as a surrogate mother. The result, 21 weeks later, was the birth of the lamb Dolly, who went on to give birth to two sets of her own lambs.

Every cell in Dolly had DNA in its nucleus that was an exact copy of the DNA in the six-year-old donor sheep. Thus Dolly was a clone of that sheep. There was no mixing of genetic material from two parents to produce Dolly—just a copying of one individual from another.

Cloning and Recombinant DNA

Once Dolly was born, science was off to the races in terms of producing clones of common mammals. To date, horses, mules, cows, pigs, cats, mice, and goats have been cloned—all of them essentially through the same process that was used with Dolly. But to what end? That is, why are scientists cloning these animals? Well, to use Dolly as an example, Ian Wilmut and his colleagues were not so much interested in cloning a sheep as they were in coupling the power of reproductive cloning to the techniques of recombinant DNA that you just looked at. The workers at Wilmut's research institute went on to produce Polly the sheep, who was not only a clone, but a clone who had had a human gene inserted into her genome and was thus able to produce a blood-clotting protein useful to hemophiliacs. A *group* of such clones would be valuable because they all would have the same starting genetic set—the same genome—and thus would all be able to produce the blood-clotting protein. Cloning's power stems in part from its ability to produce one organism after another that has the same traits. Through cloning, it is possible to sidestep the *mixing* of traits that comes with sexual reproduction.

Human Cloning

Reproductive cloning may interest scientists because of its potential to yield products such as medicines. It has captured the attention of the average person, however, because it raises the possibility of *human* cloning. How could a person be cloned? Pretty much through the same method used with Dolly—take a cell from an adult human being and fuse it with an egg cell whose nucleus has been removed. Then implant the fused cell in the uterus of a woman willing to bring the resulting embryo to term.

The prospect of a human clone is so dizzying that it's worthwhile to think about what such a person would represent in biological terms. He or she would be a genetic replica of the person who provided the donor cell with the DNA in it. One helpful way to think of this person's biological status is in terms of a more familiar concept: that of an identical twin. As it happens, identical twins also are genetic replicas of one another. Early in a human pregnancy, separate cells from an embryo can start forming as two separate human beings, and the result is identical twins. Cloning is a *different* means of producing genetically identical individuals, but the critical point is that a donor and a clone would be no more alike than two identical twins.

Be this as it may, the prospect of a human clone still has the power to stun us. The overwhelming majority of scientists and laypeople are opposed to the idea, but such a thing does seem technically feasible—more so now than just a couple of years ago, as you'll see in section 15.5, on stem cells.

Saving Endangered Species

Could the ability to clone mammals be used by scientists to save species of mammals that are threatened with extinction? Perhaps. If you look at **FIGURE 15.8**, you can see a picture of a Siberian ibex (*Capra sibirica*), a rare, goat-like animal that lives in the mountains of central Asia. In early 2004, the Chinese government announced that it had cloned the ibex as part of a species preservation effort. One problem in trying to do this for any threatened animal involves one of cloning's starting ingredients: the egg cell that fuses with the donor DNA cell. Since endangered species are rare by definition, there are few females who can serve as egg donors. But scientists now seem to have found an answer to this problem. Donor DNA cells can fuse with eggs from *closely related* species, thus bringing about a clone. This is what China did with the Siberian ibex; it used common goats as both egg-cell donors and as surrogate mothers. Hopes have been raised that this same process can be used to bring back a species that is completely extinct, a Spanish mountain goat called the bucardo. Before the last burcardo died, in 2000, researchers had the foresight to take tissue samples from it. The hope is to use cells from these samples and fuse them with egg cells from a compatible species. Note that, if this works, researchers will have brought a species back to the world *following* its extinction. So why couldn't we do this with dinosaurs, as in *Jurassic Park*? Because DNA degrades over time, and dinosaurs have been gone a *long* time—the last of them died 65 million years ago. Meanwhile, the oldest intact animal DNA we've been able to retrieve came from an individual who died a mere 30,000 years ago—a Neanderthal man.

FIGURE 15.8 Endangered and Cloned The animal on the right, a Siberian ibex, was cloned from an adult ibex, but the egg cell used in the procedure was donated by a common goat like the one at left. The goat in the picture served as the surrogate mother in the procedure, giving birth to the ibex in 2004.

15.4 Forensic Biotechnology

After New York's World Trade Center (WTC) was destroyed on September 11, 2001, public officials had to undertake the grim task of trying to identify all the victims of the attack, and to do so quickly, so as to provide some sense of closure to the victims' loved ones. In the weeks following the attack, therefore, officials collected samples of DNA not only from WTC victims, but from the toothbrushes and razors of persons who were missing since the tragedy and hence *presumed* to be victims. They also collected DNA from relatives of these presumed victims. With this DNA database in hand, workers then made an effort to match one kind of DNA with another—to match the DNA found on a toothbrush, for example, with the DNA from a WTC victim.

High-profile cases such as this one are the most visible manifestation of a huge change that has come about in the way society *identifies* persons—criminals, biological fathers, and victims, most notably. In the criminal arena, the most highly valued type of physical evidence today is not fingerprints; it is tiny bits of human tissue from which DNA can be extracted. But how do police use DNA to get a "match" in a criminal case? Let's find out.

The essence of "forensic DNA typing," as it is known, is to get two sets of physical samples—first, a sample left by the perpetrator at a crime scene in such forms as blood or semen; and second, a blood sample from a suspect. By comparing DNA sequences in both the crime-scene sample and the suspect sample, forensic scientists can establish whether the suspect was present at the crime scene. If he or she was, but had no good reason to be there, this is strong evidence that the suspect committed the crime. Another way of gathering DNA evidence is to use a victim's blood sample and test it against, for example, blood found on a suspect's clothing.

The Use of PCR

When forensic DNA typing first came into use in the 1980s, a significant limitation on it was that relatively *large* samples of human tissue had to be obtained in order to get enough DNA to work with. Eventually, however, this problem was all but eliminated by something called the **polymerase chain reaction (PCR)**: a technique for quickly producing many copies of a specific segment of DNA. Thanks to PCR, if a criminal so much as touches a telephone, police can retrieve enough DNA to get a match.

The PCR process is very simple in outline. Four kinds of materials are mixed together to carry it out. The first of these is a quantity of DNA itself. Then there is a collection of individual DNA nucleotides and DNA polymerase (which, remember, goes down the line on single-stranded DNA, affixing nucleotides to available bases). Finally, there are two DNA "primer" sequences—short sequences of single-stranded DNA that act as signals to DNA polymerase, saying "start adding nucleotides here."

The first step in the process is that the starting quantity of DNA is heated until the two strands of DNA's double helix separate, resulting in two single strands of DNA (see **FIGURE 15.9**). As the heated DNA mixture cools, primers will attach to each of the now-separate strands of DNA. Then the DNA polymerases go down the line, starting from the primers, linking available nucleotides to the template DNA and thus producing strands that are complementary to the original strands. The result is *two* double-stranded lengths of DNA, both identical to the original double strand. In short, the DNA sample has been doubled in one copying "cycle." Then the entire process of heating and cooling is repeated. Since each copying cycle takes only 1–3 minutes, by the time 90 minutes have passed, millions of copies can be created. Moreover, thanks to the specificity of primers, the sequence that is copied can be limited to a small sequence of interest within the larger genome. So useful is PCR that it is employed not just in forensic biotechnology, but in countless facets of biological research. We are focusing here on one specific use for PCR, but it has so many applications that in 1993, its inventor, Kary Mullis, was awarded the Nobel Prize in Chemistry for his achievement.

Finding Individual Patterns

Once PCR has done its job, the police have enough DNA to begin looking for individual patterns within it—one set of patterns found in the crime-scene DNA, one set of patterns seen in the suspect's DNA. But what are these patterns that each of us carries? Human genomes are filled with short sequences of DNA that are repeated over and over, from 3 to 50 times, like this:

TCATTCATTCATTCATTCAT

This TCAT example is not hypothetical. It is an actual *short tandem repeat* (or STR) that police departments use in today's DNA typing. The usefulness of STRs is that, at a given location in the genome, one person will have one number of tandem repeats, while another person is likely to have a *different* number of repeats, like so:

Individual 1: TCAT TCAT TCAT TCAT
Individual 2: TCAT TCAT TCAT TCAT TCAT TCAT

FIGURE 15.9 DNA Copying Machine The polymerase chain reaction (PCR) makes copies of a given length of DNA very quickly.

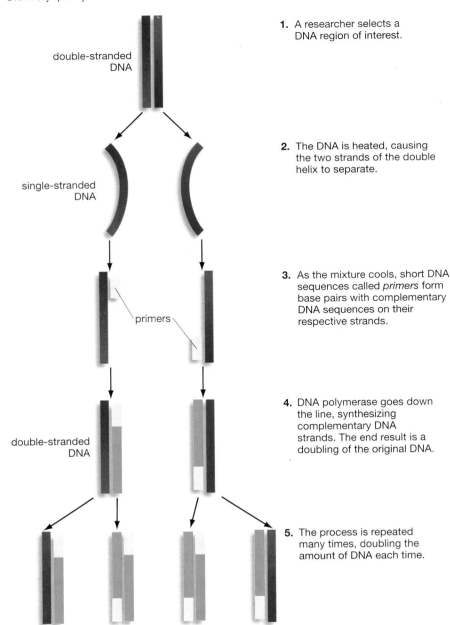

double-stranded DNA

single-stranded DNA

primers

double-stranded DNA

1. A researcher selects a DNA region of interest.

2. The DNA is heated, causing the two strands of the double helix to separate.

3. As the mixture cools, short DNA sequences called *primers* form base pairs with complementary DNA sequences on their respective strands.

4. DNA polymerase goes down the line, synthesizing complementary DNA strands. The end result is a doubling of the original DNA.

5. The process is repeated many times, doubling the amount of DNA each time.

Now, the chances are small that two unrelated individuals will have an identical number of repeats at even one location in the genome. But modern forensic DNA typing looks at STR repeats at 13 different locations. Imagine yourself, then, as a forensic investigator who is reviewing both DNA from a crime scene and DNA from a suspect in the crime. At genomic location 1, the STR patterns are as follows:

Crime scene: TCAT TCAT TCAT TCAT
Suspect: TCAT TCAT TCAT TCAT

So you have a match at this one location. If the STR patterns continue to line up this way at the other 12 locations, then what once was probable becomes certain. Your suspect was certainly at the crime scene, because the odds of any two people having identical STR patterns at 13 locations are astronomically small.

It's important to recognize that this kind of DNA evidence can help establish innocence as well as guilt. Indeed, DNA testing regularly exonerates not only suspects, but people who have been *convicted* of crimes. In the past 15 years, more than 130 prisoners who were jailed for various offenses have been freed on the basis of DNA testing. If you look at FIGURE 15.10, you can see one way that short, repeating DNA sequences can be visualized in criminal cases.

Beyond matching a known suspect to a crime scene or victim, forensic DNA typing has another power. Increasingly, it is allowing police to make so-called cold hits. All 50 states now require criminals convicted of violent crimes to provide DNA samples. Capitalizing on this, the FBI has put together a national database of the DNA profiles of these criminals. Given this, imagine a situation in which a crime is committed, and no suspect was identified, but in which a DNA sample was retrieved. A DNA profile will be produced from this sample—a profile that can be matched against all those that are stored in a DNA database. If the profile from the crime scene matches a criminal's stored profile, that in itself is sufficient evidence to make an arrest. And this is happening all the time—cold hits result in hundreds of arrests across the nation each year.

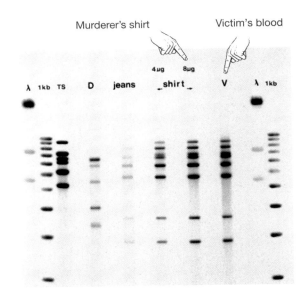

FIGURE 15.10 **Telltale Stains** A defendant in this real-life murder case claimed that blood stains found on his clothing (see the rows under "shirt") came from his own blood rather than from the blood of a young woman who was stabbed to death. The DNA evidence said otherwise. Look at the pattern in the row under V (for victim) and see how closely it matches the pattern in the shirt rows. The probability that the blood on the defendant's clothing came from anyone but the victim was 1 in 33 billion. (Courtesy of Cellmark Diagnostics, Inc., Germantown, Maryland)

So Far...

1. To "clone" means to make an exact _____.
2. True or false: Dolly was a transgenic organism.
3. PCR has been described as a copying machine for _____.

15.5 Stem Cells

Nobody who lived through the presidential campaign of 2004 could have missed the term *stem cells*, given the intense political fight over research into them. But what are stem cells, and how could they be used to combat human disease? Let's see.

Cell Fates: Committed or Not

In one way, human cells are like human beings: They start off with the potential to be lots of things, but eventually they commit themselves to being just one thing. This course of events takes place during human "development," meaning the process by which an embryo becomes a fully formed human being. Early in this process, cells need to be adaptable enough to go down different developmental pathways. Why? Because the embryo is developing in so many basic ways. For example, it needs to create both muscle cells and bone cells. Fortunately, there is a type of embryonic cell that is flexible enough to give rise to either type of cell. As time goes on, however, more and more tissues become fully specialized, which means there is less and less need for cells that are developmentally flexible. By the

A BROADER VIEW

Checking on Fidelity

It is now easy to check not just for paternity, but for fidelity between partners. Suspected semen stains left on bedding or undergarments can yield DNA that commercial laboratories will run a DNA profile on. A man can send in a sample of his own DNA as well and match its profile against the one from the stain.

So Far... Answers:

1. *genetic copy of*
2. *False; no genes from any other species were spliced into Dolly's genome. Instead, she was a clone of the sheep who provided the donor DNA cell.*
3. *DNA*

time humans reach adulthood, specialization has almost completely trumped flexibility. The vast majority of cells in our bodies have undergone what is known as *commitment*: a developmental process that results in cells whose roles are completely determined. Almost all your muscle cells, for example, are committed in that they can be nothing but muscle cells and can give rise to nothing but muscle cells.

From a human perspective, the problem with this sequence of events is that the tissues of fully developed human beings can become damaged. They break down or become diseased; they are injured in car accidents or fires. And many damaged tissues will not regenerate themselves. A person whose spinal cord is severed does not generate a new spinal cord. Thus, for medicine, the question for decades has been: Could we use developmental processes to generate new tissues? One obvious way to go about this would be to harness the power of the cells that, in nature, generate new kinds of tissue—early embryonic cells. It turns out, however, that there is also another possibility. A small proportion of *adult* cells retain the ability to produce specialized cells. Both the adult and the embryonic cells that have this generative capacity are known as **stem cells**: cells with the capacity to produce more cells of their own type, along with at least one type of specialized daughter cell.

Stem Cells from Embryos

So what is the nature of this first kind of stem cell, the type derived from an embryo? The very early embryo is a fertilized egg that has begun to undergo cell division. If we go forward about five days from the first division, the embryo has developed into a hollowed-out, fluid-filled ball of about 200 cells—a **blastocyst**. One section of the blastocyst's cells becomes the placenta that will help nurture a growing embryo, while another section, called the inner cell mass, constitutes the embryo itself (**see** FIGURE 15.11). It is these latter cells that interest us, for among the flexible, early embryonic cells we talked about, these cells are *very* flexible. Indeed, these are the cells you've heard so much about in the news, **embryonic stem cells**. How flexible are they? We can define them as cells from the blastocyst stage of a human embryo that are capable of giving rise to all the cells or tissues in the adult body.

Adult Stem Cells

Breakthroughs with embryonic stem cells in the late 1990s led to a flurry of activity regarding stem cells from adult tissues. Initial research seemed to indicate that, like embryonic stem cells, adult stem cells could be located and then give rise to many other kinds of cells. This may still turn out to be the case, but over time, doubts have increasingly been raised about the potential of adult cells. Some early, promising results with them could not be reproduced when attempted by other researchers. And in some cases, it has not been clear whether they have differentiated into desired cell types or have simply fused with existing cells of those types. The upshot is that, at the moment, scientists are unclear about the role adult stem cells will be able to play in treating disease.

FIGURE 15.11 **Cells That Can Become Anything** Human development starts with the fertilization of egg by sperm and then proceeds through several days of cell division, such that by day 3 the embryo has become a tiny, solid ball of cells. By day 5, however, it has developed into a hollow ball of about 200 cells called a blastocyst. One segment of the blastocyst, the inner cell mass, will develop into the baby. Some of the cells in the inner cell mass are the embryonic stem cells that can give rise to all the different types of cells and tissues in the adult body. Pictured on the right is a human blastocyst on day 6 of development. It has recently undergone the process of "hatching" or breaking out of the protective covering that surrounds early embryos, which is why it no longer has the spherical shape of the day-5 blastocyst.

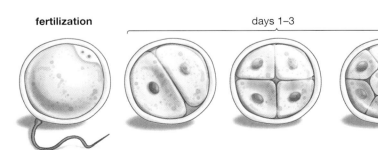
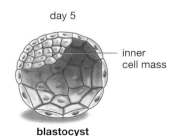

fertilization

days 1–3

day 5

inner cell mass

blastocyst

The Potential of Embryonic Stem Cells

But what about embryonic stem cells (ESCs)? No human disease has yet been cured with them, and so far they have demonstrated therapeutic potential mostly in mice and rats. Nevertheless, most parties seem to agree that they hold enormous promise. Consider an affliction such as Parkinson's disease, whose symptoms of shaking and stiffness are brought about because certain neurons in the brain quit making a substance called dopamine. The question is: Can we get ESCs to differentiate into dopamine-producing neurons? As it turns out, researchers have already done this in rats. They took stem cells, injected them into the brains of rats who had half their dopamine-making neurons destroyed, and then watched as the injected cells were transformed into functional neurons, complete with the ability to secrete dopamine. Another disease of the nervous system, multiple sclerosis, does its damage by destroying a kind of insulation, called myelin, that wraps around nerves. In 2004, researchers from California announced that in the laboratory, they coaxed ESCs into becoming myelin-producing cells. Better yet, when these cells were injected into the spinal cords of mice engineered to lack myelin, the new cells migrated to spinal cord neurons and started producing a myelin covering around them.

Encouraging as these results are, they don't guarantee anything about the ability of ESCs to regenerate damaged tissue inside human beings. Researchers have some daunting hurdles to get over before stem cells can even be tested in connection with most human illnesses. Can ESCs be developed that will differentiate *only* into the desired type of cell? (Groups of undifferentiated ESCs can spontaneously start forming tumors.) Can implanted cells be counted on to stay where they are needed, rather than "migrating" to other areas? Can we rest assured that implanted cells will be free of any harmful viruses? Given these issues, most researchers think that it may be five or even ten years before a large number of human ESC trials can get under way. This is not to say, however, that all human trials are this far off. The California researchers who did the work with the spinal cords of mice say they expect to be ready for the first human trials of their therapy in less than two years.

Stem Cells Meet Cloning

One of the additional challenges to stem cell therapy is the same challenge posed by organ transplants: Any cells put into a human body are likely to be regarded as invading *foreign* cells by the immune system, meaning these cells can set off an immune system attack. Thus, ESCs could be "rejected" by a body in the same way a transplanted kidney can be. If you think about it, one way to get around this would be for each of us to be able to generate our *own* ESCs and get them to differentiate into the type of cell we need. But how could such a thing be possible? ESCs exist only in early stage embryos, which none of us has been for a long time. Well, what about using one of our own adult cells as a donor DNA cell in the cloning procedure we talked about earlier? Then fuse this cell with an egg cell whose nucleus has been removed and let the resulting embryo divide to the blastocyst stage. Finally, "harvest" the ESCs from the blastocyst. This may seem too far-fetched to be believed, but it has already been done. In 2004, a Korean team led by researchers Woo Suk Hwang and Shin Yong Moon announced that their team had achieved it. Sixteen women volunteered to provide the team with donor DNA cells as well as eggs for the experiment. Then, in each cloning attempt, a single woman's DNA was inserted into one of her own egg cells. The result was dozens of blastocysts and a single ESC "line" that was allowed to keep dividing through 70 generations. In 2005, the same team announced dramatic improvements in the efficiency of its process.

This is another example of a biotech development so startling that we have to stop and think about what it means. Remember earlier, when it was noted that human cloning seems more technically feasible now than it did a few years ago? The Korean cloning experiments are the reason for the change. All the early steps that were carried out in cloning Dolly took place with human beings in the Korean work. Yet the goal of the Korean experiment was not *reproductive* cloning—nobody contemplated inserting one of the cloned embryos into a woman's uterus, which would have allowed the embryo to keep developing. Instead, the goal was to test the feasibility of **therapeutic cloning**: the use of cloning to produce human embryonic stem cells that can be used to treat disease. It is too early to say what future there is for therapeutic cloning. But it has attracted great interest worldwide, and its goals are clear. The idea is that each of us could help generate an embryonic clone of ourselves whose cells could be used in our own repair.

15.6 Controversies in Biotechnology

As may be apparent by now, biotechnology has a great ability to generate not only products and processes, but controversies as well. What follows is a look at a few of the issues society is grappling with as a result of biotech.

Embryonic Stem Cells and Cloning

Among the ethical controversies that biotech has generated, none is more intense than the dispute over the two areas we've just gone over: embryonic stem cells and human cloning. To be clear, scarcely any controversy exists over the idea of human *reproductive* cloning, since nearly everyone—scientists and laypersons alike—thinks it is a repulsive idea. (No one could ensure that a cloned person would be perfectly healthy, mentally or physically. Who among us would be willing to accept the idea of producing "test" human beings in cloning experiments?) Still, a religious sect based in Canada has expressed an interest in providing everything necessary for human cloning, including women willing to bring the embryos to term. And a few scientists from around the world have said they are ready to carry out the procedure. If we assume, then, that anything that can be done by human beings will be done by them eventually, we get to the ethical dilemma: Is it acceptable to advance the *techniques* necessary to bring about human reproductive cloning in the way that the Korean researchers have? Until they announced their results, most scientists believed that some technical obstacles stood in the way of human cloning. Now it is clear those obstacles can be surmounted. As noted, the benefits of *therapeutic* cloning may be significant. Yet because the techniques used in it can be employed to bring about reproductive cloning, should therapeutic cloning work be discouraged or even forbidden?

The Korean research also raised, once again, a question that figured strongly in the 2004 presidential campaign: Is it inherently unethical to do any research with embryonic stem cells? To state the obvious, embryonic stem cells come from human embryos—from collections of cells that have the potential to become fully formed human beings. Where do researchers get these embryos? In most cases from fertility clinics, which each year discard thousands of excess embryos that have been produced for in vitro fertilizations. (A couple undergoing fertility treatments generally will produce many more embryos than they need to bring about a pregnancy.) With the couples' consent, excess embryos that would have been discarded are used instead in scientific research. When scientists extract stem cells from these early stage embryos, however, the embryos are destroyed, and this is a matter of great concern to many people. For them, the destruction of an embryo is a destruction of human life. Given such objections, you might think one alternative would be to simply stop using embryonic stem cells and instead work solely with adult stem cells. As noted, however, there are serious questions about the potential of adult cells. Meanwhile, research done in the last few years has confirmed the powerful potential of embryonic stem cells.

In the United States, the practical consequence of this issue has to do with research funding. For years the federal government—the largest single source of biological research funds—did not fund any research in which embryos were destroyed. Given new interest in stem cell research in 2001, the question before President Bush was: Should the federal government fund research that merely *involves* embryonic stem cells, even if no embryos were destroyed in this research?

A BROADER VIEW

Therapeutic Cloning

If therapeutic cloning succeeded, would each of us go through a cloning procedure and then "bank" the blastocyst that resulted, thus keeping our stem cells in reserve for a time when we are ill or injured?

Could a researcher obtain embryonic stem cells from, for example, a private-sector scientist, and then receive government funding for research on the embryonic cells received?

In August 2001, the president decided that the answer is a qualified yes. Federal funding is now provided to researchers who work with cells from any of the 71 stem cell lines developed before the date of the president's announcement. The effect of this decision was to allow federal funding for some research linked to the destruction of embryos, but to restrict the number of cell lines eligible for this funding. Three years later, however, only 15 or so of the "presidential lines" were available to researchers, and some of those were not robust or were feared to have been contaminated by non-human viruses. Meanwhile, countries like Great Britain have imposed no restrictions on stem cell research funding. In the view of many American experts, it is as if the United States is trying to develop the latest computer technology with the stipulation that no microchips produced after 2001 be used in the effort. In response, private institutions and state governments have started to take matters into their own hands. Harvard University has announced a $100-million stem cell initiative, Stanford University has announced something similar, and—using private money—university researchers are developing new stem cell lines that can be distributed to the research community. In the biggest response of all, Californians voted in 2004 to spend $3 billion of state money over the succeeding 10 years on embryonic stem cell research.

Genetically Modified Foods

Worldwide, no biotechnology issue has raised more furor than that of genetically modified foods. Consider that the "golden rice" reviewed earlier had to be grown in a greenhouse in Switzerland that was made *grenade-proof* because of fear it might be attacked by opponents of GM foods. What is this controversy about? In a nutshell, proponents of GM foods see in them the potential to feed a hungry world and to lessen the environmental damage caused by such human practices as pesticide application. Opponents of GM foods see the potential to harm human health and wreak havoc on Earth's ecosystems.

If you look at FIGURE 15.12, you can see a case in point in this battle. Pictured are two sets of cotton plants, one set genetically modified, the other not. The

FIGURE 15.12 **Built-in Resistance to Pests** The cotton plants on the left have had genes spliced into them from the bacterium *Bacillus thuringiensis*. These genes code for proteins that function as a pesticide. Meanwhile, the cotton plants on the right are not transgenic in this way—they have not had genes from another organism spliced into them. Both stands of cotton were under equal attack from insects, but the *Bt*-enhanced plants fared much better.

modified plants are transgenic. They contain genes from a bacterium, called *Bacillus thuringiensis*, that is found naturally in the soil and that produces proteins that are toxic to a number of insects. Collectively, these proteins constitute a natural insecticide known as Bt, which has been sprayed on crops for years, mostly by "organic" farmers, meaning those who do not use human-made pesticides or herbicides. Enter biotechnology firms, which in the 1990s spliced *Bt* genes into crop plants, with the result that these plants—corn, cotton, and potatoes—now produce their *own* insecticide. The results have in some ways been an environmentalist's dream. In one survey conducted in the American Southeast, farmers who planted *Bt* cotton reduced the amount of chemical insecticides they applied to their fields by 72 percent. They did this, moreover, while increasing cotton yields by more than 11 percent.

Despite such benefits, we can also see, in the use of *Bt* seeds, the qualities that make environmentalists uneasy about GM crops in general. Those few insects that survive in *Bt*-enhanced fields are likely to give rise to populations resistant to the natural toxin. This raises the prospect of a valued natural insecticide losing its effectiveness against insects over the long run. So alarming is this possibility that the U.S. Environmental Protection Agency requires that at least 20 percent of any given farmer's crops must be non-*Bt* plants. The idea is to create non-*Bt* "refuges" near the *Bt* fields that will harbor bugs that have not built up a resistance to Bt. These bugs will then mate with *Bt*-resistant bugs, thus helping to ensure that *Bt* resistance does not spread.

A more general concern about GM crops is that some of them will take on the role of "super-plants" that will spread by means of their pollen fertilizing nearby plants. Then there is the idea that the foods made from GM crops could set off allergic reactions in consumers or have other adverse health effects on them.

There is no evidence that GM foods have caused either health or environmental problems so far. But public and government wariness about GM foods has had a profound effect on this area of biotech. Looking at GM acreage worldwide, it turns out that almost all of it is devoted to just four crops: cotton, corn, soybeans, and canola. Meanwhile, very few *new* GM crops are even being developed now, to say nothing of brought to market. In 1999, 120 biotech fruits and vegetables were being tested in U.S. field trials. By 2003, that number had dropped to about 20. What is driving this trend? In large part, fear of consumer resistance to GM foods. In 2004 the Monsanto corporation gave up on a program it had started seven years earlier, aimed at producing GM wheat. The company worried that few American or Canadian farmers would plant the wheat because *they* feared resistance to it from European and Japanese consumers. Such resistance, the farmers thought, might effectively put a "do not import" label on all U.S. and Canadian wheat, whether genetically modified or not. This same chain of reasoning is operating in connection with other GM foods. The upshot is that, while the reach of GM foods is very deep for a few crops, it is not broad at present and may never be.

A BROADER VIEW

Biotech on the Farm

The phrase "Roundup ready," often heard in farm states, refers to crops that have been genetically engineered to allow them to tolerate the herbicide Roundup (which kills the weeds in the vicinity of crop plants). This one aspect of biotech has had an enormous effect on farming in America. Eighty-five percent of all American soybean crops are bioengineered, essentially meaning they are bioengineered to be tolerant of Roundup.

On to Evolution

In touring the strands of DNA's double helix over the last few chapters, you've had many occasions to see how this microscopic molecule can profoundly affect our macroscopic world. DNA replicates, it mutates, it is passed on from one generation to the next. And some organisms will have more *success* in passing on their DNA—in reproducing, in other words—than will others. This latter fact has been critical in bringing about the huge variety of life-forms that Earth houses, from bacteria to bats to trees. Over billions of years, genetics has interacted with Earth's myriad environments to produce the grand story of life. That story is called evolution, and it is the subject of the next four chapters.

So Far...

1. To qualify as a stem cell, a cell must be able to produce more cells of its own type along with at least one type of _____.

2. Embryonic stem cells are derived from embryos when they are in the blastocyst stage, which occurs about (a) five days (b) five weeks (c) five months following conception.

3. The ultimate product of reproductive cloning is an _____ mammal, while the ultimate product of therapeutic cloning is _____.

So Far... Answers:

1. specialized cell (e.g., nerve cell, bone cell)

2. a

3. adult; embryonic stem cells

Chapter Review

Summary

15.1 What Is Biotechnology?

- **Biology and technology**—Biotechnology can be defined as the use of technology to control biological processes as a means of meeting societal needs. Innovations in biotechnology result from insights that are both biological and technological in nature.   (page 234)

15.2 Transgenic Biotechnology

- **Transgenic organisms**—A transgenic organism is an organism whose genome has stably incorporated one or more genes from another species. Many biotechnology products are produced within transgenic organisms. Human growth hormone, for example, is produced within a bacterium that has been made transgenic by means of incorporating a human gene.   (page 234)

- **Transgenic procedures**—Restriction enzymes are proteins derived from bacteria that can cut DNA in specific places. Plasmids are small, extra-chromosomal rings of bacterial DNA that can exist outside of bacterial cells and that can move into these cells.   (page 235)

- Human DNA can be inserted into plasmid rings by a process in which scientists use the same restriction enzyme on both the human DNA of interest and the plasmids. Complementary "sticky ends" of the fragmented human and plasmid DNA will then bond together, thus splicing the human DNA into the plasmid. This produces recombinant DNA: two or more segments of DNA that have been combined by humans into a sequence that does not exist in nature.   (page 236)

- **Transformation and protein production**—Once plasmids have had human DNA spliced into them, the plasmids can then be taken up into bacterial cells through transformation. As these cells replicate, producing many cells, the plasmid DNA inside them replicates as well. These plasmids are producing the protein coded for by the human DNA that has been spliced into them. The result is a quantity of the human protein of interest.   (page 237)

- **Transgenic products**—A large number of medicines and vaccines are produced today through transgenic biotechnology. Transgenic organisms that are used for this purpose include not only bacteria but also yeast, hamster cells, and mammals such as goats. Transgenic food crops are planted in abundance today in the United States. These genetically modified or GM crops are a subject of great controversy. (page 237)

Web Tutorial 15.1 Producing Bovine Growth Hormone

15.3 Reproductive Cloning

- **The nature of cloning**—A clone is a genetically identical copy of a biological entity. Genes can be cloned, as can cells and plants. Reproductive cloning is the process of making fully developed clones of animals. Dolly the sheep was a reproductive clone. Today, reproductive cloning of mammals is carried out through variants of the process that was used with Dolly: An egg cell has its nucleus removed and is fused with an adult cell containing a nucleus and, therefore, DNA. The fused cell then starts to develop as an embryo and is implanted in a surrogate mother.   (page 238)

- **Cloning applications**—Reproductive cloning can work in tandem with various recombinant DNA processes. Reproductive clones may be made transgenic and produce proteins of interest to humans. (page 240)

- **Human cloning**—In biological terms, two human clones would be no more alike than two identical twins. Human cloning is more technically feasible now than at any time in the past.   (page 240)

- **Cloning and extinction**—It appears possible that severely threatened animal species could have their numbers increased through reproductive cloning and that species recently made extinct might be brought back from extinction through cloning.   (page 240)

15.4 Forensic Biotechnology

- **Establishing identities**—Identities of criminals, biological fathers, and disaster victims often are established today through the use of DNA. Forensic DNA typing is the use of DNA to establish identities in connection with legal matters, such as crimes.   (page 241)

- **PCR**—The polymerase chain reaction (PCR) is a technique for quickly producing many copies of a segment of DNA. It is useful in situations, such as crime investigations, in which a large amount of DNA is needed for analysis, yet the starting quantity of DNA is small. (page 242)

- **DNA fingerprinting**—Forensic DNA typing works through comparisons of short tandem repeat (STR) patterns that are found in all human genomes. Police will, for example, compare the STR pattern in a suspect's DNA with the STR pattern in DNA extracted from a crime scene.   (page 242)

Web Tutorial 15.2 The Polymerase Chain Reaction (PCR)

15.5 Stem Cells

- **Commitment**—Most cells in the adult human body have undergone commitment: a developmental process that results in cells whose roles are completely determined.   (page 243)

- **The nature of stem cells**—Cells exist in the early embryo that have not yet undergone commitment and can give rise to various kinds of cells. A relatively small number of cells in the adult body have this same capability. Both the embryonic and adult cells that have this capability are called stem cells.   (page 244)

- **Embryonic and adult stem cells**—One source of stem cells is the early embryonic structure known as the blastocyst. Cells from the blastocyst's inner cell mass can give rise to all the different cells or tissues in the adult human body. These embryonic stem cells have thus far demonstrated good therapeutic potential in animal experiments. In contrast, the therapeutic potential of adult stem cells is uncertain. (page 244)

- **Therapeutic cloning**—Embryonic stem cells introduced into a human body may set off an immune system attack. Such an attack could be avoided if individuals could produce their own embryonic stem cells. This may be possible through the procedure known as therapeutic cloning. Many of the initial steps in therapeutic cloning were carried out by a Korean research team in 2004.   (page 245)

15.6 Controversies in Biotechnology

- **Stem cells and cloning**—One argument against therapeutic cloning is that, through it, techniques are being perfected that could be used in the reproductive cloning of human beings. A more general ethical controversy concerns the use of embryonic stem cells for any research purposes. To harvest embryonic stem cells, researchers must destroy the embryos the cells are a part of. Most of these embryos, however, come from fertility clinics and would be destroyed in any event.   (page 246)

- **Funding stem cell research**—Federal funding of embryonic stem cell research has been sharply limited by order of President Bush. In consequence, an increasing portion of this research is being funded by states and private sources.   (page 246)

- **GM foods**—No aspect of biotechnology has been as controversial as the production of genetically modified food (GM food). Critics charge that GM foods are potentially dangerous to human beings and to the environment. Proponents of GM foods say these foods are of enormous potential to benefit humanity and the environment. (page 247)

- **GM food economics**—Research into the commercialization of GM foods has been sharply curtailed in recent years, in large part because of fears that consumers will not accept these foods.   (page 248)

Key Terms

biotechnology p. 234	embryonic stem cells p. 244	recombinant DNA p. 237	stem cells p. 244
blastocyst p. 244	plasmid p. 236	reproductive cloning p. 239	therapeutic cloning p. 246
clone p. 238	polymerase chain reaction (PCR) p. 242	restriction enzyme p. 235	transgenic organism p. 235

Testing Your Understanding

In Your Own Words *(answers in the back of the book)*

1. Why is the polymerase chain reaction (PCR) so valuable?

2. Why are restriction enzymes that cut to produce sticky ends useful? If the following piece of DNA were cut with *Bam*HI, what would the fragment sequences be?

   ATCGGATCCTCCG

   TAGCCTAGGAGGC

3. If embryonic stem cells derived from fertility clinic embryos work as hoped in fighting disease, what is the point of using therapeutic cloning to produce embryonic stem cells?

Thinking about What You've Learned

1. One of the motives put forth for human cloning is that people want to replace children or other loved ones who have died. To what extent could a clone of a loved one be a replacement for that person? If the technique had then been available, should doctors in the nineteenth century have preserved the DNA of Abraham Lincoln for cloning?

2. Should society demand that there be no risks to genetically modified foods before it allows them to be developed? Does any significant technology have no risks associated with it?

3. What limits should be placed on modifications of animals to suit human interests? If you could bioengineer a chicken that was low cost because it had no eyes and no legs, would you?

How Life Was Shaped: An Introduction to Evolution

Caught in the Middle
Students leave Dover Area High School in southeastern Pennsylvania. In 2004, the Dover school board voted to require that a statement be read to Dover biology students stating that Darwin's theory of evolution is "not a fact" and that alternative accounts of life's origins exist. The result was a lawsuit challenging the legality of the action.

SCIENCE IN THE NEWS

THE SEATTLE TIMES

'EVOLUTION' TO REAPPEAR IN GEORGIA CURRICULUM

By the Los Angeles Times
February 6, 2004
ATLANTA—Georgia's superintendent of schools yesterday said she would restore the word "evolution" to the public schools' proposed science curriculum.

Kathy Cox said she originally had believed that including the word in the new teaching plan was more controversial than eliminating it. But "I am here to tell you," she said in a statement, "that I misjudged the situation, and I want to apologize for that."

SCIENCE IN THE NEWS

THE WASHINGTON POST

FOSSIL MAY SHOW APE-MAN ANCESTOR;
BONES FOUND IN SPAIN ARE CALLED LANDMARK IN EVOLUTION

By Guy Gugliotta
November 19, 2004
Scientists have discovered the remains of a tree-climbing creature that lived 13 million years ago in what is now north-eastern Spain and may be the last common ancestor of modern great apes and humans, according to a report published today.

16.1 Evolution and Its Core Principles

Why should a political ruckus like the one noted in the first newspaper story above be prompted by scientific discoveries such as the one noted in the second story? Why would an education superintendent agonize over whether to include the word *evolution* in a state teaching plan when she wouldn't have thought twice about including the words *DNA* or *ecology*? The difference is that neither DNA nor ecology

speaks directly to the question of how people and other living things came to exist. There are varying religious accounts of how this happened, but there is only one scientific account, and it can be summed up with a single word: evolution. To some religious believers, this isn't a problem. For them, it's possible to embrace both their religion and scientific findings about how life developed. For other believers, however, religion and science cannot be reconciled when it comes to this issue. To the extent that religion gives a true account of life's history, then the scientific account must be false. And with this, we get situations like the one in Georgia.

In some ways, this is a curious fight. The evidence for evolution was amassed in the same way that scientific evidence is in any field: through rigorous measurement, the running of experiments, and so forth. Work in, say, cancer research involves these same tests. And just as cancer research has informed scientists about the elements that produce cancer (DNA mutation, cell division, etc.), so evolutionary research has informed scientists about the elements that have produced the natural world (natural selection, gene flow, etc.). Both cancer researchers and evolutionary biologists have simply reported on what they've found, but the cancer findings are almost never questioned by nonscientists while the evolutionary findings are constantly challenged. Why the difference? Because cancer research tells us about a small-scale process within the living world, while evolution tells us about the origins and development of that world.

In consequence, evolution is unlike many other domains of science in that it carries considerable philosophical and even religious meaning. This alone makes it an intriguing subject for study. Setting this aside, if we look just within the discipline of biology, it is hard to overstate the importance of evolution. If biology were a house, evolution would be the mortar holding its stones together. No other concept in biology has the explanatory power of evolution, simply because no other factor has so strongly shaped life. Decades of research have told us that, in their *details*, all living things are a product of evolution. Why does the zebra have its stripes or the oak leaf its shape? Because of the channeling effects of evolution.

In sum, evolution is important, both to science and to society. Indeed, it's important enough that the next four chapters of this book are aimed at giving you some sense of how it works. In this chapter, you'll get some of the basics of evolutionary theory, along with an account of the work of evolution's founding father, Charles Darwin. In Chapter 17, you'll see how evolution operates on its smallest scale; then in Chapter 18, you'll see how new species develop. Finally, in Chapter 19, you'll go over the outcome of all these processes—you'll see how life has evolved on Earth from its beginnings to the present.

Evolution's Core Principles

Stand in a coastal redwood grove in California and look up at the trees stretching hundreds of feet into the air, and you might wonder: Why are they so tall? Watch a nature program on television and observe the exquisite plumage of a peacock, and you might think: How could such a *cumbersome* display exist on a creature that has to be wary of predators? For as long as humans have existed, people have no doubt asked questions such as these. Yet for most of human history, even the most informed individuals had no way to make *sense* of the variations that nature presented. Why does the redwood have its height or the peacock its plumage? Through the middle of the nineteenth century, the answer was likely to be either a shrug of the shoulders or a statement such as: "The creator gave unique forms to living things for reasons we can't understand."

Jumping ahead to our own time, biologists would give different answers to such questions. In the case of the redwoods, their reply would go something like this: Redwoods (and trees in general) are so tall because, over millions of years, they have been in competition with one another for the resource of sunlight. Individual trees that had the genetic capacity to grow tall got the sunlight, flourished, and thus left relatively many offspring, which made for taller trees in successive generations.

Comparing this answer to the mid-nineteenth-century answer, you might well ask: What is it that came *in between*, such that we can now make sense of the diversity of nature? What intervened was a set of intellectual breakthroughs that today go under the heading of the *theory of evolution*. Two principles lie at the core of this theory.

Common Descent with Modification   One of these principles has been labeled **common descent with modification**. It holds that particular groups, or species, of living things can undergo modification in successive generations, with such change sometimes resulting in the formation of new, separate species—one species separating into two, these two then separating further. If you had a videotape of this "branching" of species and then ran it *backward*, what you'd eventually see is a reduction of all these branches into one, meaning that all living things on Earth ultimately are descended from a single, ancient ancestor. Evolution itself can have several definitions. In this chapter, we will think of it as synonymous with common descent with modification. In Chapter 17, you'll get a more technical definition of it.

Natural Selection   Descent with modification is joined to a second evolutionary principle, **natural selection**: a process in which the differential adaptation of organisms to their environment selects those traits that will be passed on with greater frequency from one generation to the next. In the redwood example, trees that grew taller were better adapted to their environment than shorter trees, in that the taller trees got more sunlight. Accordingly, the taller trees left more offspring than the shorter trees. These offspring would likewise tend to inherit the genetic capacity to grow tall, meaning they too would have more offspring than other trees of *their* generation. In this way, the trait of tallness is being *selected* for transmission to ever-greater proportions of the tree population. As a result, the population evolves in the direction of greater tallness. As you will see, there are other processes that shape evolution, but none is as important as natural selection.

The Importance of Evolutionary Principles

With these two principles at its core, the theory of evolution provides a means for us to understand nature in all its buzzing, blooming complexity. It allows us to understand not just things that are familiar to us, such as the forms of the redwoods and peacocks, but all manner of *new* natural phenomena that come our way. When physicians observe that strains of infectious bacteria are becoming resistant to antibiotics, they don't have to ask: How could this happen? General evolutionary principles tell them that organisms evolve; that bacteria are capable of evolving very quickly (because they can produce many generations a day); and that antibiotic resistance is simply an evolutionary adaptation.

Beyond this, so far-reaching is the theory of evolution that its importance stretches outside the domain of biology and into the realm of basic human assumptions about the world. We've touched on this a little already, but consider some other ways in which this is true. Once persuasive evidence for the theory of evolution had been amassed, it meant that human beings could no longer view themselves as something *separate* from all other living things. Common descent with modification tells us that we are descended from other species, just like all organisms. Thus we sit not on a pedestal, viewing the rest of nature beneath us, but rather on one tiny branch of an immense evolutionary tree.

Evolution also meant the end of the idea of a *fixed* living world, in which, for example, birds have always been birds and whales have always been whales. Birds are the descendants of dinosaurs, while whales are the descendants of medium-sized mammals that walked on land. Life-forms only *appear* to be fixed to us, because human life is so short relative to the time frames in which species undergo change.

Finally, in evolution human beings are confronted with the fact that, through natural selection, life has been shaped by a process that has no mind and no

A BROADER VIEW

Dogs and Evolution

The process of one species giving rise to another is thought to take at least thousands of years with animals. But to see how rapidly change can come about within a species, consider the jaw-dropping diversity of dogs, whose breeds have mostly come about just in the last 400 years. The rapid evolution of dogs has been made possible by *artificial* selection—by the ability of humans to select the mating partners of their dogs as a means of producing desired traits.

So Far... Answers:

1. species; common ancestor

2. adapted; frequency

3. No; the theory supports the idea that, like all species, human beings are descended from other species. Thus, human beings sit on one ordinary branch of the evolutionary tree.

goals. Like a river that digs a canyon, natural selection shapes, but it does not design; it has no more "intentions" than does the wind. Trees that received more sunlight left more offspring, and thus trees grew taller over time. This is not a "decision" on the part of nature; it is an outcome of impersonal process. And if evolution has no consciousness, it follows that it can have no morals. This force that has so powerfully shaped the natural world is neither cruel nor kind, but simply indifferent.

So Far...

1. The principle of common descent with modification holds that groups of living things can undergo modifications that, over time, can lead to the formation of new _____ and that all living things on Earth are descended from a _____.

2. The principle of natural selection holds that organisms will vary with respect to how well they are _____ to their environment and that this difference will determine the _____ with which traits will be passed on from one generation to the next.

3. Does evolutionary theory support the idea that human beings sit at the apex of an evolutionary tree?

FIGURE 16.1 **Scientist of Great Insight** Charles Darwin, late in his life. (Painting by John Collier, 1883. London, National Portrait Gallery. © Archiv/Photo Researchers.)

16.2 Charles Darwin and the Theory of Evolution

How did such a far-reaching theory come into existence? Those who have read through the unit on genetics in this book know that the basic principles of that field were developed over the course of about a century by a large number of scientists. In contrast, a single person—nineteenth-century British naturalist Charles Darwin—is credited with bringing together the essentials of the theory of evolution (see FIGURE 16.1).

Darwin's Contribution

Darwin's contribution was twofold. First, he developed existing ideas about descent with modification while providing a large body of evidence in support of them. Second, he was the first to perceive that descent with modification is primarily driven by the process of natural selection.

To be sure, the study of evolution has been refined and expanded since Darwin. And it should be noted that before Darwin had ever published anything on his theory, a contemporary of his, fellow Englishman Alfred Russel Wallace, independently arrived at his insight about the importance of natural selection to evolution. (This makes Wallace the co-discoverer of this principle.) Yet, there is universal agreement that it is Charles Darwin who deserves primary credit for providing us with the core ideas that exist in evolutionary biology even today.

Darwin's Journey of Discovery

Darwin was born on February 12, 1809, in the country town of Shrewsbury, England. He was the son of a prosperous physician, Robert Darwin, and his wife, Susannah

Wedgwood Darwin, who died when Charles was eight. Young Charles seemed destined to follow in his father's footsteps as a doctor, being sent away to the University of Edinburgh at age 16 for medical training. But he found medical school boring; and his medical career came to a halt when, in the days before anesthesia, he found it unbearable to watch surgery being performed on children. His father then decided that he should study for the ministry. At the age of 20, Darwin set off for Cambridge University to spend three years that he later recalled as "the most joyous in my happy life." Darwin's happiness came in part from the fact that theology at Cambridge took a backseat to what had been his true passion since childhood: the study of nature. From his early years, he had collected rock, animal, and plant specimens and was an avid reader of nature books. His studies at Cambridge did yield a divinity degree, but they also gave Darwin a solid background in what we would today call life science and Earth science.

An Offer to Join the Voyage of the *Beagle*

Darwin's training, and the contacts he made at school, came together in one of the most fateful first-job offers ever extended to a recent college graduate. One of Darwin's Cambridge professors arranged to have him be the resident naturalist aboard the HMS *Beagle*, a ship that was to undertake a survey of coastal areas around the world (see FIGURE 16.2). If you look at FIGURE 16.3, you can trace the *Beagle*'s journey. In addition to numerous stops on the east coast of South America (the ship's primary survey site), the *Beagle* also stopped briefly at the remote Galápagos Islands, about 600 miles west of Ecuador—a visit you'll learn more about later.

Thus did Darwin spend time on a research vessel—five years in all—beginning in England two days after Christmas 1831 and ending back there in October 1836. Just 22 when he left, he was prone to seasickness; he had to share a 10-by-15-foot

FIGURE 16.2 **What the *Beagle* Looked Like** A reconstruction of the HMS *Beagle*, sailing off the coast of Tierra del Fuego in South America.

FIGURE 16.3 **Journey into History** The main mission of HMS *Beagle*'s 5-year voyage of 1831–1836 was to chart some of the commercially promising waters of South America. Charles Darwin served on board the ship as a resident naturalist and companion to ship's captain Robert FitzRoy. Darwin observed nature and collected specimens throughout the ship's journey, but for purposes of the theory of evolution, the ship's most important stops came in 1835 on the Galápagos Islands, west of South America. The *Beagle* had a complex itinerary; for example, it stopped twice at the Falkland Islands off the southeast coast of South America, once in 1833 and again in 1834.

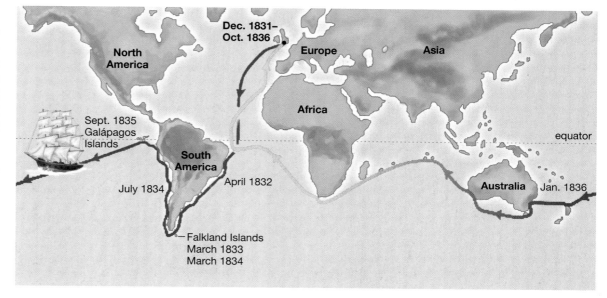

room with two other officers; he was not a traveler by nature; and the journey was dangerous (three of the *Beagle*'s officers died of illness during it). Yet Darwin was happy because of the work he was doing: looking, listening, collecting, and thinking about it all. Here he is writing about a part of coastal Brazil:

> A most paradoxical mixture of sound and silence pervades the shady parts of the wood. The noise from the insects is so loud that it may be heard even in a vessel anchored several hundred feet from the shore; yet within the recesses of the forest a universal silence appears to reign. To a person fond of natural history such a day as this brings with it a deeper pleasure than he can ever hope to experience again.

16.3 Evolutionary Thinking before Darwin

Darwin beheld these sights and sounds at a time when change was in the air with respect to ideas about life on Earth.

Charles Lyell and Geology

The single book Darwin took with him when he boarded the *Beagle* was the first volume of Charles Lyell's *Principles of Geology*, published in 1830 and bearing a message that even Lyell's fellow scientists found hard to accept: that geological forces *still operating* could account for the changes geologists could see in the

FIGURE 16.4   **Geologic Strata and Their Contents**

(a) Strata of sedimentary rock with fossils embedded

(b) Fossilized sea urchin, at least 65 million years old

Earth's surface. (As Darwin put it, "that long lines of inland cliffs had been formed, the great valleys excavated, by the agencies which we still see at work.") Under this view, Earth had not been put into final form at a moment of creation, but rather was steadily undergoing change. If such a thing were possible for the Earth itself, why not for the creatures that lived on it? (**See** FIGURE 16.4.)

Jean-Baptiste de Lamarck and Evolution

An idea along this line *had* been proposed, by the French naturalist Jean-Baptiste de Lamarck, in a book published in 1809. Lamarck believed that organisms changed form over generations through what has been termed the inheritance of acquired characteristics. Ducks or frogs did not originally have webbed feet, he said, but in the act of swimming they stretched out their toes in order to move more rapidly through the water. In time, membranes grew between their toes, effectively becoming webbing; critically, this characteristic was passed along to their offspring (**see** FIGURE 16.5). Over time, he believed, an animal would acquire enough changes that one species would diverge into two, with this branching extending all the way to human beings. In his work, Lamarck lent support to a *means* of evolution that we know today is false. (Animals don't pass along traits in accordance with the activities they carry out.) Yet note what he got right: that organisms can evolve; that one kind of organism can be ancestral to a different kind of organism.

Georges Cuvier and Extinction

Another French scientist, Georges Cuvier, in examining the fossil-laden rocks of the Paris Basin early in the nineteenth century, provided conclusive evidence of the extinction of species on Earth. This was a radical notion at the time, because many Christians believed that the creator would never allow one of his creatures to perish. (One amateur fossil collector who believed this was Thomas Jefferson.) This much about extinction Cuvier got right. However, seeing in his rock layers seeming "breaks" in the sequence of animal forms—a layer of simple forms would be followed by a layer of more complex forms as he went from older to newer layers—he held that there had been a series of catastrophes, such as floods, that wiped out life in given areas, after which the creator had carried out *new* acts of creation, bringing more complex life-forms into being with each act.

If you were to survey a broad swath of scientific thought leading up to Darwin, you'd repeatedly find what you've seen with Lamarck and Cuvier: a given scientist getting things *partly* right and partly wrong (which is almost always the way in science). The result was a rich mix of scientific findings and fanciful speculation that existed in Darwin's world as he boarded the *Beagle*. He himself believed at the time that species were fixed entities; that they did not change over time.

16.4 Darwin's Insights Following the *Beagle*'s Voyage

Darwin observed and collected wherever he went, but the most important stop on his journey took place nearly four years into it, when the *Beagle* stopped for a scant five weeks at the remote series of volcanic outcroppings called the Galápagos Islands (**see** FIGURE 16.6). There, in the dry landscape, amidst broken pieces of black lava, he saw strange iguanas and tortoises and mockingbirds

FIGURE 16.5   The Duck's Feet Jean-Baptiste de Lamarck proposed that new species evolve through a branching evolution, which is correct. He also proposed, however, that one of the forces driving this evolution is the inheritance of acquired characteristics, which does not take place. Under Lamarck's view, common at the time, a duck got its webbed feet because of activities it carried out during its lifetime, and this change was then passed on to its offspring.

FIGURE 16.6   The Galápagos Islands *Galápagos* is Spanish for *tortoise*. A part of Ecuador, the islands are still an important site of evolutionary research. Here a giant tortoise moves through the Alcedo Volcano area on one of the islands, Isabela.

FIGURE 16.7 **Three of Darwin's Finches** Thirteen species of finch evolved on the Galàpagos Islands, all of them descendants of a single species of finch native to South America. Shown, left to right, are *Geospiza fuliginosa, G. fortis*, and *G. magnirostris.*

that varied from one island to another. He shot and preserved several of these mockingbird "varieties" for transport back home, along with a number of small birds that he took to be blackbirds, wrens, warblers, and finches.

Perceiving Common Descent with Modification

Arriving back in England in October of 1836, Darwin soon donated a good deal of his *Beagle* collection to the Zoological Society of London. If there was a single most important flash of insight for him regarding descent with modification, it came when one of the Society's bird experts gave him an initial report on the birds he had collected on the Galápagos. Three of the mockingbirds were not just the "varieties" that Darwin had thought; they were separate species, as judged by the standards of the day. Beyond that, the small birds he had believed to be blackbirds, finches, wrens, and warblers were all finches—separate species of finches, each of them found nowhere but the Galápagos (**see** FIGURE 16.7).

With this, Darwin began to see a pattern: The Galápagos finches were related to an ancestral species that could be found on the mainland of South America, hundreds of miles to the east. Members of that ancestral species had come by air to the Galápagos and then, fanning out to separate islands, had diverged over time into separate species. Thus it was with Galápagos tortoises and iguanas and cactus plants as well; they had common mainland ancestors, but on these islands, had diverged into separate species. Darwin began to perceive the infinite branching that exists in evolution.

Perceiving Natural Selection

But what drives this branching? In England and set on a career as gentleman naturalist, Darwin married and settled in London for a time before moving to a small village south of London, where he would live out his days (**see** FIGURE 16.8). Two years after his homecoming, inspiration on evolution's key process came to him not from looking at nature, but from reading a book on human population and food supply—*An Essay on the Principle of Population,* by T. R. Malthus. As Darwin later wrote in his autobiography:

> I happened to read for amusement Malthus on Population, and being well prepared to appreciate the struggle for existence which everywhere goes on . . . it at once struck me that under these circumstances favourable variations would tend to be preserved and unfavourable ones to be destroyed.

A BROADER VIEW

Recognizing Dinosaurs

When did we first learn that at one time there were immense reptiles—the dinosaurs—that had flourished on Earth but then became extinct? The first dinosaur fossil to be recognized as such was a tooth found in England in 1822.

Thus did natural selection occur to Darwin as the driving force behind evolution. *Organisms* that had "favorable variations" were preserved (they lived and left many offspring) while those that had unfavorable variations were destroyed (they left fewer offspring or none at all). As a result, over generations, organisms evolved in the direction of the favorable variations.

You might think that, with two major insights in place—common descent with modification and natural selection—Darwin would have rushed to inform the world of them. In fact, more than 20 years were to elapse between his reading of Malthus and the 1859 publication of his great book, *On the Origin of Species by Means of Natural Selection*. In between, though incapacitated much of the time with a mysterious illness, Darwin published on geology, bred pigeons, and spent eight years studying the variations in barnacles—activities that each had relevance to his theory, and that yet constituted a kind of holding pattern for him as he contemplated informing the world of a theory that he knew would be controversial. Darwin finally got to work on what he thought would be his "big species book" in 1856.

16.5 Alfred Russel Wallace

Two years after Darwin began this labor, half a world away and unbeknownst to him, a fellow English naturalist lay in a malaria-induced delirium on an Indonesian island. Alfred Russel Wallace made a living collecting bird and butterfly specimens from the then-exotic lands of South America and Southeast Asia, selling his finds to museums and collectors. He had thought long and hard (and even published) about how species originate. Now, shivering with malarial fever in a hut in the tropical heat, with the question on his mind again, Wallace came upon the very insight that had come to Darwin 20 years earlier: Natural selection is the process that shapes evolution. Recovering from his illness, Wallace wrote out his ideas over the next few days and sent them to a scientific hero of his in England—Charles Darwin! He asked Darwin to read his manuscript and submit it to a journal if Darwin thought it worthy.

Darwin was stunned as he faced the prospect of another man being the first to bring to the world an insight he thought was his alone. Nevertheless, he would not allow himself to be underhanded with the younger scientist. He informed some of his scientist friends about his plight, and (without informing Wallace) they arranged for both Wallace's paper and some of Darwin's letters, sketching out his ideas, to be presented at a meeting of a scientific society in London on July 1, 1858. The readings before the scientific society turned out to have little immediate effect, but they did prod Darwin into finishing what would become *On the Origin of Species*, published some 16 months later. This event had a great effect. Indeed, it set off a thunderclap whose reverberations can still be heard today.

16.6 The Acceptance of Evolution

Sparking both scorn and praise, Darwin's ideas were fiercely debated in the years after 1859. Within scientific circles, however, it did not take long for common descent with modification to be accepted. Fifteen years or so after *On the Origin of Species* was published, almost all naturalists had become convinced of it, and it's not hard to see why. With evolution as a framework, so many things that had previously seemed curious, or even bizarre, now made sense. For example, years before *Origin* was published, scientists had known that, at a certain point in their *embryonic* development, species as diverse as fish, chickens, and humans all have

FIGURE 16.8 **Where *On the Origin of Species* Was Written** In this study in his house in rural England, Charles Darwin conducted scientific research and wrote his groundbreaking work, *On the Origin of Species by Means of Natural Selection*.

A BROADER VIEW

Wallace and Australian Animals

Where do "Asian" animals end and "Australian" animals begin? Look just east of the islands Bali and Borneo on some maps and you will see a line of division, first perceived by Alfred Russel Wallace and now known as "Wallace's Line," that marks this boundary. Marsupial mammals such as the kangaroo are only found east of the line.

Pharyngeal slits exist in these five vertebrate animals . . .

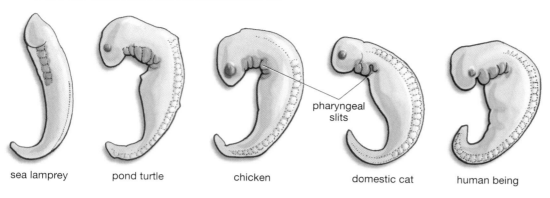

sea lamprey   pond turtle   chicken   domestic cat   human being

. . . evidence that all five evolved from a common ancestor.

FIGURE 16.9   Different Embryos, Same Structure (Adapted from M. K. Richardson, 1997.)

structures known as pharyngeal slits (**see** FIGURE 16.9). In fish, these structures go on to be *gill* slits; in humans they develop into the eustachian tubes, among other things. The question was: Why would land-dwelling and sea-dwelling organisms share a common embryonic structure? For that matter, why would organisms as different as a human and a chicken share this structure? With evolution, the answer became clear. All these organisms shared a common vertebrate *ancestor*, who had the slits. The vertebrate ancestral line had evolved into various species—and the slits into various structures—yet this element of the common ancestor persisted in the embryos of all these species.

In bringing together countless "loose ends" such as this, evolution became the mortar that unified the study of the living world. (It's interesting to think of biologists from this period as readers who start going over a detective novel a *second* time, saying, "Of course!" and "Why didn't I see that?") Given this, scientists had little trouble accepting the *fact* of evolution—that is, the occurrence of descent with modification. But it took longer for them to accept the primary process that Darwin said was driving evolution, natural selection. Indeed, the twentieth century was almost halfway over before most scientists became convinced that Darwin had been correct about natural selection's place in evolution. What settled the issue was a "modern synthesis" in evolutionary theory that brought together evidence from genetics, the fossil record, and the distribution of organisms throughout the world.

A BROADER VIEW

The Origin as Best-seller

Long and dense though it may be, Darwin's *Origin of Species* was read not just by scientists of its time, but by a broad swath of the educated public. Indeed, the entire first printing of the book sold out on the first day it was published, November 24, 1859.

So Far...

1. True or false: Charles Darwin was the first person to perceive common descent with modification.

2. The co-discoverer of natural selection was _____.

3. In his voyage around the world, Darwin gained the most insight from the time he spent on the _____ Islands, which lie off the coast of _____.

So Far... Answers:

1. *False; common descent with modification was an idea that predated Darwin. Darwin was, however, the first person to perceive natural selection as an evolutionary mechanism.*

2. *Alfred Russel Wallace*

3. *Galápagos; South America*

16.7 Opposition to the Theory of Evolution

So, why are Darwin's essential insights so widely believed to be correct? We'll now review some of the evidence for these insights. This is desirable because, as noted earlier, evolution has an unusual status among major fields of biology: Almost alone among them, its findings are regularly challenged as being unproven or simply wrong. For the average person, who can't be bothered with the details of such a controversy, these attacks make it appear that the theory of evolution is not a body of knowledge, solidly grounded in evidence, but rather a kind of scientific guess about the history of life on Earth.

What Is a Theory?

Several factors are critical in allowing the continued appearance of a "scientific debate" on the validity of the theory of evolution, when in fact none exists. The first of these has to do with a simple misunderstanding regarding terminology. You saw in Chapter 1 that, to the average person, the word *theory* implies an idea that certainly is unproven and that may be pure speculation. In science, however, a theory is a general set of principles, supported by evidence, that explains some aspect of the natural world. Accordingly, we have Isaac Newton's theory of gravitation, Albert Einstein's general theory of relativity, and many others. Lots of principles go into making up the theory of evolution, some of them on surer footing than others. Over time, however, some of these principles have achieved the status of established fact—we are as sure of them as we are sure that the Earth is round. Among these principles are the fact that evolution has indeed taken place and the fact that this has occurred over billions of years.

The Nature of Historical Evidence

A second factor that provides an opening for the opponents of evolution is the nature of the evidence for it. If you were called upon to provide the "evidence for cells," a look through a microscope at some actual cells might be enough to stop this "debate" in its tracks. Conversely, one of evolution's most important manifestations—radical transformations of life-forms—can never be observed in this way, because these transformations take place over vast expanses of time. Asking to "see" the equivalent of a dinosaur evolving into a bird before you'll believe it is like asking to see the European colonization of North America before you'll acknowledge it.

Evolution is taking place right now, all around us, but the evidence for evolution is *historical* evidence to a degree that is not true of, say, the study of genetics or of photosynthesis. Any ancient historical record is fragmentary, and evolution's historical record is ancient indeed. The fossils that scientists analyze are what remain after hundreds of millions of years of weathering and decay. Such an incomplete record leaves room for a great deal of interpretation among scientists—far more than is the case in purely experimental science. This interpretation has to do, however, with the *details* of evolution, not with its core principles. It has to do with what group descended from what other; with the rate at which evolution has proceeded; or with the role that pure chance has played in evolution. It does not have to do with *whether* evolution has occurred.

A BROADER VIEW

Misunderstandings of "Theory"

The idea that evolution is "just" a theory has great currency in America. In the 1980s, Ronald Reagan, then running for president, said that evolution was "a scientific theory only," and in 2002 the Cobb County, Georgia, school board voted to place stickers inside district biology textbooks stating that "Evolution is a theory, not a fact, regarding the origin of living things."

16.8 The Evidence for Evolution

So, what makes scientists so sure about the core principles of evolution? In any search for truth, we are more comfortable when lines of evidence are internally consistent and then go on to agree with *other* lines of evidence. The evidence for evolution satisfies both criteria. If we were to find even a single glaring inconsistency within or between lines of evidence, the whole body of evolutionary theory would be called into question. But no such inconsistencies have turned up yet, and scientists have been looking for them for almost 150 years.

Radiometric Dating

One claim of evolutionary theory, as you have seen, is that evolution proceeds at a leisurely pace, with billions of years having elapsed between the appearance of life and the present. Yet how do we know that Earth isn't, say, 46,000 years old, as opposed to the 4.6 billion that scientists believe it to be? The conceptual bedrock upon which we have determined the age of the Earth and its organisms is **radiometric dating**: a technique for determining the age of objects by measuring the decay of the radioactive elements they contain. As volcanic rocks are formed, they incorporate into themselves various elements that are in their surroundings. Some of these elements are radioactive, meaning they emit energetic rays or particles and "decay" in the process.

With such decay, one element can be transformed into another, the most famous example being uranium-238, which becomes lead-206 through a long series of transformations. The critical thing is that this transformation proceeds at a fixed *rate*; it is as steady as the most accurate clock imaginable. It takes 4.5 billion years for half a given amount of uranium-238 to decay into lead-206 (hence the term *half-life*). When such a transformation takes place in a cooled rock, the original or "parent" element is trapped within the rock, as is the "daughter" element, and the atoms of both elements can be counted. Therefore, comparing the *proportion* of a parent element to the daughter element in a rock sample provides a date for the rock with a fair amount of precision. There are now more than 40 different radiometric dating techniques used, each of them employing a different radioactive isotope. Given the range of such "radiometric clocks," scientists have been able to date objects from nearly the formation of the Earth to the present, though there are some gaps in the picture.

Fossils

Today, one line of evidence for evolution is the similarity of fossil types by sedimentary layers. Across the globe, looking at the same geologic layers of sediment, scientists find similar types of fossils, with a general movement toward more complex organisms as they go up through the newer layers. In Chapter 1, it was noted that scientific claims must be *falsifiable*; they must be open to being proved false upon the discovery of new evidence. The fossil record presents a falsifiable claim. For example, creatures called trilobites (**see** FIGURE 16.10) had a long run on Earth, existing in the ancient oceans from about 500 million years ago until some 245 million years ago, when they became extinct. By contrast, the evolutionary lineage that humans are part of, the primates, *began* about 80 million years ago. If scientists were to find, in a single fossil bed, fossils of trilobites existing side by side with those of early primates, our whole notion of evolutionary sequences would be called into question. No such incompatible pairing has happened, however, and the strong betting is it won't. Given this, the line of fossil evidence is internally consistent, but there's more. When we compare fossil placement with the dates we get from radiometric dating, we get excellent agreement *between* these two lines of evidence. We don't find trilobites embedded in sediments that turn out to be 80 million years old.

FIGURE 16.10 **Ancient Organism, Now Extinct**
Shown is a fossilized trilobite, a kind of arthropod that became extinct long before the animals known as primates came to exist. This fossilized trilobite dates from about 370 million years ago and was found in present-day Morocco.

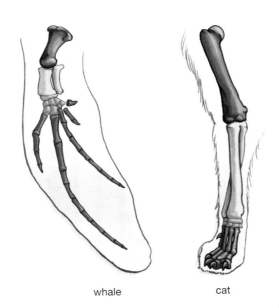

whale   cat

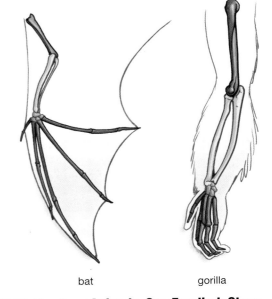

bat   gorilla

FIGURE 16.11 **Four Animals, One Forelimb Structure** Whales, cats, bats, and gorillas are all descendants of a common ancestor. As a result, the bones in the forelimbs of these diverse organisms are very similar despite wide differences in function. Four sets of homologous bones are color-coded for comparison. Note that in each case there is one upper bone, joined to two intermediate bones, joined to five digits.

Morphology and Vestigial Characters

Morphology is the study of the physical forms that organisms take. In comparative morphology, some classic evidence for evolution is seen in the similar forelimb structures found in a very diverse group of mammals—in a whale, a cat, a bat, and a gorilla, as seen in FIGURE 16.11. Look at what exists in each case: one upper bone, joined to two intermediate bones, joined to five digits. Evolutionary biologists postulate that the four mammals evolved from a common ancestor, adapting this 1-2-5 structure over time in accordance with their environments. Such features are said to be **homologous**, meaning the same in structure owing to inheritance from a common ancestor.

Along similar lines, if you look in detailed human anatomy textbooks, you'll see listings for the muscles of the human head. One of these, for example, is a muscle that retracts the tongue (the hyoglossus) while another is a muscle that helps us swallow (the aryepiglottic). In any such list, however, you'll also find the auricular muscles—three of them, which lie on each side of the head. And what do these muscles do? *They wiggle our ears.* That's it; no other natural use has ever been found for them. And, of course, in most people, they are useless even for this. Now, muscles are tissues that are specialized to do one thing—contract—and relative to, for example, fat tissue, they burn up a lot of energy. Given all this, we have to ask: Why do we humans have auricular muscles? If we imagine that our bodies were designed, then wouldn't it be strange for a designer to put in place these sets of costly, specialized, and useless structures? Against this, evolutionary theory makes sense of their presence. What we think of as our ears are really our outer ears or "pinna." Almost all mammals have pinna, but these usually are *mobile* pinna. The common house cat swivels its pinna up to 180 degrees as a means of homing in on sounds of predators or prey. We don't have this swiveling ability, however, and neither do any of the other "anthropoid" primates, which is to say monkeys and apes. Now, note that primates are among the latest evolving mammals. So, a likely explanation for human auricular muscles is that they are *vestiges* of our mammalian evolutionary history. Such muscles had a real function in our mammalian ancestors: They moved the pinna of these animals. In the anthropoid primates, however, movable pinna became less useful for some reason (the 3-D vision we have?) and over time these muscles fell into their current, functionless state.

Auricular muscles are one example among many of a **vestigial character**, meaning a structure in an organism whose original function has been lost during the course of evolution. The ostrich is a bird and it has wings, but it cannot fly. Ostrich wings do have several modern-day functions; they help balance the bird as it runs, for example. But, like muscles, wings are structures *specialized* for something—lifting animals off the ground. So is it reasonable to assume that a designer fashioned wings for the ostrich just to help it with balance? To paraphrase chemist Douglas Theobald, this would be like designing a computer keyboard that is meant to be used in hammering tacks. Evolution has a different explanation: The ostrich came from birds that could fly, but its own line *evolved* into a flightless condition. Looked at one way, the wings of the ostrich and the auricular muscles of human beings are like calling cards left by our ancestors; they tell us where we came from.

Evidence from Molecular Clocks

In the genetics unit you saw that every living thing on Earth employs DNA and utilizes an almost identical "genetic code" (*this* triplet of DNA bases specifies *that* amino acid). At the very least, this means there is a unity running through all earthly life; it is also consistent with the idea that all life on Earth ultimately had a single starting point—the single common ancestor mentioned earlier.

In recent years, molecular biology has also provided another check on what scientists have long believed to be true about evolutionary relationships—about which species are more closely related and about how long it's been since they've shared a common ancestor. All genes amount to a sequence of A, T, G, and C bases along DNA's double helix. And there are some genes that do very similar things in different organisms. There is, for example, a gene called *hedgehog* that helps regulate embryonic development in the *Drosophila* fruit fly, and a gene called *sonic hedgehog* that helps regulate embryonic development in mice.

Though the *hedgehog* and *sonic hedgehog* genes are similar, we would not expect them to be identical. This is so because gene base sequences *change* over time through the process of mutation. If the *rate* of this change is constant, then mutations are like a molecular clock that's ticking, and with the passage of a given amount of time, you get a set number of mutations. Given this, the longer it has been since two organisms shared a common ancestor, the greater the number of base differences we should see in the sequence of genes like *hedgehog* and *sonic hedgehog*.

This gene-modification hypothesis has been put to the test in connection with a gene that codes for an enzyme called cytochrome *c* oxidase, which exists in organisms as different as humans, moths, and yeasts. Going into this experiment, evolutionary theory predicted that there should be fewer DNA base-pair differences between the cytochrome *c* oxidase genes of, say, a human and a duck than between a human and a moth, because all the *other* evidence we had told us that humans and ducks share a more recent common ancestor than do humans and moths. If you look at FIGURE 16.12, you can see that evolutionary theory was fully borne out by the DNA sequencing. There are 17 sequence differences between a human and a duck, but 36 between a human and a moth. Indeed, all the differences between the species pictured fall into line with evolutionary theory. Thus we have another confirmation for evolution between lines of evidence—between DNA sequencing *and* the fossil record *and* radiometric dating *and* comparative morphology.

On to How Evolution Works

Given the abundant evidence for evolution, biologists long ago stopped asking whether it occurred. The really interesting questions for decades have been: Through what means has evolution proceeded? At what pace? In what direction, if any? These are the questions we'll look at in the next three chapters.

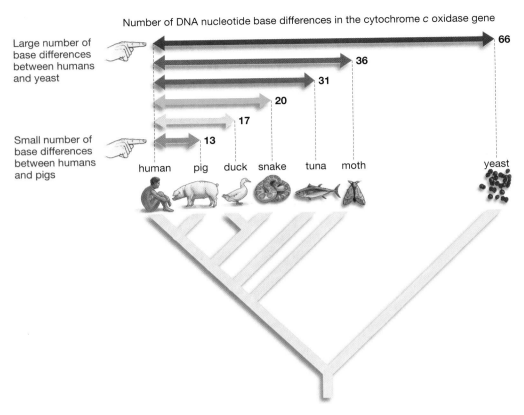

Number of DNA nucleotide base differences in the cytochrome *c* oxidase gene

Large number of base differences between humans and yeast

Small number of base differences between humans and pigs

66
36
31
20
17
13

human   pig   duck   snake   tuna   moth   yeast

FIGURE 16.12 Using Molecules to Track Evolution Diverse organisms—such as yeast, moths, and pigs—all have genes that code for an enzyme called cytochrome *c* oxidase. These organisms inherited cytochrome *c* oxidase genes from a common ancestor many millions of years ago. Over time, however, the cytochrome *c* oxidase genes have undergone mutations that have altered the sequence of their DNA "building blocks," called bases. The longer it has been since any two species shared a common ancestor, the more differences there should be in their cytochrome *c* oxidase bases. There are 13 differences between the bases found in human cytochrome *c* oxidase genes and those found in pigs, but there are 66 differences between the human and yeast genes. (Data from Whitfield, Philip. 1993. *From So Simple a Beginning: The Book of Evolution.* New York: Macmillan: Maxwell Macmillan International.)

So Far...

1. The age of the Earth has been established through the use of _____, in which scientists test for the _____ of parent and daughter elements in a given sample of rock or other material.

2. _____ features are features in different species that are the same in structure owing to inheritance from a _____.

3. The auricular muscles of human beings and the wings of ostriches are examples of _____ characters.

So Far... Answers:

1. *radiometric dating; ratio*

2. *Homologous; common ancestor*

3. *vestigial*

Chapter Review

Summary

16.1 Evolution and Its Core Principles

- **Common descent with modification**—Within the theory of evolution, a key principle is that of common descent with modification. This principle describes the process by which species of living things can undergo modification over time, with such change sometimes resulting in the formation of new, separate species. All species on Earth have descended from other species, and a single, common ancestor lies at the base of the evolutionary tree. (page 255)

- **Natural selection**—A second key principle in the theory of evolution concerns natural selection: a process in which the differential adaptation of organisms to their environment selects those traits that will be passed on with greater frequency from one generation to the next. (page 255)

- **General importance of the theory**—The theory of evolution has an importance beyond the domain of biology. Through it, human beings have become aware that (1) they are descended from other varieties of living things, and (2) the organisms that populate the living world are not fixed entities, but instead are constantly undergoing change. (page 255)

Web Tutorial 16.1 Principles of Evolution

16.2 Charles Darwin and the Theory of Evolution

- **Darwin's contributions**—Charles Darwin deserves primary credit for the theory of evolution. He developed existing ideas about descent with modification while providing a large body of evidence in support of them. And he was the first to perceive that natural selection is the primary process that drives evolution. (page 256)

- **The *Beagle*'s voyage**—Darwin's insights were inspired by the research he carried out during a five-year voyage he took around the world on the ship HMS *Beagle*, beginning in 1831. (page 257)

16.3 Evolutionary Thinking before Darwin

- **Darwin's predecessors**—Some of Darwin's ideas can be traced to the work of Charles Lyell, Jean-Baptiste de Lamarck, and Georges Cuvier, who respectively noted the dynamic geological nature of the Earth, the possibility of descent with modification, and the extinction of species on Earth. (page 258)

16.4 Darwin's Insights Following the *Beagle*'s Voyage

- **Perceiving natural selection**—Darwin understood descent with modification for several years before he comprehended that natural

selection was the most important process driving it. It was his reading of a work by Malthus that sparked his realization about natural selection. (page 259)

16.5 Alfred Russel Wallace

- **Natural selection's co-discoverer**—English naturalist Alfred Russel Wallace is the co-discoverer of the principle that evolution is shaped by natural selection. (page 261)

16.6 The Acceptance of Evolution

- **Early and later acceptance**—Descent with modification was accepted by most scientists not long after publication of Darwin's *On the Origin of Species by Means of Natural Selection* in 1859. Scientists accepted it because it explained so many facets of the living world. The hypothesis that natural selection is the most important process underlying evolution was not generally accepted until the middle of the twentieth century. Its acceptance hinged on a modern synthesis in the theory of evolution that brought together lines of evidence from several fields. (page 261)

16.7 Opposition to the Theory of Evolution

- **Scientific theories**—Even today, the theory of evolution is regularly challenged as being unproven or simply wrong. One factor leading to the appearance of a "scientific debate" over evolution is confusion about the meaning of the word *theory*. Though the average person may equate "theory" with speculation, in science, a theory is a general set of principles, supported by evidence, that explains some aspect of the natural world. (page 262)

16.8 The Evidence for Evolution

- **Lines of evidence**—Many lines of evidence are consistent with the theory of evolution. For example, radiometric dating indicates that the Earth is greater than 4 billion years old; the location of fossils is consistent with the theory of evolution and with radiometric dating; the theory of evolution explains the common occurrence of homologous physical structures in different organisms; the theory provides an explanation for otherwise puzzling vestigial characters, such as the auricular muscles in humans; and the theory is consistent with variations found in the DNA sequences of various organisms. (page 263)

Key Terms

common descent with modification p. 255	homologous p. 265 morphology p. 265	natural selection p. 255 radiometric dating p. 264	vestigial character p. 266

Testing Your Understanding

In Your Own Words *(answers in the back of the book)*

1. How does the evidence presented in Figure 16.12 support the conclusion that humans are more closely related to pigs than to yeast?

2. Describe two examples in which agreement among different lines of evidence provides evidence for evolution.

3. What is one observation that Darwin made in the Galápagos that influenced his thinking about evolution, and why was it important? Would he have been as likely to formulate his theory had he not had the opportunity to sail on the *Beagle*? Why or why not?

Thinking about What You've Learned

1. Using evolutionary principles, explain why large ears might be expected to evolve in a terrestrial plant-eater such as a rabbit or deer, but not in an aquatic mammal such as a seal.

2. Explain the evolutionary steps by which bacteria may become resistant to antibiotics, using the core requirements for evolution by natural selection.

3. Critics of the theory of evolution say it leaves no room for human purpose, since the theory asserts that human beings evolved through a process—natural selection—that has no goals or intentions. Does the theory undercut the idea of purpose in human life?

The Means of Evolution: Microevolution

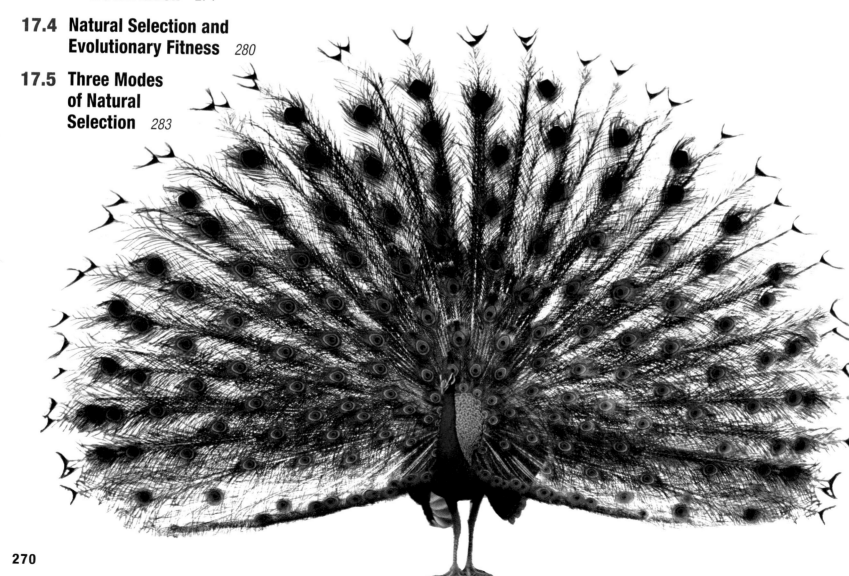

HOUSTON CHRONICLE

IT'S SURVIVAL OF FITTEST IN BROILING NASDAQ-100 OPEN

By the Associated Press
March 21, 2003
KEY BISCAYNE, Fla.—Serves turned shaky and willpower wilted Thursday in the scorching South Florida sun, which turned the Nasdaq-100 Open into a test of subtropical stamina. Old-timers Vince Spadea and Jonas Bjork-man passed, notching first-round victories. Teen prodigy Richard Gas-quet failed, quitting in the third set with a heat-related headache.

THE SAN DIEGO UNION-TRIBUNE

BIOTECH MEETS DARWIN;
IT'S SURVIVAL OF THE FITTEST FOR MANY PRIVATE SAN DIEGO BIOTECHNOLOGY COMPANIES AS THEY WAIT—AND HOPE—FOR MORE FINANCING

By Penni Crabtree
February 25, 2003
Kleanthis Xanthopoulos has always preached the doctrine of flexibility— the ability to adapt to any challenge or reversal—particularly in tough eco-nomic times. But the chief executive of Anadys Pharmaceuticals, one of about 200 privately held biotechnology com-panies in San Diego, admits the role of financial Gumby is wearing a little thin these days.

17.1 Evolution at Its Smallest Scale

Has Darwinian evolution entered the public's consciousness? Well, ask yourself: how many times have you heard the phrase "survival of the fittest"? There is some irony in this phrase being so closely associated with Darwin, in that he didn't invent it (though he did use it in later editions of *On the Origin of Species*). The real problem in linking "survival of the fittest" with Darwin, however, is that the public under-standing of *fittest* has almost nothing to do with what

Darwin meant by the term. As you'll see, Darwinian fitness has more to do with how an organism fits *in*—with its environment—rather than with any ability it has to run fast or jump high.

A concept like survival of the fittest concerns evolution's *mechanisms*, which is to say the processes through which it works. It may seem strange at first to think about evolution having underlying mechanisms. From the time we are young, what we hear about is evolution's *outcomes*—the emergence of the first life-forms, the story of how dinosaurs came and went, the relatively late appearance of human beings. But these transformations were brought about by events as small as genes being passed on within a group of living things. The purpose of this chapter is to give you some insight into the ways in which these small-scale processes lie at the root of evolution's large-scale transformations. What drives evolution? In this chapter, you'll find out.

What Is It That Evolves?

In approaching the topic of how organisms become modified over generations, the first question is: What is it that evolves? If you think about it, it's pretty clear that it is not individual organisms. A tree may inherit a mix of genes that makes it slightly taller than other trees, but if this tree is considered in isolation, it is simply one slightly taller tree. If it were to die without leaving any offspring, nothing could be said to have evolved, because no persistent quality (added height) has been passed on to any group of organisms.

In Chapter 16 you were introduced to the idea of a **species**, a group of organisms who can successfully interbreed with one another in nature, but who don't interbreed with members of other such groups. You might think that it is species—such as horses or American elm trees—that evolve. Indeed, scientists often speak this way when they refer, for example, to an amphibious mammal having evolved into today's whales. But this is a kind of shorthand whose inaccuracy becomes apparent when species are considered as they live in the real world.

Populations Are the Essential Units That Evolve   Think of a hypothetical species of frogs living in, say, equatorial Africa in a single expanse of tropical forest. Suppose that a drought persists for years, drying up the forest such that this single lush range is now broken up into two ranges separated by an expanse of barren terrain (**see** FIGURE 17.1). When the separation occurs, there is still only a single species of frog; but that species is now divided into two *populations* that are geographically isolated and hence no longer interbreeding. Each population now faces the natural selection pressures of its *own* environment—environments that may differ greatly even though they are nearly adjacent. One area may get more sun, while the other has a larger population of frog predators, for example. Thus, each population stands to be modified individually over time—to evolve. What is it that evolves? The essential unit that does so is a **population**, which can be defined as all the members of a species that live in a defined geographic region at a given time.

The question then becomes, *how* do populations evolve? The population of frogs that has more predators may, over many generations, evolve a coloration that makes them less visible to predators. But how does this slightly different coloration come about? Through genes. A given frog may be unable to change its spots, but *generations* of frogs can have theirs changed through the shuffling, addition, and deletion of genetic material.

Genes Are the Raw Material of Evolution   Recall from Chapter 11 that the genetic makeup of any organism is its **genotype**—and that a genotype provides an underlying basis for an organism's **phenotype**, meaning any observable traits that an organism has, including its physical characteristics and behavior. Many genes are likely to be involved in producing a phenotype such as coloration. In sexually reproducing organisms, *each* of these coloration genes will likely come in

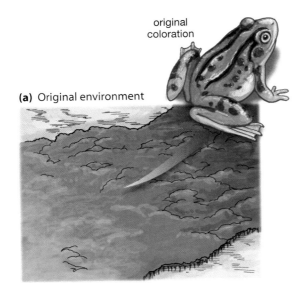

(a) Original environment

original
coloration

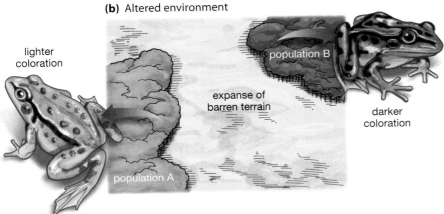

(b) Altered environment

lighter
coloration

population B

expanse of
barren terrain

darker
coloration

population A

FIGURE 17.1 **Evolution within Populations**

(a) A hypothetical species of frog lives and breeds in one expanse of tropical forest.

(b) After several years of drought, the forest has been divided by an expanse of barren terrain. The single frog population has thus been divided into two populations, separated by the barren terrain. The two frog populations now have different environmental pressures, such as the kinds of predators each faces, and they no longer breed with each other; after many generations their coloration diverges as they adapt to the different pressures.

two variant forms, called **alleles**, with offspring inheriting one allele from their father and one from their mother (**see** FIGURE 17.2). Though each of the alleles an individual inherits helps code for coloration, one may result in slightly lighter or darker coloration than the other.

Genes don't just come in two allelic variants, however; they can come in many. In most species, no one organism can possess more than two alleles of a given gene, but a *population* of organisms can possess many allelic variants of the same gene (which is one reason human beings don't just come in one or two heights, but in a continuous *range* of heights).

When the concept of a population is put together with that of genes and their alleles, the result is the concept of the **gene pool**: all

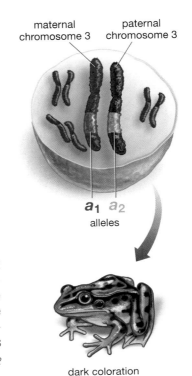

maternal
chromosome 3

paternal
chromosome 3

a_1 a_2
alleles

dark coloration

maternal
chromosome 3

paternal
chromosome 3

a_2 a_4
alleles

light coloration

FIGURE 17.2 **Genetic Basis of Evolution** Many genes can produce a trait such as body coloration, and each gene often has many alleles or variants. Each individual in a population, however, can possess only two alleles for each gene, one allele inherited from its mother and one inherited from its father. The two frogs in the figure both have maternal and paternal copies of chromosome 3 that house genes for coloration. The chromosomes of the two frogs may differ, however, in the allelic variants they have of these genes. The frog with dark coloration may possess alleles a_1 and a_2 while the light-colored frog may possess alleles a_2 and a_4 of this same gene.

the alleles that exist in a population. This gene pool is the raw material that evolution works with. If evolution were a card game, the gene pool would be its deck of cards. Individual cards (alleles) are endlessly shuffled and dealt into different "hands" (the genotypes that individuals inherit), with the strength of any given hand dependent upon what game is being played. (Survival in a drought? Survival against a new set of predators?)

17.2 Evolution as a Change in the Frequency of Alleles

Thinking of genes in terms of a gene pool gives us a new perspective on what evolution is at root: a change in the frequency of alleles in a population. This may sound a little abstract until you consider the frog coloration example. A frog inherits, from its mother and father, a coloration that allows it to evade predators slightly more successfully than other frogs of the same generation. It thus lives longer and leaves more offspring than the other frogs. It is successful in this way because of the advantageous set of alleles it inherited. Because this frog is more successful at breeding, its alleles are being passed to the *next* generation of frogs in relatively greater numbers than the alternative alleles carried by less successful frogs. Thus, this frog's alleles are increasing in frequency in the frog population. In looking at any example of evolution in a population over time, you would find this kind of change in allele frequency as its basis, assuming a stable environment.

With this perspective in mind, you're ready for a definition of **microevolution**: A change of allele frequencies in a population over a relatively short period of time. Why the *micro* in microevolution? Because evolution *within* a population is evolution at its smallest scale. This conception of microevolution allows you to understand the formal definition of evolution promised to you last chapter. **Evolution** can be defined as any genetically based phenotypic change in a population of organisms over successive generations. (Phenotypic change? Think of coloration in our frogs, or height in a tree.) Taking a step back, the large-scale *patterns* produced by microevolution eventually become visible, as with the evolution of, say, mammals from reptiles. This is **macroevolution**, defined as evolution that results in the formation of new species or other groupings of living things.

So Far...

1. The essential unit that evolves is the _____, which can be defined as all the members of a _____ living in a defined geographic region at a given time.

2. Genes come in variant forms called _____. A gene pool is defined as all the _____ that exist in a _____.

3. Microevolution is defined as a _____ in _____ in a population over a relatively short period of time.

17.3 Five Agents of Microevolution

So, what is it that causes microevolution? Put another way, what causes the frequency of alleles to change in a population? In the frog coloration example, a familiar process was at work: natural selection. The frog's coloration allowed the frog to evade predators, thus helping the frog to produce more offspring than did other frogs. This particular coloration was thus *selected* for greater transmission to future generations. With this selection, allele frequencies began to be altered in the frog population. Over generations, the alleles for protective coloration increased relative to other sets of alleles.

So Far... Answers:

1. *population; species*
2. *alleles; alleles; population*
3. *change; allele frequencies*

Natural selection is not the only process that can change allele frequencies, however. There are five "agents" of microevolution that can alter allele frequencies in populations. These are mutation, gene flow, genetic drift, sexual selection, and natural selection. You can see these processes summarized in **Table 17.1**. We'll now look at each of these agents of change in turn.

Mutations: Alterations in the Makeup of DNA

As you saw in the genetics unit, a mutation is any permanent alteration in an organism's DNA. Some of these alterations are heritable, meaning they can be passed on to future generations. Mutations can be as small as a change in a single base pair in the DNA chain (a point mutation) or as large as the addition or deletion of a whole chromosome or parts of it. Whatever the case, a mutation is a change in the informational set an organism possesses (**see FIGURE 17.3**). Looked at one way, it is a change in one or more alleles.

The rate of mutation is very low in most organisms; during cell division in humans, it might be just one DNA base pair per billion. And of the mutations that do arise, very few are beneficial or "adaptive." Most do nothing, and many are harmful to organisms. Thus mutations usually are not working to *further* survival and reproduction, as the frog coloration alleles did. Given this, they generally are not likely to appear with greater frequency in successive generations. The upshot is that mutations are not likely to account for much of the change in allele frequency that is observed in any population.

But a few mutations do occur that are adaptive. These genetic alterations are something like creative thinkers in a society: They are rare but very important. Such mutations are the only means by which *new* genetic information comes into being—

Table 17.1 Agents of Change: Five Forces That Can Bring about Change in Allele Frequencies in a Population

Agent	Description
Mutation	Alteration in an organism's DNA; generally has no effect or a harmful effect. But beneficial or "adaptive" mutations are indispensable to evolution.
Gene flow	The movement of alleles from one population to another. Occurs when individuals move between populations or when one population of a species joins another, assuming the second population has different allele frequencies than the first.
Genetic drift	Chance alteration of gene frequencies in a population. Most strongly affects small populations. Can occur when populations are reduced to small numbers (the bottleneck effect) or when a few individuals from a population migrate to a new, isolated location and start a new population (the founder effect).
Sexual selection	Occurs when some members of a population mate more often than other members.
Natural selection	Some individuals will be more successful than others in surviving and hence reproducing, owing to traits that give them a better "fit" with their environment. The alleles of those who reproduce more will increase in frequency in a population.

FIGURE 17.3 Basis of New Genetic Information
A mutation is any permanent alteration in an organism's DNA. Examples of mutations include **(a)** point mutations, in which the nucleotide sequence is incorrect; and **(b)** deletions, in which part of a chromosome is missing.

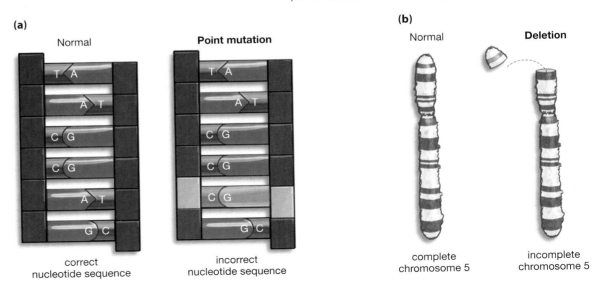

(a)

Normal   Point mutation

correct nucleotide sequence   incorrect nucleotide sequence

(b)

Normal   Deletion

complete chromosome 5   incomplete chromosome 5

by which new proteins are produced that can modify the form or capabilities of an organism. The evolution of eyes or wings had to involve mutations. No amount of shuffling of *existing* genes could get the living world from no eyes to eyes. Of course, no mutation can bring about a feature such as eyes in a single step; such changes are the result of many mutations, followed by rounds of genetic shuffling and natural selection, generally over millions of years.

Gene Flow: When One Population Joins Another

Allele frequencies in a population can also change with the mating that can occur after the arrival of members from a *different* population. This is the second microevolutionary agent, **gene flow**: the movement of genes from one population to another. Such movement takes place through **migration**, which is the movement of individuals from one population into the territory of another. Some populations of a species may truly be isolated, such as those on remote islands, but migration and the gene flow that goes with it are the rule rather than the exception in nature. It may seem at first glance that migration would be limited to animal species, but this isn't so. Mature plants may not move, but plant seeds and pollen do; they are carried to often-distant locations by wind and animals (**see** FIGURE 17.4). Of course, for a migrating population to alter allele frequencies of another population, its gene pool must be different from that of the population it is joining.

FIGURE 17.4 **Migration in Plants** Brought about by volcanic eruptions, the Hawaiian Islands have always been surrounded by the Pacific Ocean. Therefore all the plant species existing on the islands today are descended from species that were introduced to the islands through one means or another—human activity, wind currents, water currents, or animal dispersal of seeds and pollen. **(a)** Hawaiian silverswords are derived from **(b)** a lineage of California plants commonly known as tarweeds.

(a) Hawaiian silversword

(b) Tarweeds in California

North America

gene flow

Hawaiian Islands

Pacific Ocean

Genetic Drift: The Instability of Small Populations

To an extent that may surprise you, evolution turns out to be a matter of chance. You can almost see the dice rolling in the third microevolutionary agent, genetic drift. Imagine a hypothetical population of 10,000 individuals. An allele in this gene pool is carried by one out of ten of them, meaning that 1,000 individuals carry it. Now imagine that some disease sweeps over the population, killing half of it. Say that this allele had nothing to do with the disease, so the illness might be expected to decimate the allele carriers in rough accordance with their proportion in the population. If this were the case, 5,000 individuals in the population would survive, and 1 in 10—or about 500 of them—could be expected to be carriers of this allele. Let us say, however, that just by chance, *550* of the allele carriers were killed, thus leaving the surviving population of 5,000 with only 450 allele carriers. In that scenario, the frequency of the allele in this population would drop from 10 percent to 9 percent (**see** FIGURE 17.5a).

Now for the critical step. Imagine the same allele, with the same 1-in-10 frequency, only now in a population of *10*. There is now but a single carrier of the allele, and that individual may not be one of the 5 members of the population to survive the disease. In this case, the frequency of this allele drops from 10 percent to zero (**see** FIGURE 17.5b). It can be

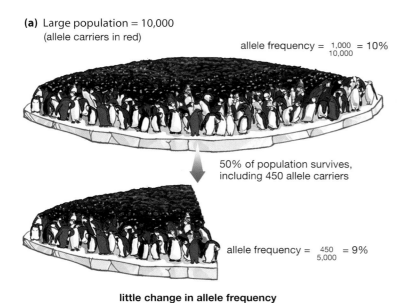

(a) Large population = 10,000
(allele carriers in red)

allele frequency = $\frac{1,000}{10,000}$ = 10%

50% of population survives,
including 450 allele carriers

allele frequency = $\frac{450}{5,000}$ = 9%

**little change in allele frequency
(no alleles lost)**

(b) Small population = 10
(allele carriers in red)

allele frequency = $\frac{1}{10}$ = 10%

50% of population
survives, with no allele
carrier among them

allele frequency = $\frac{0}{5}$ = 0%

**dramatic change in allele frequency
(potential to lose one allele)**

FIGURE 17.5   **Genetic Drift**

(a) In a hypothetical population of 10,000 individuals, 1 in 10 carries a given allele. The population loses half its members to a disease, including 550 individuals who carried the allele. The frequency of the allele in the population thus drops from 10 percent to 9 percent.

(b) A population of 10 with the same allele frequency likewise loses half its members to a disease. Because the one member of the population who carried the allele is not a survivor, the frequency of the allele in the population drops from 10 percent to zero.

replaced only by a mutation (which is extremely unlikely) or by migration from another population. Assuming that neither happens, no matter how this population grows in the future, in genetic terms it will be a different population than the original in that it lost this allele. The allele might be helpful or harmful, but its adaptive value doesn't matter. It has been eliminated strictly through chance. This is an example of **genetic drift**: the chance alteration of allele frequencies in a population, with such alterations having the greatest impact on small populations. It is true that some genetic drift has taken place in the larger population, but consider how small the effect is. The allele simply went from a 10 percent frequency to a 9 percent, with no loss of allele. Chance events can have much greater effects on small populations than on large ones.

Two scenarios, common in evolutionary history, produce the small populations that are most strongly affected by genetic drift. Populations can be greatly reduced through disease or natural catastrophe; or a small subset of a population can migrate elsewhere and start a new population. The first of these scenarios is called the *bottleneck effect*; the second is called the *founder effect*. Let's have a look at both of them in turn.

The Bottleneck Effect and Genetic Drift   The **bottleneck effect** is a change in allele frequencies in a population due to chance following a sharp reduction in the population's size. Real populations, or even species, can go through dramatic reductions in numbers. For example, northern elephant seals, which can be found off the Pacific coast of North America, were prized for the oil their blubber yielded and thus were hunted so heavily that by the 1890s, fewer than 50 animals remained. Thanks to species protection measures, the seals' numbers have rebounded somewhat in recent decades. But genetically, all the members of this species are very similar today, because they all descended from the few seals who

A BROADER VIEW

Gene Flow into Mount St. Helens

The 1980 Mount St. Helens explosion gave us some idea of how surprisingly well-traveled some species can be. In one decimated area, 43 species of spiders were blown in by the wind in the years following the blast.

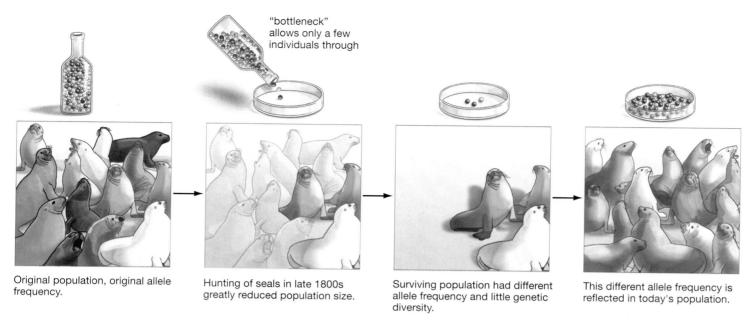

"bottleneck" allows only a few individuals through

Original population, original allele frequency.

Hunting of seals in late 1800s greatly reduced population size.

Surviving population had different allele frequency and little genetic diversity.

This different allele frequency is reflected in today's population.

FIGURE 17.6 **The Bottleneck Effect** Northern elephant seals were hunted so heavily by humans that in the 1890s, fewer than 50 animals remained. The population has grown from the few survivors (represented here by the three seals in the second frame), but the resulting genetic diversity of this population is very low. (Seal coloration for illustrative purposes only.)

A BROADER VIEW

The Bottleneck Effect on an Island

A fascinating account of the bottleneck effect can be found in Oliver Sacks' book *The Island of the Colorblind*, which tells the story of the people of the South Pacific island of Pingelap, some 8 percent of whom are completely color-blind because of a very rare genetic condition. The population in this instance was at one point reduced to 20 individuals who survived a typhoon in the late eighteenth century. Though the allele that brings about the color-blindness is harmful, it continues to exist at a high frequency within the population because of the isolation of this population over time.

made it through the nineteenth-century bottleneck. What occurs in these reductions is a "sampling" of the original population—the "sample" being those who survived the devastation (**see** FIGURE 17.6).

The reason allele frequencies change in such an event has to do with the nature of probability and sample size. Imagine a box filled with M&M's candies, with equal numbers of red, green, and yellow M&M's inside. You close your eyes and grab a handful of M&M's and pull out 12. With such a small sample, you might get, say, six reds, four greens, and two yellows, rather than the four-of-each-kind that would be expected from probability. If you pulled out *120* M&M's, conversely, your reds, greens, and yellows would be much more likely to approach the 40-of-each-kind that would be expected. In just such a way, a small sample of *alleles* is likely to yield a gene pool that's different from the distribution found in the larger population.

The Founder Effect and Genetic Drift Genetic drift can also result from the **founder effect**, which is simply a way of stating that when a small subpopulation migrates to a new area to start a new population, it is likely to bring with it only a portion of the original population's gene pool. This is another kind of sampling of the gene pool, in other words; but in this case it's caused by the migration of a few individuals rather than the survival of only a few. This sample of the gene pool now becomes the founding gene pool of a new population. As such, it can have a great effect; whatever genes exist in it become the genetic set that is passed on to all future generations, as long as this population stays isolated.

The power of the founder effect can be seen most clearly when a founding population brings with it the alleles for rare genetic diseases. There is, for example, a very rare genetic affliction of the eyes called *cornea plana* that results in a misshapen cornea—the first structure of the eye through which light passes. The result can be impaired close-range vision and a general clouding of eyesight. Cornea plana is known to affect only 113 people worldwide. The strange thing is that 78 of these people are in Finland, most of them in an area in northern Finland. Current research indicates that about 400 years ago, a small population arrived in this isolated area, with at least one member of this population carrying the recessive allele that causes this affliction. Since then, this allele has continued to profoundly affect subsequent generations in the area. This will always be the case if the descendants of a founder population stay relatively isolated over time—that is, if the descendents breed mostly among themselves.

FIGURE 17.7 **Sexual Selection** Individuals in some species choose their mates based on appearance or behavior. Female sage grouses prefer to mate with males who put on superior "displays," which include sounds, a kind of strutting, and a puffing up of their chests. Here, two males display before females on a sage grouse breeding ground.

Sexual Selection: When Mating Is Uneven across a Population

You've looked at mutation, gene flow, and genetic drift as agents of change with respect to allele frequencies in a population. Now let's look at a fourth agent, **sexual selection**, defined as a form of natural selection that produces differential reproductive success, based on differential success in obtaining mating partners. In practice, this is mating based on *phenotype*, which you may recall is any observable trait in an organism, including differences in appearance and behavior. It is the appearance of particularly nice plumage on a male peacock that makes female peacocks choose it for mating, rather than another male, while it is the behavior of strutting and chest-swelling in a male sage grouse that causes a succession of female grouse to mate with it rather than nearby competitors (**see FIGURE 17.7**). It's easy to see that if one male mates four times as much as the average male of his generation, his alleles stand to increase proportionately in the next generation.

Differential mating success among members of one sex in a species often is based on choices made by members of the opposite sex in that species. In general, it is females who are doing the choosing in these situations. Differential mating success can also, however, be based on differences in combative abilities that give individuals of one sex (generally males) greater access to members of the opposite sex. Sexual selection is a form of natural selection, in that some alleles—those that help bring about attractiveness, for example—are being preferentially selected for transmission to future generations. But while natural selection has to do with differential survival and reproductive capacity (and hence reproduction), sexual selection has to do with differential *mating* (and hence reproduction).

Natural Selection: Evolution's Adaptive Mechanism

You've already gone over a good deal about natural selection in this review of evolution, but it's time now to look at it as the last agent of change in microevolution. First, what is meant by natural selection? Here is a short definition: **Natural selection** is a process in which the differential adaptation of organisms to their environment selects those traits that will be passed on with greater frequency from one generation to the next. What does this mean in practice? Here is what biologist Julian Huxley had to say about it almost 60 years ago.

> Since there is a struggle for existence among individuals, and since these individuals are not all alike, some of the variations among them will be advantageous in the struggle for survival, others unfavorable. Consequently, a higher proportion of individuals with favorable variations will on the average survive, a higher proportion of those with unfavorable variations will die or fail to reproduce themselves.

A BROADER VIEW

Tail of the Peacock

The tail of the peacock is so exaggerated that it is usually regarded as a result of "runaway" sexual selection in action. Note that *two* traits are being selected for here; not only a male's longer tail, but a female's preference for such a tail. In each succeeding generation, then, female preference for longer tails is maintained, males with the longest tails are once again preferentially rewarded in mating, and the whole process is repeated, thus ratcheting up the size of tails.

By this process, the traits of those who are more successful in reproducing will become more widespread in a population—as the alleles that bring about these traits increase in frequency from one generation to the next.

A key concept here is that of **adaptation**, meaning a modification in the structure or behavior of organisms over generations in response to environmental change. Environmental change may come to a population (through streams drying up, for example), or a population may come to environmental change (through migration). Either way, natural selection is the process that pushes populations to adapt to new environmental conditions. The frog population noted at the outset of the chapter adapted to a new predator, for example, by evolving a darker coloration.

Among the five agents of microevolution, natural selection is the only one that consistently works to adapt organisms to their environment. Genetic drift is random; it could as easily work against an adaptive trait as for it. Although mutations can have an adaptive effect, they more often have no effect or even a negative effect. Gene flow doesn't necessarily bring in genes that are better suited to a given environment. And sexual selection has to do with the ways mating partners are selected, not with matching individuals to environment.

Natural selection is, however, constantly working to modify organisms in accordance with the environment around them. As such, it is generally regarded as the most important agent in having shaped the natural world—in having given zebras their stripes and flowers their fragrance. Because it is so important, we'll now look at it in a little more detail.

A BROADER VIEW

Natural Selection and Aging

Why do we age? Because of the way natural selection works. Think of a gene that tended to cause death among toddlers. Natural selection would remove it from the gene pool—toddlers who carried it would, on average, fail to leave children because they would tend to die young. Now think of a gene that tends to cause arthritis in 60-year-olds. People who have this gene have already had children. The arthritis gene didn't affect their ability to reproduce, so it gets passed on from one generation to the next. Natural selection works on those genes that allow us to reproduce. But it is powerless to work on any gene that asserts its effects after we reproduce.

So Far...

1. What the agents of microevolution all have in common is that each of them can _____ in a population.

2. Genetic drift is a _____ alteration of _____ in a population that has its greatest effect on _____ populations.

3. In sexual selection, differential reproductive success comes about because of differential success in obtaining _____.

So Far... Answers:

1. *alter allele frequencies*
2. *chance; allele frequencies; small*
3. *mating partners*

17.4 Natural Selection and Evolutionary Fitness

Even the strongest supporters of natural selection as a shaping force would not maintain that it is working to produce perfect organisms. The concept of "fitness" is helpful here, if only to clear up some misconceptions. To a biologist, **fitness** means the success of an organism in passing on its genes to offspring *relative* to other members of its population at a particular time. An organism cannot be deemed "fit," even if it has 1,000 offspring. It can only have *more or less* fitness relative to other members of its population (who might have 900 or 1,100 offspring). This has to do, once again, with allele frequencies in a gene pool. No matter how many offspring an individual has, its allele frequencies will increase in a population only if it has *more* offspring than do other members of its generation. Thus, fitness is a measure of impact on allele frequencies in a population.

This concept then gets us to the notion of "survival of the fittest," discussed at the start of the chapter. The phrase can be taken to imply the existence of superior beings; that is, organisms that are simply "better" than their counterparts, with images of being faster, more muscular, or smarter coming to mind. In fact, however, evolutionary fitness tells us nothing about organisms being *generally* superior, and it certainly tells us nothing about the value of any particular capacity, be it brawn or brain. All it tells us about are organisms who are better than others in their population at passing along their genes at a given time in a given environment. And environments can change, as the frog example at the start of the chapter showed. Thus, a more accurate phrase would not be "survival of the fittest," but rather "survival of those who fit—for now." Let's look at a real-life example of natural selection to see how this works out.

Galápagos Finches: The Studies of Peter and Rosemary Grant

When Charles Darwin stopped at the Galápagos Islands in 1835, some of the animal varieties he collected were various species of finches. Over the years, biologists kept coming back to "Darwin's finches" because of the very qualities Darwin found in them: They seemed to present a textbook case of evolution, with their 13 species having evolved from a single ancestral species on the South American mainland. Yet for more than 100 years after Darwin, it was a puzzle for scientists to figure out how Darwin's posited mechanism of evolution, natural selection, could have been at work with the finches. This changed beginning in the 1970s, when the husband-and-wife team of Peter and Rosemary Grant began a painstakingly detailed study of the birds.

Natural selection in the finches came into sharp focus in 1977, when a tiny Galápagos island, Daphne Major, suffered a severe drought. Rain that normally begins in January and lasts through July scarcely came at all that year. This was a disaster for the island's two species of finches; in January 1977 there were 1,300 of them, but by December the number had plunged to fewer than 300. One of the species, Daphne's medium-sized ground finch, *Geospiza fortis*, lost 85 percent of its population in this calamity. The staple of this bird's diet is plant seeds. When times get tough, as in the drought, the size and shape of *G. fortis* beaks—their beak "morphology"—begins to define what one bird can eat as opposed to another (**see FIGURE 17.8**). In *G. fortis*, larger body size and deeper beaks turned out to make all the difference between life and death in the drought of 1977. Measuring the beaks of *G. fortis* who survived the drought, the Grant team found they were larger than the beaks of the population before the drought by an average of some 6 percent.

FIGURE 17.8 **A Product of Evolution on the Galápagos** A male of the finch species *Geospiza fortis*, which is native to the Galápagos Islands.

FIGURE 17.9 **Who Survives in a Drought?** A large percentage of the population of *Geospiza fortis* died on a Galápagos island, Daphne Major, during a drought in 1977. Peter and Rosemary Grant observed in 1978 that individuals who survived the drought had a greater average beak depth than average individuals surveyed before the drought, in 1976. Individuals with larger beaks were better able to crack open the large, tough seeds that were available during the drought. The offspring of the survivors likewise had larger average beak size than did the population before the drought. Thus, evolution through natural selection was observed in just a few years on the island.

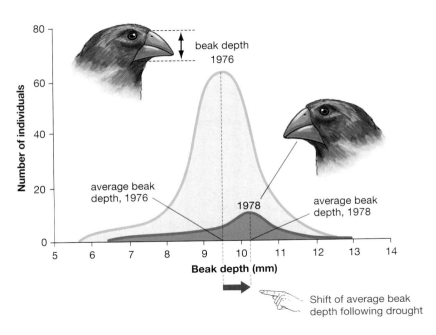

A BROADER VIEW

Dinosaurs and Fitness

Environments can change on a grand scale as well as a small scale. Dinosaurs were the dominant land animals on Earth for more than 150 million years, a time in which mammals were mostly small, night-feeders. When an asteroid struck the Earth 65 million years ago, however, the mammals survived while the dinosaurs perished. Why were the mammals evolutionarily fit relative to the dinosaurs? Perhaps pure size; small animals seem to survive extinction events better than large ones.

This was a difference of about half a millimeter, or roughly two-hundredths of an inch; such a slight difference allowed the survivors to be able to get into large, tough seeds and make it through the catastrophe, eventually to reproduce (**see** FIGURE 17.9).

This is natural selection in action, but there is more to be learned from the Grant study, which is to say *evolution* made visible. The Grant team knew that beak depth had a high "heritability" in the finches, meaning that beak depth is largely under genetic control. As it turned out, the *offspring* of the drought survivors had beaks that were 4 to 5 percent deeper than the average of the population before the drought. In other words, the drought had preferentially preserved those alleles from the starting population that brought about deeper beaks, and the result was a population that evolved in this direction.

But the Grant study yielded one more lesson. In 1984–1985 there was pressure in the opposite direction: Few large seeds and an abundance of small seeds provided an advantage to *smaller* birds, and it was they who survived this event in disproportionate numbers.

Lessons from *G. fortis* So, where is the "fittest" bird in all this? There isn't any. Evolution among the finches was not marching toward some generally superior bird. Different traits were simply favored under different environmental conditions. Secondly, there is no evolutionary movement toward combativeness or general intelligence here. Survival had to do with beak size—and not necessarily *larger* size at that. Looking around in nature, it's true that some showcase species, such as lions and mountain gorillas, gain success in reproduction by being aggressive. And it's true that our own species owes such success as it has had to intelligence. But in most instances, it is not brawn or brain that make the difference; it is something as seemingly benign as beak size and its fit with the environment.

Finally, consider how imperfect natural selection is at the genetic level. Suppose the smaller *G. fortis* that disproportionately died off in 1977 also disproportionately carried an allele that would have aided just slightly in, say, long-distance flying. In the long run this might have been an adaptive trait, but it wouldn't matter. The flight allele would have been *reduced* in frequency in the population (as the smaller birds died off) because flight distance didn't matter in 1977, beak depth did. Evolution operates on the phenotypes of whole organisms, not individual genes. As such, it does not work to spread *all* adaptive traits more broadly. Instead, the destiny of each trait is tied to the constellation of traits the organism possesses. Genes are "team players," in other words, that can only do as well as the team they came in on.

17.5 Three Modes of Natural Selection

In what directions can natural selection push evolution? As noted in Chapter 11, a character such as human height is under the control of many different genes and is thus *polygenic*. Such polygenic characters tend to be "continuously variable." There are not one or two or three human heights, but an innumerable number of them in a range. (See section 11.7, "Multiple Alleles and Polygenic Inheritance," on p. 179.) When natural selection operates on characters that are polygenic and continuously variable, it can proceed in any of three ways. The essential question here is: Does natural selection favor what is average in a given character, or what is extreme? (**See FIGURE 17.10.**)

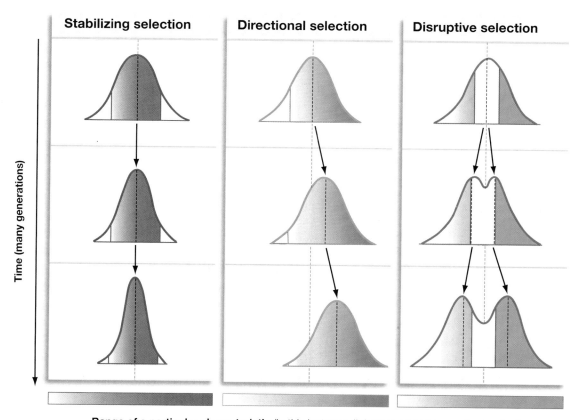

Range of a particular characteristic (in this instance, lightness or darkness of coloration)

In stabilizing selection, individuals who possess extreme values of a characteristic–here, both the lightest and the darkest colors–are selected against and die or fail to reproduce. Over succeeding generations, an increasing proportion of the population becomes average in coloration.

In directional selection, one of the extremes of a characteristic is better suited to the environment, meaning that individuals at the other extreme are selected against. Over succeeding generations, the coloration of the population moves in a directon–in this case toward darker coloration.

In disruptive selection, individuals with average coloration are selected against and die. Over succeeding generations, part of the population becomes lighter, while part becomes darker, meaning the range of color variation in the population has increased.

FIGURE 17.10 Three Modes of Natural Selection

FIGURE 17.11 Stabilizing Selection: Human Birth Weights and Infant Mortality Note that infant deaths are more prevalent at the upper and lower extremes of infant birth weights.

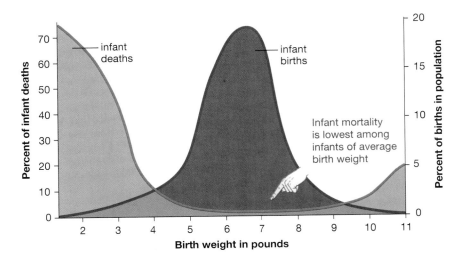

FIGURE 17.12 Directional Selection: Evolving toward Increased Brain Size Modern humans are part of an evolutionary group, called the hominins, that includes all the upright-walking or "bipedal" primate species. Cranial capacity is a measure of the volume of brain tissue that can fit inside a skull. Directional selection for cranial capacity can clearly be seen in the ever-increasing capacities of this group of hominins (all of whom are now extinct, except for ourselves). The line in the figure does not trace ancestry, but merely shows the trend toward increased brain size. Going from left to right, the full names of the species shown are *Australopithecus afarensis* (the "Lucy" species), *Australopithecus africanus*, *Homo habilis*, *Homo ergaster*, *Homo erectus*, and *Homo sapiens* (ourselves).

Stabilizing Selection

In **stabilizing selection**, intermediate forms of a given character are favored over extreme forms. A clear example of this is human birth weights. If you look at FIGURE 17.11, you can see, first, the weights that human babies tend to be. Notice that there are not only relatively few 3-pound babies, but relatively few 9-pound babies as well. A great proportion of birth weights fall at the average or "mean" of a little less than 7 pounds. Now look at the infant-mortality curve. Infant deaths are highest at both extremes of birth weight; low-birth-weight babies *and* high-birth-weight babies are more at risk than are average-birth-weight babies (though low birth weight poses the greater risk). Put another way, the children most likely to survive (and reproduce) are those carrying alleles for intermediate birth weights. Thus, natural selection is working to make intermediate weights even more common. It is not working to move birth weights toward the extremes of higher or lower weights. We think of evolution as "moving" species in one direction or another. But stabilizing selection is probably the most common type of selection operating in the natural world. In a sense, however, this is not surprising. Why? Because most organisms are well adapted to their environments.

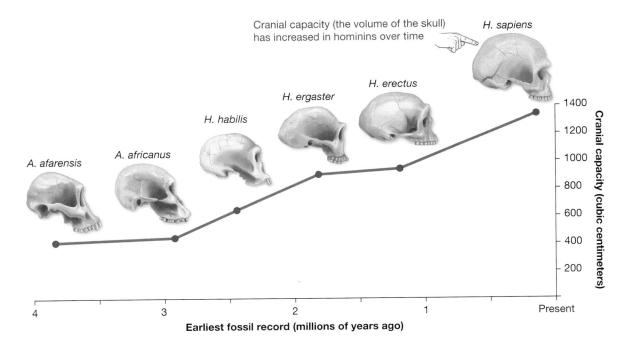

Directional Selection

When natural selection does move a character toward one extreme, **directional selection** is in operation—the mode in which we most commonly think of evolution operating. If you look at FIGURE 17.12, you can see an example of directional selection that took place over a very long period of time—about 4 million years—involving evolution toward larger brain size in some close, but now extinct, relatives of ours, a group called the hominins.

Disruptive Selection

When natural selection moves a character toward *both* of its extremes, the result is **disruptive selection**, which appears to occur much less frequently in nature than the other two modes of natural selection. This mode of selection is visible in the beaks of yet another kind of finch. A species of these birds found in West Africa (*Pyrenestes ostrinus*) has a beak that comes in only two sizes. Thomas Bates Smith, who has studied the birds, has observed that if human height followed this pattern, there would be some Americans who are 4 to 5 feet tall, and some who are 6 to 7 feet tall, but no one who is 5 to 6 feet tall (**see** FIGURE 17.13).

The environmental condition that leads to this mode of selection in the finches is, once again, diet. When food gets scarce, large-billed birds specialize in cracking a very hard seed, while small-billed birds begin feeding on several soft varieties of seed, with each type of bird being able to outcompete the other for its special variety. Given how these birds have evolved, it's probably safe to assume that a bird with an intermediate-sized bill would get less food than one with a bill of either extreme type. Bill size seems to be under the control of a single genetic factor, so that bills are able to come out either large or small. This genotype was presumably shaped by natural selection over generations, such that any alleles for intermediate-sized bills were weeded out.

FIGURE 17.13 Disruptive Selection: Evolution That Favors Extremes in Two Directions Finches of the species *Pyrenestes ostrinus*, found in West Africa. Birds of this species have beaks that come in two distinct sizes—large and small.

So Far... Answers:

1. False; evolutionary fitness exists only in a relative sense; an organism is more or less fit than another depending on whether it has more or fewer offspring than another. So, having 1 offspring or 1,000 does nothing to qualify an organism as fit or not.

2. intermediate; extreme

3. one extreme

On to the Origin of Species

Stabilizing selection does what it says: It stabilizes given traits of a population, thereby keeping it a single entity. However, both disruptive and directional selection can serve as the basis for speciation—for bringing about the transformation of a single species into two or more *different* species. How is it, though, that such speciation works? And how do we classify the huge number of species that are the outcome of evolution's countless branchings? These are the subjects of the next chapter.

So Far...

1. True or false: If an insect has 1,000 offspring, we can label it as "fit" in an evolutionary sense.

2. In stabilizing selection, forms that are _____ are favored over forms that are _____.

3. We normally think of evolution working through directional selection, in which a character moves toward _____.

Chapter Review

Summary

17.1 Evolution at Its Smallest Scale

- **Populations**—A population is all the members of a species living in a defined geographical area at a given time. Populations are the fundamental unit that evolves.   (page 272)

17.2 Evolution as a Change in the Frequency of Alleles

- **The gene pool**—Genes may be found in variant forms, called alleles. In most species, no individual will possess more than two alleles for a given gene, but a population may possess many such allelic variants. The sum total of alleles in a population is referred to as that population's gene pool.   (page 272)

- **Allele frequency change**—The basis of evolution is a change in the frequency of alleles within a population (a phenomenon that includes the appearance of new alleles through mutation). Evolution at this level is called microevolution.   (page 274)

17.3 Five Agents of Microevolution

- **The agents**—Five evolutionary forces can result in changes in allele frequencies within a population. These agents of microevolution are mutation, gene flow, genetic drift, sexual selection, and natural selection.   (page 274)

- **Mutation**—Mutation happens fairly infrequently, and most mutations either have no effect or are harmful; yet rare adaptive mutations are vital to evolution in that they are the only means by which entirely new genetic information comes into being.   (page 275)

- **Gene flow**—Gene flow, the movement of genes from one population to another, takes place through migration, meaning the movement of individuals from one population into the territory of another.   (page 276)

- **Genetic drift**—Genetic drift, the chance alteration of allele frequencies in a population, has its greatest effects on small populations. Genetic drift works on small populations in two ways. The first of these is the bottleneck effect: a change in allele frequencies due to chance during a sharp reduction in a population's size. The second is the founder effect: the fact that when a small subpopulation migrates to a new area to start a new population, it is likely to bring with it only a portion of the original population's gene pool.   (page 276)

- **Sexual selection**—Sexual selection is a form of natural selection that can affect the frequency of alleles in a gene pool. It occurs when differences in reproductive success arise because of differential success in mating. A given male in a population may, for example, sire many more offspring than the average male in the population. If so, this male's alleles will increase in frequency in the next generation of the population.   (page 279)

- **Natural selection**—Some individuals will be more successful than others in surviving and reproducing, owing to traits that better adapt them to their environment. This phenomenon is known as natural selection. Natural selection is the only agent of microevolution that consistently acts to adapt organisms to their environments. As such, it is generally regarded as the most powerful force underlying evolution.   (page 279)

Web Tutorial 17.1 Agents of Change

17.4 Natural Selection and Evolutionary Fitness

- **Fitness and environment**—The phrase "survival of the fittest" is misleading, because it implies that evolution works to produce generally superior beings who would be successful competitors in any environment. Evolutionary fitness, however, has to do only with the relative reproductive success of individuals in a given environment at a given time. One individual is said to be more fit than another to the extent that it has more offspring than another. But individuals are not born with invariable levels of fitness; instead, fitness can change in accordance with changes in the surrounding environment.   (page 280)

17.5 Three Modes of Natural Selection

- **Change and status quo**—Natural selection has three modes: stabilizing selection, directional selection, and disruptive selection. Stabilizing selection moves a given character in a population toward intermediate forms and hence tends to preserve the status quo; directional selection moves a given character toward one of its extreme forms; and disruptive selection moves a given character toward two extreme forms.   (page 283)

Web Tutorial 17.2 Three Modes of Natural Selection

Key Terms

adaptation p. 280	**fitness** p. 280	**genotype** p. 272	**phenotype** p. 272
allele p. 273	**founder effect** p. 278	**macroevolution** p. 274	**population** p. 272
bottleneck effect p. 277	**gene flow** p. 276	**microevolution** p. 274	**sexual selection** p. 279
directional selection p. 285	**gene pool** p. 273	**migration** p. 276	**species** p. 272
disruptive selection p. 285	**genetic drift** p. 277	**natural selection** p. 279	**stabilizing selection** p. 284
evolution p. 274			

Testing Your Understanding

In Your Own Words *(answers in the back of the book)*

1. Explain the statement, "Individuals are selected, populations evolve."

2. What is a gene pool, and why do changes in gene pools lie at the root of evolution?

3. What are the five causes of allele frequency changes (microevolution), and how does each work?

Thinking about What You've Learned

1. The text notes that natural selection is the process that pushes populations to adapt to environmental change. Is natural selection still going on in human populations? Why or why not?

2. Explain why genetic drift may be important when captive populations of animals or plants are started with just a few individuals.

3. Two moth populations of the same species utilize different host plants. One rests on leaves and has evolved a green color that allows it to escape predation by blending in with the leaves. The other population rests on tree trunks and is brown in color. The colors of these moths are genetically determined by different alleles of the same gene. What would be some consequences of migration of moths leading to gene flow between these two populations?

Macroevolution: The Origin of Species

Evolution's Branching
An ancestor of the California plant on the left, the Muir's tarweed (*Carlquistia muirii*) gave rise to 30 separate species of Hawaiian plants, including the plant on the right, the Haleakala silversword (*Argyroxiphium sandwicense*).

YAHOO! NEWS SAT, JAN 29, 2005
SCIENCE AP

JUMPING MOUSE LOSES FEDERAL PROTECTION

By JOHN HEILPRIN, Associated Press Writer

WASHINGTON—The Preble's meadow jumping mouse, once seen as a costly impediment to development, is now viewed by the government as a critter that never really existed—and is no longer in need of federal protection under the Endangered Species Act.

The Interior Department said Friday that new DNA research shows the 9-inch mouse, which can launch itself a foot and a half into the air and switch direction in mid-flight, is probably identical to another variety of mouse common enough not to need protection.

18.1 What Is a Species?

The Preble's meadow jumping mouse, featured in the story above, may not be nationally famous in the way that spotted owls or redwood trees are; but if you live along the corridor that runs from Colorado Springs to Cheyenne, Wyoming, you may have heard of the Preble's because of the fight over whether it is a unique variety of living thing. In 1998, the U.S. Fish and Wildlife Service listed the Preble's as "threatened," a status that brought restrictions on the development of property that was close to the mouse's habitat. And its habitat was not a matter of a small patch of land here or there. Almost 31,000 acres were designated as "critical" for the mouse, with most of this land lying

along streams in Colorado and Wyoming. In 2003, however, researchers from a Denver museum concluded that the Preble's does not deserve its status as a separate subspecies, *Zapus hudsonius preblei*, because it is in fact a more common bear lodge mouse, *Zapus hudsonius campestris*. In 2005, the Interior Department endorsed these findings when it proposed taking the Preble's off the endangered species list.

(a) Endangered species

(b) Not endangered

FIGURE 18.1 **Separate Entities** The concept of a species can have very practical effects.

(a) This bird is a northern spotted owl (*Strix occidentalis*), which is an endangered species protected by federal law.

(b) This bird is a barred owl (*Strix varia*), which is not endangered.

The fight over the Preble's mouse illustrates the importance of being able to make *distinctions* among living things. If we allow that some bacteria are harmful to human beings while others are helpful, that some varieties of rice carry disease-resistance genes while others do not, that some types of animals are threatened while others are not, then it follows that there must be some means of distinguishing one of these types from another (**see** FIGURE 18.1). The whole notion of knowing something about the natural world begins to break down if we can't say *which* type it is that is threatened or harmful.

As you'll see in this chapter, there are lots of different ways to group organisms—into units called classes, and families, and so forth. But the fundamental unit of categorization for living things is the species. Why should this be so? Because life is so powerfully shaped by breeding behavior, and species are the fundamental units within which breeding occurs. Will an elephant breed with a zebra? Of course not. But will an eastern bluebird from New York mate with an eastern bluebird from Ohio? Almost certainly, because they are members of the same species (*Sialia sialis*). By definition, members of a species breed with each other, but they do not breed with members of other species. And breeding, of course, is the way that genes are passed on. This simple fact has practical consequences. If we ask which strains of rice carry disease-resistant genes, we are likely to specify such strains *by species*, because these genes will be passed along only within a species.

Beyond such practical issues, however, if we ask how it is that the living world came to be such a tremendously diverse place, the answer is: through countless instances in which populations of a single species *stopped* breeding with one another. What this results in, as you'll see, are speciation events—events in which one species divides into two. How often has this happened? The lowest estimate for the number of species on Earth is about 4 million, though estimates commonly range up to 10 or 15 million. Suppose, though, that the number is "only" 4 million. Isn't this number staggering? This seems particularly true given its starting point. A single type of organism arose perhaps 3.8 billion years ago, branched into two types of organisms, and then the process continued—branches forming on branches—until at least 4 million different types of living things came to exist on Earth. Now here's the real surprise: These are just the *survivors*. The fossil record indicates that more than 99 percent of all species that have ever lived on Earth are now extinct. Branching indeed.

The questions to be answered in this chapter are: How does this branching work? How do we go from the microevolutionary mechanisms explored in Chapter 17 to the actual divergence of one species into two? And how do we classify the species that result from this branching?

The Biological Species Concept

To begin to answer these questions, let's look first, in a more formal way, at what a species is. In biology today, the most commonly accepted definition of a species stems from what is known as the **biological species concept**. Here is that concept, as formulated by the evolutionary biologist Ernst Mayr:

> Species are groups of actually or potentially interbreeding natural populations which are reproductively isolated from other such groups.

Note that the breeding behavior Mayr talks about can be real or potential. Two populations of finch may be separated from one another by geography, but if, upon being reunited in the wild, they began breeding again, they are a single species. Note also that Mayr stipulates that species are groups of *natural* populations. This is important because breeding may take place in captivity that would not in nature. No one doubts that lions and tigers are separate species; yet they will mate in zoos, producing little tigons or ligers, depending on whether the father was a tiger or a lion (**see** FIGURE 18.2). In natural surroundings, however, they apparently have never interbred, even when their ranges overlapped centuries ago.

FIGURE 18.2   **Hybrid Animal**   Pictured is a liger—a big cat whose father was a lion and whose mother was a tiger. These animals are produced only in captivity, never in nature.

You might think that, with the biological species concept in hand, scientists would be able to study any organism and, by discerning its breeding behavior, pronounce it to be a member of this or that species. Nature is so vast and varied, however, that this doesn't always work. We can't look at the breeding behavior of the single-celled bacteria or archaea, for example, because they don't *have* any breeding behavior; they multiply instead by simple cell division. (Microbiologists define bacterial and archaeal groupings by sequencing their DNA or RNA and looking for identifying patterns among these sequences.) Then there are separate species that sometimes interbreed in nature, producing so-called hybrid offspring as a result. Such mixing between species happens more in the plant world than the animal, but it does take place in both. If species are supposed to be "reproductively isolated" from one another, what are we to make of these crossings? (Mayr points out that it is *populations* in his definition that are reproductively isolated, not individuals, and that whole populations do tend to stay within their species confines.)

Despite these difficulties, the biological species concept provides a useful way of defining the basic category of Earth's living things. Most multicellular species carry out sexual reproduction at least part of the time, and relatively few species outside the plant kingdom regularly produce hybrids. Moreover, as noted, this species concept is rooted in a critical behavior of organisms themselves—mating, which controls the flow of genes. And as you saw in Chapter 17, it is the change in the genetic make-up of a population (a change in its allele frequencies) that lies at the root of evolution.

18.2 How Do New Species Arise?

Having defined a species, let's now see how one variety of them can be transformed into another in the process called **speciation**: the development of new species through evolution. As noted last chapter, a single species can diverge into two species, the "parent" species continuing while another branches off from it. A key question for scientists is: What brings this branching evolution about? The answer has to do with the flow of genes reviewed in Chapter 17. As you saw then, evolution within a population means a change in that population's allele frequencies. Now, however, imagine *two* populations of a single species of bird, with one being separated from the other—say, by one of them having flown to a nearby location. To the extent that they continue to breed with one another, with individuals moving between the locations, each population will *share* in whatever allele frequency changes are going on with the other population. Hence the two populations will evolve together, remaining a single species. Conversely, imagine that the migration stops between the two populations. Each population continues to undergo allele frequency changes, but it no longer shares these changes with the other population. Alterations in form and behavior may accompany allele changes, and these alterations may pile up over time. Coloration or bill lengths may change; feeding habits may be transformed. After enough time, should the two populations find themselves geographically reunited, they may no longer freely interbreed. At that point, they are separate species. Speciation has occurred.

The critical change here came when the two populations quit interbreeding. For scientists, the key question thus becomes: Why would this happen? What could drastically reduce interbreeding, and hence gene flow, between two populations of the same species?

The Role of Geographic Separation: Allopatric Speciation

In the preceding example, a very clear factor reduced gene flow between the bird populations: They were separated geographically. And geographical separation turns out to be the most important starting point in speciation.

Populations can become separated in lots of ways. On a large scale, glaciers can move into new territory, cutting a previously undivided population into two. Rivers can change course, with the result that what was a single population on one side of the river may now be two populations on different sides of it. On a smaller scale, a pond may partially dry up, leaving a strip of exposed land between what are now two ponds with two separate populations. Such environmental changes are not the only ways that populations can be separated, however. Part of a population might *migrate* to a remote area, as did the bird population, and in time be cut off from its larger population. For a real-life example of how migration can bring about speciation, see FIGURE 18.3.

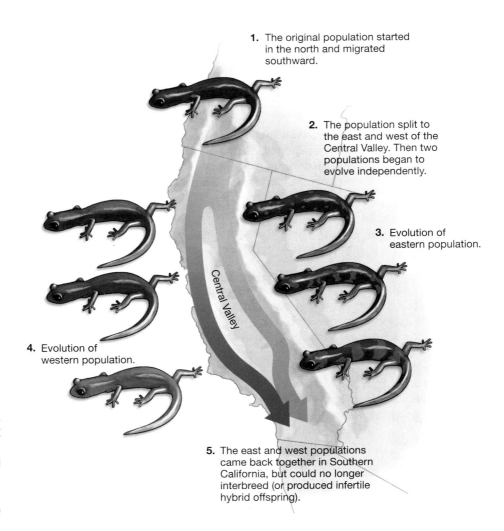

1. The original population started in the north and migrated southward.

2. The population split to the east and west of the Central Valley. Then two populations began to evolve independently.

3. Evolution of eastern population.

4. Evolution of western population.

Central Valley

5. The east and west populations came back together in Southern California, but could no longer interbreed (or produced infertile hybrid offspring).

FIGURE 18.3 **Speciation in Action** Millions of years ago, the salamander *Ensatina eschscholtzii* began migrating southward from the Pacific Northwest. When this species reached the Central Valley of California—an uninhabitable territory for it—some of its populations branched west (to the coastal range) and others went east (to the foothills of the Sierra Nevada mountains). Over time, the populations took on different colorations as they moved southward. By the time the two populations were united in Southern California, they differed enough, genetically and physically, that they either did not interbreed or produced infertile hybrid offspring when they did. (Salamanders are falsely colored in drawing for illustrative purposes.)

(a) Abert squirrel, south rim of Grand Canyon

(b) Kaibab squirrel, north rim of Grand Canyon

FIGURE 18.4 **Geographical Separation—Leading to Speciation?** These two varieties of squirrel had a common ancestor that at one time lived in a single range of territory. Then the Grand Canyon was carved out of land in Northern Arizona, leaving populations of this species separated from one another, about 10,000 years ago. In the area of the Grand Canyon, **(a)** the Abert squirrel lives on the Canyon's south rim, while **(b)** the Kaibab squirrel lives on the north rim. It's unclear whether these two varieties of squirrel would be reproductively isolated if reunited today—that is, whether they are now separate species—but the geographical separation they have experienced is the first step on the way to allopatric speciation.

When geographical barriers divide a population and the resulting populations then go on to become separate species, what has occurred is **allopatric speciation**. *Allopatric* literally means "of other countries" (**see** FIGURE 18.4). Look along the banks on either side of the Rio Juruá in western Brazil, and you will find small monkeys, called tamarins, that differ genetically from one another in accordance with how wide the river is at any given point. Where it is widest, the members of this species do not interbreed, while at the narrow headwaters they do. Are the nonbreeding populations on their way to becoming separate species? Perhaps, but only if gene flow is drastically reduced between them, probably for a very long time.

Reproductive Isolating Mechanisms Are Central to Speciation

While geographical separation is the most important factor in getting speciation going, it cannot bring about speciation by itself. Following geographical separation, two populations of the same species must then undergo physical or behavioral changes that will *keep* them from interbreeding, should they ever be reunited. Allopatric speciation thus operates through a one-two process: first the geographic separation, then the development of differing characteristics in the two resulting populations. These are characteristics that will *isolate* them from each other in terms of reproduction.

Thus arises the concept of **reproductive isolating mechanisms**, which can be defined as any factor that, in nature, prevents interbreeding between individuals of the same species or of closely related species. Geographic separation is itself a reproductive isolating mechanism, because it is a factor that prevents interbreeding. But, because the mountains or rivers that are the actual barriers to interbreeding are outside of or *extrinsic to* the organisms in question, geographic separation is called an extrinsic isolating mechanism. For allopatric speciation to take place, what also must occur is the second in the one-two series of events: the evolution of *internal* characteristics that keep organisms from interbreeding. Such characteristics are referred to as **intrinsic isolating mechanisms**: evolved differences in anatomy, physiology, or behavior that prevent interbreeding between individuals of the same species or of closely related species. In sum, allopatric speciation takes place when the extrinsic isolating mechanism of geographic separation is followed by the development of intrinsic isolating mechanisms.

So Far... Answers:
1. *breeding*
2. *gene flow; populations*
3. *geographic separation; intrinsic isolating mechanisms*

So Far...

1. Under the biological species concept, species are defined by their _____ behavior.

2. The critical factor leading to speciation is reduced _____ between two _____ of the same species.

3. Two factors always play a part in allopatric speciation: first, the development of _____ and second, the development of one or more _____.

Six Intrinsic Reproductive Isolating Mechanisms

So, what are these intrinsic mechanisms? As presented in **Table 18.1**, they are:

Ecological Isolation   Two closely related species of animals may overlap in their ranges and yet feed, mate, and grow in separate areas, which are called *habitats*. If the animals use different habitats, this means they may rarely meet up. If so, gene flow will be greatly restricted between them. Lions and tigers *can* interbreed, but they never have in nature, even when their ranges overlapped in the past. One reason for this is their largely separate habitats: Lions prefer the open grasslands, tigers the deep forests.

Temporal Isolation   Even if two populations share the same habitat, if they do not mate within the same time frame, gene flow will be limited between them. Two populations of the same species of flowering plant may begin releasing pollen at slightly different times of the year. Should their reproductive periods cease to overlap altogether, gene flow would be cut off between them.

Behavioral Isolation   Even if populations are in contact and breed at the same time, they must choose to mate with one another for interbreeding to occur. Such choice is often based on specific courtship and mating displays, which can be thought of as passwords between members of the same species. Birds must hear the proper song, spiders must perform the proper dance, and fiddler crabs must wave their claws in the proper way for mating to occur.

Mechanical Isolation   Reproductive organs may come to differ in size or shape or some other feature, such that organisms of the same or closely related species can no longer mate. Different species of alpine butterfly look very similar, but their genital organs are different enough that one species cannot mate with another.

Gametic Isolation   Even if mating occurs, offspring may not result if there are incompatibilities between sperm and egg or between sperm and the female reproductive tract. In plants, the sperm borne by pollen may be unable to reach the egg lying within the plant's ovary. In animals, sperm may be killed by

Table 18.1 Reproductive Isolating Mechanisms

Extrinsic Isolating Mechanism		**Geographic isolation** Individuals of two populations cannot interbreed if they live in different places (the first step in allopatric speciation).
Intrinsic Isolating Mechanisms		**Ecological isolation** Even if they live in the same place, they can't mate if they don't come in contact with one another.
		Temporal isolation Even if they come in contact, they can't mate if they breed at different times.
		Behavioral isolation Even if they breed at the same time, they will not mate if they are not attracted to one another.
		Mechanical isolation Even if they attract one another, they cannot mate if they are not physically compatible.
		Gametic isolation Even if they are physically compatible, an embryo will not form if the egg and sperm do not fuse properly.
		Hybrid inviability or infertility Even if fertilization occurs successfully, the offspring may not survive, or if it survives, may not reproduce (e.g., mule).

FIGURE 18.5 **Mules Are Infertile Hybrids** Even though mules cannot themselves reproduce, humans frequently cross horses and donkeys to produce mules in order to take advantage of their exceptional strength and endurance. Chromosomal incompatibilities between horses and donkeys leave the mules sterile.

the chemical nature of a given reproductive tract, or may be unable to bind with receptors on the egg. There is a form of gametic isolation, called *polyploidy*, that is very important in plants. You can read about it in Essay 18.1, "New Species through Genetic Accidents," on page 298.

Hybrid Inviability or Infertility   Even if offspring result, they may develop poorly—they may be stunted or malformed in some way—or they may be infertile, meaning unable to bear offspring of their own. A well-known example of such infertility is the mule, which is the infertile offspring of a female horse and a male donkey (**see** FIGURE 18.5).

18.3 Sympatric Speciation

So far, you have looked at the development of intrinsic reproductive isolating mechanisms strictly as a second step in speciation—one that follows a geographic separation of populations. As it happens, however, intrinsic reproductive isolating mechanisms can develop between two populations in the *absence* of any geographic separation of them. If these isolating mechanisms reduce interbreeding between the populations sufficiently, speciation can take place. What occurs in this situation is not allopatric speciation, however; it is **sympatric speciation**, which is defined as any speciation that does not involve geographic separation. (*Sympatric* literally means "of the same country.")

Sympatric speciation has been a contentious subject in biology for years. One form of sympatric speciation is the polyploidy reviewed in Essay 18.1. But setting polyploidy aside, most biologists thought for years that if sympatric speciation took place at all, it was of trivial importance. By the early twenty-first century, however, new evidence had convinced most evolutionary biologists that sympatric speciation certainly has operated in a number of instances and may be more widespread than previously thought.

Sympatric Speciation in a Fruit Fly

To see how sympatric speciation works, consider one of its best-studied examples, a species of fruit fly named *Rhagoletis pomonella*. Prior to the European colonization of North America, *R. pomonella* existed solely on the small, red fruit of

A BROADER VIEW

Singing the Right Song
Behavioral isolating mechanisms can be fascinating. As evolutionary biologist Mark Ridley has noted, a single habitat in the United States may contain up to 40 different species of breeding crickets. The males "sing" in these habitats and are approached by females, but females approach only those males that sing their own species' song.

FIGURE 18.6 **A Species Undergoing Sympatric Speciation?** The fruit-fly species *Rhagoletis pomonella*, pictured here on the skin of a green apple, may be undergoing speciation.

A BROADER VIEW

Spectacular Speciation

Africa's Lake Victoria provides a spectacular example of sympatric speciation. Though only 200,000 years old, it nevertheless harbors several hundred species of a small- to medium-sized fish, called cichlids, all of which are thought to have originated side by side in the lake.

hawthorn trees (see FIGURE 18.6). The Europeans brought *apple* trees with them, however, and by 1862 some *R. pomonella* had moved over to them. It seemed to at least one mid-nineteenth-century observer, however, that with the introduction of apples there had arisen separate varieties of the flies—what are sometimes called apple flies and haw flies—with each variety courting, mating, and laying eggs almost exclusively on its own type of tree. A modern researcher named Guy Bush led the way, beginning in the early 1960s, in investigating whether this was the case, and the answer turned out to be yes. The two varieties of flies are separated from each other in all these ways. In one study undertaken on them, only 6 percent of the apple and haw flies interbred with one another. The apple and haw *R. pomonella* are not separate species yet, but they certainly give indication of being in transition to that status.

For our purposes, the first thing to note is that this separation has not come about because of *geographic* division. The apple and hawthorn trees that the two varieties of flies live on may scarcely be separated in space at all. Given this, how did this single species move toward becoming two separate species? Bush offers a likely scenario, based on a critical difference between the hosts the two species live on. Apples tend to ripen in August and September, while hawthorn fruit ripens in September and October. In the summer, all fruit flies emerge as adults after wintering underground as larvae, after which they fly to their host tree to mate and lay eggs in the fruit.

Bush believes that about 150 years ago (when hawthorns were hosts to all the flies), some individuals in a population of *R. pomonella* experienced either a mutation or perhaps a chance combination of rare existing alleles. In either event, this change did two things: It caused these flies to emerge slightly *earlier* from their underground state than did most flies; and it drew these flies to the smell of *apples* as well as hawthorns. Because apples mature slightly earlier than the hawthorn fruit, these flies had a suitable host waiting for them. More important, these flies would have bred *among themselves* to a high degree. Recall that the other flies were emerging later, and the adult fly only lives for about a month. Thus, the variant alleles were passed on to a selected population in the next generation. Today, the apple *R. pomonella* flies indisputably do emerge earlier than the hawthorn flies. This is one thing that ensures reproductive isolation between the two groups; their periods of mating don't fully overlap.

Such a lack of overlapping mating periods may sound familiar, because it is one of the intrinsic reproductive isolating mechanisms you looked at earlier. (It is temporal isolation.) You also can see ecological isolation in operation with these flies. Though the two types of flies live in the same area, they meet up with relatively little frequency, because they have different habitats (hawthorn vs. apple trees). In short, these populations have developed intrinsic reproductive isolating mechanisms without ever having been separated from one another geographically. They are headed toward speciation, but in this case it is sympatric speciation.

To sum up, the most common form of speciation, allopatric speciation, takes place when a geographic separation is followed by the development of intrinsic reproductive isolating mechanisms. A less common form of speciation, sympatric speciation, occurs when these isolating mechanisms develop in the absence of geographic separation. One special means of sympatric speciation is speciation through polyploidy.

So Far...

So Far... Answers:

1. *ecological, temporal, behavioral, mechanical, and gametic isolation and hybrid inviability or infertility*

2. *geographic separation*

3. *plants*

1. Name two intrinsic isolating mechanisms.

2. Sympatric speciation is speciation that does not involve _____.

3. Polyploidy is a special form of sympatric speciation that is most important in _____.

(a) Modern horseshoe crab

(b) Fossilized horseshoe crab

(c) Galápagos finch

FIGURE 18.7 **Different Groups of Organisms Undergo Different Rates of Speciation** Because horseshoe crabs look essentially the same today as they did 300 million years ago, some scientists have called them "living fossils." Compare the modern organism, on the left, with the fossil at center, which is at least 145 million years old. By contrast, the Galápagos finches have diversified tremendously, forming 13 species in as little as 100,000 years.

18.4 When Is Speciation Likely to Occur?

The horseshoe crab (see **FIGURE 18.7**) is not really a crab at all, but instead is distantly related to land-dwelling arthropods such as spiders and scorpions. (It lacks the antennae that real crabs possess, but it *has* the pincers around the mouth that all spiders and scorpions do.) Horseshoe crabs have been around in something like their modern form for more than 300 million years. They exist today in a scant four or five species, each one of which is pretty much like the others. The crabs live in the shallow oceans off North America and Asia, pushing through sand or mud to feed on everything from algae to small-bodied invertebrate animals.

Now consider the Galápagos (or Darwin's) finches you looked at in Chapter 17. There are *thirteen* of these species on the small Galápagos archipelago, all of them derived from a single species of South American finch that arrived on one of the islands perhaps 100,000 years ago.

Think of the difference between horseshoe crabs and Darwin's finches. The former have remained almost unchanged for more than 300 million years throughout the world, while the latter diverged into 13 different species in the last 100,000 years in a small cluster of islands off South America.

Specialists and Generalists

Why the great differences in rate of speciation here? Two general principles are at work. First, as evolutionary biologist Niles Eldredge has pointed out, horseshoe crabs are *generalists*: Their diet is extremely diverse; they will eat plants and small animals but will also scavenge for debris on the ocean bottom. By contrast, you saw in Chapter 17 how *specialized* some of the Galápagos finches are in their feeding behavior—particularly when food has become scarce because of drought conditions, as happens on the islands. In such times, species might exist on a single variety of plant seed. Species that are tied in this way to a particular food or environmental condition must adapt in connection with changes in them or face extinction. Think of how quickly the bills of the *Geospiza fortis* population on Daphne Major evolved toward greater depth when having a deeper bill meant the difference between life and death (see Chapter 17, page 281). By definition, this kind of adaptation means change—and change is what speciation is all about. By contrast, the horseshoe crab shifts from one food to another, depending on what is available, not adapting greatly in response to changes in any one food source.

Essay 18.1 New Species through Genetic Accidents: Polyploidy

In the chapter so far, speciation has been portrayed as something that takes place over many generations. But plants (and some animals) have a means of speciating in a single generation. It is called *polyploidy*, and it is very important in the plant world; about 35 percent of all flowering plant species are thought to have come about through it. Here is one way in which it works. From the genetics unit, you may recall that human beings have 23 pairs of chromosomes, *Drosophila* flies four pairs, and Mendel's peas seven. Whatever the *number* of chromosomes, the commonality among all these species is that their chromosomes come in *pairs*, which is the general rule for species that reproduce sexually.

As noted at the start of the chapter, plants are more adept than animals at producing "hybrids," with the gametes—the eggs and sperm—from two *separate* species coming together to create an offspring. Often these offspring are sterile, however, because when it comes time for them to produce *their* gametes, the sets of chromosomes they inherited from their different parental species may not "pair up" correctly in meiosis, owing to differences in chromosome number or structure.

Now comes the accident. Suppose that, back when it was first created as a single-celled zygote, the hybrid offspring carried out the usual practice of doubling its chromosome number in preparing for its initial cell division. Now, however, the cell fails to actually divide; whereas it was supposed to put half its complement of chromosomes into one daughter cell and half into another, it doesn't do this. It doubles the chromosomes, but keeps them all in one cell (see **FIGURE E18.1.1**). Equipped with *twice* the usual number of chromosomes, this single cell then proceeds to undergo regular cell division, meaning that every cell in the plant that follows will have this doubled number of chromosomes. Critically, this plant will have doubled both its *sets* of chromosomes—the set it got from parental species A and the set it got from species B. If you double a set, by definition every member of that set now has a partner to pair with in gamete formation. Thus the roadblock to a hybrid producing offspring has been removed; the chromosomes can all pair up.

Many of our most important food crops are polyploid.

This pairing up yields gametes (eggs and sperm), which can then come together and fuse thanks to another capability of plants. Recall that many plants can *self-fertilize*. A single plant contains male gametes that can fertilize that same plant's female gametes. Thus the plant with the doubled set of chromosomes can fertilize itself, theoretically beginning an unending line of fertile offspring. This line of organisms is "reproductively isolated" from either of its parental species because it has a different number of chromosomes than either parental species. With reproductive isolation, we have a *separate species*; and with self-fertilization, we have a species that can perpetuate itself.

New Environments: Adaptive Radiation

The second lesson offered by horseshoe crabs and finches concerns the kinds of environments that induce speciation. The Galápagos Islands were formed by volcanic eruptions that brought the islands above the ocean's surface only about 5 million years ago. At that time, these volcanic outcroppings gave new meaning to the word *barren*. For a brief time, the islands were utterly sterile, with no life on them. Very quickly, however, life did come to the islands, with bacteria, fungi, plant seeds, and tiny animals landing on the islands, all being borne by air or ocean currents. The South American finches did not arrive until much later; but by then, with lots of large plant species well established, what these birds encountered was an environment rich with possibility, for there were *no birds of their kind* on the islands. Imagine that you are a graphic artist, working in a big city with lots of other graphic

A multiplication of the normal two sets of chromosomes to some other set number is known as **polyploidy**; here we have speciation by polyploidy, which is one type of sympatric speciation. The importance of this in the plant world is immense. Many of our most important food crops are polyploid, including oats, wheat, cotton, potatoes, and coffee. Indeed, the type of polyploid speciation we've looked at—which begins with a hybrid offspring—often produces bigger, healthier plants. As a result, breeders have developed ways to artificially induce polyploidy.

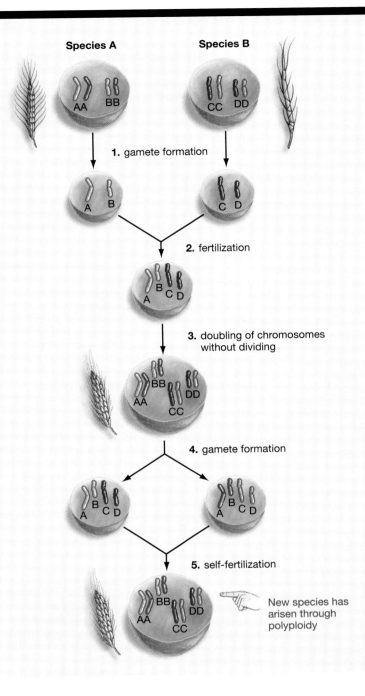

FIGURE E18.1.1 Polyploidy in Wheat Two different species of wheat exist in nature, with slightly different "genomes" or complements of DNA.

1. Gametes (eggs and sperm) are formed in the different species.
2. These gametes fuse, in fertilization, to form a zygote—a single cell that will develop into a new plant. Such a mixed-species or "hybrid" zygote generally develops into a sterile plant, because its chromosomes cannot pair up correctly when the plant produces its own gametes during meiosis.
3. In this case, however, the zygote doubles its chromosomes in preparation for cell division, but then fails to divide. With this doubling, each chromosome now has a compatible homologous chromosome to pair with during meiosis. This is the polyploidy event.
4. Gamete formation then takes place in the plant.
5. These gametes from the same plant then fuse, because this is a self-fertilizing plant. With this, a new generation of wheat plant has been produced—one that is a different species from either parent generation, because each of the parent species has two pairs of chromosomes, while this new hybrid species has four pairs.

artists, most of whom specialize in this or that (magazines, Internet websites, etc.). Now you and a few other graphic artists move to a new city in which there are few graphic artists, but a good number of *possibilities* for graphic arts work. You would thus be able to specialize fairly easily—filling a *niche* or working role in this new environment—because many of these niches would not yet have been taken. Just so did Darwin's finches rush in to fill previously unoccupied niches on the Galápagos Islands—eating this plant seed or this insect, or adopting this habitat.

Such a situation is ripe with possibilities for change (meaning speciation) because, while niches are in flux, there is a good deal of shaping of species to environment. But more of this occurred on the Galápagos, because this niche filling was taking place on 25 separate *islands*. The birds can fly from one island to another, but the water between the islands nevertheless represents a geographic barrier to bird interbreeding, and you know what follows from this: allopatric speciation.

A BROADER VIEW

Seedless Watermelons through Polyploidy

Ever have a seedless watermelon? They come about through human-induced polyploidy. Here's the recipe: Take a normal watermelon plant, induce polyploidy in it chemically, and then cross the resulting polyploid plant with a normal plant. The result is a plant that has *three* sets of chromosomes—two from the polyploid plant, one from the normal plant—which means its chromosomes cannot "pair up" in reproduction. The result is a sterile plant, which is to say one that can't produce true seeds.

A BROADER VIEW

Adaptive Radiation in Hawaii

Hawaii is not only a paradise for tourists; it is a paradise for biologists studying evolution. Isolated and with habitats ranging from desert-like to tropical, these islands were a perfect location for adaptive radiation to take place. An estimated 50 species of birds known as honeycreepers evolved from a single finch-like bird that arrived on the islands, and some 800 species of flies called drosophilids evolved from a single ancestral species. But this ancestral fly species may not have arrived as a swarm on the islands. As author Steve Olson has observed in *Evolution in Hawaii*, it is possible that Hawaii's 800 varieties of drosophilids evolved from a single pregnant female fly who was carried to the islands by wind or other accidental means millions of years ago.

The Galápagos finches exemplify something known as **adaptive radiation**: the rapid emergence of many species from a single species that has been introduced to a new environment. The finches radiated out to fill new niches on the islands, with populations adapting to the environments over time. In summary, two conditions that are conducive to speciation are specialization (of food source or environment) and migration to a new environment, particularly when there are no closely similar species in that environment.

18.5 The Categorization of Earth's Living Things

This chapter began by noting the importance of the species concept—of being able to say that this organism is fundamentally separate from that. To have a species concept means, of course, that there has to be some means of *naming* separate species. You have probably noticed that most of the species considered in this chapter have been referred to by their scientific names; you didn't just look at a fruit fly, but rather *Rhagoletis pomonella*. To the average person, such names may be regarded as evidence that scientists are awfully exacting—or just plain fussy. Why the two names? Why the Latin? Can't we just say fruit fly?

To take these questions one at a time, scientists can't just say "fruit fly" for the same reason a person can't say "the guy in the white shirt" while trying to identify a player at a tennis tournament. There are lots of guys in white shirts at tennis tournaments, and there are lots of different kinds of flies in the world. It is important to be able to say *which* fly we are talking about.

The Latin that is used stems from the fact that this naming convention was standardized by the Swedish scientist Carl von Linné in the eighteenth century, at a time when Latin was still used in the Western world in scientific naming. In fact, Linné is better known by the Latinized form of his name—Carolus Linnaeus.

Linnaeus recognized the confusion that can result from having several common names for the same creature, and thus devoted himself to giving specific names to some 4,200 species of animals and 7,700 species of plants—all that were known to exist in his time. Many of the names he conferred are still in use today.

The fact that there are *two* parts to scientific names—a binomial nomenclature—points to the central question of groupings of organisms on Earth. Consider the domestic cat, which has the scientific name *Felis domestica*. The first part of its name is its genus, *Felis*, which designates a group of closely related but still separate species. It turns out that, worldwide, there are five other species of small cat within the *Felis* genus, such as the small cat *Felis nigripes* ("black-footed cat") found in Southern Africa.

Taxonomic Classification and the Degree of Relatedness

The practical importance of the genus classification is that, if we know that two organisms are part of the same genus, then to know something about one of them is to know a good deal about the other. But what is the basis for saying that two organisms are part of the same genus? To put this another way, what is the scientific basis for putting organisms in a category such as genus?

Modern science classifies organisms largely—though not entirely—in accordance with how closely *related* they are. In this context, "related" has the same meaning as it does when the subject is extended families. If people have the same mother, they are more closely related than if they shared only a common grandmother, which in turn makes them more closely related than if they had only a common great-grandmother, and so on.

In the same way, species can be thought of as being related. Domestic dogs (*Canis familiaris*) are very closely related to gray wolves (*Canis lupus*), with all dogs being descended from these wolves. Such differences as exist between them are the product of a mere 15,000 years or so of the human domestication of dogs. It may be, then, that 15,000 years ago there was but *one* species (the gray wolf) that gave rise to both today's wolf and today's dog. A close relation indeed.

Domestic dogs are also related, however, to domestic *cats*; if we look far back enough in time, we can find a single group of animals that gave rise to both the dog and cat lines. This does not mean going back 10,000 years, however; it means going back perhaps 60 million years. So there is a big difference in how closely related dogs and wolves are, as opposed to dogs and cats. Establishing such *degrees* of relatedness is the most important task of the scientific classification system. There is a field of biology, called systematics, that is concerned with the diversity and relatedness of organisms; part of what systematists do is try to establish the truth about who is more closely related to whom. They study the evolutionary history of groups of organisms.

Setting aside for the moment the difficulty in *determining* what such evolutionary histories are, there obviously is a tremendous cataloguing job here, given the number of living and extinct species mentioned earlier. Given this diversity, a method of classifying organisms, called a *taxonomic system*, is employed in order to classify every species of living thing on Earth. There are eight basic categories in use in the modern taxonomic system: **species**, **genus**, **family**, **order**, **class**, **phylum**, **kingdom**, and **domain**. The organisms in any of these categories make up a grouping of living things, a taxon.

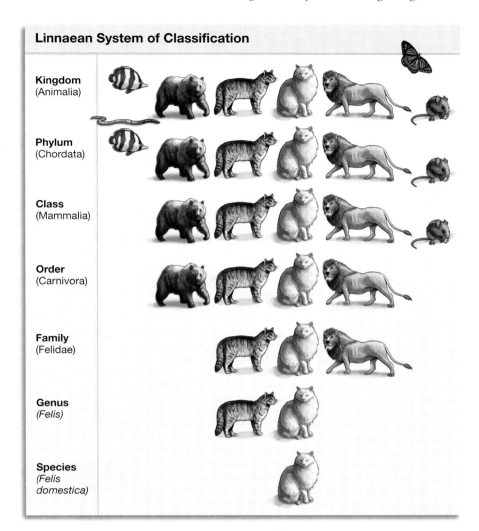

Linnaean System of Classification

Kingdom (Animalia)

Phylum (Chordata)

Class (Mammalia)

Order (Carnivora)

Family (Felidae)

Genus (Felis)

Species (Felis domestica)

FIGURE 18.8 **Classifying Living Things** The classification of the house cat, *Felis domestica*, based on the Linnaean system.

A Taxonomic Example: The Common House Cat

If you look at FIGURE 18.8, you can see how this taxonomic system works in connection with the domestic cat. As noted, this cat is only one species in a genus (*Felis*) that has five other living species in it. The genus then is a small part of a family (Felidae) that has 17 other genera in it (panthers, snow leopards, and others). The family is then part of an order (Carnivora) that includes not only big and small "cats," but other carnivores, such as bears and dogs. On up the taxa we go, with each taxon being more inclusive than the one beneath it until we get to the highest category in this figure, the kingdom Animalia, which includes all animals. (Later you'll look at the "supercategory" above kingdom, called domain; but for simplicity's sake, kingdom is the highest-level taxon considered here.)

Constructing Evolutionary Histories

So how do systematists go about putting organisms into these various groups? What evidence do they use in deciding who is more closely related to whom? They rely on radiometric dating, the fossil record, the distribution of organisms worldwide—all the things reviewed in Chapter 16 that are used to chart the history of life on Earth. In practice, there has been a revolution in this field in the last 20 years or so, in that most systematics work today is *molecular* work—scientists are comparing DNA, RNA, and protein sequences among modern organisms and using patterns among these sequences to make judgments about lines of descent.

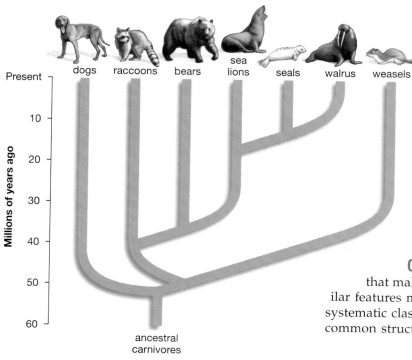

Present

Millions of years ago

dogs raccoons bears sea lions seals walrus weasels

ancestral
carnivores

FIGURE 18.9 **A Family Tree for a Selected Group of Mammalian Carnivores**

If you look at FIGURE 18.9, you can see the outcome of some systematics: an evolutionary "tree" for one group of organisms, in this case one of the major groups of mammalian carnivores. You can see that the tree is "rooted," about 60 million years ago, with an ancestral carnivore whose lineage then split two ways: to dogs on the one hand and everything else on the other. One interesting facet of this evolutionary history is how closely related bears are to the aquatic carnivores. All such histories are hypotheses about evolutionary relationships, with each such hypothesis known as a **phylogeny**.

Obscuring the Trail: Convergent Evolution   One of the things that makes the interpretation of evolutionary evidence difficult is that similar features may arise *independently* in several evolutionary lines. A bedrock of systematic classification is the existence of **homologies**, which can be defined as common structures in different organisms that result from a shared ancestry. In

(a) Homology: Common structures in different organisms that result from common ancestry

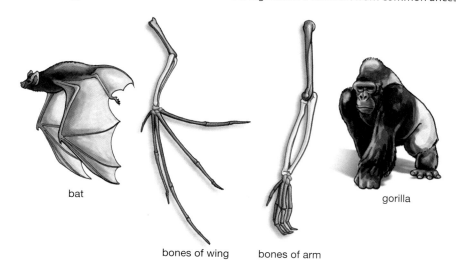

bat

bones of wing   bones of arm   gorilla

(b) Analogy: Characters of similar function and superficial structure that have *not* arisen from common ancestry

FIGURE 18.10   **Related, as Opposed to Similar**

(a) Bats and gorillas have the same bone composition in their forelimbs—one upper bone, joined to two intermediate bones, joined to five digits—because they share a common ancestor. Of course, these forelimbs are used today for very different functions.

(b) Horses and the extinct litopterns have long legs and one-toed feet that serve the same function, but these features were derived independently in the two creatures—in separate lines of descent.

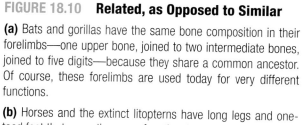

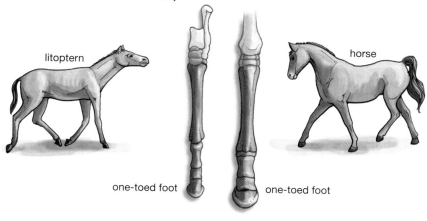

litoptern   horse

one-toed foot   one-toed foot

Chapter 16, a strong homology was noted in forelimb structure that exists in organisms as different as a gorilla, a bat, and a whale. If you look at FIGURE 18.10, you can see the gorilla and the bat again.

Now, however, consider an extinct group, the litopterns, that once lived in what is now Argentina. Like the modern-day horse, they roamed on open grasslands. Over the course of evolutionary time they developed legs that were extraordinarily similar to the horse's, right down to having an undivided hoof. In fact, litopterns are only very distantly related to the horse; but the leg similarities were enough to fool at least one nineteenth-century expert who thought he had found in litopterns the ancestor to the modern horse.

What is exhibited in the legs of the litoptern and horse is an **analogy**: a feature in different organisms that is the same in function and superficial appearance. When nature has shaped two separate evolutionary lines in analogous ways, what has occurred is **convergent evolution**. (Both the litoptern and horse lines converged in their development of similar legs.) But analogous features have nothing to do with common descent; they merely show that the same kinds of environmental pressures lead to the same kinds of designs. (There are only so many leg structures that work well for grass-eating mammals that roam over long distances, so it's not surprising that natural selection would have pushed two separate evolutionary lines toward a similar leg shape.) The problem for systematics is that analogy can be confused with homology; analogies can make us believe that organisms share common ancestry when in fact they do not.

18.6 Taxonomy and Relatedness

So, we have seen that there is a well-established taxonomic system for categorizing Earth's living things and that, within this system, relatedness is the most important factor used in placing organisms into categories. But are organisms put into categories *strictly* on the basis of relatedness? At the moment, the answer is no; modern taxonomy sometimes recognizes factors other than lines of descent in deciding on where an organism should be placed.

To appreciate this point, consider the following question. Who is more closely related: dinosaurs and lizards, or dinosaurs and birds? It may surprise you to learn that it is dinosaurs and birds. Birds split off from a dinosaur line within the last 200 million years or so, while the split between the lizard and dinosaur lines came much earlier. Despite this, conventional taxonomy says that dinosaurs and lizards belong in one class (Reptilia), while birds belong in another (Aves; **see** FIGURE 18.11). The assumption here is that, in classifying creatures, something should count *besides* their relatedness—in this case, the special qualities of birds (feathers, warm-bloodedness).

To some scientists, a categorization such as this makes no sense. Why should relatedness dictate taxonomy in some instances but not in others? Shouldn't we consistently categorize organisms on the basis of relatedness? On the other hand, aren't birds different enough from crocodiles that they deserve to be in a separate grouping? If you think that both sides in this debate are making sensible arguments, then you understand why it is so difficult to come up with a universally accepted means of classifying Earth's living things.

A BROADER VIEW

Arriving at Eyes 40 Times

Convergent evolution of the sort that produced the similar legs of the horse and the litoptern has occurred countless times during the course of evolution. The eye has evolved at least 40 separate times in various groups of animals, and even something as specialized as the echolocation that bats use to navigate has evolved three other times: in two varieties of nocturnal, cave-dwelling birds—swiftlets and oilbirds—and in the line of toothed whales and dolphins.

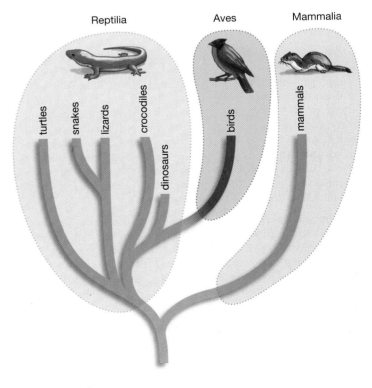

Classical view of relationships among tetrapods

FIGURE 18.11 **Are Birds Reptiles?** Conventional taxonomy employs three classes: Aves, Mammalia, and Reptilia, for birds, mammals, and reptiles. Thus, dinosaurs and birds are in separate classes—a recognition by taxonomists of the unique features of birds, such as feathers. Based on their phylogeny, however, birds are more closely related to dinosaurs than dinosaurs are to, for example, lizards—a fact that makes some systematists question the validity of putting birds in a separate class.

So Far... Answers:

1. introduced to a new environment

2. relatedness

3. genus

On to the History of Life

The millions of species that exist on Earth today took billions of years to evolve. During that time, life went from being exclusively microscopic and single-celled to being the staggeringly diverse entity it is today. How did this happen over time? And how did we go from no life to life? And what is the order of appearance for life's living things? These are the subjects of Chapter 19.

So Far...

1. Adaptive radiation is the rapid emergence of many species from a single species that has been _____.

2. Living things are classified largely in accordance with degrees of _____.

3. All species are referred to by a two-part name, the first part of which is a species' _____.

Chapter Review

Summary

18.1 What Is a Species?

- **Biological species concept**—The most accepted definition of a species is derived from the biological species concept: "Species are groups of actually or potentially interbreeding natural populations that are reproductively isolated from other such groups." Earth's organisms are so varied that this definition does not apply to all of them.   (page 290)

18.2 How Do New Species Arise?

- **Reductions in gene flow**—Branching evolution occurs when a single "parent" species diverges into two species, the parent species continuing while a second species arises from it. Branching evolution is based on reductions in gene flow between populations of the same species. To the extent that gene flow is maintained between two geographically separate populations, they will experience similar allele frequency changes, meaning they will evolve together. When gene flow is drastically reduced, however, the populations will evolve separately. When evolution has altered the physical or behavioral characteristics of the populations enough that they can no longer interbreed, even if reunited, speciation has occurred.   (page 291)

- **Allopatric speciation**—Geographical separation is the most important factor in reducing gene flow between populations of the same species. Such separation can come about due to a change in environment—a river may change course, for example—or due to the migration of part of a population. When geographic separation plays a part in the evolution of a species, allopatric speciation is said to have occurred.   (page 292)

- **Intrinsic isolating mechanisms**—Geographical separation cannot bring about speciation by itself. Following geographical separation, populations of the same species must develop internal characteristics that will keep them from interbreeding, should they be reunited. Such characteristics are referred to as intrinsic isolating mechanisms. There are six such isolating mechanisms: ecological, temporal, behavioral, mechanical, gametic isolation, and hybrid inviability or infertility.   (page 293)

Web Tutorial 18.1 Allopatric Speciation

18.3 Sympatric Speciation

- **Speciation without separation**—Speciation that occurs in the absence of geographical separation is called sympatric speciation. Like allopatric speciation, sympatric speciation always entails the development of intrinsic reproductive isolating mechanisms.   (page 295)

- **Polyploidy**—One special form of sympatric speciation is polyploidy: a doubling of the number of chromosomes in a species that can bring about speciation in one generation. Polyploidy has been especially important in plant speciation.   (page 298)

Web Tutorial 18.2 Speciation by Changes in Ploidy

18.4 When Is Speciation Likely to Occur?

- **Specialization and adaptive radiation**—Speciation is more likely to occur when a species is highly specialized with respect to a food source or other environmental condition. Speciation is also more likely to occur when a species is introduced to an environment in which few other species of its kind exist. Such an environment has many available working roles or niches available to the new species—a situation that leads to rapid specialization, which in turn leads to speciation. This series of events is known as adaptive radiation: the rapid emergence of many species from a single species that has been introduced to a new environment. (page 297)

18.5 The Categorization of Earth's Living Things

- **Biological taxonomy**—In science, a two-name or "binomial" nomenclature is used for each of Earth's species. The first name designates the genus, or group of closely related organisms, that the species is a member of; the second name is specific to the species. Genus and species fit into a larger framework of taxonomy, meaning the classification of species. Going from least to most inclusive, the eight commonly used groupings in this taxonomy are species, genus, family, order, class, phylum, kingdom, and domain. (page 300)

- **Systematics**—Species are put into these categories largely on the basis of relatedness—species that are closely related are in the same genus; species that are distantly related may only be in the same phylum or kingdom. The biological discipline of systematics is concerned with establishing degrees of relatedness among both living and extinct species. (page 300)

- **Establishing phylogenies**—Systematists establish evolutionary histories or "phylogenies" by reviewing various kinds of evidence, including radiometric dating, the fossil record, and DNA sequence comparisons. Based on this evidence, they determine the evolutionary history of a given species. (page 301)

- **Homologous and analogous**—In establishing phylogenies, one of the things systematists look for is homologous structures: common structures in different organisms that result from a shared ancestry. One problem with using these structures is that they can be confused with analogous structures: similar features that developed independently in separate lines of organisms (as with the legs of modern horses and extinct litopterns). (page 302)

18.6 Taxonomy and Relatedness

- **How should organisms be classified?** Conventional taxonomy sometimes uses factors other than relatedness in classifying organisms. For example, the special qualities of birds (feathers, etc.) have resulted in their classification in a class different from the one that they would be assigned to if lines of descent were the only criterion being utilized. This is a matter of some controversy, as some systematists hold that phylogeny (relatedness) should be the only criterion used in classifying organisms. (page 303)

Key Terms

adaptive radiation   p. 300

allopatric speciation   p. 293

analogy   p. 303

biological species concept   p. 290

class   p. 301

convergent evolution   p. 303

domain   p. 301

family   p. 301

genus   p. 301

homology   p. 302

intrinsic isolating mechanism   p. 293

kingdom   p. 301

order   p. 301

phylogeny   p. 302

phylum   p. 301

polyploidy   p. 299

reproductive isolating mechanisms   p. 293

speciation   p. 291

species   p. 301

sympatric speciation   p. 295

Testing Your Understanding

In Your Own Words (*answers in the back of the book*)

1. Why is the evolution of intrinsic reproductive isolation mechanisms required for two groups to be called separate species? Why isn't simple geographical isolation sufficient?

2. Describe one major difference between allopatric and sympatric speciation. What is one thing that they have in common?

3. What is convergent evolution, and why is it a problem for scientists who are trying to establish degrees of relatedness among organisms?

Thinking about What You've Learned

1. Living things are classified into a set of hierarchical categories. Why is this more useful to biologists than simply giving everything a one-word name (or a number) and eliminating some of the categories?

2. Populations of some kinds of organisms may be isolated by a barrier as small as a roadway, while other organisms require a much larger physical barrier to block gene flow. What features of organisms might determine how readily they become geographically isolated? What impact might these features have on how often or how rapidly speciation is likely to occur in these groups?

3. In an adaptive radiation, one species may colonize a previously empty environment, such as an island, and diversify evolutionarily into a set of closely related species that occupy a very wide range of habitats. For example, in the Darwin's finches on the Galápagos Islands, one species fills the role of woodpecker by using a stick to tap on trees and dig out insects. Do you think that such a species would have been likely to evolve in an area that already had woodpeckers? Why or why not?

The History of Life on Earth

An artist's interpretation of the appearance of the flying, feathered dinosaur, *Archaeopteryx.*

© 2002 Robert F. Walters

SAN FRANCISCO CHRONICLE

STUDY SHOWS DINOSAUR COULD FLY
WINGED CREATURE HAD BIRDLIKE SENSES, FOSSIL X-RAYS REVEAL

By David Perlman
August 5, 2004

New evidence gathered from a major advance in X-ray imaging of fossils has established that the winged dinosaur called archaeopteryx could actually fly and had much the same sense of balance and sharp vision found in today's birds.

And while its senses were a bit more primitive than its modern evolutionary descendants, the pigeon-sized archaeopteryx was certainly well equipped to navigate over land and forests looking for distant prey, scientists say. The evidence comes from a remarkable technology that allowed

University of Texas researchers to scan in three dimensions deep inside the brain case of the priceless 247-million-year-old fossil dinosaur and reveal the workings of its most critical sense organs.

19.1 Measuring the Past

"The past is never dead. It's not even the past," says one of William Faulkner's characters, in recognition of the ways in which the past stays with us. You might be able to get confirmation of this idea by just looking out your window. Do you see a bird resting on a branch or a wire? If you could get close enough to view its feet, you'd see four toes, three of them facing forward. If you could see beneath its feathers, what you'd find is an elongated, S-shaped neck. If you could look inside its bones, you'd see that they are hollow. Now trip back in time 150 million years or so and behold *Velociraptor,* made famous by *Jurassic Park.* What do you see? An elongated, S-shaped neck, hollow bones, and four toes (with the addition of a retractable "killing claw" set on top of the middle toe). Hold for a moment in your mind the image of *Velociraptor* racing across its killing grounds, and then think of a modern-day ostrich racing across the African plain. The similarities are so striking that, as far back as the nineteenth century, scientists began to think of birds as reptiles. And today? Evolutionary biologists are not joking when they refer to birds as "avian dinosaurs." So abundant is the evidence that birds evolved from two-legged dinosaurs that scientists think of birds as the last

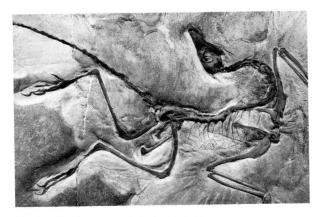

The fossilized remains of *Archaeopteryx.*

remaining dinosaurs. The scientific study written up in the *San Francisco Chronicle* story simply confirms this idea. *Archaeopteryx*, the famous "transitional" organism that had the feathers of a bird and the teeth of a dinosaur, had brain lobes enlarged in the same places as do today's flying birds. The only plausible explanation for this is that *Archaeopteryx* could fly.

If we can see dinosaurs in birds, what vestiges of our own past can we see in ourselves? Well, consider the way we walk: Our *left arm* swings forward as our *right leg* does the same. Almost all four-limbed animals walk this way. One view is that we inherited this from the elongated, four-finned fish we evolved from. They came onto the muddy land by moving in an S-pattern, the left-front fin going back as the left-rear fin went forward, then the reverse (**see** FIGURE 19.1).This is an ancient vestige of our evolutionary history, to be sure, but it manages to be relatively recent compared to the genetic code that operates within us—*this* DNA sequence specifying *that* amino acid. This critical molecular linkage represents a heritage that stretches back billions of years, to the beginning of life on Earth. We know because this code is identical in nearly every living thing. Since such a specific code is unlikely to have evolved more than once, the clear implication is that the code that operates in us is the same one that was operating in some of life's earliest microorganisms as they drifted through Earth's ancient seas.

Examples such as these make clear that evolution is not some abstract process that lies separate from us; on the contrary, it is with us in every step we take. When we look at evolution in this way, it's clear that there's quite a story of transmission here, which is to say the story of life on Earth. This chapter is aimed at giving you some sense of that story. Starting with the spark of life itself, how did evolution produce the range of creatures that inhabit the Earth? In this chapter, you'll begin to find out.

Life's Timeline

We can start this inquiry by getting a sense of life's timeline. The Earth, which is home to all the life we know of, came into being about 4.6 billion years ago. In measuring something as long as 4.6 billion years, it obviously won't do to think in terms of individual years, so scientists use something called the geological timescale, which you can see in FIGURE 19.2. It divides earthly history into broad *eras* and shorter *periods* (and shorter-yet *epochs*, which are not shown). The scale begins with the formation of the Earth, at the bottom of the figure, and runs to "historic time" at the top, meaning time in the last 10,000 years.

But what do the divisions in this timescale indicate? What, for example, is the distinction between the Permian period you can see at the middle left of Figure 19.2, and the Triassic period that followed it? The distinction largely represents a

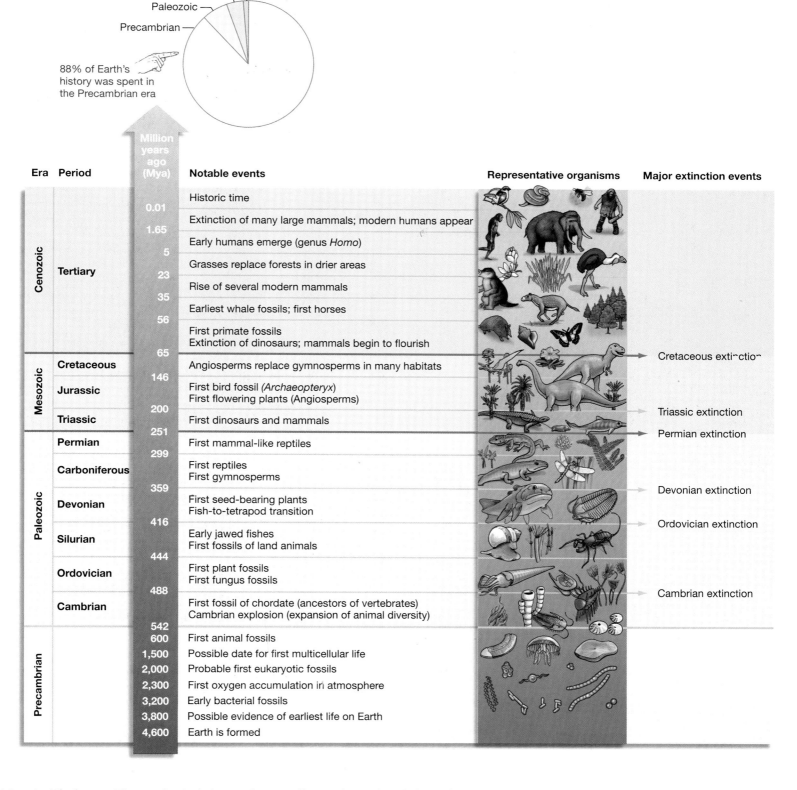

Relative Lengths of Geological Eras

88% of Earth's history was spent in the Precambrian era

FIGURE 19.2 **A Timescale for Earth and Its Living Things**

Era	Period	Million years ago (Mya)	Notable events	Representative organisms	Major extinction events
Cenozoic	Tertiary	0.01	Historic time		
		1.65	Extinction of many large mammals; modern humans appear		
		5	Early humans emerge (genus *Homo*)		
		23	Grasses replace forests in drier areas		
		35	Rise of several modern mammals		
		56	Earliest whale fossils; first horses		
		65	First primate fossils Extinction of dinosaurs; mammals begin to flourish		
Mesozoic	Cretaceous	146	Angiosperms replace gymnosperms in many habitats		Cretaceous extinction
	Jurassic	200	First bird fossil (*Archaeopteryx*) First flowering plants (Angiosperms)		
	Triassic	251	First dinosaurs and mammals		Triassic extinction
Paleozoic	Permian	299	First mammal-like reptiles		Permian extinction
	Carboniferous	359	First reptiles First gymnosperms		
	Devonian	416	First seed-bearing plants Fish-to-tetrapod transition		Devonian extinction
	Silurian	444	Early jawed fishes First fossils of land animals		Ordovician extinction
	Ordovician	488	First plant fossils First fungus fossils		Cambrian extinction
	Cambrian	542	First fossil of chordate (ancestors of vertebrates) Cambrian explosion (expansion of animal diversity)		
Precambrian		600	First animal fossils		
		1,500	Possible date for first multicellular life		
		2,000	Probable first eukaryotic fossils		
		2,300	First oxygen accumulation in atmosphere		
		3,200	Early bacterial fossils		
		3,800	Possible evidence of earliest life on Earth		
		4,600	Earth is formed		

transition in life-forms. The geological timescale actually predates the ability of scientists to assign absolute dates to the fossils they uncovered. That capability didn't come along until radiometric dating was discovered in the twentieth century. As a consequence, when the timescale was being developed—mostly in the nineteenth century—no one could say whether a fossil was 10 million or 100 million years old. Geologists of the time did know, however, that sedimentary rocks had been put down in layers (or strata), with the oldest strata on the bottom and

FIGURE 19.3 **Revealing Layers** The history of life can be traced through fossilized life-forms found in layers of sediment, seen here in America's Grand Canyon.

A BROADER VIEW

End of the Dinosaurs

How large was the asteroid that struck the Earth 65 million years ago, bringing an end to the dinosaurs? It was "only" six miles in diameter, but that was big enough to produce an impact crater 100 miles wide on Mexico's Yucatan peninsula. One result of the impact was a worldwide tsunami 150 meters high. Debate continues on whether the impact alone was responsible for wiping out the dinosaurs, or whether it simply came at the end of a long period in which these reptile behemoths were in decline.

the youngest strata on top (**see** FIGURE 19.3). And as it turned out, each stratum tended to have within it a group (or "assemblage") of fossils that was unique to it. Thus, the strata could be divided up in accordance with the *transitions* the scientists saw in life-forms. The Permian period is considered different from the Triassic period because, with movement from one to the other, a different set of life-forms appears, as recorded in the fossil record.

Features of the Timescale

Now a couple of notes on the timescale. First, it may surprise you to learn what the basis is for many of the transitions in it: death on a grand scale, in the form of "major extinction events." Six of these events are noted in Figure 19.2 (over on the right), with all of them defining the end of an era or period. The most famous extinction event, the **Cretaceous Extinction**, occurred at the boundary between the Cretaceous and Tertiary periods. This was the extinction, aided by the impact of a giant asteroid, that brought about the end of the dinosaurs, along with many other life-forms. The greatest extinction event of all, however, occurred earlier, in the boundary between the Permian and Triassic periods. This was the **Permian Extinction**, in which as many as 96 percent of all species on Earth were wiped out, perhaps because of the long-term effects of volcanic activity.

This is not to say, however, that all of Earth's eras or periods are marked off by extinctions. The transition between the Precambrian and Paleozoic eras marks not a major extinction, but a seeming explosion in the diversity of animal forms. Nevertheless, this so-called Cambrian Explosion, which you'll read about later, is just a different kind of transition in living things.

Now, when did some of these transitions come about? You might think that a transition such as a sudden increase in animal forms would have occurred very early in life's history, but the Cambrian Explosion took place 542 million years ago. This is a long time ago, to be sure, but life is thought to have begun 3.8 *billion* years ago. (Another way of writing this is 3,800 million years ago, or 3,800 Mya). There may not have been an animal on the planet for 3.2 billion years after life got going—nor any plants either. Life was single-celled to begin with and, billions of years later, it still consisted of nothing but microbes. In the Precambrian era, life evolved at such a slow pace that we scarcely have words for it. ("Glacial" greatly overstates things.) Conversely, if the fossil record is correct, all the birds, reptiles, fish, plants, and mammals that exist today came about in the last 16 percent of evolutionary time—the last 600 million years of life's 3.8-billion-year span. As you'll see, our own species, *Homo sapiens*, is an extreme latecomer in this series of events. If you look at the pie chart at the top of Figure 19.2, you can get a sense of how long life's Precambrian era was relative to all the eras that followed.

What Is Notable in Evolution Hinges on Values

One final word about the timescale and the path you will follow in this chapter. Picking out the "notable events" in evolution is inevitably an exercise in making value judgments. The notable events covered in this chapter will lead you along a line that begins with microscopic sea creatures and ends with human beings. The value judgment guiding this path is that students have a great interest in knowing about their own evolution, and that the path to humans arguably presents the broadest sweep of evolutionary development. Such a course of study may leave the impression, however, that all of evolution amounts to a march toward the development of human beings, who then get to occupy the highest branch of an evolutionary tree.

In reality, we occupy one ordinary branch among a multitude of branches. Under notable events in the Cenozoic era, the table could just as easily have listed milestones in the evolution of fish or birds. These creatures have continued evolving in the modern era, along with our own species, but in lines separate from us. Focusing on a line that leads to humans is like focusing on a railroad route that leads from, say, Baltimore to Denver, while ignoring the multitude of lines that intersect with it or that are separate from it. We regard the Baltimore–Denver route as special because we are *interested* in it, but that does not mean that it is fundamentally *different* from any other line.

19.2 How Did Life Begin?

Taking a historical approach to tracing life on Earth, the first question that arises is one of the toughest: How did life begin? Darwin himself thought about this and imagined that life began in what he called a "warm little pond" in the early Earth. Well, maybe; but the *very* early Earth had no warm little ponds. In fact, it was so hot that it was covered with a layer of molten rock, and it was periodically being slammed into by asteroids, comets, and proto-planets. This extraterrestrial bombardment seems to have slowed by about 3.8 billion years ago, however, and with this change we are left with a large question about Earth's temperature—it could have remained very hot or have become moderate to very cold, depending on its concentration of heat-trapping "greenhouse" gases, such as methane and carbon dioxide. For any of these environments to foster the development of life, however, they had to have in them the chemical raw materials that could be used in producing life's critical molecules. You saw in Chapter 3 that life's informational molecules (DNA and RNA) are composed of certain kinds of building blocks (nucleotides), while its proteins are composed of other kinds of building blocks (amino acids). The raw materials for such building blocks could have been the gases methane and ammonia. These substances certainly would have been spewed out by Earth's early volcanoes. The question is whether they would have been quickly broken down in the ultraviolet light, coming from the sun, that bathed the early Earth's atmosphere.

Life May Have Begun in Very Hot Water

Another possibility for the supply of organic materials is that they came from the methane and hydrogen sulfide that gush out even today from deep-sea vents on the floors of the world's oceans. Interestingly enough, there are creatures living near these vents today. Some of these are microscopic archaea and bacteria that thrive at temperatures of 247 degrees Fahrenheit, which is above the temperature at which water boils (see **FIGURE 19.4**). To judge by molecular sequencing, the organisms on Earth today whose roots stretch back the farthest in time are creatures such as these. This and other pieces of evidence lead to an idea that has a good

FIGURE 19.4 Hot Habitat Material pours forth from a hot-water vent on the floor of the Atlantic Ocean. The fluid being emitted is mineral-rich enough to support bacteria and archaea that in turn form the basis for a local food web.

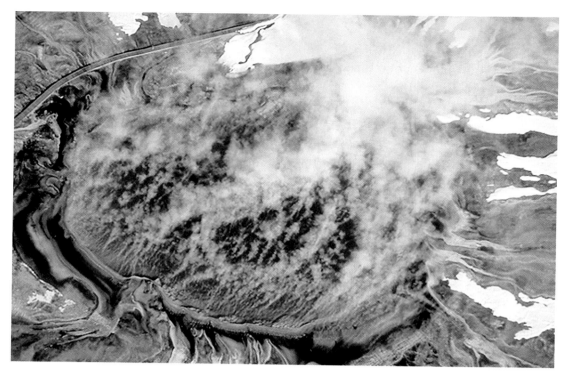

FIGURE 19.5 **Only Seemingly Inhospitable** The hot-water Grand Prismatic Pool in Yellowstone National Park, shown here in an aerial view, gets its blue color from several species of heat-tolerant cyanobacteria. The colors at the edge of the pool come from mineral deposits. The "road" at the top of the pool is a walkway for visitors.

deal of support among origin-of-life researchers. It is that life began in the "prebiotic soup" of hot-water systems—in the deep-sea vents or the kind of hot-spring pools that exist in Yellowstone Park (**see** FIGURE 19.5).

On the other hand, there are respected researchers who feel that, whatever the deep-sea vents may have had in the way of life-inducing qualities, their scalding temperatures would have obliterated any early self-replicating molecules, which were bound to be fragile. Much more likely as the first home for life, these researchers feel, are the sheltered stretches of ancient ocean beaches, where the combination of tides and the composition of shoreline rocks would have "sorted" organic materials in accordance with their adhesiveness and other qualities. In this environment, the right initial combination of materials could have come together and then been joined by fresh materials, delivered continuously by the ocean waves.

Wherever life began, it cannot have been created in a single step from basic compounds such as methane and hydrogen sulfide. There is a huge gap between these materials and organisms as complex as the archaea. Early Earth had to be a kind of natural chemical lab in which ever more complex organic molecules were produced from simpler ones. A major portion of origin-of-life research involves creating, in the laboratory, simulations of early Earth environments and then watching to see if life's building blocks will come together in them.

The Step at Which Life Begins

The critical step in life's development comes when a molecule could begin to replicate itself. Remember, from Chapter 13, how DNA and enzymes work together to replicate DNA whenever a cell divides? Imagine a single molecule that could do this on its own—a molecule that had the capability to *self*-replicate. This is the step at which life can be said to have begun. Given a molecule that can make copies of itself, a line of such molecules can come into existence. Given the *mistakes* that are bound to take place in such copying, molecules can be produced that differ slightly from one another in being, say, more resistant to the elements. Molecules that have such an advantage will leave more daughter molecules, which might just as well be called offspring. What happens, in short, is natural selection among self-replicating molecules. With this, life begins to diversify.

Most researchers are convinced that the earliest self-replicating molecule was a precursor form of today's RNA. Why RNA instead of DNA? Because certain forms of RNA can take on *two* roles: they can bear information (in the way that DNA does today), and they can catalyze chemical reactions (in the way that enzymes do today). Indeed, most researchers presume the existence of an "RNA world"—an early living world that consisted solely of self-replicating RNA molecules.

Once replication got going, life was still a long way from its modern form. Even the most primitive contemporary living things are encased in protective linings; and they can get rid of wastes, react to their environments, and so forth. Whatever the sequence of events, once elaborations such as these are joined to replication, we move from simple molecules to the cellular ancestors of today's organisms.

A BROADER VIEW

Lessons from the Deep-sea Vents

Given the importance of deep-sea thermal vents to biology, it's something of an irony that scientists didn't actually see such vents until 1977. Then again, it's not easy to spot something that exists in total darkness a mile and a half down in the ocean. (How did scientists first reach a vent? In a deep-sea submarine called *Alvin*.) Apart from the vents' possible role in the origin of life, they are important to biology because they pose so many questions about life in extreme environments. No sunlight, little oxygen, crushing water pressure, scalding temperatures: The deep-sea vents are a testament to the adaptability of living things.

So Far...

1. Each of the eras or periods within the geological timescale is characterized by a particular grouping of _____.

2. True or false: The first animals and plants evolved early in life's history.

3. Life is generally regarded as having begun with the development of _____ molecules.

So Far... Answers:

1. *life-forms*

2. *False; to judge by the fossil record, animals and plants don't appear until the last 16 percent of life's history.*

3. *self-replicating*

The Tree of Life

Once life was established, how did it evolve? If you look at **FIGURE 19.6**, you can see the biggest picture of all. At the bottom you see a "universal ancestor," which is the organism that gave rise to all current life. The evolutionary line that leads from the universal ancestor then branches out to yield three "domains" of life, which you can see at the top of the tree—Bacteria, Archaea, and Eukarya. The Eukarya domain is then further divided into four "kingdoms," which are plants, fungi, animals, and protists.

Now note the lines of descent in this family tree. The line that leads from the universal ancestor branches out to yield, on the one hand, the domain Bacteria and, on the other, a line that leads to both domain Archaea and domain Eukarya. Archaea were long thought of as simply a different kind of bacteria, but recent research has shown these organisms to be distinct from bacteria in fundamental ways. Nevertheless, there are great similarities. Bacteria and archaea are essentially single-celled, and no archaeic or bacterial cell has a nucleus. Conversely, *all* the organisms in the domain Eukarya have nucleated cells, and are thus "eukaryotes." Looking at the tree, you can see that the earliest kingdom in Eukarya is Protista—mostly one-celled, water-dwelling creatures—and it is from Protista that the other three eukaryotic kingdoms evolve. An unknown species within Protista gave rise to all

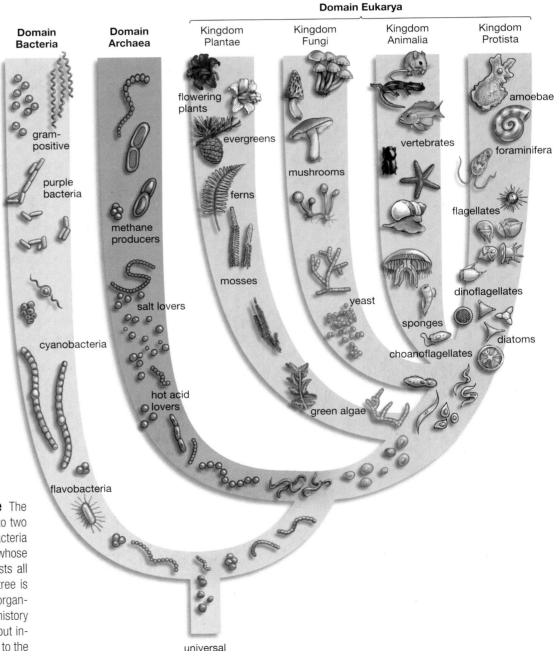

FIGURE 19.6 The Universal Tree of Life The universal ancestor at the base of the tree gave rise to two domains of organisms whose cells lack nuclei—Bacteria and Archaea. The third domain of life is the Eukarya, whose earliest kingdom was the protists. From these protists all plants, animals, and fungi evolved. The root of the tree is controversial; it may be that a large community of organisms shared genes among themselves early in the history of life. If so, there was no single universal ancestor, but instead many lines of descent that eventually gave rise to the individual lines higher in the figure.

the fungi that exist on Earth today; another grouping of protists called the choanoflagellates probably gave rise to all of the animal kingdom; and the protists we call green algae gave rise to all of today's plants. Note that, though it's common to think of plants and fungi as being alike, fungi and animals actually are more closely related than are fungi and plants.

These branches on life's tree are accepted by almost all scientists, but the *root* of the tree is another matter. Was there actually a single organism that was ancestral to all other living things? Many respected evolutionary biologists have their doubts. We are used to thinking of genes being passed *vertically*—from parent to offspring—but it is clear that single-celled organisms are adept at passing on genes *horizontally*, meaning from one "adult" to another. Such horizontal gene transfer seems to have been very common early on in life's history—so much so that many researchers imagine not a single common ancestor, but instead a community of simple organisms that promiscuously passed genes around. If so, life's root should be pictured not as the nice, straight line of one universal ancestor, but instead as a kind of spaghetti ball of gene-swapping organisms that only later gave rise to the clear lines of descent you can see higher up in Figure 19.6.

19.3 A Long First Era: The Precambrian

With this picture in mind, let's start tracing life from its single-celled beginnings. As noted, living things started small and stayed that way for a long, long time. Recall that the Earth was formed about 4.6 billion or 4,600 million years ago (Mya). The earliest evidence we have of life is not a fossil but instead a kind of chemical signature of life left in the barren, frozen Isua rocks of Greenland (see FIGURE 19.7). Carbon trapped in these 3,700 to 3,800-million-year-old rocks is of a type produced by living things. This evidence is controversial, but most authorities believe that life existed by 3,800 Mya, or at the latest by 3,500 Mya.

All of this life was either bacteria or archaea, who had the planet to themselves for a long time. Indeed, it was probably almost 2 billion years before any other life-form came into existence. When the first *multi*celled organisms appeared isn't clear. Certainly they exist by the time we have the first fossil evidence for animals, which is 600 Mya. But molecular clock studies put their evolution much further back, perhaps about 1,500 Mya. Even supposing this latter figure is right, note the big picture here. It took about 900 million years for life to appear after the Earth was formed; but then it took another 2.3 billion years to get from these single-celled life-forms to multicelled life-forms. Some authorities say this is evidence that the real hurdle in evolution was not the initiation of life, but rather the initiation of complex, multicellular life. What is certain is that evolution took its time in the Precambrian. The years rolled by in the millions, the land was barren—save for some bacteria—and the oceans were populated mostly by creatures too small to see with the naked eye.

FIGURE 19.7 Looking at the Oldest Rocks The rocks at Isua, Greenland, are believed to be the oldest in the world. In examining them, scientists have found possible chemical signatures of life that have been dated to nearly 3.8 billion years ago.

Notable Precambrian Events

None of this is to say, however, that "nothing happened" in the Precambrian era. Those of you who have read Chapter 8 know how critical photosynthesis is for life on Earth. This capability first came about in the Precambrian era—and fairly early in the era at that.

We have good evidence that photosynthesis had begun in bacteria by 3,400 Mya. This event was a turning point. Had this capacity not developed, evolution would have been severely limited, for the simple reason that there would have been so little to eat on Earth. The earliest organisms subsisted mostly on organic material in their surroundings, but the supply of this material was limited. In photosynthesis, the sun's energy is used to *produce* organic material—in our own era, the leaves and grasses and grains on which most animal life depends. Large organisms require more energy and, through photosynthesis, a massive quantity of energy-rich food was made available.

One particular kind of photosynthesis, again beginning in the Precambrian, had a second dramatic impact on evolution. One very early type of bacteria, the *cyanobacteria*, were the first organisms to produce *oxygen* as a by-product of photosynthesis. For us the word *oxygen* seems nearly synonymous with the word *life*, but until about 2,400 Mya there was almost no oxygen in the atmosphere. When it finally did arrive, through the work of the cyanobacteria, its effect was anything but life-giving. For most organisms it was a deadly gas, producing what has been termed an "oxygen holocaust" and establishing a firm rule for life on Earth: Adapt to oxygen, stay away from it, or die.

In the creatures that adapted to oxygen, we find one of the most interesting stories in evolution. Those of you who went through Chapter 7, on cellular energy harvesting, will recall that most of the energy that we eukaryotes get from food is extracted in special structures within our cells called *mitochondria*. Though mitochondria are tiny organelles within eukaryotic cells, they have characteristics of free-living *cells*, specifically free-living bacterial cells. As it turns out, that is probably what they once were. Ancient bacteria that could metabolize oxygen took up residence in early eukaryotic cells and eventually struck up a mutually beneficial relationship with them (**see** FIGURE 19.8). The bacteria benefited from the eukaryotic cellular machinery, and the eukaryotes got to survive in an oxygen-rich world. In a second merger with bacteria, algae and their plant descendants acquired a second kind of organelle, the chloroplasts in which photosynthesis is carried out.

The oxygen revolution then had one more momentous consequence. Molecules of oxygen came together to form the gas called *ozone*, which rose through the atmosphere to form the ozone layer, giving Earth, for the first time ever, protection against the ultraviolet radiation that comes from the sun. Marine creatures had been shielded from this radiation by the ocean itself, but prior to the formation of the ozone layer, there was no chance for life to develop to any great extent on *land*.

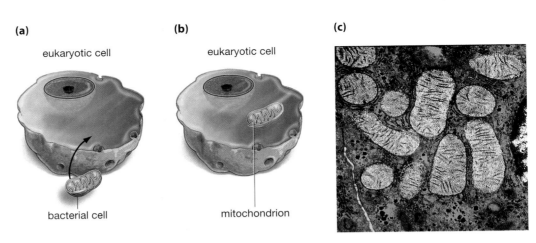

(a) eukaryotic cell — bacterial cell

(b) eukaryotic cell — mitochondrion

(c)

FIGURE 19.8 **Adapting to Oxygen** The "powerhouses" of our cells—the organelles called mitochondria—have several characteristics of free-standing bacteria.

(a) There is general agreement among scientists that our mitochondria originally were bacteria that invaded eukaryotic cells many millions of years ago. These bacteria benefited from the eukaryote's cellular machinery, and the cell benefited from the bacteria's ability to metabolize oxygen.

(b) Over time, the bacteria came to be integrated into the host cells, replicating along with them. Two organisms had now become a single organism.

(c) Today, the mitochondria in your cells turn the energy from food into a form that allows you to run, think, and turn the pages of this book. These artificially colored mitochondria functioned in a liver cell of a mouse.

19.4 The Cambrian Explosion

As the Precambrian was coming to a close, life consisted of bacteria, archaea, and many kinds of protists, most of which lived in the ocean. By about 600 Mya, however, we get the first fossil evidence of *animals* in the seas. The term *animal* here fits our common-sense definition: They all are multicelled organisms, and they get their nutrition from other organisms or organic material.

There currently is a debate about whether the earliest animal fossils represent a line of creatures that died out, or are instead related to today's animals. Whatever the case, these early animals can be seen as a kind of early stirring; a first blip on the screen bringing news of something big to come. To judge by the fossil record, about 60 million years after the emergence of these animals, we get it. A tidal wave of new animal forms, the like of which has never been seen before or since in evolution. This is the **Cambrian Explosion**, which began about 542 Mya.

Recalling the taxonomic system you went over in Chapter 18, you may remember that just below the "domain" and "kingdom" categories there is the **phylum**,

A BROADER VIEW

Life Influences Earth

We commonly think of Earth's environment influencing life, but note that the Precambrian provides some sterling examples of life influencing Earth's environment. Living things were responsible for increasing the oxygen content of the atmosphere—from near zero to today's 21 percent—and their activity likewise provided the ozone that shields Earth from ultraviolet radiation.

FIGURE 19.9 Life on the Ancient Sea Floor The Cambrian Explosion resulted in a multitude of new animal forms. In this artist's interpretation of the fossil evidence, a number of now-extinct Cambrian animals are shown in their ancient sea-floor habitat. Two of these are Anomalocaris, rising up with the hooked claws, and Hallucigenia, on the sea floor with the spikes extending from it.

which can be defined informally as a group of organisms that share the same body plan. There are at least 36 phyla in the animal kingdom today, some of these being Porifera (sponges), Arthropoda (insects, crabs), and Chordata (ourselves, salamanders). With one exception, the fossil record indicates, every single one of these came into being in the Cambrian Explosion. This seeming riot of sea-floor evolution was actually more extensive than even this implies, in that a good number of phyla that *don't* exist today—phyla that have become extinct—also seem to have first appeared in the Cambrian. Some of these creatures are so bizarre, it's surprising that Hollywood hasn't mined the Cambrian archives for new ideas on monsters (**see FIGURE 19.9**). As if all this weren't enough, the fossil record indicates the Cambrian Explosion took place in a very short period of time relative to evolution's normal pace—perhaps as little as 6 million years.

But did it? The Cambrian Explosion is so extreme that one of the primary challenges to evolutionary biology has been to explain why such a thing came about—or why it came about only once. Note the series of events: only a few animal forms before the Cambrian, then the explosion, then almost no new animal forms since. In the 1990s, several lines of evidence suggested something else: that animal forms actually began to diverge well back in the *Pre*cambrian. In this view, all the Cambrian Explosion amounted to was an explosion of forms big and hard enough to leave *fossils*.

If the Cambrian Explosion did in fact occur, what could have sparked it? Lots of ideas have been proposed; one of the best-received is that the explosion was triggered by the rise in atmospheric oxygen. To get bigger creatures, you need more oxygen, and levels of atmospheric oxygen may have reached a critical threshold about this time.

So Far...

1. Life has three domains, and one of these domains has four kingdoms within it. Name both the domains and the kingdoms.

2. Life is thought to have begun as early as _____ years ago. Name some important evolutionary events that took place during life's first era, the Precambrian era.

3. The Cambrian Explosion is a seeming explosion in the number of _____ life-forms.

So Far... Answers:

1. *Domains: Bacteria, Archaea, Eukarya. Kingdoms within Eukarya: Protista, Plantae, Fungi, Animalia.*

2. *3.8 billion. The development of photosynthesis; the production of atmospheric oxygen by photosynthesizers; the production of the ozone layer from atmospheric oxygen; the union of oxygen-using bacteria with eukaryotic cells (resulting in mitochondria in the eukaryotic cell) and the union of photosynthesizing bacteria with eukaryotic cells (resulting in chloroplasts within algae and plant cells).*

3. *animal*

19.5 The Movement onto the Land: Plants First

The teeming seas of the early Cambrian period stand in sharp contrast to what existed on land at the time, which was no life at all except for some hardy bacteria. Earth was simply barren—no greenery, no birds, no insects. When multicelled life did come to the land, the first intrepid travelers were plant-fungi combinations. (Even today, most plants have a mutually beneficial relationship with fungi, the plants providing fungi with food through photosynthesis, the fungi providing plants with water and mineral absorption through their filament extensions.) Exactly when this transition onto land took place is a matter of debate. The earliest undisputed fossil evidence we have of plants and fungi on land dates from 460 Mya, meaning these organisms had become abundant by that time. But when did they first appear? Molecular work estimates that land (or "terrestrial") fungi have been around for 1.3 billion years. These fungi couldn't have existed without plants—since they need them to live—which would mean that plants were on

Evolution of Plants

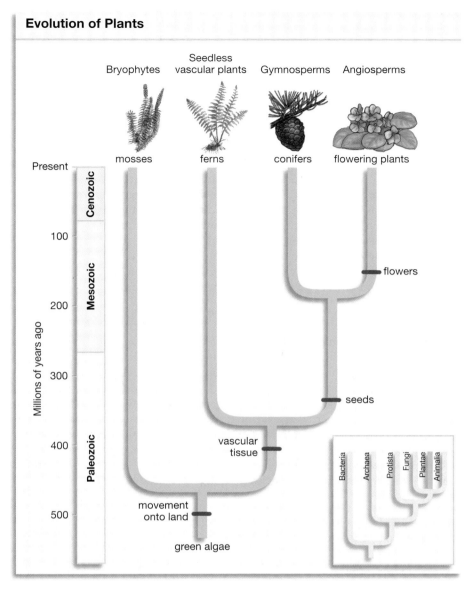

land at this time too. The general transition here began with algae in the ocean, then continued with freshwater algae, then freshwater algae that came to exist in *shallow* water, living partially above the waterline, then plants that lived in damp environments.

If you look at FIGURE 19.10, you can see how the major divisions of plants evolved, all of them from algae. The most primitive land plants, the bryophytes (represented by today's mosses), had no vascular structure—no system of tubes that transports water and nutrients. Plants without such a system have very limited structural possibilities. They can grow out, but they lack ability to grow *up* very far, against gravity (see FIGURE 19.11). By contrast, the ancestors of today's ferns developed a vascular system, and over the ensuing 100 million years, a variety of seedless vascular plants evolved, including huge seedless trees such as club mosses, some of them 130 feet tall.

FIGURE 19.11 **Lying Low** Moss covering rocks along a stream in Tennessee's Great Smoky Mountains National Park. Mosses are members of the earliest-evolving group of plants, the bryophytes.

FIGURE 19.12 **A Possible Early Flowering Plant** In 2002, a team of Chinese and American scientists announced the discovery of *Archaefructus sinensis*, whose fossilized, 125-million-year-old impression can be seen at left. The plant clearly had angiosperm-like traits—for example, the male reproductive structures known as anthers. Yet it had no petals. One hypothesis is that it represents a transitional form in the evolution of angiosperms. On the right is an artist's conception of what *A. sinensis* looked like.

A BROADER VIEW

The Plants in the Dinosaur Paintings

Ever notice how, in paintings of the dinosaur era, ferns and plants that look something like palm trees abound? The artists who render these drawings are trying to be true to life by showing the seedless vascular plants and early gymnosperms that dominated the landscape during the Triassic and Jurassic periods. The ferns and club mosses that were abundant then later gave way to ginkgos and palm-like cycads, which are gymnosperms.

Plants with Seeds: The Gymnosperms and Angiosperms

Even as the seedless vascular plants were dominating the landscape, a revolution in plants was well under way as some of them developed the reproductive packages we know as seeds. With the first seed plants, offspring no longer had to develop within a delicate plant on the forest floor. Instead they developed within a seed, which can be thought of as a reproductive structure that includes not only an embryo (brought about when sperm fertilizes egg), but food for this embryo, and a tough outer coat.

The living descendants of the first seed plants are gymnosperms—today's pine and fir trees, for example—whose seeds are visible in the well-known pinecone. Gymnosperms also represent the final liberation of plants from their ocean past. Mosses and ferns had sperm that could make the journey to eggs only through water. Not much water was needed; a thin layer would do, as with the last of a morning's dew on a fern leaf. With seed plants, however, the water requirement is gone entirely. Sperm are encased inside pollen grains, which can be carried by the *wind*. Think how far a windblown pollen grain could disperse compared to a sperm moving across a watery leaf. Several types of gymnosperms developed beginning about 350 Mya. Today, gymnosperm trees cover huge stretches of the Northern Hemisphere.

The Last Plant Revolution So Far: The Angiosperms

At the time the dinosaurs reigned supreme among land animals, the first flowering on Earth occurred, with the development of flowering plants, also known as angiosperms. Evolving about 165 Mya, the angiosperms eventually succeed the gymnosperms as the most dominant plants on Earth. Today there are about 700 gymnosperm species, but some 260,000 angiosperm species, with more being identified all the time. Angiosperms are not just more numerous than gymnosperms, they are vastly more diverse as well. They include not only magnolias and roses, but oak trees and cacti, wheat and rice, lima beans and sunflowers. If you look at FIGURE 19.12, you can see the fossilized impression of what may be an early flowering plant, dating from 125 Mya, along with an artist's conception of what it looked like.

19.6 Animals Follow Plants onto the Land

The movement of plants onto the land made it possible for animals to follow. With plants came food and shelter from the sun's rays. Recalling the division of the animal kingdom into various phyla, it was the phylum of arthropods— today, the phylum of insects and spiders, among others—that first moved to land. About 20 million years after the first plant fossils were laid down, a creature similar to a modern centipede laid down the oldest set of terrestrial animal tracks we know of. It makes sense that arthropods were the first animals to come onto land, because the hallmark of all of them is a tough external skeleton (an "exoskeleton"), which can prevent water loss and guard against the sun's rays. For millions of years, the arthropods were the only animals on land—but this would change.

Vertebrates Move onto Land

You'll recall that back at the time of the Cambrian Explosion, almost all animals existed on the ancient ocean floor. Within just a few million years, however, animals with backbones, known as vertebrates, were moving throughout the ocean; several of them were primitive, now extinct, orders of fishes. About 450 Mya, we get the rise of the family of fishes, the gnathostomes, that are the ancestors of nearly all the fish species still alive today, from sharks to gars to goldfish. Critically, gnathostomes are the first creatures to have jaws. With the development of jaws, these vertebrates and their descendants (one of whom is us) could securely grasp all kinds of things; big chunks of food, to be sure, but also offspring and inanimate objects.

The new foods that became available to those with jaws allowed the gnathostomes to outgrow their jawless vertebrate relatives.

Primitive Fish First It was a particular, later-arriving type of gnathostome, the "lobe-finned" fishes, that represent the transitional organisms between sea life and land life. The lobed *fins* these creatures possessed are the precursors to the four *limbs* that are found in land vertebrates such as ourselves (see FIGURE 19.13). It is a line of such four-limbed or "tetrapod" vertebrates—the amphibians, the reptiles, and the mammals—that you'll follow for the rest of your walk through evolution. Here is the order of emergence among these forms. The lobe-finned fish give rise to amphibians, and early, salamander-like amphibians give rise to reptiles. Early reptiles then diverge into several lines, one leading to the dinosaurs and another leading to mammals. Dinosaurs eventually die out, of course, but before they do, a lineage that will live on—the birds—diverges from them (see FIGURE 19.14).

(a) Ray-finned fish **(b)** Lobe-finned fish **(c)** Amphibian

FIGURE 19.13 **Vertebrates onto Land**

(a) Our ray-finned fish ancestors lacked bones in their fins.

(b) The lobe-finned fish that evolved from ray-finned fish had small bones in their fins that enabled them to pull out of the water and support their weight on tidal mudflats and sandbars.

(c) Finally, the early amphibians lost ray fins altogether, and had only bones in their limbs.

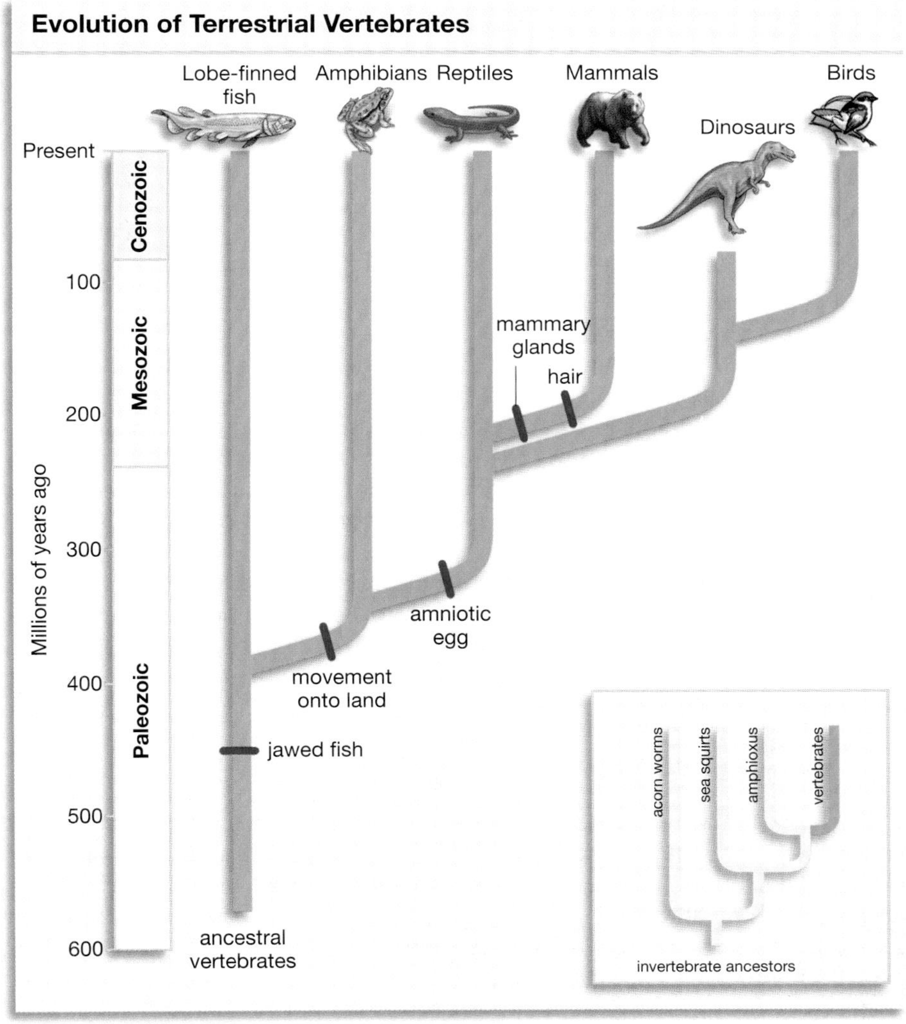

Evolution of Terrestrial Vertebrates

Lobe-finned fish
Amphibians
Reptiles
Mammals
Dinosaurs
Birds

Present

Cenozoic

Mesozoic

Paleozoic

100

200

300

400

500

600

Millions of years ago

mammary glands
hair
amniotic egg
movement onto land
jawed fish
ancestral vertebrates

acorn worms
sea squirts
amphioxus
vertebrates

invertebrate ancestors

FIGURE 19.14 How Terrestrial Vertebrates Evolved A tree of life for lobe-finned fish, amphibians, reptiles, mammals, and birds.

FIGURE 19.15 **Transitional Animal** An artist's conception of *Ichthyostega*, one of the earliest-known amphibians, seen at upper left climbing on the log and at center. Sometimes thought of as a "four-legged fish," *Ichthyostega* spent most of its time in water and probably used its back legs mostly for paddling. All land creatures with backbones, including human beings, ultimately trace their ancestry to tetrapod fishes such as *Ichthyostega*.

FIGURE 19.16 **Primate Characteristics** These modern-day lemurs in Madagascar exhibit several characteristics common to most primates: Opposable digits that enable grasping, front-facing eyes that allow binocular vision, and a tree-dwelling existence.

Amphibians and Reptiles   In FIGURE 19.15, you can see an artist's rendering of *Ichthyostega*, one of the earliest-known amphibians, which has a tail much like that of its lobe-finned ancestors. *Amphibian* literally means "double life," which is appropriate because in these creatures we can see the pull of both the watery world from which they came and the land onto which they moved. Many species of modern-day frogs spend their adult lives as air-breathing land dwellers, but must return to the water to reproduce. The young they give rise to then live as swimming tadpoles, complete with gills and a tail, before developing into land-dwelling adults. A double life indeed.

A Critical Reptilian Innovation: The Amniotic Egg   In the transition from amphibian to *reptile*, there is a severing of the amphibian ties to the water through a new kind of protection for the unborn. Amphibian eggs require a watery environment; lacking any sort of outer shell, they dry out if taken out of the water, thus killing the embryo inside. The early reptiles solved this problem with the remarkable adaptation of the **amniotic egg**: an egg that has not only a hard outer casing, but an inner padding in the form of egg "whites" and a series of membranes around the growing embryo. These membranes help supply nutrients and get rid of waste for the embryonic reptile. With this hardy egg, the tie to the water had been broken; reptiles could move *inland*.

One line of reptiles evolved, about 220 Mya, into the most gigantic creatures ever to walk the Earth—the dinosaurs. For 155 million years, no other land creatures dared challenge them for food or habitat. Not all dinosaurs were huge, and many did not eat meat. Yet even a leaf eater such as *Apatosaurus* must have commanded a good deal of respect, because it measured more than 85 feet from head to tail and weighed more than 30 tons. By comparison, a modern African elephant may weigh about 7 tons.

From Reptiles to Mammals   Scurrying about in the underbrush when the dinosaurs reigned supreme were several species of insect-eating animals, few of them bigger than a rat. These animals fed their young on milk derived from special female mammary glands, however. In addition, they probably possessed fur coats over their skins—an important feature for animals restricted to feeding at night, when the dinosaurs were less active. These were the mammals, the first of which appeared about 210 million years ago—about 10 million years after the dinosaurs arose. In an example of being "born at the wrong time," these small mammals then lived in the shadow of the reptilian behemoths for more than 100 million years. Toward the end of the Cretaceous period, they began to radiate out into more environments and consequently evolved into more forms. At that point fate intervened: The asteroid that helped bring the reign of the dinosaurs to a crashing close 65 Mya also brought the mammals out of hiding. They radiated into niches far and wide, becoming the largest form of land animal, a status they retain to this day. Some of their members even returned to the sea, as with seals and whales. (Who is the whale's closest living relative? It turns out to be the hippopotamus.) An obvious question here is: Why did the mammals survive the Cretaceous Extinction while the dinosaurs died out? Alas, we have no clear answers. Pure size may have had something to do with it; small creatures seem to survive extinction events better than large ones.

The Primate Mammals

A species of mammals gave rise to the order of mammals called *primates*, though exactly when this happened is not clear. The oldest primate fossils we have date from 55 Mya, but it will not surprise you to learn that molecular work places their origins much further back—at about 90 Mya. If you look at FIGURE 19.16, you can see, in the lemur of Madagascar, a modern-day descendant of these earliest primates. Three things characteristic of most primates are apparent in this picture: large, front-facing eyes that allow for binocular vision (which enhances depth

FIGURE 19.17   **How Primates May Have Evolved**

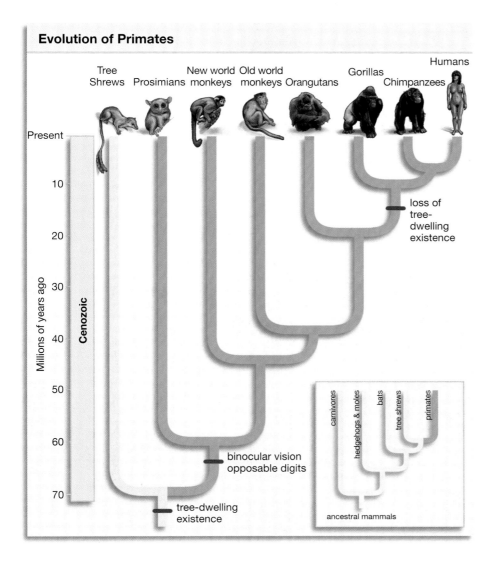

Evolution of Primates

perception); limbs that have an opposable first digit, like our thumb (which makes grasping possible); and a tree-dwelling existence. FIGURE 19.17 sets forth a possible primate family tree.

So Far...

1. All plants are thought to have evolved from water-dwelling _____.

2. Amphibians evolved from _____, and in turn gave rise to _____, who gave rise to _____.

3. The amniotic egg that came about in reptile evolution was important because it could be laid _____.

19.7 The Evolution of Human Beings

If you look to the upper right in Figure 19.17, you can see a fork that leads to chimpanzees on the one hand and human beings on the other. Down at the branching point of this fork is a presumed "common primate ancestor"—a single species that gave rise to both the chimpanzee family tree and the human family tree. By working backward through a set of primate molecular clocks, geneticists have determined

So Far... Answers:

1. *algae*

2. *lobe-finned fish; reptiles; mammals and birds*

3. *out of water*

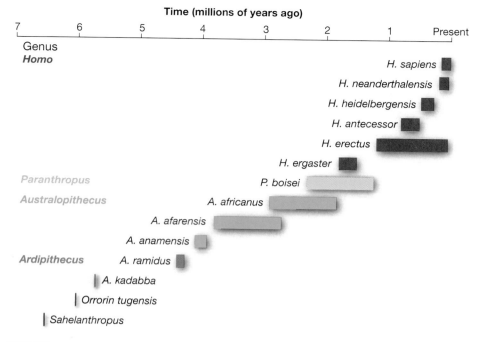

Time (millions of years ago)

Genus
Homo

H. sapiens
H. neanderthalensis
H. heidelbergensis
H. antecessor
H. erectus
H. ergaster

Paranthropus
P. boisei

Australopithecus
A. africanus
A. afarensis
A. anamensis

Ardipithecus
A. ramidus
A. kadabba
Orrorin tugensis
Sahelanthropus

FIGURE 19.18 **Selected Members of Hominini, the Human-like Primates**

that this branching probably occurred between 6 and 7 million years ago.

This split, of course, raises the question: Who was the common primate ancestor? We know that it was not the equivalent of a modern-day chimpanzee; today's chimpanzees did not arise until about the same time modern humans did. All we can say is that it was an ape-like creature whose closest living relatives include not only ourselves and chimpanzees, but the chimp-like primates known as bonobos.

Figure 19.17 might leave the impression that it only took a single speciation event to go straight from the primate ancestor to modern human beings. But the line extending up to humans is something like a satellite image of a river—it shows only the river's general outline, instead of its small-scale turns. To give you some idea of the number of branches in the human family tree, look at FIGURE 19.18. You'll notice that this is not a family tree per se, in that it lacks lines of descent—lines that tell us who evolved from whom. Instead, it's a partial inventory of the human-like species discovered so far, with the dates corresponding to the time in which they lived, as indicated by their fossils. It's possible to construct a plausible tree from these individuals; indeed, a number of these trees have been drawn up. But for purposes of our own ancestry, the only aspect of such trees that seems agreed upon by all experts is that the *Australopithecus afarensis* species (which you can see in green toward the middle of the figure) gave rise to all the species of our own genus, *Homo*, including ourselves. (These species are color-coded red in the figure). Other than this, there is great disagreement about the line of human descent, both in the early and late time frames covered in the figure.

Despite this lack of agreement, we still know a great deal about the human family tree. All the species in Figure 19.18 are members of a taxonomic grouping called Hominini, which is to say the group of human-like primates. Every species you see in Figure 19.18 is a hominin, including human beings.

What are some of the notable features of the hominin family tree? Well, for one thing, every species in it is extinct except for our own. Numerous species arose in this tree over the past 6 or 7 million years, but all of them are gone except us. If you could take a time machine back 50,000 years, you would find, in Africa and elsewhere, fully modern human beings, which is to say hominins who physically are indistinguishable from us. But in Europe you would also find stocky Neanderthals (species *Homo neanderthalensis*), and over in the Far East you would find the dwindling remnants of a once far-flung species, *Homo erectus*. Jump forward another 10,000 years and all the *H. erectus* are gone; jump forward another 13,000 and all the Neanderthals are gone. Today, only we *Homo sapiens* remain among the hominins.

Another way to think of this tree is in terms of relative time frames. An "ancient" civilization, such as that of Egypt, stretches back perhaps 5,000 years. Yet the Neanderthals disappeared from the face of the Earth 22,000 years before Egyptian civilization began, and they had been in existence for 200,000 years. Even so, the Neanderthals were latecomers among the hominins. Recall that the hominin line stretches back as far as 7 million years. Thus, human evolution took a great deal of time, even when dated just from the last primate ancestor. On the other hand, when we consider the entire sweep of evolution, all hominins are latecomers. If we assume that life on Earth began 3.8 billion years ago, and think of all evolution as taking place in one 24-hour day, then the most ancient of the hominins evolved in the last $2\frac{1}{2}$ minutes of that day. And modern human beings? We arose at most 200,000 years ago, which means we have only been around for the last 5 seconds of Earth's evolutionary day.

Yet another way to think of human evolution is in terms of *where* it took place. The message here is quite straightforward: For perhaps 70 percent of the time we hominins have been around, we lived and died solely in Africa. This is clear because the earliest hominin fossils found *outside* of Africa date to only 1.75 Mya. Meanwhile, all the early and mid-period hominin fossils have been unearthed on the African continent, as you can see in FIGURE 19.19.

As little as 10 years ago, a finer point could have been put on this: It was not just in Africa, but in *East* Africa that all hominin evolution took place. Until the mid-1990s, all early- and middle-period hominin fossils were found in a swath of Africa that runs from the eastern portion of South Africa up through the Great Rift Valley that traverses modern Kenya, Tanzania, and Ethiopia. But the range of

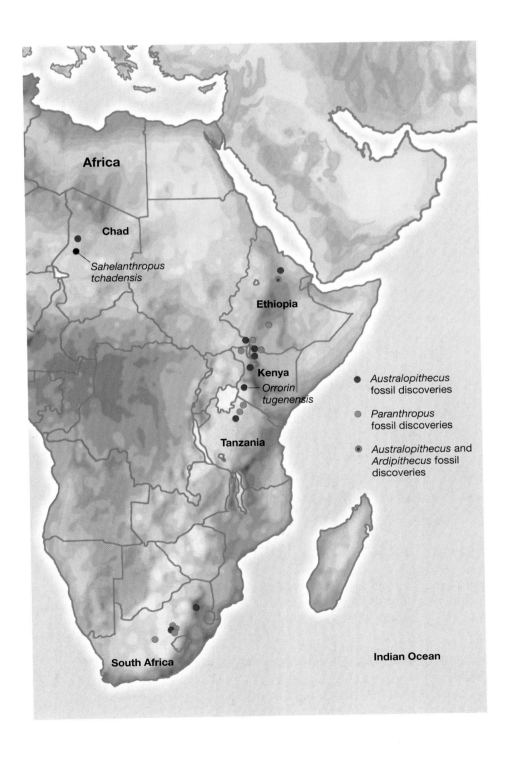

FIGURE 19.19   Hominin Fossil Sites in Africa All of the oldest hominin fossils have been found on the African continent.

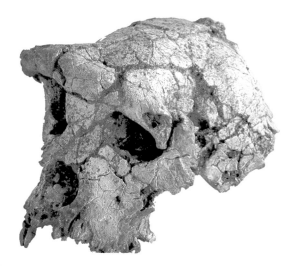

FIGURE 19.20 **An Ancient Human Ancestor?** This cranium of *Sahelanthropus tchadensis*, or Toumaï, was discovered in the deserts of Chad, and has been determined to be 6 or 7 million years old. Toumaï's discoverers believe it is a very ancient hominin, but other researchers believe it should be grouped with apes.

likely hominin fossils has now been expanded some 2,500 kilometers west of the Great Rift Valley, to Chad. There, in 2002, Michel Brunet and his colleagues announced that they had found a very primitive hominin, a creature they dubbed *Sahelanthropus tchadensis* but who they informally refer to as Toumaï (**see** FIGURE 19.20).

The Toumaï fossils actually were remarkable in another way. They capped a period of two years in which the root of the hominin family tree was pushed back by more than 2 million years. Prior to 2000, the oldest hominin fossil was dated at 4.4 million years old. Then, in rapid succession, fossil discoveries for all three of the earliest species you can see in Figure 19.18 were announced by different teams of researchers. With the Toumaï find, hominin fossils were pushed back to between 6 and 7 Mya. As you may recall, the human and chimpanzee lines only split at about 7 Mya at most. One possibility, then, is that Brunet's Toumaï is a creature that lies right at the base of the hominin family tree. Another possibility, however, is that Toumaï is not a hominin, but belongs in another line of primate ancestors, specifically the apes.

Snapshots from the Past

It shouldn't surprise us that a controversy exists over whether Toumaï belongs in the hominin or the ape family tree. After all, Toumaï lived right about the time that the hominin and ape lines were parting company. If we looked at each of the species noted in Figure 19.18, moving from most ancient to most recent, what we would see is a transition that reflects this split—a transition from creatures that are very much like apes to those that look very much like ourselves. We don't need to examine all the species noted in the figure to get a sense of this transition, however. Here are three snapshots from the past that will provide some idea of the nature of our hominin relatives over time. An artist's interpretation of the appearance of three hominin species—two of which are featured in our snapshots—can be seen in FIGURE 19.21.

(a) *Paranthropus boisei*

(b) *Homo ergaster*

(c) *Homo neanderthalensis*

FIGURE 19.21 **Three Extinct Hominins** What did now-extinct members of the hominin grouping look like? Working from fossil and other evidence, artist Jay Matternes has produced drawings that provide some idea of their appearance.

(a) *Paranthropus boisei* lived about 2 million years ago. Though it was not a member of our genus (*Homo*), and not an ancestor of modern humans, it was a hominin.

(b) *Homo ergaster*, who lived about 1.6 million years ago, clearly has a form more human-like than that of *P. boisei*—not surprising, since *H. ergaster* probably was a human ancestor. This is the species represented by the Turkana Boy fossils.

(c) *Homo neanderthalensis* lived in Europe and elsewhere in a period running from about 200,000 years ago to 27,000 years ago. These were the Neanderthals, a species that lived in proximity to modern humans for thousands of years before becoming extinct.

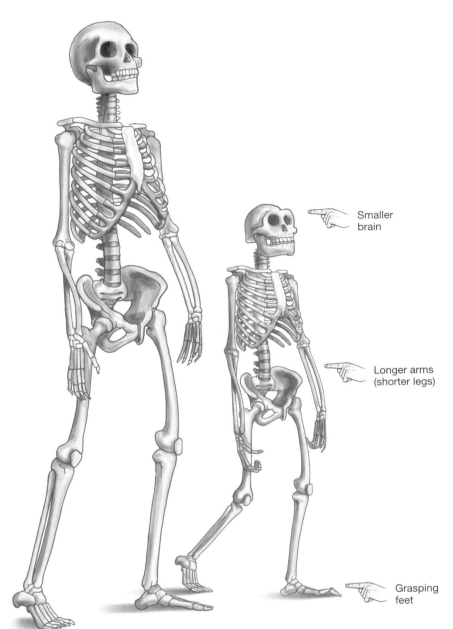

FIGURE 19.22 **"Lucy" and Modern Humans Compared** What was the hominin Lucy like compared to a modern human female? The figure gives an idea. Lucy stood about three-foot-seven and had a much smaller brain than modern humans, even allowing for her smaller stature. Lucy's hip and pelvic bones make clear that she was bipedal, but note the longer arms and grasping feet common to tree-dwelling primates. Lucy's combination of features have prompted some researchers to label her a "bipedal chimpanzee." (Adapted from "Lucy's skeleton compared to modern human" by Owen Lovejoy)

Smaller
brain

Longer arms
(shorter legs)

Grasping
feet

Australopithecus afarensis: **Lucy and Her Kin**   If we jump forward 3 million years from Toumaï—to about 3.5 Mya—we find a hominin whose features illustrate the transition from ape-like to human-like. The species, *Australopithecus afarensis*, is best known through its most famous individual, dubbed Lucy, whose fossilized remains were discovered in Ethiopia in the 1970s. In Lucy, we can see the full development of a hallmark of every hominin: bipedalism, which is to say walking on two legs rather than four. We know Lucy was a biped because of the structure of her pelvis. Nevertheless, she possessed long arms, short legs, and feet that were built for grasping—features consistent with a species that no longer lived exclusively in the trees, but that probably still spent considerable time in them. Lucy also differed from us in the size of her brain; she had a cranial volume of about 450 cubic centimeters—about that of a chimpanzee—while ours is about 1,400 cubic centimeters. To get some idea of Lucy's anatomy compared with ours, see **FIGURE 19.22**. As noted earlier, Lucy's species is generally agreed to have given rise to our own genus (*Homo*). In Lucy, therefore, we are seeing not just a relative, but an ancestor.

A BROADER VIEW

Why Stand Upright?

Why did we hominins start standing upright? One of the ideas proposed is that it freed us to move about while carrying objects (such as the kill from a hunt). Another is that an upright posture minimized sun exposure. Consistent with this, note that, though we have lost most of the hair our primate relatives possess, we have retained hair on one part of our bodies—the tops of our heads, which would have continued to receive plentiful sun once we started standing upright.

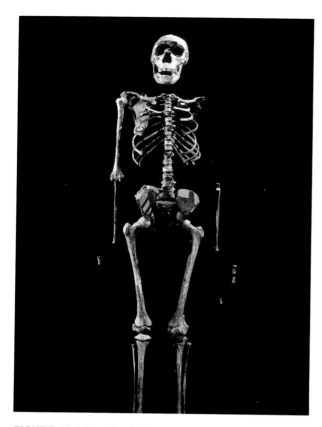

FIGURE 19.23 **Much More Like Us** In the remarkably complete skeleton of "Turkana Boy," we can see the evolution of hominins to a form more like our own. This member of the species *Homo ergaster* was tall and had a much larger brain capacity than the earlier hominin "Lucy." His remains, dating from 1.6 million years ago, were found near Kenya's Lake Turkana.

A BROADER VIEW

Neanderthals in Novels

You've heard of historical novels. What about prehistorical novels? Beginning in 1980, Jean M. Auel wrote an "Earth's Children" series whose most popular book, *The Clan of the Cave Bear*, imagined an orphaned *Homo sapiens* girl, Ayla, being raised by a Neanderthal clan. Robert Sawyer's 2002 novel *Hominids* also imagined life in a Neanderthal group, though his group lived in a universe parallel to our own.

Homo ergaster: More Like Us

Changes to physical forms that are more like ours come at several steps in the hominin line, but the most dramatic change comes with the rise of *Homo ergaster*, whose best-preserved remains come from a boy who died 1.6 million years ago on the shores of Lake Turkana in Kenya. Experts estimate that, had this 9-year-old "Turkana Boy" grown to maturity, he would have reached a height of 6 feet. His brain was more than half the size of the average modern *Homo sapiens* brain. He had a much more modern face, long limbs typical of those humans who dwell today in arid climates in Africa, and advanced tool technology. If you look at FIGURE 19.23, you can see Turkana Boy's skeleton, which was found in amazingly complete form.

Homo ergaster is regarded by many researchers as being ancestral not only to our own species, but to another species—*Homo erectus*—that was one of the earliest to migrate out of Africa, beginning at least 1.6 million years ago. *H. erectus* ended up very far from Africa indeed, as its remains have been found in both China and the Indonesian island of Java, where the last remaining *H. erectus* individuals died out about 40,000 years ago. Keep *H. erectus* and its extinction date in mind, as they will be important for what's coming up.

The Neanderthals: Relatives, Not Ancestors

Of all the lost members of the hominin family, it is the Neanderthals that have most strongly captured the modern human imagination. This is so partly because the Neanderthals were the first extinct hominins whose fossils were discovered. When Neanderthal remains were uncovered in 1856, at a limestone quarry in Germany's Neander Valley, humanity got a wake-up call about human evolution. Though the notion was resisted at first, the Neanderthal bones made clear that there once had been a species on Earth that was very much like humans, but that was not *exactly* human. In the years that followed, the Neanderthals became the best known of the extinct hominins, simply because they left so many traces of themselves behind. In Europe alone, there are perhaps 20 major Neanderthal excavation sites, and others have been found as far away as Uzbekistan. Note that the Neanderthals are very recent hominins. The first evidence we have of them dates from only 200,000 years ago, which is recent indeed compared to the 1.6 million years ago when *H. ergaster* flourished, to say nothing of the 3.5 Mya when *A. afarensis* lived.

Critically, the Neanderthals (*Homo neanderthalensis*) were not our ancestors, but instead were more like distant cousins who lived side-by-side with us for an extended period of time. We and the Neanderthals may have shared a common ancestor in *Homo ergaster*. But recent DNA sequencing work has convinced most scholars that Neanderthals and modern human beings did not interbreed, even during thousands of years when they lived in close proximity. Hence, modern humans cannot be the descendants of Neanderthals.

So, what were the Neanderthals like? Figure 19.21c shows you an artist's interpretation of their physical stature: short, stocky bodies—the average man was about five-foot-six—powerfully built, with a heavy "double arched" brow and a receding chin; big-boned, with large joints to match. It's easy to see how such features were turned into a *caricature* that has been with us since the nineteenth century: that of the caveman. It is the Neanderthals who provided us with the image of dumb, prehuman brutes, grunting their way through the Stone Age. But does this image fit with the reality? Well, consider that the Neanderthals had a cranial capacity slightly larger than ours; that some put up shelters in their campsites; and that one reason we have so many artifacts from them is that they took the trouble to bury their dead.

Modern *Homo sapiens*

The Neanderthals arrived in Europe about 200,000 years ago, but beginning about 40,000 years ago they had hominin company in the form of ourselves—modern human beings (*Homo sapiens*). Where had we come from? The short answer is Africa. DNA sequencing has convinced most scholars that we humans evolved into our modern anatomical form in Africa before ever migrating to Europe and the rest of the world. This same kind of molecular testing has yielded approximate dates for the first appearance of modern human beings—between 100,000 and 200,000 years ago, with 160,000 years ago being a best estimate.

Who Lives, Who Doesn't?

Now let's think about extinction and survival among three of the hominin species we've looked at. The arrival of modern humans in Europe 40,000 years ago was followed by the extinction of the Neanderthals 13,000 years later. Meanwhile, it's clear that modern humans had reached Australia and New Guinea (via Indonesia) by 46,000 years ago. In connection with this, do you remember the extinction date for the *Homo erectus* species that made its way to China and Indonesia? The last of them died out in Indonesia about 40,000 years ago. One obvious interpretation of these dates is that modern humans fanned out across the globe, from 100,000 to 50,000 years ago, and proceeded to "replace" both the Neanderthals and *H. erectus*. This is not to say that we simply killed them outright. We have no evidence of violent encounters between ourselves and the Neanderthals or *H. erectus*. But similar species are bound to compete for the same kinds of resources, such as food and shelter. And we have every reason to believe that modern humans could *outcompete* both the Neanderthals and *H. erectus*, thanks to our superior brain power. (Neanderthals thrust spears at prey, but modern humans developed the valuable technique of throwing spears from a distance. Neanderthals had relatively primitive tools, while *Homo sapiens* constructed much finer implements.) So, one possibility is that the hominin family is now a family of one—ourselves—in part because our success brought about the death of our near relatives.

Next-to-Last Standing: *Homo floresiensis*

In 2004, a fascinating twist was added to this story. Researchers working on the Indonesian island of Flores reported finding skeletons of a hominin species whose members stood just three feet tall and who lived on Flores as recently as 18,000 years ago (see FIGURE 19.24). This discovery was stunning for three reasons. First, anthropologists have long believed that the last hominins other than ourselves died out not 18,000 years ago, but 27,000 years ago, with the passing of the Neanderthals. Second, if modern humans really were spanning the globe replacing other hominin species 50,000 years ago, how did the Flores people, dubbed *Homo floresiensis*, manage to avoid this fate for some 30,000 years? Third, the Flores people used fire and had sophisticated tools. Yet they had a cranial capacity of only 380 cubic centimeters—less than that of a chimpanzee! How could they do so much in the way of culture with so little in the way of brain capacity?

The answers to these questions may be a while in coming. One possibility is that much of the hoopla about *H. floresiensis* has been undeserved. Some researchers think it likely that *H. floresiensis* represents not a new hominin species, but a modern human who had a medical condition—microcephaly, which results in a very small head and brain. The researchers who made the Flores discovery rejected this assertion, however, and a later examination of the *H. floresiensis* fossils seemed to rule this possibility out. Indeed, the examination suggested that *H. floresiensis* had a brain structure that would allow it to do more with less—that would allow it to build sophisticated tools, for example, even with a very small brain.

Assuming that these Flores "Hobbit people" really are a new species, how do they fit into the greater hominin family? They appear to be a later-evolving offshoot of *Homo erectus*, the species that migrated from Africa 1.6 million years ago and ended up in China and Indonesia. The supposition is that a group of *H. erectus* migrated along the Indonesian archipelago and made a crossing to the island of Flores where, in isolation, they evolved into *H. floresiensis*. (Sound familiar? This would be allopatric speciation in action.) If this account is true, then the Flores people are in some ways like the Neanderthals: They are not our ancestors but instead are relatives of ours who lived at the same time as our ancestors. Their small stature can be explained by a phenomenon seen elsewhere in the animal kingdom: "island dwarfism," in which large mammal species who end up isolated on islands evolve into small sizes, perhaps because of inbreeding and the "resource poverty" of their environment. More research will hopefully sort out the truth about the Flores people. Even with what we know now, however, they represent a fascinating addition to the story of human evolution.

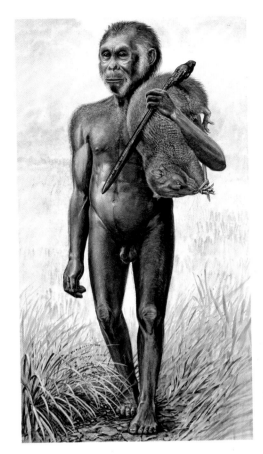

FIGURE 19.24 **Small Hominin, Big Discovery** An artist's interpretation of the appearance of *Homo floresiensis*, the three-foot-tall hominin whose skeletal remains were found on the Indonesian island of Flores.

A BROADER VIEW

Last of the Hominins

A classic moment in old Western movies comes when one cowboy says to another: "This town ain't big enough for the both of us." Is it possible that this world wasn't big enough for two hominin species? Given the similar resources that we and other hominins were bound to compete for, was it inevitable that only one species of our type would survive? Or could two or more hominin species have survived, providing that one of them was not us?

So Far... Answers:

1. *6 and 7 million; chimpanzee*
2. *Africa*
3. *bipedalism; on two legs rather than four*

On to the Diversity of Life

In reviewing the history of life in this chapter, you necessarily looked, in some small way, at almost all of Earth's major life-forms—bacteria, animals, plants, and so forth. In the three chapters coming up, you'll take a more detailed look at each of these life-forms. This is a tour that will make plain the astounding diversity that exists in the living world.

So Far...

1. The taxonomic group of human-like primates, the hominins, is thought to have originated between _____ years ago with a common primate ancestor who gave rise to both the human and _____ family trees.

2. In what part of the world did all early and mid-period hominin evolution take place? _____.

3. All hominins share the trait of _____, which is walking _____.

Chapter Review

Summary

19.1 Measuring the Past

- **Life's timeline**—Earth's 4.6 billion years of history are measured in the geological timescale, which is divided into broad eras and shorter periods. These time frames were defined mostly in the nineteenth century, based in part on the different life-forms that existed within each of them, as reflected in the fossil record. Many of the major transitions in the fossil record have come about because of extinction events. (page 308)

- **Starting small**—Life is thought to have begun on Earth about 3.8 billion years ago; from that time until perhaps 2 billion years ago, it was strictly microscopic and single-celled. According to the fossil record, all the animals and plants that exist today came about in the last 16 percent of this 3.8-billion-year period. (page 310)

- **The shape of evolution's tree**—Evolution can be conceptualized as a branching tree, with humans sitting on one ordinary branch. It is not shaped as a pyramid with humans at the top. The story of evolution recounted in the chapter leads to human beings, but it could just as easily have led to any other modern organism. (page 310)

19.2 How Did Life Begin?

- **Where did life begin?** There currently is a debate regarding not only the kind of environment in which life is likely to have arisen but also how this environment could have accumulated the necessary chemical raw materials to allow the construction of life's complex organic molecules. One hypothesis is that life began in a "prebiotic soup" of hot-water systems—in thermal deep-sea vents or in hot-spring pools. (page 311)

- **Self-replication**—There is agreement that life began with the origin of self-replicating molecules. Through mistakes in replication, such molecules will begin to differ, meaning that some will leave more offspring than others. This is the beginning of Darwinian evolution and the diversification of living things. (page 312)

- **The tree of life**—Under one hypothesis, life today can be traced to a universal ancestor that gave rise to the domains Archaea, Bacteria, and Eukarya. The earliest kingdom in Eukarya, the protist kingdom, went on to give rise to the other kingdoms in Eukarya: animals, plants, and fungi. Some scientists believe that a community of gene-sharing organisms, rather than a single common ancestor, gave rise to life's three domains. (page 313)

Web Tutorial 19.1 The Tree of Life

19.3 A Long First Era: The Precambrian

- **The earliest life**—The earliest evidence we have for life is a chemical signature of it, carbon utilization, found in rocks in Greenland dated to about 3.8 billion years ago. Carbon signatures indicating photosynthesis (and hence life) are abundant beginning 3.5 billion years ago. All of this life was either bacterial or archaeal, and all of it existed in the oceans. (page 314)

- **Precambrian events**—Oxygen came to exist in the Earth's atmosphere through the activity of photosynthesizing organisms, originally photosynthesizing bacteria. Photosynthesis meant an abundance of food on the Earth. However, atmospheric oxygen was toxic to many existing species. Some early eukaryotes adapted to oxygen by establishing a union with bacterial cells that were able to metabolize oxygen. Algal cells struck up a similar relationship with other bacterial cells that could perform photosynthesis. (page 314)

19.4 The Cambrian Explosion

- **A real explosion**? The fossil record indicates a tremendous, rapid expansion in the number of animal forms in a "Cambrian Explosion" that began about 542 Mya, but this evidence has been challenged. It may be that the Cambrian Explosion was merely an explosion in the number of animal forms big and hard enough to leave fossils, and that the surge in animal diversity took place over a relatively long period of time beginning well back in the Precambrian era. (page 315)

19.5 The Movement onto the Land: Plants First

- **Plants and fungi**—Plants, which evolved from green algae, made a gradual transition to land in tandem with their partners, the fungi. Plant-fungi fossils exist from about 460 Mya, but molecular studies indicate they were on land 1.3 billion years ago. A major transition in plant life came with the development of a "vascular" or fluid-transport system.   (page 316)

- **Later-evolving plants**—Later plants went on to develop seeds, which can be thought of as packages containing an embryo and food for it, encased in a tough outer covering. The first seed plants were the grouping called the gymnosperms, represented today by conifers. The flowering plants, called angiosperms, developed about 165 Mya and eventually succeeded the gymnosperms as the most dominant plants on Earth. Today, angiosperms include many food crops, cactus, and tree varieties.   (page 318)

19.6 Animals Follow Plants onto the Land

- **Arthropods onto land**—The first land animals were arthropods; a centipede-like creature laid down the oldest terrestrial animal tracks we know of. The arthropods were the only land animals for millions of years thereafter.   (page 318)

- **Vertebrates onto land**—Four-limbed vertebrates (tetrapods) moved onto land in the form of lobe-finned fishes, which gave rise to amphibians (represented by today's frogs and salamanders). Reptiles later branched off from amphibians, and mammals branched off from reptiles. Birds are the living descendents of the reptile branch that included the dinosaurs.   (page 318)

- **Amphibians and reptiles**—Amphibians inhabit two worlds, water and land, in keeping with their evolutionary transition from one to the other. Reptiles evolved a protective amniotic egg that allowed their offspring to develop away from water, freeing reptiles to migrate inland.   (page 320)

- **The rise of mammals**—Mammals appear as a group of small, insect-eating animals about 210 million years ago. They began to evolve into the varied forms seen today shortly before the reign of the dinosaurs was ended with an asteroid impact 65 million years ago; after this they radiated into many niches, becoming the largest land animals.   (page 320)

19.7 The Evolution of Human Beings

- **Humans and chimpanzees**—A common primate ancestor is believed to have given rise to both the chimpanzee and the human family trees between 6 and 7 million years ago. The structure of the human family tree is a matter of considerable debate among researchers.   (page 321)

- **The hominin family tree**—Human evolution is the study of the taxonomic grouping called hominins or human-like primates. Every member of this grouping is referred to as a hominin, including human beings. All the members of the hominin grouping are extinct except for *Homo sapiens*, the human species. All early- and mid-period hominin evolution took place in Africa.   (page 322)

- *Australopithecus* **and** *Homo ergaster*—Bipedalism is clearly seen in the hominin genus *Australopithecus*, which is thought to be ancestral to our own genus, *Homo*. The most famous Australopithecine is "Lucy," whose remains were discovered in Ethiopia in the 1970s. A change to a physical form and mental capacity much closer to ours comes with the evolution of *Homo ergaster*, exemplified by "Turkana Boy," who lived 1.6 Mya.   (page 325)

- **Neanderthals**—The best known of the extinct hominins are the Neanderthals (*Homo neanderthalensis*), who populated Europe as well as parts of Asia and North Africa for nearly 200,000 years. The last of them died 27,000 years ago. Modern humans may share a common ancestor with the Neanderthals (*Homo ergaster*), but they appear not to have interbred with them.   (page 326)

- **Modern humans**—Modern human beings fully evolved in Africa between 200,000 and 100,000 years ago, with a best-estimate date being 160,000 years ago. Their arrival in Europe 40,000 years ago was followed by the extinction of the Neanderthals 13,000 years later, while their arrival in Australia and New Guinea 46,000 years ago was followed by the extinction of the *Homo ergaster* species, 6,000 years later. It is commonly assumed that modern humans "replaced" both species, perhaps by outcompeting them.   (page 326)

- *Homo floresiensis*—In 2004, researchers reported finding a species of a previously unknown hominin, *Homo floresiensis*, who stood only three feet tall and apparently survived on the Indonesian island of Flores until at least 18,000 years ago.   (page 327)

Key Terms

amniotic egg p. 320	**Cretaceous Extinction** p. 310	**Permian Extinction** p. 310	**phylum** p. 315
Cambrian Explosion p. 315			

Testing Your Understanding

In Your Own Words (answers in the back of the book)

1. What is the amniotic egg, and why was it a breakthrough for terrestrial animals? In which vertebrate class did the amniotic egg evolve?

2. What are two physical characteristics of primates that are not found in other mammals? Why are they important for the evolution of hominins?

3. Were Neanderthals the ancestors of modern humans?

Thinking about What You've Learned

1. How would Earth be different if photosynthesis had never developed in any organism?

2. Why does it make sense that life started so small?

3. Suppose an isolated population of Neanderthals had survived into modern times. How do you think society would treat them? As a species more akin to chimpanzees, which is to say subject to being kept in zoos? Or as regular, if somewhat different, human beings?

Microbes:
The Diversity of Life 1

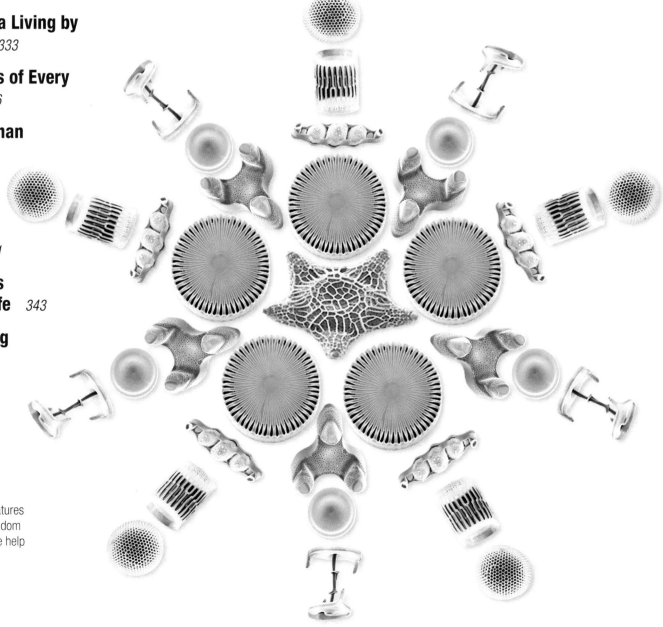

Diatoms are single-celled, aquatic creatures who are part of a mostly microbial kingdom in the living world, the protists. With the help of a microscope, these diatoms were arranged into this artistic array.

LONG BEACH PRESS-TELEGRAM

STAPH GETTING RESISTANT

By Linda A. Johnson, Associated Press
September 30, 2004

Flesh-eating bacteria cases, fatal pneumonia and life-threatening heart infections suddenly are popping up around the country, striking healthy people and stunning their doctors.

The cause? Staph, bacteria better known for causing skin boils easily treated with standard antibiotic pills. No more, say infectious disease experts, who increasingly are seeing these "super bugs"—strains of Staphylococcus aureus unfazed by the entire penicillin family and other first-line drugs.

THE HARTFORD COURANT

BACTERIA PLAY VITAL ROLE IN HEALTHY INTESTINES, YALE STUDY SHOWS

By William Hathaway, Courant Staff Writer
July 27, 2004

Scientists at Yale University School of Medicine have discovered new evidence that millions of bacteria in the gut are indispensable allies in keeping our intestines healthy.

The good gut bacteria, known as commensal bacteria, activate proteins called Toll-like receptors, or TLRs, that maintain the health of intestinal cells and help them respond to injury, according to a study published in the current issue of the journal *Cell*. Previously, TLRs were thought to only direct immune system response to dangerous bacteria.

20.1 Life's Categories and the Importance of Microbes

Well, which is it for bacteria: human flesh-eaters or human digestive partners? To judge by the newspaper stories above, the answer is "both." If we ask, however, which role any of us is *likely* to see bacteria playing in our lives, then flesh-eater isn't even in the running. The odds of being attacked by flesh-eating bacteria are probably about those of being struck by lightning. Meanwhile, the "good" bacteria

The White Cliffs of Dover are the limestone remains of countless generations of an ocean-dwelling microbe called a foram.

noted above operate every day in almost all of us. We're in trouble only if we *don't* have these bacteria living in our digestive tract. Still, the exceptional bacteria deserve our attention because they are very dangerous indeed. The staph bacteria mentioned in the story are like an army we can defeat only in individual battles—that is, in individual patients—but that we can never hope to eradicate entirely. Worse yet, this is an army whose soldiers are constantly becoming immune to the weapons we use against them. (Imagine if human soldiers from a given country could evolve over the course of a few years so that they could walk right through clouds of poison gas, and you begin to get the idea.)

Interacting Forms of Life

The broader message from these examples is that, while human beings and bacteria are very different forms of life, they are *interacting* forms of life—relating to each other sometimes as combatants, sometimes as partners, often just as joint tenants of a common space. The purpose of this chapter is to give you some sense of the nature of bacteria and their fellow *microbes*: creatures so small they can't be seen with the naked eye.

Of course, the interactive web of life is much broader than just microbes and humans; it includes all the broad categories of living things (see FIGURE 20.1). Accordingly, in this chapter and the two that follow, we'll be looking at life as it exists across all its categories. Those readers who went through Chapter 19 know that all living things can be placed into one of three "domains." Two of these are Archaea and Bacteria, whose members are all single-celled and microscopic. The third domain, Eukarya, encompasses such incredible diversity that it is further divided into four separate "kingdoms." These are plants, animals, fungi, and a group of

(a) Mouse beneath the foot of an elephant

(b) Aquatic snails moving across kelp

(c) Fungus growing on a fallen log

(d) Water flea swimming near green algae

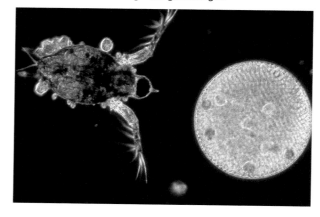

FIGURE 20.1 **An Amazing Diversity in the Living World**

mostly microscopic organisms called protists. We will begin our tour of the living world by starting small, looking in this chapter at Domains Bacteria and Archaea and at the protists within Domain Eukarya. Then in Chapter 21, we'll look at plants and fungi; and in Chapter 22 we'll review animals. FIGURE 20.2 shows you the evolutionary relationships among all these categories of life.

Microbes and Life

Because this chapter is devoted to mostly unfamiliar microbes, it might be helpful to say a word at the start about their significance. Microbes are something like the foundation of a house: seldom thought of but critically important. To the extent that they cross our minds at all, we are aware of them mostly for the diseases they cause. This is an important topic—one we'll go into extensively—but thinking of microbes strictly as disease-causing organisms is like thinking of cars strictly as wreck-causing machines. As noted earlier, bacteria, whose very name sounds creepy to us, are partners with us in our digestion of food. Beyond this, where does the oxygen we breathe come from? If you answered "plants," that's partly correct. But the photosynthesis carried out by plants supplies less than half our oxygen; the balance comes from microscopic algae and bacteria—mostly drifters on the ocean's surface. How about the nitrogen that is indispensable to all plants and animals? Among living things, only bacteria are capable of plucking nitrogen out of the air and turning it into a form that plants can use. If there were no bacteria, there would be no plants—and consequently no land animals. And how is it that dead tree branches or orange peels or the bones of animals are broken down and recycled into the Earth? Almost entirely through the work of bacteria and fungi. Without these organisms, Earth would long ago have become a garbage heap of dead, organic matter. Animals cannot decompose a dead tree branch, but bacteria and fungi can.

The upshot of all this is that life on Earth would grind to a halt without microbes; they are an essential underpinning to all forms of life. They come to this status in part because of the environments they live in, which is to say all environments on Earth that support life of any kind. Wherever a plant or animal can live, microbes can be found too. Just as impressive is the sheer number and tonnage of these tiny organisms. A common denominator for measuring the amount of living material is something called *biomass*, defined as the total dry weight of material produced by a given set of organisms. Each bacterium or microscopic alga has an infinitesimally small weight, of course—but then again, there are so many of them! It has been estimated that the number of bacteria living in your mouth right now is greater than the number of people who have ever lived, and there seem to be more bacterial cells in the human body than there are human cells. A typical gram of fertile soil—an amount about the size of a sugar cube—has about 100 million bacteria living in it, and a mere drop of water near the ocean's surface contains thousands of the tiny algae called phytoplankton. All this adds up. When the calculations are done, we find that microbes probably make up more than half of Earth's biomass. Put another way, the microbes of the world outweigh all the plants and animals of the world combined.

In sum, microbes matter, and not just in ways that are harmful. We'll now start to look at the portion of the living world that is made up of them. Our first stop on this tour, however, will be a category of tiny replicating entities that most biologists would say lie just outside of life, though they have a great effect on living things.

20.2 Viruses: Making a Living by Hijacking Cells

If a living thing is too small to be seen and can cause disease, most people would put it in the vague category known as "germs." But there is a fundamental distinction to be made between *two* kinds of infectious organisms. On the one hand there are bacteria, which may be very small but nevertheless are *cells*, complete with a

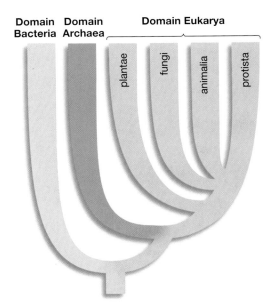

FIGURE 20.2 **The Tree of Life** Life is presumed to have developed from a single common ancestor (the base of the tree), which gave rise to the three domains of the living world: Bacteria, Archaea, and Eukarya. Domain Eukarya is then further divided into four "kingdoms," plants, fungi, animals, and protists.

A BROADER VIEW

Small but Tenacious

Who will inherit the Earth? Bet on microbes. We can imagine an environmental catastrophe—such as an asteroid striking the Earth—that would be severe enough to wipe out all large life-forms (plants, animals). But it's hard to imagine anything that could wipe out all microbial life short of something that turned Earth into a waterless cinder. Certainly, whatever catastrophe occurred would have to be a long-lasting one. In the 1990s bacterial spores were found *living* inside a bee that had been encased in amber for 25–40 million years.

(a) Human immunodeficiency virus (HIV)

(b) Life cycle of HIV

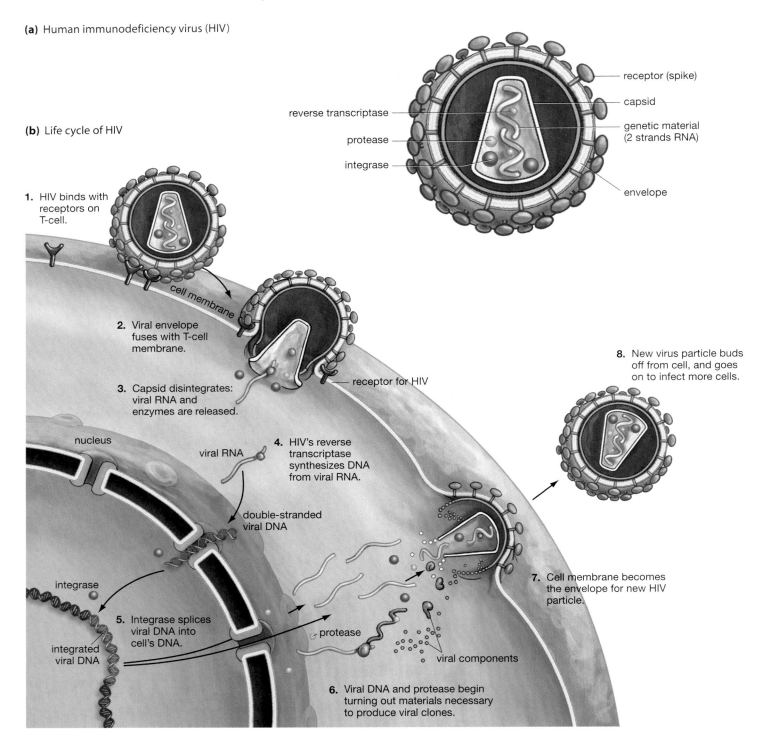

reverse transcriptase

protease

integrase

receptor (spike)

capsid

genetic material
(2 strands RNA)

envelope

1. HIV binds with receptors on T-cell.

cell membrane

2. Viral envelope fuses with T-cell membrane.

3. Capsid disintegrates: viral RNA and enzymes are released.

receptor for HIV

nucleus

viral RNA

4. HIV's reverse transcriptase synthesizes DNA from viral RNA.

double-stranded viral DNA

integrase

5. Integrase splices viral DNA into cell's DNA.

integrated viral DNA

protease

viral components

6. Viral DNA and protease begin turning out materials necessary to produce viral clones.

7. Cell membrane becomes the envelope for new HIV particle.

8. New virus particle buds off from cell, and goes on to infect more cells.

FIGURE 20.3   **The Virus That Causes AIDS**   The anatomy and life cycle of HIV, the human immunodeficiency virus.

protein-producing apparatus, a mechanism for extracting energy from the environment, a means of getting rid of waste—in short, complete with everything it takes to be a self-contained living thing. On the other hand there are viruses, which by themselves possess none of these features. Indeed, viruses can be likened to a thief who arrives at a factory he intends to rob possessing only two things: the tools to get inside and some software that will make the factory turn out items he can use. The factory being broken into is a living cell, and the software that the virus brings is its DNA or RNA, which it puts inside this "host" cell. Once this is accomplished, viruses employ different tactics, as you'll see, but the common result is that viruses make more copies of themselves. Viruses are an integral part of the living world, and not all of them cause harm. But unlike the case with bacteria, it's right to think of them first and foremost in terms of the diseases they cause, not only in humans, but in all kinds of living things. **Viruses** can be defined as noncellular, replicating entities that must invade living cells to carry out their replication.

HIV: The AIDS Virus

If you look at FIGURE 20.3a, you can see a simplified rendering of a particular virus, the human immunodeficiency virus (HIV), which is the cause of AIDS. Almost all viruses have two of the three large-scale structures you can see in HIV: the genetic material at the core (in HIV's case, two strands of RNA), and a protein coat, called a *capsid*, that surrounds the viral genetic material. Many viruses also have a third major element—a fatty membrane, called an *envelope*, which you can see surrounding the HIV capsid. Protruding from the HIV envelope is a series of receptors, often referred to as spikes, which are proteins capped with carbohydrate chains. These serve the critical role of binding with receptors that protrude from the target cell, thus giving the virus a way to get in.

If you look at FIGURE 20.3b, you can see how the life cycle of HIV proceeds from this initial binding. The target cell in this case is the one most commonly invaded by HIV, an immune-system cell called a helper T-cell. As you can see, once HIV binds with two receptors on the T-cell's surface, the viral envelope fuses with the T-cell membrane. With this, HIV's capsid, with the genetic material enclosed, is inside the cell. Once there, the capsid disintegrates. Now two HIV enzymes that had been enclosed in the capsid get to work. If you look at Figure 20.3a, you can see these enzymes, integrase and reverse transcriptase. The reverse transcriptase gets busy first, turning HIV's RNA strands into double-stranded DNA. Then integrase does just what its name implies: It integrates the viral DNA into the *cell's* DNA through a cut-and-paste operation.

At this point, the viral DNA might simply stay integrated in the T-cell's DNA, causing no great harm but getting copied, along with the cell's DNA, each time the cell divides. The effect of this, however, is that every "daughter" cell of this infected cell will be infected too; each new cell will have viral DNA spliced into its own DNA. Each of these cells thus has a time bomb ticking within it, because at some point the HIV DNA will change course. It will no longer simply go on replicating within the cell's DNA. Instead, it will start turning out the materials necessary to make whole new copies of the virus. As you can see in the figure, virus construction takes place just inside the cell's outer membrane (helped along by a third enzyme that HIV brought with it, protease). This location is important, because the cell's membrane ends up serving as the final structural part of each new HIV "particle," as the viral clones are called. When the capsid is completed, it buds off from the cell, taking with it part of the cell's membrane, which now becomes the capsid's envelope. When enough viral particles have budded off from the cell in this way, it dies. The viral particles that emerge from the cell then go on to infect other cells.

Viral Diversity

In the HIV replication cycle, you can see four steps that are common to most viruses: getting genetic material inside the host cell, turning out viral component parts, constructing new particles from these parts, and moving these particles out of the cell. HIV uses RNA as its genetic material, but many viruses use DNA. Viruses that lack an envelope don't get their genetic material inside the cell by fusing with the cell's membrane in the way HIV does. If you look at FIGURE 20.4, you can see a spacecraft-shaped virus called T-4 that infects bacteria through an alternate means. T-4 keeps its capsid outside the target cell, but gets its DNA inside by injecting it through the bacterium's cell wall and membrane.

Viruses are small things compared to the organisms they infect—thousands of them typically fit inside a bacterial cell—and they are very simple things as well.

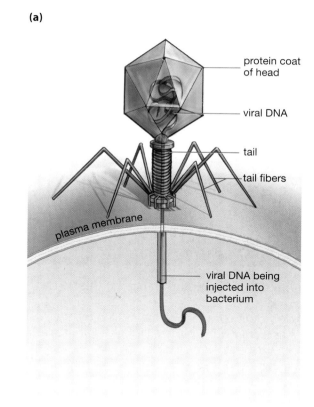

(a)

protein coat of head

viral DNA

tail

tail fibers

plasma membrane

viral DNA being injected into bacterium

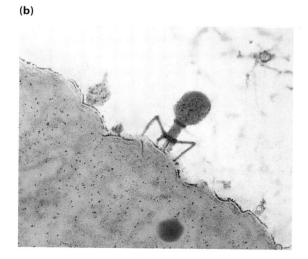

(b)

FIGURE 20.4 Another Variety of Virus

(a) The T4 virus looks like a spacecraft that has landed on the surface of a bacterium.

(b) A viral invader lands. An artificially colored T4 virus injects its DNA into an *E. coli* bacterium (in blue).

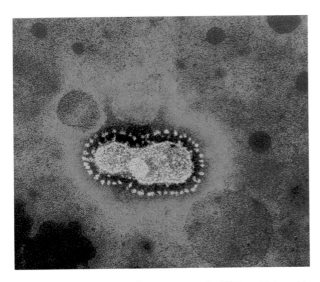

FIGURE 20.5 **From Annoyance to Killer** Pictured is a corona virus, which for years was a cause of nothing more serious than the common cold. Then, in 2003, one strain of corona virus underwent a mutation that made it the cause of a deadly human respiratory infection. The result was worldwide alarm over SARS—severe acute respiratory syndrome. The virus is named for the crown-like appearance of the spikes extending from its envelope.

A BROADER VIEW

HIV in America and Africa

Where do we stand with HIV infections? In the United States in 2003, about 0.6 percent of the adult population was HIV-positive—just under 1 million people were in this category. By contrast, in sub-Saharan Africa, about 7.5 percent of the adult populace was HIV-positive—25 million people were in this category. In at least two African countries, however, the adult infection rate tops 33 percent.

So Far... Answers:

1. Bacteria, Archaea, Eukarya; protists, plants, animals, fungi
2. production of oxygen; transformation of atmospheric nitrogen into a form plants can use; decomposition of dead organic matter
3. living cells

For starters, they lack the usual machinery that cells have, such as a protein assembly line, a waste disposal system, and so forth. But even when we look at the one thing all of them do have, which is genetic material, they are still simple entities. Human beings are thought to have 20,000–25,000 genes, and even a primitive organism like baker's yeast has 6,000 genes. But the AIDS virus? It has nine genes. To be sure, some viruses are more complex than HIV, but in no case is a virus complex enough to replicate by itself. Because viruses can carry on scarcely any of life's basic functions by themselves, most scientists don't classify them as living things. (For a look at another agent that is infectious, but not living, see Essay 20.1, "Prions and Mad Cow Disease," at right.)

The Effects of Viruses

But what trouble is caused by something that isn't even alive! Apart from AIDS, viruses are responsible for smallpox, chicken pox, measles, rabies, polio, herpes, rubella, and some forms of cancer, hepatitis, and pneumonia, to say nothing of common colds and flus. In 2003, parts of the world were thrown into turmoil by the appearance of a new disease dubbed SARS (severe acute respiratory syndrome) whose cause was quickly identified as a so-called corona virus (so named because its club-like spikes make it look as though it has a crown, as you can see in FIGURE 20.5). Normally corona viruses cause nothing more than colds, but the deadly SARS came about because of a common occurrence in viruses: an existing virus, which previously infected various animals, mutated and "jumped" to human beings. What changed in the corona virus, allowing it to make this jump? The spikes that bind with cells. Once those changed form through mutation, the virus had a way to get into human cells.

Viruses have no special affinity for human hosts, however. They invade every life-form, from protists to fungi to plants to bacteria, not to mention all varieties of animals. In early 2001, hundreds of thousands of cattle and sheep had to be destroyed in Europe because they had been exposed to a virus that causes an affliction known as foot and mouth disease. In humans, viruses are combated by the immune system and by vaccination. You can read about both things in Chapter 27.

So Far...

1. The three domains of life are _____, _____, and _____. The four kingdoms within one of these domains are _____, _____, _____, and _____.

2. Microbes perform several activities that are critical to the maintenance of all life on Earth. Name three of these activities.

3. In order to replicate, viruses must invade _____.

20.3 Bacteria: Masters of Every Environment

Domain Bacteria is made up entirely of the microbes called bacteria. Keeping in mind that almost all the organisms we're looking at are microbes, what separates bacteria from any of the others? Readers who went through Chapter 4 will recall that if we set a bacterial cell and, say, a fungal cell side by side, anyone peering through a microscope could see a fundamental difference between the two. The fungal cell would have a large, circular-shaped structure near its center while the bacterial cell would not. The circular structure is the fungal cell's nucleus, which contains almost all of its DNA. But bacteria are **prokaryotes**: organisms whose DNA is not contained within a cell nucleus. Only bacteria and archaea lack a nucleus. All other

Essay 20.1 Prions and Mad Cow Disease

When we think of a disease that is transmissible, we normally assume that some sort of tiny living thing—a microorganism—is involved in transmitting it. We have, for example, salmonella, meningitis, and tuberculosis: three diseases, each caused by a different bacterium that can move from one living "host" to another. And this is the case with most transmissible diseases; at root, they are caused either by microscopic living things, or by viruses, which always have the DNA or RNA that living things possess.

But then there is the illness that has come to be called mad cow disease. Almost alone among transmissible human diseases, it has a *component part* of a living thing as its cause—a protein. Recall from Chapter 3 that much of the scaffolding, or structure, of living things is made from proteins; that the chemical tools called enzymes are made from proteins; and that some hormones are made from them. Clearly, proteins have many functions in living things. But proteins themselves are not living.

Mad cow disease spread among British cattle because prions were transmitted through a kind of recycling.

So, how could a protein cause a transmissible disease? A variety of protein called a prion protein exists in living tissue in a normal, harmless form. Such prion proteins also can come to exist, however, as "rogue" proteins that are *misshapen*. In their normal form in humans, prion proteins reside in the outer or "plasma" membranes of various cells, notably in nerve cells. For reasons not yet understood, a prion protein will become deformed and then go on to deform *more* prion proteins when it comes into contact with them. The result is a chain reaction in which more and more proteins become deformed. Collections of these deformed proteins destroy brain cells; the ultimate result is a brain with sponge-like holes in it—and death for the victim. Rogue prion proteins, referred to as just-plain prions (PREE-ons), are the infectious agent in mad cow disease.

A number of rare human illnesses are believed to be caused by prions, among them an affliction, identified in the 1920s, called Creutzfeldt-Jakob disease. However, even in its most common form, Creutzfeldt-Jakob disease is very rare, striking only about 1 in a million adults, for reasons that are still unknown. In contrast, the late 1990s saw a small flurry of prion-linked human illness across Western Europe, peaking with 28 deaths in Great Britain in the year 2000. This was something new, and it was linked to a series of incidents involving farm animals.

In 1984, a single cow in England began behaving erratically and then dropped dead of a disease that affected its nervous system. This incident turned out to be the start of an epidemic of prion-caused disease in British cattle—the actual mad cow disease, officially known as bovine spongiform encephalopathy (BSE). Scientists now believe that mad cow disease spread among British cattle because *prions* were transmitted through a kind of recycling: Cattle may have originally been infected by eating the remains of sheep that had developed a prion-caused disease called scrapie. Some of these cattle then themselves became feed, as parts of them were processed or "rendered" into a dietary supplement for other cattle. In this way, prions were spread among an ever-growing population of British cattle.

This cycle was ended in Great Britain in 1996 with the passage of laws banning the use of rendered parts for animal feed. But the move came too late to stop transmission of the disease to humans. People who ate prion-contaminated beef products likewise ingested prions, and these prions are active in human beings. How is this possible? Mad cow disease has been transformed into a human variant, dubbed variant Creutzfeldt-Jakob disease or vCJD, which, as of early 2004, had stricken 146 people within Great Britain and 10 people in other countries.

The cattle-based British outbreak of vCJD is not expected to result in a large number of additional victims, neither in the U.K. nor elsewhere. And restrictions on using animal renderings as feed—now in effect in Canada and the United States as well—should help close off one of the main avenues by which mad cow disease can spread. But so great is the fear of the disease that the slightest evidence of it in a country's beef supply can have tremendous consequences. In 2003, a single confirmed instance of mad cow disease in the United States caused Japan to ban the importation of U.S. beef, thus shutting down a $1.2 billion market for U.S. ranchers. To guard against mad cow disease, what's needed is a means to test for the affliction in cattle that are headed for slaughter. Several varieties of such tests are in development; but so far, no one test has been shown to be fast, accurate, and cheap.

FIGURE E20.1.1 **A Mad Cow Quarantine** Cattle in Washington State that were quarantined in 2003 due to fears of mad cow infection.

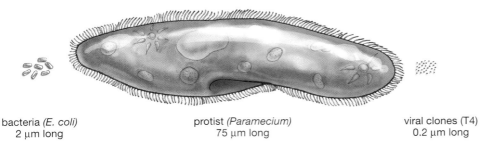

bacteria (E. coli)
2 μm long

protist (Paramecium)
75 μm long

viral clones (T4)
0.2 μm long

FIGURE 20.6 **Small Is a Relative Thing** The bacteria, protist, and viral particles shown in the figure are all microscopic. Nevertheless, they differ greatly among themselves in terms of size. A typical human cell might be about a third the size of the *Paramecium*.

organisms—all plants, animals, fungi, and protists—are composed of cells that have a nucleus, which makes them **eukaryotes**. The lack of a cell nucleus is, however, just one defining feature of bacteria. Here is a list of others.

- *No membrane-bound organelles.* A nucleus is an example of an *organelle*: a highly organized structure within a cell that carries out specific cellular functions. You may remember from Chapter 4 that, in addition to a nucleus, eukaryotic cells have such organelles as mitochondria and lysosomes, which are surrounded by a membrane. Bacteria have only a single kind of organelle, the ribosome, which differs from other organelles in that it is not surrounded by a membrane.

- *Single-celled organisms with small cell size.* Bacteria often come together in groupings that have an organization to them, but each bacterial cell is a self-contained living thing—each one can live and reproduce without the aid of other cells. Bacterial cells are also very small, relative to eukaryotic cells; thousands of them could typically fit into a single eukaryotic cell. (If you look at **FIGURE 20.6**, you can see a size comparison of bacteria, a T4 virus, and a protist you'll be looking at called a paramecium.)

- *Asexual reproduction.* Bacteria reproduce by a simple cell-splitting, or **binary fission**: One cell splits into two, with both "daughter" cells being exact replicas of the parental cell. In contrast, most eukaryotes are capable of sexual reproduction, in which the genetic material of two separate organisms produces offspring, usually through the fusion of egg and sperm.

If you look at **FIGURE 20.7**, you can see micrographs of the three forms bacterial cells most often take: round, rod-shaped, and spiral-shaped. A round bacterium is known as a *coccus*; a rod-shaped bacterium is a *bacillus*; and a spiral-shaped bacterium goes under a variety of names, the best known of which is a *spirochete*. Different species of bacteria fit into all three groups.

FIGURE 20.7 **Different Shapes of Bacteria**

(a) The round bacterium *Staphylococcus epidermidis* is a normal part of the bacteria that cover our skin.

(b) The rod-shaped bacterium *Bacillus anthracis* is the cause of the deadly disease anthrax.

(c) The spiral-shaped bacterium *Leptospira interrogans* often infects rodents, sometimes infects dogs, and occasionally infects human beings.

(a) Round bacteria (cocci)

(b) Rod-shaped bacteria (bacilli)

(c) Spiral-shaped bacterium (spirochete)

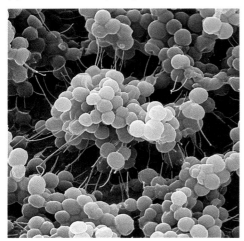

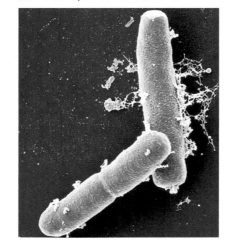

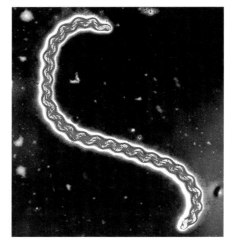

Almost all bacteria have a structure called a cell wall at their periphery. This may seem like a technical detail, but the fact that bacteria have cell walls is of great importance to humans because, as you'll see, it represents a *difference* between bacterial cells and human cells that matters in human health.

Bacterial Diversity

Scientists have taken small soil samples, looked for DNA sequence differences among the bacteria in these samples, and concluded that about 10,000 different species of bacteria exist in such handfuls of earth. Extrapolating from these samples, it's safe to say that millions of species of bacteria exist. Surprisingly, perhaps, only 4,500 species of bacteria and archaea have been identified to date.

An incredible diversity exists among the various kinds of bacteria. It may seem strange to think of these tiny creatures as "diverse," given that they are all microscopic and single-celled. But physical form is only one kind of diversity. Imagine that only some human beings got their nutrition in the conventional way—by ingesting food—while others could *make* their own food by carrying out photosynthesis. Bacteria do things both ways. Imagine that only some human beings needed oxygen to live, while others could take it or leave it, and still others were poisoned by it. Bacteria fit into all three categories. Animals and plants have diverse forms, but bacteria have diverse metabolisms, which is to say diverse means of carrying out the basic chemical reactions that support life.

20.4 Bacteria and Human Disease

We saw earlier that bacteria play an important role in our digestion, that they produce much of our oxygen, that they are responsible for passing on vital nitrogen to plants, and so forth. It goes without saying, however, that not all bacteria are beneficial to human beings. Even within our digestive tract, there are bacteria that are **pathogenic**, which is to say disease causing. These are not bacteria that routinely inhabit any part of our body in large numbers. Instead, they are opportunists that invade specific tissues. About 200 species of bacteria are known to cause human diseases, and the roster of bacterial afflictions reads like a chronicle of human misery: tuberculosis, syphilis, gonorrhea, cholera, tetanus, botulism, leprosy, typhoid fever, and diphtheria are a few of the diseases caused by bacteria, not to mention many kinds of food poisonings and blood-borne infections. You may recall a substance referred to as "anthrax" that was maliciously mailed to various offices after the September 11, 2001 attacks. Anthrax actually is a bacterium (*Bacillus anthracis*) that, when not in the hands of terrorists, occasionally infects farm animals.

How do pathogenic bacteria cause illness? A few invade human cells and reproduce there. More often, however, bacterial damage comes from substances bacteria either secrete or leave behind when they die. These compounds are *toxins*: substances that harmfully alter living tissue or interfere with biological processes. In the case of respiratory anthrax, *Bacillus anthracis* secretes a trio of toxins that cause blood vessels in the lungs and brain to "hemorrhage" or leak. Of course, all pathogenic bacteria must have a way to get into the body in the first place. The bacterium that causes bubonic plague, *Yersinia pestis*, enters on the bite of a flea, while the tetanus bacterium, *Clostridium tetani*, may hitch a ride in on a rusty nail. Meanwhile, the tuberculosis bacterium, *Mycobacterium tuberculosis*, can simply be inhaled.

Killing Pathogenic Bacteria: Antibiotics

Prior to the 1930s, doctors were nearly powerless to stop any of these bacterial invaders. Some progress was made in the late 1930s with the development of the synthetic chemicals called sulfa drugs. But modern medicine was born in the 1940s with the development of **antibiotics**, meaning chemical compounds produced by one microorganism that are toxic to another microorganism. The first antibiotic, penicillin, was a substance produced by a fungus. Like all antibiotics, it

A BROADER VIEW

What Is Botox?

What is the nation's *best-selling* bacterial toxin? Botox. It's a purified form of a toxin produced by the bacterium *Clostridium botulinum*. In nature, the *C. botulinum* toxin can cause death by interfering with the nerve signals that allow us to breathe. In the offices of plastic surgeons, Botox is used to interfere with nerve signals that put age lines on middle-aged foreheads.

Essay 20.2 The Persistence of Herpes

The first outbreak usually is the worst, but the real problem is that the first outbreak usually isn't the last. Herpes comes back, as herpes sufferers well know. A few weeks or months after the initial outbreak, everything seems fine, but then there may be a tingling—nothing that can be seen yet, but instead just a feeling. Not much later, however, it's back: the same unsightly "cold sore" on the mouth or lesions on the genitals, in the same place that the first outbreak occurred.

This series of events is familiar to lots of people. An estimated one in five Americans over the age of 12 is infected with the herpes virus, though this statistic overstates the trouble caused by it. The vast majority of herpes carriers do not even realize they harbor the disease, because such outbreaks as they have are so mild as to go unnoticed. This very thing facilitates the spread of herpes, however. People who have no symptoms can nevertheless be "silent shedders": They can pass herpes along to others, who may get the full-blown form of the affliction.

Why does herpes come back in the way it does?

But why does herpes come back in the way it does? Where does it go between outbreaks, and why can't we just get rid of it permanently? The term *herpes* actually is shorthand for "herpes simplex viruses," a pair of viruses that belong to a group of 100 known viruses in the herpes family. (Others in the family cause chicken pox, shingles, and the Epstein-Barr disease.) The herpes sores that can be so distressing are the result of newly formed virus particles breaking out of skin cells, thereby de-

stroying them. Fever, swollen glands, and other symptoms often accompany these lesions, particularly in an initial infection. Within a couple of weeks or so, however, the immune system has killed all the virus particles in and around the skin cells, and the symptoms are gone. The problem is that, by this time, the virus has also infected *nerve* cells in the area. Such nerve cells (or neurons) have long extensions of themselves called axons. With the initial infection, herpes virus particles start making a slow migration up these axons to the *body* of the nerve cells, specifically to the nucleus inside the cell body. In the case of genital herpes, this means a journey from the vagina or penis to cell bodies that sit right at the base of the spinal cord. And these cell bodies are where the virus stays— forever. Once there, it does next to nothing; it's not causing any nerve damage or replicating. The nerve cells it is in never divide, so it isn't passed on to a larger group of cells. But neither is it eliminated. It just resides in the nerve cell nuclei until one day—often a stressful day—it gets activated by an unknown mechanism and begins a journey back up the very axons it came in on. Once it reaches the original nerve-cell endings it spreads to skin cells, and the result is new lesions in the same old place.

As noted, some people with herpes can shed virus particles from time to time even when they have no lesions, meaning they are infectious during these periods. This, of course, puts these carriers at terrible risk of passing the virus on to their uninfected partners. The good news about herpes is that there is a drug, called valacyclovir, that has shown significant ability to suppress the virus at all times. Meanwhile, in 2004, researchers from Harvard reported finding a compound that keeps the herpes virus from replicating within an infected cell—not a cure by any means, but a clear advance in developing one.

had two critical qualities: It could kill microorganisms within human beings while leaving the humans themselves relatively unharmed.

How can antibiotics be toxic to bacteria but not to people? They exploit the differences between bacterial cells and human cells. Remember earlier when we noted that almost all bacteria have cell walls, while human cells don't? Penicillin and several other antibiotics work by blocking cell-wall construction, which bacteria have to undertake when they divide. More specifically, penicillin confiscates a construction tool; it binds with a bacterial enzyme that helps stitch the cell wall together. A bacterium without a cell wall is doomed: It ruptures, floods, and dies. But the *human* cells around it—lacking cell walls—are left unharmed.

The Threat of Antibiotic Resistance

This power of antibiotics ought to spark alarm in us about something that is happening today, which is that antibiotics are steadily *losing* their ability to kill bacteria. How could this happen? An estimated 30 percent of antibiotic prescriptions written today are thought to be needless, in the sense that the person taking the antibiotic doesn't even have a bacterial infection, but rather is suffering from a viral or other illness. (So why do people get such prescriptions? They ask their doctors for them and the doctors go along.) Even when people do have bacterial infections, they often take their antibiotics only until they start feeling better, rather than following through with the full course of recommended treatment, which can last weeks or even months.

Those of you who have read through the chapters on evolution may be able to see what this leads to. Normal bacteria may be wiped out by a course of antibiotics foolishly cut short by a patient. But a few bacteria will survive these inadequate treatments because they have sets of genes that allow them to do so. In the *absence* of antibiotics, these survivors might not have had anything special going for them, relative to others in their population. (They aren't *generally* superior bacteria; they just resist antibiotics better.) But with antibiotics these survivors have been left without bacterial competitors for food and habitat. As a result, they flourish, giving rise to antibiotic-resistant *lines* of bacteria. Moreover, they will do this rapidly, since one generation of bacteria can give rise to another in as little as 20 minutes.

Other factors also enter into this picture. It may surprise you to learn that nearly half the antibiotics used in the United States don't go into people at all; they are put into animal feeds because they serve as growth stimulants. In a similar vein, farmers spray huge tracts of fruit trees with antibiotics to ward off bacterial infections. The upshot of this massive overuse of antibiotics is that bacterial infectious diseases are making a comeback, as noted in the newspaper story at the start of the chapter. The problem is most pressing in hospitals, where physicians are seeing one formerly useful drug after another fail to cure infections. If we cannot stay ahead of bacterial evolution by developing new antibiotics—while cutting back on the needless use of existing antibiotics—we may find ourselves returning to the days of our grandparents, when a simple cut could be life threatening.

20.5 Archaea: From Marginal Player to Center Stage

As little as 10 years ago, any biology textbook you picked up would likely have referred to the archaea as archae*bacteria*, and most would have mentioned the archaea as a kind of marginal life-form, existing in only a few extreme habitats, such as hot springs or salty lakes. Today, the archaea are recognized as constituting their own domain in life—standing alongside Domain Bacteria and Domain Eukarya—and existing in large numbers in many environments. One importance of the archaea is that, of all the organisms living today, they may have been the first to exist. Whether this means that archaea are ancestral to all other existing life-forms remains to be seen—many researchers have their doubts—but it is certain that archaea lie near the trunk of life's family tree. In particular, the *thermophiles*, or heat-loving archaea, appear to be a very old branch on life's family tree, which is one piece of evidence leading many scientists to suspect that life began in or near hot-water vents of the type seen on the ocean floor today.

It took a long time to recognize the unique qualities of archaea because, superficially, they are very similar to bacteria. They are microscopic, single-celled, and their cells lack a nucleus (making them prokaryotes, along with the bacteria). But the sequencing of the genome of an archaean named *Methanococcus jannaschii* in 1996 revealed that an amazing 56 percent of its genes were completely unknown to science—they were unlike anything seen in either bacteria or in eukaryotes. As for the remaining 44 percent of the genome, some *M. jannaschii* genes worked like those of eukaryotes, while others worked like those of bacteria. It makes sense, then, that archaea lie between bacteria and eukaryotes on life's family tree, as you can see in Figure 20.2 (page 333).

(a)

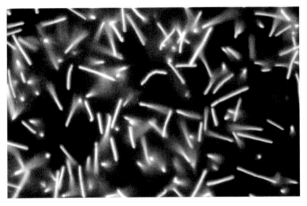

(b)

FIGURE 20.8 **Life in Extreme Environments**

(a) Deep-sea archaeans *Methanopyrus kandleri* is an archaean that lives in the ocean at temperatures near those of boiling water. It has been found in geothermal vents on the seafloor at depths of about 1.2 miles down. One of many archaeans that produce the gas methane as part of their metabolism, it can only live in environments in which there is no oxygen.

(b) Prospecting for extremophiles Researchers looking for archaean extremophiles in the Obsidian Pool in Yellowstone National Park.

A BROADER VIEW

Extremophiles and Police Work

In TV shows such as *CSI*, the "PCR" process police use to get usable DNA samples is dependent on an enzyme that came originally from an extremophile, the hot-spring-dwelling bacterium *Thermus aquaticus*. PCR requires heating DNA samples to a temperature above the range at which human DNA enzymes can operate. *T. aquaticus* has no problem with PCR-level heat, however, since it lives in temperatures above 170 degrees Fahrenheit.

So Far... Answers:

1. proteins

2. one microorganism; toxic to another microorganism

3. Both are single-celled life-forms whose cells lack a nucleus (making them prokaryotes).

Prospecting for "Extremophiles"

The species of archaea that inhabit extreme environments have in recent years become the target of a kind of new-age prospecting, with "miners" being scientists from chemical and biotechnology firms. The archaea they are looking for are "extremophiles"—archaea or bacteria that flourish in conditions that would kill most organisms, such as high heat, high pressure, high salt, or extreme pH (**see** FIGURE 20.8). One of the most extreme of the extremophiles was reported on in 2000, when researchers revealed the existence of an archaean they found thriving in the acidic wastes of an abandoned copper mine. Dubbed *Ferroplasma acidarmanus*, this hardy microbe can live in liquid that has a pH of zero—a habitat more acidic than battery acid. To live in such environments, extremophiles ("extreme-lovers") must produce *enzymes* that function in these conditions. And, through modern biotechnology processes, enzymes can be isolated from an organism and then turned out in quantity in factories. These enzymes can then be put to use in *human-made* extreme environments, such as the inside of a washing machine or the confines of a laboratory container. The use of archaea in industry is a project still in its infancy, but scientists hope to make extensive use of the archaean extremophiles in coming years.

The extremophile archaeans get a great deal of attention in science, but archaea are now recognized as living in large numbers even in environments that are not so extreme. Almost a third of the microscopic organisms living in the surface waters off Antarctica turn out to be archaea, for example, and huge populations of them inhabit deeper-ocean waters.

So Far...

1. Prions, the cause of mad cow disease, are not living things but instead are _____.

2. Antibiotics are chemical compounds produced by _____ that are _____.

3. Name two ways that bacteria and archaea are similar life-forms.

20.6 Protists: Pioneers in Diversifying Life

We've thus far looked at two of life's domains, Bacteria and Archaea. As noted, the third domain, Eukarya, is so diverse we will look at only one of its four kingdoms in this chapter, the mostly microscopic organisms known as protists.

What's a protist? We don't really have a satisfactory definition. They are something like the objects in a drawer marked "miscellaneous": The primary thing they have in common is that they don't belong in any other drawer. Domain Eukarya has within it three well-defined kingdoms—plants, animals, and fungi—but protists are defined in terms of what they are *not* relative to these life-forms: Protists are eukaryotic organisms that do not have all the defining characteristics of either a plant or an animal or a fungus.

You may recall from Chapter 19 that animals, plants, and fungi *evolved* from protists—animals from protists called choanoflagellates, plants from green-algae protists, and fungi from an unknown variety of protist. Not surprisingly, then, there are protists that are *like* plants, animals, and fungi. Some protists ingest food, as animals do; others make their own food, as plants do; and still others send out slender extensions of themselves and digest their food externally, as fungi do. However, protists themselves don't actually stay confined within these categories. To give one well-known example of a category-breaker, there is a microscopic protist called *Euglena gracilis* that is bright green and, when at rest, takes on a cigar shape (see FIGURE 20.9). The green color of *Euglena* comes from the chloroplasts within it, which you might remember are the organelles that perform photosynthesis. So, when swimming around its home, which tends to be a freshwater pond, *Euglena* is making its own food through photosynthesis, which would presumably make it plant-like. Yet when sunlight becomes scarce, *Euglena* simply switches over to an animal-like nutritional mode—it ingests organic matter. Then, when sunlight returns, it goes back to photosynthesis.

Protists, then, are difficult to categorize. The only things we can say about all of them is that they are eukaryotes and that they live in environments that are at least moist, if not fully aquatic—damp forest floors as well as oceans and lakes. Almost all are microscopic, though as you'll see, some are very large. About 100,000 species are known to exist. Though only a small proportion are parasites that affect humans, the rogue's gallery of protists is a formidable one. A microscopic protist, *Plasmodium falciparum*, causes one of the world's most widespread diseases, malaria, by passing between human and mosquito hosts. Campers in the United States are aware of the intestinal parasite *Giardia*, which contaminates water. Other protists cause sleeping sickness and amoebic dysentery. Pathogenic protists also affect humans in less direct ways. The cause of the great Irish Potato Famine

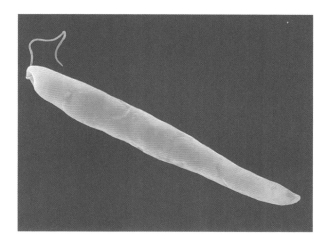

FIGURE 20.9 **More Like a Plant or an Animal?**
The protist *Euglena gracilis* switches back and forth between the plant-like behavior of making its own food and the animal-like behavior of capturing food from its external environment.

of the 1840s was a fungus-like protist, *Phytophthora infestans*, which devastated the Irish potato fields. A close relative of this pest, *Phytophthora ramorum*, has now wiped out tens of thousands of oak and bay laurel trees in California and Oregon.

When we look at the living world as a whole, the protists in it are fascinating because they are the organisms that represent the break with prokaryotic life—with life as nothing but single-celled bacteria and archaea. After life first appeared on Earth, it consisted solely of bacteria and archaea for more than 2 billion years. The first life-forms to appear other than bacteria and archaea were the protists. And in these organisms, we can see how the living world branched out; we can see living things making transitions to life as we know it today. You'll see examples of this now, as we look at a few representatives of the protists. We'll review them within the categories of those that perform photosynthesis, and those that don't.

20.7 Photosynthesizing Protists

The protists that perform photosynthesis are perhaps the only large group of protists known to the average person. The catchall name for them is **algae**. If you look at FIGURE 20.10, you can see three varieties we'll be reviewing.

FIGURE 20.10a shows a microscopic "golden" alga called *Synura scenedesmus* that can be found swimming in freshwater ponds in the United States. In this species, we can see a transition between the single-celled life carried out by bacteria and the multicelled life carried out by more complex organisms. All varieties of golden algae come in species that exhibit **colonial multicellularity**: a form of life in which individual cells form stable associations with one another but do not take on specialized roles. Each of the three objects you see in the picture of *Synura* is a cluster of *Synura* cells—a colony of them. Each member of the colony keeps its two whiplike flagella pointed to the outside of the group, the flagella beat back and forth, and the colony literally rolls through the water. None of these cells, however, performs a function different from any other cell, which is why this *Synura* is referred to as colonial and not truly multicellular.

If you look at FIGURE 20.10b, you can see a form of *green* algae, called *Volvox*, that represents an evolutionary step beyond *Synura*. The beautiful, translucent circles that outline *Volvox* are a gelatinous material that initially encloses a single layer of anywhere from 500 to 60,000 *Volvox* cells, all of them spaced around the periphery of the sphere. Almost all of these cells have flagella that they move back and forth in unison, so that *Volvox* moves slowly through the water in a given direction, spinning along an axis as it goes (something like a torpedo). For our purposes, the key thing about *Volvox* is that not all of the cells in this colony will reproduce. Only a select group of cells without flagella, generally at the rear pole of the axis, will divide to produce new colonies. These colonies in turn give rise to new spheres that exist inside the original parent sphere. Eventually the gelatinous material around the parent sphere ruptures and the daughter spheres emerge,

A BROADER VIEW
Algae and Art
Any list of nature's most beautiful creations would have to include the algal protists known as diatoms, which incorporate the mineral silica into their cell walls. The result is tiny creatures who take on an astounding array of colors and geometric shapes. England's Victorians carried this natural beauty a step further by using microscopes to arrange diatoms on glass slides, thus producing "diatom art," a tradition still carried on today. For one example, see the photo on this chapter's page 330. For other examples, see: http://thalassa.gso.uri.edu/flora/arranged.htm

FIGURE 20.10 **Photosynthesizing Protists—the Algae**

(a) Three colonies of the golden algae *Synura*, each of which is composed of a cluster of individual cells. *Synura* is a colonial protist, as none of the cells in a colony takes on any specialized function.

(b) Colonies of the green algae *Volvox* with daughter colonies growing inside. *Volvox* colonies can be regarded as truly multicellular organisms, in that they have specialized reproductive cells. The largest *Volvox* colonies can be seen with the naked eye.

(c) A frond of the brown algae known as sea kelp. These unusually large protists have collections of cells that conduct food throughout the organism.

(a) Golden algae

(b) Green algae

(c) Brown algae

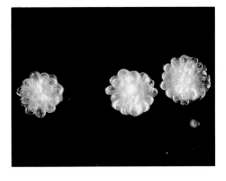
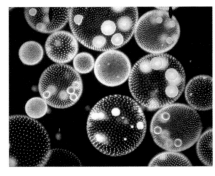

each of them then continuing the process. The important point is that *Volvox* has cells that specialize. As such, *Volvox* has arguably achieved **true multicellularity**: a form of life in which individual cells exist in stable groups, with different cells in a group specializing in different functions. In human beings, of course, we can see a great specialization in, for example, nerve and muscle cells. *Volvox* shows us specialization in a rudimentary form.

Finally, if you look at FIGURE 20.10c, you can see some representative *brown* algae—the giants of the protist world. One variety of brown algae, the sea kelp, has individuals that may be 100 meters long and that often grow in enormous "kelp forests." For our purposes, brown algae are notable because their leaf-like "blades" and stem-like "stalks" represent another leap up in complexity: They contain organized groups of cells that conduct food throughout the organism.

One other thing is important about the algae. Note that the golden algae and *Volvox* share three characteristics: They are microscopic, aquatic, and carry out photosynthesis. If we were to go through all the algae, we would find these same qualities in countless species. Creatures in this category, including some bacteria, are known as **phytoplankton**: small photosynthesizing organisms that float near the surface of water. Such organisms are extremely important to life on Earth, for two reasons. First, they produce most of the Earth's oxygen. Second, they sit right at the base of most aquatic food chains. All large life-forms on Earth ultimately depend on photosynthesizers for their food. On land, lions may eat zebras, but zebras eat grass. On the ocean, whales may eat the shrimp-like creatures called krill, but krill eat phytoplankton.

A BROADER VIEW

Phytoplankton

When we think of abundant living things, trees or grass may come to mind, but remember that more than 70 percent of Earth's surface is covered with ocean water. Since phytoplankton are abundant both there and in fresh water, it may be that they cover three-quarters of the planet's surface.

20.8 Heterotrophic Protists

Lots of species of protists do not get their nutrients from photosynthesis, as algae do, but instead get them from consuming other organisms or bits of organic matter that they find. This makes them heterotrophs (or "other-eaters").

Heterotrophs with Locomotor Extensions

Some of these protists have evolved slender extensions, called cilia and flagella, that allow them to move toward their prey or away from danger. If you look at FIGURE 20.11, you can see two varieties of these highly mobile heterotrophic protists.

FIGURE 20.11a shows a *Paramecium*, which is known as a ciliate because of all the hair-like cilia that cover its exterior. These cilia beat in unison, allowing paramecia to move forward or back in an exacting fashion—toward or away from objects in its environment. *Paramecium* may be single-celled, but it is complex. It takes food in through a mouth-like passageway called a gullet, digests it in a special organelle called a food vacuole, and then empties the resulting waste into the outside world through a special pore on its exterior. Some species even have tiny dart-like structures (called trichocysts) that they can shoot out when threatened.

FIGURE 20.11b shows a famous human parasite that enters human beings through contaminated water. So widespread is this "camper's parasite," *Giardia lamblia,* that health authorities have advised Americans never to drink stream or lake water without purifying it. *Giardia* lives in the small intestines of

FIGURE 20.11   Movement through Cellular Extensions

(a) A green *Paramecium*, whose many cilia both help it move and bring food to its gullet, visible in the center.

(b) The parasite *Giardia lamblia* within the human small intestine. Seen here as the pear-shaped green and white objects, *Giardia* use sucking discs to attach themselves to finger-like extensions of the intestine's inner lining.

(a) A *Paramecium*

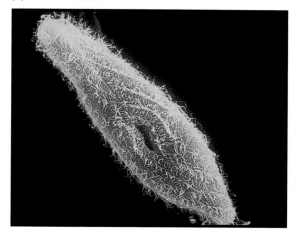

(b) The *Giardia* parasite

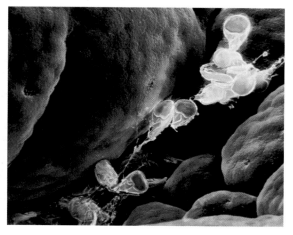

humans and other animals, where it causes symptoms that include nausea, diarrhea, and vomiting. It gets around not by the cilia of *Paramecium*, however, but by pairs of whiplike flagella, as we saw in the algae protists.

Heterotrophs with Limited Mobility

FIGURE 20.12a shows a so-called amoeboid protist, meaning a protist that gets around by means of a pseudopod or "false foot." The amoeba sends out a slender extension of itself (the foot), and then the rest of its body flows into this extension. It takes in bits of food by encircling them with extensions and then fusing these extensions together. What once had been outside the amoeba is now inside, being digested. Many amoeboid protists are parasites, and one of them, *Entamoeba histolytica*, is a serious human parasite that causes an estimated 100,000 deaths per year, mostly in the tropics, by entering the human digestive system through contaminated water. When a parasite invades the human intestinal wall, the resulting condition is known as dysentery. When that parasite is an amoeba, the result is amoebic dysentery.

FIGURE 20.12b shows a protist known as a plasmodial slime mold. This organism assumes several forms during its life cycle. In the form you see in the picture, its name is appropriate, because it is essentially moving slime. ("Slime" is actually one of the nicer common names for these organisms. Gardeners refer to one variety as "regurgitated cat breakfast.") If you looked at this slime mold under a microscope, you would see lots of cell nuclei, but almost none of the *boundaries* (the plasma membranes) that normally separate one cell from another. The result is a free-flowing, but slowly moving mass of cytoplasm—the interior of a single cell spread out over a large area, with only one plasma membrane at its periphery. This material moves by the same "cytoplasmic streaming" you saw in the amoeboids. As it rolls over rocks or forest floor, it consumes underlying bacteria, fungi, and bits of organic material. Should an object stand in its path, it simply flows around it, or even through it, if the object is porous enough.

By now, you may be getting the impression that protists could be star contestants in a show called "Can You Top This for Strangeness?" If you look at FIGURE 20.13, you can see another contestant. Pictured is a much-studied organism, *Dictyostelium discoideum*, a so-called cellular slime mold that has a multiple personality. At one point in its life cycle, it exists in the form of individual amoeba cells that crawl along the forest floor feeding on bacteria. Each "Dicty" is on its own at this stage, but if the bacterial food runs low, these individual cells begin "aggregating"

FIGURE 20.12 **Life as a Blob**

(a) An amoeba bends itself around food, which it will soon ingest.

(b) This plasmodial slime mold, *Physarium polycephalum*, is growing on and around a tree trunk in Pennsylvania.

(a) Amoeba

(b) Slime mold

(a) *Dictyostelium* slug

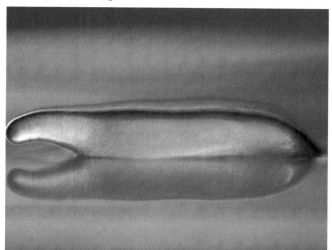

(b) *Dictyostelium* reproductive structure

or coming together in an organized group. Eventually, they develop into a barely visible "slug" of up to 100,000 cells that has front and back ends and that migrates to a more fertile area of the forest. Then at a certain point, they change form again. They arrange themselves into a tower-like reproductive structure that has, at its top, spores that will be dispersed and develop into new individual cells that once again crawl along the forest floor. The whole aggregation process will not take place unless starvation is imminent, however; without hunger as a motivator, the individual cells stay separate.

Now, once you get past the shock of realizing that such a thing could happen—that individual cells could just collect themselves and create a larger, moving organism—you can begin to see what interesting questions it raises. What's the signaling mechanism that brings 100,000 separate cells together? How do they begin to cooperate to form these various structures? (To put this another way, how does multicellularity begin to work?) And, perhaps most intriguingly, why should some cells sacrifice themselves by serving on the *stalk* of the reproductive structure—from which location they will not reproduce themselves—while others get to be at the top and disperse to a new life in better hunting grounds? If evolution is all about passing on genes, why should some cells give up the chance to pass on theirs? Scientists study Dicty in hopes of learning more about all these questions.

FIGURE 20.13 Changing Forms

(a) A *Dictyostelium* in its slug stage, composed of individual cells that have come together to produce an organism capable of moving across the forest floor. Though the slug is composed of tens of thousands of cells, it is no bigger than a grain of sand.

(b) Cells in the slug eventually arrange themselves into this tower-like reproductive structure. Reproductive cells called spores will be dispersed from the top of the structure to the forest floor. There they will begin life as individual *Dictyostelium* cells.

On to Fungi and Plants

The bacteria, archaea, and protists reviewed in this chapter represent only a portion of the living world. Coming up next are two life-forms that most people think of as related and similar, though in fact they are only distantly related and are quite dissimilar. These are fungi and plants.

So Far...

1. Protists tend to be _____-celled organisms that live in _____ environments. In contrast to bacteria and archaea, however, all protist cells have a _____, which makes protists eukaryotes.

2. The organisms known as phytoplankton are important to life in general for two reasons. What are they?

3. The catchall term for protists that perform photosynthesis is _____.

So Far... Answers:

1. *single; aquatic; nucleus*

2. *Phytoplankton produce much of the world's oxygen, and they sit at the base of aquatic food chains.*

3. *algae*

Chapter Review

Summary

20.1 Life's Categories and the Importance of Microbes

- **Classification**—All living things on Earth can be classified as falling into one of three domains of life: Bacteria or Archaea, whose members are all single-celled and microscopic; or Eukarya, whose diversity is so broad it is further divided into four kingdoms: plants, animals, fungi, and protists. (page 331)

- **Microbes and life**—Microbes are living things so small they cannot be seen with the naked eye. These organisms are indispensable to all life on Earth. They produce more than half of Earth's atmospheric oxygen; the bacteria among them are responsible for putting nitrogen into a form plants can use; and bacteria and fungi are the most important "decomposers" of the natural world—they break down dead organic matter, such as tree branches, and recycle the resulting elements back into the Earth. (page 333)

20.2 Viruses: Making a Living by Hijacking Cells

- **The nature of viruses**—Viruses are noncellular, replicating entities that invade living cells to carry out their replication. Because viruses cannot carry out replication by themselves, most scientists do not classify them as living things. (page 333)

- **HIV as an example**—The human immunodeficiency virus (HIV), which causes AIDS, has two structures common to almost all viruses: genetic material and a protein coat, called a capsid, surrounding this material. HIV also has one other structural element that many viruses possess: a fatty membrane, called an envelope, which surrounds the capsid. (page 335)

- **Viral replication**—Most viruses carry out four steps in their life cycle. They get their genetic material inside a "host" cell; turn out viral component parts, construct new virus clones or "particles" from these parts; and move the new particles out of the cell, at which point they go on to infect more cells. (page 335)

- **Range of infections**—Viruses cause a host of human illness. New viruses harmful to human beings periodically come along by means of viruses that infect other animals mutating and then "jumping" to human hosts. All forms of life are vulnerable to viral infections. (page 336)

Web Tutorial 20.1 HIV: The AIDS Virus

20.3 Bacteria: Masters of Every Environment

- **The nature of bacteria**—Bacteria are microscopic, single-celled organisms that are prokaryotes: organisms whose genetic material is not contained within a nucleus. Other defining features of bacteria are that they have only a single organelle (the ribosome) and reproduce asexually through a simple cell-splitting called binary fission. Millions of species of bacteria exist. Bacteria are metabolically far more diverse than plants or animals. (page 336)

20.4 Bacteria and Human Disease

- **Pathogenic bacteria**—Only a small number of bacteria are pathogenic or disease causing, but these bacteria are responsible for some of humanity's worst diseases. A few pathogenic bacteria cause harm by invading human cells, but bacteria generally do their damage by releasing or leaving behind harmful substances called toxins. (page 339)

- **Antibiotics**—The primary human defense against pathogenic bacteria is the class of drugs known as antibiotics: substances produced by one microorganism that are toxic to another. The first antibiotic, penicillin, was developed in the 1940s. Antibiotics work by exploiting the differences between bacterial and human cells, such that they kill bacteria while leaving human cells relatively unharmed. (page 339)

- **Antibiotic resistance**—The power of antibiotics is being threatened by the emergence of antibiotic-resistant strains of bacteria. These bacteria are evolving in greater numbers because of an overuse of antibiotics in medicine and agriculture. (page 340)

20.5 Archaea: From Marginal Player to Center Stage

- **The nature of archaea**—Archaea were once thought to be a form of bacteria, but have been shown to differ greatly from bacteria in their genetic makeup and metabolic functioning. Like bacteria, they are single-celled, microscopic, and are prokaryotes (cells without nuclei). They appear to be among the earliest-evolving organisms on Earth. (page 341)

- **Products from extremophiles**—The extreme environments that so many archaea live in have prompted biotechnology firms to "prospect" for novel archaeans (and bacteria) with the intent of developing commercial products from the enzymes these "extremophiles" produce. (page 342)

20.6 Protists: Pioneers in Diversifying Life

- **The nature of protists**—A protist is a eukaryotic organism that does not have all the defining features of a plant, an animal, or a fungus. Protists are mostly single-celled, and all of them live in environments that are at least moist, if not aquatic. Plants, animals, and fungi evolved from them. Thus, there are protists that are plant-like, animal-like, and fungi-like, but many protists do not fit neatly into any of these categories. (page 343)

20.7 Photosynthesizing Protists

- **The algae**—Protists that get their nutrition by performing photosynthesis are known as algae. Some algal species provide examples of colonial multicellularity, defined as a form of life in which individual cells form stable associations with one another but do not take on specialized roles. Other algal protists provide examples of true multicellularity: a form of life in which individual cells exist in stable groups, with different cells specializing in different functions.   (page 344)

- **Phytoplankton**—Microscopic algae are important members of the group of organisms known as phytoplankton: small photosynthesizing organisms that float near the surface of water. Phytoplankton are very important to life in general, because they produce most of Earth's oxygen and form the base of so many aquatic food chains. All phytoplankton are either algae or bacteria.   (page 345)

20.8 Heterotrophic Protists

- **Heterotrophs with cilia or flagella**—Heterotrophic protists do not get their nutrients by performing photosynthesis, but instead get them from consuming either other organisms or bits of organic matter. Some have evolved tiny slender extensions, cilia and flagella, with which they move toward prey or away from danger.   (page 345)

- **Heterotrophs with limited mobility**—The protists called amoeba move through use of a pseudopod or "false foot"—a slender extension of the amoeba into which the rest of the body flows. Likewise, the protists called plasmodial slime molds and cellular slime molds move by means of this "cytoplasmic streaming." The cellular slime mold called *Dictyostelium discoideum* exists as a collection of individual amoeboid cells that come together to form a unitary "slug" and then a reproductive stalk in times of low food.   (page 346)

Key Terms

algae p. 344	colonial multicellularity p. 344	phytoplankton p. 345	true multicellularity p. 345
antibiotics p. 339	eukaryote p. 338	prokaryote p. 336	virus p. 335
binary fission p. 338	pathogenic p. 339		

Testing Your Understanding

In Your Own Words   (answers in the back of the book)

1. Why do most biologists classify viruses as nonliving?

2. Which of the following classes of living things evolved from protists? Animals, fungi, plants.

3. What is an amoeba? What class of diseases is associated with them?

Thinking about What You've Learned

1. What does it mean to be a "successful" organism? Bacteria are certainly the most numerous organisms on Earth, they live in nearly all environments, and they are the least susceptible to being eliminated through environmental catastrophe. Are they Earth's most successful organisms?

2. What does it mean to be an organism? The cellular slime mold *Dictyostelium discoideum* exists initially as a collection of separate cells moving over the forest floor. In times of starvation, these cells come together to form a single "slug" that moves to more fertile territory. Eventually it will change into a new structure that will throw off separate cells again. Is the slug "Dicty" an organism or a group of temporarily cooperating cells?

3. What would life on Earth be like if there were no decomposers?

Fungi and Plants: The Diversity of Life 2

The mushroom *Amanita muscaria* (a fungus) surrounded by moss (a plant).

CHICAGO TRIBUNE

IT'S HARD TO BEAT MOLD—BUT YOU CAN TRY TO CONTAIN IT

By Bill Hendrick, Cox News Service
November 19, 2004

Like the melon-sized pods in the cult movie "Invasion of the Body Snatchers" that consumed every person they touched, millions of green mold spores are wafting unseen around our homes, offices and movie theaters, silently attacking our lungs, sinuses and eyes.

THE RECORD

LONDON RESTAURANT FORKS OVER RECORD $52,000 FOR WHITE TRUFFLE

By The Associated Press
November 26, 2004

ROME—A London restaurant dished out a record $52,000 for a 2.4-pound Italian white truffle during a charity auction in Tuscany. Zafferano restaurant outbid other buyers to take home the delicacy, said Daniel Pescini, a town official from San Miniato, the village west of Florence where the truffle was found.

21.1 The Fungi

So, which is it for fungi: Invading spores or pinnacles of gourmet eating? The answer, of course, is "both." Mold, noted in the first newspaper story above, is a type of fungus that has gained great notoriety not only for damaging human health, but for damaging human houses. On the other hand, the antibiotic penicillin is the product of a fungus—once again a mold—while the much-prized truffle is an edible fungus.

The white truffle, weighing 2.4 pounds, that was bought at auction for $52,000.

Like the bacteria we looked at last chapter, the vast group of living things called fungi are both helpful and harmful. So distinctive are the fungi in the way they undertake basic processes of living that they constitute one of the four "kingdoms" of the living world that we've been reviewing. Another kingdom is the plants, but with them the helpful/harmful dichotomy doesn't really apply. Why not? Because it's so hard to find harmful plants. Aside from occasionally growing in more abundance than we might like—think of kudzu or common weeds—can you name anything about plants that isn't positive for human beings? And what a set of positives they present us with! If we made up a list, the first thing on it might be: providing us, either directly or indirectly, with all the food we eat. We began a tour of Earth's life-forms last chapter with microbes; in this chapter we will continue our tour with plants and with fungi.

At first glance, plants and fungi may seem to be alike. After all, they're fixed in one spot and tend to grow in the ground. But as you'll see, the main connection between plants and fungi is that many of them have struck up underground partnerships. When it comes to evolution, the surprising thing is that fungi and *animals* are more closely related than fungi and plants. There was a single evolutionary line (of protists) that gave rise to fungi, plants, and animals, but plants branched off from this line first.

The Nature of Fungi

So if a fungus is not a plant, what is it? Critically, fungi do not make their own food as plants do. Instead, fungi are *heterotrophs*—like us, they consume existing organic material in order to live. They make a living by sending out webs of slender, tube-like threads to a food source. This food source could be an old piece of bread or a fallen log in a forest (**see** FIGURE 21.1). Whatever the case, fungi cannot immediately take in or "ingest" their food in the way that animals do. Fungal cells have cell walls, which means that only relatively small molecules can pass into the fungal threads. The fungus' solution to this problem is to digest its food *externally*, through digestive enzymes it releases. When the food has been broken down sufficiently by these enzymes, the resulting molecules are taken up into the filaments. This is a defining characteristic of fungi. We can state this characteristic and several others in a more formal way:

- Fungi largely consist of slender, tube-like filaments called **hyphae**.

- Fungi get their nutrition by dissolving their food externally and then absorbing it into their hyphae.

- Collectively, the hyphae make up a branching web, called a **mycelium**. If you look at FIGURE 21.2, you can see the nature of this structure. You may have thought of a mushroom as the main part of a fungus, but in reality it is merely a reproductive structure, called a *fruiting body*, that sprouts up from the

A BROADER VIEW

An Enormous Living Thing

What is the world's largest living thing? You might guess it would be a blue whale or redwood tree, but a case can be made that it is a fungus. In 1992, researchers reported finding an underground fungus in Washington State that runs through an area of 1,500 acres and that arguably is a single organism.

FIGURE 21.1 **Fungal Diversity**

(a) Fungi serve with bacteria as the major "decomposers" in nature. Here a honey fungus, *Armillaria mellea*, is breaking down a dead tree in a forest in Poland.

(b) Fungi are also human pests. Pictured is a human nail with fungal spores on it that are the cause of athlete's foot.

(c) Fungi distribute their spores in various ways. Pictured are spores being expelled from some puffballs in Costa Rica. What brings about the expulsion? Drops of rain falling on the puffballs.

(a)

(b)

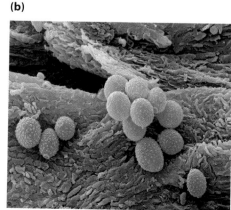

(c)

larger mycelium below. A fruiting body amounts to a large, folded-up collection of hyphae.

- Fungi are **sessile** or fixed in one spot, but their hyphae grow. The fungal mycelium does not move, but its individual hyphae grow toward a food source. This growth comes, however, only at the tips of the hyphae, not throughout their length.

- Fungi are almost always multicellular. The exception to this is the yeasts, which are unicellular and thus do not form mycelia.

21.2 Roles of Fungi in Society and Nature

Like bacteria, fungi are mostly hidden from us—and mostly known to us by the trouble they cause. This trouble is considerable, as fungi are responsible for mildew, dry rot, vaginal yeast infections, bread molds, general food spoilage, toenail and fingernail infections, athlete's foot, "jock itch," and ringworm. The agricultural blights of corn smut and wheat rust are fungal infections. Indeed, fungi probably are the single worst destroyers of crop plants. One survey in Ohio indicated that only 50 plant diseases there were caused by bacteria and 100 by viruses, but 1,000 by fungi. If you look at FIGURE 21.3, you can see an example of what happened to American elm trees in cities across the United States beginning in the 1930s because of a fungus called *Ophiostoma ulmi*—the cause of Dutch elm disease. In recent years, molds have caused great damage to housing in such states as Texas and California. In Texas alone, insurance companies paid more than $1 billion in mold insurance claims in 2001–2002.

Even given all this fungal pestilence, however, we could say of the fungi what we said of bacteria: While we could do without some of them, we could not live without others. If you know someone who has high cholesterol, he or she may control it with one of the so-called statin drugs (Pravastatin, Lovastatin, or Simvastatin). All these drugs contain a compound produced by the fungus *Aspergillus terreus*. In the realm of food production, brewer's yeast makes bread rise and beer ferment, and blue cheese is blue because of the veins of mold within it. Most soft drinks contain citric acid that is produced from a fungus, and we eat fungi outright in the form of mushrooms. Out in nature, fungi join bacteria as the major "decomposers" of the living world, breaking down organic material such as garbage and fallen logs and turning it into inorganic compounds that are recycled into the soil. In fact, the final breakdown of woody material is almost entirely the work of fungi. And, as noted, fungi are involved in a critical association with plants, about which you'll learn more later.

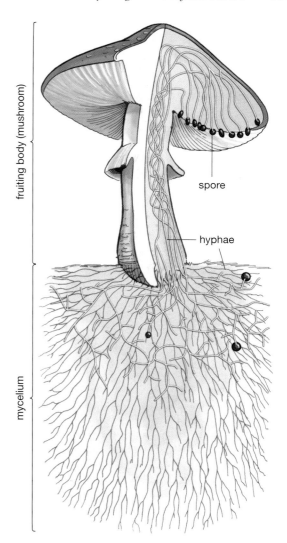

FIGURE 21.2   Structure of a Fungus Fungi are composed of tiny slender tubes called *hyphae*. The hyphae form an elaborate network called a *mycelium*. The same hyphae also form a reproductive structure called a fruiting body, which in this case is the familiar mushroom.

(a) Gillet Avenue, Waukegan Illinois, 1962   **(b)** Gillet Avenue, Waukegan Illinois, 1972

FIGURE 21.3   The Effects of Dutch Elm Disease

So Far... Answers:

1. hyphae; mycelium
2. (b)
3. breaking down organic materials, thus recycling the compounds that make them up

Some 77,000 species of fungi have been identified so far, but this is just a fraction of the number that actually exist. Because fungi tend to grow in inaccessible places, most of them are unknown to us. By one estimate, the total number of fungal species may be about 1.6 million.

So Far...

1. The essential structure of most fungi is a set of slender filaments called _____ that collectively make up a structure called a _____.

2. Which of the following is not a characteristic of fungi? (a) fixed in one spot (b) make their own food (c) tend to be multicellular

3. Fungi are indispensable to the natural world because of their role in _____.

21.3 Reproduction in Fungi

Let's now look at just one type of fungus as a means of learning something about fungi in general. In FIGURE 21.4, you can see a so-called club fungus that we know as the common mushroom. If you look down at the figure's lower right, you can see that the mushroom's hyphae amount to thin lines of cells—something that is true of all hyphae.

FIGURE 21.4 The Life Cycle of a Fungus

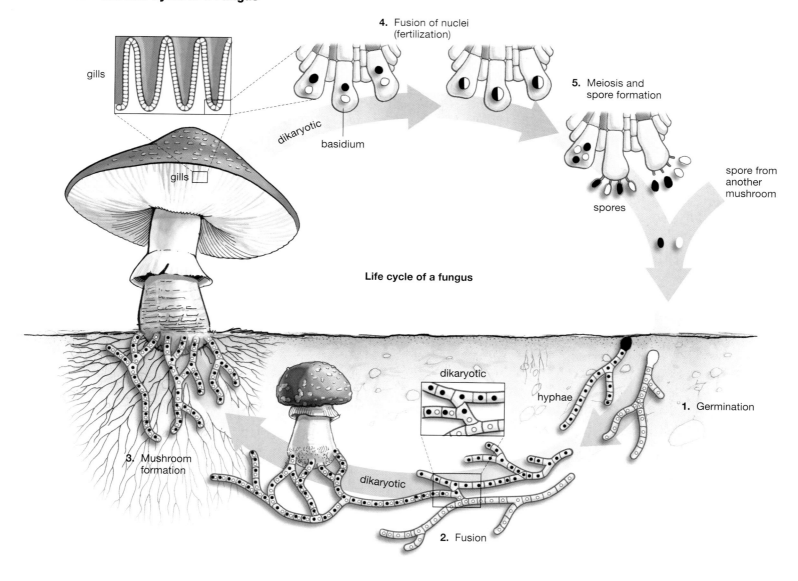

4. Fusion of nuclei (fertilization)

gills

dikaryotic

basidium

5. Meiosis and spore formation

spore from another mushroom

spores

gills

Life cycle of a fungus

dikaryotic

hyphae

1. Germination

3. Mushroom formation

dikaryotic

2. Fusion

If you look at the upper left of the figure, you can see how the mushroom cap is structured. The underside of each cap is made up of accordion-like folds called gills. Each gill is simply a collection of hyphae. And right at the tip of some of these hyphae is a reproductive structure called a basidium. Reproductive cells called spores are produced on the basidium's surface and then ejected from it. Caught by the wind, these spores are then carried away to new ground on which they can "germinate" or sprout new hyphae. Fungal spores of this sort are, however, something like lottery tickets: multitudes are made, but only a few will be winners. Such a tiny fraction of spores end up germinating that huge numbers must be released to ensure that all nearby environments receive some. Tens of millions will be ejected each hour from the underside of a fairly small mushroom cap.

The mushroom cap is one example of a large fungal reproductive structure, but most fungi don't have this kind of size. Think of a flat circle of blue-gray mold on some spoiled bread. Reproductive structures are present, topped by spores, but magnification is necessary to make out any of this detail, as you can see in FIGURE 21.5. To a great extent, fungi can be regarded as microbes.

The Life Cycle of a Fungus

We've noted the role that fruiting bodies play in releasing mushroom spores, but how do these spores get produced in the first place? For that matter, what's a spore? These questions can best be understood within the context of the life cycle of a fungus, which is set forth in Figure 21.4. Taking the life cycle from the beginning, over at step 1, you can see that it starts with spores that have been released from two separate fruiting bodies (in this case mushrooms). These spores germinate, generate hyphae, and in step 2, two of these hyphae fuse—two hyphal cells come together and become one cell, from which a single hypha grows. Note, however, that the cells in this new hypha have *two* nuclei each (symbolized by the little black and white dots within them). This is an unusual condition for a cell; in the normal case, one cell has one nucleus. This condition has come about because, while two cells fused in step 2, their nuclei remained separate. Thus begins a phase of life that is unique to fungi: the **dikaryotic phase**, in which cells in a fungal mycelium have two nuclei. Once a mushroom mycelium enters dikaryotic phase, it can stay there for a long time. The phase can go on for years, in fact, and ends only when the most visible manifestation of the mycelium, the mushroom cap, sprouts above-ground as seen in step 3.

In step 4, you can see again the structure from which new spores will be released, the basidium. Note that, in the first drawing in step 4, cells at the tip of the basidium are still in dikaryotic phase; but in the second drawing, there is a fusion of the two nuclei in these cells. Now think about where each of these nuclei came from. One of them came originally from the black spore over at step 1, while the other came from the white spore. They existed separately in a single cell for a while, but now they have fused. With this, we have had a fusion of genetic material from two separate organisms—sex has taken place. It's not the merger of egg and sperm we're used to in human reproduction, but it's sex nevertheless.

Finally, as you can see in step 5, the cells whose nuclei fused now divide—in the process called meiosis, reviewed in Chapter 10—and the result is a set of individual spores. These are the spores that will be released by the millions, with a few of them germinating into new hyphae. Note, however, the difference between the spores of the mushroom and the *gametes* of a human being—eggs or sperm. A human being cannot develop from an egg or a sperm; a fusion of both cells is required. Meanwhile, the spores that mushrooms develop from do not undergo fusion. A fungal spore is produced and, once it comes into being, it is dispersed and can develop into a whole new mycelium by itself. Thus, a **spore** can be defined as a reproductive cell that can develop into a new organism without fusing with another reproductive cell. The black and white spores of step 5 represent different "mating types." Just as only opposite sexes can come together to create offspring in human beings, so only opposite mating types can come together (in hyphae fusion) to create offspring in fungi that are reproducing sexually.

FIGURE 21.5 Bread Mold Up Close The dark spots on moldy bread are composed of a mycelium that forms stalks tipped by spores, visible in this micrograph as tiny black balls. The spores will be dispersed to produce new mycelia elsewhere—perhaps on more bread. These are spores of the mold *Rhizopus nigricans*, growing on the surface of some bread.

A BROADER VIEW

Deadly Mushrooms

Hardly a year goes by without a newspaper story recounting a tale of a hiker who has either died or become desperately ill after eating poisonous mushrooms. Such mushrooms are relatively rare—of the thousands of mushroom species, only about 50 are even toxic—and a single species, *Amanita phalloides*, may account for 90 percent of fatalities worldwide. Even so, it's good to recall an old saying: There are old mushroom hunters, and there are bold mushroom hunters, but there are no old, bold mushroom hunters.

(a) Spores on the basidium of a club fungus

(b) One of the cup fungi, the edible morel mushroom

FIGURE 21.6 **Club and Cup Fungi**

21.4 Categories of Fungi and Fungal Associations

Fungi are placed into different large categories or "phlya" based on the sexual reproductive structures they have. As you've seen, mushroom fungi have a reproductive structure called a basidium; accordingly, mushrooms are in the fungal group known as basidiomycetes. Likewise, there is a phylum called the ascomycetes and one called the zygomycetes—both named for the reproductive structures they have. The basidiomycetes are often called *club fungi* (because of the club-like shape of the basidium); the ascomycetes are sometimes called *cup fungi* (because many of their fruiting bodies have cup shapes); and the zygomycetes are called *bread molds* (because they are commonly found growing on old bread). Figure 21.5 shows a representative bread mold. FIGURE 21.6 shows representatives of the club and cup fungi.

Several types of fungi can enter into associations with life-forms outside the fungal kingdom; indeed, these associations are widespread, as we'll now see.

Lichens

If you've ever taken a walk in the woods, you've probably seen the thin, colorful coverings called *lichens* that seemingly can grow anywhere—on rocks as well as on trees; in Antarctica as well as in a lush forest (**see** FIGURE 21.7). As it turns out, lichens are not a single organism. A **lichen** is a composite organism composed of a fungus and either algae or photosynthesizing bacteria. If you look at FIGURE 21.8, you can see that this association is structured as a kind of sandwich: an upper layer of densely packed fungal hyphae on top; then a zone of less-dense hyphae that includes a layer of algal or bacterial cells; then another layer of densely packed hyphae on the bottom that sprout extensions down into the material the lichen is growing on (such as a rock). The example here is an association between

A BROADER VIEW

Smut in Two Forms

The corn- and wheat-attacking pathogens called smuts are one variety of club fungi. Given this harmful role, isn't it interesting that the English word *smut* refers both to a variety of fungus and to pornography?

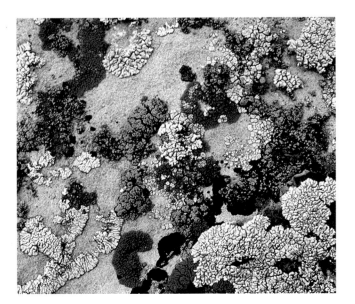

FIGURE 21.7 **Life on the Rocks** Lichens growing on sandstone rocks, near the Montana-Wyoming border.

a fungus and an alga, which is the case in about 90 percent of lichens. In such a relationship, the fungal hyphae either wrap tightly around the algal cells or actually extend into them.

A lichen is mostly a fungus, then, but it is a fungus that could not grow without its algal partner. Why? Because the alga makes its own food through photosynthesis and then proceeds to supply the fungus with most of the nutrients it needs. For years, it was assumed that what fungi do for algae is keep them from drying out. Increasingly, however, scientists have come to believe that most fungi are doing nothing at all for the algae, in which case the fungi should be considered an algal parasite, rather than a partner.

Mycorrhizae

By some estimates, 90 percent of seed plants live in a cooperative relationship with fungi. The linkage here is one of plant roots and fungal hyphae, with both the plants and the fungi benefiting. What the fungus gets from the association is food, in the form of sugars that come from the photosynthesizing plant; what the plant gets is minerals and water, absorbed by the profusion of fungal hyphae. So important is this relationship to plants that some species of trees, such as pine and oak, cannot grow without fungal partners. Other plants will have stunted growth without a fungal partner.

Fungal hyphae generally grow into the plant roots, but in some instances they wrap around the root without penetrating it. In either case, the root-hyphae associations of plants and fungi are known as **mycorrhizae** (**see** FIGURE 21.9). The mycorrhizal relationship is one of the oldest and most important in nature. It appears that at least 460 million years ago, when the very first plants made the transition to life on land, they did so with fungal partners. Research done recently indicates that mycorrhizal relationships are broader than one-plant/one-fungus. A given fungus, receiving sugars from a plant, may go on to transmit some of this food to a second plant. Indeed, mycorrhizae sometimes seem to function like an underground food network for a broad plant/fungus community.

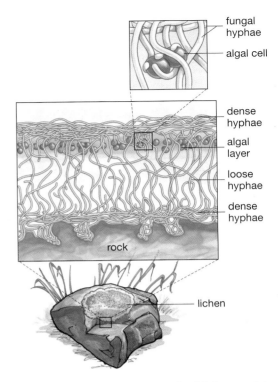

FIGURE 21.8 **The Structure of a Lichen** A lichen is not a single organism. It is composed of a fungus and an alga (or sometimes a photosynthesizing bacterium). The fungal hyphae form a dense layer on the top and the bottom and a loose layer in the middle, within which the alga is nestled. The fungus may provide a moist, protective environment for the alga, and the alga, which performs photosynthesis, provides food for the fungus.

Figure 21.8 labels: fungal hyphae, algal cell, dense hyphae, algal layer, loose hyphae, dense hyphae, rock, lichen

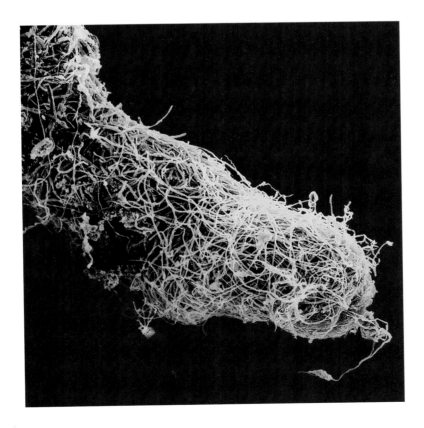

FIGURE 21.9 **Underground Partners** Mycorrhizae are associations between plant roots and fungal hyphae that benefit both the plant and the fungus. Shown is an association between the root of a eucalyptus tree and thread-like hyphae that are wrapped around it. The hyphae help bring water and minerals to the trees, while the tree provides the fungus with food.

So Far... Answers:

1. *fusing with another reproductive cell*

2. *fungus; algae; bacteria*

3. *fungal; plant*

So Far...

1. A spore is a reproductive cell that can develop into a new organism without _____.

2. A lichen is a composite organism, always composed of a _____ and either _____ or _____.

3. The cooperative associations called mycorrhizae provide the _____ partner with sugars, while providing the _____ partner with water and minerals.

A BROADER VIEW

Slow but Steady

The lives of lichens bear some comparison to French movies: They tend to be long and slow. Some lichens have been alive for thousands of years, but at most they spread out at a rate of about 3 centimeters a year, which is a little over an inch. Indeed, they have been observed to grow as little as 1 millimeter per year, at which rate they would grow 4 inches in a century.

21.5 Plants: The Foundation for Much of Life

It was noted earlier that, while we human beings have reason to feel ambivalent about fungi and bacteria, our relationship with plants is not so complicated. Indeed, we might ask why it is that dogs own the title of "man's best friend" when a strong case can be made that this honor should belong to plants. For starters, we humans are utterly dependent upon plants for food. Try naming something you eat that isn't a plant or that didn't itself eat plants, and you will end up with a very short list. Then there is the atmospheric oxygen that plants produce as a by-product of photosynthesis. Beyond this, plants stabilize soil, provide habitat for animals, and lock away the "greenhouse" gas carbon dioxide. They yield lumber and medicines, and certain varieties of them are so beautiful that human beings spend hours tending them. Against all this, what's the worst thing we could say of them? That poison ivy is a problem, or that mowing the grass can be a pain? What other form of life helps us so much and harms us so little?

The Characteristics of Plants

We all have an intuitive sense of what plants are, though not all plants match these intuitions. A few are parasites (on other plants) and some, like the Venus flytrap, consume animals. Allowing for such exceptions, plants share the following characteristics.

FIGURE 21.10 **Characteristics of Plants** All cells have an outer membrane, but plant cells have a wall external to this membrane. The compounds cellulose and lignin, which can help make up the cell wall, impart strength to it. Plant cells have a higher proportion of water in them than do animal cells, with much of this water located in an organelle called a central vacuole. The sites of photosynthesis in plants are the organelles called chloroplasts.

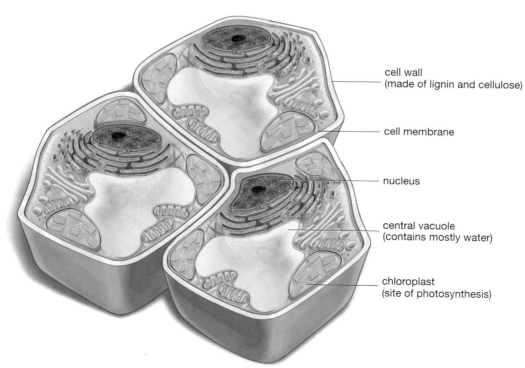

cell wall
(made of lignin and cellulose)

cell membrane

nucleus

central vacuole
(contains mostly water)

chloroplast
(site of photosynthesis)

- *Plants are photosynthesizing organisms that are fixed in place, multicelled, and mostly land-dwelling.* We know that plants are fixed in one spot (sessile), and that they make their own food through photosynthesis. That they are multicelled follows from the fact that they develop from embryos, which are multicelled by definition. During the course of their evolution from green algae, plants made a transition from water to land, and land is where most of them are found today—though some have made the transition *back* to water, as with water lilies.

- *Plant cells have cell walls.* Both plant and animal cells are enclosed by a plasma membrane, but plant cells have something just outside the membrane: a **cell wall**, defined as a relatively thick layer of material that forms the periphery of plant, bacterial, and fungal cells (**see** FIGURE 21.10). The plant cell wall is composed in large part of a tough, complex compound called cellulose. In woody plants, cellulose is joined by a strengthening compound called lignin. Together, cellulose and lignin allow plant cells to support the massive weight of trees.

- *Plant (and algae) cells have organelles called chloroplasts that are the sites of photosynthesis.* If there is one feature that has fundamentally shaped plant evolution, it is the chloroplast. Why are plants *able* to be fixed in one spot? Because they make their own food in chloroplasts and thus don't need to move toward food, as animals do.

- *Successive generations of plants go through what is known as an alternation of generations.*

One way to conceptualize this characteristic is to compare plant reproduction with human reproduction. The eggs and sperm that human beings produce are so-called gametes that are *haploid* cells, meaning they have but a single set of chromosomes in them. When these haploid cells fuse in the moment of conception, what's produced is a *diploid* fertilized egg, meaning an egg with two sets of chromosomes—one set from the father and one from the mother. This egg gives rise to more diploid cells through cell division, and the result is a whole new human being (**see** FIGURE 21.11a).

A typical *plant* in its diploid phase will, like human beings, produce a specialized set of haploid reproductive cells. Instead of these cells being egg or sperm, however, what's produced is a *spore*: a reproductive cell that can develop into a new organism without fusing with another reproductive cell. Like the spores we saw in fungi, plant spores have the ability to grow into a new generation of plant all on their own. In some plants, this spore lands on the ground, starts dividing, and after a time a mature plant exists. When *this* plant reproduces, however, it does not do so through more spores. It produces haploid gametes—eggs and sperm. These go on to fuse, and the result is a diploid plant just like the one we started with.

The upshot of all this, as you can see in FIGURE 21.11b, is that plants move back and forth between two different kinds of generations, one of them spore-producing, the other gamete-producing. In this cycle, the generation of plant that produces spores is known as the **sporophyte generation**, while the generation of plant that produces the gametes is the **gametophyte generation**. The fern shown in Figure 21.11b gives you some idea of how physically different these generations can be, but the disparities actually can be much greater than this. In a massive organism such as a redwood tree, the tree that is familiar to us is the sporophyte generation. But what do its spores develop into? On the male side, barely visible pollen grains. On the female side, small collections of cells, hidden deep within a redwood "pinecone." The fact that these gametophyte individuals are tiny, however, doesn't mean that they are unnecessary. The eggs and sperm they produce are required to bring about a whole new redwood tree. In sum, all plants go

A BROADER VIEW

Plants and Human History

In his best-selling book *Guns, Germs, and Steel,* Jared Diamond argues that Europe ended up colonizing the New World—rather than the other way around—in part because Europe got lucky in what might be called the plant sweepstakes. The European environment was especially conducive to the growth of highly nutritious cereal plants. The result was a surplus of food that allowed people in European societies to specialize in terms of vocation, becoming craftsmen and soldiers, for example, rather than subsistence hunters and farmers.

(a) Human reproduction

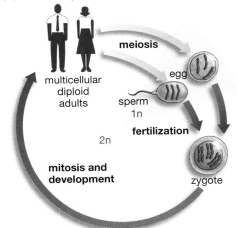

(b) Plant alternation of generations

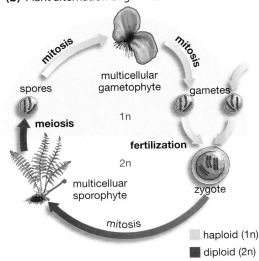

FIGURE 21.11 **The Alternation of Generations in Plants, Compared to the Human Life Cycle**

(a) Humans Almost all cells in human beings are diploid or 2n, meaning they have paired sets of chromosomes in them. The exception to this is human gametes (eggs and sperm), produced through meiosis, which are haploid or 1n, meaning they have a single set of chromosomes. In the moment of conception, a haploid sperm fuses with a haploid egg to produce a diploid zygote that grows into a complete human being through mitosis.

(b) Plants In plants, conversely, diploid (2n) plants—the multicellular sporophyte fern in the figure—go through meiosis and produce individual haploid (1n) spores that, without fusing with any other cells, develop into a separate generation of the plant. This is the multicellular gametophyte shown. This gametophyte-generation plant then produces its own gametes, which are eggs and sperm. Sperm from one plant fertilizes an egg from another, and the result is a diploid zygote that develops into the mature sporophyte generation. The alternation between the sporophyte and gametophyte forms is called the alternation of generations.

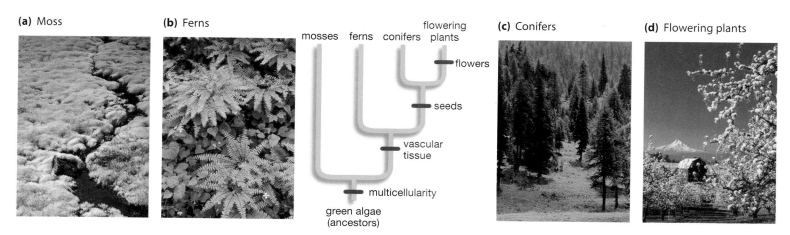

(a) Moss **(b)** Ferns **(c)** Conifers **(d)** Flowering plants

FIGURE 21.12 **Four Main Varieties of Plants** The most primitive type of plant, **(a)** the bryophytes (moss in the figure) lack a fluid-transporting vascular system. **(b)** Ferns, representing the seedless vascular plants, have a vascular system but do not have seeds. **(c)** Gymnosperms (conifers in the picture) do utilize seeds. **(d)** Flowering plants (blossoming pear trees in the picture) produce seeds and are responsible for several other plant innovations, among them fruit.

FIGURE 21.13 **Three Kinds of Bryophytes** **(a)** Mosses, **(b)** liverworts, and **(c)** hornworts. These plants have no vascular tissue and thus tend to be small. The sperm of bryophytes must travel through water to get to eggs. For this reason bryophytes are found most often in moist environments.

(a) Mosses

(b) Liverworts

(c) Hornworts

through an **alternation of generations**: a life cycle practiced by plants, in which successive plant generations alternate between the diploid sporophyte condition and the haploid gametophyte condition.

21.6 Types of Plants

For our purposes, it is convenient to separate the members of the plant kingdom into just four types (**see** FIGURE 21.12). These are the *bryophytes*, which include mosses; the *seedless vascular plants*, which include ferns; the *gymnosperms*, which include coniferous ("cone bearing") trees; and the *angiosperms*, a vast division of flowering plants—by far the most dominant on Earth today—that includes not only flowers such as orchids but also oak trees, rice, and cactus. We'll look at all four types briefly here, starting with the relatively primitive plants that were the first to move on to land from the water.

Bryophytes: Amphibians of the Plant World

If you visit low-lying, wet terrain, you are likely to see a carpet-like covering of moss. Mosses are the most familiar example of a primitive type of plant that falls under the classification of **bryophyte**: a type of plant that lacks a true vascular system (**see** FIGURE 21.13). What's a vascular system? A network of tubes within an organism that serves to transport fluids. We can get a good idea of what the bryophytes are like by looking at the mosses.

Mosses are representatives of some of the earliest plants that made the transition from water to land. As such, they can be thought of as plants that made only a partial break with the aquatic living of their evolutionary ancestors, the algae. In making this transition, they had to deal with something the algae didn't, which is the effect of gravity. Lacking a vascular system, mosses cannot transport water and other substances very *far* against the force of gravity; instead they must lie low, hugging the surface to which they are attached while spreading out horizontally to maximize their exposure to sunshine. In keeping with their aquatic origins, they have sperm that can get to eggs only by swimming through water. Not much water is necessary; a thin film left over from the morning dew will suffice. But some water must be present. Not surprisingly, mosses and other bryophytes usually are found in moist environments.

Seedless Vascular Plants: Ferns and Their Relatives

As noted, all plants except the bryophytes have a *vascular system*, or network of fluid-conducting tubes, which transport both food and water. When we begin to look at vascular plants, we find that the most primitive variety of them are the **seedless vascular plants**: plants that have a vascular system, but that do not produce seeds as part of reproduction.

Easily the most familiar representatives of the seedless vascular plants are ferns, with their often beautifully shaped leaves, called fronds (**see** FIGURE 21.14). These plants have moved a step further in the direction of separation from an aquatic environment. Their vascular system allows them to grow *up* as well as out. Despite this evolutionary innovation, the sperm of the seedless vascular plants is like that of the bryophytes; it needs to move through water to fertilize eggs.

A BROADER VIEW

Antiseptic Peat Moss

The dead, submerged portions of peat moss are very acidic—about as acidic as vinegar—with the result that not much except peat moss can remain alive in the "bogs" that these bryophytes help create. This quality preserved the remains of the 2,000-year-old "Lindow man," found in a bog in England in 1984. Fungi and bacteria that would normally decompose such a body couldn't live in the bog's acidic environment. During World War I, the antiseptic quality of peat moss combined with its absorptive power to make it a widely used dressing for war wounds.

(a) Ferns

(b) Horsetails

(c) Club mosses

FIGURE 21.14 **Three Kinds of Seedless Vascular Plants (a)** Fall-colored ferns in New Hampshire, **(b)** horsetails, and **(c)** club mosses. Because these plants have vascular tissue, they are able to grow taller than most bryophytes. Like bryophytes, however, they do not produce seeds and are tied to moist environments.

(a) Spruce tree

(b) *Ginkgo biloba*

FIGURE 21.15   **Gymnosperms**

(a) This spruce tree is a member of the grouping of gymnosperms known as conifers, which account for about three-quarters of all gymnosperm species. The conifers also include redwood, pine, juniper, and cypress trees.

(b) There are other types of gymnosperms as well. Pictured are the leaves and seeds of the maidenhair tree, *Ginkgo biloba*, which are used today in herbal preparations.

The First Seed Plants: The Gymnosperms

The first variety of seeded plant to evolve were the gymnosperms, which began to replace seedless vascular plants about the time the dinosaurs came to dominance on Earth. The dominance of the gymnosperms was relatively short-lived, however, as these plants were in turn succeeded by the angiosperms. Today there are only about 700 gymnosperm species compared to 260,000 angiosperm species. Nevertheless, the gymnosperm presence is considerable, especially in the northern latitudes, where they exist as vast bands of coniferous trees, including pine, fir, and spruce. Conifers such as these provide most of the world's lumber (**see** FIGURE 21.15).

Why were the gymnosperms able to outcompete so many of the seedless vascular plants back in the dinosaur days? Part of the answer lies in sperm transport. In gymnosperms, sperm are contained in tiny spheres called pollen grains. Critically, these grains are transported to the female eggs through the *air*, rather than being limited to *swimming* to them, as is the case with the sperm of seedless plants. With their pollination innovation, gymnosperms could propagate over great distances, which gave them a competitive advantage over the seedless plants.

Another advantage the gymnosperms had were their seeds, which can be thought of as tiny packages of food and protection for developing embryos. If you look at FIGURE 21.16, you can see one example of a gymnosperm seed, this one from a pine tree. FIGURE 21.17 then shows you how this seed fits into the life cycle of the pine. As you can see, pollen grains are carried on the wind to the female reproductive structure, which resides in the familiar pinecone. Then the sperm in the pollen grain comes together with an egg in the reproductive structure, and the result is a fertilized egg that begins to develop as an embryo. This is no different in principle than human egg and sperm coming together to create a human embryo. In the pine tree, however, the embryo is developing inside a seed, a structure that is unique to plants. Seeds have a tough coat, inside of which is not only the embryo but a food supply for it—a kind of sack lunch of stored carbohydrates, proteins, and fats. In sum, a **seed** is a plant structure that includes a plant embryo, its food supply, and a tough, protective casing.

Gymnosperm seeds turn out to differ from the seeds produced by angiosperms. Angiosperm seeds come wrapped in a layer of tissue—called fruit—

FIGURE 21.16   **A Gymnosperm Seed** Seeds are tiny packages of food and protection. They come in many shapes and sizes, but they all contain an embryo, some food, and a protective seed coat.

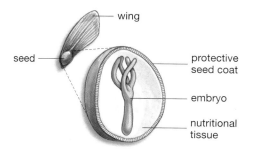

seed — wing

protective seed coat

embryo

nutritional tissue

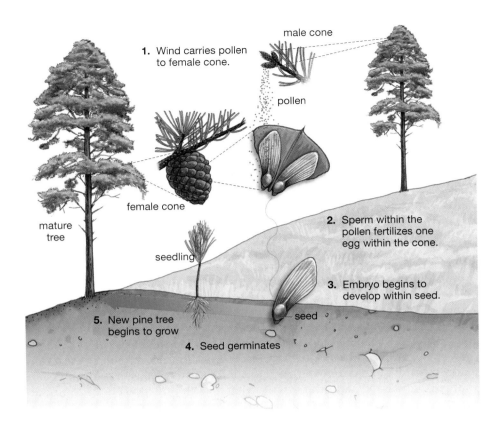

1. Wind carries pollen to female cone.

male cone

pollen

female cone

mature tree

seedling

2. Sperm within the pollen fertilizes one egg within the cone.

3. Embryo begins to develop within seed.

5. New pine tree begins to grow

seed

4. Seed germinates

FIGURE 21.17 The Life Cycle of a Gymnosperm
Pine trees have two kinds of cones. Pollen is produced within the smaller male cones, while eggs are produced within the larger female cones. When the wind carries pollen onto the female cone, the sperm within the pollen fertilizes one of the eggs within the cone. An embryo then begins to develop inside a seed, which falls to the ground. Once conditions are suitable, the seed germinates and a whole new pine tree begins to grow.

that gymnosperm seeds do not have. With this in mind, a gymnosperm can be defined as a seed plant whose seeds are not surrounded by fruit. The very name *gymnosperm* comes from the Greek words *gymnos*, meaning "naked," and *sperma*, meaning "seed."

Angiosperms: Earth's Dominant Plants

The term *flowering plants* may bring to mind roses or tulips and, indeed, these flowers are angiosperms. But the angiosperm grouping includes all manner of other plants as well—almost all trees except for the conifers, all our important food crops, cactus, shrubs, common grass: The list is very long (**see** FIGURE 21.18).

FIGURE 21.18 Angiosperm Variety

(a) Calla lilies on the California coast

(b) Cholla cactus in Arizona

(c) Corn in a field

(a) Pollen-covered honeybee on a dandelion **(b)** Carib martinique pollinating a flower **(c)** Lesser long-nosed bat pollinating a cactus

FIGURE 21.19 Animals Help Pollen Get from Here to There Flowering plants were the first to take advantage of the mobility of animals to transport pollen from one plant to another. Relationships between flowers and pollinators can be species-specific, which helps to ensure that the pollen will be delivered to the right address.

As noted, **angiosperms** are defined by an aspect of their anatomy; they are plants whose seeds are surrounded by the tissue called fruit. With respect to the sperm transport we've been looking at, angiosperms were responsible for the innovation you can see in FIGURE 21.19: by and large, angiosperms rely on *animals* to move sperm-bearing pollen from one plant to another. This means of transport is unique to angiosperms and very important to them. Why do flowers have nectar and their fragrances and their colors? Because all of these features serve to attract animal pollinators.

Some flowering plants also add an inducement for the dispersal of their *seeds*: the tissue called fruit. This is not fruit in the technical sense, but fruit as that term is commonly understood—the flesh of an apricot or cherry, for example. To understand how the production of fruit benefits plants as well, imagine a bear who consumes a wild berry, which consists not only of the fruit flesh, but of the seeds inside this flesh. The fruit is digested by the bear as food, but the *seeds* are tough; they are passed through its system intact to be deposited, with bear feces as fertilizer, at a location that may be very remote from the place where the bear *ate* the berry. The result is seed dispersal at what may be a promising new location for a berry plant (**see** FIGURE 21.20).

FIGURE 21.20 Seed Carriers Angiosperms have taken advantage of the mobility of animals not only to transfer pollen but also to disperse seeds.

(a) Some seeds are wrapped in tasty fruit that is consumed by animals, such as bears.

(b) Other seeds come wrapped in burrs and spines that stick to the fur of animals and are carried away.

(a) Wild berries induce seed dispersal **(b)** Burrs help seeds travel

On to a Look at Animals

In this chapter and the one that preceded it, you've looked at bacteria and archaea, and at three of the kingdoms within the eukaryotic domain—protists, fungi, and plants. This leaves one more eukaryotic kingdom to go. It is a kingdom that is arguably the most familiar of all, because we human beings are part of it. Yet it is a kingdom so diverse that it holds a good many surprises as well. It is the kingdom of animals.

So Far... Answers:

1. *spores; eggs; sperm*

2. *spruce tree; orchid*

3. *wind*

So Far...

1. All plants go through an alternation of generations in which one generation produces the reproductive cells called _____, while the alternate generation produces the reproductive cells _____ and _____.

2. Of the following plants, which produce(s) seeds? moss, fern, spruce tree, orchid

3. The pollen of gymnosperms generally is transported by _____.

Chapter Review

Summary

21.1 The Fungi

- **Nutrition in fungi**—Fungi do not make their own food, as plants do, but instead are heterotrophs—they consume existing organic material in order to live. Fungi consist largely of webs of slender tubes, called hyphae, that grow toward food sources. Fungi cannot immediately ingest the food the hyphae reach, since only small molecules can pass through the cell walls of fungal cells. Fungi thus digest their food externally, through release of digestive enzymes, and then bring the resulting small molecules into the hyphae.   (page 352)

- **The structure of fungi**—Collectively, hyphae make up a branching web, called a mycelium, that forms the bulk of most fungi. Many fungi have a reproductive structure called a fruiting body that produces and releases reproductive cells called spores. Fungi are sessile or fixed in one spot, though their hyphae grow toward food sources.   (page 352)

- **Multicellular**—Almost all fungi are multicellular. Single-celled yeasts are the sole exception to this.   (page 353)

21.2 Roles of Fungi in Society and Nature

- **Harmful and helpful fungi**—Fungi are the cause of many crop diseases, human infections, and forms of damage to buildings and homes. They are also sources of medicines, and they are used extensively in food processing. In nature, fungi join bacteria as major "decomposers" of the living world, breaking down organic material such as fallen trees and recycling the resulting products into the soil. (page 353)

21.3 Reproduction in Fungi

- **Fruiting bodies and spores**—The fruiting body of the mushroom fungi, the mushroom cap, is made up on its underside of accordion-like gills that both produce spores and release them into the wind, which carries them to new locations. Only a tiny fraction of fungal spores will successfully germinate, or sprout new mycelia. Because of this, fruiting bodies release huge quantities of spores as a means of ensuring new fungal growth.   (page 354)

- **Dikaryotic phase**—Within their life cycle, many fungi have a so-called dikaryotic phase of life, in which the hyphae from two separate fungi fuse to produce a single hypha whose cells each have two nuclei (one from each of the original two organisms).   (page 355)

- **Sexual union**—When the below-ground mycelium in a mushroom fungus sprouts into the above-ground mushroom cap, all the cells in the cap initially are dikaryotic. The nuclei of certain cells in the cap's reproductive structure, the basidium, then undergo fusion, however, meaning a union of genetic material from the two original separate organisms has occurred—that is, sex has taken place.   (page 355)

- **Spores**—The cells that undergo fusion of their nuclei produce spores that are attached to the tips of the basidium. These spores are then released and blown by the wind to new locations, where they may germinate and result in new mycelia. A spore is a reproductive cell that can develop into a new organism without fusing with another reproductive cell.   (page 355)

Web Tutorial 21.1 The Life Cycle of a Mushroom

21.4 Categories of Fungi and Fungal Associations

- **Three fungal phyla**—There are three major categories or "phyla" of fungi: basidiomycetes (also known as club fungi), ascomycetes (also known as cup fungi), and zygomycetes (also known as bread molds). Fungi are placed in these categories in accordance with the structures they use in sexual reproduction.   (page 356)

- **Lichens**—Lichens are composite organisms, made up of both fungi and algae (or fungi and photosynthesizing bacteria). Within a lichen, the relationship between fungi and algae may be mutually beneficial to both kinds of organisms: Fungi derive food from the photosynthesizing algae, and the algae are shielded by the fungi from forces that would dry them out. Many scientists have concluded, however, that algae derive nothing from the relationship, in which case the fungi actually are parasites.   (page 356)

- **Mycorrhizae**—Up to 90 percent of seed plants live in a cooperative association with fungi that links plant roots with fungal hyphae. In this relationship, plants supply fungi with food produced in photosynthesis, while the fungi supply plants with minerals and water, gathered by the web of fungal hyphae. Associations of plant roots and fungal hyphae are called mycorrhizae.   (page 357)

21.5 Plants: The Foundation for Much of Life

- **The importance of plants**—Plants are the foundation for much of life on Earth, because they are responsible for much of the living world's production of food and oxygen. In addition, they stabilize soil, provide habitat for animals, and lock up carbon dioxide.   (page 358)

- **Plant characteristics**—All plants are multicelled, and almost all are fixed in one spot and carry out photosynthesis. All plant cells have a cell wall and contain organelles called chloroplasts, which are the sites of photosynthesis.   (page 358)

- **Alternation of generations**—Plants reproduce through an alternation of generations: a life cycle in which successive plant generations produce either diploid spores (the sporophyte generation) or haploid gametes (the gametophyte generation). Within a given species, these two generations can differ greatly in size and structure.   (page 359)

Web Tutorial 21.2 Alternation of Generations

21.6 Types of Plants

- **Four kinds of plants**—The four principal categories of plants are bryophytes, seedless vascular plants, gymnosperms, and angiosperms. (page 360)

- **Bryophytes**—Bryophytes, such as mosses, are representative of the earliest plants that made the transition from water to land. They lack a fluid-transport or vascular system and thus tend to be low-lying. Bryophyte sperm can get to eggs only by swimming through water. Because of this, bryophytes are most commonly found in damp environments.   (page 360)

- **Seedless vascular plants**—Seedless vascular plants, represented by ferns, have a vascular system but do not produce seeds in reproduction. Their sperm must move through water to fertilize eggs.   (page 361)

- **Gymnosperms**—Gymnosperms, represented by coniferous trees, are seed-bearing plants whose seeds are not encased in tissue called fruit. There are only about 700 gymnosperm species, but their presence is considerable, particularly in northern latitudes. The sperm of gymnosperms is encased in pollen grains that are carried to female reproductive structures by the wind. Gymnosperms carry out reproduction by producing seeds: plant structures that include a plant embryo, its food supply, and a tough, protective casing.   (page 362)

- **Angiosperms**—Angiosperms, or flowering plants, are seed plants whose seeds are encased in tissue called fruit. Angiosperms are easily the most dominant group of plants on Earth, with some 260,000 species having been identified to date. Angiosperm pollen grains generally are transferred from one plant to another by animals, such as insects and birds. Such animal-assisted pollination is unique to angiosperms. The tissue called fruit, produced as a seed covering by angiosperms, can act as an incentive for animal dispersal of angiosperm seeds.   (page 363)

Key Terms

alternation of generations p. 360	dikaryotic phase p. 355	lichen p. 356	seedless vascular plant p. 361
angiosperm p. 364	gametophyte generation p. 359	mycelium p. 352	sessile p. 353
bryophyte p. 360	gymnosperm p. 363	mycorrhizae p. 357	spore p. 355
cell wall p. 358	hyphae p. 352	seed p. 362	sporophyte generation p. 359

Testing Your Understanding

In Your Own Words   *(answers in the back of the book)*

1. Fungi could be said to take a "lottery" approach to reproduction. In what respect?

2. How does a spore differ from a gamete?

3. Describe two common associations that fungi make with other organisms. What does each party in each of these associations do for the other, and what do they gain themselves?

Thinking about What You've Learned

1. Animals move because they need to—they have to get to prey and they must evade predators. These dual needs led to the evolution of the animal nervous system, which in turn led to the development of varying levels of intelligence in animals. Meanwhile plants never had a need to move, because they made their own food through photosynthesis. Do the separate evolutionary paths of animals and plants mean that there is an inherent value to struggle in the living world—in the animals' case, the struggle to obtain food?

2. Like all living things, fungi require water to begin growing. The recent scourge of mold fungi in American homes has primarily affected new homes. Why should they be affected more than old homes?

3. Plants and fungi often are thought of as being alike, but in one way they lie at opposite ends of a cycle involving organic material. Can you think of what this cycle is and what roles plants and fungi play in it?

Animals: The Diversity of Life 3

A barn owl (*Tyto alba*) swoops to capture its prey.

THE GRAND RAPIDS PRESS

A WHALE OF A FIND

By The Associated Press
November 19, 2003

Japanese scientists say they have identified a new species of whale—a remarkable discovery if confirmed. The animal is a type of baleen, the family of whales that strain tiny plankton and other food from seawater, the researchers say.

"Can you imagine? An animal of more than 10 meters was unknown to us even in the 21st century," said Tadasu Yamada of Tokyo's National Science Museum, the senior author of the study that appears in this week's issue of the journal *Nature*.

MILWAUKEE JOURNAL SENTINEL

LURKING BELOW: MYTH, MYSTERY OF THE GIANT SQUID

By Henry Fountain
May 17, 2004

With a length up to 75 feet, the giant squid, Architeuthis, is the largest invertebrate on Earth. But it is also the most elusive. It has never been seen alive in its natural habitat. As such, Architeuthis (pronounced ark-uh-TOOTH-us) has something of a mythical reputation. There has been speculation that the creatures live for decades, even a century, at depths of several thousand feet.

22.1 What Is An Animal?

Here's an idea for a quiz-style CD game that could sit alongside the popular games *You Don't Know Jack: Television* and *You Don't Know Jack: Sports*. Let's call it *You Don't Know Jack: Animals*. Even the experts can join in for these questions. For $50, how many species of the whales called roquals are there? If you answered seven, you might be right; but then again, if you answered eight or nine you might be right too. As the newspaper story above makes clear, things are still in flux with respect to how many whale species there are on Earth. OK, so how about this: Just to within the nearest thousand feet, how far down in the ocean does the giant squid

The red sea urchin *Strongylocentrotus franciscanus*.

Architeuthis dux live? Until recently, the common best estimate was 2,000–3,000 feet, but a new analysis indicates the answer may be 600–1,000 feet. No one really knows, however, because no one has ever seen a giant squid in its natural habitat. Finally, the sea urchin *Strongylocentrotus franciscanus* is found along the Pacific coast of North America. How long does it live? Until recently, the answer was 7–10 years, but new research indicates it lives at least 50 years and may live up to 200. If so, that spiky little red thing you saw in a California tide pool may have started life in the same year that General Robert E. Lee was born.

All right, it's true that questions like these amount to a stacked deck—experts actually know a great deal about animal life. But isn't it interesting to see some examples of things they don't know? You might expect biologists to be puzzling over the number of bacterial or fungal species, but whale species? We actually could conduct a little *You Don't Know Jack: Animals* quiz for non-experts as well, however. Look at the three species in FIGURE 22.1. Each one of them is an animal, but if you hadn't been told, would you ever have guessed this was the case? Of all the life-forms in the world, animals may be the most familiar to us, but all it takes is a look at a feather star or a sea squirt to let us know that only some animals are familiar. The purpose of this chapter is to give you a sense of what the word *animal* can mean when we look at the entire animal kingdom, rather than just considering those animals who are "like us" in the sense of having, say, legs or a backbone. A good place to start in this effort is to pose a question that would send most *You Don't Know Jack* teams into a brainstorming huddle. The question is: What's an animal?

Defining an Animal

Since whales, feather stars, and sea squirts are all animals, it's pretty clear that common-sense notions about animals won't get us far in defining them. It turns out, however, there is a single, rather technical feature that sets animals apart from all other living things.

- Animals pass through something called a blastula stage in their embryonic development. A *blastula* is a hollow, fluid-filled ball of cells that forms soon after an egg is fertilized by sperm. All animals go through a blastula stage, but no other living things do.

Three other characteristics are found in all animals, but they're found in other kinds of organisms as well. All animals:

- Are multicelled; there are no single-celled animals.

- Are heterotrophs; they must get their nutrition from outside themselves.

FIGURE 22.1 **Animals All** The physical forms of animals are extremely varied.

(a) Ocean-dwelling feather stars such as this one may look like plants, but they are members of the same animal phylum as sea stars. They extend their arms to catch bits of food drifting in the ocean currents.

(b) Sea squirts are members of our own phylum, Chordata. These animals, also known as tunicates, spend their adult lives attached to rocks, filter feeding. Their common name comes from their practice of suddenly squirting a spout of water from an opening near the top of their bodies.

(c) Tube worms such as this one are ocean-dwelling members of a phylum, Annelida, that also includes the common earthworm. The tube worms stay in one spot, burrowed into tubes they themselves have constructed, generally within a coral. The bushy "Christmas trees" that extend from the worm's head are structures called radioles that it uses to filter-feed from the water.

(a) Feather star

(b) Sea squirts

(c) Tube worm

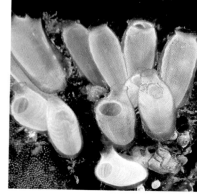

- Are composed of cells that do not have cell walls. (The outer lining of animal cells is the plasma membrane. By contrast, all plants and most other creatures have a relatively thick additional lining—the cell wall—outside their plasma membrane.)

Apart from these universals, animals tend to move; although there are animals that, for at least part of their lives, are as *sessile*—or fixed in one spot—as any plant.

Then there are characteristics of animals that don't fit our preconceived notions. While it's true that animals generally are large relative to bacteria or protists, some animals are microscopic. We often think of animals as creatures with backbones (or "vertebral columns"), but there are far more **invertebrates**, or animals without vertebral columns, than animals with them. Indeed, 99 percent of all animal species are invertebrates. We think of animals primarily as land-living or "terrestrial" creatures, but there is far more animal diversity in the ocean than on land. This makes sense because animals existed in the sea for at least 100 million years before any of them came onto land.

A BROADER VIEW

Blastulas and Stem Cells

In human beings, the blastula is known as the blastocyst, a term that might ring a bell. Blastocysts have been in the news a lot because they are the source of embryonic stem cells—the cells that scientists hope to use to generate new human tissues in connection with such conditions as Parkinson's and Alzheimer's disease.

22.2 Across the Animal Kingdom: Nine Phyla

One way to get a handle on the range of animal life is to look at the large-scale categories of animals and see how the creatures in these categories differ from one another. The animal kingdom's broadest categories are known as phyla, with each *phylum* informally defined as a group of organisms that shares a basic body structure. Depending on who is counting, there are between 36 and 41 animal phyla. What we'll do now is briefly walk through nine of these phyla—a group chosen to give you some idea of the range of animal forms. Once we've finished with this walk, we'll look at some common features of animals and note the degree to which these features are shared across the phyla.

Phylum Porifera—These are the sponges, easily the most primitive form of animal life. Almost all are ocean dwellers, though a few live in freshwater. The adult form is sessile and is a filter feeder—it feeds by drawing water through a system of pores and then filtering out the nutrients.

Phylum Cnidaria—Jellyfish are the symbols of this phylum, though sea anemones and the tiny animals that make up coral reefs are also included in it. Almost all are marine (sea dwelling), though a few live in freshwater. The hallmark of all cnidarians (nee-DAHR-ee-uns) is their use of stinging tentacles to capture prey. They tend to have two life stages: a "medusa" stage that is free-swimming (think of a jellyfish) and a "polyp" stage that is attached to a surface (think of a sea anemone).

Phylum Platyhelminthes—These are the flatworms, which include such human parasites as tapeworms and flukes. Most members of the phyla, however, are tiny worms that carry out life in marine, freshwater, or moist land environments, completely out of contact with humans.

Phylum Annelida—Annelids are the segmented worms. We are very familiar with one member of this phylum, the common earthworm, whose body seems to be made up of a series of little rings. These rings define a series of internal segments that give segmented worms their name. Earthworms live in moist, terrestrial environments; but most annelids are found in marine environments, with a large number also living in freshwater. Leeches are annelids.

Phylum Mollusca—This phylum contains three well-known classes of animals: Gastropods, which include snails and slugs; bivalves, which include oysters, clams, and mussels; and cephalopods, which include octopus, squid, and nautilus. All molluscs possess a layer of tissue, called a mantle, that secretes material that often becomes a shell. The largest and smartest invertebrates in the world are molluscs—squid and octopus.

Phylum Nematoda—These are roundworms, or nematodes, which are found in tremendous numbers in nearly all habitats in which life exists—from farmland, to deserts, to ice caps. Many are microscopic, but a few of the parasitic nematodes reach several meters in length. Several species are a major cause of damage to such crops as soybeans and corn. All nematodes have an external skeleton, called a cuticle, that they shed or "molt" during their lifetime.

Phylum Arthropoda—This is an enormously varied phylum that usually is thought of in terms of its three major groupings, called "subphyla": Uniramia, which includes insects, millipedes and centipedes; Crustacea, which includes shrimp, lobsters, crabs, and barnacles; and Chelicerata, which includes spiders, ticks, and mites. All arthropods have paired, jointed appendages (think of the legs of a grasshopper) and, like the nematodes, all arthropods have a skeleton on the *outside* of their body—an exoskeleton.

Phylum Echinodermata—This is the phylum of "star fish" (sea stars), sea urchins, and such exotic creatures as sea cucumbers. All live in marine environments; some are sessile, but most move slowly across surfaces such as rocks and algae.

Phylum Chordata—These are mostly animals with vertebral columns, though this phylum includes any animal that, at some point in its life, has a rod-shaped support structure called a notochord; a nerve cord on its back or "dorsal" side; a series of openings toward its head called pharyngeal slits; and a "post-anal" tail, meaning a tail located posterior to its anus. As vertebrates, we humans are part of a subphylum in Chordata, called Vertebrata, that also includes all other mammals as well as all fish, amphibians, reptiles, and birds.

So Far...

1. A feature common only to animals is that all animals pass through a _____ stage in their _____ development.

2. Three other features that all animals share is that they are _____, _____, and their cells do not have _____.

3. True or false: vertebrates make up a small portion of the animal kingdom.

22.3 Lessons from the Animal Family Tree

So, how can we make sense of this rich diversity of animal life? What traits do the animals in these nine phyla share? And what is the relationship between, say, a primitive sponge and a clever squid? You can see some answers to these questions in FIGURE 22.2, which sets forth the nine phyla again, this time as an animal family tree. One way to conceptualize this tree is that, over time, animals became more complex through a series of *additions* to the characteristics found in more primitive animals. The twist is that only some varieties of animals evolved to get these additions, while others retained the primitive, "ancestral" condition. To get a handle on this large-scale evolution, a good place to start is with some little red lines you

So Far... Answers:

1. *blastula; embryonic*

2. *Multicelled; heterotrophic (get their food from outside themselves); cell walls*

3. *True; 99 percent of all animal species are invertebrates.*

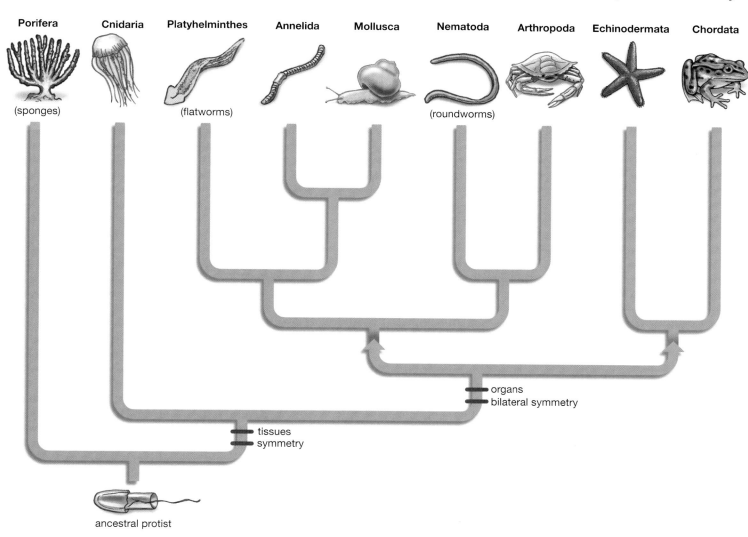

Porifera   Cnidaria   Platyhelminthes   Annelida   Mollusca   Nematoda   Arthropoda   Echinodermata   Chordata

(sponges)   (flatworms)   (roundworms)

organs
bilateral symmetry

tissues
symmetry

ancestral protist

FIGURE 22.2   A Family Tree for Animals The animal kingdom is divided into groups called phyla. The members of each phylum have in common physical features that are evidence of shared ancestry. Nine of the estimated 36–41 animal phyla are shown here. All these phyla are regarded as having evolved from an ancestral species of protist, shown at lower left in the family tree. The red horizontal lines on the tree mark the points in animal evolution at which certain structural innovations appeared, such as tissues and bilateral symmetry. Human beings are members of phylum Chordata, at upper right.

can see at the bottom of the tree. We're concerned first with one animal trait labeled there as "tissues" and another labeled as "symmetry." Later on we'll look at the significance of the trait labeled "organs."

Tissues

A **tissue** is a group of cells that performs a common function. In your own body, you have cells that perform the common function of contracting. These are muscle cells, and a group of them constitutes a muscle tissue. Likewise, you have nervous tissue and connective tissue, and other varieties. Looking down at the trunk of the tree in Figure 22.2, you can see that all animals have a common ancestor—probably a protist similar to the modern-day protists called choanoflagellates. Now notice that the tree splits above this common ancestor and yields, over to the left, the phylum Porifera, which is made up of sponges. Porifera are truly the outliers of the animal world in that, almost alone among animals, they lack tissues (and symmetry,

A BROADER VIEW

Cells of a Sponge

How independent are the cells that make up a sponge? In experiments, scientists have strained some sponge species through a filter, separating them into the individual cells they are made of. The scientists then watched as these cells came back together to make up a single sponge once again.

which we'll get to in a second). If you look again at the split above the common ancestor, this time going to the right, you can see that the animals on this branch did develop tissues. Note also that this right branch leads to all the other animal phyla. Thus, all the animals that stem from this branch will have tissues. If we looked, for example, at the Cnidarian phylum (which includes jellyfish), we'd see that its members have nervous tissue and tissue that functions like our own muscle tissue. But sponges? They are more like a collection of cells that have come together to form an organism. In a given sponge, each cell acquires its own oxygen and eliminates its own wastes, and there are no collections of cells that perform common functions. Tissues are a marker of organization, and in the animal kingdom, only sponges lack this organization.

Symmetry

So what about this second quality noted with a red line, symmetry? If you look at FIGURE 22.3a, you can see that the jellyfish that's pictured can be thought of as being divided by imaginary planes, thus yielding body sections. These sections are mirror images of one another; they have **symmetry**, meaning an equivalence of size, shape, and relative position of parts across a dividing line or around a central point. The jellyfish actually has a particular kind of symmetry. It has **radial symmetry**, meaning a symmetry in which body parts are distributed evenly around a central point. In simpler terms, the symmetry of jellyfish is the symmetry of pie sections.

To appreciate symmetry, consider what a lack of it is like by looking again at our outlier, the sponge, in FIGURE 22.3b. Where is its symmetry? There isn't any, because there is no section of the sponge that is a mirror image of any other.

Symmetry can, however, come in several forms. Look at the dog in FIGURE 22.3c, divided by what is known as an imaginary "sagittal" plane. The dog obviously has symmetry, but it is a different kind of symmetry from that of the jellyfish. As it

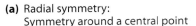

(a) Radial symmetry:
Symmetry around a central point

(b) Asymmetry:
No planes of symmetry

(c) Bilateral symmetry:
Symmetry across the sagittal plane

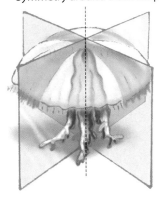

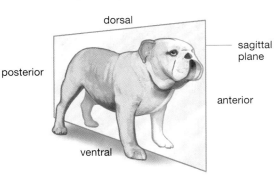

dorsal

posterior

sagittal plane

anterior

ventral

FIGURE 22.3   Symmetry Is an Equivalence in Body Sections If an imaginary plane drawn through an animal can divide that animal into sections that are mirror images of one another, then that animal has symmetry.

(a) Radial symmetry The imaginary planes drawn through the jellyfish show that it has radial symmetry: a symmetry in which body sections are distributed evenly around a central point. Radial symmetry is characteristic of the phylum Cnidaria, which includes jellyfish.

(b) Asymmetry The sponge has no symmetry—no plane drawn through the sponge body would yield sections that are mirror images of one another.

(c) Bilateral symmetry The dog has a kind of symmetry common to most animals: bilateral symmetry, in which the sides of an animal are mirror images of one another. More formally, bilateral symmetry exists when body sections on opposite sides of a sagittal plane are symmetrical to one another. Notice also the terms for different parts of the dog—dorsal and ventral, meaning "back" side and "belly" side; and anterior and posterior, meaning "head" end and "tail" end.

turns out, the dog has a kind of symmetry that unifies most of the animal kingdom. Animals usually are *different* front-to-back and top-to-bottom, but *symmetrical* side-to-side. Put another way, your head is different than your feet, and your chest is different than your back, but your left *side* is very similar to your right. This is what it means to have **bilateral symmetry**: a bodily symmetry in which opposite sides of a sagittal plane are mirror images of one another. In looking at the animal family tree, you can see that bilateral symmetry was an evolutionary innovation that affected all modern-day animals except for the sponges and the Cnidarians.

The Body Cavity

The animal family tree has another lesson for us regarding unity and diversity among animals. Your stomach expands when you've just had a big meal, but then contracts when you're busy for a few hours. Your heart expands and contracts perhaps 70 times a minute. You bend over to tie your shoes and, unbeknownst to you, many of your internal organs slide out of the way. What allows you to do all this? The answer is an internal space you have; a centrally placed, fluid-filled body cavity called a **coelom** ("SEE-lome"). We humans are not alone in having such a cavity. There are only three animal phyla in Figure 22.2 that *don't* have one: the sponges, the cnidarians, and the members of phylum Platyhelminthes (the flatworms).

What's the value of such a cavity? Well first, an expandable stomach has the same value as a gas tank: It allows you to go for a while without refueling. It also allows reproductive organs to expand to accommodate eggs or offspring. Then there is the fact that if a heart couldn't expand and contract, it couldn't work at all. Finally, a body cavity protects organs from bodily blows and provides a large part of the body with flexibility.

In most instances the coelom surrounds another physical structure, the *digestive tract*—meaning the tube, functioning in digestion, that runs from the mouth to the anus. We can therefore think of the coelom as one tube that encircles another; the coelom is generally tube-shaped, and it surrounds the tube that is the digestive tract. **FIGURE 22.4** shows you what it means to have no coelom, as in flatworms; and a *true coelom,* as in earthworms. (Not shown is another variation on this theme, a pseudocoel, which roundworms have.) You may wonder why there are no red lines on Figure 22.2 showing where the coelom evolved. So valuable is this internal space that it seems to have evolved independently several times among animals.

Having reviewed a few broad-scale features of the animal kingdom, let's look at some of the ways that varying groups of animals take care of business. How do they reproduce, how do they get oxygen, how do they circulate their bodily fluids, and how do they move?

22.4 Sexual and Asexual Reproduction

Those of you who have been through this book's chapters on microbes and fungi know that living things reproduce in a lot of different ways. You might expect that animal reproduction would be a little more familiar, but the range of reproductive "strategies" that animals employ is extremely varied, as you'll see.

Asexual Reproduction

Perhaps the most basic distinction in reproduction has to do with whether it is sexual or asexual—do eggs and sperm from two different individuals come together and fuse their genetic material (sexual reproduction), or is there reproduction

FIGURE 22.4 **An Important Space** Most animal bodies have an enclosed cavity or coelom—an internal space that surrounds their digestive tract or other internal structures. A coelom gives an animal flexibility, protects its organs from external blows, and provides space for the expansion of such organs as the stomach. Only three of the phyla covered in the chapter lack a coelom.

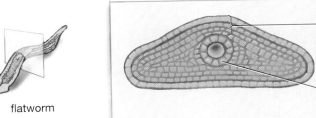

uninterrupted tissue from exterior to gut

gut (digestive tract)

flatworm

No coelom. Phylum Platyhelminthes, composed of flatworms, is one of the phyla that has no coelom. Note that from its gut to its exterior, the flatworm is composed of uninterrupted tissue.

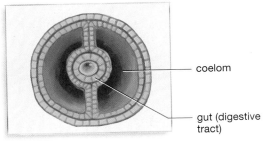

earthworm

coelom

gut (digestive tract)

Coelom. The earthworm is one of the many animals that has a coelom — a fluid-filled central cavity that usually surrounds the gut.

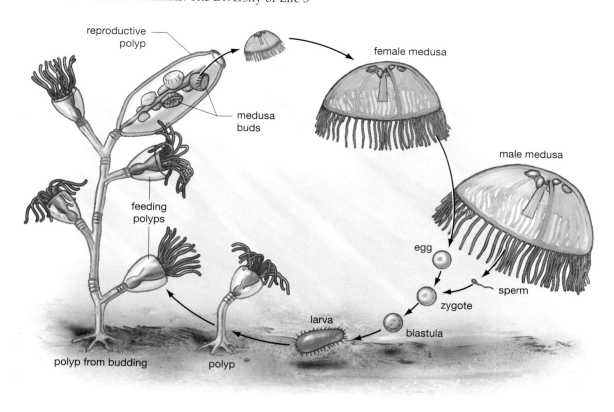

FIGURE 22.5 **Sexual and Asexual Reproducer** This cnidarian, known as *Obelia*, reproduces through both sexual and asexual means. Sexual reproduction can be observed at the right of the figure, where male and female medusae release sperm and eggs, respectively, that fuse to form the new generation of *Obelia*. The polyp that results from this fusion, however, can reproduce by asexual means. Cells bud off from the first polyp (center) and grow into another polyp, shown at left, which may be the first of many that will develop this way. Polyps within these larger colonial structures specialize. Some are involved in feeding, while others function in reproduction. The reproductive polyps have medusa developing inside them (the medusa buds), which are released to the surrounding water, starting the life cycle over again. The different stages of the life cycle are drawn at different scales.

A BROADER VIEW

Purely Sexual Reproducers

Are there any groups of animals that *never* reproduce asexually? Only two: birds and mammals.

without this fusion (asexual reproduction)? We're used to thinking of animal reproduction as being sexual because the animals we're most familiar with—the vertebrate animals—almost always reproduce sexually. But when we look over the animal kingdom as a whole, we find that the more primitive animal phyla are apt to reproduce *both* sexually and asexually. As we move up the complexity scale in the kingdom, sexual reproduction becomes more and more the norm; but asexual reproduction occasionally exists even in very complex animals, as you'll see.

If you look at FIGURE 22.5, you can see an example of a primitive animal that reproduces both sexually and asexually. The creature is *Obelia*, a tiny jellyfish-like Cnidarian, and its sexual reproduction is straightforward: Female medusa-stage *Obelia* produce eggs, the males produce sperm, both kinds of reproductive cells are ejected into the water, sperm fertilize eggs, and a new generation of *Obelia* begins development. But note that the larva that represents this new generation develops into a "polyp stage" *Obelia* that implants on a surface (perhaps some algae or a rock). Once this happens, the polyp reproduces by **budding**: a form of asexual reproduction in which an offspring is produced as an outgrowth of a parent organism. The polyp that results from this budding could be thought of as a colony that develops some polyps that are specialized for feeding and others that are specialized for the production of medusa-stage *Obelia*—the very thing our life cycle started with. These medusa *Obelia* then go on to engage in sexual reproduction, and the whole cycle repeats. (See FIGURE 22.6 for a picture of *Obelia*.)

FIGURE 22.6 ***Obelia* Up Close** A micrograph of the tiny cnidarian *Obelia* in its colonial polyp stage. The red, central stalk attaches to two kinds of polyps: those without tentacles, which are are reproductive; and those with tentacles, which function in feeding. The reproductive polyps contain medusa buds that are being released into the water as mature medusae.

Forms of asexual reproduction other than budding are possible as well. Within the flatworm phylum (Platyhelminthes), certain worms can reproduce by what is known as *fission*, which simply means they break themselves roughly in half. The posterior ("tail") portion of the worm grasps the material beneath it, while the anterior ("head") portion moves forward. The separated halves then grow whatever tissues they need to make themselves complete worms.

Next, there is a form of asexual reproduction that is found among bees and ants, and that even exists in a small number of reptile species. Among common honeybees (*Apis mellifera*), a queen bee might lay 2,000 eggs per day. Nearly all of these eggs are fertilized by sperm the queen has received during one of her "nuptial flights," and each one of these eggs will develop into one of the worker bees of the colony—every one of them a female. A queen can, however, choose to let some of her eggs go unfertilized; no sperm from a male ever fuses with these eggs, yet bees develop within them and hatch from them. Since egg and sperm do not come together in this process, this is not sexual reproduction. Instead, each of these bees has been derived through **parthenogenesis**: a form of asexual reproduction in which an unfertilized egg develops into an adult organism. Among the honeybees, all the bees derived through parthenogenesis are males—these are the few "drones" of a bee colony. But in other animals, parthenogenesis can produce females. Indeed, if you look at FIGURE 22.7, you can see a lizard whose species produces nothing *but* females by employing parthenogenesis. All the members of this species are female, and all reproduction in the species comes through parthenogenesis. Thus each new lizard develops solely from one of her mother's eggs, meaning that each is a clone of her mother.

FIGURE 22.7 **All-Female Species** Pictured is a whiptail lizard, *Aspidoscelis tigris*, that reproduces solely through parthenogenesis. Because the eggs these lizards produce are not fertilized by males, they contain only one set of chromosomes, the mother's. As a consequence, each hatchling is genetically identical to her mother—each is a clone of her mother.

FIGURE 22.8 **Hermaphrodite, with Organs**

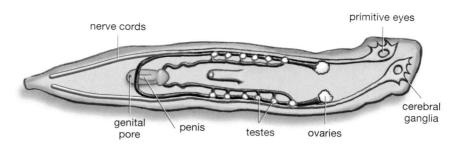

The flatworm *Dugesia* is a hermaphrodite, as it possesses both the sex organs of a male (penis and testes) and those of a female (ovaries and genital pore). When two *Dugesia* copulate, each projects its penis and inserts it in the genital pore of the other. In addition to its sexual organs, *Dugesia* has other organs—primitive eyes, for example, and clusters of nerve cells (the cerebral ganglia) that function like a brain.

Variations in Sexual Reproduction

Having looked at some of the ways animals reproduce asexually, you might think that all sexual reproduction in animals would follow the pattern we saw with *Obelia*: Sperm and eggs are produced by male and female members of a species; these reproductive cells come together and result in offspring. But the animal kingdom is more varied than this. If you look at FIGURE 22.8, you can see one variation in the form of the flatworm *Dugesia*, which is part of the phylum Platyhelminthes. Note that *Dugesia* has both testes *and* ovaries, along with both a penis and a genital pore (which is analogous to a vagina). *Dugesia* is a **hermaphrodite**, meaning an animal that possesses both male and female sex organs. When two *Dugesia* copulate, each projects its penis and inserts it in the genital pore of the other. (Can a given flatworm fertilize itself? It occasionally happens in the class of parasitic flatworms known as tapeworms.)

Several variations exist on the theme of hermaphrodism. If we look at the oysters that are part of phylum Mollusca, we find that, while they have separate sexes, a given individual can *change* its sex between mating seasons. A frequent course of events is for an oyster to mature as a male and then change to a female later on. This kind of male-to-female transition is surprisingly common in the animal kingdom; some fish, eels, snails, and shrimp go through it as well. Female-to-male transitions are less common, but they do occur.

Later on, we'll explore one more theme in animal sexual reproduction, which has to do with how eggs get fertilized and protected. For now, however, let's shift over to a different kind of lesson that the *Dugesia* flatworm has for us.

A BROADER VIEW
Pseudocopulation

Among the all-female species of lizards noted in the text, males may be gone, but they have not been entirely forgotten. Females in these species engage in what is called pseudocopulation. One female will mount another and bite her on the back of the neck—behaviors that males exhibit during copulation in other lizard species. Pseudocopulation seems to trigger a hormonal surge that allows females who have been mounted to lay a larger clutch of eggs.

22.5 Organs and Circulation

If you look again at Figure 22.8, you can see that, in addition to having testes and ovaries, *Dugesia* has two primitive eyes and two collections of nerve cells (called cerebral ganglia), which might be thought of as a primitive brain. Not shown is a highly branched intestine that delivers nutrients to all parts of its body. What all this adds up to is that *Dugesia* has **organs**: highly specialized structures that generally are formed of several kinds of tissues. (You may recall that tissues are formed of collections of similar cells.) We're so used to the idea of animals having organs—such as a heart or a liver—that *Dugesia*'s possession of organs may not seem all that impressive. But *Dugesia*'s phylum, Platyhelminthes, actually is where organs *start* in the animal kingdom, as you can see in Figure 22.2. Sponges

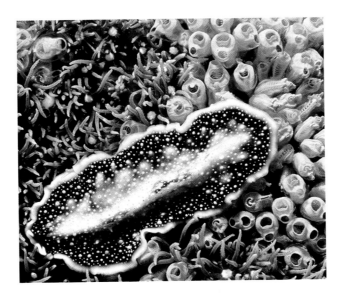

FIGURE 22.9 **Flatworm** The flatworm in this picture, *Pseudoceros ferrugineus*, was photographed in the oceans off the coast of the Philippines. Most flatworms are free-living like this one, but there are also many parasitic flatworms that live for at least part of their lives inside a host.

don't even have tissues, and jellyfish have nothing as organized as testes or ovaries. From Platyhelminthes all the way up the ladder of complexity, however, organs are the rule.

If a flatworm such as *Dugesia* is organizationally sophisticated compared to some animals, however, it is organizationally *simple* compared to others. Like all its fellow flatworms, *Dugesia* has no circulatory system; as a result, every one of its cells must get its oxygen directly from the environment around it, rather than from circulating blood. This is why flatworms are flat—they have to maximize the number of cells that are near an exterior surface, as opposed to being buried deep inside. (How flat are flatworms? About the width of a sheet of typing paper.) Moreover, flatworms lack a complete digestive tract, and they have no anus, which means waste must be expelled through their mouths. (To see what a flatworm looks like, see **FIGURE 22.9**.)

All this stands in contrast to what you can see in **FIGURE 22.10**, which is an aquatic snail that is part of phylum Mollusca. Notice that the snail has organs that are familiar to us—a heart and stomach—along with gills, which serve to move oxygen into its circulation. Among the items that have helped free the snail from the constraints of shape and size seen in *Dugesia* are its coelom or internal body cavity, which allows an expandable stomach and a beating heart; and its circulatory system, which allows nutrients to get to tissues that exist far away from its surface.

Open and Closed Circulation

The snail's circulatory system turns out to offer us another comparative lesson, this time not relative to a flatworm, but relative to ourselves. We have what is known as a **closed circulation system**—a system of circulation in which blood *stays within* the blood vessels. There is some "leakage" of the liquid portion of our blood, but the basic arrangement is that, while oxygen and nutrients diffuse out of our blood vessels, the blood itself stays confined. In contrast, the snail and most other molluscs have an **open circulation system**: a circulation system in which blood flows out of blood vessels altogether, into spaces or "sinuses" where it bathes the surrounding tissues. Veins then collect the blood (or "hemolymph") in these sinuses for a return trip to the gills and the heart. The most complex molluscs, the cephalopods (squid, octopus, etc.), have a closed circulation system like ours; but this isn't surprising, since only a closed system has the efficiency to support an animal as large as a squid. The animal kingdom as a whole is mixed with respect to open and closed systems. All vertebrates have closed systems, for example, while all arthropods have open systems.

FIGURE 22.10 **Organ Complexity** Gastropods such as this snail have a sophisticated set of organs. For example, the snail has gills to transfer oxygen into its body, a heart to pump its blood, and a stomach to hold and digest its food.

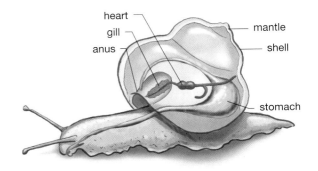

So Far... Answers:

1. *tissues; symmetry*
2. *internal body cavity*
3. *eggs; sperm; parthenogenesis*
4. *stays within; leaves the vessels*

So Far...

1. The sponges in Phylum Porifera are "outliers" in the animal kingdom in that, almost alone among animals, their bodies are not composed of _____ and their body structure lacks the quality of _____.

2. A coelom is an _____ possessed by almost all animals except sponges, jellyfish, and flatworms.

3. Sexual reproduction in animals always entails a fusion of _____ and _____ from different individuals. _____ is a form of asexual reproduction in which an unfertilized egg gives rise to an adult.

4. In closed circulation systems, blood _____ blood vessels, while in open circulation systems, it _____ to bathe surrounding tissues.

22.6 Skeletons and Molting

How do we raise our arm? Two things are required: a muscle to exert force, and an object to exert this force against—a bone, which is part of a support network called a skeleton. If we look at the human body, we take for granted that our skeleton lies interior to our muscles. But this isn't always the case in the animal kingdom. When we look at arthropods (insects, crustaceans, etc.), we find that they have an **exoskeleton**: an external material covering of the body that provides support and protection. Arthropod muscles pull against a skeleton, but this skeleton lies external to the arthropods' muscles.

Having a skeleton on the outside of the body, however, presents a problem: The arthropod body can only grow so much before it expands right into the skeleton. The solution to this problem is **molting**: a periodic shedding of an old skeleton followed by the growth of a new one. If you look at FIGURE 22.11, you can see a picture of a crab that has just slipped out of its old skeleton. Molting activity of this sort is not limited to the arthropods. You may remember from our walk through the phyla that roundworms also have an exoskeleton, and it turns out that they shed it four times during their lives. Since roundworms and arthropods make up such a large proportion of the animal kingdom, the way to think of molting is as a widespread phenomenon among animals.

The exoskeletons that are part of molting stand in contrast not only to our own *endo*skeleton, but to a third variation on animal skeletons. If you look at FIGURE 22.12, you can see some of the anatomy of an earthworm, which is part of Phylum Annelida. Now, an earthworm has muscles and everyone knows it can move; yet it has no bones for its muscles to pull against. So *how* does it move? If you look in

A BROADER VIEW

Do Crabs Feel Pain?

Lobsters and crabs, a couple of the crustaceans who have exoskeletons, are part of an ongoing animal rights controversy. The key question in this dispute is whether these creatures feel pain when humans boil them alive during meal preparation. You might think this is bound to be painful for them; but a Norwegian government report, released in 2005, concluded it probably is not. We tend to assume that any large creature can feel pain, but as the report noted, crustaceans are arthropods—in the same phylum as insects—and as such probably do not have a nervous system that allows them to feel pain.

FIGURE 22.11 **Old and New** Arthropods have an outer or exoskeleton of a fixed size. They are able to grow by periodically shedding, or molting, this stiff skeleton. The crab in the photograph has just finished molting; its old exoskeleton lies nearby.

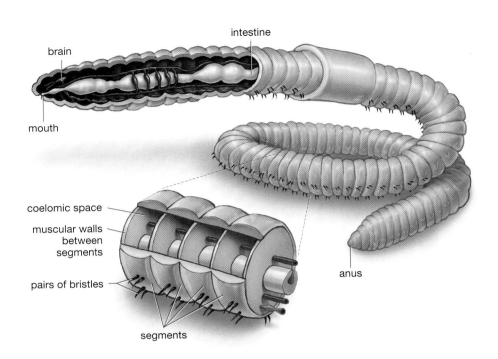

FIGURE 22.12 **Hydrostatic Skeleton** Animals such as the earthworm move through use of a fluid or "hydrostatic" skeleton. Fluid fills the coelomic spaces that can be seen in the lower drawing. When the worm squeezes its muscles, the pressure of this fluid causes segments in the worm to lengthen and narrow, just as a water balloon will lengthen and narrow when squeezed.

the lower drawing of Figure 22.12, you can see the worm's "coelomic space," which is the internal cavity noted earlier. This space is filled with fluid, and it is this fluid that serves as the earthworm's skeleton. Here's how it works. Imagine squeezing a water-filled balloon. The volume of the water doesn't change, but the *shape* of the balloon does, in response to the pressure of the water. In the same way, the earthworm can squeeze one set of muscles and *lengthen* a set of the segments you can see in the illustration. Thus, the earthworm employs what is known as a **hydrostatic skeleton**: a skeleton in which the force of muscular contraction is applied through a fluid. Hydrostatic skeletons are common in the animal kingdom. Apart from annelids like the earthworm, they exist in Cnidarians (jellyfish, etc.), flatworms, and roundworms.

22.7 Egg Fertilization and Protection

Having started our review of animal functions by going over some aspects of reproduction, we'll now finish up on functions by looking at reproduction again. In this case, the central question before us could be stated as: "Where are the eggs, and how are they being taken care of?" Note that we're now concerned only with sexual reproduction, which is to say instances in which offspring are produced by means of a sperm from a male fertilizing an egg from a female.

Egg Fertilization

If you think back to the jellyfish-like creature known as *Obelia*, you may remember where the eggs came from and went to in that instance—from the female and into the ocean. The male likewise deposited his sperm into the ocean, after which water currents brought the two kinds of reproductive cells into contact. Two facets of reproduction that interest us are on display here: the ways in which eggs and sperm can *come together* for fertilization, and the ways in which eggs might be *protected*, either before or after fertilization. On both counts, *Obelia* represents what is generally regarded as the most primitive and ancestral condition in the animal kingdom. Both males and females simply "broadcast" their eggs and sperm into the environment at the same time. Neither male nor female attempts to protect these "gametes" (eggs and sperm), the gametes themselves are fragile, and the union of gametes is a matter of water currents and chance.

A BROADER VIEW

Extremes in Reproduction

One way of thinking about reproductive strategies of animals is in a spectrum that runs from "many eggs/no care" to "few eggs/much care." So, which species lie at the extremes of this spectrum? The female Atlantic oyster (*Crassostrea virginica*) can produce up to 100 million eggs per year, which she simply broadcasts into the water and provides no care for thereafter. By contrast, a human female will release 13 eggs per year (through ovulation), but in most cases will provide years of care to that small fraction of eggs that become fertilized.

If we followed the twin issues of gamete union and egg protection across the animal kingdom, what we would see, with movement up the species complexity scale, is a general trend toward fewer eggs, more directed means of bringing these eggs together with sperm, and greater protection for the fertilized eggs that result.

With respect to bringing gametes together, a fundamental question is whether fertilization takes place outside the female (external fertilization, as in *Obelia*) or takes place inside the female (internal fertilization, as in the human female). As it turns out, external fertilization is rarely seen in animals except in aquatic environments, for the simple reason that unfertilized eggs desiccate—they perish by drying out—when they are not either in a body of water or in a female's reproductive tract.

You might expect that all instances of internal fertilization would involve something that is familiar to us from the world of mammals: **copulation**, which is to say the release of sperm within a female by a male reproductive organ. But the animal kingdom is more varied than this. The males of many species produce what are known as spermatophores (sperm packets), which in some cases must be introduced into the female reproductive tract by the females themselves. The male of a European amphibian called the great crested newt (*Triturus cristatus*) will do a courtship dance in front of a female, waving his tail, and then deposit a spermatophore on the ground in front of her and block her path in hopes that she will scoop the packet into the opening of her reproductive tract.

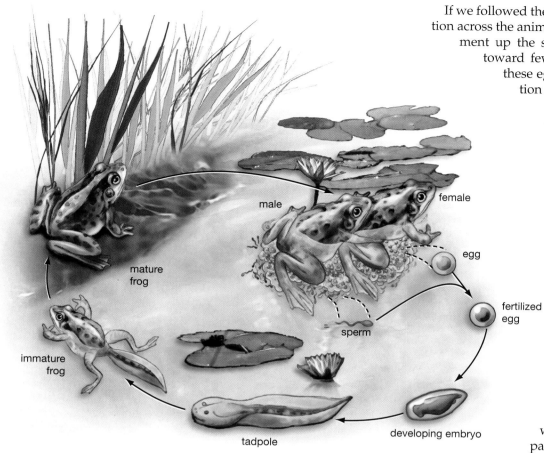

FIGURE 22.13 **Fragile Eggs, Aquatic Environment** Amphibians such as frogs utilize external fertilization, releasing both eggs and sperm into the water. The eggs must remain in the water, lest the embryos in them perish from drying out. The tadpoles that hatch out of the eggs are likewise aquatic until such time as they lose their tails, develop lungs, and begin living on land.

Protection for Eggs: The Vertebrates

With respect to the other issue we're concerned with, the protection of fertilized eggs, different strategies are followed across the animal kingdom. But we're going to focus here on just a few reproductive "snapshots" from one small part of the kingdom, the vertebrates in phylum Chordata.

If you look at FIGURE 22.13, you can see the life cycle of a familiar amphibian vertebrate, the frog, whose reproduction usually involves the kind of contact between the male and female that you can see in the figure: the male clasps the female in an embrace called amplexus, which can go on for hours or even days. This stimulates the female in such a way that she lays her eggs into the water, after which the male deposits his sperm on top of them. These eggs are fragile, however, and predators tend to be plentiful in the environments in which frogs live. A common response of frogs to this situation is to invest time and effort in keeping the eggs protected. A clear, if unusual, example of this is seen in the Central American frog *Physalaemus pustulosus*, whose females will, while locked in amplexus with the male, secrete a mucus that both frogs then proceed to beat into a froth with their legs. What results is a foam nest, resting on top of the water, that the eggs and sperm are deposited into (see FIGURE 22.14). Four days later, little tadpoles drop out of the underside of the nest and start swimming away.

Egg protection efforts of this sort are common across much of the animal kingdom. What interests us is to follow an evolutionary progression that starts with these amphibians. Note that the frogs laid their eggs in the water. They did so because their

FIGURE 22.14 **Stirring Up Protection** Two tungara frogs (*Physalaemus pustulosus*) stir up foam that will serve as a protective hiding place for eggs that will be deposited by the female and fertilized by the male. These frogs were photographed in the Panamanian rainforest.

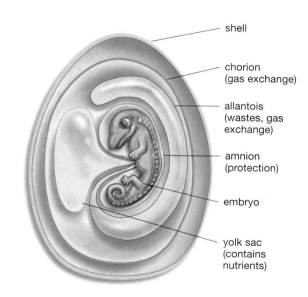

FIGURE 22.15 **An Egg for Many Environments** The amniotic egg was an evolutionary development that allowed vertebrates to move inland from the water. The egg's features meant that, in dry environments, the embryo would not "desiccate" or perish through a loss of fluids. The egg's tough shell is the first line of defense; it limits evaporation of fluids. The amnion is one of several membranes that provide some cushioning protection for the embryo. The allantois serves as a repository site for the embryo's waste products. The chorion provides cushioning protection and works with the allantois in gas exchange, meaning the movement of oxygen to the embryo and the movement of carbon dioxide away from it. The yolk sac provides the nutrients the embryo will need as it develops.

eggs are vulnerable not just to predators, but to the drying out we talked about. Those of you who went through Chapter 19, on the history of life, know that, while amphibians were the first true land vertebrates, they gave rise to another vertebrate line that was able to move *inland*, away from the water. How did these later-evolving creatures preserve their eggs out of the water? They employed a different kind of egg. If you look at FIGURE 22.15, you can see the **amniotic egg**: an egg produced primarily by reptiles and birds whose shell and system of membranes provide protection and life support to a developing embryo. Protection? The shell maintains a watery environment *inside* the egg, while several of the egg's membranes provide cushioning. Life support? Other membranes in the egg allow for oxygen to flow to the embryo, for carbon dioxide to flow from it, and for waste products to be stored away.

Protection of the Young

Modern reptiles generally lay their eggs and then abandon them before the young ever hatch out of them. If we ask why reptiles do this, one answer might be: because they can. The amniotic egg provides such a protective environment that parental guarding of the eggs isn't necessary. There are, however, some fascinating exceptions to this rule. If you look at FIGURE 22.16, you can see a reptile family tree that starts, at the base, with an ancestral reptile. Note that, early on, mammals evolved from this reptile line (over on the left). What really interests us, however, are the three evolutionary lines at upper right: dinosaurs, birds, and crocodilians (meaning both crocodiles and alligators). Now,

Evolutionary Relationships of Reptiles, Birds, and Mammals

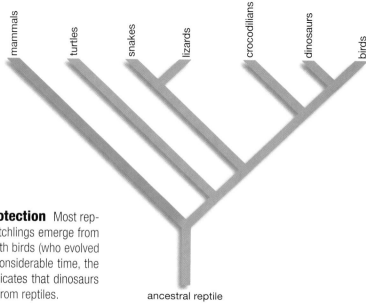

FIGURE 22.16 **The Evolution of Offspring Protection** Most reptiles do not stay with their eggs through the time that hatchlings emerge from them. This is not the case, however, with crocodiles or with birds (who evolved from dinosaurs). They guard both their eggs and, for a considerable time, the young that hatch out of them. Some recent evidence indicates that dinosaurs may have done this as well. Note that mammals evolved from reptiles.

FIGURE 22.17 **Parent and Offspring?** Researchers in China found the fossilized remains of a parrot-like dinosaur, *Psittacosaurus*, surrounded by 34 juveniles. The 125-million-year-old fossils provide evidence that dinosaurs provided care to their young after they were hatched.

A BROADER VIEW
From Conception to Birth

In the placental mammals, the period of gestation—the time from conception to birth—can be extremely long. In humans, of course, it is about nine months; but in elephants it can be about two years. In marsupial animals, such as the kangaroo and opossum, this period is much shorter because marsupial young develop largely outside their mother's bodies. The gestation time for a kangaroo is about a month, but it then embarks on six months of "pouch life" after being born.

among contemporary creatures who employ an amniotic egg, which of them also *guard* not only the eggs, but the young who hatch out of them? The answer is birds and crocodiles. Birds are well known for their protective behavior toward their eggs and "hatchlings," but female crocodiles also turn out to be fierce protectors of their eggs and young. (In some crocodile species, the mother stays with the young for over a year after they hatch.) Now, what about the third line at upper right in the reptile family tree, the dinosaurs? Recent fossil evidence indicates that dinosaurs may have stayed with their newly hatched too! The phrase "attentive parents" may not readily come to mind in connection with dinosaurs, but this may have been the case. In 2004, fossils of a plant-eating dinosaur were found in China surrounded by 34 juvenile dinosaurs—a configuration that is hard to explain unless the adult had a close association with the juveniles (**see** FIGURE 22.17). Taking a step back from this, the interesting thing is that protective behavior toward the young may have evolved in an evolutionary line that stretches all the way from ancient crocodilians through modern birds.

Protection in Mammals

So, thus far we've seen protective behavior on the part of amphibians, a more protective egg that evolved in reptiles, and this egg combined with protective behavior among a select group of reptiles. What else could vertebrate evolution come up with in the way of protection for offspring? A dandy of an innovation, as it turned out. When we look at *mammals*, we find that their embryos are protected and nurtured by the mother's entire body. An egg that is fertilized in a mammalian female's reproductive tract then *stays* in it, implanting in the female's uterus and developing there. Most mammals are thus **viviparous**: Reproduction occurs by means of fertilized eggs developing inside the mother's body. In contrast, most reptiles and all birds are **oviparous**: Reproduction occurs by means of the young developing in fertilized eggs that are laid outside the mother's body.

In most mammals, implantation of the embryo triggers the formation of a network of blood vessels and membranes, called the placenta, through which nutrients and oxygen flow to the embryo, and through which wastes and carbon dioxide flow from it. Most mammals are thus said to be "placental," but there are some well-known exceptions to this rule. A small group of early-evolving mammals lay eggs, just as reptiles do. These are the monotremes, represented by the duck-billed platypus (**see** FIGURE 22.18). And there are mammals in which the young develop only to a limited extent inside the mother's body. These are the marsupials, which include kangaroos. But the vast majority of mammals today are placental.

Mammalian care doesn't stop with the birth of the young, of course. Indeed, mammals are the supreme caregivers of the living world. If we look, for example, at elephants, we find that an elephant calf might be nurtured by its mother and a larger kinship group for the first 10 years of its life. Part of the care that the mammalian young get is based on physiology: All female mammals provide milk for their young through a system of mammary glands.

It's important to note that mammals do not occupy some special *moral* place in the animal kingdom because of the way they care for their young. Intensive protection of the young developed in mammals for the same reason that any trait evolves in a group of living things: because it "works" in an evolutionary sense. In the mammal line, additional time that parents spent with their offspring tended to bring about greater survival of those offspring. If our mammalian ancestors could have produced more offspring by abandoning their young, then that's what mammals would be doing today. But parental care has worked for mammals. To see a stellar example of it, we need look no further than a species that is very familiar to us, the primate known as *Homo sapiens*. Where does this species stand with respect to parental care? It lavishes more care on its offspring than any other creature in nature.

(b) Marsupial

(a) Monotreme

(c) Placental mammal

FIGURE 22.18 **Three Types of Mammals** All mammals have mammary glands that provide milk for their young. However, mammals differ in their reproductive strategies.

(a) This duck-billed platypus is a monotreme—a mammal that lays eggs.

(b) Kangaroos are marsupials—they give birth to physically immature young, which then develop further within the mother's pouch. Pictured are a mother and her fairly mature "joey" in eastern Australia.

(c) The young of placental mammals, such as this grizzly bear (*Ursus arctos*), develop to a relatively advanced state inside the mother, with nutrients supplied not by an egg, but by a network of blood vessels and membranes—the placenta.

On to Ecology

Over the last three chapters, we've reviewed the incredible diversity of life-forms in the living world—bacteria and archaea, protists and fungi, plants and animals. In a sense, however, we have reviewed each life-form in isolation from any of the others. In reality, no variety of living thing is self-contained. Rather, living things are constantly *interacting*, with members of their own species, with members of other species, and with the nonliving world around them. In the next two chapters, you'll be looking at what might be called interactive life. What you'll be studying is ecology.

So Far...

1. Insects and crustaceans have a skeleton different than ours. Theirs is an _____: a skeleton that forms an _____ on an animal body.

2. One of the problems with having an exoskeleton is that it can limit the size of an animal. Animals with an exoskeleton deal with this problem through the practice of _____, which can be defined as _____.

3. Most reptiles and all birds are oviparous, meaning their young develop in _____. Most mammals are viviparous, meaning their fertilized eggs develop _____.

So Far... Answers:

1. *exoskeleton; external covering*

2. *molting; a periodic shedding of an old skeleton followed by the growth of a new one*

3. *fertilized eggs laid outside the mother's body; inside the mother's body*

Chapter Review

Summary

22.1 What Is An Animal?

- **The blastula**—A single physical feature is sufficient to define animals. All animals pass through a blastula stage in embryonic development, but no other living things do. A blastula is a hollow, fluid-filled ball of cells that forms once an egg is fertilized by sperm. (page 370)

- **Other common features**—Three other features are characteristic of all animals, but they are also shared, to varying degrees, by members of other kingdoms in the living world. All animals are multicelled, are heterotrophs (they cannot make their own food), and are composed of cells that do not have cell walls. (page 370)

22.2 Across the Animal Kingdom: Nine Phyla

- **Animal phyla**—The animal kingdom is divided into large-scale categories called phyla. There are between 36 and 41 animal phyla, each phylum being informally defined as a group of organisms that shares a basic body structure. (page 371)

- **Across the animal kingdom**—Nine phyla, reviewed in the text, provide a good idea of the range of animal forms. These are Porifera (sponges), Cnidaria (jellyfish), Platyhelminthes (flatworms), Annelida (segmented worms), Mollusca (snails, oysters, squid), Nematoda (roundworms), Arthropoda (insects, lobsters, and spiders), Echinodermata (sea stars), and Chordata (vertebrates). (page 371)

22.3 Lessons from the Animal Family Tree

- **Tissues**—Except for the sponges in Porifera, the members of all the phyla reviewed have bodies composed of tissues: groups of cells that perform a common function (for example, muscle cells performing the function of contracting). (page 373)

- **Symmetry**—All phyla reviewed except Porifera exhibit symmetry: an equivalence of size, shape, and relative position of parts across a dividing line or around a central point. Except for Porifera and Cnidaria, all phyla reviewed exhibit bilateral or "side-to-side" symmetry. Cnidaria exhibits radial or "circular" symmetry. (page 374)

- **The coelom**—The members of all phyla reviewed other than Porifera, Cnidaria, and Platyhelminthes possess a coelom: a centrally located, fluid-filled body cavity that provides protection and allows for the expansion of internal organs. (page 375)

Web Tutorial 22.1 The Architecture of Animals

22.4 Sexual and Asexual Reproduction

- **Modes of reproduction**—Sexual reproduction always entails the fusion of eggs and sperm from two individuals, while asexual reproduction does not. Members of the less-complex animal phyla commonly reproduce both sexually and asexually. Sexual reproduction is the norm with more-complex animals, but asexual reproduction is sometimes seen even in them. (page 375)

- **Asexual reproduction**—Asexual means of reproduction among animals include budding, in which offspring are produced as outgrowths of parents; fission, in which individuals split themselves in two and grow new body parts; and parthenogenesis, in which an unfertilized egg develops into an adult organism. (page 375)

- **Hermaphrodites**—Sexual reproduction can be practiced by hermaphrodites: individual animals that possess both male and female sex organs. Some animals change sex over the course of their lifetimes. (page 378)

22.5 Organs and Circulation

- **Organs**—Organs are complex bodily structures composed of several types of tissues. Organs are found in all the animal phyla reviewed except Porifera and Cnidaria. (page 378)

- **Circulation systems**—Vertebrates have closed circulation systems, in which blood stays within blood vessels. Arthropods and small molluscs have open circulation systems, in which blood flows out of blood vessels and into spaces in which it bathes surrounding tissues. (page 378)

22.6 Skeletons and Molting

- **Types of skeletons**—Three kinds of skeletons are found in animals: endoskeletons, which lie interior to muscles; exoskeletons, which form an external covering on an animal; and hydrostatic skeletons, in which the force of muscular contraction is applied through a fluid. Vertebrates have endoskeletons, arthropods and roundworms have exoskeletons, and annelid worms have hydrostatic skeletons. (page 380)

- **Molting**—To continue their growth, animals with exoskeletons must periodically shed or molt their old skeletons. (page 380)

22.7 Egg Fertilization and Protection

- **Strategies across the phyla**—Animals employ a range of strategies for bringing eggs and sperm together and for protecting eggs following fertilization. The most primitive, ancestral strategy is that of parents "broadcasting" eggs (along with sperm) into the environment (generally water) and thereafter providing no care for these eggs. With movement up the scale of animal complexity, species tend to produce fewer eggs, to employ more directed means of bringing these eggs together with sperm, and to provide greater protection to the fertilized eggs that result. (page 381)

- **Means of fertilization**—Fertilization of eggs can be external (outside the mother's body) or internal (inside it). Only some species that employ internal fertilization also employ copulation: the release of sperm within a female by a male reproductive organ. Other species rely on females to insert sperm packets (spermatophores) into their own reproductive tracts. (page 381)

- **Vertebrate care of offspring**—Vertebrate evolution produced a range of sophisticated strategies for protecting eggs and the offspring that develop from them. Amphibians frequently spend time and effort protecting fertilized eggs; reptile evolution resulted in the amniotic egg, whose shell and system of membranes allowed eggs to be laid in dry environments; and mammalian evolution produced the placental system, in which fertilized eggs develop inside the mother's body, aided by a network of blood vessels shared by both mother and embryo—the placenta. (page 382)

Key Terms

amniotic egg   p. 383

bilateral symmetry   p. 375

budding   p. 376

closed circulation system   p. 379

coelom   p. 375

copulation   p. 382

exoskeleton   p. 380

hermaphrodite   p. 378

hydrostatic skeleton   p. 381

invertebrate   p. 371

molting   p. 380

organ   p. 378

open circulation system   p. 379

oviparous   p. 384

parthenogenesis   p. 377

radial symmetry   p. 374

symmetry   p. 374

tissue   p. 373

viviparous   p. 384

Testing Your Understanding

In Your Own Words   *(answers in the back of the book)*

1. A coelom amounts to internal space in an animal. Why is such a feature valuable?

2. Why is it fair to characterize sponges as "primitive" animals?

3. In what group of animals did the amniotic egg first evolve? What are some of its features, and what did it make possible?

Thinking about What You've Learned

1. Animals reign supreme in the living world in their ability to sense various sorts of stimuli and to move toward or away from these stimuli in different ways. (Hearing and true vision are unique to animals, as is flying.) What system do animals possess that makes these various capabilities possible?

2. Plants can have elaborate defense systems, and they can respond in sophisticated ways to their environment—for example, in preserving resources, they can go into a dormant state in winter. Given this, can plants have intelligence? Or is it only animals that have this trait, to one degree or another?

3. The text notes that heads in animals—a concentration of sensory capability at one of their ends—evolved in tandem with the trait of bilateral symmetry. Why does it make sense that sponges never developed either trait?

Populations and Communities in Ecology

THE PHILADELPHIA INQUIRER

WILDLIFE GROUP BACKS DEER KILLS—THE N.J. AUDUBON SOCIETY SAID LETHAL MEANS COULD HELP PRESERVE FORESTS FROM THE GROWING HERDS

By The Associated Press
March 15, 2005

The New Jersey Audubon Society is endorsing the use of hunts, hired guns, and other lethal measures to thin out the state's 200,000-strong white-tailed deer herd in the interest of preserving the state's forests.

In a 17-page report, the society said yesterday that "lethal control" was a legitimate option for a problem that had gone from gardening nuisance to ecological menace as New Jersey's record-high deer population continues to grow.

23.1 The Study of Ecology

The Audubon Society is generally known for a kindly attitude toward wildlife. So how did it happen that in November 2004, Kristen Berry of the National Audubon Society ended up writing in an editorial, "If you love birds, shoot deer." Four months later, the New Jersey Audubon Society called for reducing its state's white-tailed deer population by hunting—by bringing in sharpshooters if necessary. And this wasn't some hastily taken position. The group studied the problem for a year and concluded that the threat deer pose to other wildlife is so severe that increased hunting is the best solution to the state's deer-driven "disaster."

What disaster is this? Well, for the Audubon Society, one important issue is that deer eat small-plant species that birds live in or feed on. Indeed, the National Audubon Society reports that bird species reliably get decimated in accordance with the number of deer per square mile. Beyond the problem with birds, when deer mow down native plant species, they leave an opening for invading alien plant species. In New Jersey, the native witch-hazel and spicebush forest shrubs are being replaced by plants such as Japanese stiltgrass. In some places, deer are eating so many saplings that forests themselves are threatened—

FIGURE 23.1 **An Ecologist on the Job** Ecologists spend a good deal of time going out into nature to study the conditions that exist in various ecosystems—however inaccessible those ecosystems may be. Here a researcher is studying tree-branch growth in the rain-forest canopy in Costa Rica.

A BROADER VIEW

Ecology and Public Disputes

Is it any surprise that, among all the disciplines of biology, ecology is the one that figures most often in public disputes? We may read more about health and medicine in the newspaper, but most of the coverage of these subjects concerns research findings, not contentious public policy issues. Many of ecology's concerns are inherently public concerns because they have to do with what is sometimes called "the commons," meaning the air that everyone breathes, the oceans that everyone may use, and so forth. Moreover, since ecology has to do with the linkage between nature's component parts, touching on one ecological issue often means touching on many, each of which can evoke strong feelings. Some bird lovers might not mind seeing deer shot, but animal rights activists are likely to have other ideas.

no new trees are coming along to replace old trees as they die out. And, of course, it's not just wildlife that deer impact. Food crops, timber, and cultivated flowers all get damaged by them. In 2003, there were an estimated 1.5 million collisions between motor vehicles and deer in the United States.

How did things come to this pass? How did the nation's deer population go from half a million in the early 1900s to at least 25 million today? Well, here's a recipe for how to explode a national deer population. First, get rid of a key deer predator, the wolf, by driving it to extinction in most states. Then, build houses farther and farther out into the suburbs, on land adjacent to fields and forests. Now the deer have plenty of land to graze on (suburban lawns) that doesn't even have *human* predators on it. Then couple a growing sensibility about animal rights with a decline in hunting nationally. (Hunters are a potent force in this issue; in Wisconsin alone, they killed 484,000 deer in 2003.) The result? Deer right in your back yard, on your streets, ruining your forest preserve. You might think that the silver lining to this is that the *deer* are at least coming out of it fat and content, but consider that in Ohio's Cuyahoga Valley National Park, deer have been starving to death in recent winters because their habitat has been picked clean by ... deer.

For our purposes, the story of deer in the United States drives home the message that living things do not live in isolation from one another. What happens with one species affects another. What happens with wolves affects deer, then deer affect plants, and plants affect birds. Predators and prey, native and alien species, plants and animals—all these things are part of an *interrelated* web of life. Moreover, this is a web of life that exists in a larger context yet: the nonliving world, which is to say Earth's climate, its sea currents and volcanoes, its atmosphere, and the energy it gets from the sun.

In your walk through biology so far, you have made it through the molecular world (DNA, for example) and through the world of whole organisms, such as fungi and plants. Now it is time to take the next steps up; in this chapter and the one that follows, you will consider biology's largest realm of all, **ecology**, defined as the study of the interactions that living things have with each other and with their environment. What's coming up is a tour of what might be called the interactive living world.

Starting Out: What Ecology Is Not

In common speech the word *ecology* has become synonymous with the conservation of natural resources or with the "environmental movement" in all its political dimensions. Under this view, you might assume that the job of an ecologist is to help preserve the natural world. In fact, however, the job of an ecologist is to *describe* the natural world in its largest scale; to say what is, rather than what should be. Are ecologists environmentalists? Most are, but this shouldn't be surprising. We would no more expect ecologists to be indifferent to the environment than we would expect art historians to be indifferent to paintings. Moreover, there is a branch of biology, conservation biology, that is directly concerned with preserving natural resources. But the business of ecology per se is to tell us about the large-scale interactions that go on in the natural world. This work has an important use; it provides the *information base* that society can use to make decisions about the environment. Does species diversity make for healthier natural communities? How many gray whales are there in the world? How important are prairie dogs to the prairie? To answer these questions, ecologists end up working in some places you might expect—in tents near the Arctic Circle or at treetop level in tropical rain forests (FIGURE 23.1). They also work in some locations you might not expect, however, such as in offices, hunched over computers, seeing how mathematical models of global warming work out.

Path of Study

In this chapter and the next one, you'll follow a bottom-to-top approach by moving from smaller ecological units to larger ones. What are the scales of life that concern ecology? The smallest is that of the physical functioning, or physiology, of given organisms. At this level, ecologists are examining individual living things, but at all other scales they are looking at *groups* of organisms or species. For the ecologist, life is organized into:

Populations

Communities

Ecosystems

The Biosphere

From Individuals to a Population   The smallest level of group organization is a **population**, defined in Chapter 17 as all the members of a single species that live together in a specified geographic region. The North American bullfrog (*Rana catesbeiana*) is a species that can be found from southern Canada to central Florida, but all the *R. catesbeiana* in a given pond constitute a single *population* of this species. Of course it's possible to define a population to be all the bullfrogs in *two* ponds, or in two states, or two countries—whatever geographic region is most useful for the question an ecologist is asking.

From Populations to a Community   Going up the scale, if you take the populations of *all* species living in a single region, you have a **community**. Often the term *community* is used more restrictively, to mean populations in a given area that potentially interact with one another.

From a Community to an Ecosystem   If you add to the community all the *non*living elements that interact with it—rainfall, chemical nutrients, soil—you have an ecosystem. More formally, an **ecosystem** is a community of organisms and the physical environment with which they interact. Ecosystems can be of various sizes; you'll be looking at a range that goes from fairly small (a single field, for example) up through the enormous ecosystems called *biomes* that may take up half a continent.

From Ecosystems to the Biosphere   The largest scale of life is the **biosphere**, which can be thought of as the interactive collection of all the Earth's ecosystems. Given what you've reviewed about ecosystems, this means all life on Earth and all the nonliving elements that interact with life. Sometimes, however, the term *biosphere* is used purely in a territorial sense—to mean that portion of the Earth that supports life. If you look at FIGURE 23.2, you can see a graphic representation of ecology's scales of life. We'll now begin to look at ecology through its levels of organization, starting with populations.

FIGURE 23.2   The Scales of Life That Concern Ecology In this chapter, you will consider how organisms of one species are associated in populations, and how populations of different species are associated in communities. In Chapter 24, you will look at ecosystems, which include not only living community members but also the nonliving factors that interact with them (such as the rainfall in the ecosystem panel above). Chapter 24 also reviews the large-scale ecosystems called biomes and the biosphere, which is the interactive collection of all the Earth's ecosystems.

Biosphere

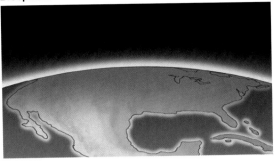

Ecosystem

Community

Population

Organism

23.2 Populations: Size and Dynamics

So what is it ecologists want to know about a population? Well, they need to know how to count it, how and why it is distributed over its geographical area, and how and why its size *changes* over time—what its population dynamics are, to use the term employed by ecologists.

Estimating the Size of a Population

The reason ecologists want to count the members of a given population is straight-forward: Without such a count, there is no way to answer a question such as how much territory a group of cheetahs must have in order to flourish, or how fast a population of finches recovers from a drought. (How would you know about the latter unless you could compare the population after the drought with the population before it?)

With large, immobile species such as trees, taking a census can sometimes be easy; just mark off an area and count. Things become more difficult when the area under consideration is so large that not all individuals in it can be counted, but rather must be estimated based on a population counted in a smaller representative area. Estimating becomes more difficult yet when the individuals are numerous and mobile, as with birds. Ecologists employ various means to estimate such populations; they count animal droppings within a defined area or survey bird populations as they migrate, for example.

Growth and Decline of Populations over Time

How is it that populations *change* size? As you begin to think about this question, it's worth going over a more general concept, which is the way a population of *anything* might increase in number. If you look at, say, the number of cars coming off the end of a production line, there might be 1,000 on Monday and then 1,000 on Tuesday, and so forth. The important thing to note is that the number of cars produced on Tuesday is not related to the number produced on Monday. The increase in the number of cars is thus an **arithmetical increase**: Over each interval of time (a day in this case), an unvarying number of new units (cars) is added to the population.

Now contrast this with what happens to living things. In most cases, each new unit (each living thing) is capable of playing a part in giving rise to *more* units, which certainly is not the case with cars. Population increases for living things are thus *proportional to* the number of organisms that already exist. Thus the increase in a population of organisms comes about through a different sort of increase, an **exponential increase**, which occurs when, over an interval of time, the number of new units added to a population is proportional to the number of units that exist. FIGURE 23.3 gives an example of the difference between the two kinds of growth over a period of weeks, using cars for arithmetic growth and the tiny water flea *Daphnia* for exponential growth. Let's assume we start out at the end of day 1 with the 1,000 cars produced that day and with 1,000 water fleas existing in an optimal laboratory environment. As you can see, *Daphnia* is a relatively slow starter, but its population quickly overwhelms that of the car population—not surprising, because the *Daphnia* population doubles every 3 days.

Population Growth in the Real World
As noted, the *Daphnia* were growing in an "optimal" laboratory environment—plenty of food, habitat kept clean, no predators. Thanks to human intervention, this was a kind of paradise for water fleas, in other words. But could any population ever grow like this in the real world? In all cases the answer is "not forever," and in most cases the answer is "not for long." If you look at FIGURE 23.4a, you can see what the *Daphnia* population's growth

FIGURE 23.3 Arithmetic and Exponential Growth
When the same number of objects is produced in a given interval of time—in this case, cars from a factory each day—arithmetic growth is at work. By contrast, the populations of organisms, such as the water flea *Daphnia*, can exhibit exponential growth, at least for a time, meaning the population grows in proportion to its own size.

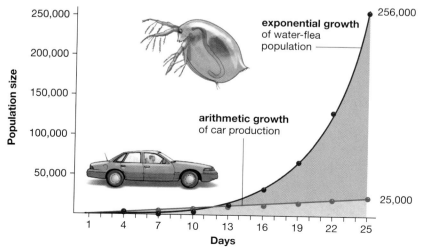

exponential growth
of water-flea
population

256,000

arithmetic growth
of car production

25,000

Population size

250,000
200,000
150,000
100,000
50,000

Days
1   4   7   10   13   16   19   22   25

looks like when plotted on a graph. An exponential increase such as *Daphnia's* creates what is called the *J-shaped growth curve* because it can be so extreme that when plotted it looks like the letter J. Note that there is relatively slow growth at first, then faster growth, then faster yet. This is population growth, in other words, in which the *rate* of increase keeps accelerating—it is **exponential growth**.

Now look at Figure 23.4b to see a second model of growth for natural populations, the *S-shaped growth curve*. This model tracks **logistic growth**: growth of a population in which the rate of growth slows and finally ceases altogether due to outside influences. Note that this growth stabilizes at a certain level, denoted in the figure as *K* (which you'll learn the significance of shortly).

What has intervened to account for the difference between the dizzying increase of the J-shaped curve and the moderate increase of the S-shaped curve? All the forces of the environment that act to limit population growth; in ecology, they are known as **environmental resistance**. Organisms will run out of food or have their sunlight blocked; greater numbers of predators will discover the population; the wastes produced by the organisms will begin to be toxic to the population. There are populations whose dynamics over time look something like either the J- or S-shaped curve, but in the real world, population size is likely to vary in more complex patterns. In Figure 23.4c, you can see one example of such a pattern, with the curve moving above and below the line denoted as *K*.

But what is *K*? What is this point around which the population hovers? It is the maximum population density of a given species that a defined geographical area can support over time. The term for this measure is an area's carrying capacity. Though we didn't name it as such, we actually touched on **carrying capacity** earlier, in connection with deer in Ohio's Cuyahoga Valley National Park. In recent years, average densities of deer in the park have been about 40 per square mile, though in some areas the density reaches as high as 130. Meanwhile, a healthy density for deer is considered to be certainly no more than 30 per square mile and perhaps no more than 20. So, what happens when these numbers are exceeded? When winter comes, deer starve to death, as they did in the park in the winter of 2005, because their habitat no longer contains enough food to support them. It's also possible to see what factors led to this extreme outcome. A 1997 analysis of deer who died in the park identified the effects of malnutrition on them. Biologists found that the deer evidenced not only low body fat, but a high prevalence of lungworms and a high prevalence of a respiratory virus. Effects such as these are examples of what happens when carrying capacity is exceeded. To put this another way, effects such as these are the means by which environmental resistance asserts itself.

The carrying capacity *K* line can generally be established for a given population by estimating its numbers over an extended period of time. For many species, their population may exceed or "overshoot" *K* at times, but this excess will be offset by periods during which the population is under *K*. Once *K* is known, it is valuable in predicting the change in a population's size, because it often serves as a factor that can be used in mathematical equations to calculate the limits to population growth. The higher a population is relative to *K*, the more severely its growth will be limited.

(a) Exponential growth

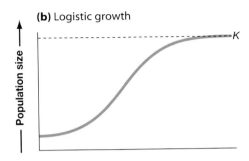

(b) Logistic growth

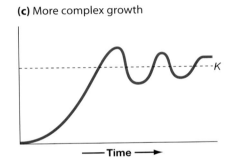

(c) More complex growth

Time →

Population size →

FIGURE 23.4   Three Models of Growth for Natural Populations

So Far...

1. A population is all the members of a species that live together in a _____. A community is the populations of all _____ living in a single region. An ecosystem is a community of organisms and the _____ with which they interact.

2. Environmental resistance can be defined as all the forces of the environment that act to _____.

3. Carrying capacity can be defined as the maximum _____ of a given species that a defined geographical region can _____.

So Far... Answers:

1. *specified geographic region; species; physical environment*
2. *limit population growth*
3. *population density; support*

A BROADER VIEW
Rabbits Gone Wild

Human tampering with the environment has resulted in several instances in which exponential growth of a population has gone largely unchecked for a time. In 1859, Thomas Austin released 24 wild European rabbits (*Oryctolagus cuniculus*) onto his estate in southern Australia, near Melbourne, in order to provide more game for sport hunting. This species was alien to Australia and, with no natural predators to keep its numbers down, it spread like wildfire; within 16 years it had expanded its range north and west through the entire latitude of the Australian continent, a distance of almost 1,100 miles. This scourge was finally brought under control only in the mid-1990s through use of a rabbit-killing virus.

FIGURE 23.5 **K-Selected and r-Selected Species**
The elephants on the left are a *K*-selected species. Here, a mother stays close to her calf, providing the kind of careful attention typical of *K*-selected species. The pond flies on the right are an *r*-selected species, bearing many young but giving no attention to them after birth.

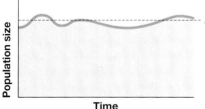

K-selected
equilibrium species → r-selected
opportunist species

Population size:
• limited by carrying capacity (*K*)
• density dependent
• relatively stable

Organisms:
• larger, long lived
• produce fewer offspring
• provide greater care for offspring

Population size:
• limited by reproductive rate (*r*)
• density independent
• relatively unstable

Organisms:
• smaller, short-lived
• produce many offspring
• provide no care for offspring

23.3 *r*-Selected and *K*-Selected Species

It's well known that for human beings, about 9 months will elapse between conception and birth; but it's less well known that for an elephant this same process takes almost 2 years. For any given human or elephant female, then, births are relatively few and far between. And once the offspring come, the amount of attention lavished on them is great indeed. It's not unusual for an elephant calf to be nurtured by its mother and a larger kinship group for at least the first 10 years of its life. In contrast, the common housefly can produce a new generation once a month, and the parental generation provides no care whatsoever to the offspring.

Houseflies and elephants lie at opposite ends of a continuum of *reproductive strategies*—characteristics that have the effect of increasing the number of fertile offspring an organism bears. So how does the strategy of elephants differ from the strategy of houseflies? This may be apparent from what you've observed already. For elephants the strategy is to bring forth few offspring, but lavish attention on them. For houseflies it is to bring forth a multitude of offspring, but give no attention to any of them. These strategies are in turn related to other characteristics of these species, as you'll now see (see FIGURE 23.5).

K-Selected, or Equilibrium, Species

Elephants will not seek out a totally new environment; they experience their environment as a relatively stable entity, and compete among themselves and with other species for resources within it. Given this stability, species like elephants are known as an *equilibrium species*. In line with this, the elephant population stays relatively stable compared to a fly population; its numbers will fluctuate in a relatively narrow range above and below the environment's carrying capacity. This is another way of saying that the population regularly bumps up against carrying capacity (*K*), which is the density of population that a unit of living space can support. Two things follow from this. First, elephants are said to be a **K-selected species**: one whose population sizes tend to be limited by carrying capacity. The second point, which follows from this definition, is that the pressures on the elephant population are **density dependent**: As the density of the population goes up, the factors that limit the population—food supply, living space—assert themselves ever more strongly.

r-Selected, or Opportunist, Species

In contrast to elephants, houseflies are known as an *opportunist* species—a species whose population size tends to fluctuate greatly in reaction to variations in its environment. Should favorable weather suddenly arrive, or a food supply suddenly appear, the fly population in the area will skyrocket. When the food is gone, or if the temperature suddenly changes, this same population will plunge in number. In short, environment tends to be highly variable for fly populations, and the flies have a high population growth rate that allows them to take advantage of environmental opportunities. For these animals, the strategy is to produce a multitude of offspring very fast. Ecologists denote a population's potential growth rate by the symbol *r*. Flies are therefore said to be an **r-selected species**, meaning one whose population sizes tend to be limited by reproductive rate.

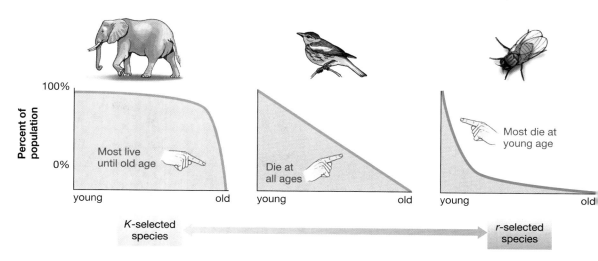

FIGURE 23.6 **When Is Death Likely to Come?** An organism's chance of living a long life is related to the reproductive strategy of its species, as reflected in these survivorship curves. Elephants and humans (late-loss species) produce few young, but most survive until old age. At the other extreme, flies (early-loss species) produce many young, but most die young. In between are constant-loss species; they produce a moderate number of young, which die at a fairly constant rate during a typical life span.

This definition implies that fly population numbers have nothing to do with fly population density, at least for a time. The pressures on the population are thus said to be **density independent**. The forces at work in these populations tend to be *physical* forces—frost, temperature, rain—as opposed to the *biological* forces that operate in density-dependent control (disease, competition for food). Given this, the *r*-selected strategy of many offspring/little attention is understandable. When you exist at the whim of environmental forces, "life is a lottery and it makes sense simply to buy many tickets," as R. M. May and D. I. Rubinstein have phrased it.

Survivorship Curves: At What Point Does Death Come in the Life Span?

K- and *r*-strategies are related to the concept of "survivorship curves," which, as you can see in FIGURE 23.6, are thought of in terms of three ideal types: late loss, constant loss, and early loss (sometimes referred to as types I, II, and III, respectively). Humans and most of our fellow *K*-selected mammals fall into type I, because we tend to survive into old age (our lives are lost late). Insects and many amphibians are type III, because their death rates are very high early in life, but level out thereafter. Other types of living things, such as birds, fall into type II, because they die off at a nearly constant rate through their life span.

23.4 Thinking about Human Populations

We'll look at the final elements in population dynamics in connection with the human population. Some of the concepts involved could be applied to any population of living things. But here the focus is on human beings, so that you can consider population principles along with the real-world issue of human population growth, which figures so prominently in environmental issues.

Population Pyramids: What Proportion of a Population Is Young?

You've seen that population growth in the natural world can be exponential (for a time) because a population of living things grows in proportion to its own size. However, all members of a population do not count equally in calculating this growth. If scientists want to peer into the future of a given population, what they want to know is: What proportion of the population is *past* the age of reproduction

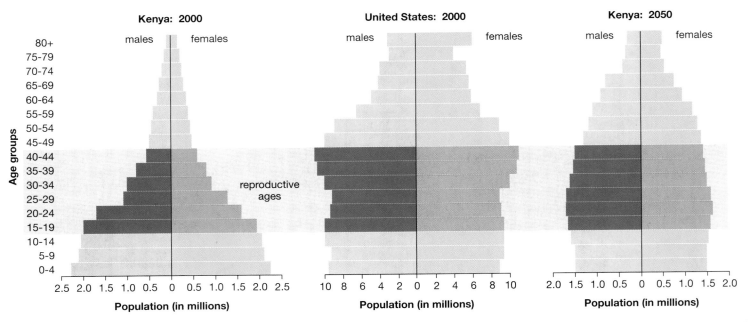

FIGURE 23.7 Age Structure in Populations: Kenya and the United States Each bar on the graphs represents a five-year age grouping of the population. Note the greater proportion of the Kenya 2000 population that is of "pre-reproductive" age—an indication of greater population increases in the future than is the case with the United States. The graph on the right shows the change expected to come about in Kenya's age structure by mid-century. (Data from the United States Census Bureau.)

as opposed to the proportion that is, or will be, of reproductive age? If you look at FIGURE 23.7, you can see the answer to this question, expressed as a "population pyramid" for two countries in the year 2000. Each bar on the graph represents a 5-year age grouping of the population (those who are 0–4 years old, 5–9, and so on), with the length of each bar representing the size of the grouping. The reproductive age-range for humans generally is considered to be between 15 and 45. If you look at the graph for Kenya, you can see that it has many more individuals of reproductive and pre-reproductive age than of post-reproductive age. What does the near future hold for Kenya? In all probability, a large growth in the population. In contrast, the age-structure diagram for the United States shows that a much greater proportion of its population is at or beyond reproductive age.

The World's Human Population: Finally Stabilizing

You probably noticed that there is a third graph in Figure 23.7, which displays Kenya's expected population pyramid for the year 2050. It's easy to see that this projected age distribution for Kenya looks something like the current age distribution for the United States. As a result, further into the future, Kenya should have slower population growth. This Kenyan projection actually is just one example of a momentous shift that is going on with human population around the world today. Only in the last few years has it become clear that, following centuries of explosive growth, the human population finally appears to be stabilizing.

To put this change into a larger context, look at FIGURE 23.8, which shows you how human population has grown during "historical time," which is to say the last 10,000 years. For centuries human numbers grew relatively little, but then began an upward climb about 1700. Improved sanitation, better medical care, and increases in the food supply came together to produce the rate of growth you see. The Earth's human population didn't pass the 1-billion mark until 1804; it then took 123 years to double to 2 billion (in 1927), then 48 years to double to 4 billion (in 1974), and now it has exceeded 6.3 billion. The curve you see that resulted from this growth may look familiar, because it is a more extreme version of the J-shaped curve, though greatly elongated on its left side. Were we looking at any species other than our own and saw this kind of growth curve, we would predict that this kind of increase could not go on for long; that environmental resistance would assert itself and that the population might even crash in one catastrophic sense or another.

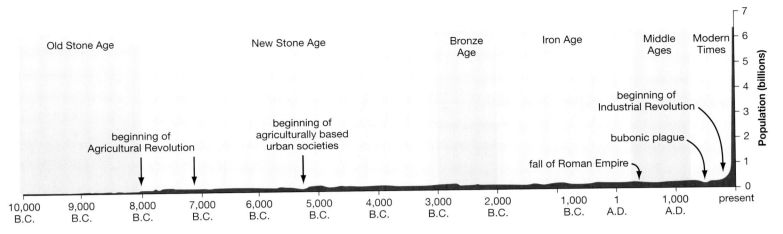

FIGURE 23.8 **Human Population Increase through the Centuries**

But something has changed. In 1992, demographers in the United Nations were predicting that, by 2050, the Earth would have 12 billion people on it. But by 2004, the UN was predicting that the human population will peak at about 9.2 billion people in 2075. To be sure, 9.2 billion represents a big increase from today's 6.3 billion. But the *rate* of increase is slowing. From 1990 to 1995, world population was increasing by 82 million people per year; by 2015–2020, the increase should be down to 69 million people a year. Should this trend continue, the world's human population will at least stabilize in this century, and it may even decline slightly from its peak by century's end.

What has made the difference? The key measurement used by demographers to predict human population change is **female fertility**—the average number of children born to each woman in a population. In more-developed countries, the "replacement" fertility level is 2.1. If fertility is above 2.1, the population will grow; if it's below 2.1, the population will shrink. (Why 2.1 instead of 2.0? Some children will not live to have children of their own.) With this in mind, look at **FIGURE 23.9**. Pictured are female fertility levels for a selected group of countries in two periods: 1960–65 and 2000–2005. The dramatic declines you can see are not being repeated in all countries. But when all the world's countries are averaged together, it's clear there has been a major change in fertility. Women are having fewer babies, and this is why the world's population seems to be stabilizing.

The global average in fertility, however, masks some tremendous differences between more-developed and less-developed countries. Think of it this way. In 2002, the world's population was increasing by 151 people per *minute*. And of these 151 additional people, 149 were born in less-developed countries. The South Asian country of Bangladesh—which is about the size of Iowa—is expected to have a population increase of more than 100 million people between 2003 and 2050. Meanwhile, in 15 countries of the European Union—among them France, Germany, and Sweden—the fertility rate is now 1.5. If this persists for another 20 years, the current population of these countries will be reduced by 88 million people by century's end.

Where does the United States stand in this spectrum? In the coming decades, it is expected to have more population growth than any other large, developed country, both in proportional and absolute terms. The U.S. population is expected to go from 292 million people in 2003 to 422 million people in 2050—a 45 percent increase. What makes the United States different from the other developed countries? Partly

FIGURE 23.9 **Big Changes in Fertility** Pictured are female fertility rates in six countries during two periods: first during the 1960s and then projected for the period 2000–2005. Changes such as these have brought about a remarkable decline in average female fertility worldwide. (From Population Reference Bureau website, 2003. Reprinted with permission.)

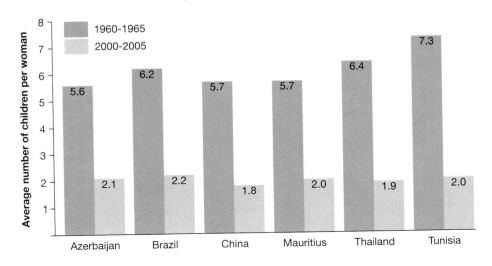

the American fertility rate, which at 2.03 is not quite at replacement level, but is much higher than those of European countries. The biggest single factor driving the U.S. population increase, however, is immigration. More than 1 million people immigrate to the United States each year.

Human Population and the Environment

Almost everyone is aware of a linkage between the quality of the environment and the size of human populations. Some scientists argue that there is no greater single environmental threat than the continued growth of the human population. The basis for this conclusion is that population affects so many environmental issues: the use of natural resources, the amount of waste that is pumped into the environment daily, the reduction of species habitat, the decimation of species through hunting and fishing. Look at almost any environmental problem, and you are likely to find human population growth playing a part in it.

Given that the human population is still growing—but that stabilization of it is on the horizon—it's possible to see the twenty-first century as a crucial moment in time for the environment. The biologist Edward O. Wilson has characterized our century as an environmental *bottleneck*: as a period in which the Earth's ecosystems will be under the greatest assault they are ever likely to face. On the other side of the bottleneck, with human population stabilized or even declining, the pressures on the environment should ease. The question is: How many species and habitats can make it *through* the bottleneck? How many primate species can be preserved? How many acres of rain forest can be kept intact? How many acres of American wetlands can remain as they are?

Some experts argue that it is not population, but rather the use of resources *per person* that is of most pressing concern. Consider that the United States used less water in 1995 than it did in 1980, even though the U.S. population grew by 16 percent during the period. Another perspective on this is provided by the use of resources in more-developed, as opposed to less-developed, countries. If you look at FIGURE 23.10, you can get a sense of the differences that exist. Carbon dioxide (CO_2) is a gas that is found naturally in the atmosphere, but that is also put into the atmosphere when human beings burn "fossil fuels," such as coal or gas, to power car engines or run power plants. CO_2 produced by these human activities is now regarded by most scientists as a cause of global warming—the rise in Earth's surface temperature that has occurred over the past century. Figure 23.10a shows the amount of CO_2 produced per person by human activity in five countries. Note how little CO_2 is produced per person in less-developed countries, such as India or China, compared to that produced in the more-developed countries, such as the

A BROADER VIEW

A Population in Free-fall

The French mathematician Auguste Comte is credited with coining the phrase "demography is destiny." By this, he meant that a nation's future is inevitably shaped by the characteristics of its human population. So steep have the declines in fertility been in almost all of the more-developed countries—and many of the less-developed countries—that some demographers are now writing about the problems that will come with having too *few* people. The future may already have arrived for Russia, whose fertility rate at one point dropped to an astoundingly low 1.17 per woman. Couple this with a simultaneous rise in Russian mortality, and the result is that Russia's population is expected to drop from 146 million in 2000 to 104 million in 2050.

FIGURE 23.10   Use of Resources by Country The use of natural resources per person (or "per capita") varies greatly from one country to the next. The average resident of a developed country, such as the United States, uses far more resources than the average resident of a less-developed country, such as India, and this has environmental consequences.

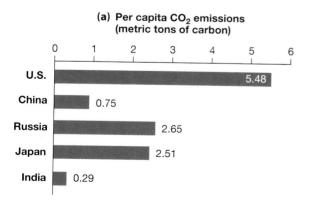

(a) Per capita CO_2 emissions
(metric tons of carbon)

U.S.	5.48
China	0.75
Russia	2.65
Japan	2.51
India	0.29

One measure of resource use is per capita carbon dioxide (CO_2) emissions: the amount of CO_2 a country releases into the atmosphere through human-caused activities, divided by the number of people that country has. The graph displays per capita CO_2 emissions for five countries in 1997, as measured in metric tons of carbon.

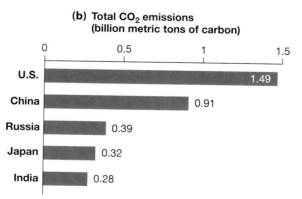

(b) Total CO_2 emissions
(billion metric tons of carbon)

U.S.	1.49
China	0.91
Russia	0.39
Japan	0.32
India	0.28

The differences in per capita emissions have significant consequences for total emissions within a country. In 1997, the United States released 64 percent more CO_2 into the atmosphere than did China, even though the U.S. had only one-fifth as many people as China.

United States. Now, in Figure 23.10b, note the effect this per capita difference has on total CO_2 emissions. The United States put 64 percent more carbon into the atmosphere than did China in 1997, even though China had nearly five times as many people as the United States (1.3 billion compared to 265 million in the U.S.). With this, you can begin to see why environmentalists are worried about the strain that economic development will put on the Earth's environment *irrespective* of population growth. What will happen when people in China start driving cars and using air conditioners at the rate that Americans do?

So Far... Answers:

1. *few; abundant; many; little or no*

2. *greater*

3. *average number of children born to each woman; 2.1 births per woman*

So Far...

1. *K*-selected or equilibrium species tend to have _____ offspring to whom they provide _____ care. Conversely, *r*-selected or opportunist species tend to have _____ offspring to whom they provide _____ care.

2. A country whose population pyramid is widest at the bottom can expect to have _____ population growth in the near future than a country whose population pyramid is of uniform width throughout.

3. A country's human population changes can be predicted by looking at female fertility, defined as the _____. The "replacement" level for fertility in more-developed countries is _____.

23.5 Communities

In this tour of ecology, you've so far been looking at species in isolation—at one population or another as it increases or decreases in size. In nature, of course, species live together in a rich mix of combinations. How rich? To take the high end of diversity, which is the tropical rain forest, Smithsonian biologist Terry Erwin once found an estimated 1,700 species of beetle in a single tree in Panama. Not 1,700 beetles, mind you; 1,700 *species* of beetle.

In beginning to think about the interactions of these diverse populations, a helpful concept is that of the community. As noted earlier, a **community** generally is defined as the populations of all species that inhabit a given area; all the plants, animals, fungi, bacteria, protists—every living thing, in other words. The term also is used, however, to mean a collection of populations in a given area that potentially *interact* with each other.

So what is it ecologists would like to know about any community? For starters, they'd like to know what at least some of the species are and what their relative numbers are. They'd also like to know about the importance to the community of some of its individual members.

Large Numbers of a Few Species: Ecological Dominants

There's a tremendous variability in the mix of species found in different communities, but many communities tend to be dominated by only a few species. Forests tend to be populated by certain kinds of trees, stretches of prairie by certain kinds of grasses. The few species that are abundant in a given community are called **ecological dominants** (see FIGURE 23.11). These

FIGURE 23.11 **Ecological Dominants**

(a) German forest dominated by single species of tree

(b) Kansas prairie dominated by tallgrass

FIGURE 23.12 **Keystone Species** When the predatory sea star *Pisaster ochraceus* was removed from a small area of rocky shore along the Pacific coast of the United States, the species composition of the community changed drastically, more so than if one of the other species had been removed. Pictured are several *Pisaster* sea stars on the Pacific Northwest coast.

generally are plants, but they can be other life-forms, as is the case with the tiny animals known as corals that are the ecological dominants in coral reefs.

Importance beyond Numbers: Keystone Species

Ecologists have long recognized that there are also species that may not be numerous in a given area, but whose *absence* would bring about significant change in the community. Each of these is a **keystone species**: a species whose impact on the composition of a community is disproportionately large relative to its abundance within that community. The concept of the keystone species was introduced in the 1960s by marine ecologist Robert Paine, who went with his students to a shallow-water zone of the Pacific Ocean in Washington State and for six years regularly removed all sea stars in the genus *Pisaster* from a small area. Such a sea star may sound harmless enough, but *Pisaster* was, in fact, the **top predator** in the area. It preyed on other species, but no species preyed upon it (**see** FIGURE 23.12). The impact of the *Pisaster* removal was big: Before the change there had been 15 species in the area, and after it there were 8. One species of mussel, freed from its former predator's control, took over much of the attachment space in the area and crowded out other animals, such as barnacles.

Over time, the keystone species concept has undergone some modification. Where once scientists thought of keystones as always sitting at the top of food chains, they now recognize that organisms in other positions can take on a keystone role—beavers building dams, for example, or even lichens that are critical in getting communities going in the desert. Beyond this, it turns out that there are communities without keystones; remove any one species from such a community, and its role will be taken over by another species.

Variety in Communities: What Is Biodiversity?

Apart from ecological dominants and keystones, a third element that ecologists pay attention to in any community is the *range* of species it has in it. This touches on the more general concept of **biodiversity**, which can be defined as variety among living things. In everyday speech, what biodiversity means is a diversity *of species* in a given area. This is an important measure of diversity, but it is only one among several. You can see why from an imaginary experiment that has been noted by ecologist Paul Ehrlich.

Suppose that you could get a few members of every species on Earth, but that you restricted each species to a single population housed somewhere (in a single zoo or an aquarium or botanical garden). Species *diversity* would not drop at all in such a scenario—you'd still have some of each kind of creature, after all—but the Earth quickly would become barren, because what's needed is a rich *distribution* of species in populations across the planet. This geographical distribution of populations is the second measure of biodiversity. The third measure exists *within* populations or species. It is genetic diversity, meaning a diversity of "alleles" or variants of genes in a population. Without such diversity, populations are vulnerable to disease; their members may die young or suffer from a variety of inherited mental and physical afflictions. In summary, biodiversity means species diversity, geographic diversity, and genetic diversity (**see** FIGURE 23.13).

23.6 Types of Interaction among Community Members

Having considered some general issues regarding communities, let's now turn to the subject of how members of a community might interact with each other. Here's a short list of ways:

Competition

Predation and parasitism

Mutualism and commensalism

Before continuing with the exploration of these modes of interaction, let's go over a couple of concepts that will apply to all of them.

High biodiversity   **Low biodiversity**

(a) Species diversity

many different species   few species

(b) Geographic diversity

broad distribution of species   narrow distribution of species

(c) Genetic diversity

high genetic diversity within population   low genetic diversity within population

FIGURE 23.13 **Three Types of Biodiversity**

Two Important Community Concepts: Habitat and Niche

A **habitat** is the physical surroundings in which a species can normally be found. Though two populations of a species may be widely separated, they can generally be found dwelling in similar natural surroundings. A habitat is sometimes described as a species' "address," but a more accurate metaphor might be a species' preferred type of neighborhood.

The word **niche** has been defined in several ways, but it is useful to think of it in terms of a simple metaphor: A niche is an organism's occupation. How and where does the organism make a living? What does it do to obtain resources? How does it deal with competition for these resources? The horseshoe crab has found a niche walking on the bottom of shallow coastal waters, feeding on food items that range from algae to small invertebrates. Note that this is about more than what the horseshoe crab eats. It includes specific surroundings (shallow ocean waters), specific behaviors (ocean-floor crawling), and perhaps seasonal or daily feeding times, among other things. If you were to specify all the things that define the horseshoe crab's niche, the odds are that no other organism would exactly fit into it.

Competition among Species in a Community

With these two concepts in mind, let's look at the ways organisms interact in communities, starting with a familiar type of interaction—competition.

Even though niches tend to be specific to given organisms, some species—particularly closely related species—have niches that *overlap* to some degree in a community. A large proportion of both species' diet may be made up of a given organism, or both species may occupy similar kinds of spaces on rocks or branches or pond surfaces.

(a) Competitive exclusion

When two species compete for the same limited, vital resource, one will always drive the other to local extinction—as the paramecium *P. aurelia* did to the paramecium *P. caudatum*. This is the competitive exclusion principle at work.

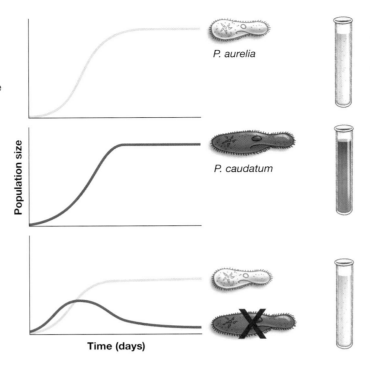

(b) Resource partitioning

Conversely, when Gause put *P. aurelia* together with another paramecium, *P. bursaria*, the two species divided up the habitat and both survived. This is a demonstration of resource partitioning.

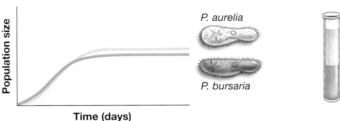

FIGURE 23.14 Competition for Resources among Species Laboratory experiments by G. F. Gause showed that competition for resources between two species can have two possible outcomes.

A BROADER VIEW

Competitive Exclusion within Us

If we want to see competitive exclusion in nature, we need look no further than our own digestive tracts. By the time we are two years old, each of us has a well-established community of bacterial species living in our intestines. (Thousands of species live in the human gut, but each of us has an individualized community of about 100 varieties.) One of the reasons that we suffer from intestinal infections so rarely is competitive exclusion: Invading bacteria can't gain a foothold because they are outcompeted by the established bacteria, which are well adapted to their environment.

What may come to mind with competition between two species is a never-ending series of physical battles, but things seldom work like this. For one thing, competition tends to be indirect. It often is a competition for *resources*, in which the winner generally triumphs not by fighting but by being more efficient at doing something, such as acquiring food.

Competitive Exclusion When the niches of two species greatly overlap, it's unlikely that you'd see *long-standing* competition among them, for the simple reason that one species or the other is likely to win such a competition in fairly short order. Because the competition is for nothing less than vital resources, the result is that the losing species will be driven to local extinction.

It was a laboratory experiment, performed by the Russian biologist G. F. Gause in the 1930s, that pointed the way on this latter principle. Gause grew two species of *Paramecium* protists, *P. caudatum* and *P. aurelia*, in culture and found that in time, *P. aurelia* was always the sole survivor. True to what was just noted, *P. aurelia* wasn't eating or wounding *P. caudatum*; instead, it grew faster and thus used more of the surrounding resources. Nevertheless, the outcome was always the death of all the *P. caudatum*. This led Gause to formulate what came to be called the **competitive exclusion principle**: When two populations compete for the same limited, vital resource, one will always outcompete the other and thus bring about the latter's local extinction (**see** FIGURE 23.14a).

Ecologists wouldn't expect to witness the competitive exclusion principle operating much in nature, however. This is because, as noted, the *Paramecium* scenario

FIGURE 23.15 **Kudzu Vines Run Wild** The kudzu plant was introduced in the American South in the 1930s and has since spread at a rapid rate, locally eliminating many plants in its path. This is competitive exclusion in action. Here kudzu has overgrown an abandoned house in Mississippi.

is likely to be played out quickly. Nevertheless, competitive exclusion has been observed in nature, often when humans have a hand in things. For example, humans introduced a Southeast Asian vine, the kudzu, to the American South on a large scale in the 1930s. Growing at up to 1 foot per day, kudzu has now taken over millions of acres in the South, locally eliminating many plants in its path (**see FIGURE 23.15**).

Resource Partitioning Is Common in Natural Environments

Competitive exclusion notwithstanding, there are instances in nature in which two related species will use the same kinds of resources from the same habitat over a long period of time. So, why isn't one of them eliminated? The answer is contained in another experiment conducted by Gause. He took the successful species from the earlier experiment, *P. aurelia*, and placed it in a test tube with a different paramecium, *P. bursaria*. This time, neither species was eliminated (see Figure 23.14b). Instead of competing, the two species divided up the habitat, *P. aurelia* feeding in the upper part of the test tube and *P. bursaria* flourishing in the lower part. (*P. bursaria* had an advantage in the lower, oxygen-depleted water because it has symbiotic algae that grow with it, producing oxygen.) In a nutshell, this result describes a situation that often exists in nature: **coexistence** (a sharing of habitat) through a practice called **resource partitioning**, which can be defined as a dividing up of scarce resources among species that have similar requirements. If you look at FIGURE 23.16, you can see how this works among some species that are a little more familiar—several varieties of warbler.

Other Modes of Interaction: Predation and Parasitism

It is one thing for two species to compete for resources; it is another for one species to *be* a resource for another—to be eaten or used by another species. The difference between the two things distinguishes the first mode of interaction in communities, competition, from the second, predation.

Predation can be defined as one organism feeding on parts or all of a second organism. The prey here can be plants, protists, animals—whatever is preyed upon. Predation is generally thought of in terms of animals killing other animals, but note that the definition given includes such things as animals consuming whole plants or their seeds. **Parasitism** is a variety of predation in which the predator feeds on prey, but does not

FIGURE 23.16 **Resource Partitioning** Ecologist Robert MacArthur spent long stretches of time over several years in the 1950s observing the feeding patterns of several species of warblers. All of them ate caterpillars, but from substantially different, though overlapping, parts of the tree.

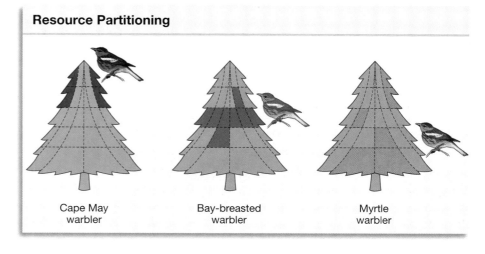

Resource Partitioning

Cape May warbler

Bay-breasted warbler

Myrtle warbler

FIGURE 23.17 **Plants Parasitizing Plants** The orange tendrils in the picture belong to a species of dodder, *Cuscuta pentagona*, a parasitic plant that survives by tapping into the food reserves of other plants, thus feeding off them. The dodder is orange, rather than green, because it lacks the chlorophyll that is used in photosynthesis.

kill the prey immediately—and may not ever kill it. Parasites that come to mind tend to be animals, but there are an estimated 3,000 species of parasitic plants, one of which can be seen in **FIGURE 23.17**.

The Value of Predation and Parasitism Predation and parasitism are the features of nature that many nature lovers can't stand, perhaps because these practices seem cruel or unfair from a human point of view. Arguably, however, predation and parasitism have had a clear value, in that they have spurred on evolution by stimulating the "arms race" that has resulted in evolutionary adaptations such as vision and flight. (One organism's predatory adaptation spurs the development of another organism's defensive adaptation, and vice versa.) Whatever we may think of it, predation is simply a fact of nature, and it is impossible to imagine life without it. But it's also true that to admire the beauty of nature is in significant part to admire the handiwork of predation, since it has done so much to shape the living world.

Predator-Prey Dynamics It may be obvious that although predators attack prey, predators are also *dependent* on prey as a food source. An ongoing question in ecological research is how tight this linkage is. To what extent do predator and prey population sizes tend to move up and down together? It's clear that the population dynamics of some predator and prey species are very tightly linked. The small lemmings of northeast Greenland (*Dicrostonyx groenlandicus*) are fed upon by only four predators: the stoat, the arctic fox, the snowy owl, and another bird called the long-tailed skua. Of these, the stoat is a "specialist" predator, in that it feeds almost entirely on lemmings. If you look at **FIGURE 23.18**, you can see that both the lemming and stoat populations predictably go through four-year up-and-down population cycles. Moreover, note that the movement up or down in stoat population *lags* the change in lemming population, generally by a few months. What happens is that a higher lemming population means more food for the stoats, which then increase in population. This increase, of course, then means more predation of the lemmings, but the fall you can see in lemming population is not brought about solely by the increase in stoat numbers. When the lemmings' density reaches high levels, the arctic fox, snowy owl, and long-tailed skua begin to prey on them. It is this predation, in combination with the increased stoat predation, that then drives the lemming numbers down. Once the lemmings exist in low densities again, the stoat becomes the only predator, lemming numbers go up, and the cycle repeats.

The Finnish and German researchers who reported these findings found that declines in lemming numbers could be explained *entirely* by predation; neither the food nor the space available to the lemmings, for example, played a part in reductions of their population. Such a crystal-clear connection between a prey species and its predators naturally raises the question of how common such a linkage is in nature. What we can say for certain is that in most species, population dynamics involve much more than just predator-prey relationships. Environmental resistance can take many forms, of which increased predation is only one. You might say that the connection between this population of lemmings and their predators serves as an example of how strong predator-prey linkage *can* be, but it is not representative of how

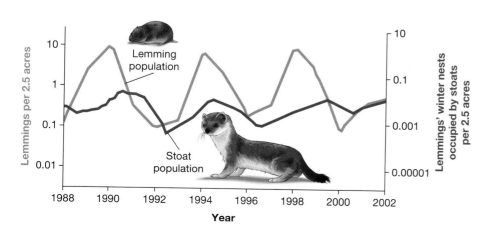

FIGURE 23.18 **Predator and Prey Populations** The population dynamics of a prey species (the lemming) and one of its predators (the stoat) in Northeast Greenland over a period of 14 years. (Reprinted with permission from "The Population Dynamics of a Prey Species (the lemming)" and "One of its Predators (the stoat) in Northeast Greenland Over a Period of 14 Years" by O. Glig et al, *Science*, 302:867, 2003.)

Modern Myths

Lemmings Commit Suicide

Like lemmings to the sea. The idea behind this well-worn metaphor is, of course, that the small northern-latitude rodents called lemmings march periodically toward the sea and hurl themselves into it from cliffs, thereby plunging to their deaths. The only problem with this idea is that there's not a shred of truth to it. As the text notes, some lemming populations do go through "boom and bust" cycles; and when the population is in what might be called full boom, the lemmings may *migrate* in large numbers. But they are not jumping off cliffs together.

So how did the idea of lemming suicide get started? Riley Woodford of the Alaska Department of Fish and Game has written that a 1958 Walt Disney documentary, *White Wilderness*, is to blame. The makers of this film, Woodford says, simply faked scenes of lemmings running off of cliffs. Lemming experts in both the United States and Europe agree that lemmings never undertake the self-destructive behavior shown in *White Wilderness*. The curious thing is how persistent the idea of their mass suicide is, given that the only basis for this idea is a "documentary" made almost 50 years ago.

strong such linkage is likely to be. Finally, one of the enduring myths about lemmings is that they control their own population densities by committing suicide en masse—something you can read more about in *Modern Myths,* above.

Parasites: Making a Living from the Living

Let's look now at a special variety of predation, *parasitism*. Some familiar parasites, such as leeches and ticks, have a straightforward strategy: Get on, hold on, and consume material from the prey, known as the **host**. But the relationship between parasites and hosts can be more sophisticated than this. Often a parasite makes a host part of its reproductive cycle while feeding on it. Many species of wasps, for example, deposit their eggs directly into the adult (or larval) bodies of caterpillars. When the wasp eggs hatch into larva within the caterpillar, they begin to use the caterpillar's body not only as an environment in which to mature, but as a food source. Eating away at the caterpillar body from within, they crawl out of the animal about the time it dies from its internal losses. For an example of parasitism more complex (and chilling) than this, see Essay 23.1, "Why Do Rabid Animals Go Crazy?" on page 406.

The Effect of Predator-Prey Interactions on Evolution

Over evolutionary time, predator-prey interactions can shape the physical forms of both predators and prey. To see an example of this on the predator side, look at the warm-water "frogfish" in FIGURE 23.19. These fish have evolved a spine on their dorsal fins that is tipped with a piece of flesh that looks for all the world like a small worm floating free in the ocean. When a would-be predator tries to snatch up this

A BROADER VIEW

Parasitism and Evolution

Some scientists believe that parasitism has had the profound evolutionary effect of bringing about sexual reproduction. Microbial invaders often work by means of latching onto specific molecules on the surface of a target cell—molecules that will stay the same, generation after generation, in organisms that reproduce asexually, since each "daughter" cell produced this way is a clone of its parental cell. Sexual reproduction presents potential invaders with an ever-changing target; each generation differs from its parents because each results from a mixture of its parents' genetic material.

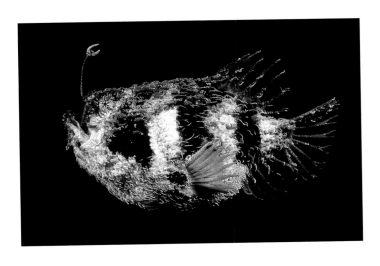

FIGURE 23.19 **Fooling Predators about Prey** The predatory frogfish uses a modified spine resembling a tasty worm to lure its prey. Here a tasseled frogfish clearly shows its lure while swimming near Edithburg, South Australia.

Essay 23.1 Why Do Rabid Animals Go Crazy?

There is a tiny worm, called *Plagiorhynchus cylindraceus*, that lives its adult life as a parasite in the intestines of starlings and other songbirds. While in a starling, this worm produces eggs that pass out of the bird in its feces. After this, both eggs and feces may be eaten by pillbugs, which are the familiar little arthropods that "roll up into a ball" (see **FIGURE E23.1.1**). The worm eggs, however, are not digested by the pillbugs. As a result, a worm *hatches* inside a pillbug, which makes the bug the second "host" for this parasite. And here is where things get interesting, as researcher Janice Moore has shown.

The pillbug is dark, but once infected by a worm, it starts spending more time on light-colored surfaces. What's more, it stops sheltering itself under overhanging objects such as leaves. And the females among the bugs start moving around more and resting less. The effect of all this is that worm-infected pillbugs are more visible to their predators and thus are more likely to be eaten by them. And who are these predators? Some of them are starlings.

Rabid raccoons or dogs are not simply "going crazy" because of their disease.

Now, when a starling eats a pillbug, the pillbug dies; but the worm inside the bug remains unharmed. Indeed, this is the way the worm gets back inside a starling host (where it will develop into an adult, lay eggs, and so forth). And that's just the point. It appears that the worm *engineers* this outcome by bringing about the self-destructive behavior of the pillbug. Through some unknown mechanism, this parasite prompts the pillbug to become an accomplice in its own death. When the pillbug is infected, its behaviors are such that it might as well have a sign on it saying, "Please eat me." The starling is happy to oblige, but the real winner is the *P. cylindraceus* worm, which gets to move from one of its hosts to another.

It would be nice to report that such behavioral takeovers by parasites are rare, but in fact they are common. "Horsehair" worms need to mate in the water; but they spend most of their lives inside arthropods such as ants, whose abdomens they feed on. Just before an ant dies from this, however, it aids the worm in one final way—it makes a journey to water. Indeed, a worm-infected ant seemingly has an uncontrollable urge to get wet, as it will return repeatedly to water if blocked. Once it wades into a pool or puddle, however, things proceed quickly. In a matter of minutes, the worm inside the ant will emerge, ready to mate with another worm.

Now, think about all this in connection with one of the things children are taught to fear from an early age: a rabid animal. Rabies is caused by a virus that spreads through animal saliva. In the "furious" form of the disease, the virus is passed on by means of one animal biting another, thus injecting its saliva into the second animal. Not surprisingly, one of the things the virus does is affect salivary tissue in an infected animal's head and neck. This ensures that there is plenty of saliva to be transmitted. The virus also inhibits swallowing in an infected animal, which ensures more saliva. Finally, the virus affects the animal's central nervous system in such a way as to bring about the famous *behavioral* change in infected animals—frenzied, unprovoked attacks on other animals. Thus, rabid raccoons or dogs are not simply "going crazy" because of their disease. Their behavior has a function. To be sure, it's a *perverse* function from a human perspective, but then again, aren't the behaviors of infected pillbugs or ants disturbing in a similar sort of way? The general message here is straightforward but hard to take: Living things can be forced to act in ways that aid the very organisms that have infected them.

FIGURE E23.1.1 Accomplice to Its Own Death The common pillbug *Armadillidium vulgare*, whose behavior can be manipulated by the parasitical worm *P. cylindraceus*.

"worm," however, it finds *itself* the prey. On the prey side, look at **FIGURE 23.20** to see the amazing kinds of camouflage that animals use to keep themselves hidden from the eyes of predators. Needless to say, predator-prey evolution goes on within all the kingdoms of life. Plants have developed not only spines, thorns, and other protective structures but also a vast array of chemical compounds—more than 15,000 have been characterized so far—that protect them from predators.

Mimicry Is a Theme in Predator-Prey Evolution

The ability to fool an opponent by faking is as valuable in nature as it is in sports, as you've seen with the frogfish. Another way such deception takes place in nature is through **mimicry**, a phenomenon through which one species has evolved to assume the appearance of another. One form of mimicry involves three participants: a model, a mimic, and a dupe. If you look at FIGURE 23.21, you can see a model on the left, the yellowjacket wasp, which obviously can provide a painful lesson to any animal that tries to eat it. On the right you can see the mimic, the clearwing moth, which is harmless but *looks* a great deal like a wasp. For the dupe, you could select any predator species that sees a clearwing moth and passes it by, believing it to be a dangerous yellowjacket. This is so-called **Batesian mimicry**: the evolution of one species to resemble a species that has a superior protective capability.

In a second type of mimicry, **Müllerian mimicry**, several species that *have* protection against predators come to resemble each other (**see** FIGURE 23.22). This creates a visual warning that becomes known to an array of predators. Given this, a would-be predator learns, by interacting with an individual from one species, to keep away from all individuals in any look-alike species. The result is that fewer individuals in these species will be disturbed or killed.

(a) A spanworm looking like a twig on a maple tree

(b) A casque-head chameleon against a tree in Kenya

FIGURE 23.20 **Avoiding Predation through Camouflage**

So Far...

1. As a keystone species, the sea star *Pisaster* has an impact on the composition of a community that is _____ relative to its _____ within the community.

2. There are three types of biodiversity: _____, _____, and _____.

3. In the interaction known as _____, one species feeds on all or parts of a second organism; in the interaction known as _____, a predator feeds on prey but does not kill it immediately, if ever.

So Far... Answers:

1. *disproportionately large; abundance*
2. *species; geographic; genetic*
3. *predation; parasitism*

FIGURE 23.21 **Harmless, but Looking Dangerous** In an example of Batesian mimicry, the clearwing moth (on the right) has no sting but has evolved to look like an insect that does, the yellowjacket wasp (left). The yellowjacket is the model and the moth is the mimic.

FIGURE 23.22 **Müllerian Mimicry** The South American butterflies *Heliconius cydno* (top) and *Heliconius sapho* are different in some ways, but they share the characteristic of tasting bad to predators. Over time, they evolved to look like each other, with their appearance serving as a warning to would-be predators. Natural selection favors this Müllerian mimicry, because predators can learn about the butterflies' unpalatable taste from both species. The result is that fewer members of either species are likely to be killed or bothered by predators.

Beneficial Interactions: Mutualism and Commensalism

The competitive and predatory modes of interaction you've reviewed have an each-organism-against-all quality to them, but other kinds of interactions exist in nature as well. Consider **mutualism**, meaning an interaction between individuals of two species that is *beneficial* to both individuals. There's a great example of this in Chapter 21 (page 357), with the linkage between plant roots and the slender, below-ground extensions of fungi called *hyphae*. What the fungi get from linking to plant roots is food that comes from the photosynthesizing plant; what the plants get is minerals and water, absorbed by the network of hyphae. By some estimates, 90 percent of seed plants have this relationship with fungi. Perhaps the most famous example of mutualism, however, is between the rhinoceros and oxpecker birds. Sitting on the back of the rhino, the oxpecker removes parasites and pests (ticks and flies), thus getting food and a safe place to perch, while the rhino gets relief from its tiny adversaries (FIGURE 23.23).

There is also **commensalism**, meaning an interaction between two species in which one benefits while the other is neither harmed nor helped. Birds can make their nests in trees, and benefit from this, but generally don't affect the trees at all.

Coevolution: Species Driving Each Other's Evolution

When different types of organisms are tightly linked in any way over long periods of time, they are likely to shape one another's evolution. Perceiving this, scientists Paul Ehrlich and Peter Raven developed the concept of **coevolution**, meaning the interdependent evolution of two or more species. Some of the clearest examples of coevolution involve flowering plants and the animals that pollinate them. Many flowers have both fragrances that attract insects and ultraviolet color patterns (invisible to us) that guide the insects to the proper spot for pollination. These features are nature's own homing signals and landing lights, communicating the messages "Food lies this way" and "Land here" (**see** FIGURE 23.24). Now, one of the organisms linked with flowering plants is the honeybee. Consider that the color vision of honeybees is most sensitive to the colors that exist in the very flowering plants they pollinate. What seems likely is that honeybee eyesight evolved in response to plant coloration, and that plant coloration evolved to be maximally attractive to the pollinating insects. These groups of organisms coevolved, in other words.

Having considered how community members interact with one another, let's now go on to consider how communities change over time.

23.7 Succession in Communities

When Washington State's Mount St. Helens volcano erupted in May of 1980, it first collapsed inward, thus sending most of the mountain's north face sliding downhill in the largest avalanche in recorded history. Then came the actual eruption, which sent a huge volume of rock and ash hurtling not straight up but out at an angle, toward the north. Some stands of forest in the path of this blast were instantly incinerated down to bare rock; another 86,000 acres of trees were snapped in two like so many twigs. Then came the mudslides caused by the vast expanse of snow and ice melted in the explosion; then came the fall of hundreds of millions of tons of ash, some of it landing as far away as Wyoming. After viewing the devastated area around the blast, President Jimmy Carter said that it "makes the surface of the moon look like a golf course."

But today? No one would claim that the Mount St. Helens area has returned to anything

A BROADER VIEW

Ants as the First Farmers

The world's first farmers were not human beings, but ants, specifically the Amazonian leaf-cutter ants of the genus *Atta*. These ants do not eat the leaves they carry off. Instead they bring them down into their nests, where they process the leaves into a pulp, and then spread the pulp onto a fungus that develops only in their nests. The fungus then feeds on the food that has been brought to it, and grows, after which the ants harvest its knob-like "stalks" and consume them. There is clearly a mutualism between ant and fungus, with both parties getting a continuous food supply out of the arrangement, but there's coevolution as well: The fungus has evolved to exist only under the care of the ants. Meanwhile, the ants have evolved glands that produce secretions harmful to alien fungi, but not to the fungi they tend.

FIGURE 23.23 **Mutually Beneficial**

(a) Rhinoceros and oxpecker birds

(b) Snapping shrimp and shrimp goby

Several oxpecker birds sit atop a black rhinoceros, ridding the rhino of ticks and other pests while the rhino provides a safe habitat for the birds. This is a demonstration of mutualism—an interaction between two species that is beneficial to both.

The orange–spotted shrimp goby, on the right, and the snapping shrimp at left also exhibit mutualism. Here the shrimp goby stands guard near the snapping shrimp, which digs out the burrow on the sea floor that the two creatures share.

(a) What we see

(b) What a bee sees

FIGURE 23.24 **Coevolution of Plants and Their Pollinators** This flower has an ultraviolet color pattern that is not normally visible to humans, but that attracts bees. It is likely that these color patterns evolved in the flowers because they aided in attracting the bees. Meanwhile, the bees developed vision that was sensitive to the colors exhibited by these same plants.

like its former state, but look at the pictures in FIGURE 23.25 to get an idea of the transition that has occurred at one location around Mount St. Helens.

The exact form of this rejuvenation may surprise us, but don't we intuitively expect something like this? Just from our everyday experience of watching, say, an abandoned urban lot becoming progressively weed-filled, don't we expect this would generally be the case? As it turns out, our intuition is right, because almost any parcel of land or water that has been either abandoned by humans or devastated by physical forces will be "reclaimed" by nature, at least to some degree.

The question is, how does this reclamation proceed? Areas that are rebounding don't just get one type of vegetation and then retain it. Instead, the vegetation changes through time—one type of growth *succeeds* another—with a general movement toward more and larger greenery as time goes by. Ecologists call this phenomenon **succession**, meaning a series of replacements of community members at a given location until a relatively stable final state is reached. Within this framework, there are two kinds of succession. The first is **primary succession**: a succession in which the starting state is one of little or no life and no soil. Parts of the Mount St. Helens area underwent primary succession following the blast. Then there is **secondary succession**: a succession in which the final state of a habitat has been disturbed by some force, but life remains and soil exists. The classic example of this is land that has first been cleared for farming but later abandoned by the farmer.

The relatively stable community that develops at the end of any process of succession is called a **climax community**. There will be some shifts over time in a climax community, just as there are in any other. A prolonged drought will cause some change in the mix of the community's animal and plant life, for example. But, with some exceptions, what does not happen is a shift to a *fundamentally* different mix of life-forms, barring a major change in climate. Grassland stays grassland, and forest stays forest.

Succession can be fairly predictable within small geographical areas. Two abandoned farm plots that lie close together will have very similar kinds of succession—so much so that local farmers or naturalists can sometimes tell to within a few years how long it's been since a field was abandoned. As you may be able to tell, succession generally is thought of in terms of the various *plant* communities that succeed one another, though modern ecology is attempting to bring animals into the picture as well.

An Example of Primary Succession: Alaska's Glacier Bay

One of the best examples of primary succession comes from Alaska's 65-mile-long Glacier Bay, whose lower-lying areas were covered by a glacier until about 240 years ago, when the glacier began retreating. Such withdrawal initially leaves behind a rocky terrain that is completely devoid of

FIGURE 23.25 **Rebound from Disaster** Mount St. Helens, with its crater in the distance, as it appeared in June 1980, one month after the blast and as it appeared in 1998. This is an example of ecological succession in action.

organic matter—no fallen leaves or branches, no decomposing animals. Indeed, what's left is not soil, but a pulverized rock (called till) that has a high pH and no available nitrogen—not a promising environment for life to establish itself. But establish itself it does. If you look at FIGURE 23.26, you can see how succession proceeded in Glacier Bay. Some of the types of changes that occurred there are common to primary succession in general. We can say that succession is likely to include:

- An arrival of photosynthesizing pioneer species—in Glacier Bay's case, the bacteria and lichens you can see in the figure—which can establish themselves in the most barren environments. Photosynthesis is the foundation of almost all life on Earth; as such, organisms who can perform photosynthesis must begin to flourish in order for a community to develop.

- An increase in "biomass" or the dry weight of material produced by living things. By just glancing at the figure, you can see how biomass increased in Glacier Bay over time.

- A general movement toward longer-lived species. The *Dryas* shrub you see in the figure can live for 50 years, but alders can live for 100 years, and spruce trees can live for 700.

- The facilitation of the growth of some later species through the actions of earlier species. The *Dryas* shrubs had bacteria in their roots that could *fix nitrogen*—that could take nitrogen from the air and convert it into a form that plants can use. This is how a community of larger plants could get started in Glacier Bay, which *lacked* nitrogen until the *Dryas* appeared.

- The competitive driving out of some species by the actions of later species. The spruce trees in the figure "shaded out" the smaller alder bushes, thus depriving them of the sunlight they needed for photosynthesis.

FIGURE 23.26 **One Example of Primary Succession: Glacier Bay, Alaska**

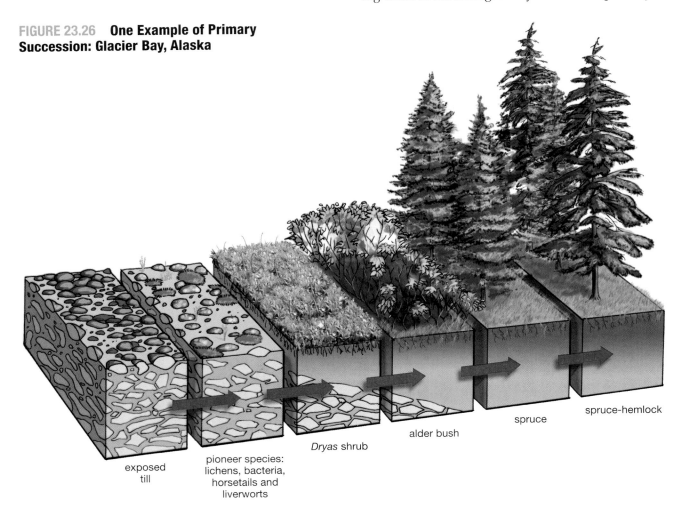

exposed till

pioneer species: lichens, bacteria, horsetails and liverworts

Dryas shrub

alder bush

spruce

spruce-hemlock

On to Ecosystems and Biomes

In this chapter, you have looked at populations and communities. In populations, you saw the building blocks of communities. In communities, you looked at the interactions of different species. But it's obvious that communities interact with *nonliving* forces and factors, such as weather and soil composition. When living things and nonliving factors are considered together, the resulting whole is called an ecosystem—the subject of the first part of Chapter 24. Very large-scale ecosystems are called biomes; a review of them forms the concluding portion of Chapter 24 and will bring to a close this book's coverage of ecology.

So Far... Answers:
1. *mutualism; commensalism*
2. *replacements; stable state*
3. *increase; longer-lived*

So Far...

1. In the species interaction known as _____, both species benefit; in the interaction known as _____, one species benefits while the other is neither helped nor harmed.

2. Succession can be defined as a series of _____ of community members at a given location until a relatively _____ is reached.

3. In primary succession, over time, biomass tends to (pick the correct term from the two) increase/decrease; and species tend to be shorter-lived/longer-lived.

Chapter Review

Summary

23.1 The Study of Ecology

- **Defining ecology**—Ecology is the study of the interactions living things have with each other and with their environment. (page 390)

- **Scales of life**—There are five scales of life that concern ecology: physiology, populations, communities, ecosystems, and the biosphere. A population is all the members of a single species that live together in a specified geographical area; a community is all the populations of all species living in a single area; an ecosystem is a community and all the nonliving elements that interact with it; the biosphere is the interactive collection of all the Earth's ecosystems. (page 390)

23.2 Populations: Size and Dynamics

- **Estimating populations**—Ecologists employ several means to estimate the size of populations of living things, among them counting animal droppings or surveying bird populations as they migrate. (page 392)

- **Arithmetical and exponential**—An arithmetical increase occurs when, over a given interval of time, an unvarying number of new units is added to a population. An exponential increase occurs when the number of new units added to a population is proportional to the number of units that exists. Populations of living things are capable of increasing exponentially, because living things are capable of giving rise to more living things. (page 392)

- **J- and S-shaped curves**—The rapid growth that sometimes characterizes living populations is referred to as exponential growth, which is illustrated in the J-shaped growth curve. Populations that initially grow, but whose growth later levels out because of external forces, have experienced logistic growth, which is illustrated in the S-shaped growth curve. (page 392)

- **Environmental resistance**—The size of living populations is kept in check by environmental resistance, defined as all the forces of the environment that act to limit population growth. (page 393)

- **Carrying capacity**—Carrying capacity, denoted as K, is the maximum population density of a given species that can be sustained within a defined geographical area over an extended period of time. (page 393)

Web Tutorial 23.1 Population Growth

23.3 *r*-Selected and *K*-Selected Species

- ***K*-selected species**—Different species have different reproductive strategies, meaning characteristics that have the effect of increasing the number of fertile offspring they bear. *K*-selected, or equilibrium, species tend to be large, to experience their environment as relatively stable, and to lavish a good deal of attention on relatively few offspring. The pressures on *K*-selected species tend to be density dependent, meaning that as a population's density goes up, factors that limit the population's growth assert themselves ever more strongly. (page 394)

- ***r*-selected species**—*r*-selected or opportunist species tend to be small, to experience their environment as relatively unstable, and to give little or no attention to the numerous offspring they produce. The pressures on *r*-selected species tend to be density independent, meaning pressures that are unrelated to the population's density. (page 394)

- **Survivorship curves**—Survivorship curves describe how soon species members tend to die within the species' life span. There are three idealized types of survivorship curves: late loss, constant loss, and early loss, also referred to as types I, II, and III. (page 395)

23.4 Thinking about Human Populations

- **Population pyramids**—An important step in calculating the future growth of human populations is to learn what proportion of the population is at or under reproductive age. A population pyramid displays this proportion. Populations whose pyramids are heavily weighted toward younger age groups are likely to experience relatively large near-term growth. (page 395)

- **Human population trends**—Following centuries of explosive increase, the world's human population is projected to stabilize in the coming decades, going from about 6.3 billion now to a maximum of about 9.2 billion by 2075. This stabilization is being brought about by a decrease in female fertility, defined as the number of children born, on average, to each woman in a population. (page 396)

- **Differences across countries**—The global reduction in female fertility masks enormous, ongoing differences between fertility in more-developed and less-developed countries. Fertility in less-developed countries tends to be much higher than that in more-developed countries. The fertility in most European nations is now so low that the continent's population stands to shrink significantly by mid-century. The population of the United States, however, is projected to grow significantly during this same period. Primary factors bringing about the U.S. increase are immigration and relatively high female fertility. (page 397)

- **Population and environment**—Some scientists believe that there is no greater single threat to the environment than the continued growth of the human population. Others argue that a more important concern is the use of natural resources per person. (page 398)

23.5 Communities

- **Communities**—An ecological community is the populations of all species that inhabit a given area, though the term can be used to mean a collection of populations in a given area that potentially interact with one another. (page 399)

- **Dominants and keystones**—Most communities tend to be dominated by only a few species; the few species that are abundant in a given area are called ecological dominants. A keystone species is a species whose impact on the composition of a community is disproportionately large relative to its abundance within that community. (page 399)

- **Biodiversity**—Biodiversity is variety among living things. It takes three primary forms: a diversity of species in a given area; a distribution of species across the Earth; and genetic diversity within a species. (page 400)

23.6 Types of Interaction among Community Members

- **Types of interaction**—There are three primary types of interaction among community members: competition, predation and parasitism, and mutualism and commensalism. (page 401)

- **Habitat and niche**—Habitat is the physical surroundings in which a species can normally be found. Niche can be defined metaphorically as an organism's occupation, meaning what the organism does to obtain the resources it needs to live. (page 401)

- **Competitive exclusion**—The competitive exclusion principle states that when two populations compete for the same limited, vital resource, one always outcompetes the other and thus brings about the latter's local extinction. (page 402)

- **Resource partitioning**—There are numerous instances in nature in which two related species do not compete for resources from a given habitat but instead divide them up, such that neither of the species undergoes local extinction. This phenomenon is called coexistence through resource partitioning. (page 403)

- **Predation and parasitism**—Predation is defined as one free-standing organism feeding on parts or all of a second organism. Parasitism is a variety of predation in which the predator feeds on prey, but does not kill it immediately, if ever. (page 403)

- **Predator-prey dynamics**—The population dynamics of a predator and its prey can be linked, but predator-prey interaction generally is only one of several types of environmental resistance controlling the population level of either group. (page 404)

- **Mimicry**—Mimicry is a phenomenon by which one species has evolved to assume the appearance of another. A mimic species evolves to match the appearance of a model species. In general, the value of mimicry is that the mimic species suffers less predation because of its resemblance to the model species. Batesian mimicry is the evolution of one species to resemble a species that has a superior protective capability. In Müllerian mimicry, several species that have protection against predators come to resemble each other. (page 407)

- **Mutualism**—Mutualism is an interaction between individuals of two species that is beneficial to both individuals. Commensalism is an interaction in which an individual from one species benefits while an individual from another species is neither harmed nor helped. (page 408)

- **Coevolution**—Coevolution is the interdependent evolution of two or more species. Flowers have evolved colors and fragrances that attract bees, for example, while bees have evolved vision that is most sensitive to the colors in the flowers they pollinate. (page 408)

23.7 Succession in Communities

- **Nature of succession**—Parcels of land or water that have been abandoned by humans or devastated by physical forces will almost always be reclaimed by nature to some degree. The process by which this takes place is called succession: a series of replacements of community members at a given location until a relatively stable final state is reached. (page 408)

- **Primary and secondary succession**—Primary succession proceeds from an original state of little or no life and no soil. Secondary succession occurs when a final state of habitat is first disturbed by some outside force, but life remains and soil exists. The final community in any process of succession is known as the climax community. (page 409)

- **Succession processes**—A common set of developments occurs in most instances of primary succession, including the arrival of "pioneer" photosynthesizers, facilitation of the growth of some later species through the actions of earlier species, and the competitive driving out of some species by the actions of later species. (page 409)

Web Tutorial 23.2 Primary Succession

Key Terms

Testing Your Understanding

In Your Own Words   *(answers in the back of the book)*

1. Name the four levels of group organization of living things, in order of increasing complexity.

2. Explain what is meant by environmental resistance and its relationship to population growth.

3. The red-spotted newt *Notophthalmus viridescens* secretes toxins from its skin that make predators avoid it. The red salamander *Pseudotriton ruber* has no such secretions but resembles the red-spotted newt and gains protection from this resemblance. What is this an example of? Which is the model, the mimic, and the dupe?

Thinking about What You've Learned

1. Species in general cannot continue exponential growth for very long, because environmental resistance will assert itself. However, the human population has been growing exponentially for about 200 years. Do you think that human populations are truly without an upper bound, or do you think that carrying capacity (*K*) is just always rising? What evidence do you see that humans might eventually reach an absolute upper limit (other than that of pure physical space)? Does the emergence of disease factor into this discussion?

2. What general principle about environmental resistance can you deduce from the explosive growth of such species as the kudzu in the United States?

3. Some scientists argue that no single factor poses a greater threat to the environment than the continued growth of the human population, while others believe the use of resources per person is a more critical factor. What arguments can you think of for or against either proposition?

Ecosystems and Biomes

SCIENCE IN THE NEWS

THE DALLAS MORNING NEWS

STUDIES: GLOBAL WARMING TO BE FELT FOR CENTURIES—TEMPERATURES, OCEANS TO RISE EVEN IF GAS LEVELS STEADY, RESEARCH SHOWS

By Alexandra Witze
March 18, 2005

The effects of global warming will be felt for several centuries even if the world's nations could somehow immediately stabilize the amount of heat-trapping "greenhouse" gases in the atmosphere, two new computer modeling studies suggest.

Were gas levels held constant, worldwide temperatures would still rise about 1 degree Fahrenheit by 2100, while sea level would rise more than 4 inches, according to one of the studies.

SCIENCE IN THE NEWS

THE DENVER POST

OZONE DECLINE STUNS SCIENTISTS—THINNING IN ARCTIC—SOLAR STORMS, BITTER COLD AND MANMADE CHEMICALS ARE BELIEVED TO BE BEHIND THE PHENOMENON

By Katy Human and Kim McGuire,
Denver Post Staff Writers
March 2, 2005

Solar flares and frigid temperatures are believed to be working with human chemicals to eat away at the protective ozone layer above the North Pole, surprising scientists who have been looking for evidence that the planet's ozone layer is healing.

24.1 The Ecosystem and Ecology

The news about global warming has been unrelentingly bad in recent years, but the news about the Earth's ozone layer has been positive on the whole. So it was a surprise when, in 2005, the atmosphere above the North Pole suffered its greatest ozone thinning ever. Actually, this turn of events was a surprise in two ways. First, it's the *South* Pole that's been the area of greatest concern over the years with respect to

Telephone poles in Alaska are tilted because the "permafrost" that underlies them is no longer permanently frozen, perhaps because of global warming.

ozone. Second, the thinning over the Arctic left scientists wondering if they really understand the forces that affect this protective layer of gas. One possibility was that the North Pole thinning was not brought about by the human-made chemicals that break down ozone. Instead, some scientists speculated that the culprit may have been . . . global warming. Oddly enough, warmer surface temperatures on Earth mean a colder stratosphere, and a colder stratosphere means less ozone.

Global warming and Arctic ozone thinning underscore, once again, how *interactive* the living world is. Global temperature and the ozone layer are *abiotic*, to use the scientific term—they are not themselves alive. Yet look what profound effects these things stand to have on Earth's *biotic* realm, which is to say, Earth's living things. Last chapter, when you went over ecological communities, you actually were studying the interactions of living things in isolation from any nonliving entities. To get a complete picture of how the living world functions, however, it's necessary to understand how living things interact not only with each other, but with nonliving elements.

What do such elements add to life? All communities need an original source of energy, generally meaning the sun; they all need water and a supply of nutrients, such as nitrogen and phosphorus; and all organisms need to exchange gases with their environment (think of your own breathing). The sun's energy is making a one-way trip through the community, ultimately being transformed into heat; but water and nutrients are being *cycled*—from the earth into organisms and then back again into the earth.

Looking at this, we can begin to perceive the reality of ecological *systems*: of working units that tightly link both biotic and abiotic elements. Put another way, we can begin to perceive the reality of the fundamental unit of ecology, the **ecosystem**, which can be defined as a community of organisms and the physical environment with which they interact. The goal of this chapter is to introduce you to some of the elements that make up ecosystems, and then to look at some very large-scale ecosystems—the geographical regions known as biomes. In today's world, it's impossible to note the ways in which ecosystems function without also noting the ways in which human activity is damaging them.

24.2 Abiotic Factors in Ecosystems

In overview, abiotic factors fall into two categories. First are the resources that exist in the ecosystem, such as water and nutrients; second are the conditions in which an ecosystem exists, such as average temperature. Resources will be considered first in this chapter and conditions later.

The Cycling of Ecosystem Resources

Those of you who went through Chapter 2, on chemistry, know what a chemical **element** is: a substance that is "pure" in that it cannot be broken down into any other component substances by chemical means. The 92 stable elements include such familiar substances as gold and helium, but only 30 or so of these elements are vital to life and are thus called **nutrients**. Some nutrients, such as iron or iodine, are needed only in small or "trace" quantities by living things, but other nutrients are needed in large quantities, among them carbon, oxygen, nitrogen, and phosphorus. Living things also have a great need for water, which is not an element but a molecule composed of two elements, hydrogen and oxygen. Water and nutrients move back and forth between the biotic and abiotic realms, with the term for this movement being a mouthful: **biogeochemical cycling**. Let's look now at the cycling of two elements, carbon and nitrogen, and then at the cycling of water.

Carbon as One Example of Ecosystem Cycling A certain amount of cosmic debris makes its way to Earth's surface in the form of meteorites, but this material is pretty sparse, relatively speaking. What else comes to Earth from space? The sun's rays, to be sure, as well as light from distant stars, and some other forms of radiation.

But when we're talking about chemical elements, the Earth really is a spaceship: It carries a fixed amount of resources with it. Nothing comes to Earth *from* the outside, and relatively little leaves Earth *for* the outside. For an ecologist, this self-containment simplifies things: What we possess in terms of elements is all we'll ever possess. Thus the question becomes not so much what we have, but where we have it.

In recent years there has been considerable concern about the buildup of the heat-trapping gas carbon dioxide CO_2 in the Earth's atmosphere, as this increase is now regarded as a cause of global warming. Well, if Earth has this fixed quantity of elements—in this case the carbon and oxygen that make up CO_2—how could there be a buildup of carbon dioxide in the atmosphere? The answer is that carbon has been *transferred* from one place on Earth to another. It has moved from the "fossil fuels" of coal and oil, where it was stored, into the atmosphere.

Storage and transfer; these two concepts are fundamental to biogeochemical cycling. When we look a little deeper into carbon cycling, we find other players involved in carbon's storage and transfer. Plants need CO_2 to perform photosynthesis, and they take in great quantities of it for that purpose, producing their own food as a result. While the plants are alive, they use part of the carbon they take in to grow—to build up leaves and stems and roots. And as long as these structures exist, they will store carbon within them, just as coal and oil do. (Thus, one of the proposed solutions to global warming is to grow a huge number of trees, each of which could be thought of as a kind of piggy bank for carbon.) Eventually the plants die, of course, after which their decomposition by bacteria and fungi releases carbon into the soil and atmosphere, again as CO_2. With this, you can begin to see the whole of the carbon cycle in the natural world (see **FIGURE 24.1**).

Some plants will, of course, be eaten by *animals*, who need carbon-based molecules for tissues and energy, just as plants do. The difference is that animals get these molecules *from plants*—in the form of seeds and leaves and roots—rather than from the air. These animals will likewise lock up carbon for a time, only to return it to the

FIGURE 24.1 The Carbon Cycle

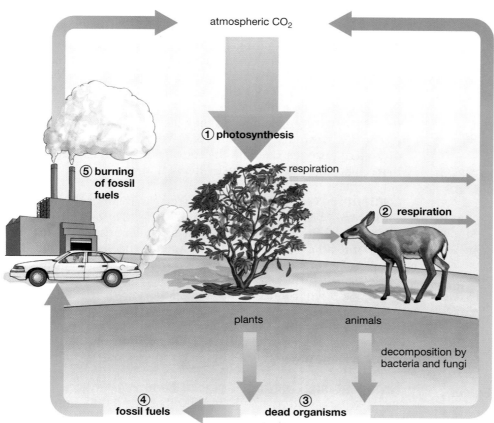

The carbon cycle

1. Plants and other photosynthesizing organisms take in atmospheric carbon dioxide (CO_2) and incorporate the carbon from it into their own tissues.

2. The physical functioning or respiration of organisms converts some of the carbon in their tissues back into CO_2.

3. Plants and animals die and are decomposed by fungi and bacteria. Some CO_2 results, which moves back into the atmosphere.

4. Some of the carbon in the remains of dead organisms becomes locked up in carbon-based compounds such as coal or oil.

5. The burning of these fossil fuels puts this carbon into the atmosphere in the form of CO_2.

soil with their death and decomposition. Note, however, the critical difference between animals and plants: Carbon comes into the living world only through plants (and their fellow photosynthesizers algae and some bacteria). An animal might take in some carbon dioxide in breathing, but nothing life-sustaining happens as a result. Conversely, a plant that takes in carbon dioxide can *incorporate* it into a life-sustaining molecule, in this case the sugar that plants produce through photosynthesis. With this sugar as a food source, plants grow, and this bounty ultimately is passed along to animals and all other life-forms. Look at this simplified cycle:

$$\text{atmospheric } CO_2 \longrightarrow \text{plants} \longrightarrow \text{animals} \longrightarrow \text{decomposers} \longrightarrow \text{atmospheric } CO_2$$

What does it track, the path of carbon or the path of food? The answer is both. This is one of the reasons that biologists refer to life on Earth as "carbon based."

How Much Atmospheric CO_2? In following the carbon trail, it's probably apparent that, just as money is a medium of exchange in our economy, CO_2 is the living world's medium of exchange for carbon; photosynthesizers take *in* carbon as CO_2, and decomposers yield *up* carbon as CO_2. (All organisms also release CO_2 as a by-product of carrying out life's basic processes, as with your own exhalations in breathing.) Given how much **biomass** or material produced by living things there is in the world, you might think there would have to be a great deal of CO_2 in the atmosphere. But as a proportion of atmospheric gases, CO_2 is a bit player, making up about 0.035 percent of the atmosphere. But what an important player it is! It's critical in feeding us all, and changes in its atmospheric concentration are now leading to significant changes in global temperature.

The Nitrogen Cycle Like carbon, nitrogen is an element; and like carbon, nitrogen cycles between the biotic and abiotic domains. Nitrogen makes up only a small proportion of the tissues of living things, but it is a critical proportion because it is required for DNA, RNA, and all proteins. All living things need this element, then. The question is, how do they get it?

The source for all the nitrogen that enters the living world is atmospheric nitrogen, which exists in great abundance. You've seen that carbon dioxide makes up less than 1 percent of the atmosphere, and it turns out that oxygen makes up about 21 percent. But nitrogen makes up 78 percent of the atmosphere. So rich is the atmosphere in this element that a typical garden plot has tons hovering above it. The problem? All this nitrogen is in the form of pairs of nitrogen atoms (N_2) that have a great tendency to stay together, rather than combining with anything else. Thus, in a case of so near and yet so far, plants have no direct access to atmospheric nitrogen. So how do they take it up? In the natural world, the answer essentially is, through the actions of bacteria. It is bacteria that are carrying out the process of **nitrogen fixation**: the conversion of atmospheric nitrogen into a form that can be taken up and used by living things.

Bacterial Fixation of Nitrogen You can see nitrogen fixation and the rest of the nitrogen cycle diagrammed in FIGURE 24.2. The essence of the cycle is that several types of nitrogen-fixing bacteria take in atmospheric nitrogen (N_2) and convert it into ammonia (NH_3). This ammonia then is converted (either in water or by other bacteria) into two types of nitrogen-containing compounds that plants can *assimilate*—can take up and use. And as you've seen, what comes into the plant world will come into the animal world, when animals eat plants. The two usable nitrogen compounds produced in the cycle are the ammonium ion (NH_4^+) and nitrate (NO_3^-). Some of the nitrate that results from this process is used by yet another kind of bacteria, denitrifying bacteria, which can convert nitrate back into atmospheric nitrogen, and the cycle is complete.

Nitrogen as a Limiting Factor in Food Production If you were looking at Earth's nitrogen story 100 years ago, there would be little more to it than what you've just gone over. In other words, through most of human history, nitrogen

---

A BROADER VIEW

Carbon Cycling Times

So, how long does it take for carbon to move from organisms, into the atmosphere, and then back again? The time-frames are extremely variable, as researcher Vaclav Smil notes in his book *Cycles of Life*. The carbon dioxide that bacteria liberate from, for example, a decomposing animal can be taken up by a tree almost immediately upon being released (and then used by the tree for photosynthesis). However, those CO_2 molecules that manage to rise up above a tree canopy can be lifted high into the atmosphere and remain there, circling the planet, for more than a century.

A BROADER VIEW

Fritz Haber and WMDs

Who invented the process of manufacturing a usable form of nitrogen, thus helping to feed the world? German chemist Fritz Haber. Who invented the first weapon of mass destruction? Arguably Fritz Haber, who conceived of and led a German military program during World War I to develop poison gas, which was first used by the Germans in April 1915. Indeed, the first use of Haber's nitrogen-fixing process was not the manufacture of fertilizers, but the manufacture of compounds that could keep the German munitions factories running. Haber was branded a war criminal by the allies at the end of the war but was never prosecuted. In 1919, he received the Nobel Prize in chemistry for the nitrogen process and its contributions to agriculture.

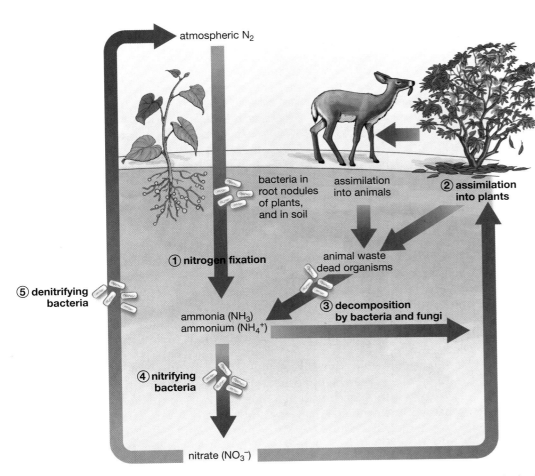

The nitrogen cycle

1. Nitrogen-fixing bacteria convert N_2 into ammonia (NH_3), which converts in water into the ammonium ion (NH_4^+). The latter is a compound that plants can assimilate into tissues. In the diagram, bacteria living symbiotically in plant root nodules have produced NH_4^+, which their plant partners have taken up and used. Meanwhile, free-standing bacteria living in the soil have likewise produced NH_4^+.

2. Other plants take up NH_4^+ that has been produced by soil-dwelling bacteria and assimilate it. Animals eat plants and assimilate the nitrogen from the plants.

3. Animal waste and the tissues of dead animals are decomposed by fungi and by other bacteria, which turn organic nitrogen back into NH_4^+.

4. Other "nitrifying" bacteria convert NH_4^+ into nitrate (NO_3^-), which likewise can be assimilated by plants.

5. Some nitrate, however, is converted by "denitrifying" bacteria back into atmospheric nitrogen, completing the cycle.

FIGURE 24.2   The Nitrogen Cycle Some bacteria "fix" nitrogen, meaning they convert atmospheric nitrogen (N_2) into an organic form that can be used by other living things.

was fixed almost solely by bacteria. Human beings had a problem with this single route to nitrogen, however, because nitrogen is critical for all plant growth, including *agricultural* plant growth. Indeed, for centuries a lack of nitrogen was a primary limiting factor in getting more food from a given amount of land. Long before anyone knew what nitrogen was, farmers were trying to get more of it to crops in two ways. One was by applying organic fertilizer, such as rotting organic material, to their fields. (Remember how the Indians taught the Pilgrims to bury dead fish around their corn plants?) The second was by planting crops that carried their own nitrogen-fixing bacteria within them. This is the case with certain legumes, such as soybeans, which carry bacteria in their root nodules.

Early in the twentieth century, however, a momentous change came about in the use of nitrogen when the German chemist Fritz Haber developed an *industrial* process for turning atmospheric nitrogen into ammonia. In essence, human beings became nitrogen fixers and, by the 1960s, they had taken up this activity on a grand scale. In 1990 about 80 million metric tons of biologically active nitrogen was manufactured for use as fertilizer; in addition, human beings planted crops that resulted in the production of another 40 million tons (**see** FIGURE 24.3). This human-driven production was roughly equal to the amount of active nitrogen that nature fixed on its own.

It is hard to overstate the importance of the industrial fixation of nitrogen. Put simply, the world's human population could not be fed without such fixation. Only with synthetic nitrogen can the Earth's farmland produce enough food to feed the Earth's burgeoning human population.

FIGURE 24.3   Nutrients Beyond What Nature Provides A helicopter applies fertilizer to a sugar-beet crop in California.

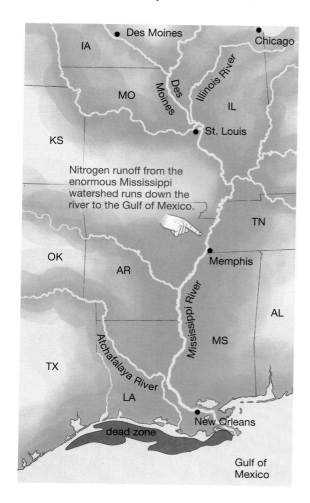

Nitrogen runoff from the enormous Mississippi watershed runs down the river to the Gulf of Mexico.

dead zone

Gulf of Mexico

FIGURE 24.4 **Map of the Gulf of Mexico "Dead Zone"** In 2002, the seasonal "dead zone" in the Gulf of Mexico stretched from the mouth of the Mississippi River to eastern Texas. The waters in the zone are oxygen depleted, particularly at their lower depths. The primary cause is the tremendous amount of nitrogen carried into the Gulf from both the Mississippi and Atchafalaya rivers. This nitrogen originates as fertilizer that is applied to farmland all the way up the Mississippi River basin.

This massive human intervention in the nitrogen cycle has a downside as well, however. Remember that human-manufactured nitrogen ends up being used in a very *concentrated* way—it is poured onto farmland. A good deal of this nitrogen does not end up in the crops it was intended for, however, but instead departs as runoff. The amount of runoff generated by each farm adds up, and the end result can be large amounts of nitrogen flowing into creeks, ponds, and rivers. On a regional basis, such nitrogen flows can be enormous. The Mississippi River carries an estimated 1.5 million metric tons of nitrogen into the Gulf of Mexico each year.

You might think that, since nitrogen is a nutrient, this would be a good thing for the Gulf waters, but the opposite is true. The nitrogen brings on a seasonal "bloom" of algae, the algae result in a huge bacterial population, and the bacteria use up most of the available oxygen at the water's lower depths. The result is a giant "dead zone" that drives away mobile fish and that kills immobile, bottom-dwelling sea creatures (**see** FIGURE 24.4). This problem of *nutrient pollution* is not limited to the Gulf of Mexico, however; it is played out on scales both large and small in bodies of water around the world.

FIGURE 24.5 **The Water Cycle** More than 95 percent of Earth's water is stored in the oceans. When water evaporates from the ocean, 90 percent returns to the ocean directly by way of precipitation. The other 10 percent falls on land. There, the water either runs back into the ocean, moves into groundwater storage, is stored in such structures as glaciers, or is moved by transpiration and evaporation back into the atmosphere.

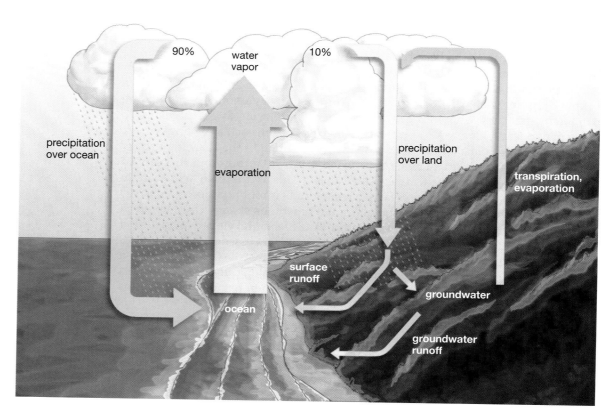

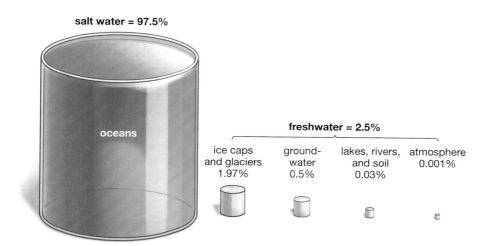

salt water = 97.5%

oceans

freshwater = 2.5%

| ice caps and glaciers 1.97% | ground-water 0.5% | lakes, rivers, and soil 0.03% | atmosphere 0.001% |

FIGURE 24.6 **Earth's Water** Although there is a tremendous amount of water on our planet, only a small fraction of it is available to us as freshwater.

The Cycling of Water

Now let's turn from nutrient to water cycling. Those who went through the material on water in Chapter 3 know how important water is to life. The human body is about 66 percent water by weight, and any living thing that does not have a supply of water is doomed in the long run. As with nitrogen or carbon, all water is either being cycled or is locked up. Carbon may be stored in coal or trees, but water can be stored in ice—in glaciers or in polar ice, for example.

The oceans hold more than 95 percent of Earth's water, but thankfully for us, when the oceans' salt water evaporates, it falls back to Earth as freshwater. We can also be grateful that not all ocean water falls back to Earth on the oceans; instead, about 10 percent falls on land (see FIGURE 24.5). This water represents about 40 percent of the precipitation that land areas get; the other 60 percent comes from a form of cycling called *transpiration*, meaning the process by which water is taken up by the roots of land plants, then moved up through their stems and out through their leaves as water vapor. After evaporating into the atmosphere, water returns to Earth as rain, fog, snow—all the forms of precipitation possible. The driving force behind the water cycle is the energy of the sun; its heat powers both evaporation and transpiration.

With more than 95 percent of Earth's water in the oceans, perhaps as little as 2.5 percent of the Earth's water is freshwater at any given time. Of this freshwater, more than 75 percent is locked up in glaciers and other forms of ice (see FIGURE 24.6). Thus, any snapshot we would take of Earth would show that as little as 0.5 percent of its water is available as liquid, freshwater; but even here we have to qualify things. About 25 percent of this freshwater is **groundwater**, meaning water that moves down through the soil until it reaches porous rock that is saturated with water. In line with this, about 20 percent of the water used in the United States comes from groundwater, and at least a quarter of the world's population depends on groundwater to satisfy basic needs. Groundwater is stored in porous rock called an **aquifer**, an example of which is pictured in FIGURE 24.7. Major aquifers can be enormous, as you can see from the map of

FIGURE 24.7 **Aquifers Store Groundwater** When water seeps into the ground, it moves freely through layers of sand and porous rock but moves either extremely slowly or not at all through layers of impermeable rock. Water thus becomes trapped in different underground layers, from which humans can draw water. Unconfined aquifers receive water directly from a large surface area and therefore tend to be more vulnerable to pollution by chemicals such as fertilizers and pesticides. Confined aquifers are located between layers of impermeable rock. They are replenished more slowly but tend to contain purer water.

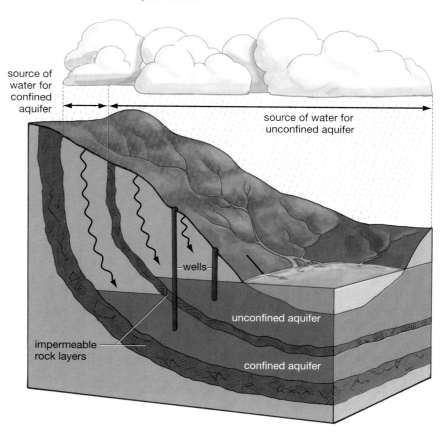

source of water for confined aquifer

source of water for unconfined aquifer

wells

impermeable rock layers

unconfined aquifer

confined aquifer

FIGURE 24.8 **Enormous Store of Underground Water** The High Plains Aquifer, perhaps the largest aquifer in the world, underlies an expanse of land stretching across eight states. Billions of gallons of water are withdrawn from it each day, mostly for agriculture. Since the 1940s, more than 200,000 wells have been drilled into it for this purpose. (Modified from E. D. Gutentag, F. J. Heimes, N. C. Krothe, R. R. Luckey, and J. B. Weeks, *Geohydrology of the High Plains Aquifer in Part of Colorado, Kansas, Nebraska, New Mexico, Oklahoma, South Dakota, Texas, and Wyoming.* U.S. Geological Survey Professional Paper 1400-B, 1984.)

the High Plains Aquifer in FIGURE 24.8. Even an aquifer of this size, however, can lose more water through pumping than it gains through replenishment from precipitation. In Texas, the depth of the High Plains Aquifer declined by more than a foot per year from 1940 to 1980 and then by a little less than half a foot per year between 1980 and 1998.

Global Human Use of Water Given the sheer number of people on Earth and the importance of water to each of them, it will not surprise you to learn that human beings are laying claim to a great deal of Earth's freshwater; civilization now uses more than half the world's accessible supply. But from what you've seen so far, think of what this means. There is no less *water* than there ever was. Like nitrogen or carbon, all the water we have is either being recycled or stored. What's changed is that the human population is using an ever-greater proportion of the water that's available. Human beings fight among themselves for water, of course, because it is a scarce commodity in locations throughout the world. In fact, *sanitary* water is so scarce that an estimated 1 billion people worldwide lack access to it (see FIGURE 24.9). A large part of the problem is that there is a mismatch between the location of freshwater and the location of human populations. Twenty-six percent of the world's precipitation falls in South America, but only 6 percent of the world's people live there. Meanwhile, 36 percent of the world's precipitation falls in Asia, while 60 percent of the world's people live there.

One part of the human water problem, water expert Peter Glieck has noted, is that humans make such poor use of the water they capture. Mexico City's leaky water system *loses* enough water to meet the needs of a city the size of Rome. If the toilet in your house is more than 10 years old, chances are it uses six gallons of water with each flush. If it was built since the 1990s, following imposition of new federal standards for low-flow toilets, it probably uses less than a third of that—a mere 1.6 gallons.

When we think of the human use of water, however, household use actually is relatively small. About 70 percent of the water we withdraw from rivers, lakes, and groundwater goes to agricultural irrigation. People need food, of course, but the *way* irrigation is carried out in much of the world is extremely wasteful: water moves from, say, a river to crops by means of an open channel. En route, much of this water either evaporates or is taken up by land that isn't being irrigated. An efficient alternative is "micro-irrigation" in which water gets piped to crops.

Human beings are, of course, capable of diverting water from species that have no say in this division of resources. More than 20 percent of all freshwater fish species are estimated to be threatened or endangered precisely because their water

FIGURE 24.9 **Freshwater Is a Limited Resource** Women carry drinking water through a polluted slum in Port-au-Prince, Haiti.

(a) Atmospheric CO_2 concentration

76% naturally occurring 24% human-caused

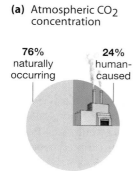

(b) Terrestrial nitrogen fixation

42% naturally occurring 58% human-caused

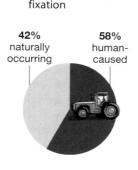

(c) Accessible surface water

46% available 54% used

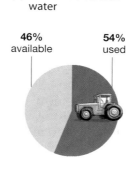

FIGURE 24.10 Human Impact on Resources

(a) Nearly a quarter of the current atmospheric carbon dioxide concentration is produced by human activity, primarily through the burning of fossil fuels.

(b) Over half of terrestrial nitrogen fixation comes about because of human activity, including the manufacture of fertilizer for agriculture.

(c) Over half of the Earth's accessible surface water is now diverted for use by humans, mostly for agriculture. Percentages are approximate.

has been dammed or diverted for human use. California's natural populations of coho salmon have been reduced by some 94 percent since the 1940s, in significant part through diversions of the freshwater streams that the salmon swim up in order to spawn. If you look at FIGURE 24.10, you can get a quick idea of humanity's impact on the three resources just reviewed—carbon, nitrogen, and water.

Human Beings Are Not Separate from the Earth on Which They Live

One final thing about nutrient and water cycling. The average person, who has no knowledge of these cycling processes, may regard human beings as a kind of independent line that stretches back in time. Under this view, water and food may pass through us, but they are as separate from us as gasoline is from a car engine.

From your reading of this section, however, it may now be apparent to you that this is not true. Some of the carbon in the bread you ate this morning could have been contained in a tree that decomposed in China 100 years ago. After many cycles, carbon from this tree was taken up by a wheat plant in America whose seeds were ground up, yielding flour that ended in the bread on your shelf. This carbon may be burned by your body to supply the energy it needs, it's true, but it may also become part of your bones or your nerves. Then, when you die, the material you are made of will go *back* to the Earth to be recycled in the ways you've been reading about, perhaps to be part of a tree or a person in a future generation. We are intertwined with the Earth in some basic, physical ways.

So Far...

1. An ecosystem includes not only a community of organisms, but the _____ with which they _____.

2. The carbon that helps make up the living world moves into it by means of organisms that perform _____. Carbon then returns to the soil and atmosphere through means of the _____ of organic material by the organisms called _____ and _____.

3. Almost all the nitrogen that enters the living world by natural means does so through the actions of _____. Nitrogen is then returned to the atmosphere by the actions of _____.

24.3 Energy Flow in Ecosystems

Having looked at the abiotic elements of nutrients and water, let's now turn to another important abiotic component of any ecosystem, energy. Those of you who read through Chapter 6, on energy, know that one of energy's inflexible laws—the first law of thermodynamics—is that energy is never gained or lost, but only transformed. The sun's energy is not used up by green plants; rather, some of it is *converted* by plants into a chemical form—initially into the chemical bonds of the sugar that plants produce in photosynthesis.

So Far... Answers:

1. *physical environment; interact*

2. *photosynthesis; decomposition; bacteria; fungi*

3. *bacteria; bacteria*

The second law of thermodynamics is that energy spontaneously flows in only one direction: from more ordered to less ordered. Sugar's chemical bonds are very ordered things—atoms of carbon, oxygen, and hydrogen in a precise spatial relationship to one another. Contrast this with another form of energy, heat, which is the *random* motion of molecules—clearly a disordered form of energy compared to the chemical bonds in sugar. Indeed, heat is the least-ordered form of energy. And, because all energy spontaneously moves toward less order, heat is the ultimate *fate* of all energy. The chemical bonds in gasoline can be broken through combustion and thus be used to power a car engine, but not all the energy released from the combustion drives the engine. Some dissipates as heat, and this happens *every* time energy is used, whether we're talking about a piston firing or a cell dividing.

All of this provides a framework to conceptualize an important part of ecosystems, which is how energy flows through them. Think of the sun's energy, with the energetic rays that leave it as a starting point. Now think of the ultimate fate of all this energy, which is heat, randomly dispersed in the universe. Looked at one way, life on Earth *intervenes* in this flow. Life is an enormous energy collection and storage enterprise, gathering some of the sun's energy and locking it up for a time in the form of chemical bonds. These bonds can then be broken—think of digesting a muffin—which *releases* this stored energy so that an organism can grow and reproduce. Life does not stop the march of the sun's energy into heat, but it does intervene in it by transforming it into chemical bonds that have order and stability.

Once we see life in these terms, we can begin to think of the *flow* of energy through it. That's what you'll be looking at in the sections that follow.

Producers, Consumers, and Trophic Levels

One way to look at food and energy is in terms of production and consumption. Plants and other photosynthesizers are an ecosystem's **producers**, while the organisms that eat plants are one kind of **consumer**: an organism that eats other organisms, rather than producing its own food. But as you know, there are consumers *of* consumers—grass is eaten by zebras, but zebras are eaten by lions. Thus do we get to the concept of feeding *levels* or, as ecologists put it, trophic levels. More formally, a **trophic level** is a position in an ecosystem's food chain or web, with each level defined by a transfer of energy between one kind of organism and another. Producers are one level, and then there are several levels of consumers. You may remember that an animal that eats only plants or algae is a **herbivore**; then there are **carnivores**, which eat only meat, and **omnivores**, which eat both plants and meat. With this in mind, here's a list of trophic levels through four stages.

First trophic level: **Producers** (photosynthesizers)
Second trophic level: **Primary consumers**—plant predators (herbivores)
Third trophic level: **Secondary consumers**—herbivore predators (carnivores)
Fourth trophic level: **Tertiary consumers**—organisms that feed on secondary consumers (carnivores)

It is the sun's energy that is locked up in the leaves and stems generated by producers, and it is this energy that is then being passed along through the other trophic levels (see FIGURE 24.11). Of course, many organisms cannot be assigned to just a single trophic level. Most human beings, for example, are primary, secondary, and tertiary consumers.

FIGURE 24.11 **Trophic Levels** Plants, algae, and some bacteria are called producers because they use the sun's energy to produce their own food through photosynthesis. Organisms at all other trophic levels ultimately derive their energy from these photosynthesizers.

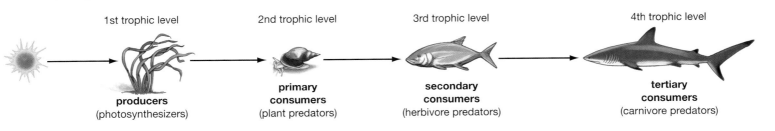

1st trophic level	2nd trophic level	3rd trophic level	4th trophic level
producers (photosynthesizers)	**primary consumers** (plant predators)	**secondary consumers** (herbivore predators)	**tertiary consumers** (carnivore predators)

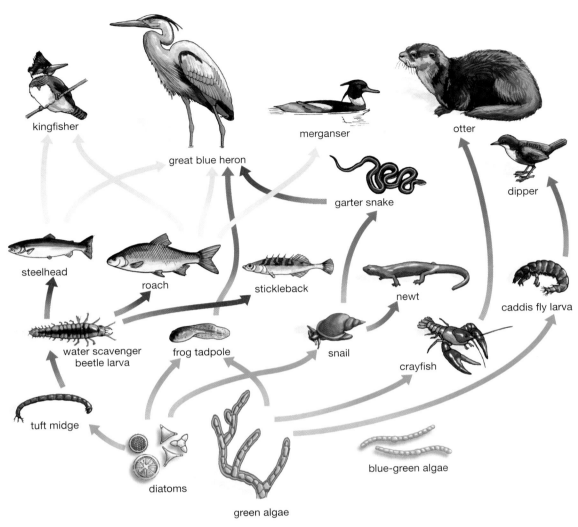

FIGURE 24.12 **A River Food Web** The diagram indicates who eats whom along a portion of the Eel River in Northern California. The arrows have been color-coded by trophic level. Note that there are up to five trophic levels in the web. (Adapted with permission from an original drawing by Mary E. Power, University of California, Berkeley.)

You have thus far thought in terms of food chains, which is to say trophic *lines* in which a single organism follows another. In reality, of course, nature is much more complex than this; what really exists are feeding patterns called *food webs*, one example of which you can see in FIGURE 24.12.

A Special Class of Consumers: Detritivores

To this scheme of trophic levels, ecologists add a special class of consumers, the **detritivores**. These are consumers that feed on *detritus*, which in normal usage simply means a collection of debris. In ecology, however, detritus is the remains of dead organisms or cast-off material from living organisms. A fallen branch from a living tree is detritus, for example, and a worm or a dung beetle feeding on this material is a detritivore.

Of particular interest is a special kind of detritivore, a **decomposer**: an organism that, in feeding on dead or cast-off organic material, breaks it down into inorganic components that are recycled back into an ecosystem. (*Inorganic* can be thought of as chemical building blocks.) The most important decomposers are

Chain of Detritivores

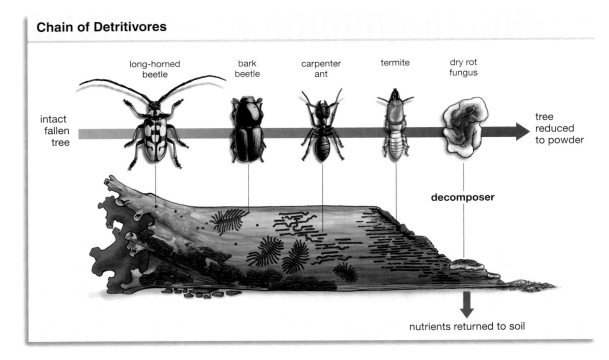

FIGURE 24.13 How Organic Material Is Broken Down Different kinds of detritivores consume the organic material in a branch that has fallen from a tree. Most of these animals are consuming pieces of the log, but the detritivore called a decomposer (a fungus in the figure) is breaking organic materials into their inorganic components, thus returning nutrients to the soil.

fungi and bacteria. If you look at FIGURE 24.13, you can see the role that decomposers normally play—as the last links in a chain of detritivores that break down organic material.

Stepping back from this, it's worth noting that if a factory made such complete use of its materials as nature does, that factory would be considered a wonder of the world. Think of it this way: What is thrown away in the feeding and decomposing chain you've been looking at? Nothing, which is all to the good, because if there were any natural garbage of this sort, spaceship Earth would eventually be filled up with this useless material. This very thought ought to give us some pause about the compounds human beings currently are putting into the environment that are not **biodegradable**, meaning capable of being decomposed by living organisms. Of particular concern here are plastics, many of which will remain intact for hundreds of years following their production.

Accounting for Energy Flow through the Trophic Levels

All organic material eventually is recycled in ecosystems, but this is not true of energy, which is just passing through, on its way to heat. One interesting question that arises from this is: How much of the sun's energy does the living world collect? The answer is, very little: Only about 2 percent of the sun's energy that falls around plants is taken in for photosynthesis. Now, of the energy that is collected by plants, how much is then available to the *next* trophic level up? To get a handle on this question, you need to understand something about what might be called energy costs in living things.

Plants don't get to retain all the material they produce through photosynthesis. Why? Because they incur energy costs in just staying alive. It takes energy to perform photosynthesis, to move sap from one end of a plant to the other, and so forth. Such costs are subsumed under the heading of "cellular respiration," but you could think of them as the plant's overhead; they are the price of doing business. Apart from this, we know that, in every energy transaction, some of the energy is lost as heat. After subtracting these various costs, the plant is left with the amount of material it *accumulates* as a result of photosynthesis—its leaves and stems and roots. And it turns out that, of the solar energy that plants initially receive, no more than 85 percent—and perhaps as little as 30 percent—will be transferred into this accumulated material. In other words, at the low end of this range, only 30 percent of the solar energy a plant takes in is transferred into material that is even potentially available to the next trophic level up.

Now, what happens when you go up to that trophic level—to those primary consumers? They can't even get *to* a lot of the tissue that's been produced in plants, and they then have their own energy costs to take care of. When we take all these factors into account, the result is what you see in FIGURE 24.14: a drastic reduction in available energy with the transition from one trophic level to the next. The amount that is passed on between levels varies by community, but there is a rule of thumb in ecology that for each jump up in trophic level, the amount of available energy drops by 90 percent. Take all the energy converted into tissue by second-level herbivores (rabbits, mice) and of this, only 10 percent ends up being converted into tissue by third-level carnivores (foxes, weasels). There is a great variability in conversion rates, however, and 10 percent probably is more like a maximum passed on.

The Effects of Energy Loss through Trophic Levels
So, what is the consequence of this energy reduction at trophic levels? Nothing less than the character of the living world. Walk through a forest and what do you see? Innumerable small plants and trees, a fair number of small animals—birds, insects, squirrels—and, on very rare occasions, a predator such as a bobcat or a fox. The energy reduction that comes at each trophic level dictates this reduction in the number of living things at each trophic level. If you had a field that could feed 100 field mice, how many weasels could the field support, if those weasels were existing solely on the mice? Remembering that, at most, only 10 percent of a trophic level's energy is likely to be passed along to the next level, you might guess that 10 weasels could make a living in the field; but even this number would be too high. For 10 predators to make a living from 100 mice, these predators would have to be no *bigger* than the mice, which is certainly not true of weasels. Because weasels are not about to start eating the same things as mice, weasels are locked into a trophic level with a more limited energy supply. When there's only so much energy, there can be only so many weasels. With really large predators, such as wolves, the numbers fall even more drastically.

Productivity Varies across the Earth by Region

As you've seen, photosynthesizers are the producers of Earth's biomass or living material. Hence, ecologists think of a region as being *productive* to the extent that it has plants and other photosynthesizers working away in it. But not all regions of the Earth are equally productive. If you look at FIGURE 24.15, you can get an idea of productivity differences around the globe. Much of what you see there is intuitive. Productivity is very low in such desert areas as North Africa and very high in such tropical rain-forest areas as equatorial Africa. Yet why should one part of the Earth be a rain forest and another part a desert? The answer has to do with Earth's large-scale physical environment.

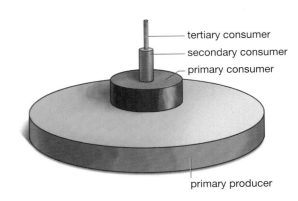

FIGURE 24.14   Energy Pyramid Only a small fraction (10 percent or less) of energy at each trophic level is available to the next higher trophic level.

A BROADER VIEW
Livestock as a Trophic Level

About 70 percent of the grain produced in the United States doesn't go to feed people directly. It goes to an intermediate trophic level—the livestock that end up as the beef, chicken, and pork on our grocery shelves. Since roughly 90 percent of the energy at one trophic level is lost to the level above it, the results are predictable: it takes 9 pounds of feed to produce 1 pound of beef ready for human consumption. What does this mean in practice? A harvest of 2,000 pounds of grain will feed five people for a year. Take roughly this same harvest, however, feed it to livestock, and then feed the *livestock* products to people, and only one person will be fed for a year.

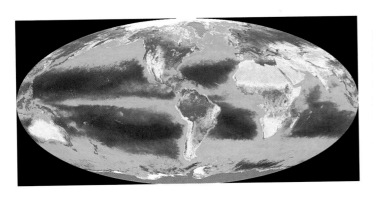

FIGURE 24.15   Productivity by Geographic Region This composite satellite image shows how the concentration of plant life varies on both land and sea. On land, the areas of lowest productivity are tan-colored, as with the enormous swath of land that begins with the eastern Sahara Desert in North Africa. More productive land areas are yellow, more productive yet are light green, and the most productive of all are dark green. On the oceans, the gradient runs from dark blue (the least productive) through lighter blue, green, yellow, and orange-red (the most productive). The ocean measurements are of concentrations of phytoplankton—the tiny, photosynthesizing organisms that drift on the water and form the base of the marine food web. Note the high productivity of the oceans in both the far northern and southern latitudes. Antarctica itself is gray because the imaging technique did not work well for it.

So Far... Answers:

1. *photosynthesis; producers; herbivore*
2. *decomposers; biodegradable*
3. *energy; trophic level*

So Far...

1. All the organisms at the first trophic level perform _____ and thus are referred to as _____. Any organism that feeds solely on first-trophic-level organisms would have to be a (choose one) carnivore/herbivore/omnivore.

2. All material produced by living things is capable of being broken down into its inorganic components by the class of living things called _____. This material is thus said to be _____.

3. There are fewer predator than prey animals due to drastic reductions of _____ at each _____.

24.4 Earth's Physical Environment

Earth's Atmosphere

Earth itself exists within an environment, called the **atmosphere**, which is the layer of gases surrounding the Earth. One of the main things to realize about Earth's atmosphere is how it differs from outer space. There's something *to* Earth's atmosphere (a mix of gases), while outer space really is *space*; there's almost nothing in it.

The atmosphere is divided into several layers, but we will concern ourselves with only two of them, which you can see in FIGURE 24.16. The lowest layer of the atmosphere, called the **troposphere**, starts at sea level and extends upward about 7.4 miles, and it contains the bulk of the gases in the atmosphere. Nitrogen and oxygen make up 99 percent of the troposphere, but carbon dioxide exists there in small amounts, as do some other gases, such as argon and methane. After a transitional zone, the next layer above the troposphere is the **stratosphere**. Of greatest concern in it is the gas we've already touched on, ozone, that reaches its greatest density at about 13 to 21 miles above sea level.

The oxygen we breathe comes in the form of two oxygen molecules bonded together (O_2). When ultraviolet sunlight strikes O_2 in the stratosphere, however, it can put it in a form of three oxygen molecules bonded together (O_3). This is the gas known as **ozone**. Created initially as a *product* of life—since atmospheric oxygen came from photosynthesis—this ozone layer ultimately came to protect life by blocking some 99 percent of the ultraviolet (UV) radiation the sun showers on the Earth. It was this blockage that allowed life to come onto land some 460 million years ago, and it is this blockage that protects life now. The UV radiation that pours from the sun can wreak havoc on living tissue, bringing about cancer and immune-system problems in people, and damaging vegetation as well.

The Worrisome Issue of Ozone Depletion

Given the importance of the ozone layer, it is sobering to contemplate how fragile it is. Some human-made chemical compounds have the effect of destroying stratospheric ozone, and such destruction went on unchecked for years until atmospheric chemists Sherwood Rowland and Mario Molina revealed, in 1974, that compounds called chlorofluorocarbons posed a direct threat to the ozone layer. Chlorofluorocarbons or CFCs—found at one time in spray cans, refrigerators, and plastic foams—are undoubtedly the most famous of the ozone-depleting compounds, but they are by no means the only chemicals

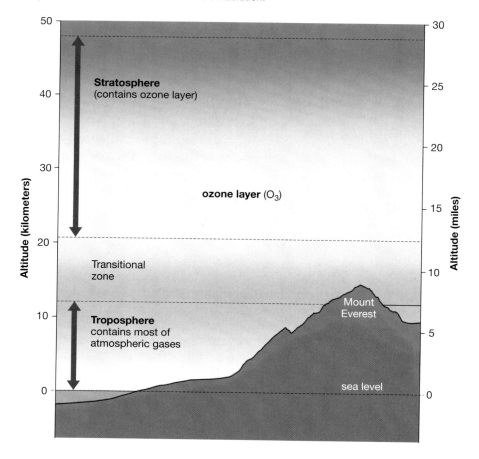

FIGURE 24.16 Earth's Atmosphere The two lowest layers of the atmosphere are shown, with Mount Everest (the tallest mountain) included for vertical scale. The troposphere contains most of the atmosphere's gases—mostly nitrogen and oxygen, with some carbon dioxide, methane, and other gases. The stratosphere contains the ozone layer, which blocks 99 percent of the sun's harmful ultraviolet radiation.

that have this effect. Other harmful compounds include methyl bromide, which is used as a pesticide, notably on tomato and strawberry crops; and bromine, which is found in a group of fire-extinguishing chemicals. The damage that these compounds have done to stratospheric ozone has brought about the spectacle of annual reports on how big the "ozone hole" is over the South Pole, along with a concern about a general thinning of the ozone layer.

Despite the ominous nature of this news, the long-term outlook on ozone depletion actually is good. In an agreement signed in Montreal in 1987, many of the world's nations pledged to phase out production of ozone-depleting chemicals, and this agreement is working. Production of CFCs was banned in the United States even before the Montreal agreement was signed and has dropped dramatically throughout the world in recent years. Methyl bromide continues to be used on U.S. agricultural crops, though the amount used in 2005 was less than two-thirds the amount used in 1991.

Critically, the reduction in the use of these compounds has brought about a stabilization or reduction in the levels of ozone-depleting compounds in the atmosphere. As a consequence, the *rate* of ozone thinning is certainly slowing and may have stopped altogether. Should this trend continue, it seems likely that the ozone layer will be restored, though this process may take decades. One wildcard in this issue, however, is something mentioned at the start of the chapter: global warming, which has the potential to cool the stratosphere and hence thin the ozone layer. Scientists will be looking carefully to see if the Arctic ozone thinning that occurred in 2005 was a one-time event or a first sign of long-term trouble.

The Worrisome Issue of Global Warming

It's to global warming itself that we now turn. The essential questions regarding this issue are straightforward: Is the Earth getting warmer? If so, to what degree is human activity responsible for this warming? And, if the Earth is warming, what consequences will this have? Put another way, what will a warmer Earth be like?

Is the Earth Getting Warmer?
At one time, all of these questions were contentious within the scientific community. In the last few years, however, unanimity seems to have been reached on *whether* the Earth is warming—it is. Earth's surface temperature probably has increased by about 0.6° Celsius (or about a degree Fahrenheit) in the past century, with much of this increase coming in just the past 20 years.

Is Human Activity Responsible?
This leads to the second question, which is whether human activity has caused this increase. Here, disagreement has almost disappeared. There is a strong consensus that human activity is at least partly to blame for global warming. The most authoritative conclusion we have on this issue comes from a group called the United Nations Intergovernmental Panel on Climate Change (IPCC). After receiving input from hundreds of scientists around the globe, this group issued a report in 2001 stating "there is new and stronger evidence that most of the warming observed over the past 50 years is attributable to human activities." In the period since the IPCC released its findings, evidence has continued to mount indicating that human activities are a cause of global warming. What activities are these? The burning of fossil fuels, such as coal and oil, and the deforestation of the Earth. Both put more carbon dioxide and other "greenhouse" gases into the atmosphere. Warming has resulted from this increase because greenhouse gases trap heat, as you'll see.

What makes scientists think that human activity and global warming are linked? If you look at FIGURE 24.17, you can see that a strong correlation exists between increasing temperatures and

FIGURE 24.17 Global Warming Atmospheric carbon dioxide concentration is increasing, and global temperature is as well. It is difficult to be certain that the CO_2 increase is driving the increase in temperature because, over long periods of time, there are great natural fluctuations in global temperature. Most experts have now concluded, however, that rising levels of greenhouse gases such as CO_2 are making the world warmer.

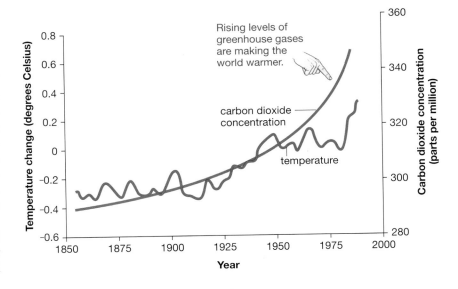

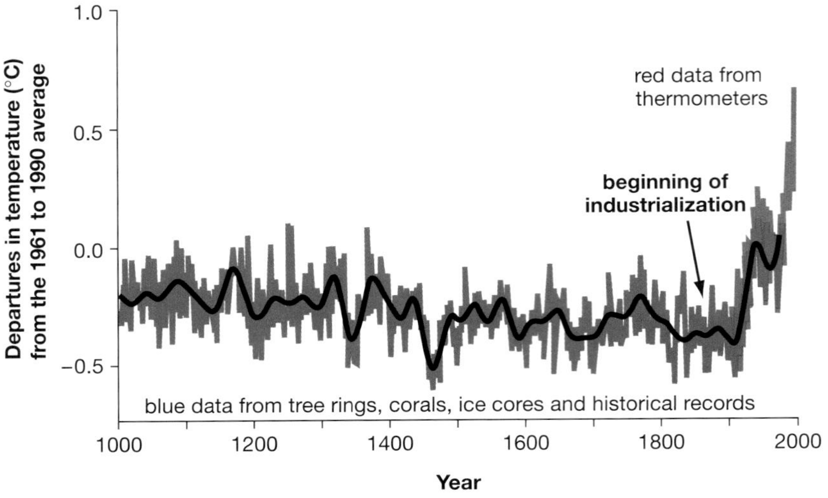

FIGURE 24.18 **Earth's Temperature during the Past 1,000 Years** Using data from such sources as tree rings and ice cores, scientists have estimated average temperatures in the Northern Hemisphere over the last 1,000 years. They concluded that the increase in Earth's temperature in the twentieth century seems to have been greater than the increase that occurred in any other century during the last 1,000 years. Further, the 1990s appear to have been the warmest decade—and 1998 the warmest single year—over the past 1,000 years. (Adapted from *The Third Assessment Report of Working Group I of the Intergovernmental Panel on Climate Change* (IPCC) 2001, *Summary for Policymakers*, Figure 1b.)

rising atmospheric carbon dioxide (CO_2) levels over the last 150 years. If you then look at FIGURE 24.18, you can see temperature trends in the Northern Hemisphere for a much longer period, the last thousand years. Note the strong increase in temperature that took place beginning after 1900, as civilization was becoming more industrialized, and the even more prominent upturn in the past few decades. This longer-term temperature trend is likewise consistent with longer-term CO_2 trends. The IPCC report notes that the atmospheric concentration of CO_2 has increased by 31 percent since 1750 and that Earth's current CO_2 concentration "has not been exceeded during the past 420,000 years and likely not during the past 20 million years."

A firm rule in science is that correlation is not causation; the fact that two trends are correlated does not mean that one caused the other—that rising CO_2 levels caused global warming, in this case. But by constructing theoretical models of what might be at play, IPCC scientists concluded that the most plausible explanation for the rise in Earth's temperature is that human activity has combined with natural forces, such as solar variation and volcanic activity, to produce the warming we have experienced.

What Is the Greenhouse Effect? Why should the concentration of gases such as CO_2 have anything to do with a warmer planet? Sunlight that is not filtered out by ozone comes to Earth in the form of very energetic, short waves that can easily pass through the atmosphere (**see** FIGURE 24.19). Once this energy reaches the land and ocean, most of it is quickly transformed into heat that does all the things you've just read about: warms the planet, drives the water cycle, and so forth. This heat ultimately is radiated back toward space. But heat is not short-wave radiation; it is *long-wave* radiation, and it can be *trapped* by certain compounds, among them carbon dioxide and methane. And when heat is trapped in the atmosphere, higher global temperatures result.

What Are the Likely Consequences of Global Warming? All of this leads to the third question that frames the issue of global warming: What will its consequences be? The 1-degree Fahrenheit rise recorded so far in Earth's temperature may seem tiny, but consider the effects that *already* appear to have been caused by this small increase. Arctic sea-ice has shrunk by about 6 percent since 1978; ice cover on lakes and rivers in some northern latitudes now lasts about two weeks less per year than it did 150 years ago; in the European Alps, some plant species have been

FIGURE 24.19 **The Greenhouse Effect** The high-energy rays of the sun can easily penetrate the layer of gases in the troposphere. However, the lower-energy radiation (heat) that reflects from Earth's surface cannot penetrate the layer of gases as easily. Carbon dioxide and methane thus take on the role of glass panes in a greenhouse: They let solar energy in, but they retain a good deal of heat.

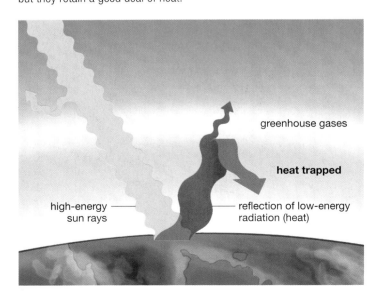

greenhouse gases

heat trapped

high-energy sun rays

reflection of low-energy radiation (heat)

migrating to higher altitudes at the rate of 13 feet per decade; in Europe and North America migratory birds are now arriving earlier in the spring and leaving later in the fall; and general snow coverage in the Northern Hemisphere appears to have shrunk by 10 percent since 1972. Glaciers around the world, from the Himalayas to the Andes, are shrinking because of higher temperatures (**see** FIGURE 24.20). Meanwhile, the ice cap on top of Africa's Mount Kilimanjaro is melting at such a rapid rate that it is expected to disappear in the next 15 to 20 years.

Given such changes, if the temperature increase of the past hundred years were merely repeated in the *next* hundred years, we might worry about what is in store for the Earth. But IPCC research predicted an increase not of 1 degree Fahrenheit for the coming century, but of 2.7 to 10 degrees. In the time since the panel's report was released "climate modelers"—scientists who make predictions about the climate based on computer simulations—have arrived at a single number that is a consensus best-estimate for how much temperature will rise. A doubling of CO_2 levels, which may happen by the end of the twenty-first century, is likely to lead to a 4.7 degree Fahrenheit rise in Earth's surface temperature.

What would changes of this magnitude mean? One likely result is a worldwide rise in sea levels. In fact, recent studies indicate that such a rise will inevitably take place, even if CO_2 emissions could be stabilized at levels reached back in the year 2000. Two factors are at work in sea-level changes. First, as seawater warms, it expands, which raises its level. Second, remember how a great proportion of Earth's freshwater is locked up in glaciers and ice sheets? Well, if the great ice sheets of Greenland and the Antarctic start to melt, this water is no longer locked up on land; it becomes part of the world's oceans. The lowest estimate for rise in sea levels by 2100 is about 5 inches, while other estimates range up to about 1.5 feet.

The prospects for the melting of polar ice sheets seem more likely because, as it turns out, the Earth is not warming uniformly. The average temperature in the Arctic has been rising at almost twice the rate of global average temperature, and parts of the Antarctic Peninsula have experienced a whopping 9-degree Fahrenheit increase in temperature over the past 50 years. Under some scenarios, by the end of this century, the sea-ice that now exists year-round over the Arctic will melt away almost entirely during summer months. One likely outcome of this would be the extinction of polar bears, who move from one area to another on sea-ice.

The picture on global warming may not be totally negative, however. A second IPCC report notes that crop yields in some midlatitude regions probably will increase because of the greater warmth, as would timber yields in some areas. More water might become available in areas, such as parts of Southeast Asia, where water is scarce now. Conversely, insect-borne diseases, such as malaria, can be expected to move both north and south from equatorial regions into formerly temperate areas. Australia and New Zealand are expected to experience a drying trend. Crop yields in current warm-weather regions can be expected to fall.

Many of the predictions about the effects of global warming are very tentative because—as you have come to know by now—no environmental change happens in isolation from other changes, and we have so little experience with changes of this sort. The interactions of the Earth's biotic and abiotic realms are simply too complex to make firm predictions about what global warming will mean. To look at this issue one way, humanity has embarked on an unplanned experiment with planet Earth and its temperature. And as with any experiment, we don't know what the results will be.

(a) The Qori Kalis glacier 1978

(b) The Qori Kalis glacier 2000

FIGURE 24.20 **Warming Planet, Disappearing Glaciers** Global warming is melting glaciers around the world. Peru's Quelccaya ice cap, in the southern Andes, has shrunk by at least 20 percent since 1963. One of the main glaciers flowing from the cap, Qori Kalis, is shown here as it existed in 1978 and then in 2000. A 10-acre lake now exists where the glacier once extended. Its rate of retreat has reached 155 meters or 509 feet per year. This is three times greater than its rate of shrinkage from 1995 to 1998.

A BROADER VIEW

Mountaintop Warming

As you might expect, some of the largest effects of global warming are being felt in cool climates. In the high-mountain or "alpine" regions of the world, native plants and animals are being driven to ever-higher elevations as temperatures warm. The treeless alpine ecosystems are, in effect, islands of cool habitat sitting above an "ocean" of warmer habitat at lower elevations. But as the ocean rises, the alpine habitats shrink. One study of New Zealand's alpine islands predicts that 80 percent of them will be eliminated in the next 100 years.

So Far... Answers:

1. oxygen; stratosphere; ultraviolet radiation
2. heat; carbon dioxide; methane
3. expansion; Greenland; Antarctica

So Far...

1. Ozone is a gas composed entirely of the element _____ that exists in the layer of the atmosphere known as the _____ and that has the effect of keeping _____ from reaching Earth's surface.

2. Most of the energetic short-wave radiation from the sun is transformed into _____, which can be trapped by greenhouse gases such as _____ and _____.

3. Global warming may increase sea levels through two means: the _____ of sea water as it warms and the melting of the large ice sheets of _____ and _____.

Earth's Climate: Why Are Some Areas Wet and Some Dry, Some Hot and Some Cold?

Did you ever look at a globe of the Earth and wonder why it is tilted? Earth exists in space, after all, so what could it be tilted against? When we say that a rod stuck in the ground is tilted, we mean that it is not *perpendicular* to the ground. Here, of course, we're thinking of the ground as a flat surface—a plane. Earth, too, can be viewed as existing on a plane, in this case the plane of its orbit around the sun. (We generally think of this orbit as looking like a large hula hoop with the sun as a yellow ball in the middle.) Now think of the spherical Earth on this plane as having a rod sticking through it in the form of its north-south axis—the imaginary line that runs from the North Pole straight through to the South Pole. The critical thing is that this rod does not stick straight up and down with respect to the plane of the Earth's orbit; it is *tilted*, at an angle of 23.5°. This tilt dictates a good deal about Earth's climate, and Earth's climate dictates a great deal about life on Earth. Once again, in other words, the sun determines the basic conditions for life.

You can see the effects of Earth's tilt in FIGURE 24.21: In June, our Northern Hemisphere tilts toward the sun, while in January it tilts away from it. Thus the sun's rays strike us more directly in June, and the days are warmer (and longer) than days in January. The angle at which the sun's rays strike a given portion of the Earth is very important. Relative to, say, Brazil, sunlight strikes the Arctic and Antarctic at an indirect or "oblique" angle. Thus, the far north and south get less warmth and less of the solar energy that powers photosynthesis.

The Circulation of the Atmosphere and Its Relation to Rain

The variation in temperature caused by these sunlight differences is the most important factor in the *circulation* of Earth's atmosphere. Near the equator Earth simply gets more warmth and, critically, warm air *rises*. Warm air also can retain more moisture than cold air. You've seen an example of this whenever you've looked at

A BROADER VIEW

Earth's Driest Place

What's the driest place on Earth? Probably the Atacama Desert, in northern Chile, which is scarcely more than 100 miles wide at most, but almost 1,000 miles long. In some parts of the Atacama, rain has never fallen in recorded history. Not surprisingly, vegetation tends to occur only in patches, and animal life is sparser yet. One source of moisture, however, is the fog that regularly rolls in along the coast.

FIGURE 24.21 Sunlight and Seasons Regions of the Earth are warmed to the extent that sunlight strikes them more directly. Sunlight strikes the Earth's polar regions at an indirect or "oblique" angle, but strikes the Earth's equatorial regions more directly. Thus, the equatorial regions are warmer and receive more of the sunlight that drives photosynthesis. Because of the Earth's tilt, the Northern Hemisphere gets more sunlight in June and less in January, while the reverse is true for the Southern Hemisphere. This is the reason for seasonal climate variations over large portions of the globe.

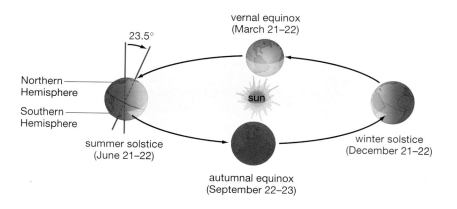

the beads of "sweat" that form on the outside of a cold glass in summer. Moisture-laden warm air comes into contact with the cooler air immediately around the glass; thus cooled, the air cannot hold as much moisture and releases it onto the glass.

This same thing happens with the warm air that rises on both sides of the equator. Because of the heat of the tropics, air is rising in quantity from the tropical oceans; because this air is warm, it is carrying with it a great deal of moisture. The air cools as it rises, however, and then drops much of its moisture on the tropics, which is why they're so wet.

Following this a step further, this volume of air is now cooler, drier, and moving toward the poles in both directions from the equator (see FIGURE 24.22). At about 30° north and south of the equator, it descends, warming as it drops and actually absorbing moisture *from* the land. The land will be dry at these latitudes because this is where the dry, hot air descends.

The air that has descended now flows in two directions from the 30° point. In the Northern Hemisphere this means north toward the North Pole and south toward the equator. Traveling at a fairly low altitude, this air eventually picks up moisture, rises, and deposits its moisture, this time at about 60° north and near the equator, with the same events happening at the same locations in the Southern Hemisphere.

What you get from this rising and falling is what you see in Figure 24.22: a set of interrelated "circulation cells" of moving air, each existing all the way around the globe at its latitude, and each acting like a conveyor belt that is dropping rain on the Earth where it rises but drying the Earth where it descends.

The Impact of Earth's Circulation Cells

The full meaning of this becomes obvious only when you look at a map such as the one in FIGURE 24.23 (or better yet, a globe). Draw your finger across the equatorial latitude on the map and look at how much green there is—Southeast Asia, equatorial Africa, the Amazon. This is no surprise, because we expect the equator to be wet and green. Now, however, do the same thing at the dry 30° north latitude. Look at how much desert there is at this latitude all around the globe—it's where we find the Sahara, the Arabian, and the Sonoran deserts, among others. And there's more desert territory at 30° south, with the Australian desert and Africa's Namib. From this you can see that the climatic fate of Earth's various regions is largely written in our globe's large-scale wind patterns.

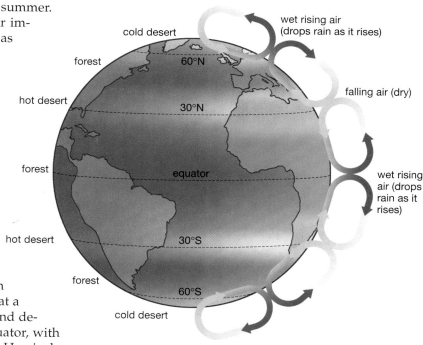

FIGURE 24.22 Earth's Atmospheric Circulation Cells Because the Earth receives its most direct sunlight at the equator, this region is warm—and warm air holds moisture and rises. This air moves north and south from the equator, cooling as it rises, and then loses its moisture as rain, with the result that the land and ocean immediately north and south of the equator get lots of rain. As the now-dry air moves farther north and south, it cools and sinks again at about 30° in both hemispheres. Thus, the land at these latitudes tends to be dry. Two other bands of circulation cells exist at higher latitudes, operating under the same principles.

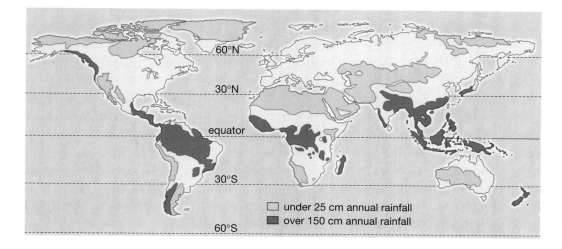

under 25 cm annual rainfall
over 150 cm annual rainfall

FIGURE 24.23 Earth's Atmospheric Circulation Cells and Precipitation The wettest regions of the Earth lie along the equator, while the driest regions lie along the latitudes 30° N and S. Compare these regions to the circulation cells shown in Figure 24.22. The variations of rainfall patterns within cells are explained by factors such as the presence of mountain ranges.

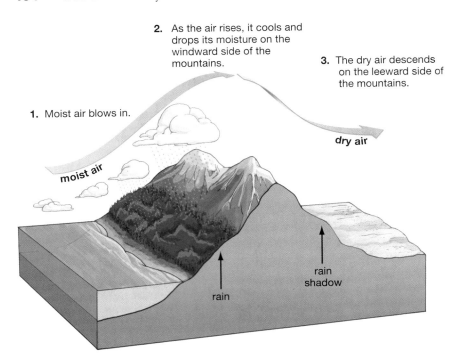

1. Moist air blows in.

2. As the air rises, it cools and drops its moisture on the windward side of the mountains.

3. The dry air descends on the leeward side of the mountains.

moist air

dry air

rain

rain shadow

FIGURE 24.24 **Rain Shadows** Mountain ranges force moving air to rise. This cools the air, causing it to lose its moisture on the windward side of the range. This creates a rain shadow, meaning a lack of precipitation, on the leeward side of the range.

Mountain Chains Affect Precipitation Patterns

Latitude doesn't tell the whole story about precipitation on Earth, however. Mountain ranges force air upward and it too cools as it rises, dropping its moisture on the windward side of the range. Then the air descends on the opposite (leeward) side of the range, only this time it is dry air and is thus *picking up* moisture from the ground, rather than depositing it (**see** FIGURE 24.24). A dramatic example of this "rain shadow" effect can be seen on the Pacific Coast of the United States in southern Oregon, where the western side of the Cascade Mountains is lush with greenery all year round, while the basin below the eastern slopes is a desert.

Beyond this, we know that it's possible to be in a warm latitude, such as that of Ecuador in South America, and still be very cold—if you go to the top of the Ecuadorian Andes. As altitude increases, temperature drops, in other words. The changes in climate that occur as you move from flatland to mountain peak are similar to the changes that occur as you move from the equator to one of the poles.

The Importance of Climate to Life

The factors you have just reviewed are among the most important in shaping the large-scale **climates** or average weather conditions we find on Earth. It's obvious that climate has a great deal of influence on life, since we never see polar bears in Panama or monkeys in Montana, but this influence is even greater than you might think. Looking over the globe, you see different forms of dominant vegetation in different areas—grassland in America's Great Plains, rain forest in northern Brazil. So strongly does climate affect this pattern that Earth's vegetation regions essentially are defined by Earth's climate regions. Far in Canada's north, there is a fairly distinct boundary at which Canada's vast coniferous forest gives way to the mat-like vegetation called *tundra*. North of this line there is tundra; south of it are the forests. What is this line? In summer it is the far southern reach of a mass of cold air called the arctic frontal zone (**see** FIGURE 24.25).

FIGURE 24.25 **Line of Transition** In Alaska's Denali National Park, a region of forested taiga vegetation gives way to a region of grassy tundra vegetation in a fairly abrupt way.

Vegetative formations generally overlap one another to a greater extent than is seen with Canada's tundra and coniferous forest, so that the closer we get to a vegetation boundary the less distinct it looks. Nevertheless, there are discrete regions of vegetation on Earth whose boundaries are largely determined by climate—particularly by temperature and rainfall. These distinct vegetative regions of the Earth are the subject to which we now turn.

24.5 Earth's Biomes

When large regions of land vegetation are looked at as *ecosystems*, complete with animals, plants, nutrient cycling, and all the rest, they go by the name **biomes**. Grasslands, whether found in Russia or in the American Midwest, are ecologically very similar to one another and thus constitute one type of biome. The tropical rain forests found in Africa, South America, and Asia constitute another type of biome. The world can be divided into any number of different biome varieties— this is a human classification system overlaid on nature—but six are recognized as the minimum: tundra, taiga, temperate deciduous forest, temperate grassland, desert, and tropical rain forest. Polar ice and mountains often are recognized as separate biomes as well, as is another grassland variation, the savanna, found most famously in equatorial Africa. There is also chaparral, a shrub-dominated vegetation formation found in "Mediterranean" climates such as those on California's coast. If you look at FIGURE 24.26, you can get an overview of locations of these biome types throughout the globe. Let's look briefly now at each of Earth's major biome types, after which we'll look at Earth's aquatic ecosystems.

Cold and Lying Low: Tundra

Along with polar ice, **tundra** is the biome of the far north, stretching in a vast ring around the northern rim of the world. So inaccessible is tundra that the average person may never have heard of it, yet it occupies about a fourth of Earth's land surface. The word *tundra* comes from a Finnish word that means "treeless plain," and the description is apt. Its flat terrain stretches out for mile after mile with little

A BROADER VIEW

Why Are Mountaintops Colder?

Why does the air get colder as we go up a mountain? The answer has to do with atmospheric pressure, which is simply the weight of the air that exists above a given location. At sea level, atmospheric pressure is 14 pounds per square inch, but in mile-high Denver it is only 12.2 pounds per square inch (because Denver has less atmosphere above it than an ocean location has). As a given quantity of air rises, therefore, the pressure on it decreases; this allows its molecules to expand, and when air expands, it cools. Think of taking the lid off a boiling pan of water and allowing the steam inside to escape. As it expands in volume, it cools.

FIGURE 24.26 **Distribution of Biomes across the Earth**

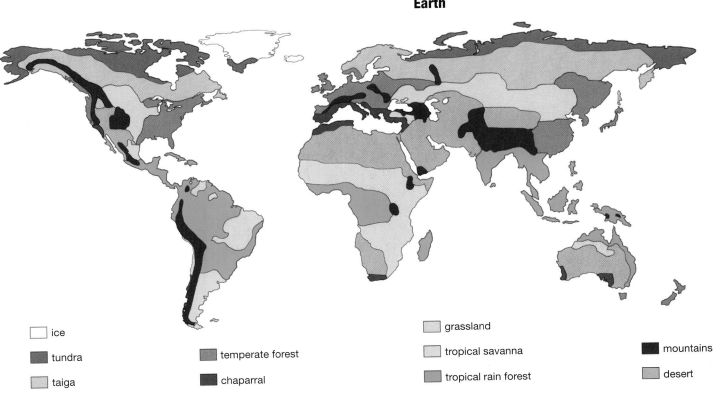

☐ ice	☐ grassland	
☐ tundra	☐ temperate forest	☐ tropical savanna	■ mountains
☐ taiga	☐ chaparral	☐ tropical rain forest	☐ desert

FIGURE 24.27 **Tundra** Grizzly bears forage for fall berries in Alaska in the shrub-and-grass biome known as the tundra.

change in the vegetational pattern of low shrubs, mosses, lichens, grasses, and the grass-like sedges (**see** FIGURE 24.27). No more than three feet below the surface, the tundra ground is permanently frozen into *permafrost*, or permanent ice. Tundra is, however, a biome of very little rainfall—about 10 inches per year.

Northern Forests: Taiga

The **taiga** or boreal (northern) forest includes the enormous expanse of coniferous trees mentioned earlier that juts up against the tundra. Conditions are still dry and cold, but in both respects the taiga is less severe than the tundra. Note in Figure 24.26 that, as with tundra, taiga is found almost solely in the Northern Hemisphere. Taiga is a model of species uniformity as only a few types of trees—spruce, fir, and pine—are the ecological dominants, but they are present in great numbers. The taiga's vegetation supports lots of animals, including large mammals such as moose, bear, and caribou (**see** FIGURE 24.28).

Hot in Summer, Cold in Winter: Temperate Deciduous Forest

Temperate deciduous forests are biomes characterized by seasonal growth patterns of both trees and an "understory" of smaller plants, and relatively abundant rainfall—between 30 and 78 inches per year. This is a biome familiar throughout much of the United States, since it exists roughly from the Great Lakes south, nearly to the Gulf of Mexico; and from the Great Lakes east, to the Atlantic Ocean. The "deciduous" in the name refers to trees that exhibit a pattern of loss of leaves in the fall and regrowth of them in the spring (FIGURE 24.29).

Dry but Sometimes Very Fertile: Grassland

There is an irregular line in western Indiana that forms the boundary between deciduous forest and prairie. What is "prairie" in the United States is known as "steppes" in Russia, "pampas" in Argentina, and "veldt" in South Africa. These are all names for the biome that ecologists call **temperate grassland**, a biome whose ecological dominants tend to be seasonal grasses, rather than trees. (FIGURE 24.30).

Looked at one way, grasslands are what often literally lie between forests and deserts. Whereas the precipitation range for deciduous forests is from 30 to 78 inches per year, for grassland the range is from 10 to 39 inches. The North American

FIGURE 24.28 **Taiga** Alaskan caribou in winter in the northern-forest biome known as taiga.

FIGURE 24.29 Temperate Deciduous Forest
Plenty of water and seasonally warm temperatures lead to abundant plant and animal life in the biome known as temperate deciduous forest. This forest is in the Great Smoky Mountains National Park, which straddles Tennessee and North Carolina.

grassland has its own internal division based on the amount of rainfall. It ranges from tall-grass prairie in Illinois to a drier mixed-grass prairie that begins at about the middle of Kansas and Nebraska—this is where the Great Plains begin—to a short-grass prairie that lies in the rain shadow of the Rocky Mountains. The soil of temperate grasslands is some of the most fertile in the world.

The Challenge of Water: Deserts

Deserts can be defined as biomes in which rainfall is less than 10 inches per year and water evaporation rates are high, relative to rainfall. Around the globe, some 20 major deserts collectively cover about 30 percent of Earth's land surface. Note that deserts are not defined by temperature. There are cold deserts (the Gobi in China), temperate deserts (the Mojave in Southern California), and hot-weather deserts (the Sahara in North Africa). Rainfall is not only sparse in deserts, but sporadic as well; most of the moisture a desert gets will come in just a few days during

FIGURE 24.30 Temperate Grassland A white-tailed deer fawn stands among the grass and flowers in a tall-grass prairie in Missouri. Tall-grass is one kind of vegetation found in the biome known as temperate grassland.

FIGURE 24.31 **Desert** When water gets sparse, life does too. Shown is a desert biome in Monument Valley, Arizona.

the year. Most deserts harbor a variety of water-conserving life-forms, such as small rodents, reptiles, and cacti (**see** FIGURE 24.31). What we think of as desert can transition imperceptibly into dry grassland.

Tropical Rain Forests: Habitat Destruction and Biodiversity

We'll look at our last biome type, the tropical rain forest, in a little more detail than the others, because of several issues that this biome puts into sharp relief: productivity, species diversity, and habitat destruction. **Tropical rain forests** can be defined as biomes characterized by heavy rain and stable, warm temperatures (**see** FIGURE 24.32). The norm for rainfall is between 78 and 175 inches a year, but some areas get up to 390 inches annually.

Many people have an idea that the Earth's tropical rain forests are important, but they're not quite sure why. There are two main reasons. First, along with coral reefs, they are the most productive biome type in the world, meaning they produce more biomass per square meter of territory than any other biome. Second, no other ecosystem can touch them in terms of species diversity. As noted earlier, for mile after mile in the taiga, trees might be limited to one or two species. In the tropical forests of the Amazon, more than *three hundred* species of trees have been identified in about 2.5 acres. Indeed, half of all the identified plant and animal species on Earth reside in tropical rain forests. This despite the fact that, compared to the other biomes you've looked at, they don't occupy a lot of territory: about 7 percent of the Earth's surface.

The problem, as most people know, is that the rain forests are disappearing—significant portions of them are either being burned down or cut down. Several periods of temperate forest loss came before the twentieth century, but we are living in the period of great tropical forest loss. How significant is the reduction? Satellite imaging of humid tropical forests indicates that, between 1990 and 1997, about 14 million acres

FIGURE 24.32 **Tropical Rain Forest** Abundant rain and warm weather mean abundant growth and a great diversity in life-forms in the biome known as the tropical rain forest. Shown is a lowland rain forest on the Segama River in Borneo.

were lost each year—an area that is about twice the size of Maryland. South America has what is by far the world's largest tropical rain forest, the Amazon Basin Forest that runs from Colombia through Bolivia and at one point nearly from the Atlantic to the Pacific Ocean. Significant rain forests also can be found in Central America, equatorial Africa, and regions of Southeast Asia near the equator.

So Far...

1. The warm surface air of the tropics _____ as it rises, causing it to _____ moisture. At about 30° north and south of the tropics, however, air descends and _____ moisture from the land.

2. Earth's large vegetation regions are essentially defined by _____.

3. Starting at the North Pole and going south, list the order in which you would be likely to find the following biomes: temperate deciduous forest, tundra, tropical rain forest, taiga.

So Far... Answers:
1. *cools; lose; absorbs*
2. *climate*
3. *tundra, taiga, temperate deciduous forest, tropical rain forest*

24.6 Life in the Water: Aquatic Ecosystems

Marine Ecosystems

We turn now from life on land to life in the water. Because we humans are land creatures, it might be good to get our bearings by thinking about how one kind of aquatic environment, the ocean, can be divided up in terms of the life in it. The ocean's **coastal zone** extends from the point on the shore where the ocean's waves reach at high tide to a point out at sea where an ocean-floor formation called the continental shelf drops off (**see** FIGURE 24.33). Beyond this point there is the *open sea*. Within the coastal zone, there is an area bordered on one side by the ocean's low-tide mark and on the other by its high-tide mark. This is the **intertidal zone**.

More Productive Near the Coasts, Less Productive on the Open Sea

Each of these zones marks off an area that is meaningful in relation to life. Start first at the intertidal zone. This is a world *in between*, we might say, because the creatures in it live part of the time submerged in the ocean and part of the time exposed to the air. It turns out that the intertidal zone is extremely productive for a reason that you might think would make it *unproductive*: The ocean waves, which bring in nutrients, carry away wastes, and expose more of the surface area of photosynthesizers to sunlight.

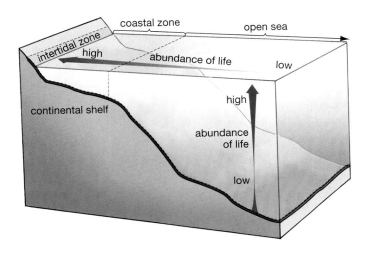

FIGURE 24.33 **Ocean Zones**

(a) A man and two children investigate tidepools

(b) King penguins swim underwater

(c) Sweepers and fairy basslets feed in currents

Ocean life continues to be abundant out past the intertidal zone and in general remains so all the way to the end of the coastal zone. Go past the continental shelf and into the open sea, however, and productivity can drop off to levels that are less than those of a desert (see FIGURE 24.34). Why should this be so? Because sunlight strong enough to power photosynthesis penetrates ocean waters only to a depth of 100 yards or so. Meanwhile, the average depth of the ocean is about 1.8 miles. Thus there is a vast depth of open ocean that simply isn't very productive. In fact, life at greater ocean depths essentially depends on organic particles, referred to as "marine snow," that rain down from above. In short, the most productive parts of the ocean are the shallower parts of the ocean. One measure of the coastal zone's importance is that it is the location of all the world's great fisheries. Fish do exist in the open ocean as well, however, and concern is increasingly being raised about the health of fish populations in both kinds of areas, as you can see in Essay 24.1, "Our Overfished Oceans," on page 442.

More Productive Near the Poles, Less Productive Away from Them Apart from the shallow-to-deep transition, the ocean also has one other gradient with respect to the amount of life it harbors: There is relatively more ocean life toward the poles, with this concentration decreasing with movement toward the equator—the exact opposite of what we find with land life. The oceans that surround Antarctica are filled with life. **Phytoplankton**, meaning floating, microscopic photosynthesizers such as algae and cyanobacteria, are the base producers for it all. These are consumed by the tiny floating animals known as zooplankton. Shrimp-like crustaceans called krill feed mostly on the phytoplankton and serve as the principal source of food for the whales and seals we associate with this area.

Cities of Productivity: Coral Reefs Despite the richness of the colder oceans, the tropical oceans do boast marine communities that are as productive as any in the ocean—indeed, as any on Earth. These are the coral reefs, which lie in shallow waters at the edge of continents or islands at tropical latitudes. Coral reefs actually are the piled-up remains of many generations of coral animals, with each individual animal, known as a polyp, being a tiny relative of the more familiar sea anemones. Coral polyps secrete calcium carbonate, better known as limestone, a practice they share with a type of red algae with whom they share their habitat. The limestone forms an external skeleton for the corals and, when each coral dies, its limestone skeleton remains. Thus, coral reefs are composed largely of the stacked-up remains of countless generations of coral polyps, along with the remains of their neighboring red algae. But the thin outer veneer of each reef is composed of *living* algae and the latest generation of pinhead-sized corals.

FIGURE 24.34 **Diversity in Ocean Life**

(**a**) A man and two children investigate tide pools, a part of the ocean's intertidal zone. The deeper water beyond them can be seen both above and below the surface.

(**b**) King penguins swim underwater near a breeding colony in the fertile waters off the coast of Antarctica.

(**c**) Fish called sweepers and fairy basslets feed in the currents above a coral plate covered with animals called crinoids. The structure is off the coast of Papua New Guinea.

You can get an idea of how many generations of coral are piled up by considering that most reefs are between 5,000 and 8,000 years old—and some are millions of years old. What the reef as a whole creates is *habitat*: nooks and crannies and tunnels and surfaces that are home to an amazing array of living things. So rich is this habitat that it covers only 2 percent of the ocean's floor, but is home to perhaps 25 percent of the ocean's species.

Sadly, coral reefs are now imperiled cities of biological productivity. By one estimate, 27 percent of the world's coral reefs have now been lost as functioning ecosystems, though with coral reefs, loss of activity is not always permanent. The greatest single source of this destruction was the El Niño-caused ocean warming of 1997–98, which brought about a phenomenon known as *coral bleaching*, meaning a discoloration of the reefs due to the death of symbiotic algae that live within the coral animals. Temporary weather events aside, the longer-term threat to coral reefs is human activity. Oil is dumped into the ocean; sewage and runoff from cities harm the reefs; divers may trample on them. Perhaps most disturbing, thousands of pounds of the poison sodium cyanide are dumped each year on coral reefs in the Philippines and other parts of Asia as a means of capturing colorful aquarium fish. (The cyanide stuns the fish, but kills the coral animals.) Global warming has now been tied to the threat to coral reefs as well; both ocean warming and increased ocean CO_2 concentrations are harmful to the reefs.

Freshwater Systems

Whereas the oceans cover almost three-quarters of Earth's surface, freshwater ecosystems—inland lakes, rivers, and other running water—together cover only about 2.1 percent of Earth's surface. Like the oceans, lakes can be divided into biological zones, which you can see described in FIGURE 24.35.

Nutrients in Lakes
Lakes can be naturally nutrient rich, or nutrient poor. A lake that has few nutrients will have relatively few photosynthesizers, and this in turn means few animals. Nutrient-poor lakes generally are clear (and deep), while nutrient-rich lakes often have an abundant algal cover.

Nutrient enrichment can happen naturally over time, or it can take place because of human intervention. There is such a thing as having too *many* nutrients in a body of water, however. We touched on this earlier in connection with the Gulf of Mexico's "dead zone," which results from the tons of nitrogen that flow into it. But smaller bodies of water can also get too many nutrients and thus undergo a process called *eutrophication*. An overabundance of nutrients such as phosphorus or nitrogen brings on an overabundance of algae—an "algal bloom"—and this eventually means an overabundance of dead algae falling to the bottom of the lake

A BROADER VIEW

The Mighty Great Lakes

An astonishing one-fifth of all the world's fresh surface water is contained in the five Great Lakes of North America. Lake Superior is the largest single lake in the world in terms of surface area—it is about the size of South Carolina—but it does not hold the most water. That honor belongs to Russia's Lake Baikal, whose maximum depth reaches more than 5,000 feet, compared to Superior's 1,300.

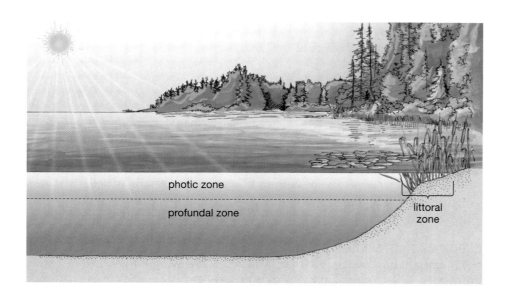

FIGURE 24.35 **Zones in a Lake** A lake can be divided into zones that support different kinds of life. A lake's littoral zone starts at water's edge and extends out to the point at which rooted plants can no longer grow. The photic zone starts at the lake's surface and extends down to the point at which sunlight no longer drives photosynthesis. The profundal zone is the area beneath the photic zone in which photosynthesis cannot be performed.

Essay 24.1 Our Overfished Oceans

At one time, it was thought that the bounty of fish in the world's oceans was so immense it could never be exhausted. But that idea began to be challenged in the late 1980s and by the mid-1990s it was effectively dead, as it became clear that the numbers of certain kinds of fish had not just dwindled, but had dropped to alarmingly low levels. The news has not gotten better since then. In 2003, researchers reported that the global ocean has lost 90 percent of its big predatory species—bluefin tuna, blue marlin, Antarctic cod, and sharks among them (**see** FIGURE E24.1.1). One of the ways the researchers arrived at this conclusion was to count the number of fish caught per baited hook on the "longlines" that stretch out for miles behind oceangoing fishing vessels. Just after World War II, Japanese fishermen commonly caught 10 fish per hundred longline hooks; now that figure is often down to 1 fish per hundred hooks.

> In many commercial species, the trend is toward depletion, if not endangerment.

About the time this study came out, the Canadian government banned the fishing of all Atlantic cod in three large regions off the Canadian coast. The move came after an independent research group declared Canadian cod an endangered species. Meanwhile, across the ocean in Britain's North Sea, cod populations have dropped to 15 percent of their 1970s levels, with the result that scientists are calling for similar reductions in fishing there. To be sure, not all populations of commercial fish have dropped by this amount; but in many commercial species, the trend is toward depletion, if not endangerment.

How did this happen? In essence, world fishing fleets got more efficient at scouring the Earth for large populations of fish. By using ever-more sophisticated technology, the fleets have managed to find fish no matter what kind of terrain they are hidden in. But in some instances—as with Canada's Atlantic cod—there are no large populations left to find.

Fisheries scientist Daniel Pauly and his colleagues have pointed out that this turn of events actually should not be surprising. We would not expect industrialized *hunting*—of, say, deer or buffalo—to be sustainable over the long run if left unregulated. Why should we expect anything different with industrialized fishing? Many scientists believe that a three-pronged approach has to be taken if global fish populations are to bounce back. First, global fishing must be reduced in general. Second, countries need to restrict the use of "bottom trawling" techniques in which nets and accompanying gear—sometimes weighing thousands of pounds—are dragged across the ocean floor, destroying habitat as well as fish. Third, "no take" zones need to be created so that specific fish populations can have a chance to establish themselves.

FIGURE E24.1.1 **One of a Dwindling Number** A blue marlin (*Makaira nigricans*), one of the large predatory fishes whose numbers have been reduced by 90 percent in recent decades, according to a global assessment.

(**see** FIGURE 24.36). When this happens, decomposing bacteria flourish, and they are using *oxygen*—so much of it that the fish in a lake can suffocate. How do these harmful levels of phosphorus or nitrogen get into freshwater aquatic ecosystems? Fertilizers from lawns or agriculture, sewage, and detergents are some of the sources. This is one form of human environmental impact that can be reversible in a straightforward way: Reduce the nutrients that are flowing into the ecosystem.

Estuaries and Wetlands: Two Very Productive Bodies of Water There is one important type of aquatic water system that always straddles the line between a freshwater and saltwater habitat. This is the **estuary**, an aquatic area, partly enclosed

FIGURE 24.36 **Too Many Nutrients** A pond with an overabundance of nutrients has experienced an overgrowth of algae—an algal bloom—that may be detrimental to other life-forms in the pond.

by land, where a stream or river flows into the ocean. Estuaries are very productive ecosystems, with the cause of this productivity being the constant movement of water—the same force at work in the ocean's intertidal zone. Ocean tides and river flow are constantly stirring up estuary silt that is rich in nutrients. Plants and algae thus get an abundance of these substances, grow in abundance, and pass their bounty up the food chain. In **FIGURE 24.37** you can see an image of an important estuary in the United States, the Chesapeake Bay.

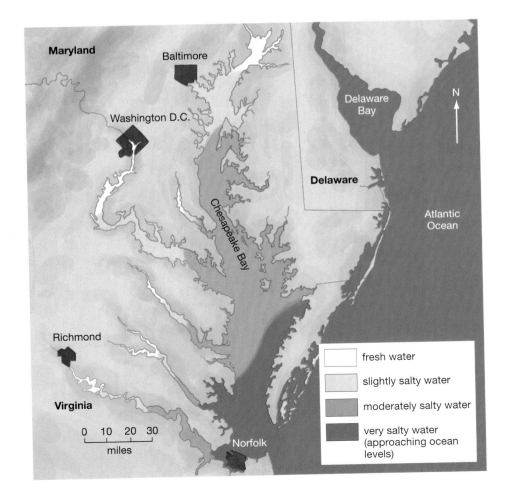

FIGURE 24.37 **Where Salt Water and Freshwater Meet** The Chesapeake Bay is one of the largest and most productive estuaries in the world. In it, the freshwater of 19 principal rivers meets the salt water of the ocean, forming zones of different degrees of saltiness. Each zone supports specific communities of organisms. (Map adapted from "Where Salt and Fresh Water Meet, Chesapeake Bay" from *Life in the Chesapeake Bay* by A.J. Lippson and R.L. Lippson, 1984.)

So Far... Answers:

1. decreases

2. phytoplankton; perform photosynthesis

3. a stream or river flows into the ocean

Another important type of aquatic ecosystem is **wetlands**, which are just what they sound like—lands that are wet for at least part of the year. "Wet" here covers a variety of conditions, from soil that is merely temporarily waterlogged—a prairie "pothole" in Minnesota, for example—to a permanent cover of water that may be several feet in depth (the bayous of Louisiana). Wetlands go by such specific names as swamps, bogs, marshes, and tidal marshes; the overwhelming majority of them are freshwater *inland wetlands*, rather than the ocean-abutting *coastal wetlands*, which can be freshwater or salt water (**see** FIGURE 24.38). Wetlands are sites of great biological productivity, and they serve as vital habitat for migratory birds. It's been estimated that the amount of wetlands in the United States has been reduced by 55 percent since the arrival of Europeans—120 million of the 215 million acres that were originally wetlands have been drained or paved over.

Life's Largest Scale: The Biosphere

You've been reviewing life on a very large scale, which is to say the aquatic ecosystems and biomes of the Earth. But there is one larger scale of life yet, which is the collection of *all* the world's ecosystems, the **biosphere**, which is also thought of as that portion of the Earth that supports life. Much could be said about the biosphere, but here's just one observation about life on this global scale. The Earth is about 24,000 miles in circumference, with life existing along any line you could draw all around the globe. If we look at life in its *vertical* dimension, however, it has been found to exist only about 1.8 miles below Earth's surface and about 5.2 miles above sea level (in the upper Himalayas). This means that all of life on Earth exists in a band that is at most about 7 miles thick, with the vast bulk existing in an even narrower span. Actually, this is all the life that we know of in the *universe*, so that if we could imagine ourselves traveling by cosmic elevator, we could potentially come up from the molten center of Earth, pass through the bacteria lodged between underground rocks, and then move through a tissue-thin layer of a green living world, after which we might never see anything like it again, no matter how far out into the universe we traveled. Taking this perspective, the really remarkable thing is not how much life there is, but how little.

FIGURE 24.38   **Inland Wetland and Coastal Wetland**

(a) A bird's nest in prairie wetland in the United States

(b) An overhead view of the sawgrass prairie in Florida's Everglades National Park

On to Human Physiology

In this chapter and the one that preceded it, we've been taking a very broad view of the living world, looking at the relationships between its various parts and the physical environment with which they interact. In what's coming up, however, we're going to narrow our focus to just one life-form: human beings. In the next five chapters, the focus will be on the anatomy and physiology of humans—how the human heart works, what digestion is, what the kidneys do, how the immune system fights off infections, and more. Though our focus will have narrowed, you'll see that, when it comes to the human body, there is no end of pathways to explore. Coming up: what makes the human body tick.

So Far...

1. With increasing distance from the shoreline toward open sea, the abundance of marine life _____.

2. The tiny organisms collectively called _____ lie at the base of most marine food webs, thanks to their ability to _____.

3. An estuary is an area where _____.

Chapter Review

Summary

24.1 The Ecosystem and Ecology

- **Nature of the ecosystem**—An ecosystem is a community of organisms and the physical environment with which these organisms interact. Very large-scale ecosystems are known as biomes, which are defined by the types of vegetation within them. (page 415)

24.2 Abiotic Factors in Ecosystems

- **Cycling in ecosystems**—The chemical elements that are vital to life are known as nutrients. Along with water, nutrients move back and forth between abiotic (nonliving) and biotic (living) domains on Earth in a process called biogeochemical cycling. (page 416)

- **The carbon cycle**—Carbon comes into the living world through the organisms that perform photosynthesis. Animals obtain their carbon from these photosynthesizing organisms. Carbon moves back into the atmosphere in the form of carbon dioxide, which plants and animals give off as a product of their respiration, and as a product of their decomposition after death. Carbon dioxide makes up a small but critical proportion of the Earth's atmosphere. It is vital to life and may greatly affect global temperature. (page 416)

- **The nitrogen cycle**—Prior to the twentieth century, nitrogen entered the biotic domain almost entirely through the action of certain bacteria that have the ability to convert atmospheric nitrogen into forms that can be used by living things. Other bacteria have the ability to convert this organic nitrogen back into atmospheric nitrogen, completing the cycle. Human beings now use industrial processes to convert atmospheric nitrogen into a biologically usable form, thus greatly expanding agricultural productivity. Nitrogen runoff from agriculture can be a form of nutrient pollution that can harm both small and large aquatic ecosystems. (page 418)

- **The water cycle**—As with carbon or nitrogen, all of Earth's water is either being cycled or being stored—in such forms as glaciers or polar ice. As little as 0.5 percent of Earth's water is available as fresh, liquid water, with about 25 percent of this being groundwater. (page 421)

- **The scarcity of water**—Fresh, sanitary water is a scarce commodity even for human beings, despite the fact that civilization now uses more than half the world's accessible water. The scarcity of water can be traced in significant part to the inefficient ways in which humans use it. Human diversion of water from natural environments is having harmful impacts on species such as fish. (page 422)

24.3 Energy Flow in Ecosystems

- **Trophic levels**—Plants and other photosynthesizers are an ecosystem's producers, while the organisms that eat plants are its consumers. Every ecosystem has a number of feeding or trophic levels, with producers forming the first trophic level and consumers forming several additional levels. (page 424)

- **Detritivores**—A detritivore is a class of consumer that feeds on the remains of dead organisms or cast-off material from living organisms. A decomposer is a special kind of detritivore that breaks down dead or cast-off organic material into its inorganic components, which can then be recycled through an ecosystem. (page 425)

- **Flow through trophic levels**—Very little of the energy that a given trophic level receives is passed along to the next trophic level. A rule of thumb in ecology is that for each jump up in trophic level, the amount of available energy drops by 90 percent. This explains why large, predatory animals are rare: Very little of the solar energy assimilated by plants is available to the higher trophic levels the predators occupy. (page 426)

24.4 Earth's Physical Environment

- **Composition of the atmosphere**—The atmosphere is the layer of gases surrounding the Earth. The lowest layer of the atmosphere, the troposphere, contains the bulk of the atmosphere's gases. Nitrogen and oxygen make up 99 percent of the troposphere, with carbon dioxide and small amounts of other gases making up the rest. (page 428)

- **Ozone depletion**—The gas called ozone screens out 99 percent of the sun's potentially harmful ultraviolet radiation. Human-made compounds used in various consumer and industrial products can destroy ozone. International action taken on this issue in the 1980s seems to be bringing the ozone back to health. (page 428)

- **Global warming**—There is near unanimity among scientists that global warming is taking place, and a strong consensus that human activity is partly responsible for it. The greenhouse effect is one mechanism that underlies global warming. In it, higher atmospheric concentrations of gases such as carbon dioxide trap additional quantities of the heat that comes to the Earth from the sun. (page 429)

- **Effects of global warming**—The Earth appears to have warmed by 1° Fahrenheit in the past century. The effects of this warming can be seen in such phenomena as the shrinkage of Arctic sea-ice and the migration of mountain plant species to higher altitudes. The best current estimate for additional warming by the end of the twenty-first century is 4.7° Fahrenheit. One likely consequence of this warming is a rise in global sea levels. (page 430)

- **Global rainfall**—Sunlight strikes the equatorial region of the Earth more directly than it strikes the polar regions. The differential warming that results produces a set of enormous interrelated circulation cells of moving air, each existing all the way around the globe at its latitude. Each of these cells drops rain on the Earth where it rises, but dries the Earth where it descends. This is why some regions of the Earth get so much more rainfall than others. (page 432)

- **Rain-shadow effect**—More rain will be deposited on the windward side of a mountain range than on the opposite leeward side. This rain-shadow effect can cause opposite sides of a mountain range to have dramatically different vegetation patterns. (page 434)

- **Climate and vegetation**—A climate is an average weather condition in a given area. Large vegetative formations essentially are defined by climate regions. (page 434)

Web Tutorial 24.1 Earth's Atmosphere and Climate

Web Tutorial 24.2 Rain Shadows

24.5 Earth's Biomes

- **Types of biomes**—A biome is an ecosystem defined by a large-scale vegetation formation. Six types of biomes are recognized at a minimum: tundra, taiga, temperate deciduous forest, temperate grassland, desert, and tropical rain forest.   (page 435)

- **Tundra**—Tundra is the biome of the far north, frozen much of the year, but with a seasonal vegetation formation of low shrubs, mosses, lichens, grasses, and grass-like sedges.   (page 435)

- **Taiga and temperate deciduous forest**—Taiga is another biome of the north; it includes the enormous expanse of coniferous trees that lies south of the tundra at northern latitudes. The taiga exhibits a great deal of species uniformity because only a few types of trees—spruce, fir, and pine—are the ecological dominants. Temperate deciduous forests grow in regions of greater warmth and rainfall than is the case with tundra or taiga. These forests, existing over much of the eastern United States, are composed of an abundance of trees, complemented by a robust understory of plants.   (page 436)

- **Temperate grassland**—Temperate grassland is characterized by less rainfall than that of temperate forest and by grasses as the dominant vegetation formation. Such regions can be very fertile agricultural land.   (page 436)

- **Desert**—Deserts are characterized by rainfall that is low and evaporation rates that are high. Deserts may be hot, cold, or temperate, but all desert life is shaped by the need to collect and conserve water. (page 437)

- **Tropical rain forest**—The tropical rain forest biome is characterized by warm, stable temperature, abundant moisture, great biological productivity, and great species diversity. Found in Earth's equatorial region, tropical rain forests are being greatly reduced in size through cutting and burning.   (page 438)

24.6 Life in the Water: Aquatic Ecosystems

- **Aquatic productivity**—Ocean or marine ecosystems are most biologically productive near the coasts, with the deep open oceans having a productivity that can be less than that of deserts. Coastal areas benefit from wave actions that bring in nutrients, carry away wastes, and expose more of the surface area of photosynthesizers to sunlight.   (page 439)

- **Species distribution and decimation**—There is relatively more ocean life toward the poles, with this concentration decreasing as one moves toward the equator. The food webs in the oceans surrounding Antarctica are based on photosynthesizing phytoplankton. Modern fishing techniques have seriously depleted many once-abundant commercial fish species.   (page 440)

- **Coral reefs**—Coral reefs are warm-water marine structures composed of the piled-up remains of generations of coral animals. Coral reefs provide a habitat that results in a rich species diversity.   (page 440)

- **Freshwater**—Freshwater ecosystems, which include inland lakes, rivers, and other running water, cover only about 2.1 percent of Earth's surface. Freshwater lakes are most productive near their shores and near their surface. Lakes can be naturally nutrient rich or nutrient poor. Human activity can sometimes introduce too many nutrients into a lake, a process known as eutrophication.   (page 441)

- **Estuaries**—Estuaries are partially enclosed areas where streams or rivers flow into the ocean. They are characterized by high biological productivity due to the constant movement of water within them.   (page 442)

- **Wetlands**—Wetlands are lands that are wet for at least part of the year. Wetland soil may merely be waterlogged for part of the year or under a permanent, relatively deep cover of water. Wetlands are very productive and are important habitats for migratory birds. The amount of wetlands in the United States is estimated to have been reduced by 55 percent since the arrival of the Europeans.   (page 442)

Key Terms

aquifer   p. 421

atmosphere   p. 428

biodegradable   p. 426

biogeochemical cycling   p. 416

biomass   p. 418

biome   p. 435

biosphere   p. 444

carnivore   p. 424

climate   p. 434

coastal zone   p. 439

consumer   p. 424

decomposer   p. 425

desert   p. 437

detritivore   p. 425

ecosystem   p. 416

element   p. 416

estuary   p. 442

groundwater   p. 421

herbivore   p. 424

intertidal zone   p. 439

nitrogen fixation   p. 418

nutrient   p. 416

omnivore   p. 424

ozone   p. 428

phytoplankton   p. 440

primary consumer   p. 424

producer   p. 424

secondary consumer   p. 424

stratosphere   p. 428

taiga   p. 436

temperate deciduous forest   p. 436

temperate grassland   p. 436

tertiary consumer   p. 424

trophic level   p. 424

tropical rain forest   p. 438

troposphere   p. 428

tundra   p. 435

wetlands   p. 444

Testing Your Understanding

In Your Own Words *(answers in the back of the book)*

1. In a large reserve in Idaho, Forest Service wildlife scientists measured wolf populations as well as elk populations, and found that the wolf population was much lower than that of the elk population. If wolves feed on elk, why aren't the population sizes similar?

2. Explain why most deserts across the globe occur at about 30° north and south of the equator.

3. A great many dead fish are found floating on the surface of a pond. What might be the cause of such a fish kill?

Thinking about What You've Learned

1. Bacteria and fungi are to the living world as a foundation is to a house: not often thought of, but vitally important. If all bacteria and fungi were somehow instantly eliminated, almost no life-forms would survive for long. This is so because of the role bacteria and fungi play in ecological processes. Can you name two of these processes?

2. Increasing levels of carbon dioxide in Earth's atmosphere are likely to increase the rate of a basic, critically important biological process. Can you think of what that process is?

3. Tundra and desert are in some ways more fragile biomes than are grasslands or forests. Why should this be so?

The Human Body and Three of Its Systems

SCIENCE IN THE NEWS

AUSTIN AMERICAN-STATESMAN

WITH ACL TEARS, IT'S A LONG ROAD TO RECOVERY

By Pamela LeBlanc, American-Statesman Staff
February 21, 2005

When it goes, you know it. Usually, there's a telltale pop. A stab of pain. Suddenly, your knee feels loose.

You've torn your ACL, or anterior cruciate ligament. It's one of the most common knee injuries in America, especially among people who play football, basketball, soccer and other sports that involve pivoting or jumping.

SCIENCE IN THE NEWS

THE TENNESSEAN

TORN ACL ENDS YEAR FOR SPENCER

FOURTH LADY VOLS PLAYER TO HURT KNEE THIS SEASON

By Tim Vacek for the Tennessean
February 25, 2005

KNOXVILLE—The season of the injured knee just gets worse for the Lady Vols.

UT sophomore Sidney Spencer is out for the rest of the season with a torn right ACL she suffered in practice Wednesday. Spencer, the fourth Lady Vol sidelined with a knee injury this season, came down on her knee wrong during practice and had to be carried off the floor.

25.1 Organization of the Body

I f you watch sports on TV, it won't take more than a game or two to hear the term *ACL*, as in: "She's recovering from that ACL tear," or "Oh, boy, it looks like the left knee is the trouble; hope it's not an ACL injury." Scarcely anyone ever says "anterior cruciate ligament," of course, partly because it's such a mouthful of a term and partly because ACL is such a familiar acronym. But here's a question even for those who know that ACL

New York Giants defensive end Keith Washington lies on the field after tearing his ACL in a game against the Chicago Bears.

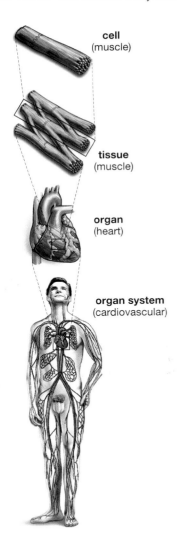

cell
(muscle)

tissue
(muscle)

organ
(heart)

organ system
(cardiovascular)

FIGURE 25.1 **Four Levels of Organization in the Human Body** A group of cells that performs a common function is a tissue; here muscle cells have formed heart muscle tissue (whose common function is contraction). Two or more tissues combine to form an organ such as the heart. The heart is one component of the cardiovascular system; other components are blood and blood vessels. All the organ systems combine to create an organism, in this case a human being.

stands for anterior cruciate ligament: What's a ligament? If you have trouble coming up with an answer right away, the mystery may deepen when *ligament* is paired with another term. An athlete might also get a strained or ruptured *tendon*, in the knee or elsewhere. So, what's the difference between a tendon and a ligament? You'll get details on both terms soon, but here's the short answer. A tendon is a tissue that attaches bone to *muscle*, while a ligament is a tissue that attaches bone to *bone*. So your ACL attaches your thigh bone (femur) to your shin bone (tibia) and the result is a good, tight linkage that allows quick "cuts" on the football or soccer field.

Terms like *ligament* and *tendon* serve to remind us that our bodies manage to be both familiar and foreign to us at the same time. Our heart starts to pound and we know it; sweat pours out of us after a workout and we can feel it; our stomach is upset and there's no mistaking it. Yet how does our heart beat, where does sweat come from, and how does our stomach operate? So sealed off is our body from us that these processes might as well be taking place on the moon for all the access we have to them. Ironically, to find out what's going on within the confines of our own skin, we must turn to books and CDs and television programs. A world of information is available from these sources, however, thanks to the generations of scientists who pieced together the story of how the human body operates. The purpose of this chapter, and the four that follow, is to walk you through a portion of this story. How do we breathe? What is blood made of? How does a newly formed embryo implant in a mother's womb? In the chapters coming up, you'll find out.

Levels of Physical Organization

We'll start our tour of the human body by thinking about how it is organized. Just as, in our everyday world, several individuals may make up an office, several offices a department, and several departments a working company, so in nature each living thing is built up from a series of individual units that go on to make up a working organism. Here's how these units are organized if we consider just the animals among life's creatures (one of which is us). Recall from Chapter 4 that the basic unit of all life is the cell. A group of cells that performs a common function is known as a **tissue**. There is, for example, muscle tissue—a group of cells that performs the common function of contracting. Tissues in turn are arranged into various types of **organs**, meaning complexes of several kinds of tissues that perform a special bodily function. The stomach is an organ that has both muscle and nerve tissue within it. Organs and tissues then go on to form **organ systems**, which are groups of interrelated organs and tissues that serve a particular function. Contractions of the heart push blood into a network of blood vessels. The heart, blood, and the blood vessels form the cardiovascular system, which is one of 11 systems in the human body. FIGURE 25.1 shows you these levels of structural organization within the body, using the cardiovascular system as an example.

25.2 The Human Body Has Four Basic Tissue Types

Having seen how the body's levels of organization fit together, let's now look in somewhat greater detail at each of the levels, starting with tissues. What kinds of tissues are there in the human body? It turns out that there are only four fundamental tissue types: epithelial, connective, muscle, and nervous. All other tissue varieties are subsets of these fundamental types. We'll now look briefly at the characteristics of these four varieties.

Epithelial Tissue

The key word in epithelial tissue is *surfaces*, because **epithelial tissue** is tissue that covers an internal or external surface. One such surface is the exterior of our body and, as such, the outer layer of our skin is composed of epithelial tissue. But there are other surfaces as well. Think of a pipe; it has not only an external surface but also an internal surface—the portion that comes into contact with the fluid that runs through it. Similarly, your body has arteries and veins that have the fluid called blood running through them. And the tissues that line the internal blood-vessel surfaces are epithelial tissues. Likewise, the lining of your stomach is a surface composed of epithelial tissues.

Given their location, epithelial tissues often serve as a barrier, as in the case of our skin; but many epithelial tissues are "transport" tissues that materials are moved *across*. (The food that enters your bloodstream does so by being transported across the epithelial tissue that lines your small intestines.) Epithelial tissues can be several layers of cells thick, or as thin as a single layer, depending on their function (see FIGURE 25.2 on page 452).

Epithelial tissues often produce substances that aid in various bodily processes. The skin has epithelial tissue that produces the water-resistant protein keratin, for example, while the stomach has epithelial tissue that produces digestive juices. This substance-producing function can be carried out by isolated epithelial cells, but it is sometimes undertaken by concentrations of cells. Such concentrations are known as **glands**: organs or groups of cells specialized to secrete one or more substances.

Connective Tissue

Connective tissue is tissue that stabilizes and supports other tissue, and it is very different from epithelial tissue. Whereas epithelial tissue always covers a surface, connective tissue never does. Whereas epithelial tissue is almost completely composed of cells, connective tissue usually is composed of cells that are separated from each other by an extracellular material—a material that is secreted by the connective-tissue cells themselves. Indeed, the prime function of many connective tissues is to produce this kind of material. A good example of this is bone tissue, which is composed only in small part of bone *cells*. Most of the material we think of as bone is a mixture of protein fibers and calcified material that bone cells secrete.

Muscle Tissue

Our third variety of tissue, **muscle tissue**, is tissue that is specialized in its ability to contract, or shorten. There are three kinds of muscle tissue—skeletal, smooth, and cardiac—which differ in their structure and in the way they are prompted to contract. **Skeletal muscle** is muscle that is attached to bone and is contained in, for example, our biceps. It is under our conscious control and has a striped or "striated" appearance when looked at under a microscope, owing to the parallel orientation of the cells that make it up. **Cardiac muscle** exists only in the heart and likewise is striated, but it contracts under the influence of its own pacemaker cells. **Smooth muscle**, which gets its name from its lack of striated appearance, is muscle responsible for contractions of blood vessels, the uterus, the passageways of the lungs, and other structures. As you might guess from this list of tasks, smooth muscle is not under voluntary control.

Nervous Tissue

Nervous tissue is tissue that is specialized for the rapid conduction of messages, which take the form of electrical impulses. There are two basic types of nervous-tissue cells: neurons, which actually carry the nervous-system messages; and glial cells, which perform support functions for the neurons.

A BROADER VIEW

Deep Inside, but Vulnerable

Epithelial tissue can lie deep inside us and yet be exposed to substances from the environment outside us, which means that this tissue needs to protect itself. The epithelial cells that line our lungs have hair-like extensions on their surface, called cilia, that constantly move back and forth like so many tiny brooms, working to get rid of any harmful material that comes in with the air we inhale. Likewise, our stomach, which receives food from the outside world, secretes acids that can kill most of the harmful microbes that hitch a ride in on the food.

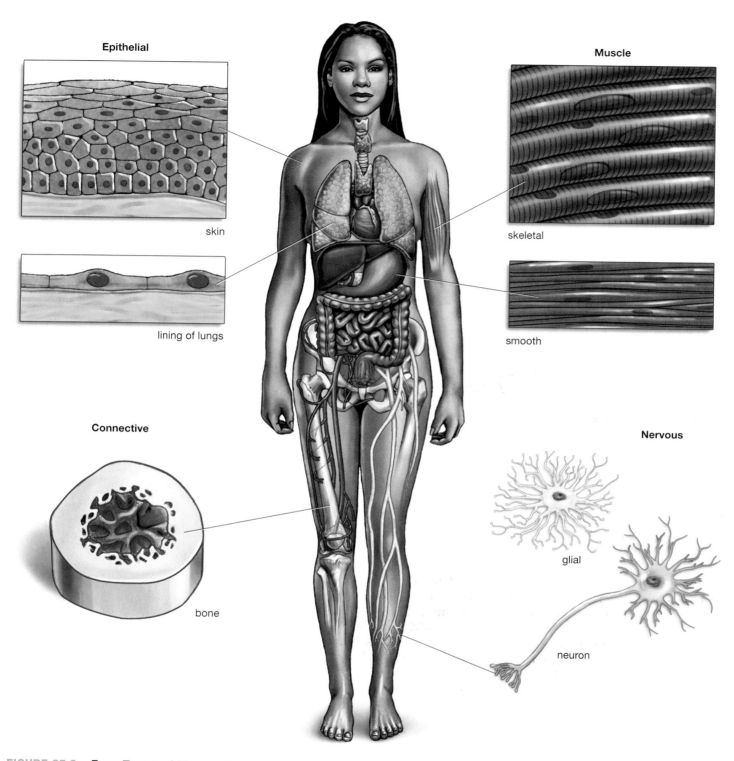

Epithelial

skin

lining of lungs

Muscle

skeletal

smooth

Connective

bone

Nervous

glial

neuron

FIGURE 25.2 **Four Types of Tissue** There are four basic tissue types in the human body: epithelial, connective, muscle, and nervous. Pictured are examples of each of the four types in some representative locations. Epithelial tissue, which always covers a surface, can take the form of many layers of cells (as with the skin) or a single layer of cells (as with the lining of lungs). Connective tissue, which supports and stabilizes other tissue, can take such diverse forms as bone (pictured) or fat. Muscle tissue, which is specialized for contraction, comes in a striped or "striated" form (in both skeletal and cardiac muscle) and in a smooth form that lines the digestive tract, blood vessels, and other structures. Nervous tissue is specialized for electrical signal transmission. The two types of nerve cells are neurons, which conduct nervous-system signals, and glial cells, which support the neurons.

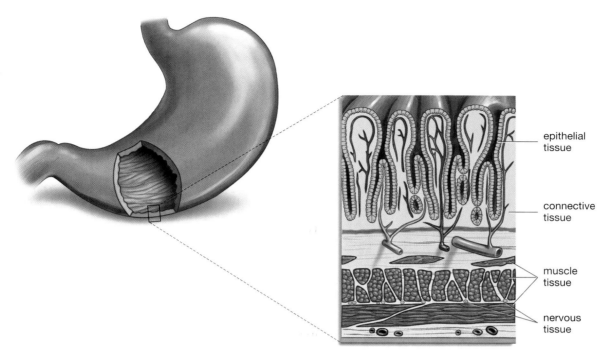

FIGURE 25.3 **Organs Are Composed of Tissues** The stomach is a typical organ in that it is composed of various tissue types. Nervous tissue controls the release of digestive juices into the stomach and also controls the contractions of the stomach's digestive muscle tissues. Connective tissue helps support and stabilize the other tissues around it. Epithelial tissue lines the stomach's surface, which comes into contact with the food we eat. The individual epithelial cells pictured secrete digestive acids and enzymes.

25.3 Organs Are Made of Several Kinds of Tissues

Having looked at the four tissue types, let's now see how they combine to form an organ. FIGURE 25.3 shows you the anatomy of the human stomach, with an emphasis on the kinds of tissues that go into making it up. The muscle tissue noted in the figure is one of several layers of smooth muscle that bring about contractions that help digest food. Likewise, there is nervous tissue, which controls the release of digestive juices and the contractions of the stomach's muscles. Supportive connective tissue exists as well. And, of course, since the inner lining of the stomach is a surface, there is epithelial tissue. True to what you saw earlier, some of this epithelial tissue is secretory—that is, it secretes substances that help with digestion.

25.4 Organs and Tissues Make Up Organ Systems

You've now seen how cells make up tissues, and how tissues make up organs. Let's take the final step up and see how tissues and organs together make up organ systems. What follows is a brief look at each of the body's 11 organ systems. Once you have finished reviewing them, the balance of this chapter will be devoted to a more detailed look at the first three of these systems. In Chapters 26 through 29, you'll be taking a look at the other eight.

(a) The integumentary system

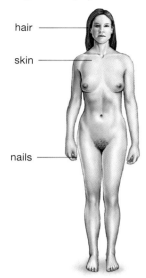

hair

skin

nails

(b) The skeletal system

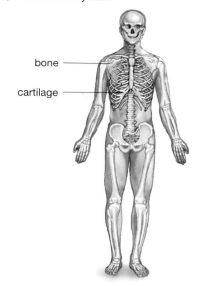

bone

cartilage

(c) The muscular system

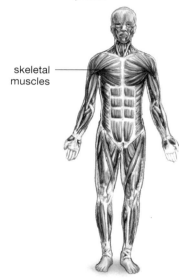

skeletal
muscles

(d) The nervous system

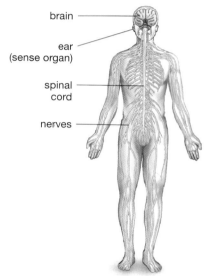

brain

ear
(sense organ)

spinal
cord

nerves

(e) The endocrine system

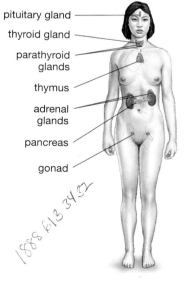

pituitary gland

thyroid gland

parathyroid
glands

thymus

adrenal
glands

pancreas

gonad

(f) The immune system
(lymphatic network)

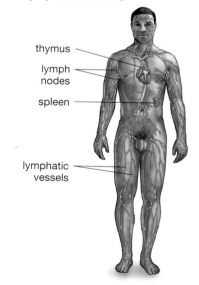

thymus

lymph
nodes

spleen

lymphatic
vessels

Organ Systems 1: Body Support and Movement—the Integumentary, Skeletal, and Muscular Systems

- The **integumentary system** (FIGURE 25.4a) is sometimes thought of simply as skin. But a number of structures are associated with skin, including hair and nails and the secretory glands we noted earlier. In addition to serving as an outer, protective wrapping, skin also helps keep body temperature stable.

- The **skeletal system** (FIGURE 25.4b) is composed not only of bones, but of cartilages and ligaments. This is a complex framework, as human beings have 206 bones in their body, along with a large number of cartilages. Bones provide support and protection, and they can store fat and minerals.

- The **muscular system** (FIGURE 25.4c) includes all the skeletal muscles of the body—about 700 of them. These muscles provide for movement, posture, and support, and their contraction warms our bodies. The smooth and cardiac muscle tissues noted earlier are not thought of as part of the muscular system. Cardiac muscle is part of the cardiovascular system, and smooth muscle is part of several systems—in the stomach, for example, it is part of the digestive system.

Organ Systems 2: Coordination, Regulation, and Defense—the Nervous, Endocrine, and Immune Systems

- The nervous tissue reviewed earlier goes on to make up the **nervous system** (FIGURE 25.4d), a rapid communication system in the body that includes the brain, spinal cord, all the nerves outside the brain and spinal cord, and sense organs such as the eye and ear.

- The endocrine system (FIGURE 25.4e) bears comparison with the nervous system, in that it is also in the business of sending signals throughout the body. The **endocrine system**, however, is a communication system that works through the chemical messengers called **hormones**. Many of these hormones are produced in specialized glands (such as the thyroid gland).

- The **immune system** consists of a collection of cells and proteins whose central function is to rid the body of invading microbes. In (FIGURE 25.4f) you can see a key component of the immune system, the body's lymphatic network, which serves a dual function. First, it acts as a kind of drainage system, collecting fluid that lies outside of cells and routing it back to the bloodstream. Second, it acts as an *inspection* system. Lymphatic organs contain heavy concentrations of the

FIGURE 25.4 **The Organ Systems of the Human Body**

body's defenders—its immune-system cells—and these cells monitor lymphatic fluid as it passes by for signs of invading microbes. Lymphoid organs include the well-known lymph nodes located under the arms and elsewhere, along with the thymus gland, tonsils, and spleen. Immune-system cells and proteins don't reside solely within the lymphatic network however, but exist throughout the body.

Organ Systems 3: Transport and Exchange with the Environment—the Cardiovascular, Respiratory, Digestive, and Urinary Systems

- The **cardiovascular system** (FIGURE 25.4g) consists of the heart, blood, and blood vessels and an inner, "marrow" portion of bones where red blood cells are formed. The cardiovascular system is the body's mass transit system. It carries nutrients, dissolved gases, and hormones *to* tissues throughout the body, and it carries waste products *from* tissues to the sites where they are removed: the kidneys and lungs.

- The **respiratory system** (FIGURE 25.4h) includes the lungs and the passageways that carry air to and from the lungs. Through this system, oxygen comes into the body and carbon dioxide is expelled from it.

- The central feature of the **digestive system** (FIGURE 25.4i) is the digestive tract, a long tube that begins at the mouth and ends at the anus. Along its course, there are the organs of the system—the stomach, small intestines, and large intestines—and accessory glands, such as the pancreas and liver.

- The **urinary system** (FIGURE 25.4j) has two major functions, one of them well known, the other not so much appreciated. The well-known function is the elimination of waste products from the blood through the production of urine. The underappreciated function is conservation. The system is very good at retaining what is useful to the body—water, proteins, and so forth—even as it eliminates what is useless. The primary organs that carry out this screening process are the kidneys, but the system also includes various other parts, such as the bladder.

The Reproductive System

- Males and females have different **reproductive systems** (FIGURES 25.4k and 25.4l). The two are linked, of course, in that together they produce offspring. Both systems are discussed in detail in Chapter 29.

(g) The cardiovascular system

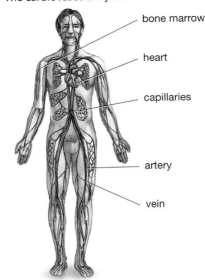

bone marrow
heart
capillaries
artery
vein

(h) The respiratory system

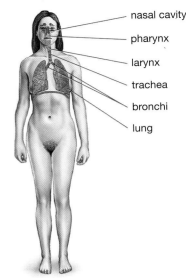

nasal cavity
pharynx
larynx
trachea
bronchi
lung

(i) The digestive system

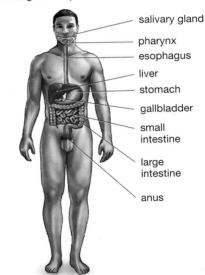

salivary gland
pharynx
esophagus
liver
stomach
gallbladder
small intestine
large intestine
anus

(j) The urinary system

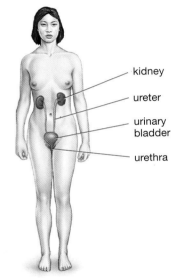

kidney
ureter
urinary bladder
urethra

(k) The male reproductive system

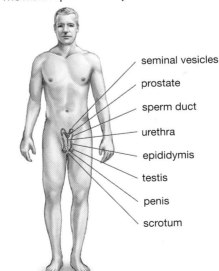

seminal vesicles
prostate
sperm duct
urethra
epididymis
testis
penis
scrotum

(l) The female reproductive system

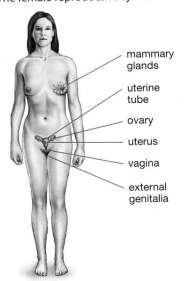

mammary glands
uterine tube
ovary
uterus
vagina
external genitalia

FIGURE 25.4 *(continued)*

With this brief review of the organ systems completed, you're ready to look in more detail at the first three organ systems noted earlier—the integumentary, skeletal, and muscular systems. They are batched together in this chapter because each of them has to do with body support and movement. We'll start with the integumentary system.

So Far...

1. Going from least inclusive to most inclusive, how would the following levels of organization be ordered in an animal? organ system, cell, tissue, organ

2. What are the four basic tissue types in a human being?

3. _____ tissue is specialized in its ability to contract, while _____ tissue is specialized in its ability to conduct electrical impulses.

25.5 The Integumentary System

The primary component of the integumentary system is the organ known as skin. It may seem strange to think of skin as an organ, since it is not a compact, clearly defined structure like the heart or the liver. But remember the definition of an organ: a complex of several kinds of tissues that performs a special bodily function. Clearly, that's what skin is. Joining skin to make up the integumentary system are the accessory structures of hair, nails, and a variety of glands.

The primary function of skin is easy to see: It covers the body and seals it off from the outside world. Skin also protects underlying tissues and organs, however, and helps regulate our temperature. Meanwhile, glands that are accessories to the skin excrete materials such as sweat, water, oils, and milk; and specialized nerve endings in the skin detect such sensations as pressure, pain, and temperature.

The Structure of Skin

Skin is an organ that is organized in two parts: a thin outer covering, the epidermis, and a thicker underlying layer, the dermis, that is composed mostly of connective tissue. Beneath the dermis—and not part of the skin proper—is another layer, called the hypodermis, made up of connective tissue that attaches the skin to deeper structures such as muscles or bones. You can see a cross section of human skin in FIGURE 25.5.

The Outermost Layer of Skin, the Epidermis

The outer layer of our skin, the epithelial tissue called the **epidermis**, would seem to be facing an impossible task. On the one hand, it has to serve as a permanent, protective barrier, keeping out everything from water to invading microorganisms; on the other it has to continually renew itself, given that it is cut, scraped, and simply worn away. How does it manage to do both things? The answer lies in the many *layers* of cells the epidermis is organized into. If we could take a micro-elevator down to the innermost layer of the epidermis, we would see a group of rounded cells rapidly dividing and, in the process, pushing the cells on top of them *up*, toward the surface of the skin. As we move toward the surface, we notice that the epidermal cells are flattening out. Furthermore, they're getting less active; at a certain point they cease dividing altogether. What is happening is that they are now so far from the underlying blood vessels (down in the dermis) that they are losing their blood supply. They are still carrying out a task, however, which is the production of **keratin**—a flexible, water-resistant protein, abundant in the outer layers of skin, that also makes up hair and fingernails. By the time epidermal cells reach the surface, they are scarcely cells at all anymore. They are dead and don't even have the remnants of internal organelles within them. Instead, they

So Far... Answers:
1. *cell, tissue, organ, organ system*
2. *epithelial, connective, muscle, nervous*
3. *Muscle; nervous*

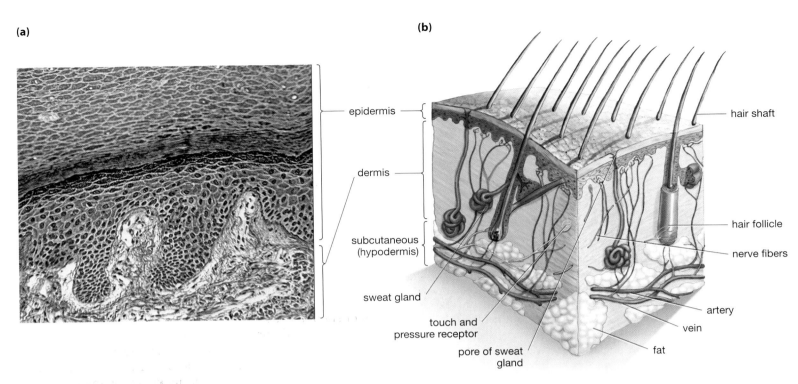

(a)

(b)

epidermis

dermis

subcutaneous (hypodermis)

sweat gland

touch and pressure receptor

pore of sweat gland

hair shaft

hair follicle

nerve fibers

artery

vein

fat

FIGURE 25.5 **Our Outer Wrap: What Is Skin Made Of?**

(a) A micrograph of human skin, showing the epidermal and dermal layers. Note how the cells of the epidermis flatten out about halfway toward the top.

(b) A cross section of the skin's epidermis and dermis layers, and the hypodermis layer beneath them (which is not part of the skin).

have become a series of tightly interlocked keratin sacs. In another two weeks they will be scraped or washed away, replaced by the next cells that have pushed up from the inner epidermis.

Beneath the Epidermis: The Dermis and Hypodermis

Whereas the epidermis is epithelial tissue, the inner layer of skin, the **dermis**, is mostly connective tissue. But if you look again at Figure 25.5, you can see that nerves and muscles exist in the dermis as well, along with sweat glands, hair follicles, and more. This is just another example of how all the various kinds of tissues tend to exist side by side in a typical organ.

Many of the structures in the dermis are serving the epidermis. For example, the nervous system touch and pressure receptors shown in Figure 25.5 allow us to know when something makes contact with our skin. Note also, in Figure 25.5, the nature of hair. Each of the 5 million or so hairs on the human body grows out of one of the tube-like hair follicles, seen in the figure, that begins fairly deep in the dermis. Each follicle is a tissue complex that has not only its own blood vessels and muscles, but its own nerve endings as well. If you rub your hand with just the slightest stroke over the hair on your head, you can get a sense of how exquisitely sensitive hair is to our sense of *touch*. The reason for this is the structure of the hair follicles: when you touch a hair, you stimulate the nerve endings of its follicle.

As noted earlier, the "hypodermis," the layer beneath the dermis, is not actually part of the skin. But it deserves special mention here because of one of the aspects of it you can see in Figure 25.5—its abundant fat cells (which technically are known as adipocytes). These fat cells make up the body's *subcutaneous fat*, which can be thought of as a layer of fat that varies in thickness throughout the body. The body also has what is known as *visceral fat*, however, which is fat that lies deeper in the abdomen, surrounding organs such as the liver. In the past 10 years, our conception of fat cells has changed radically, as you can see in Essay 25.1, "Fat Cells: Why Weight Matters," on page 458.

Finally, here's one more thing about the skin: If you want to make it age fast, one sure way to do it is to spend a lot of time in the sun. You can read more about this in Essay 25.2, "How Sun Exposure Ages the Skin," on page 460.

A BROADER VIEW
Why Tattoos Require Needles

Ever wonder why a tattoo must be made with a needle sunk into the skin? The cells in the skin's deeper layer, the dermis, are fixed in place, while those of the epidermis are constantly moving toward the skin's surface. For a tattoo to be permanent, it needs to be deposited into fixed cells, which means those of the dermis. The term *hypodermis* may ring a bell because it sounds like the more familiar term hypoder*mic*, as in needles. The fatty tissue of the hypodermis doesn't have much of a blood supply and thus is a good place to inject medicines that need a relatively slow, steady entry into circulation.

Essay 25.1 Fat Cells: Why Weight Matters

At one time, scientists thought of the body's billions of fat cells as little more than fuel depots—as storage sites for the calories we ingest but don't burn up. This model began to fall apart in 1994, however, with the discovery of a hormone called leptin that plays a role in regulating hunger and that is *produced* by fat cells. In the years that followed, scientists identified 11 more hormones whose source is fat cells, and the betting is that there are more of these substances yet to be discovered. The upshot is that fat cells are no longer regarded as passive warehouses; instead, they are thought of as active chemical factories. Indeed, many scientists now look at the body's collection of fat cells as a single hormonal *organ*, on a par with, say, the pancreas (**see FIGURE E25.1.1**). The problem is that the products of this organ are mostly bad news when a person has an excess of fat cells, which is to say, when a person is overweight. And who is overweight? Perhaps two-thirds of the adults in the United States. This proportion is so high—and the afflictions linked to fat cells so serious—that research into these cells is now a top public health priority.

Fat cells are no longer regarded as passive warehouses.

Consider type 2 diabetes, which comes about not because the body has stopped producing insulin, but because the body has become resistant to its effects. Insulin helps move glucose from the bloodstream into various cells in the body. Thus, when cells become resistant to insulin, glucose accumulates in the bloodstream. This is where the trouble starts, because elevated levels of glucose can result in nerve damage, heart disease, blindness, and more. Now, why is it that being overweight is the most important risk factor for developing type 2 diabetes? Well, it turns out that, of the dozen known hormones produced by fat cells, seven have the effect of increasing insulin resistance. One of these hormones, the well-named resistin, actually is produced in greater quantity in a type of immune-system cell, called a macrophage; but macrophage numbers turn out to be linked to . . . fat-cell numbers. Macrophages seem drawn to collections of excess fat cells, as if these cells were foreign invaders. When we become overweight, therefore, the result is not just more (and larger) fat cells, but more macrophages as well, which means increased production of resistin.

The connection between fat cells and the immune system then carries over into an equally serious health threat, heart disease. The old conception of heart disease was essentially one of a steady buildup of fatty tissue in one of the coronary arteries of the heart. When a sufficient amount of fatty issue accumulated, this view held, the result was a complete blockage of a coronary artery—a heart attack. We now know, however, that most heart attacks come about through a fatty buildup *coupled with* the immune-system response known as inflammation. And compounds that fat cells produce help bring about inflammation. Indeed, fat cells contribute to heart disease in several ways, as they also produce a compound (called angiotensinogen) that raises blood pressure.

The conception of fat cells as chemical factories has seemingly answered the question of why being overweight has so many serious health consequences. It's not the weight itself. It's that fat cells change character when they grow too large and too numerous. The set of chemical reactions they engender—their metabolism—goes from being benign to being harmful. So far, the only practical consequence of this knowledge is that it fills us in on the *details* of why it's desirable to lose excess pounds. In the long run, however, research on fat cells may allow us to intervene in fat-cell metabolism in such a way that we can control *it*, rather than having it control us.

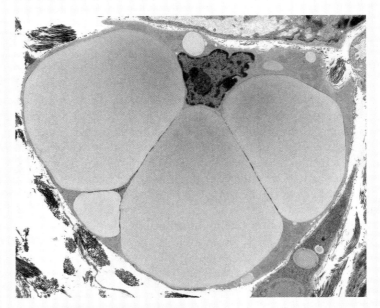

FIGURE E25.1.1   **The Fat Cell** Fat cells, or adipocytes, often contain several lipid droplets whose size dwarfs the cell's other components. In this case, the three large central droplets do just this. The cell's nucleus is the purple object near the top.

So Far...

1. The primary component of the integumentary system is an organ called _____, which is composed of two primary layers, an outer layer called the_____ and an inner layer called the _____.

2. The outermost cells of our skin are essentially a set of water-resistant sacs composed mostly of the protein _____.

3. Each hair that we have grows out of a tissue complex called a _____.

25.6 The Skeletal System

We now shift gears, going from a consideration of our outer covering to a consideration of the framework that underlies it—our skeletal system. This system has only three structural elements to it: bone, ligaments, and cartilage. Most of us are quite familiar with **bone**, which can be defined as a connective tissue that provides support and storage capacity, and that is the site of blood-cell production. Ligaments are fairly inaccessible to us; but if you look at FIGURE 25.6, you can see some typical ligaments carrying out a typical task—linking bones together. You can also see the third element of the skeletal system, cartilage. Note that cartilage is located *in between* bones in the foot, and this turns out to be one of its common functions. **Cartilage** is a connective tissue that serves as padding in most joints. It is more familiar to us, however, in other parts of the body. It forms our larynx (voice box), the C-shaped "rings" on our trachea (windpipe), our outer ears, and the tip of our nose.

It's a good idea to look at Figure 25.6 to get a sense not only of the difference between bone, ligaments, and cartilage but also of the difference between all of these skeletal elements and another type of tissue mentioned at the start of the chapter, tendons. Though not part of the skeletal system, tendons are always associated with the bones of the system. Looking at all four elements in the figure, here's one way to think of them. Bones are the basic framework; cartilage often serves as padding between bones; **ligaments** are tissues that join bone to bone; and **tendons** are tissues that join bone to muscle.

Function and Structure of Bones

It's tempting to think of bones as being nothing more than support beams, but this conception is wide of the mark in two ways. First, support beams are not alive, but bone is very much a living, dynamic tissue, as you'll see. Second, bones do more than just support us. Once we are past infancy, all of our blood cells (both red and white) are created in the interior of our bones. Some bones store fat, in the form of yellow marrow, as a kind of energy reserve; and bones serve as the storage sites for important minerals such as calcium and phosphate. Beyond this, the only reason we can move at all is that our muscles are attached to bones. When we contract our bicep, for example, our forearm is raised by means of the bicep pulling on the bones of the forearm.

The Structure of Bone Each bone in our body actually is considered a separate organ, because each bone is composed of several kinds of tissue. There is the calcified or "osseous" tissue we think of as bone, to be sure, but each bone also has its own blood vessels and nerves. (Cartilage, meanwhile, is composed mostly of a single kind of protein—a type of collagen—and it has no nerves and no blood vessels.)

As noted earlier, bones are a great example of connective tissue, in that bone *cells* don't make up much of bone tissue; cells may account for as little as 2 percent of a bone's mass. What is the other 98 percent of bone made of? It is overwhelmingly a cell-secreted substance that is composed of hard, calcium-containing crystals and

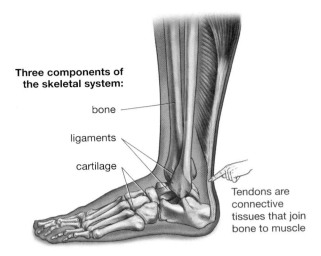

Three components of the skeletal system:

bone

ligaments

cartilage

Tendons are connective tissues that join bone to muscle

FIGURE 25.6 Elements of the Skeletal System
Bones are the basic element of the skeletal system. Ligaments link one bone to another, and cartilage often serves as padding between bones. Tendons, while not part of the skeletal system, always link bone to muscle.

So Far... Answers:

1. *skin; epidermis; dermis*

2. *keratin*

3. *follicle*

Essay 25.2 How Sun Exposure Ages the Skin

The French clothing designer Gabrielle "Coco" Chanel once remarked that "fashion changes, but style remains." Decades ago, however, Chanel herself helped launch a fashion that seems stubbornly resistant to change. The fashion is that of getting a suntan, but this practice might as well be called the fashion of damaging your skin so that in a couple of decades it will be saggy, wrinkled, and spotted.

Throughout the early part of the twentieth century, it was chic to have pale skin, rather than a tan, as a tan was a sign of someone who *had* to be outdoors—working a job. According to one popular account, Gabrielle Chanel (of Chanel No. 5 fame) got some sun on an ocean cruise in the 1920s, stepped off the boat tanned, and conveyed the message that it was chic to look that way. People in industrial countries were ready to hear this message because it came at a moment when their leisure time was increasing, meaning they had more opportunities to be out in the sun playing, rather than working.

These days, the constant message from health professionals is to *avoid* the sun, but this is a message society has not really taken to heart. Consider that there are an estimated 15,000 businesses in the United States devoted to *giving* people tans through exposure to the ultraviolet light that the sun delivers.

But in what ways does sun exposure harm the skin? The most serious thing it does is help cause skin cancer—three varieties of it, actually—but let's set this aside and focus on what it does to the appearance of the skin. By some estimates, 90 percent of the aging that comes about in the skin is attributable to sun damage. What does this damage look like? Well, a fair-skinned 50-year-old who has spent time in the sun is likely to have brown marks on the skin, commonly known as "liver spots." These marks, however, don't have anything to do with

the liver; they're purely caused by sun exposure. Then there are likely to be small white spots—patches of skin where it seems the pigment has just given up. This too is sun exposure. Finally, there is likely to be sagging and wrinkling of the skin. It turns out that just below the surface of the skin, in the layer called the dermis, there is a protein called elastin that does pretty much what it sounds like: It springs back to its original shape after having been bent or twisted in some way. Unfortunately, ultraviolet light breaks down elastin, so that after enough sun exposure, the skin loses that full, robust look—it sags and wrinkles.

> This practice might as well be called the fashion of damaging your skin so that in a couple of decades it will be saggy, wrinkled, and spotted.

But what about sunblocks? They are indeed recommended by dermatologists, with the understanding that they need to be SPF 15 or higher, applied in nice thick layers, and reapplied after activities such as swimming. The problem is that many people seem to stay in the sun long after their sunblock has worn off. The general advice of dermatologists is simply to avoid lengthy exposure to direct sunlight, particularly between the hours of 10 a.m. and 2 p.m. People who stay out of the sun may not end up with the most fashionable-looking skin over the next two weeks, but over the next two decades, they will be the clear winners.

tough, yet flexible collagen fibers. This combination means that bones are hard without being brittle—they can support our weight and yet can bend or twist a little without breaking.

A BROADER VIEW

Simple and Compound Fractures

Bones do break, of course. When they break "cleanly," without any bone protruding from the skin, a simple fracture has taken place. When bone does protrude through the skin, a compound fracture has taken place. Once a simple fracture has been set, it generally takes about six to eight weeks to heal.

Large-Scale Features of Bone   The typical features of a long bone such as the humerus (in the arm) are shown in FIGURE 25.7. A long bone has a central shaft and expanded ends. Within such a bone, there are two types of bone tissue. **Compact bone**, which is relatively solid, forms the outer portion of the bone (all the way around it, as you can see in Figure 25.7b). **Spongy bone** then fills the expanded ends, and is well named. It is porous enough that it contains another type of tissue, called marrow, that comes in two forms: *red marrow*, which gives rise to blood cells; and *yellow marrow*, made up of energy-storing fat cells. Yellow marrow also exists in the so-called *marrow cavity* of a long bone, which you can see in Figure 25.7a.

Dynamic Bone   *Dynamic* probably is not a word that most of us would associate with bones, but bone is, in fact, continually remaking itself. Three kinds of cells operate on bone tissue, one type building it up, one type maintaining it, and one type breaking it down. The bone "remodeling" that results is so extensive that it replaces our entire skeleton every 10 years. Moreover, this dynamism is affected by our activities. To the extent we are physically active, bone density increases; to the extent we are inactive, bone density decreases. People who take up an exercise such as jogging get denser leg bones. Conversely, people who are using crutches to take the weight off a broken leg bone may temporarily lose up to a third of the mass in this bone.

Interestingly, though we develop almost all our height by late adolescence, we are still gaining in bone density up until about the age of 30. Then, beginning in middle age, the body may start removing more bone than it adds. This phenomenon particularly afflicts women, in the form of a condition known as *osteoporosis*, meaning a thinning of bone tissue that can result in bone breakage and deformation. One way we can guard against osteoporosis is to reach the age of 30 with as much bone density as we can muster—the equivalent of putting money in the bank as a hedge against a later withdrawal. And there are a couple of proven ways to get more bone density before 30: Take up moderate programs of "weight-bearing" exercise such as jogging or walking, and modestly increase calcium intake.

Joints

Joints exist wherever two bones meet. What qualities are required of a joint? Well, some only need strength. We think of our skull as being a single bone; but in fact, it's composed of several bones that interlock with each other in joints that are immovable. For other joints, flexibility is the key requirement. The ball-and-socket joint at our shoulder permits a range of motion so extensive that movement of our upper arms is limited more by our shoulder muscles than by our bones.

Highly movable joints are typically found at the ends of long bones, such as those in the legs as well as the shoulder. You can see a view of such a joint—the human knee joint—in FIGURE 25.8. A joint such as this must serve two competing functions. On the one hand, it must keep the massive femur (thigh bone) and large tibia (shin bone) in a stable arrangement. On the other, it serves as a hinge that allows the tibia to swing as free as a screen door off the femur. How does it do both things? Well, the two bones are partly held in place by four main ligaments, some of which are visible in the figure. Recall that ligaments link bone to bone, but there are also tendons, which attach bone to muscle; the knee joint has these in place as well.

With respect to the motion that such a joint allows, a critical element is that one bone does not actually touch another. Instead, an extensive network of padding and lubricating

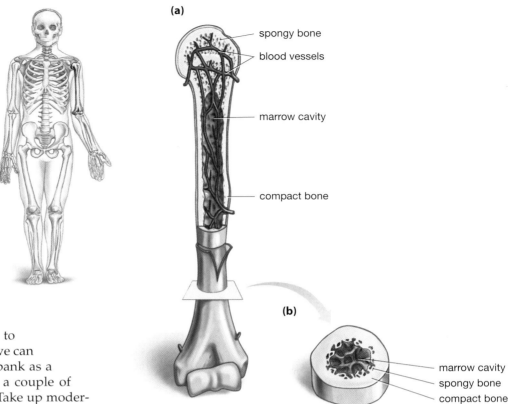

(a)

- spongy bone
- blood vessels
- marrow cavity
- compact bone

(b)

- marrow cavity
- spongy bone
- compact bone

FIGURE 25.7 **Large-Scale Features of a Typical Bone**

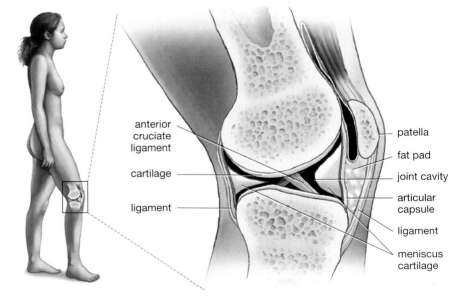

- anterior cruciate ligament
- cartilage
- ligament
- patella
- fat pad
- joint cavity
- articular capsule
- ligament
- meniscus cartilage

FIGURE 25.8 **A Highly Movable Joint** A simplified view of the human knee joint. Note that the two bones never touch each other, because they are covered with cartilage. The joint has shock-absorbing cartilage in the form of the meniscus and a fat pad that protects the cartilage.

A BROADER VIEW

Avoiding Arthritis

One of the biggest health problems in America is a problem with joints: the affliction known as osteoarthritis, which is a slow, steady degeneration of hip, knuckle, or knee joints. Osteoarthritis generally appears later in life, but young people can take steps to avoid it. Risk factors for the condition include being overweight and injuring the joints through such contact sports as football or soccer.

tissues in between the bones performs the same function as the "gel" found in running shoes. Note that the top of the tibia and bottom of the femur are surrounded by cartilage, and that more padding lies on each side of the joint in the form of the meniscus cartilage (which is a frequent site of sports injuries). The yellow tissue shown is fat padding that fills in spaces that are created when the bones move.

25.7 The Muscular System

Bones may provide a scaffolding for the body, but how is it that this scaffolding is capable of movement? The answer is muscles, specifically the skeletal muscles that were introduced earlier, which are numerous and large enough that they account for about 40 percent of our body weight. Like bones, skeletal muscles are organs in that they contain a variety of tissues. At each end of a muscle, fibers of the outer muscle layer come together to form the tendons that attach muscle to bone.

The Makeup of Muscle

As you can see in FIGURE 25.9, each muscle has within it a number of oval-shaped bundles called fascicles. Inside each fascicle there is a collection of muscle cells; but the term *cell* here may be misleading, because any one of these cells can be as long as the muscle itself—a gigantic length relative to the strictly microscopic dimensions of most cells. Because skeletal muscle cells are so elongated, they are referred to as muscle fibers. Along their length, muscle fibers are divided into a set of about 10,000 repeating units, called **sarcomeres**, that are the fundamental units of muscle contraction. But what is it that's contracting in a muscle? A look inside a single muscle fiber reveals that it is composed of more strands—perhaps a thousand long, thin structures called myofibrils that run the length of the cell. Each myofibril, in turn, has a large collection of two kinds of strands inside it that alternate with one another; these are thin filaments made of the protein *actin* and thick filaments made of the protein *myosin*. It is these filaments that contract, as you'll now see.

FIGURE 25.9 **Structure of Skeletal Muscle** Skeletal muscles have oval-shaped bundles within them, called fascicles, that are composed of several muscle cells, each one of which is so elongated it is referred to as a "fiber." Each fiber is composed of many thin myofibrils, and each myofibril has within it a collection of alternating thick and thin protein strands called filaments. The thin filaments are made of the protein actin, the thick filaments of the protein myosin. Each myofibril is divided up lengthwise into a series of repeating functional units called sarcomeres.

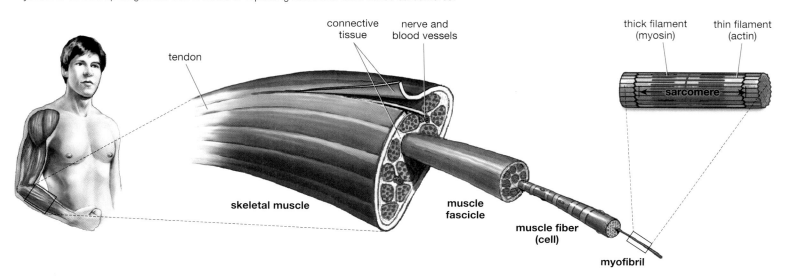

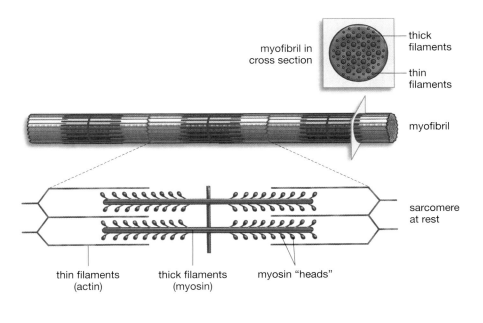

(a) Muscle at rest

The myofibril contains several sarcomeres, one of which is shown at rest. The thin filaments within the sarcomere, made of actin, overlap with adjacent thick filaments, made of myosin. The thick filaments have club-shaped outgrowths, called heads. The myofibril is also shown in cross section, at a location where thick and thin filaments overlap.

myofibril in cross section

thick filaments

thin filaments

myofibril

sarcomere at rest

thin filaments (actin)

thick filaments (myosin)

myosin "heads"

(b) Muscle contraction

Muscle contraction is a matter of the thin actin filaments sliding together within a sarcomere. This comes about when the myosin heads attach to the actin filaments and pull them toward the center of the sarcomere. The myosin heads pivot at their base and are capable of alternately binding with, or detaching from, the thin filaments.

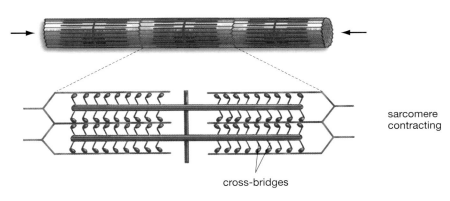

sarcomere contracting

cross-bridges

FIGURE 25.10 How a Skeletal Muscle Contracts

How Muscles Work

To understand contraction, note first that the thick myosin filaments lie in the *center* of a given sarcomere. Meanwhile, the thin actin filaments are attached to either *end* of the sarcomere and extend toward the center, where they overlap with the thick filaments (**see FIGURE 25.10**). In contraction, the thin filaments slide toward the center, causing the unit to shorten—which is to say, causing the muscle to contract. To understand this, place just the tips of the fingers of your right hand in between the tips of the fingers of your left hand. Now slide your right hand as far as it will go to the left. Note that neither set of *fingers* shortened in length, but that the total distance made up by both sets of fingers did shorten—exactly what happens within a sarcomere.

But what enables the thin filaments to slide? The thick myosin filaments have numerous club-shaped "heads" extending from them. These myosin heads are capable of alternately binding with or detaching from the adjacent thin filaments, creating so-called *cross-bridges*. The myosin heads pivot at their base, as if they were on a hinge. Here's the order of their activity: Attach to the actin filament, pull it toward the center of the sarcomere, detach from the actin, reattach to it, and pull again. In actual muscle contraction, this process happens many times in the blink of an eye; at any one point in a contraction, some myosin heads will be pulling an actin filament while others are detaching from it and still others binding to it.

The cycle that runs from myosin binding, through pulling, and then detaching is known as a *twitch*. The muscle fibers that go through this cycle come in two varieties—fast-twitch and slow-twitch—whose names describe how long it takes

A BROADER VIEW

Is Stretching Helpful?

Stretching muscles would seem to be a common-sense way of avoiding muscle injuries. But a 2004 analysis of 361 studies on stretching concluded that there is no reason to believe that stretching—either before or after exercise—does anything to ward off injuries. As the authors of the analysis noted, stretching elongates muscles beyond their normal range of motion, but most exercise injuries happen while a muscle is within its normal range of motion.

Modern Myths

Staying Slim by Replacing Fat with Muscle

It's an attractive idea, and it's based on something that is true about metabolism: that muscle tissue, while at rest, burns up more calories than fat tissue does. And it's also true that people who engage in regular, long-term weight-lifting programs tend to add muscle tissue as a *replacement* for fat tissue. The idea that stems from this is that, if we lift weights, the muscle tissue we add should help us keep excess pounds off, around the clock, simply by virtue of burning more calories while at rest than the fat tissue it replaced.

One problem with this idea is that any amount of muscle we can add to our bodies is small compared to the amount of muscle we have. Suppose a woman weighs 120 pounds. On average, about 48 pounds of this weight will be muscle. Now, how much muscle is she likely to add through regular, long-term weight training? To be generous, maybe 4.5 pounds—about 9 percent of her

original amount. And how many calories is this added muscle likely to burn per day when at rest? About 26, which is about the same amount our woman could get from a couple gulps of a soft drink.

The problem, as you can see, is that muscles at rest just don't burn that many calories. A kilogram (2.2 pounds) of skeletal muscle burns about 13 calories per day when at rest. This is almost 10 times the calories that a kilogram of fat tissue will burn, but it's not enough to make much of a difference in a person's resting metabolic rate unless huge amounts of muscle are being added.

None of this is to say, however, that there's no benefit to weight lifting. In fact, there are four benefits: Weight lifting burns up a lot of calories during the course of exercise, it increases bone density, it increases strength, and it appears to increase the body's sensitivity to insulin, thus helping people avoid type 2 diabetes.

A BROADER VIEW

How We Bulk Up

So, how does a muscle grow bigger and stronger when we exercise it? Vigorous contraction of a muscle fiber creates a series of microscopic tears in it. Once this happens, satellite cells lying outside a fiber get a "repair needed" signal and begin to multiply rapidly. Eventually, some of the new cells fuse with the existing fiber. Unlike most cells, muscle fibers contain not one cell nucleus, but many. With the process of fusion, new nuclei are added to the existing fiber. These additional nuclei allow the fiber to ramp up its production of proteins, which are then used to create new myofibrils within the fiber. Thus, the fiber bulks up because of its new myofibrils, and it strengthens because these myofibrils enhance its ability to contract.

them to complete a cycle. Fast-twitch fibers are not just faster than slow-twitch fibers, however; they also can generate more power. But there is a catch to this, in that fast-twitch fibers can generate power only for brief periods. In contrast, slow-twitch fibers generate less power but can do it over a sustained period of time. Thus, fast- and slow-twitch fibers are something like the hare and tortoise of muscle tissue. Most muscles contain a mixture of fast- and slow-twitch fibers. We simply utilize different proportions of each kind of fiber, depending on what we are doing. As you might guess, a long-distance runner primarily utilizes slow-twitch fibers, while a sprinter primarily uses fast-twitch. Finally, the fact that muscle tissue burns more calories than fat has led to a common misconception about the ability of weight lifting to keep a person slim, something you can read about in the "Modern Myths" box, above.

On to the Nervous and Endocrine Systems

Having looked at the basic characteristics of the body and at three of its organ systems, we're now ready to move on to two other systems, both of which function in communication. These are the nervous and endocrine systems.

So Far...

1. The three elements of the skeletal system are _____, _____, and _____.

2. Name two functions of bone other than support.

3. Muscle contraction works through the process of a _____ of two kinds of muscle filaments past each other. One of these filaments is a relatively thin variety called _____, while the other is a relatively thick variety called _____.

So Far... Answers:

1. *bones; cartilage; ligaments*
2. *production of blood cells; storage of energy reserves as fat*
3. *sliding; actin; myosin*

Chapter Review

Summary

25.1 Organization of the Body

- **Cells to organ systems**—Animal bodies function through a series of smaller-to-larger working units whose order runs as follows: cells, tissues, organs, and organ systems. Tissues are groups of cells that have a common function. Organs are complexes of several kinds of tissues that perform a special bodily function. Organ systems are groups of organs and tissues that perform a given function.   (page 450)

25.2 The Human Body Has Four Basic Tissue Types

- **Epithelial tissue**—There are four fundamental tissue types in the human body: epithelial, connective, muscle, and nervous. Epithelial tissue covers surfaces and forms glands, which are organs or groups of cells that secrete one or more substances.   (page 451)

- **Connective tissue**—Connective tissue supports and protects other tissues. A prime function of most connective-tissue cells is to produce extracellular material. For example, bone cells secrete a connective material that largely forms the solid bones we are familiar with. (page 451)

- **Muscle tissue**—Muscle tissue is specialized in its ability to contract. There are three primary types of muscle tissue: skeletal, cardiac, and smooth. Skeletal muscle is always attached to bone and is under voluntary control. Cardiac muscle exists only in the heart and contracts under the control of its own pacemaker cells. Smooth muscle is "smooth" because it lacks the striations that skeletal muscle has; it is not under voluntary control and is responsible for contractions in blood vessels, air passages, and organs such as the uterus.   (page 451)

- **Nervous tissue**—Nervous tissue is specialized for the rapid conduction of electrical impulses. It is made up of two basic types of cells: neurons, which actually transmit the impulses; and glial cells, which are support cells for neurons.   (page 451)

25.3 Organs Are Made of Several Kinds of Tissues

- **Tissues and organs**—The stomach provides an example of how tissue types come together to form a working organ. A typical cross section of the stomach is made up of muscle, nervous, connective, and epithelial tissue.   (page 453)

25.4 Organs and Tissues Make Up Organ Systems

- **Integumentary, skeletal, muscular**—There are 11 organ systems in the body. The first three of these—the integumentary, skeletal, and muscular systems—function in body support and movement. The integumentary system is made up of skin and such associated structures as glands and hair. The skeletal system is made up of bones, ligaments, and cartilage. The muscular system includes all the skeletal muscles, but not the cardiac or smooth muscles.   (page 454)

- **Nervous, endocrine, immune**—The nervous, endocrine, and immune systems function in communication, regulation, and defense of the body. The nervous system is a rapid communication system that includes the brain, spinal cord, all the nerves, and sense organs such as the eye and ear. The endocrine system is a communication system that works more slowly, through the substances called hormones, many of which are produced in specialized glands. The immune system is a network of cells and proteins whose central function is the elimination of microbial invaders. The immune system works in part through the body's lymphatic network, whose vessels collect extracellular fluid and deliver it to the blood vessels. Immune cells within lymph glands inspect lymph as it passes by to identify invading microbes.   (page 454)

- **Cardiovascular, respiratory, digestive, urinary**—The cardiovascular, respiratory, digestive, and urinary organ systems function in transport and in exchange with the environment. The cardiovascular system consists of the heart, blood, blood vessels, and the inner "marrow" portion of the bones where red blood cells are formed. The respiratory system includes the lungs and the passageways that carry air to them. The central feature of the digestive system is the digestive tract, a tube that begins at the mouth and ends at the anus. Along its course, there are digestive organs of the system: the stomach, small intestines, and large intestines and such accessory glands as the liver and pancreas. The urinary system functions to eliminate waste products from the blood through the formation of urine and to retain substances that are useful to the body. It is made up of the kidneys and various other structures, such as the bladder.   (page 455)

25.5 The Integumentary System

- **Skin**—The primary component of the integumentary system is the organ called skin. Integumentary structures associated with skin are hair, nails, and a variety of glands.   (page 456)

- **Functions of skin**—In addition to covering the body, skin protects underlying tissues and organs and regulates heat loss. Glands associated with skin excrete such materials as water, oils, and milk. Specialized nerve endings in the skin detect touch, pressure, pain, and temperature.   (page 456)

- **The structure of skin**—Skin has a thin outer layer, the epidermis, composed of epithelial tissue; and a thicker underlying layer, the dermis, composed mostly of connective tissue. Beneath the dermis—and not part of the skin—is a third layer of tissues, the hypodermis. Epidermal skin cells are constantly worn away and replaced by new epidermal cells being pushed up from the inner epidermis. As they move toward the surface, these cells are transformed into a series of dead, tightly linked sacs made of the protein keratin. The dermis is filled with accessory structures, such as hair follicles and sweat glands. (page 456)

25.6 The Skeletal System

- **Elements of the system**—The human skeletal system is composed of bone, ligaments, and cartilage. Bone is a connective tissue that provides support and storage capacity, and that is the site of blood-cell production. Cartilage serves as padding in most joints, and forms our larynx (voice box) and trachea (windpipe). Ligaments are tissues that join bone to bone. Tendons link bone to muscle but are not part of the skeletal system.   (page 459)

- **Bone support**—Each bone of the skeleton is an organ that contains not only connective tissue, but blood vessels and nervous tissue as well.   (page 459)

- **Structure of bone**—The typical bone features a long central shaft and expanded ends. There are two types of bone tissue: compact bone, which is relatively solid and forms the outer portion of the bone; and spongy bone, which is less dense and fills the ends. Spongy bone is filled with marrow, which comes in two forms: red marrow, which gives rise to blood cells; and yellow marrow, which is composed of energy-storing fat cells. Long bones have a central marrow cavity that is filled with yellow marrow.   (page 459)

- **Bone dynamics**—Three different types of cells continually build up, maintain, and break down bone tissue. The bone remodeling that results is sensitive to our activities; weight-bearing use of a bone increases its density, while inactivity decreases its density.   (page 460)

- **Joints**—Joints exist wherever two bones meet. Some joints provide a strong linkage between bones, while others provide great flexibility. Ligaments (joining bone to bone) and tendons (joining bone to muscle) help provide the stability that large-bone joints need. Normally, the bony surfaces of highly movable joints do not meet directly, but are instead covered with special cartilages and pads that provide shock absorption and lubrication.   (page 461)

25.7 The Muscular System

- **Muscle fibers**—A given skeletal muscle cell can be as long as the muscle itself; because of their elongation, these cells are called muscle fibers. Along their length, muscle fibers are divided into a set of repeating units called sarcomeres, which are the basic units of muscle contraction. Each muscle fiber is composed of thin structures, called myofibrils, that are in turn composed of assemblies of two kinds of protein strands that alternate with one another: thin filaments made of the protein actin, and thick filaments made of the protein myosin. (page 462)

- **Muscle contraction**—Actin filaments attached to the end of each sarcomere overlap with the myosin filaments that lie in the center of each sarcomere. The myosin filaments bring about contraction by attaching to the actin filaments, pulling them toward the center of the sarcomere, detaching, and then pulling again. As the actin filaments slide toward the center of the sarcomere, the sarcomere shortens. (page 463)

Web Tutorial 25.1 Contraction of Skeletal Muscle

Key Terms

bone p. 459	**epidermis** p. 456	**muscular system** p. 454	**skeletal system** p. 454
cardiac muscle p. 451	**epithelial tissue** p. 451	**nervous system** p. 454	**skin** p. 456
cardiovascular system p. 455	**gland** p. 451	**nervous tissue** p. 451	**smooth muscle** p. 451
cartilage p. 459	**hormone** p. 454	**organ** p. 450	**spongy bone** p. 460
compact bone p. 460	**immune system** p. 454	**organ system** p. 450	**tendon** p. 459
connective tissue p. 451	**integumentary system** p. 454	**reproductive system** p. 455	**tissue** p. 450
dermis p. 457	**keratin** p. 456	**respiratory system** p. 455	**urinary system** p. 455
digestive system p. 455	**ligament** p. 459	**sarcomere** p. 462	
endocrine system p. 454	**muscle tissue** p. 451	**skeletal muscle** p. 451	

Testing Your Understanding

In Your Own Words *(answers in the back of the book)*

1. What role does the lymphatic network play in immune function?

2. What makes our skin water-resistant?

3. People who have torn their anterior cruciate ligament report that their knee feels "wobbly" until they get it repaired. Why would this be?

Thinking about What You've Learned

1. Cells called melanocytes that lie deep in the epidermis produce a pigment, called melanin, that gives skin its color. Exposure to ultraviolet light—from the sun or a tanning lamp—causes melanocytes to produce more melanin, which they pass along to the skin cells above them, yielding a suntan. But why would sun exposure prompt this increased melanin production? What is the function of a tan, in other words?

2. Bone density increases in response to the stresses on bone that are provided by exercises such as running. Imagine two fairly inactive people who decide to start exercising. One becomes a runner, the other becomes a swimmer. Who experiences the greater increase in bone density: the swimmer (in the arm bones) or the runner (in the leg bones)?

3. We think of hair as material that primarily serves to keep us warm. If so, then why should hair be so finely tuned to our sense of touch? Given this sensitivity, what other function might hair have served for our evolutionary ancestors? (Hint: Think of a cat's whiskers.)

The Nervous and Endocrine Systems

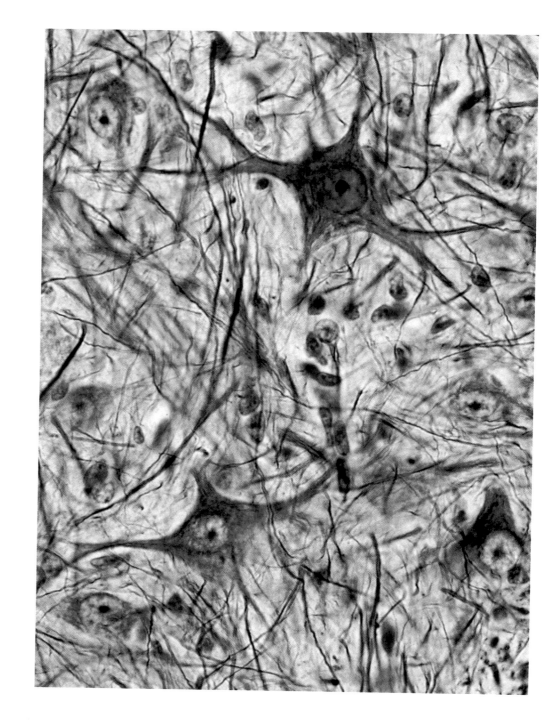

A micrograph of a portion of the human brain, showing the nerve cells called neurons (in brown) and their associated support cells (in purple).

THE NEW YORK TIMES

MCGWIRE OFFERS NO DENIALS AT STEROIDS HEARINGS

By Duff Wilson
March 18, 2005

Washington, March 18—Mark Mc-Gwire, one of the top home run hitters in baseball history, refused repeatedly during a Congressional hearing Thurs-day to say whether he used steroids while he played. Two other star players testified that they had not used steroids.

THE DES MOINES REGISTER

YOUNG ATHLETES' STEROID ABUSE CALLED PERILOUS

By Jane Norman
Register Washington Bureau
July 14, 2004

Washington, D.C.—Abuse of steroids and dietary supplements that have the same effect is becoming a widespread problem among college, high school and even middle school athletes, wit-nesses testified at a congressional hear-ing Tuesday.

26.1 Communication within the Body

If the drugs called steroids were being abused only by athletes, we might count the nation's problem with these substances as serious enough. But it appears that steroids also appeal to "vanity users," meaning people who simply want their bodies to take on a super-mus-cular appearance. We have pretty good information on who uses steroids, thanks to a number of surveys that have been completed in recent years. More problematic is

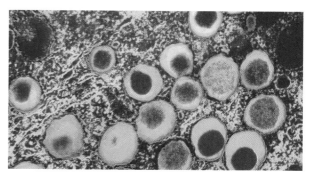

Hormone-filled granules (the red circles) lie within one of the insulin-secreting cells of the pancreas.

469

finding out what these drugs *do* to users over the long run. It would be unethical to give steroids to, say, a group of high school athletes as a means of seeing how their health holds up over time. So, with respect to steroids' consequences, what we mostly have are anecdotes—personal stories from steroid users and their doctors. It's a little risky to take anecdotes at face value, but to the extent that the ones about steroids can be believed, some of the effects of these drugs seem . . . strange. Though steroids are intended to mimic the "male" hormone testosterone, they sometimes end up giving female-sized breasts to men. Though they are intended to strengthen muscles, they can weaken tendons. Though they are intended to enhance growth, they can bring growth to a halt in adolescents.

These seeming paradoxes don't come as a surprise to the scientists who study steroids, however. These researchers are familiar with muscle-building steroids as a general class of hormones. And it's been well understood for decades that some hormones are like dollar bills: once they're in circulation, they can be put to many different uses. We shouldn't be surprised, then, that some of the other reported effects of steroid abuse are clitoral enlargement and acne.

But how is it that hormones are able to bring about effects such as these? In essence, hormones are *communication* molecules. We could think of them as couriers who not only travel some distance to complete their work, but who then flip a switch upon arrival, thus setting in motion some process (such as building up muscle tissue). Lots of different hormones function within us, of course, and together they make up a hormonal *system*—the endocrine system. Complex in its own right, this system then works hand in hand with an even more complex communication network in the body, the nervous system, which could be thought of as the hare to the endocrine system's tortoise. Nervous system communication is instantaneous, while hormonal communication can unfold over minutes, or even hours. Why do we need two systems? Because we're so complex. Some messages need to be delivered right away (the plate is hot) while others can be delivered more slowly (put more calcium into circulation). The purpose of this chapter is to show you how the body handles its various communication needs through the interlinked workings of its two communication networks, the endocrine and nervous systems. We'll start with the nervous system, which holds a special place of honor among the human organ systems we've been reviewing, in that it is both our means of perceiving the world and our means of understanding it.

Structure of the Nervous System

Taking the broadest view, the nervous system includes all the cells in the body that can be defined as nervous tissue, plus the sense organs we have, such as the eye and ear. Recall from Chapter 25 that two types of cells make up nervous tissue. These are neurons, which actually transmit nervous-system messages; and glial cells, which support the neurons.

One helpful way to think about the nervous system is to consider three essential tasks it has to perform. It first has to *receive* information, both from outside and inside our bodies. (For example, what taste sensations are coming in?) It next has to *process* this information. (Do I like this?) Then it has to *send* information out that allows our body to deal with this input. ("Lift your hand for another spoonful.")

The nervous system in which these activities take place can be divided into two fundamental parts. The first of these is the **central nervous system**: that portion of the nervous system consisting of the brain and the spinal cord. The second is the **peripheral nervous system**: that portion of the nervous system outside the brain and spinal cord, plus the sensory organs (see FIGURE 26.1a).

The central nervous system's brain and spinal cord are not physically separate entities; the spinal cord simply expands greatly at its top, and we call this flowering the brain. Because the brain and spinal cord are so different in terms of structure and function, however, they are thought of as individual units.

Meanwhile, the peripheral nervous system, or PNS, can be pictured as a group of nerves and related nerve cells that fan out from either the brain or spinal cord.

(a) The nervous system has two components.

(b) How these two components interact

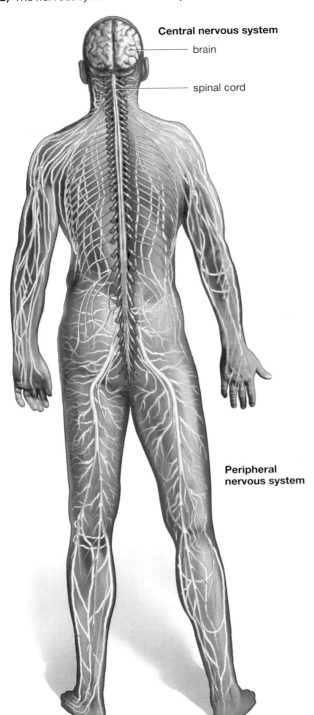

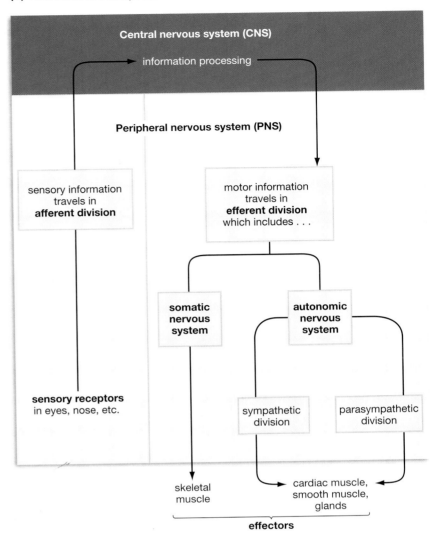

FIGURE 26.1 **Divisions of the Nervous System**

(a) The two fundamental branches of the nervous system are the central nervous system (CNS), which consists of the brain and the spinal cord; and the peripheral nervous system (PNS), which consists of all the nervous tissue outside the brain and spinal cord, plus the sensory organs.

(b) Information about the body and its environment comes to the nervous system from its sensory receptors—for example, cells in the eyes—which are part of the peripheral nervous system. This information then goes through the afferent division of the peripheral nervous system to the brain and spinal cord. After processing this information, the brain and spinal cord issue motor commands through the peripheral nervous system's efferent division. These commands go to "effectors" such as skeletal muscles and glands. The peripheral nervous system's efferent division has two systems within it: the somatic nervous system, which provides voluntary control over skeletal muscles; and the autonomic system, which provides involuntary regulation of smooth muscle, cardiac muscle, and glands. The autonomic system is further divided into sympathetic and parasympathetic divisions.

One way to think about these peripheral nerves and related cells is to consider the direction of the messages they carry. Any nerves that help carry messages *to* the brain or spinal cord are said to be part of the **afferent division** of the peripheral nervous system. Any nerves that help carry messages *from* the brain or spinal cord are said to be part of the **efferent division** of the PNS. An easy way to remember the difference between these two terms is to remember that "efferent" sounds like "effect." And the efferent division *effects change* in various organs in the body.

What kinds of change? Many of the activities the nervous system controls are voluntary activities, which often mean movement. You decide to lift your finger; you decide to stand up, and so forth. You may remember from Chapter 25 that all movement is handled by skeletal muscles—muscles, like the biceps, that are attached to bones and that are under voluntary control. So, one part of the PNS's efferent division is the **somatic nervous system**: that portion of the peripheral nervous system's efferent division that provides voluntary control over skeletal muscle.

But, there are lots of processes in the body that are not under voluntary control. When you walk from bright sunshine into a darkened movie theater, your pupils dilate, to let in more light, but you have no control over this dilation. Certain muscles allow our pupils to open up, but these are not skeletal muscles; they are the involuntary "smooth muscles." Likewise, the cardiac muscle of your heart beats in a rhythm that is largely out of your conscious control. And your glands release hormones in a way you have little control over. It is a *second* part of the PNS's efferent division that controls these operations, the **autonomic nervous system**: that part of the peripheral nervous system's efferent division that provides involuntary regulation of smooth muscle, cardiac muscle, and glands (**see** FIGURE 26.1b).

Now let's go down just one more level, this time strictly within the autonomic system. It turns out that the autonomic system is divided into two divisions, the *sympathetic* and the *parasympathetic*. These terms will be formally defined later; for now, just be aware of two characteristics they have. First, these divisions differ in that the nerves that make them up stem from different locations in the brain and spinal cord. Second, the sympathetic division generally has stimulatory effects on us—we get adrenaline going through this division, for example—while the parasympathetic division generally has relaxing effects.

26.2 Cells of the Nervous System

So, what about the cells that make up these systems? Remember first that the nervous-system cells that transmit signals are called neurons. These cells come in three varieties that neatly parallel the idea of a nervous system that receives, processes, and sends information (**see** FIGURE 26.2a). The first type of neuron is the **sensory neuron**, which does just what its name implies: It senses conditions both inside and outside the body and brings this received information to the central nervous system. (Given the direction of this information, sensory neurons are afferent neurons.) When someone brushes the top of your hand, sensory neurons just beneath the skin sense lots of things about this touch—what direction it came from, what shape the touching object was—and then convey a message containing this information that goes from your hand, into your spinal cord, then up into your brain.

The second type of neuron is the interneuron (or association neuron). Located solely in the central nervous system, **interneurons** interconnect other neurons. These can be very simple connections, but they can be complex as well. Our ability to recall events in our past amounts to a massive mobilization of interneurons. How do we remember? We process information in complex webs of interneurons.

The third type of neuron is the **motor neuron**: a peripheral-system neuron that sends instructions from the central nervous system (CNS) to such structures as muscles or glands. (Given the direction of this transmission, these are efferent neurons.) The key to understanding motor neurons is to think of them as neurons that transmit messages to organs or tissues that lie *outside* the nervous system,

A BROADER VIEW

Having to Go

How do we know when we "have to go" to the bathroom? Information about the bladder filling up is sent to the spinal cord, which sends an autonomic (involuntary) message for smooth muscle in the bladder to contract. This contraction puts a sometimes uncomfortable pressure on the bladder that lets us know about its fullness. Of course, we exercise voluntary control over the bladder as well, by means of the somatic nervous system controlling a kind of valve—the bladder's "external sphincter"—that must open before urination can occur.

(a) Three types of neurons

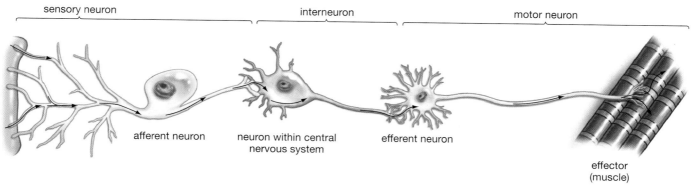

(b) Anatomy of a neuron

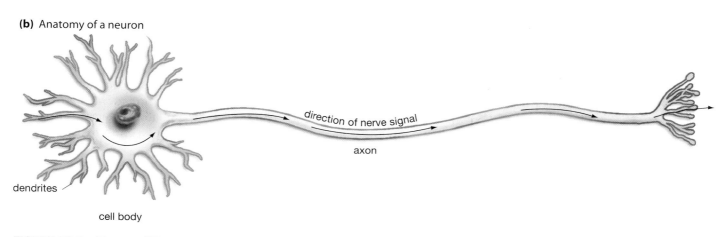

FIGURE 26.2 **Types of Neurons and Neuron Anatomy**

thus prompting some kind of action. If you look at the right side of **Figure 26.2a**, you can see how this works. A message comes from the cell body of a motor neuron, travels down an extension of it, and then gets transferred to a muscle—causing the muscle to contract. Any organ or group of cells that responds to this kind of nervous-system signal is called an *effector*. Thus, a gland prompted to release hormones through a motor neuron signal is also an effector.

To sharpen the point about the difference between sensory and motor neurons, consider the newly popular drug Botox. Physicians inject this drug into the foreheads of older people who want to get rid of the "age lines" there. Botox works by blocking the signal between motor neurons and the muscles they stimulate. The linkage works as follows: Botox blocks the nerve signal going to forehead muscles; this means there will be no forehead muscle contraction; and this means there will be no age lines. Now, one of the frequently asked questions about Botox is: Will it make my forehead numb? The answer is no, because Botox is blocking motor neuron transmission, which means it is blocking the ability to *send* a signal (to a muscle) not the ability to *receive* a signal (from a sensory neuron).

Anatomy of a Neuron

Now let's look at how a neuron is structured. Like any other cell, a neuron has a nucleus, it makes proteins, it gets rid of waste, and so forth. If you look at FIGURE 26.2b, you can see, however, that the neuron has some extensions sprouting from it that clearly separate it from other cells. Projecting from the cell body are a variable number of **dendrites**: extensions of neurons that carry signals *toward* the neuronal cell body. There is also a single, large **axon**: an extension of the neuron that carries signals *away* from the neuronal cell body. Though you can't see it, the other thing that separates the neuron from other cells is its outer or "plasma" membrane, which has an amazing ability to respond to various kinds of stimulation, as you'll see.

The Nature of Glial Cells

Now, what about the second kind of nervous-system cell, the glial cells (also known as glia)? Found in both the CNS and PNS, glia have no signal-processing ability of their own, but they provide all kinds of support to neurons. For example, some glia wrap their cell membranes around the axons of neurons in the CNS and PNS. The membranous covering that glia provide to neurons is called **myelin**, and an axon wrapped in this way is said to be myelinated. The importance of this is that axons that are myelinated carry nerve impulses faster than do those that are not. (The impulse skips from one gap or "node" in the myelin covering to the next.) If this sounds like some technical detail, consider that the disease multiple sclerosis results from a dismantling of the myelin covering in the brain and spinal cord by the body's own immune system. If you look at FIGURE 26.3a, you can see what a myelinated axon looks like.

Nerves

All this information about neurons and systems provides a way to understand a term that's been used extensively thus far, but that has not been defined. What is a **nerve**? It is a bundle of axons in the PNS that transmits information to or from the CNS. If you look at FIGURE 26.3b, you can see a representation of a nerve. Note that nerves have support tissue in the form of blood vessels and connective tissue.

A BROADER VIEW

Oxygen and Brain-death

Neurons are voracious users of energy; witness the fact that, though the brain makes up less than 3 percent of our body weight, it typically consumes about 20 percent of our calories when we are at rest. The pressing need that neurons have for energy—which oxygen helps supply—means that the brain is ground zero for any *cutoff* of our oxygen supply. Terri Schiavo, who was at the center of the famous Florida "right to die" case in 2005, collapsed in 1990 because her heart quit beating. When this occurred, her brain was deprived of oxygen for about five minutes—not a long time, but long enough to destroy most of her brain's processing ability.

So Far...

1. The brain and spinal cord make up the _____ nervous system, while all the nervous tissue outside the _____ plus the sensory organs make up the _____ nervous system.

2. The efferent division of the peripheral nervous system is made up of nerves that help carry messages _____ the brain or spinal cord, and it is divided up into a portion that provides voluntary control over skeletal muscles—the _____ nervous system—and a portion that provides involuntary regulation over smooth muscle, cardiac muscle, and glands—the _____ nervous system.

3. The cells of the nervous system that carry nervous-system messages are called _____, each of which has a variable number of _____ that carries signals toward the cell body, and a single large _____ that carries signals away from the cell body.

26.3 How Nervous-System Communication Works

With this anatomy under your belt, you're ready to see how nervous-system signaling works. We'll look at a two-step process, which we can think of as (1) signal movement down a cell's axon and (2) signal movement from this axon over to a second cell, across what is known as a synapse.

Communication within an Axon

All the action in this first step is going to take place on either side of the thin plasma membrane that constitutes the outer border of any animal cell. Inside the plasma membrane is the cell and all its contents; outside is extracellular fluid

So Far... Answers:

1. *central; central nervous system; peripheral*

2. *from; somatic; autonomic*

3. *neurons; dendrites; axon*

(a) A myelinated axon

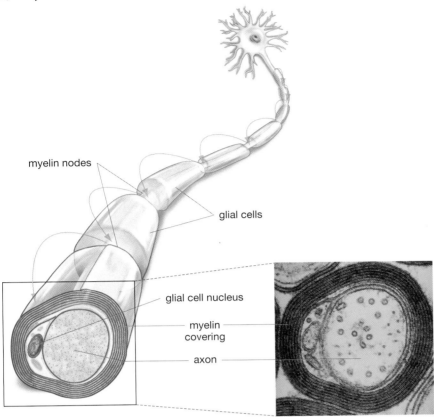

myelin nodes

glial cells

glial cell nucleus

myelin covering

axon

(b) Anatomy of a nerve

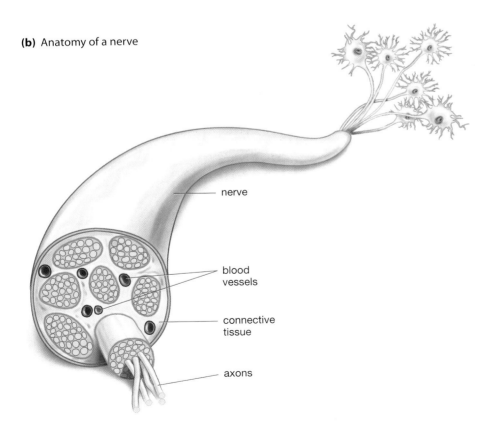

nerve

blood vessels

connective tissue

axons

FIGURE 26.3 One Function for Glial Cells; the Nature of Nerves

(a) Some glial cells produce a fatty material called myelin that wraps around the axons of neurons, as shown here. Nerve signals travel faster down myelinated axons, because the nerve signal skips from one myelin gap or "node" to the next. The micrograph at right shows a cross section of a single vertebrate axon that is wrapped in multiple layers of myelin covering.

(b) A nerve is a bundle of axons in the peripheral nervous system. It includes not only axons themselves, but supporting blood vessels and connective tissue.

(a) Resting potential

Electrical energy is stored across the plasma membrane of a resting neuron. There are more negatively charged compounds just inside the membrane than outside of it. As a result, the inside of the cell is negatively charged relative to the outside. The charge difference creates a form of stored energy called a membrane potential. Protein channels (shown in green) that can allow the movement of electrically charged ions across the membrane remain closed in a resting cell, thus maintaining the membrane potential.

(b) Action potential

1. Nerve signal transmission begins when, upon stimulation, some protein channels open up, allowing a movement of positively charged sodium ions (Na^+) into the cell. The flow of these ions is aided not only by their electrical charge, but by their concentration gradient. Since there were more of them outside the cell than inside prior to the channels opening, there is a natural net movement of them into the cell. For a brief time, the inside of the cell becomes positively charged.

2. The Na^+ gates close and the gates for positively charged potassium ions (K^+) open up, allowing a movement of K^+ out of the cell. With this, there is once again a net positive charge outside the membrane. The influx of Na^+ at one point in the cell membrane then triggers the same sequence of events in an adjacent portion of the membrane—note the influx of Na^+ next to the outflow of K^+.

3. The nerve signal continues to be propagated one way along the axon by means of this action potential.

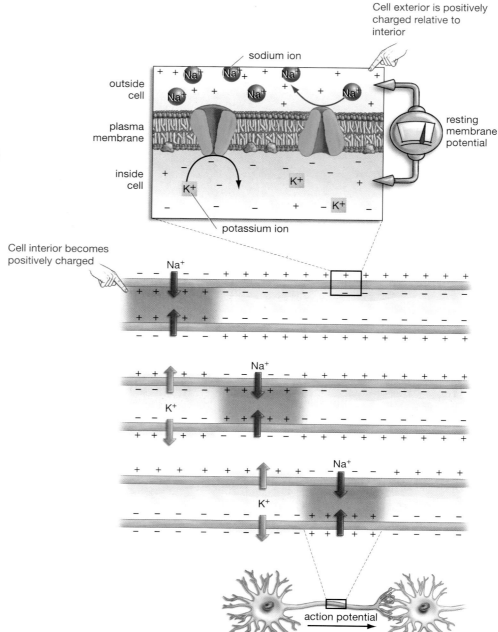

FIGURE 26.4 **Nerve Signal Transmission within a Neuron**

(**see** FIGURE 26.4). Those of you who read Chapter 5 will recall that the plasma membrane regulates the passage of substances in and out of the cell. You may remember that some of these substances are the electrically charged particles called ions, meaning atoms that have gained or lost one or more electrons. Because ions are charged, they get the little + or − signs after their chemical names, as with Na^+ for the positively charged ion sodium. Ions cannot simply pass through the plasma membrane; rather, they need to move through special passageways, called protein channels. Some of these are so-called leak channels that are always open, but others are "gated" channels that can open or close.

It turns out that a "resting" neural cell—one not transmitting a signal—has a greater positive charge outside itself than inside. This is so mostly because there are more proteins inside the cell than outside, and proteins carry a *negative* charge.

The charge difference also exists, however, because there are more of the positively charged Na^+ ions outside the membrane than there are positively charged potassium ions (K^+) inside, as you can see in Figure 26.4. As you know, in electricity, opposites attract, meaning the positively charged Na^+ ions outside the plasma membrane are attracted to the negatively charged interior of the cell. This amounts to a form of potential energy, in the same way that a rock perched at the top of a hill does. The rock would roll down the hill (releasing energy) if given a push, and the Na^+ ions would rush into the cell (releasing energy) if the proper channels were opened to them. The charge difference that exists from one side of the neuronal plasma membrane to the other is known as the **membrane potential**.

This potential exists in part because of what you've already seen: the greater abundance of Na^+ ions outside the cell. It's important to note that this abundance doesn't exist by accident. Every cell keeps this imbalance in place by constantly pumping three Na^+ ions out for every two (K^+) ions it pumps in. It takes a lot of energy to maintain this "sodium-potassium pump," but the cell does it for the same reason that you charge your cell-phone battery: to have ready access to a capability. In your case, the capability is talking on the phone; in the cell's case, the capability is that of having Na^+ ions available to rush into it. Indeed, the action of the pump "primes" the neuron for activity in two ways. In addition to the charge difference it maintains, the pump maintains a *concentration* difference of the ions. The fact that there are more Na^+ ions outside the cell than inside means that the Na^+ ions will have a natural tendency to move down their "concentration gradient"—they will naturally move from where they are more concentrated to where they are less concentrated. Since they are more concentrated outside the cell than inside, there will be a natural net movement of these ions into the cell. (For more on concentration gradients, see Chapter 5.) In sum, the sodium-potassium pump maintains both an electrical charge difference and a concentration difference that can get a nerve signal going, as you'll now see.

When a neuron is stimulated by getting a chemical signal from another neuron, the effect of this stimulation is that, up near where the neuron's axon meets its cell body, Na^+ ion gates open up. Now the Na^+ ions can act on their electrical attraction and concentration difference, and they do so, rushing into the cell. After a very brief time, this influx is great enough that the inside of the cell is more positive than the outside. This very state causes the *potassium* gates to open up more fully, however. Recall that potassium (K^+) ions exist in more abundance inside the cell than outside it. With the opening of their gates, *they* act on their electrical and concentration differences, rushing out of the cell, and the positive charge that has briefly existed on the inside of the cell begins to decline. Eventually, enough K^+ ions have moved out that the cell returns to its resting state, with a negative charge inside the membrane. This entire process takes place in a few milliseconds, or thousandths of a second.

Movement Down the Axon

All this alters the membrane potential at *one spot* along the neuron's axon. But how does a nerve signal get transmitted down the entire length of the axon? The key to this transmission is that an influx of Na^+ ions at an initial location on the membrane triggers reactions that cause the *neighboring* portion of the membrane to begin an Na^+ influx (see Figure 26.4). What occurs, in other words, is a chain reaction that moves down the entire length of the axon membrane. This is an **action potential**: a temporary reversal of cell-membrane potential that results in a conducted nerve impulse down an axon. (Temporary reversal? Remember the inside of the cell goes from negative to briefly positive, then back.) Action potentials have been compared to what happens to a lighted fuse: The heat of the spark causes the neighboring section of the fuse to catch fire, thus moving the spark along.

All action potentials are of the same strength. Once the original signal on the cell-body/axon border reaches sufficient strength to allow Na^+ ions to rush in, the

A BROADER VIEW

Ultra-long Axons

How long is an axon? Most are just a few millionths of a meter in length, but some can stretch out for a full meter or more. We have cell bodies located in our lower spine whose axons extend all the way to the muscles on the soles of our feet.

action potential will get going. Thanks to its fuse-like quality, this potential will be the same strength all the way down the axon. How fast can this signal go? At best, about 120 meters per second—very fast indeed, but much slower than the electricity that comes from a wall socket.

Communication between Cells: The Synapse

Once it has traveled the length of an axon, an action potential then reaches a tip of the axon. How does the signal then get to the *next* neuron (or muscle, or gland cell)? This question was once very intriguing to biologists, because it was clear that, with a few exceptions, an axon does not touch the downstream cell. It comes extremely close, but there is almost always a small intervening space between the two cells. There are thus three entities involved in cell-to-cell transmission: the sending neuron, the receiving cell, and the gap between them. The area where all three come together is called a **synapse** (see FIGURE 26.5).

The action potential arrives at a branch of the axon, called a *synaptic terminal*, which has stored within it small sacs (or vesicles) containing a chemical called a neurotransmitter. The arrival of the action potential causes the synaptic terminal

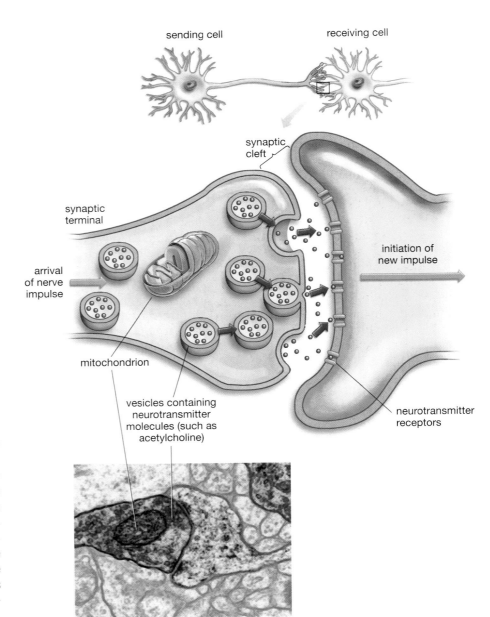

sending cell   receiving cell

synaptic cleft

synaptic terminal

arrival of nerve impulse

initiation of new impulse

mitochondrion

vesicles containing neurotransmitter molecules (such as acetylcholine)

neurotransmitter receptors

FIGURE 26.5 **Structure and Function of a Synapse**
A synapse is made up of a sending neuron, a receiving cell, and the gap between them, called a synaptic cleft. As a nerve impulse reaches the synapse, molecules of a neurotransmitter (such as acetylcholine) are released into the synaptic cleft from synaptic terminals of the sending neuron. This occurs when small membrane-bound vesicles—each filled with thousands of neurotransmitter molecules—fuse with the outer membrane of the synaptic terminal. The neurotransmitter molecules then move across the cleft and stimulate receptors in the receiving cell's membrane, thus initiating a nerve impulse within it. The lower figure is a color-enhanced micrograph of a synapse.

to release neurotransmitter molecules into the gap in the synapse, which is called a **synaptic cleft**. With this, the molecules diffuse over to the receiving cell and bind to receptors in that cell's outer membrane. This binding stimulates the opening of sodium gates there, allowing the now-familiar influx of Na^+ ions in the *receiving* neuron, and this keeps the signal transmission going.

But why doesn't the released neurotransmitter just keep on stimulating the receiving cell? Serving as one means of control are the proteins called enzymes. Released into the synaptic cleft, they break down the neurotransmitter, thus inactivating it. The sending cell may also be capable of taking the neurotransmitter back into itself, by a process known as "reuptake." In a given nerve impulse, a sending cell typically releases tens of thousands of neurotransmitter molecules, which are stored in hundreds of sacs. With all this in mind, you can understand the definition of a **neurotransmitter**: a chemical, secreted into a synaptic cleft by a neuron, that affects another neuron or an effector by binding with receptors on it.

The Importance of Neurotransmitters

It is difficult to overstate the importance of neurotransmitters. To gauge their significance, consider that one neurotransmitter, called acetylcholine, must travel from nerve-cell terminals to muscles for any skeletal muscle to work. If acetylcholine doesn't make it out of a motor neuron, we can't contract the muscles that allow us to breathe. Likewise, the shaking and stiffness of Parkinson's disease are caused when selected cells in the brain die, thus reducing the brain's supply of the neurotransmitter dopamine. On a different level, remember the reuptake process we talked about, whereby a releasing cell will take back a neurotransmitter that it has secreted into a synaptic cleft? Well, the antidepressant drugs Prozac, Zoloft, and Paxil are all called SSRIs, which stands for selective serotonin reuptake inhibitors. Serotonin is a neurotransmitter found in the brain, and SSRIs are aimed at reducing serotonin reuptake. The result is an increased amount of serotonin in synaptic clefts—and perhaps less depression. By the same token, drugs of abuse such as cocaine and the amphetamine-like ecstasy all work by altering neurotransmitter release or reuptake. What makes us feel bad, or OK, or ecstatic? A big part of the answer is: the levels of neurotransmitters in our brains.

A BROADER VIEW

Comprehending the Synapse

The first person who realized that nerve cells do not touch each other was the Spanish scientist Santiago Ramón y Cajal. Working with microscopes at the end of the nineteenth century, he not only came to understand the reality of synaptic gaps, but made a painstaking set of drawings that laid out the details of what he saw. The drawing task wasn't as difficult for Cajal as it would be for most people, however, because he was a gifted artist as well as a scientist. A selection of his scientific drawings can be seen at www.psu.edu/nasa/cajal2.htm.

So Far...

1. Transmission of a nerve signal down an axon takes advantage of something called a membrane potential, which can be defined as a difference in _____ that exists between the inside and the outside of a neuron.

2. The key to signal transmission down the length of an axon is that an influx of Na^+ ions at an initial location on the axon membrane triggers an _____ on a _____ of the membrane.

3. Communication between neurons takes place by means of substances called _____ moving across a _____ from a sending cell to a receiving cell.

26.4 The Spinal Cord

Now that you've looked at how nerve impulses are transmitted at the cellular level, it's a good time to take a step back and look at the nervous system in its larger dimensions. Let's start by examining the spinal cord, which serves two key functions. First, it can act as a communication center on its own, receiving input from sensory neurons and directing motor neurons in response, with no input from the brain. This is what a reflex amounts to. Second, most sensory impulses that go to the brain don't go *directly* to the brain; they are channeled first through the spinal cord.

Extending from the base of the brain to an area just below the lowest rib, the spinal cord is about half an inch wide and consists of 31 segments, each one having

So Far... Answers:

1. *electrical charge*

2. *influx of Na^+ ions; neighboring portion*

3. *neurotransmitters; synapse*

FIGURE 26.6   **Structure of the Spinal Cord** Thirty-one pairs of spinal nerves extend from the spinal cord. These nerves are grouped according to the body part they control. Eight cervical nerves control the head, neck, diaphragm, and arms; 12 thoracic nerves control the chest and abdominal muscles; five lumbar nerves control the legs; and five sacral nerves control the bladder, bowels, sexual organs, and feet.

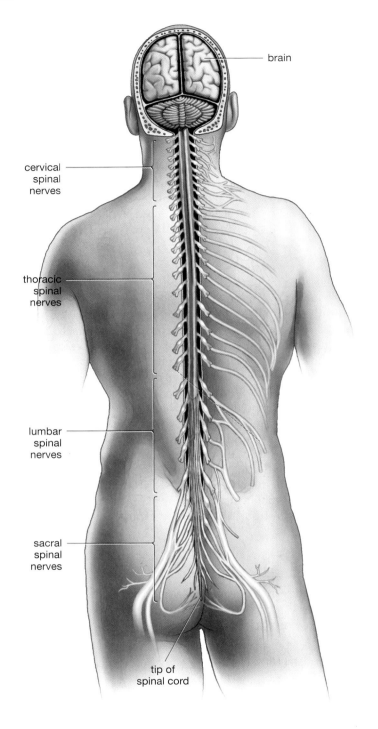

brain

cervical spinal nerves

thoracic spinal nerves

lumbar spinal nerves

sacral spinal nerves

tip of spinal cord

both a left and right spinal nerve stemming from it (**see** FIGURE 26.6). These nerves are grouped into the classes you see named in the figure, which correspond to the region of the body they serve. ("Thoracic" is the chest area; "lumbar" the lower back, and so forth. For an account of what happens when the spinal cord is severed in these areas, see "Spinal Cord Injuries" on page 482.) Remember how we noted that the peripheral nervous system could be thought of as a group of nerves and related nerve cells that fan out from the brain and spinal cord? In Figure 26.6, you can see the start of a lot of this fanning, in the form of the spinal nerves. (The other major set of nerves that do this is the so-called cranial nerves, which fan out from the brain.)

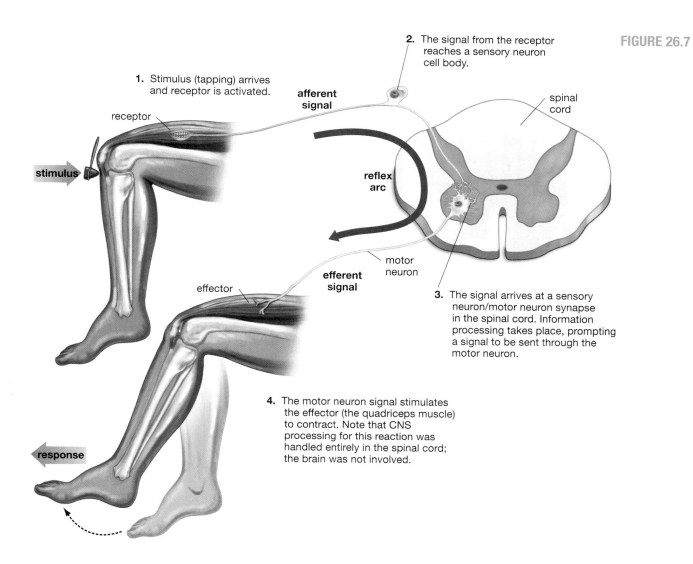

FIGURE 26.7   **Steps in a Reflex Arc**

2. The signal from the receptor reaches a sensory neuron cell body.

1. Stimulus (tapping) arrives and receptor is activated.

afferent signal

receptor

spinal cord

stimulus

reflex arc

motor neuron

effector

efferent signal

3. The signal arrives at a sensory neuron/motor neuron synapse in the spinal cord. Information processing takes place, prompting a signal to be sent through the motor neuron.

4. The motor neuron signal stimulates the effector (the quadriceps muscle) to contract. Note that CNS processing for this reaction was handled entirely in the spinal cord; the brain was not involved.

response

Quick, Unconscious Action: Reflexes

The spinal cord is a central player in body **reflexes**, which can be defined as automatic nervous-system responses that help us avoid danger or preserve a stable physical state. If you accidentally touch a hot stove, you automatically pull back your hand. No conscious thought of "Oh, I've touched a hot stove; I'd better pull my hand back," is necessary. The neural wiring of a single reflex is called a *reflex arc*. A reflex arc begins at a sensory receptor, runs through the spinal cord, and ends at an effector, such as a muscle or gland. FIGURE 26.7 shows one of the best-known examples, the knee jerk, or patellar reflex, in which a properly placed sharp rap on the knee produces a noticeable kick. Note what's at work: The receptor of a sensory neuron is stimulated, and the resulting signal moves up through the neuron's dendrite and into the neuron's cell body (which lies just outside the spinal cord). The signal then continues through the neuron's axon, into the spinal cord, to a synapse with a motor neuron. This neuron then issues a command that is carried out by an effector—in this case a skeletal muscle. This is an example of the simplest possible reflex arc, in that the sensory neuron is linked directly to the motor neuron. Because only one synapse is involved, this kind of simple reflex controls the most rapid motor responses of the nervous system. Many other reflexes have at least one interneuron placed between the sensory receptor and the motor neuron.

Essay 26.1 Spinal Cord Injuries

In the movie *Million Dollar Baby*, Maggie, the boxer played by Hilary Swank, sustains an injury that she says has made her "a complete C1–C2." What she means is that her spinal cord was severed at the level of cervical spinal nerves 1 and 2—C1 and C2. If you look at Figure 26.6 on page 480, you can see the cervical spinal nerves bracketed. There are eight of these nerves, the topmost of them being C1, the next C2, and so forth. (The thoracic spinal nerves are then called T1, T2, etc.) Maggie's condition was so terrible because she sustained an injury so high up in her spinal cord.

> Maggie's condition was so terrible because she sustained an injury so high up in her spinal cord.

People whose spinal cords are severed between the T1 and upper lumbar segments become paraplegic—they lose the use of their legs. People whose injury comes between C5 and T1 become quadriplegic—they lose the use of both their legs and arms. But people whose injury comes above C3 lose all this functionality *and* the ability to breathe on their own. Why? The nerve signals that raise our diaphragm—thus allowing us to inhale—travel through the C3–C5 spinal nerves. Since Maggie's injury was above C3, she could be kept alive only by being connected to an artificial respirator.

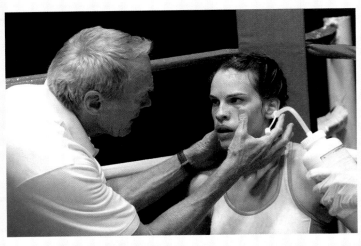

FIGURE E26.1.1 **In the Ring** Clint Eastwood and Hilary Swank in a scene from *Million Dollar Baby*.

A BROADER VIEW

Priming Fight or Flight

What do we need for fight or flight? Adrenaline helps, by providing us with a surge of energy. We need blood in our skeletal muscles, so a diversion of blood from our digestive tract to the muscles is good. And we're going to need plenty of air, so it would be helpful if our lung passages could open up further. Each one of these effects comes about by means of signals sent out by the sympathetic division of our autonomic system.

26.5 The Autonomic Nervous System

As noted earlier, the part of the peripheral nervous system's efferent or "outgoing" division over which we have no conscious control is called the autonomic nervous system. You can grasp its importance by imagining having to consciously control the beating of your heart. Recall that this system controls the involuntary regulation of smooth muscle, cardiac muscle, and glands.

Sympathetic and Parasympathetic Divisions

Also recall that, within the autonomic nervous system, there are two "divisions," the sympathetic and parasympathetic. The **sympathetic division**, often called the *fight-or-flight* division of the autonomic nervous system, usually stimulates tissue metabolism, increases alertness, and generally prepares the body to deal with emergencies. The **parasympathetic division**, often regarded as the *rest-and-digest* division of the autonomic nervous system, conserves energy and promotes activities such as digestion. To give you some idea of the division of duties here, sympathetic signals increase heart rate, but parasympathetic signals decrease it; sympathetic signals raise blood pressure, but parasympathetic signals lower it. In FIGURE 26.8, you can see which spinal and cranial nerves are part of these two divisions. You can also see the effects each division has on various bodily functions. Although some organs are connected to only one division or the other, most vital

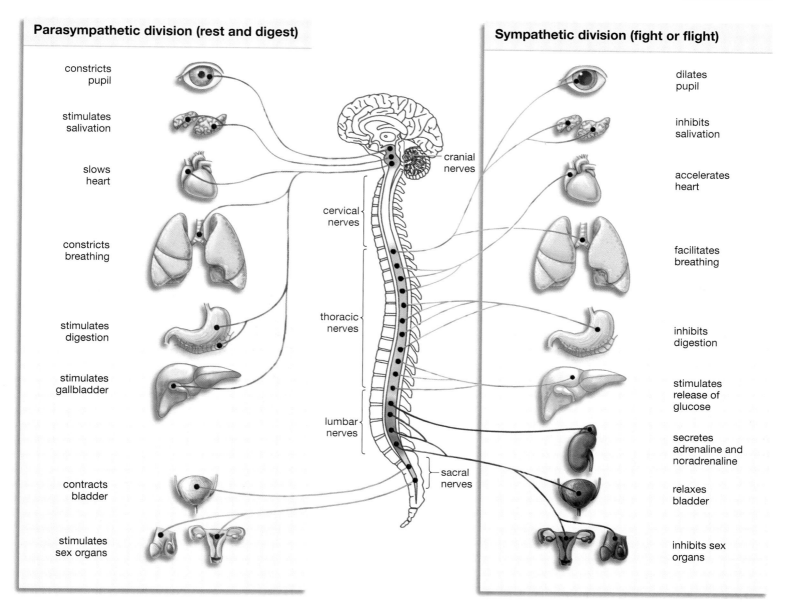

Parasympathetic division (rest and digest)

constricts
pupil

stimulates
salivation

slows
heart

constricts
breathing

stimulates
digestion

stimulates
gallbladder

contracts
bladder

stimulates
sex organs

Sympathetic division (fight or flight)

dilates
pupil

inhibits
salivation

accelerates
heart

facilitates
breathing

inhibits
digestion

stimulates
release of
glucose

secretes
adrenaline and
noradrenaline

relaxes
bladder

inhibits sex
organs

cranial
nerves

cervical
nerves

thoracic
nerves

lumbar
nerves

sacral
nerves

FIGURE 26.8 **Involuntary Control of Bodily Functions** The autonomic nervous system has two divisions, the sympathetic and the parasympathetic, which exercise automatic control over the body's organs, generally in opposing ways. Note how the parasympathetic aids in digestion, while the sympathetic controls fight-or-flight responses, such as an increase in heart rate. Axons of the parasympathetic division emerge not only from the spinal cord, but from the brain as well.

organs receive both sympathetic and parasympathetic signals. Where such dual signaling exists, the two divisions often have opposing effects, keeping the body's stability mechanisms working in balance.

26.6 The Human Brain

Now that we've reviewed the spinal cord, it's time to look at the second major part of the central nervous system—the brain, which weighs about 3.3 pounds, is about the size of a grapefruit, and has the consistency of cream cheese. Like the spinal cord with its spinal nerves, the brain has nerves extending from it that allow it to communicate directly with other body tissues and organs. These are the

FIGURE 26.9 **The Brain**

(a) View of the left surface of the brain and a cross section of the brain showing its cerebrum and cerebral cortex.

(b) Structures of the brain outside the cerebrum and cerebellum. The pituitary gland is connected to the hypothalamus, but is a part of the body's endocrine system. The brainstem is made up of structures called the midbrain, pons, and medulla oblongata.

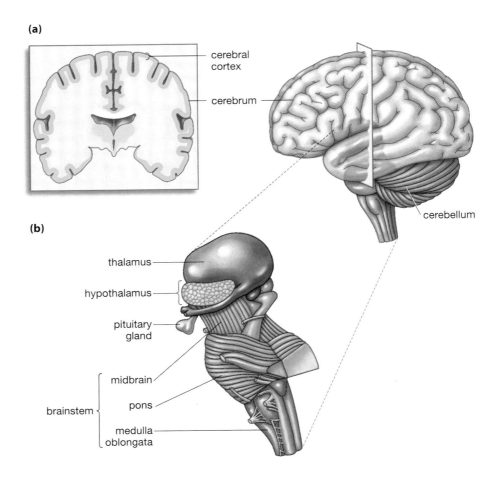

cranial nerves and, as you might expect, some of them go to nearby organs (the eyes, the nose, the face muscles); but some also go to organs that are farther away (the heart and lungs). If you look at FIGURE 26.9, you can see the major structures that make up the brain. Let's look briefly now at each of them.

Six Major Regions of the Brain

- The **cerebrum**, the largest region of the human brain, is responsible for much of our higher mental functioning. With its valley-like fissures and fatty appearance, the roundish cerebrum fills up most of our skulls and is effectively draped over many other portions of the brain. It is divided into left and right *cerebral hemispheres*. The outer layer of the cerebrum is the **cerebral cortex**, the site of our highest thinking and processing. It covers the entire cerebrum, but it amounts to a *thin* covering—only two or three millimeters thick, which means it is little more than one-tenth of an inch deep at most. Its billions of neurons are extremely varied in their activities. The cerebral cortex has, for example, areas called the visual cortex and the olfactory cortex, which are the primary processing centers for sight and smell, respectively. Likewise, there is a sensory cortex that gets touch information from our skin. Then there is a "prefrontal" association area, up toward our foreheads, that is involved in "executive" functioning, meaning it helps prioritize our behaviors. As may be apparent, these cortex areas are regions of higher processing. Out in the body, there are neurons that sense various forms of stimulation, but the cerebral cortex *makes* sense of the information these neurons provide. The cerebrum is more than the cerebral cortex, however. It also contains an area called the hippocampus, which plays a critical role in forming long-term memories; and the amygdala, which functions in the fear response and our memory of it.

- The *cerebellum* is a portion of the brain that refines our movements based on our "sense memory" of them. People who have damaged their cerebellum may find that their movements have become jerky, or that they reach too far (or not far enough) for everyday objects. The cerebellum also helps us maintain balance.

In looking at the brain, if you take away the cerebrum and the cerebellum, you are left with (a) the thalamus and hypothalamus; and (b) the brainstem, which is composed of the midbrain, pons, and medulla oblongata. All of these structures can be seen in Figure 26.9b.

- When we see, hear, taste, or touch anything, the resulting sensory perceptions are channeled through the **thalamus** before moving on to the cerebral cortex for more processing. Lying below the thalamus is the hypothalamus, which is small but extremely important in several ways. Since we'll be seeing it several times in this chapter, a formal definition might be helpful. The **hypothalamus** is a portion of the brain that is important in regulating drives and in maintaining homeostasis in the body. With respect to drives, cells in the hypothalamus are critical in telling us whether we are hungry or thirsty, for example. Likewise, the hypothalamus helps control our sleep-wake cycle, or "circadian rhythms." The hypothalamus is important in preserving homeostasis—the maintenance of a stable internal environment—because it controls a good number of the hormones that are released from our glands, as you'll see later.

- The *midbrain* helps us maintain muscle tone and posture through control of involuntary motor responses. Along with the amygdala, the midbrain is involved in our fear responses. Also located in the midbrain are the dopamine-producing cells that, as noted earlier, keep our hands steady and our movements fluid.

- The Latin term *pons* refers to a bridge, and the pons serves as one. Its primary function is to relay messages between the cerebrum and the cerebellum. It also helps control involuntary breathing.

- Located next to the spinal cord, the *medulla oblongata* contains major centers concerned with the regulation of unconscious functions such as breathing, blood pressure, and digestion. The medulla oblongata actually has a connection to a well-known term. What does it mean to be "brain dead"? Under one definition, it means that all the centers of the brain *except* the medulla oblongata have permanently ceased to function. As long as the medulla is still working, a person continues to breathe, despite the loss of all conscious ability.

So Far...

1. The spinal cord can receive input from _____ and direct _____ in response, without input from the brain. This entire process is known as a _____.

2. The autonomic nervous system is divided into the sympathetic or "_____" division and the parasympathetic or "_____" division.

3. The portion of the brain that is the site of our highest thinking and processing, the _____, is the outer portion of the brain's _____.

26.7 The Endocrine System

We now leave the nervous system to focus on the second system we talked about at the start of the chapter, the endocrine system. Like the nervous system, the endocrine system is in the communication business. And, like the nervous system with its neurotransmitters, the endocrine system works through a group of chemical messengers. The endocrine messengers are not neurotransmitters, however,

So Far... Answers:
1. *sensory neurons; motor neurons; reflex*
2. *fight-or-flight; rest-and-digest*
3. *cerebral cortex; cerebrum*

but instead are substances called hormones. In this text, we will define a **hormone** as a substance secreted by one set of cells that travels through the bloodstream and affects the activities of other cells. Such a definition sets hormones apart from other kinds of signaling molecules in the body. There are signaling molecules that do not travel through the bloodstream, but instead diffuse from one or more cells to a nearby group of cells, causing a metabolic change in them. Likewise, a cell can be affected by its own secreted chemical messenger. But here we will look only at communication molecules that are transported through the bloodstream.

This very means of distribution provides a key for getting to the heart of what separates the endocrine and nervous systems. In the nervous system, signals go from neurons A to B to C, in well-defined lines of transmission—something like a telephone call going through relay stations. By contrast, a typical hormone is "broadcast," in a sense, as it moves through the bloodstream. Like a television signal, it can be "picked up" by any cell that has the proper "receiver." What are these receivers? They are receptors that are shaped in such a way that they can latch onto the hormone. With this, we get to the concept of **target cells**: those cell types that can be affected by a given hormone (see FIGURE 26.10). Hormones differ greatly with respect to the number of target cells they affect and the location of these cells. The target cells for a hormone called antidiuretic hormone are located primarily in the kidneys, while the target cells for the hormone insulin are located not only in the liver, but in muscle and fat cells throughout the body.

Hormonal production takes place to a significant extent within specialized organs called endocrine glands, and this represents another contrast with the nervous system. You may remember that a gland is any localized group of cells that work together to secrete a substance. **Endocrine glands** are glands that release their materials directly into the bloodstream or into surrounding tissues, without using the tubes known as ducts. The endocrine system's use of glands stands in contrast to the nervous system, which has no such organs—only its individual neurons, secreting neurotransmitters. The major endocrine glands are shown in

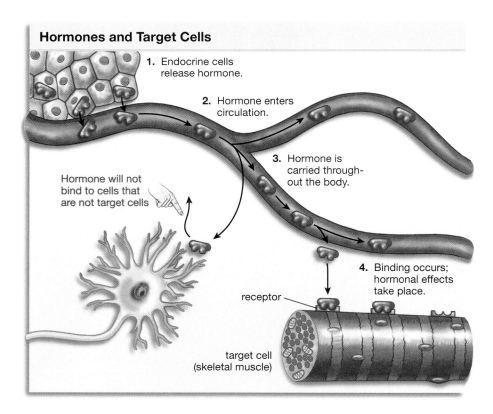

FIGURE 26.10 Hormones and Their Target Cells
For a hormone to affect a target cell, that cell must have receptors that can bind the hormone. The binding of hormone to receptor then initiates a change in the target cell's activity. The figure shows a peptide hormone that affects skeletal muscle tissue. The hormone does not affect the nerve cell, also shown in the figure, because this cell does not have the appropriate receptors to bind with the hormone.

Hormones and Target Cells

1. Endocrine cells release hormone.
2. Hormone enters circulation.
3. Hormone is carried throughout the body.
4. Binding occurs; hormonal effects take place.

Hormone will not bind to cells that are not target cells

receptor

target cell (skeletal muscle)

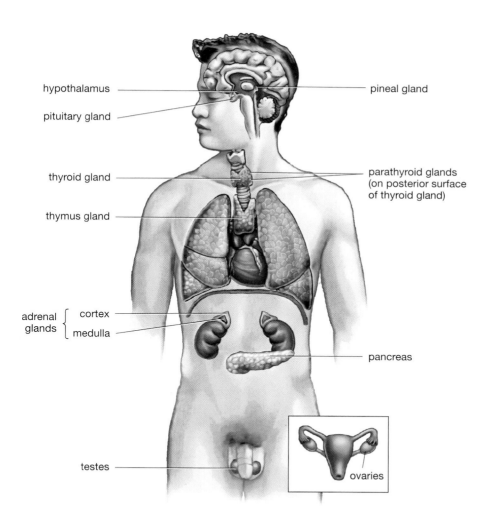

hypothalamus

pituitary gland

thyroid gland

thymus gland

adrenal glands { cortex / medulla

testes

pineal gland

parathyroid glands
(on posterior surface
of thyroid gland)

pancreas

ovaries

FIGURE 26.11 **The Major Hormone-Secreting Glands of the Body** Though the hypothalamus is part of the brain (and not an endocrine gland), it is shown because it plays a central role in hormonal regulation.

FIGURE 26.11. While many hormones are secreted by these glands, it's important to note that not *all* hormones are. The heart, kidneys, stomach, liver, small intestine, placenta, and fatty tissue all secrete hormones, yet they are not glands.

In a final point of comparison, noted earlier, the endocrine system tends to work more slowly than the nervous system, but its effects tend to be more long-lasting. The fastest-acting hormones take several seconds to work, while the slowest-acting ones take several hours. Contrast this with the almost instantaneous effects of nervous-system messages. The opposite side of this coin is that the longest-lasting hormones can keep exerting their effects for hours after they have been released. In contrast, nervous-system signals disappear as fast as they arise.

Despite their differences, the endocrine and nervous systems are tightly linked, in that signals from one system often affect the other. Beyond this, modern research actually is blurring the distinction between the two systems. Scientists have a growing list of substances that once were thought to work strictly as hormones, but that now turn out to be both hormones and neurotransmitters. For example, the hormone adrenaline works as a neurotransmitter within the brain, shuttling across synapses; but then it also works as a hormone—it is secreted into circulation by the body's adrenal glands.

26.8 Types of Hormones

Hormones come in three primary classes. The first of these is the **amino-acid-based hormones**: hormones that are derived from modification of a single amino acid. The hormone adrenaline, just mentioned, is produced through the modification of the amino acid tyrosine, for example, while the hormone melatonin is produced through a modification of the amino acid tryptophan. The second class of hormones, the **peptide hormones**, are hormones composed of *chains* of amino acids. Such chains can vary greatly in length. Human growth hormone, for example, is composed of 191 amino acids while antidiuretic hormone is composed of only nine. Insulin is a peptide hormone, as are most of the hormones secreted by an important endocrine gland we'll be looking at, the pituitary.

With a few exceptions, peptide and amino-acid-based hormones have target cells whose hormone receptors are located on their outer membranes. These receptors are the "broadcast receivers" that stand ready to bind with these hormones when they pass by, as depicted in Figure 26.10. The effects of this binding are very diverse. Ion channels may be opened in a target cell as a result, or enzymes may be activated. In all cases, the cell will change its activities as a result of the binding.

The third major class of hormones is the **steroid hormones**, which are all constructed around the chemical framework of the cholesterol molecule. (To get a sense of this linkage, see Figure 3.14 on page 45.) Most of the steroid hormones are released from just a few glands—notably the male and female reproductive glands and the adrenal glands, which sit atop the kidneys. Unlike amino-acid-based and peptide hormones, most steroid hormones pass *through* a cell's plasma membrane, and then bind with a receptor protein inside the cell. This combined hormone/receptor molecule then enters the nucleus of the cell and binds with the cell's DNA there. This binding turns on one or more of the cell's genes, thus bringing about the production of one or more proteins. The complexity of this series of events makes steroids the snails of the hormone world; hours might pass between the time a steroid hormone is produced and the time it has an effect. While testosterone is undoubtedly the best-known steroid hormone, estrogen is also a steroid hormone. From this, it's easy to see that the term *steroid*, as commonly used, actually means one *kind* of steroid—the muscle-building or "anabolic" steroid hormones.

So Far...

1. A hormone can be defined as a substance produced in one set of _____ that travels through the _____ and affects the activities of other _____.

2. Many hormones are produced by organs called _____, which release their materials directly into the _____.

3. There are three primary classes of hormones: _____, _____, and _____. Of these three, only the _____ hormones work by passing through the target-cell membrane and binding with a receptor inside the cell.

26.9 Hormone Secretion and Negative Feedback

What prompts the body to produce a given hormone? As you'll see, signals from the brain bring about the release of a number of hormones. But then again, the brain *gets* numerous signals saying, "More of this hormone needed now." And, in many instances, the brain isn't involved at all in starting up, or shutting down, the production of a hormone. So, what's at work in hormone secretion? A system of self-regulation that can be best understood by way of an analogy.

Most people are aware of how a home heating system works. Falling temperature causes a thermostat to turn on a furnace. To look at this another way, there is a stimulus (cold air) that brings about a response (furnace operation). The *product* of this response is hot air. When enough hot air circulates to raise the temperature, the thermostat senses this and shuts the furnace down.

A BROADER VIEW

Hormones and Irrational Behavior

When you begin to see how important hormones are in both long-term processes (growing to our full height) and short-term processes (taking care of blood-sugar surges), you begin to see some irony in the fact that, to the average person, "hormones" are substances that drive us to behave in irrational ways. This is not to say that there is no such thing as premenstrual syndrome or excessive, hormone-linked behavior among adolescent boys. But thinking of hormones primarily as intoxicating substances is like thinking of forks primarily as weapons.

So Far... Answers:

1. *cells; bloodstream; cells*

2. *endocrine glands; bloodstream or surrounding tissue*

3. *amino-acid-based, peptide, steroid; steroid*

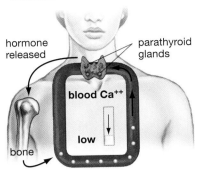

Stimulus: *low* calcium levels in blood

hormone released

parathyroid glands

blood Ca⁺⁺

low

bone

Response: release of calcium into blood

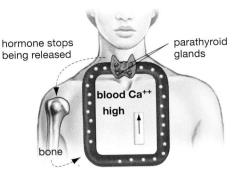

Stimulus: *high* calcium levels in blood

hormone stops being released

parathyroid glands

blood Ca⁺⁺ high

bone

Response: release of calcium is shut down

FIGURE 26.12 **Calcium Levels: An Example of Negative Feedback** Calcium levels in the bloodstream are regulated in part by the parathyroid glands. These glands can sense when calcium levels are too low (the stimulus in the drawing at left). In response, the parathyroids secrete a hormone that causes the bones to release calcium. This brings about high calcium levels, which the glands also can sense (the stimulus in the drawing at right). High calcium levels cause the parathyroids to halt their production of the calcium-liberating hormone.

Now let's think of your body which has, as an example, a certain amount of calcium circulating through it in the bloodstream. When levels of calcium in the blood fall too low (stimulus), organs called the parathyroid glands sense this and secrete a hormone that causes your bones to release calcium (response). Thus, the product of *this* stimulus-response chain is released calcium; when enough of it is circulating, the parathyroid glands sense this and stop releasing their calcium-liberating hormone (**see** FIGURE 26.12).

With both the thermostat and the parathyroid glands, we can see that their responses to a stimulus bring about a decrease in their own activity. In other words, their responses feed back on their activity in a negative way—they reduce it. This is **negative feedback**, which can be defined as a process in which the elements that bring about a response have their activity reduced by that response.

Negative Feedback and Homeostasis

Though our parathyroid example concerns negative feedback as it affects hormonal release, the negative feedback mechanism actually has a terrific *general* importance in the body. To put things simply, most large-scale processes in the human body are controlled by negative feedback. The rate of our breathing, the pH level of our blood, the production of enzymes in our cells; all these processes and many more are governed by negative feedback. And the pervasiveness of this mechanism provides us with an important insight about how the human body operates: It is *self-regulating*. There is no "boss" in the body that directs all of its various systems and subsystems. Instead, there is an endless series of stimulus-and-response reactions that either open up or shut down an activity.

The central place that negative feedback has in human functioning raises an interesting question. Why did evolution settle on *it* as the body's primary regulatory mechanism? The answer is that negative feedback produces the very thing the body needs to maintain itself: stability. The human body must guard against being too hot or too cold; too dehydrated or too hydrated; too stimulated or too relaxed. If it is to exist at all, it must avoid extremes. To put this another way, it must seek **homeostasis**, meaning the maintenance of a relatively stable internal environment. And negative feedback provides this, as you can see from the parathyroid example. Should there be too little calcium in the bloodstream, the parathyroids go to work; when the level is high enough, the parathyroids shut down. Through this mechanism, the body keeps its blood levels of calcium within a very narrow range. There is thus a stability here. And, were we to start looking at almost any major process in the body, we would find this same kind of stability—a stability provided by negative feedback.

So Far... Answers:

1. *reduced*
2. *stable*
3. *brain; pituitary*

Hormonal Secretion: The Hypothalamus

A central player in hormonal release is one of the brain structures we looked at earlier, the hypothalamus. Consistent with what you just learned about negative feedback, the hypothalamus is prompted to act based on input it gets from other sources. (Sensory nerves feed into it, for example.) But it's possible to learn a lot about the endocrine system just by focusing on signals that come from the hypothalamus, rather than the signals that come to it. To put things in a nutshell, activity in the hypothalamus prompts a good deal of the body's hormonal activity, as you can see in FIGURE 26.13. Note all the endocrine tasks that are undertaken by this small part of the brain. First, it acts as a hormonal organ itself, releasing, through one part of the pituitary gland—the posterior pituitary—two hormones that it produces. Second, it exercises control over the endocrine cells of the inner portion (or medulla) of the adrenal glands. Upon nervous-system stimulation from the hypothalamus, the adrenal medulla releases into the bloodstream the hormones adrenaline and noradrenaline. Third, the hypothalamus exercises control over a second part of the pituitary gland— the anterior pituitary—which is referred to as a "master gland" in that several of the hormones it releases go on to control *other* endocrine glands. Though lots of hormones are produced outside the control of the hypothalamus, you can see from this list that it is a central structure in the endocrine system. Note that the hypothalamus is a part of the brain, yet it is releasing some hormones and controlling the release of others. This is an example of how extensive the overlap is between the nervous and endocrine systems. **Table 26.1** provides a list of the hormones produced by most of the major endocrine glands in the body, along with the effects of these hormones.

On to the Immune System

Having looked at the capabilities of the nervous and endocrine systems, it's time to move on to another of the body's systems. This system is not involved in communication, but instead has the job of defense. Indeed, we might say it is involved in a never-ending war of defense. On one side in this combat are all kinds of microscopic organisms—bacteria, viruses, fungi—that use every opportunity to invade the body. On the other side is the body's chief defender, the immune system, whose story is the subject of Chapter 27.

So Far...

1. Negative feedback occurs when the elements that bring about a response have their activity _____ by that response.

2. Negative feedback is important because it preserves homeostasis in the body, meaning a _____ internal state.

3. The hypothalamus, a part of the _____, exercises direct control over the body's "master gland," the _____ gland.

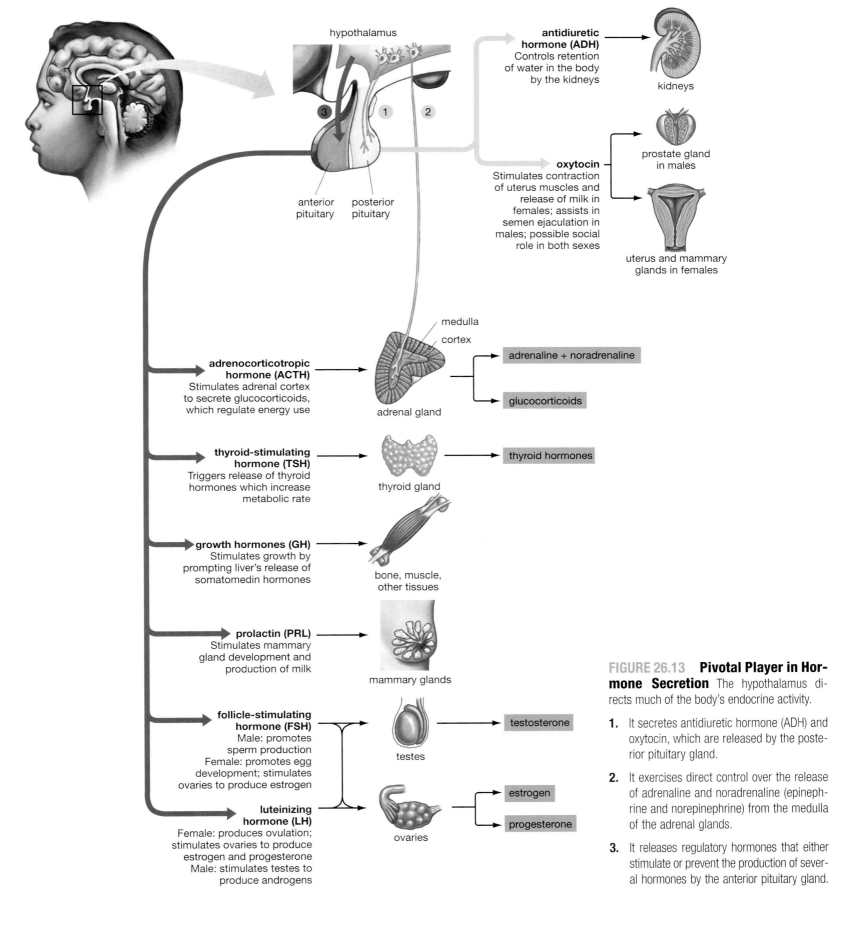

hypothalamus

antidiuretic hormone (ADH)
Controls retention of water in the body by the kidneys

kidneys

oxytocin
Stimulates contraction of uterus muscles and release of milk in females; assists in semen ejaculation in males; possible social role in both sexes

prostate gland in males

uterus and mammary glands in females

anterior pituitary   posterior pituitary

medulla
cortex

adrenocorticotropic hormone (ACTH)
Stimulates adrenal cortex to secrete glucocorticoids, which regulate energy use

adrenal gland

adrenaline + noradrenaline

glucocorticoids

thyroid-stimulating hormone (TSH)
Triggers release of thyroid hormones which increase metabolic rate

thyroid gland

thyroid hormones

growth hormones (GH)
Stimulates growth by prompting liver's release of somatomedin hormones

bone, muscle, other tissues

prolactin (PRL)
Stimulates mammary gland development and production of milk

mammary glands

follicle-stimulating hormone (FSH)
Male: promotes sperm production
Female: promotes egg development; stimulates ovaries to produce estrogen

testes

testosterone

luteinizing hormone (LH)
Female: produces ovulation; stimulates ovaries to produce estrogen and progesterone
Male: stimulates testes to produce androgens

ovaries

estrogen

progesterone

FIGURE 26.13 Pivotal Player in Hormone Secretion The hypothalamus directs much of the body's endocrine activity.

1. It secretes antidiuretic hormone (ADH) and oxytocin, which are released by the posterior pituitary gland.

2. It exercises direct control over the release of adrenaline and noradrenaline (epinephrine and norepinephrine) from the medulla of the adrenal glands.

3. It releases regulatory hormones that either stimulate or prevent the production of several hormones by the anterior pituitary gland.

Table 26.1 Hormones of the endocrine system: Their sources and effects

Gland/Hormone	Effects
Hypothalamus	
Releasing hormones	Stimulate hormone production in anterior pituitary
Inhibiting hormones	Reduce hormone production in anterior pituitary
Anterior pituitary	
Thyroid-stimulating hormone (TSH)	Triggers release of thyroid hormones
Adrenocorticotropic hormone (ACTH)	Stimulates adrenal cortex cells to secrete glucocorticoids
Follicle-stimulating hormone	Female: promotes egg development; stimulates ovaries to produce estrogen Male: promotes sperm production
Luteinizing hormone (LH)	Female: produces ovulation (egg release); stimulates ovaries to produce estrogen and progesterone Male: stimulates testes to produce androgens (e.g., testosterone)
Prolactin (PRL)	Stimulates mammary gland development and production of milk
Growth hormone (GH)	Stimulates growth by prompting liver's release of somatomedin hormones
Posterior pituitary	
Antidiuretic hormone (ADH)	Controls retention of water in the body by the kidneys
Oxytocin	Stimulates contraction of uterus muscles and release of milk in females; assists in semen ejaculation in males; possible social role in both sexes
Thyroid	
Thyroxine	Increases general rate of body metabolism
Calcitonin	Reduces calcium ion levels in blood
Parathyroid	
Parathyroid hormone (PTH)	Increases calcium ion levels in blood
Thymus	
Thymosins	Stimulate development of white blood cells (lymphocytes) in early life
Adrenal cortex	
Glucocorticoids	Includes cortisol, which stimulates glucose production and breakdown of fats. A stress-response hormone.
Mineralocorticoids	Cause the kidneys to retain sodium ions and water and excrete potassium ions
Adrenal medulla	
Adrenaline	Also known as epinephrine; stimulates release of energy stores; increases heart rate and blood pressure
Noradrenaline	Also known as norepinephrine; effects similar to epinephrine
Pancreas	
Insulin	Decreases glucose levels in blood
Glucagon	Increases glucose levels in blood
Testes	
Testosterone	Promotes production of sperm and development of male sex characteristics
Ovaries	
Estrogens	Support egg development, growth of uterine lining, and development of female sex characteristics
Progesterones	Prepare uterus for arrival of developing embryo and support of further embryonic development
Pineal gland	
Melatonin	Establishes day/night cycle

Chapter Review

Summary

26.1 Communication within the Body

- **Endocrine and nervous systems**—Communication needs within the human body are handled in large part through two communication networks, the endocrine and nervous systems.   (page 469)

- **Nervous-system structure**—The nervous system includes all the nervous tissue in the body, plus the body's sensory organs, such as the eyes and ears. Nervous tissue is composed of neurons, which transmit nervous-system messages; and glial cells, which support neurons.   (page 470)

- **Central and peripheral systems**—The two major divisions of the human nervous system are the central nervous system (CNS), consisting of the brain and spinal cord; and the peripheral nervous system (PNS), which includes all the neural tissue outside the CNS, plus the sensory organs.   (page 470)

- **Afferent and efferent divisions**—The PNS has an afferent division, which brings sensory information to the CNS; and an efferent division, which carries action commands to the body's "effectors"—muscles and glands. Within the PNS's efferent division are two subsystems: the somatic nervous system, which provides voluntary control over skeletal muscles; and the autonomic nervous system, which provides involuntary regulation of smooth muscle, cardiac muscle, and glands.   (page 472)

- **Sympathetic and parasympathetic**—The autonomic system is further divided into the sympathetic division, which generally has stimulatory effects; and the parasympathetic division, which generally has relaxing effects.   (page 472)

26.2 Cells of the Nervous System

- **Types of neurons**—There are three types of neurons: sensory neurons, which sense conditions inside and outside the body and convey information about these conditions to neurons inside the CNS; motor neurons, which carry instructions from the CNS to such structures as muscles or glands; and interneurons, which are located entirely within the CNS and interconnect other neurons.   (page 472)

- **Axons and dendrites**—Each neuron has extensions called dendrites that receive signals coming to the neuron cell body, and a single large extension, called an axon, that carries signals from the cell body. Glial cells serve to protect neurons and to facilitate communication by them. A nerve is a bundle of axons in the PNS that transmits information to or from the CNS.   (page 473)

26.3 How Nervous-System Communication Works

- **Two-step communication**—Nervous-system communication can be conceptualized as working through a two-step process: signal movement down a neuron's axon, and signal movement from this axon to a second cell across a structure known as a synapse.   (page 474)

- **Membrane and action potentials**—An electrical charge difference, called a membrane potential, exists across the plasma membrane of neurons because the inside of the neuron is negatively charged relative to the outside. This represents a form of potential energy that is put to use when channels in the neuron's membrane open up upon stimulation, thereby allowing charged particles called ions to flow into the neuron's axon. Another factor influencing this influx is difference in the concentration of ions inside and outside the membrane. A greater concentration of sodium (NA^+) ions outside the membrane means that these ions will have a natural tendency to flow into the neuron with the opening of the channels. The influx of NA^+ ions at an initial point on the axon triggers reactions that cause the adjacent portion of the axonal membrane to initiate the same influx of ions. Thus a conducted nerve impulse, called an action potential, moves down the entire axon in a set of linked reactions.   (page 474)

- **Synaptic transmission**—A nerve signal moves from one neuron to another (or to a muscle or gland cell) across a synapse, which includes a "sending" neuron, a "receiving" cell, and a tiny gap between the two cells, the synaptic cleft. A chemical called a neurotransmitter diffuses across the synaptic cleft from the sending neuron to the receiving neuron. It then binds with receptors on the receiving neuron, thus keeping the signal going. Neurotransmitters can be degraded in synaptic clefts by enzymes, or taken back into a sending cell in the process called reuptake.   (page 478)

26.4 The Spinal Cord

- **Spinal cord functions**—The spinal cord can act as a nervous-system communication center, receiving input from sensory neurons and directing motor neurons with no input from the brain. It also channels sensory impulses to the brain. Spinal nerves extend from the spinal cord to most areas of the body.   (page 479)

- **Reflexes**—Reflexes are automatic nervous-system responses, triggered by specific stimuli, that help us avoid danger or preserve a stable physical state. The neural wiring of a single reflex, called a reflex arc, begins with a sensory receptor, runs through the spinal cord, and proceeds back out to an effector such as a muscle or gland.   (page 481)

26.5 The Autonomic Nervous System

- **Sympathetic and parasympathetic**—The sympathetic division of the autonomic nervous system is often called the fight-or-flight system, because it generally activates bodily functions. The parasympathetic division is often called the rest-and-digest system, because it conserves energy and promotes digestive activities. Most organs receive input from both systems.   (page 482)

Web Tutorial 26.1 The Nervous System

26.6 The Human Brain

- **Regions of the brain**—There are six major regions in the adult brain: the cerebrum, thalamus and hypothalamus, midbrain, pons, cerebellum, and medulla oblongata. The cerebrum is divided into right and left cerebral hemispheres and is the seat of our higher thinking and processing. The cerebrum also has a thin outer layer, the cerebral cortex, that is the site of our highest thinking. The brainstem is a collective term for three brain areas—the midbrain, pons, and medulla oblongata. These brainstem structures are active in controlling involuntary bodily activities (such as breathing and digesting), in relaying information, and in processing sensory information. Most of the body's sensory perceptions are channeled through the thalamus before going to the cerebral cortex. The hypothalamus is important in sensing internal conditions and in maintaining stability or homeostasis in the body. The cerebellum is important in refining movement and maintaining balance.   (page 483)

26.7 The Endocrine System

- **The nature of hormones**—The endocrine system functions in the control and regulation of bodily processes. It works through a group of chemical messengers called hormones: substances secreted by one group of cells that travel through the bloodstream and affect the activities of other cells. Hormones stand in distinction to other signaling molecules the body uses that do not travel through the bloodstream. (page 485)

- **Target cells**—Each hormone works only on specific cells—the hormone's target cells. Hormones bind to their target cells via receptors on or in the target cells. This binding then spurs chemical reactions within the target cells.   (page 486)

- **Endocrine glands**—Hormonal production and secretion takes place to a significant extent within endocrine glands, meaning glands that secrete materials directly into the bloodstream or into surrounding tissues. Some hormones are secreted, however, not by specialized glands, but by organs such as the heart or kidneys.   (page 487)

Web Tutorial 26.2 The Endocrine System

26.8 Types of Hormones

- **Three classes of hormones**—There are three principal classes of hormones: amino-acid-based hormones, peptide hormones, and steroid hormones. Each amino-acid-based hormone is derived from a chemical modification of a single amino acid. Peptide hormones are composed of chains of amino acids. Steroid hormones are all constructed around the chemical framework of the cholesterol molecule. Amino-acid-based and peptide hormones generally link to their target cells via receptors that protrude from the target cells' outer membranes. Most steroid hormones pass through a cell's plasma membrane and bind with a receptor protein inside the cell. The combined steroid hormone/receptor molecule then binds with the cell's DNA, thus turning on one or more cell genes.   (page 487)

26.9 Hormone Secretion and Negative Feedback

- **Control of hormone secretion**—Almost all hormone secretion is controlled by negative feedback: a process in which the elements that bring about a response have their activity reduced by that response. Negative feedback controls most large-scale processes in the body. It plays this central role because it is so well suited to preserving homeostasis: the maintenance of a relatively stable internal environment.   (page 488)

- **Activity of the hypothalamus**—The brain's hypothalamus is an important part of the endocrine system in that it (1) acts as an endocrine organ, producing two hormones that are released by the posterior pituitary gland; (2) exercises control, via the nervous system, over the release of two hormones—adrenaline and noradrenaline—that are produced by the adrenal glands; and (3) controls release of six hormones secreted by the anterior pituitary gland. The anterior pituitary is known as the body's "master gland" because several of the hormones it releases go on to affect the release of hormones in other endocrine glands.   (page 490)

Key Terms

action potential   p. 477

afferent division   p. 472

amino-acid-based hormones
p. 487

autonomic nervous system
p. 472

axon   p. 474

central nervous system (CNS)
p. 470

cerebral cortex   p. 484

cerebrum   p. 484

dendrites   p. 473

efferent division   p. 472

endocrine gland   p. 486

homeostasis   p. 489

hormone   p. 486

hypothalamus   p. 485

interneuron   p. 472

membrane potential   p. 477

motor neuron   p. 472

myelin   p. 474

negative feedback   p. 489

nerve   p. 474

neurotransmitter   p. 479

parasympathetic division   p. 482

peptide hormones   p. 488

**peripheral nervous system
(PNS)**   p. 470

reflex   p. 481

sensory neuron   p. 472

somatic nervous system   p. 472

steroid hormones   p. 488

sympathetic division   p. 482

synapse   p. 478

synaptic cleft   p. 479

target cells   p. 486

thalamus   p. 485

Testing Your Understanding

In Your Own Words *(answers in the back of the book)*

1. Three functional types of neurons are found in the nervous system. What are they, and what role does each fill?

2. Describe how a simple reflex arc works.

3. What are the differences between amino-acid-based, peptide, and steroid hormones?

Thinking about What You've Learned

1. Should we step on a sharp object, we have a one-two reflex response that prompts us not only to lift the leg that stepped on the object, but to straighten out our other leg (so that we won't fall over). In a similar vein, any surprising, loud sound causes us to turn our heads in the direction of the sound (in the so-called orientation reflex). Can you think of other actions we take that are completely involuntary? In other words, what other reflexes do we have?

2. Thanks to its various sorts of sensory receptors, the body can respond to many different kinds of stimulation—vibrations, smells, pressure, light, and taste among them. Can you think of any other kinds of stimulation that our bodies did not evolve to respond to, but might have? Could the senses we have be extended in any way? (Hint: Think of the olfactory abilities of dogs.)

3. Given the duties of various regions of the brain, why is it almost literally true to speak of "higher" brain functions?

The Immune System

SCIDS patient David Vetter, inside his isolation chamber in 1976.

THE DESERET NEWS

UTAH 'BUBBLE BOY' A GIRL

By Lois M. Collins, Deseret
Morning News
June 5, 2004

Abby Flemetakis is a smiling, perky 1-year-old who would be indistinguishable from others her age were it not for the tiny tube running through her nose into her stomach and the central line in her chest. That and the fact that she only gets out, aside from occasional strolls through the neighborhood, when it's time to go to the doctor. Which happens a lot. Even there she has to wear a face mask.

Abby was diagnosed at 7 months with a form of severe combined immune deficiency syndrome (SCIDS). Her body doesn't produce its own T-cells, so she has no immune system. It's one form of the condition sometimes called "Bubble Boy" syndrome.

27.1 Defenders of the Body

Who was the "bubble boy" whose condition conferred a name on the illness that Abby Flemetakis suffers from? He was David Vetter, born in 1971, deceased in 1984, and alive for almost all of the 12 years in between in a series of isolation chambers that came to be known as "bubbles" because of the clear plastic they were made from. The water with which he was baptized had to be sterilized. The air he breathed was filtered; the clothes he wore were treated with disinfectants. His parents and doctors could handle him only with special gloves that were attached to the bubble. By the time he was 12, the prospect of continued life within the bubble seemed so bleak that his parents decided to risk a bone-marrow transplant—with tissue provided by his sister—as a means of providing him with an immune system. A virus went undetected in his sister's marrow, however, and without an immune system to fight off its effects, he died of cancer four months after getting his transplant.

The severe combined immune deficiency (SCID) that David Vetter suffered from no longer results in children being placed in isolation chambers. But that does not

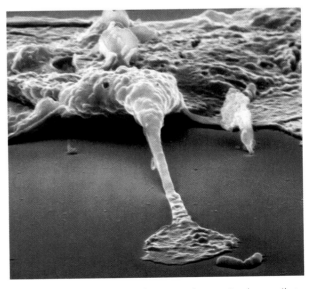

An immune-system cell called a macrophage extends a portion of itself in an attempt to ingest an invading bacterial cell.

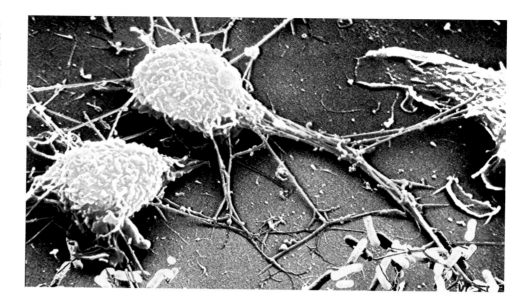

FIGURE 27.1 **Defenders and Invaders** The large, spherical cells on the left are immune-system macrophages that are in the process of attacking invading *E. coli* bacterial cells, at the bottom of the photo. The macrophages will engulf the bacterial cells, ingest them, and dissolve them. (Micrograph ×2,150)

mean that life is pleasant for the children who are afflicted with this condition. Abby Flemetakis's house and all of her belongings must be scrubbed or sterilized with great frequency; she takes a host of medications; and to provide her with an immune system, doctors have given her something called a cord-blood stem cell transplant, whose outcome is uncertain.

With children like Abby Flemetakis and David Vetter serving as examples, we don't need to wonder what our immune system does for us. It allows us to live. Our bodies are constantly under assault from all kinds of microbes—viruses, bacteria, fungi (see FIGURE 27.1). Remarkably, however, the war that we fight with these organisms takes place completely outside our awareness, except on those rare occasions when the immune system can't stop an attack quickly enough. Even then, what we generally suffer is a couple days in bed—time enough for the immune system to gain the upper hand and bring us back to normal.

The workings of this system are all the more extraordinary because of the diversity of invaders that it faces. As most people know, disease-causing bacteria and viruses don't just come in a few varieties; cold viruses alone come in hundreds of variants. How can the immune system deal with a huge number of *different* invaders? And how can it perform the neat trick of getting rid of a disease like chicken pox once and thereafter keep us free from this disease for the rest of our lives? Beyond this, how does it work with vaccinations to keep us from ever getting diseases like polio and diphtheria? The goal of this chapter is to explain how all of this and more is taken care of by the defender of the body, the immune system.

General Features of the Immune System

The human immune system is armed with two essential types of defenses. First, it has **nonspecific defenses**: immune-system defenses that do not discriminate between one invader and the next. Our tears contain bacteria-fighting enzymes, called lysozymes, that do not discriminate between one type of bacterium and another. Likewise, some immune cells will attack any cell perceived to be foreign. Working with these elements is the system's second line of defense, its **specific defenses**: immune-system defenses that provide protection against *particular* invaders. You have immune cells that will latch onto bacterium A, but not to bacterium B. And these specific defenders *remember* invaders they have faced, which enables them to generate a rapid response to any subsequent invasion by the same microbe. The result is **immunity**: a state of long-lasting protection that the immune system develops against specific microorganisms.

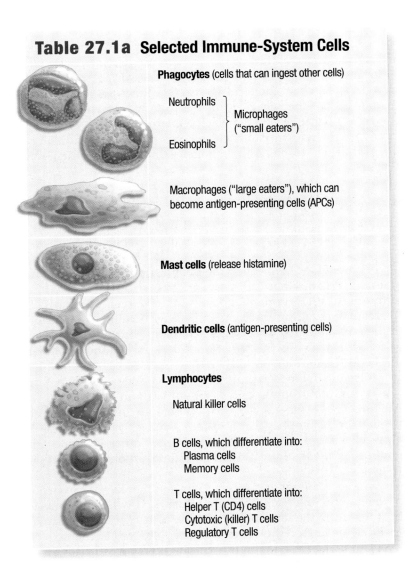

Table 27.1a Selected Immune-System Cells

Phagocytes (cells that can ingest other cells)

Neutrophils
Eosinophils
} Microphages ("small eaters")

Macrophages ("large eaters"), which can become antigen-presenting cells (APCs)

Mast cells (release histamine)

Dendritic cells (antigen-presenting cells)

Lymphocytes

Natural killer cells

B cells, which differentiate into:
Plasma cells
Memory cells

T cells, which differentiate into:
Helper T (CD4) cells
Cytotoxic (killer) T cells
Regulatory T cells

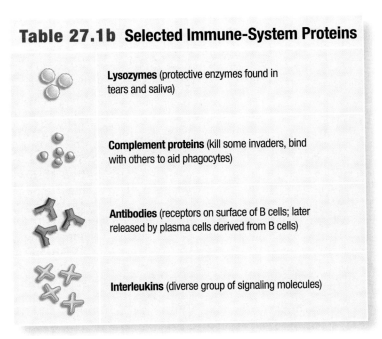

Table 27.1b Selected Immune-System Proteins

Lysozymes (protective enzymes found in tears and saliva)

Complement proteins (kill some invaders, bind with others to aid phagocytes)

Antibodies (receptors on surface of B cells; later released by plasma cells derived from B cells)

Interleukins (diverse group of signaling molecules)

As you'll see, nonspecific and specific defenses can work together in fighting off a given infection. Moreover, both kinds of defenses are prompted to act by the same thing: the presence of an **antigen**, which can be defined as any foreign substance that elicits an immune-system response. A bacterial cell in the lining of our lungs is, of course, foreign to our bodies. But it is certain proteins or carbohydrates on the surface of the bacterial cell that serve as the antigens, setting off the immune response. In general, antigens are component parts of living things.

Table 27.1 lists just a few of the many players that take part in the immune process. All the entities in the table do one of three things: They destroy antigens, they mark antigens for destruction, or they act as communication molecules within the immune system. Note that all of the immune cells listed in Table 27.1a are different kinds of white blood cells. Such cells are the key players in immunity; any cell that is active in the immune system is likely to be a variety of white blood cell.

27.2 Nonspecific Defenses

The body's first line of defense actually is not a part of the immune system per se, but instead is a set of barriers. To cause trouble, a disease-causing organism—known as a *pathogen*—must enter the body's tissues. What stands in the way? One barrier is skin, described in Chapter 25, which has multiple layers of cells, a

A BROADER VIEW

Latex as an Antigen

We can encounter antigens in the most unexpected places. A small proportion of people experience allergic reactions to latex, such as is found in the gloves used by doctors. Why should latex set off an immune reaction? Because it is the product of a living thing (the Brazilian rubber tree) and as such contains proteins that can trigger an immune response.

waterproof keratin coating, and a network of tight seams that lock adjacent cells together. These specializations create a very effective barrier that protects underlying tissues. The exterior surface of the body is also protected by hairs and glandular secretions. Secretions from sweat glands flush the surface of the skin, washing away microorganisms and chemical agents. Our lungs, which come into contact with air from the outside world, have the protective material we call mucus, along with a set of hair-like cilia that sweep their passageways clean like so many tiny brooms. Meanwhile, the stomach contains a powerful acid that can destroy pathogens; and urine, which has a pathogen-killing low pH, flushes the urinary passageways.

Nonspecific Cells and Proteins

But what happens if invaders get past these barriers? We have a line of nonspecific defense waiting to meet them. Let's look at some of the members of this team of defenders.

There is a type of white blood cell, called the **phagocyte**, that can be defined as a cell capable of ingesting another cell, parts of cells, or other materials. As you might guess from this description, phagocytes can take on a warrior role—they essentially surround invading microorganisms with cellular extensions, take them in, and then dissolve them in a set of acid baths. But phagocytes also take on a less-exalted role, which is that of janitor. They engulf and haul away the body's *own* cells and tissue fragments when they have become worn out or damaged. As you can see in Table 27.1a, phagocytes come in several varieties, two of them smaller (the neutrophils and eosinophils) and one larger (the macrophages). All three have the capacity to kill foreign invaders outright. Later, you'll be seeing the macrophage in a different role: It not only kills invaders, but then presents fragments of these invaders to other immune cells, something like a soldier who is holding an enemy's uniform out for the rest of the troops to get a look at.

When an invader is a virus, rather than a bacterium, another key player in nonspecific defense is the natural killer cell, so named because it will attack any cell that it perceives as "foreign" in the sense of having unusual surface proteins. Cells infected by viruses often are altered this way; when an "NK" cell encounters one of these infected cells, it can kill them in one of two ways. First, it can use proteins to create so many pores in the membranes of infected cells that these cells break up, like a ship that's been riddled with holes. Second, it can cause these cells to commit suicide, through a process called apoptosis.

Table 27.1a also lists an immune-system cell called the mast cell. This cell releases a substance called histamine, whose role we'll get to shortly. Finally, it is not just cells that are involved in nonspecific defense; it is proteins as well. A group of them listed in Table 27.1b, the complement proteins, will make an appearance soon. Let's now see how all the players work together to mount a type of well-orchestrated nonspecific defense, the *inflammatory response*.

Nonspecific Defense and the Inflammatory Response

Let's say you accidentally puncture your skin with a nail that's been sitting outdoors. This means you not only have a puncture, you have lots of microscopic invaders that have gotten past the barrier of your skin. How does the immune system deal with the bacteria among them? Three types of responses take place. First, blood vessels near the site of injury dilate and become more permeable—they increase in diameter and they undergo a change that allows cells and proteins to pass out of them. Second, these cells and proteins leave the blood vessels and move to the site of injury, where they kill the bacteria. Third, various compounds are released that wall off the site of injury, thus limiting the spread of the infection.

Mast cells help get the first step going by releasing a substance called **histamine**: a compound that, in the inflammatory response, brings about blood-vessel dilation and increased blood-vessel permeability. Mast cells have significant quantities of histamine to release because they store it inside themselves in tiny granules. The blood-vessel dilation that histamine prompts is important because it results in increased blood flow near the injury site—an efficient means of bringing in more immune-system cells and proteins. The increased permeability histamine causes is important because it allows these cells and proteins to move *out* of the blood vessels so that they can get to the site of the infection. Normally, blood-vessel cells are locked tightly together; histamine has the effect of opening up spaces between them.

The first immune-system cells to squeeze through these spaces probably will be one of the smaller phagocytes, the neutrophils. Slower in arriving, but longer lived, are the macrophages. Both varieties of cells are guided to the site of injury by a form of chemical signaling: Injured cells release compounds that produce a kind of trail that leads to them. Once the neutrophils and macrophages arrive, they begin ingesting bacterial invaders; later the macrophages will clean up the debris.

Flowing through the bloodstream at this time, and also activated by the attack, are the complement proteins noted earlier—about 20 of them in all. They are called complement proteins because their actions are complementary to those taken by other parts of the immune system. In this instance, one of their roles is to cut holes in the cell membranes of the bacteria. The site of infection is likewise being limited in size at this time. The area is sealed off partly through creation of a network of fibers composed of a blood protein that, appropriately enough, is called fibrin.

In looking at this simplified account of the inflammatory response, you can see that it brings more blood near a site of injury and more cells and proteins to the site. Blood is warm and red, of course, and this explains why the area around an injury site takes on these same qualities. Meanwhile, the added cells and proteins at the site cause the swelling and pain that we associate with injuries.

It's important to note that all these responses are nonspecific: Any group of invaders that entered through this route would get pretty much this same treatment by this same lineup of immune-system players. But what about instances in which the body's nonspecific defenses can't contain an infection? What happens when, say, a chicken pox virus comes into the nasal passages and starts to spread from there throughout the body? If this invader can't be defeated by nonspecific defenses, the body has another line of defense at the ready—the specific defenses, which have a marvelous ability to target specific invaders.

So Far...

1. The human body's _____ defenses provide protection against particular invaders. The memory of these defenses allows the body to generate a rapid response to any subsequent invasion by the same microbe. This long-lasting state of protection is called _____.

2. An antigen is any substance that _____.

3. All of the cells most important to immune-system function are varieties of _____ cells. One variety of these cells, the phagocyte, is a cell that is capable of _____.

So Far... Answers:

1. *specific; immunity*

2. *elicits an immune-system response*

3. *white blood; ingesting another cell, parts of cells, or other materials*

27.3 Specific Defenses

Specific defenses work through something known as acquired immunity, which is the immunity that results from what humans come into contact with during their lifetimes. One type of acquired immunity is provided to us by others. **Passively acquired immunity** is immunity gained by the administration of disease-fighting

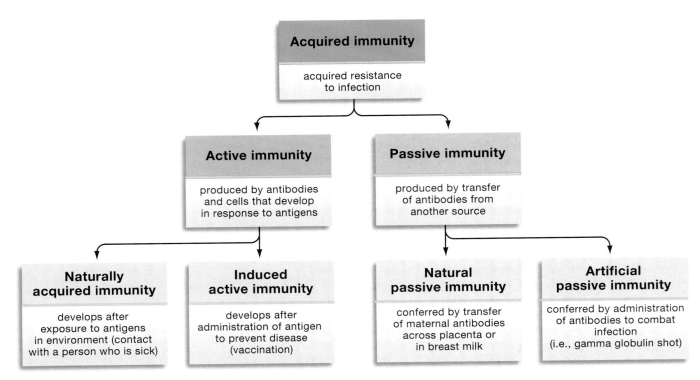

FIGURE 27.2 **Types of Acquired Immunity**

substances, called antibodies, that have been produced by another individual. Mothers pass antibodies to their children in the womb and in breast milk, for example. As an adult, you might receive antibodies in the form of a gamma globulin shot, which people sometimes get prior to taking a trip abroad (**see FIGURE 27.2**). Invader-fighting antibodies made by others are degraded rather rapidly in our system, however, which means that the immunity these antibodies provide lasts a few weeks or months. Of more interest to us is the long-lasting immunity that our bodies generate themselves. This is **actively acquired immunity**: immunity developed as a result of accidental or deliberate exposure to an antigen.

Note the critical term there: exposure to an *antigen*—a substance the body recognizes as foreign. Think about this in relation to one well-known form of actively acquired immunity, the vaccinations mentioned at the start of the chapter. In vaccinations, a person is not being treated with a substance that kills an invader. Rather, the person is being injected with at least a part of the invader itself. This is deliberate exposure to an antigen, and it results in our body mounting an attack on the antigen in ways that we'll review shortly. Of course this "invader" has been made harmless back in the laboratory. It may be a poliovirus, for example, that has been inactivated, so that it can no longer cause disease. But the immune system's attack on it is full blown; and having mounted it, the system enters a high state of readiness for any subsequent attack by the *real* invader—the live poliovirus. What vaccines do, then, is elicit readiness on the part of the immune system. Getting a disease like chicken pox does this as well—it elicits protection that keeps us from getting chicken pox again. But how does this trick of "invader memory" work? For that matter, how does the immune system defeat specific invaders the first time they attack? These are the subjects to which we now turn.

A BROADER VIEW

Slow and Fast Responses

Any bacterial invader that gets past the barrier of the skin will be met immediately by cells of the nonspecific defenses. In contrast, specific defenses take a long time to ramp up. It takes the body at least four days to produce a significant number of antibodies specific to a new invader.

Antibody-Mediated and Cell-Mediated Immunity

Actively acquired immunity has two major arms. The first arm is called **antibody-mediated immunity**: an immune-system capability that works through the production of proteins called antibodies. The second arm is called **cell-mediated immunity**: an immune-system capability that works through the production of cells that destroy *other* cells in the body—those that have been infected by an invader. Cells that are central to both arms originate in the same place, which is the marrow of our bones. It is there that all white blood cells begin development in adults. The white blood cells that are key to actively acquired immunity are called *lymphocytes*. Some lymphocytes go on to migrate to the body's thymus (located at the base of the throat), and there they specialize. They develop into the main cells of cell-mediated immunity, the **T-lymphocyte cells** or just **T cells** (with the *T* standing for "thymus" cell). Meanwhile, lymphocytes that remain in the bone marrow can develop into the central cells of *antibody*-mediated immunity, the **B-lymphocyte cells** or **B cells**. The development of T and B cells is tracked in FIGURE 27.3.

Eventually, both types of cells will migrate to the body's lymphatic system organs—the spleen and the lymph nodes, for example—where they are prepared to fight invaders. The lymphatic system is laid out for you in FIGURE 27.4. Think of it as a system of vessels that picks up fluids (called interstitial fluids) that have leaked from blood vessels. Eventually, these fluids are channeled back into the circulatory system; but while they are in the lymphatic system, they pass through the lymphoid organs, which are packed with immune-system cells. This is why, when you get sick, your "lymph glands" are swollen—they are filled with an abnormally large number of cells that are busy fighting off the infection. If you look back at Table 27.1, you can see the various types of lymphocytes listed.

From this point forward, the antibody-mediated and cell-mediated immune operations will be treated separately. The two systems work together, but for now we'll look at them in isolation, starting with antibody-mediated immunity.

27.4 Antibody-Mediated Immunity

As you might imagine, a key player in antibody-mediated immunity is the **antibody**, which can be defined as a circulating immune-system protein that binds to a particular antigen. From this definition, it may be apparent that antibodies and antigens are closely linked. (They are so closely linked, in fact, that one is named for the other; the word *antigen* is short for "antibody generating.") In general, a given antibody will bind to a specific antigen and to no other antigen. In their binding capacity, you will see antibodies in two roles: first as *receptors* for antigens on the surface of B cells, extending in a generalized Y-shape from the cells' outer membranes; second as *free-standing* molecules that are produced by B cells and then stream away from them in great numbers, moving through the bloodstream.

The Fantastic Diversity of Antibodies

As noted, there are hundreds of viruses that bring on the general set of symptoms we call a cold. Then there are all the other potential invaders, such as bacteria, fungi, and so forth. How can the immune system cope with such a variety of foes? One of its key strengths is that it produces B cells that differ from each other with respect to the type of antigen receptors they have on their surfaces. Look at one B cell and it will have receptors that latch onto an antigen on *this* virus only; look at a *different* B cell and its receptors bind with an antigen of a different virus. This mind-boggling complexity is made possible by DNA arranging that takes place back in the bone marrow, when the precursors of B cells are being formed. The result is B cells with tens of millions of variations in antigen receptors. Now, how do these B cells do their job?

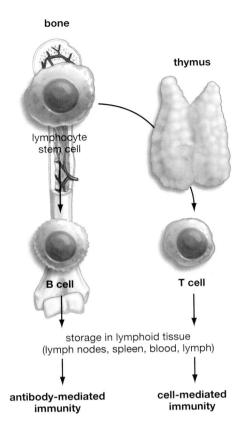

FIGURE 27.3 **Development of T Cells and B Cells**
Key players in the body's specific defense system are T cells and B cells, both of which start out as a type of white blood cell, called a lymphocyte stem cell, in the bone marrow. Some of these lymphocytes migrate to the thymus gland, where they differentiate into T cells (thymus cells). Others that remain in the bone marrow develop into B cells. Most T and B cells then migrate to lymphoid tissues, such as lymph nodes and the spleen, where they will serve their immune function.

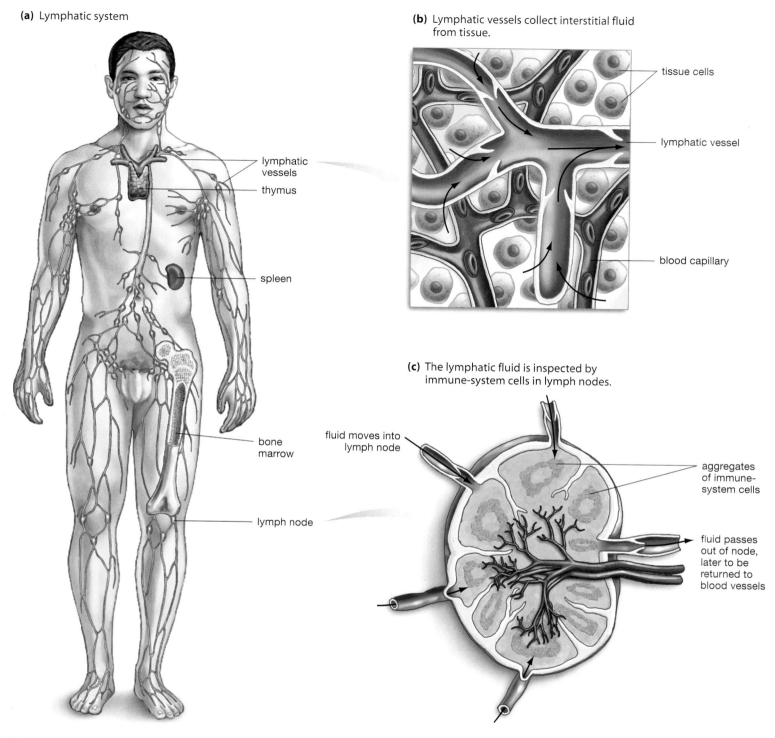

(a) Lymphatic system

(b) Lymphatic vessels collect interstitial fluid from tissue.

lymphatic vessels

thymus

spleen

bone marrow

lymph node

tissue cells

lymphatic vessel

blood capillary

(c) The lymphatic fluid is inspected by immune-system cells in lymph nodes.

fluid moves into lymph node

aggregates of immune-system cells

fluid passes out of node, later to be returned to blood vessels

FIGURE 27.4   The Human Lymphatic System

(a) Fluid moving through lymphatic vessels passes through lymphatic organs, such as lymph nodes or the spleen, where the fluid is inspected by an array of immune-system cells.

(b) Detail of interstitial fluid moving into lymphatic vessels.

(c) Detail of fluid moving through a lymph node.

504

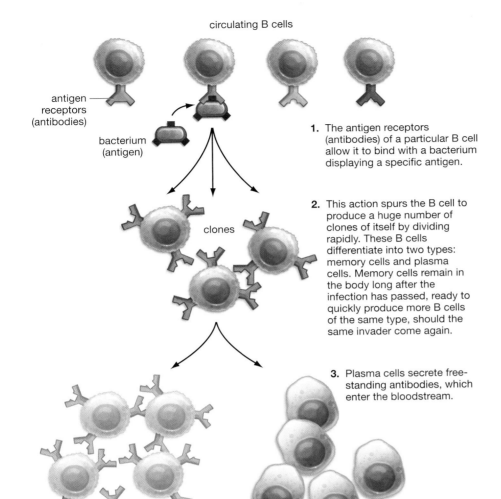

circulating B cells

antigen receptors (antibodies)

bacterium (antigen)

clones

memory cells

plasma cells

FIGURE 27.5 **B Cells and Antibody-Mediated Immunity**

1. The antigen receptors (antibodies) of a particular B cell allow it to bind with a bacterium displaying a specific antigen.

2. This action spurs the B cell to produce a huge number of clones of itself by dividing rapidly. These B cells differentiate into two types: memory cells and plasma cells. Memory cells remain in the body long after the infection has passed, ready to quickly produce more B cells of the same type, should the same invader come again.

3. Plasma cells secrete free-standing antibodies, which enter the bloodstream.

4. Antibodies produced by the plasma cells attack the invader.

The Cloning and Differentiation of B Cells

Lying in readiness in the lymph nodes or circulating through the bloodstream, a B cell encounters, say, a bacterium—one that has an antigen that can be bound by this B cell's antigen receptors. If you look at FIGURE 27.5, you can see what happens next. This very binding causes the B cell to start dividing very rapidly. Each new B cell of this type gives rise to more identical B cells, creating a selected "clone," or a huge number of identical cells of this type. Such numbers are necessary because the *bacteria* will be multiplying rapidly as well.

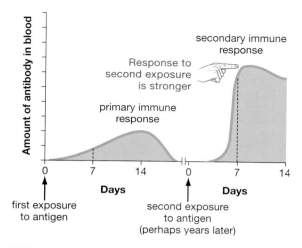

FIGURE 27.6 **Prepared for an Invasion** The memory cells produced by the body during a first attack by an invader allow it to mount a faster, more vigorous defense should the same invader attack a second time.

A BROADER VIEW

Antibodies as Guided Missiles

The exquisite specificity with which antibodies bind to antigens has long held out a tantalizing prospect: Couldn't we produce exactly the antibody we wanted in the laboratory and then use it as a kind of guided missile against, say, cancer tumors? The answer is yes; in the past few years, molecules called monoclonal antibodies have shown an ability to latch onto and disable proteins that facilitate cancer. As an example, a monoclonal antibody called Avastin has been developed that reduces the blood supply going to colon-cancer tumors. How? It binds with a signaling protein, secreted by the tumors, that normally travels to nearby blood vessels and induces them to sprout new vessels—in the direction of the tumors. When bound to the antibodies, however, the signaling molecules are "occupied" and are thus incapable of binding to the blood vessels.

So Far... Answers:

1. exposure to an antigen; readiness

2. lymphatic system; immune-system cells; circulatory

3. antigen; B; exported

After a time, these B cells then differentiate into two types of cells. One variety is called a **plasma cell**: an immune-system B cell that is specialized to produce antibodies. These antibodies are not the cell-surface receptors you've been looking at, however. They are free-standing antibodies that leave the plasma cells by the millions—they are *exported* from the plasma cells—after which they move through the bloodstream and fight the invader directly.

The other type of B cell is called the **memory B cell**, and it is the body's permanent sentry. This is one part of the long-lasting invader readiness noted earlier. After the initial war with an invader is over, this specific variety of memory B cell remains in the body. Should this invader come in a *second* time, these memory cells will be ready to divide and produce more plasma cells to mount a very quick defense. This is why being inoculated with poliovirus antigens protects us permanently from this crippler. If you look at FIGURE 27.6, you can see the typical time-course of exposure to antigen and production of antibodies. What is known as the "primary immune response" has reached a substantial level 14 days out from first exposure. Look at the second exposure, however, to see what a difference this first exposure has made. Now there is a much greater response in a much shorter time.

The Action of the Antibodies

But what is it that the circulating antibodies do to combat the invader? First, because they are binding to the invader's antigens, the antibodies can sometimes prevent the invader from attaching to anything *else*, which stops its spread. Antibodies can also come together with antigens in a clumping or "agglutination" that renders the invaders inactive. Third, antibodies work in concert with an arm of the immune system you've already reviewed—the nonspecific defenses. Remember the cell-ingesting phagocytes and the complement proteins that you saw in nonspecific defense? Many invading bacterial cells have a slick outer coating that keeps phagocytes from latching onto them. Working together, antibodies and complement proteins often *can* bind to many of these invaders, and this gives the phagocytes something to hold onto, so that they can begin ingesting the enemy.

So Far...

1. Vaccination, a deliberate _____, elicits a high state of _____ on the part of the immune system to any second attack by a given microbe.

2. The spleen and lymph nodes, parts of the body's _____, are packed with _____ that inspect lymphatic fluid as it is being returned to the _____ system.

3. An antibody is a circulating immune-system protein that can bind to a particular _____. Antibodies are seen in two forms: as receptors on the surface of _____ cells and as free-standing proteins that are _____ by these cells.

27.5 Cell-Mediated Immunity

Though antibody-mediated immunity is quite spectacular, it also turns out to be limited in that it works strictly on free-standing foreign organisms. The problem with this is that many viruses (and some bacteria) are successful in invading the body's *own cells*—they exist and multiply inside these cells, which makes the viruses inaccessible to antibodies. To deal with this threat, the body has developed another arm of acquired immunity, the cell-mediated immunity mentioned earlier.

You may recall that the central player in cell-mediated immunity is the T-lymphocyte or T cell, which comes in three main varieties. Let's now go through the actions of the cell-mediated immune system, taking as a starting point a body that has been infected with a virus.

Cells Bearing Invaders: Antigen-Presenting Cells

Any cell that has been infected by a virus puts protein fragments from the virus on its surface. Certain immune-system cells, however, don't merely display such fragments; they *present* them, as antigens, to other immune cells as part of the immune-system response. One of these **antigen-presenting cells** or **APCs** is called a dendritic cell. (Another APC is the macrophage you've already been introduced to, and still another is a B cell). Dendritic cells begin their work by engulfing and killing some of the invaders that are free-standing, and then displaying the invader's fragments. Outfitted with these antigens on display, dendritic cells migrate to lymph nodes or the spleen where, you'll recall, immune-system cells exist in abundance. There, the dendritic cells encounter our first variety of T cell, the helper T cell. Among the huge number of helper T cells that exist in a lymph node, a tiny fraction will have receptors specific to this antigen. If you look at FIGURE 27.7, you can see how the next interaction plays out. Helper T cells now lock onto these APCs. Once this interaction has taken place, the helper T cell has been "activated," and this initiates the creation of a helper T cell clone. The first activated helper T cell divides into two cells, these two cells divide into four, and so forth. Critically, each cell in this clone is specific to this invader. These T cells then leave the lymph nodes or spleen and begin circulating throughout the body.

Helper T Cells, Cytotoxic T Cells, and Regulatory T Cells

The helper T cells that have been activated now take on a central role in specific immunity. They secrete a signaling molecule, called interleukin-2, that puts several other processes in motion. First, this secretion spurs production not just of more helper T cells, but of the variety of T cell that actually will eliminate infected cells in the body, the **cytotoxic T cell**. Critically, both cytotoxic and helper T cell clones are being produced not only in active types, but in "memory" types for use in future invasions. Second, the T cell's interleukin-2 secretion spurs the activation of another type of T cell, the **regulatory T cell**, which acts as a kind of weapons-control mechanism—it limits the immune system's response so that the body's own tissues don't get damaged in the battle that ensues. Third, other interleukins secreted by helper T cells activate *B-cell* immunity by promoting the production and differentiation of B cells. Indeed, the entire process of B-cell development is dependent on signals that come from helper T cells.

Once this production process has ramped up, the system's cytotoxic T cells are primed to recognize any cell that is displaying the fragments of the viral invader. Recall that this means any cell in the body that's been infected by the invader, because all infected cells will put some invader fragments on their surface. Using their own special receptors, the cytotoxic T cells now latch onto these infected cells and kill them in one of two ways. First, they puncture the outer membranes of infected cells, causing them to burst—the same kind of lethal action that natural killer cells carry out, but now carried out by a huge number of cells that are targeting this specific invader.

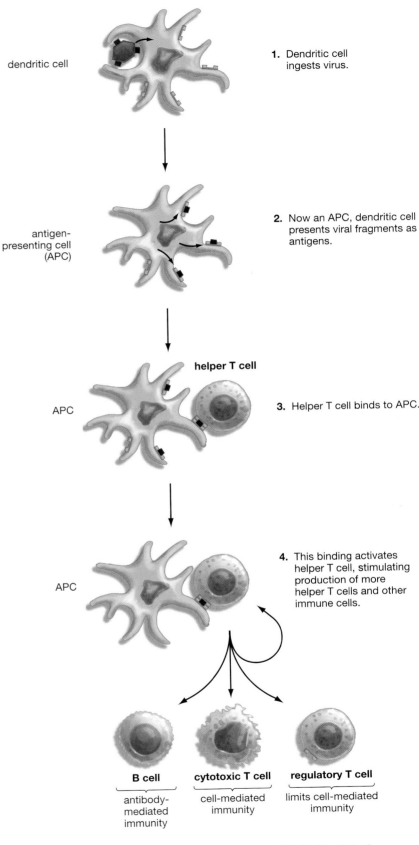

dendritic cell

1. Dendritic cell ingests virus.

antigen-presenting cell (APC)

2. Now an APC, dendritic cell presents viral fragments as antigens.

helper T cell

APC

3. Helper T cell binds to APC.

APC

4. This binding activates helper T cell, stimulating production of more helper T cells and other immune cells.

B cell
antibody-mediated immunity

cytotoxic T cell
cell-mediated immunity

regulatory T cell
limits cell-mediated immunity

FIGURE 27.7   T Cells and Cell-Mediated Immunity

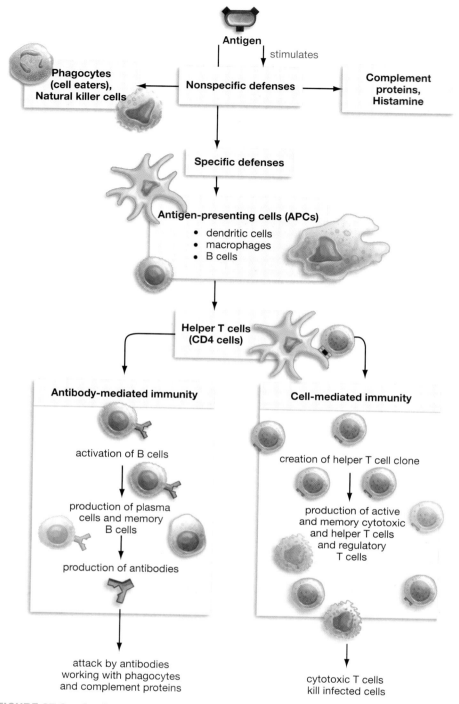

FIGURE 27.8 **An Overview of the Immune-System Process**

Second, they secrete signaling molecules that cause the infected cells to commit suicide through apoptosis. (Given such actions, it's not surprising that another name for cytotoxic T cells is the "killer" T cells.) While this destruction is taking place, the newly activated regulatory T cells are on hand to make sure that the attack doesn't spread too far.

Note that cytotoxic T cells are binding with the body's own infected cells. From this, it may be apparent why this is called *cell-mediated* immunity. In the end, it is an attack involving not the antibodies described before, but several sets of cells.

Taking a step back from cell- and antibody-mediated immunity, in **FIGURE 27.8** you can see how the two systems work together, along with nonspecific defenses. In specific defense, however, note the central role of the **helper T cell**. We can define it as an immune-system cell that helps activate both cell- and antibody-mediated immunity (since it is necessary for the production of both T and B cells).

Helper T Cells and AIDS

Once you understand the pivotal position of the helper T cell in the immune system, you are in a better position to understand one of the most devastating diseases of our time: AIDS. As you probably know, AIDS is caused by a virus—the human immunodeficiency virus, or HIV. And as it turns out, the primary target of HIV in the human body is the helper T cell. The virus invades helper T cells and then uses these cells to turn out more copies of itself. And each time a newly made virus copy "buds off" from an infected T cell, it takes a part of the cell with it, a process that eventually kills the cell. Meanwhile, the emerging virus copies go on to infect more cells, and the cycle is repeated. The result, ultimately, is a body with a very low number of helper T cells. Physicians often refer to helper T cells as "CD4" cells (from the name of one of their receptors). Not surprisingly, a critical question for any physician treating an AIDS patient is: What is the patient's CD4 count? Put another way, how many CD4 cells does the patient have left to fight off infection? Having seen all that the helper T cell does in immune activity, you can understand why this question is so critical. Without helper T cells, the whole immune system falls apart, and the AIDS patient becomes vulnerable to a host of infections that a normal immune system would make quick work of. In sum, AIDS is so devastating because its initial target is a critical player in the immune system.

27.6 The Immune System Can Cause Trouble

In the account you've read so far, the immune system has been portrayed as a kind of multipart weapons system that the body uses to protect itself. The problem with weapons, however, is that they can get pointed in the wrong direction—toward their owners, rather than their targets. And this is what happens with the

FIGURE 27.9 When the Body Attacks Itself
Rheumatoid arthritis has so disfigured this person's hands that simply signing a greeting card has become a difficult act. Rheumatoid arthritis is an autoimmune disorder.

immune system. The price we pay for having it is that it sometimes attacks *us*, rather than the organisms that invade us.

A critical factor in the immune response is the system's ability to distinguish "self" (the body's own uninfected cells) from "non-self" (infected cells or antigens). The problem is that the immune system makes mistakes; it sometimes regards "self" as "non-self." The result is an **autoimmune disorder**: an attack by the immune system on the body's own tissues.

T cells learn to distinguish self from non-self during the embryonic or newborn phases of life. Developing in the thymus during this period, some T cells may never "learn" to distinguish correctly between self and non-self. Conversely, activated B cells can make antibodies against the body's own tissues. Why would such a thing happen? Well, suppose an invading organism had protein antigens whose building-block amino acid sequences were similar to those found in the body's own proteins. Antibodies meant to defeat a virus might end up targeting the body's own tissues instead. The upshot is that the body can attack itself with devastating consequences. If you look at FIGURE 27.9, you can see the results of the disease known as rheumatoid arthritis, which seems to be a result of antibodies attacking connective tissue around joints. There is also multiple sclerosis, in which the immune system starts attacking the nerve-covering myelin noted in Chapter 26. And there is type 1 diabetes, an autoimmune disorder in which antibodies destroy the pancreatic cells that produce insulin.

In addition to autoimmune diseases, the immune system is responsible for **allergies**, which are immune-system overreactions to antigens that result in the release of histamine. Any immune-system attack is triggered by substances that the body perceives as foreign. The foreign substances that trigger allergic reactions are called **allergens**. These substances cause the body to start releasing the histamine noted earlier, and this leads to inflammatory responses that can range from small and harmless (sneezing and sniffling) to large and life-threatening (asthma or anaphylactic shock). Allergens need not be living things, but they are almost always at least derived from living things. Thus pollen, foods, fur, and dust mite excrement are common allergens.

Finally, the immune system is the single most important problem in organ transplantation. Imagine a person who receives, say, a kidney via transplantation. That person's immune system will regard the new kidney as non-self and immediately begin an attack upon it. The only exception to this rule comes when donor and recipient are identical twins. The challenge for modern medicine has been to walk a fine line: to dampen the recipient's immune response somewhat, but not to dampen it so much that the recipient is made vulnerable to serious attacks by microbes.

Given the various types of harmful immune-system responses that can occur, it's not surprising that one of the hottest topics in immune-system research today is the regulatory T cells we looked at earlier—the cells that *limit* immune-system responses. Recent studies indicate that multiple sclerosis patients may have faulty regulatory cells and that asthma and hay-fever sufferers may likewise be victims of regulatory cells that fail to restrain immune-system attacks.

On to Transport and Exchange

Having looked at the amazing capabilities of the immune system, it's time to move on to reviews of four other systems that keep the body functioning. Chapter 28 is concerned with blood, breathing, digestion, and elimination.

A BROADER VIEW

CD4 Counts and AIDS

CD4 cell counts are one of the main tools used to place persons at differing stages of HIV infection. When a person's CD4 count drops below 200 cells per cubic millimeter of blood, that person is defined as having AIDS. The 200-count threshold is used because, at that point, an infected person becomes particularly vulnerable to the kinds of infections that a healthy immune system would quickly fight off.

A BROADER VIEW

Anaphylactic Shock

The worst allergic reactions bring on a condition, called anaphylactic shock, that can be fatal. Penicillin, peanuts, and insect venom are some of the allergens that can cause anaphylactic shock (though only in persons with extreme allergies to these substances). The immediate cause of this condition is the release of histamine. Recall how, in the inflammatory response, histamine caused blood to flow more freely out of blood vessels at the site of a puncture wound? Well, when histamine is released body-wide, as can happen in anaphylaxis, the result is the disastrous drop in blood pressure known as shock. Second, histamine causes smooth muscle to contract, and it is smooth muscle that surrounds the passages of the lungs. The result can be such an extreme narrowing of lung passages that suffocation can ensue. The antidote to anaphylactic shock is an injection of adrenaline.

So Far... Answers:

1. *T; the body's own cells*

2. *helper T; cytotoxic or killer T cell*

3. *autoimmune disorders; attacks by the immune system on the body's own tissues*

So Far...

1. Cell-mediated immunity works primarily through the immune system's _____ cells and is critical in eliminating microbes that reside in _____.

2. A central cell in mediating the body's entire specific immune response is the _____ cell. The key cell in actually destroying invaders in cell-mediated immunity is the _____.

3. Multiple sclerosis and rheumatoid arthritis are examples of _____, which can be defined as _____.

Chapter Review

Summary

27.1 Defenders of the Body

- **Two strategies**—The immune system has two primary defensive strategies: nonspecific defenses, which do not discriminate between one invader and the next; and specific defenses, which provide protection against particular invaders.   (page 498)

- **Antigens and immune cells**—An antigen is any foreign substance that elicits an immune-system response. These substances generally are proteins or carbohydrates that are part of an invading organism. Most of the cells active in immune responses are different varieties of white blood cells.   (page 499)

27.2 Nonspecific Defenses

- **The body's barriers**—The body's first line of nonspecific defense is a set of barriers, such as tightly locked skin cells, and the secretions that protect these barriers, such as the acid secretions of the stomach and the mucus that lines the lungs.   (page 499)

- **The inflammatory response**—Nonspecific defenses mount a coordinated action called the inflammatory response. It involves not only cells and proteins that kill invaders, but substances that increase blood flow and blood-vessel permeability near the site of an injury. (page 500)

27.3 Specific Defenses

- **Types of acquired immunity**—Acquired immunity is the immunity that results from what humans come into contact with during their lifetimes. It exists in two forms. Passively acquired immunity is achieved through the administration of disease-fighting antibodies produced by another individual—for example, manufactured gamma globulins. Actively acquired immunity develops as a result of natural or deliberate exposure to an antigen. Actively acquired immunity can be produced by getting an immunization shot or by catching a disease such as chicken pox.   (page 501)

- **Antibody-mediated and cell-mediated immunity**—Actively acquired immunity has two major arms. The first arm is antibody-mediated immunity, which works through the production of proteins called antibodies. The second arm is cell-mediated immunity, which works through the production of cells that destroy infected cells in the body.   (page 503)

- **Lymphocytes**—White blood cells called lymphocytes are fundamental to both antibody-mediated and cell-mediated immunity. One variety of lymphocyte, the B-lymphocyte cell (or B cell) is central to antibody-mediated immunity. Another type of lymphocyte, the T-lymphocyte cell (or T cell) is central to cell-mediated immunity and helps initiate antibody-mediated immunity.   (page 503)

- **The lymphatic system**—Mature B and T cells are most often found in organs of the body's lymphatic system, which is a system of vessels that captures fluids leaked from the bloodstream; that then subjects these fluids to scrutiny by immune-system cells; and that then returns the fluids to the circulatory system.   (page 504)

27.4 Antibody-Mediated Immunity

- **Antibodies**—An antibody is a circulating immune-system protein that binds to a particular antigen. Antibodies exist (1) as receptors on the surface of B cells that bind to specific antigens, and (2) as free-standing invader-fighting proteins that are secreted by B cells and move through the bloodstream.   (page 503)

- **B cells and immunity**—Antibody-mediated immunity begins with the binding of a B-cell antigen receptor to an antigen. This binding produces a cascade of effects, including the production of a B-cell clone—a huge number of identical copies of the B cell that bound originally to the antigen. Some of these cells are plasma B cells that export free-standing antibodies that fight the invader; others are memory B cells that stay in the body to provide a quick response to a subsequent attack by the same invader.   (page 505)

- **Actions of antibodies**—Antibodies limit infectious attacks in several ways. Their binding with antigens may keep the antigen from binding with anything else, and in concert with complement proteins, antibodies bind to invaders in such a way that macrophages can bind with the invaders and kill them.   (page 506)

Web Tutorial 27.1 Antibody-Mediated and Cell-Mediated Immunity

27.5 Cell-Mediated Immunity

- **The T cell**—Cell-mediated immunity provides protection in instances in which the body's own cells have become infected by an invader. Viruses and some bacteria can infect the body's cells and remain inside them, in which case antibodies will have no access to these invaders. The central player in cell-mediated immunity is the T cell, which comes in three main varieties: helper T cells, cytotoxic (or killer) T cells, and regulatory T cells.   (page 506)

- **Antigen presentation**—Cell-mediated immunity begins with several classes of white blood cells killing invaders and displaying protein fragments from them on their cell surfaces. Any cell that plays this role is known as an antigen-presenting cell, or APC. Once APCs display fragments from an invader, they migrate to lymphatic organs and bind with helper T cells that have receptors specific to this invader. This "activation" of helper T cells prompts the production of a helper T cell clone, with all the members of this clone being specific to this invader.   (page 507)

- **Helper T cell activity**—Helper T cells from the clone then migrate out of the lymphatic organs to encounter infected cells wherever they may be in the body. Substances secreted by helper T cells activate cytotoxic T cells, which prompts the creation of a cytotoxic T cell clone. Both this clone and the helper T cell clone are being produced in active and "memory" varieties. In addition, helper T cell activity stimulates regulatory T cell activation and B-cell clone development.   (page 507)

- **Cytotoxic T cell and regulatory T cell actions**—Cytotoxic T cells bind with infected cells and bring about their destruction in two ways: They puncture the cells' membranes, and they initiate suicide or apoptosis in the infected cells. Regulatory T cells limit immune-system attacks, thus sparing the body's own tissues from damage.   (page 507)

- **Helper T cells and AIDS**—AIDS, or acquired immune deficiency syndrome, is caused by HIV, the human immunodeficiency virus. This virus is devastating to the body because its main target is the helper T cell (or CD4 cell), which is central to initiating both antibody-mediated and cell-mediated immunity.   (page 508)

27.6 The Immune System Can Cause Trouble

- **Autoimmune disorders**—The immune system can attack the body's own healthy tissues, essentially mistaking the body's own tissues ("self") for foreign substances ("non-self"). These attacks produce such autoimmune disorders as rheumatoid arthritis, multiple sclerosis, and type 1 diabetes.   (page 508)

- **Allergies**—An allergy is an immune-system overreaction to a foreign substance (an allergen) that causes the body to release infection-fighting histamine. The result is an inflammatory response that can range from small and harmless (sneezing and sniffling) to large and life-threatening (asthma or anaphylactic shock).   (page 509)

- **Organ transplantation**—The immune system is the most important problem in organ transplantation. The immune system regards any transplanted organ as "non-self" and thus initiates an attack on it. (page 509)

Key Terms

actively acquired immunity
   p. 502

allergen   p. 509

allergies   p. 509

antibody   p. 503

antibody-mediated immunity
   p. 503

antigen   p. 499

antigen-presenting cell (APC)
   p. 507

autoimmune disorder   p. 509

B-lymphocyte cell (B cell)
   p. 503

cell-mediated immunity   p. 503

cytotoxic T cell (killer T cell)
   p. 507

helper T cell (CD4 cell)   p. 508

histamine   p. 500

immunity   p. 498

memory B cell   p. 506

nonspecific defenses   p. 498

passively acquired immunity
   p. 501

phagocyte   p. 500

plasma cell   p. 506

regulatory T cell   p. 507

specific defenses   p. 498

T-lymphocyte cell (T cell)
   p. 503

Testing Your Understanding

In Your Own Words *(answers in the back of the book)*

1. Name the main actions that take place in the inflammatory response.

2. HIV infection can lead to several forms of cancer, including Kaposi's sarcoma and a type of lymphoma; it can lead to a kind of dementia and a physical wasting; it makes its victims prone to several kinds of fungal infections that are rarely seen in the general population. How can one virus have this many effects?

3. Why is the immune system an obstacle to organ transplantation?

Thinking about What You've Learned

1. Public health officials agree that, since the 1960s, asthma has been increasing among children in developed countries such as the United States. Among the tentative explanations for this increase is the "hygiene hypothesis," which holds that the problem is that many children are growing up in environments that are "pristine"—very clean and devoid of household pets. What could the linkage be between an immune-system condition such as asthma and growing up in a pristine environment?

2. The immune system is a particularly complex biological system. Why do you think it evolved to have so many different subsystems and component parts?

3. One of the key treatments for children who lack an immune system is a bone-marrow transplant. Why would this procedure help to establish an immune system in a person who does not have one?

Transport and Exchange

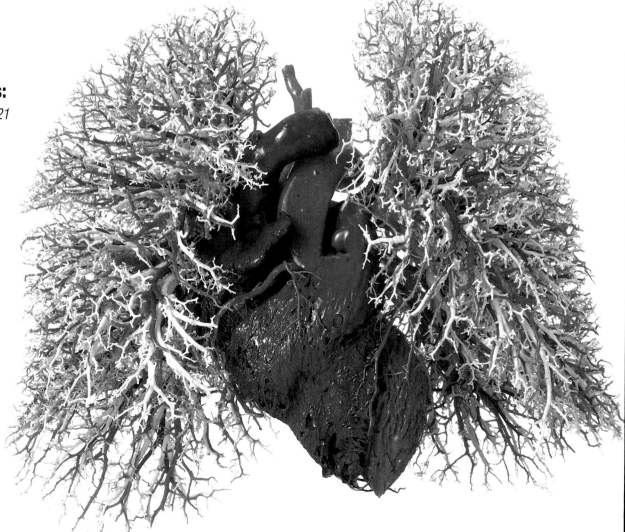

A model of the circulatory system of the human lungs, with the human heart in the center.

SCIENCE IN THE NEWS

THE WICHITA EAGLE

BEATLE GEORGE HARRISON DIES

By Eagle news services
December 1, 2001

George Harrison, the Beatles' lead guitarist and the composer of several of the group's most beautiful songs including "While My Guitar Gently Weeps" and "Something," has died. He was 58.

Harrison had surgery for throat cancer in 1998 and was treated for lung cancer and a brain tumor this year.

SCIENCE IN THE NEWS

THE DESERET NEWS

PERFORMER WARREN ZEVON DIES AFTER BATTLING CANCER

By The Associated Press
September 9, 2003

LOS ANGELES—Singer-songwriter Warren Zevon, who battled death with the same twisted sense of humor found in his songs "Life'll Kill Ya," "Werewolves of London" and "Excitable Boy," has lost his yearlong fight against lung cancer at age 56.

28.1 Transport and Exchange in the Body

Rock stars are known for excessive behavior, so who'd have predicted that, if a drug was going to bring down Beatle George Harrison and songwriter Warren Zevon, it would be the nicotine in boring old cigarettes? Then again, this turn of events isn't too surprising, given nicotine's power. Multiple-drug abusers consistently report that nicotine is the hardest drug of all to kick. Heroin and cocaine may be harder to

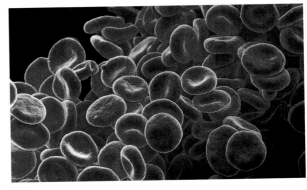

Human red blood cells (magnified ×1583).

get off initially, but nicotine is harder to *stay* off. Couple this with the fact that nicotine doesn't interfere with the ability to work, and its use just goes on and on. Eventually, of course, its use often stops suddenly, given the heart attacks that smoking helps cause. Lung cancer moves more slowly, but even with it, the time between diagnosis and death usually can be measured in months. And for anyone who thinks that smoking doesn't get you until you're so old that there's nothing left to get, consider Harrison and Zevon. They were 58 and 56 at the time of their deaths—still writing songs, still making records. It is true that no one dies of smoking at 30, and almost no one does even at 40. But if you're still smoking at 55 or so, it might be time to take a tip from the title of an album dedicated to Zevon and *enjoy every sandwich*, because the big clock of your life might be winding down.

Since smoking is so widespread, we tend to take it for granted. But it has not always been a part of every human culture. Up until the time of Columbus, Europeans had never smoked anything. The idea of inhaling smoke into the lungs was so foreign to them that, as legend has it, Sir Walter Raleigh's valet once doused him with water, thinking he was on fire—when in fact he was just smoking this newly discovered substance, called tobacco, that he'd brought back from the New World.

When you think about smoking in these terms, it is kind of strange, isn't it? Our lungs are specialized to take in the oxygen that we need to live. But we "clever" human beings discovered that, by inhaling the remains of burned leaves, our lungs can also serve as drug delivery vehicles. Indeed, the fact that our lungs are such *efficient* delivery vehicles is one of the reasons nicotine is so addictive. When a person inhales cigarette smoke, the nicotine in it reaches the brain in about eight seconds. Thus, the psychological boost that nicotine provides is an almost *instantaneous* boost, which is the kind that human beings find most rewarding.

The way nicotine gets to the brain is by moving into one of the streams that flow by our lungs. These streams are the blood vessels of our circulatory system. In the normal course of events, these vessels deliver oxygen and nutrients to every part of the body. Then, beyond delivering what's needed, they help get rid of what's *not* needed, which is to say carbon dioxide (circulated back to the lungs) and other metabolic waste products (which go to the bladder as urine). The main way the body gets rid of these latter, blood-borne waste products is by utilizing a couple of filtering plants, which we call the kidneys. Blood goes into the kidneys, and most of this blood comes back out of them; but while *in* the kidneys, the useful materials in blood get channeled one way while the waste products get channeled another—to the bladder. Coming full circle on this, a chief reason we have these metabolic waste products in the first place is because of our processing of food and water. And how is the processing of these materials handled? By our digestive system, which essentially is a long tube in which food is broken down until its building blocks can be moved into … the circulatory system.

So if you're keeping count here, we have digestion, circulation, filtration, and respiration—four critical tasks that certainly are different, but that nevertheless have something in common. They all involve *exchanges* between our body and the outside world. How does a given substance get into the body, how does it get distributed, and how are the wastes from it removed and sent out of the body? The goal of this chapter is to tell you how this overarching task is accomplished through the workings of four physical systems (**see** FIGURE 28.1). These are the respiratory, digestive, urinary, and circulatory systems (the last of which is more properly known as the cardiovascular system). We'll look at these systems individually, starting with one that is central to transport and exchange, in that it serves all the other systems that take part in this work. Indeed, we could think of this system as the body's mass transit operation.

FIGURE 28.1 **The Transport and Exchange Systems of the Body** The human body has four systems that are active in transport and exchange: the cardiovascular, digestive, urinary, and respiratory systems. The cardiovascular system is central to all the others, in that it transports materials to and from them.

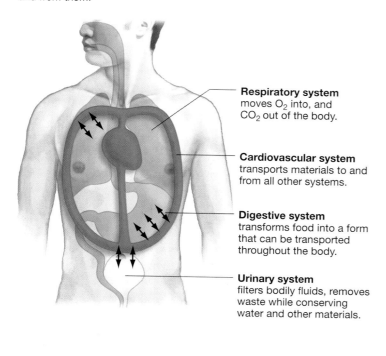

Respiratory system moves O_2 into, and CO_2 out of the body.

Cardiovascular system transports materials to and from all other systems.

Digestive system transforms food into a form that can be transported throughout the body.

Urinary system filters bodily fluids, removes waste while conserving water and other materials.

28.2 The Cardiovascular System

The human **cardiovascular system** is a fluid-transport system that consists of the heart, all the blood vessels in the body, the blood that flows through these vessels, and the bone-marrow tissue in which red blood cells are formed. This system bears a rough comparison with a car's cooling system. A car uses antifreeze and water as its circulating fluids; the cardiovascular system uses blood. A car may use a water pump to circulate its fluids; the cardiovascular system uses a pump called the heart. A car has an assortment of conducting pipes; the cardiovascular system has blood vessels. The similarities end there, however, because, while a car's cooling system transports only a couple of things, the cardiovascular system transports lots of things. It moves nutrients, vitamins, waste products, hormones, and immune-system cells and proteins, not to mention the oxygen we breathe in and the carbon dioxide we breathe out. All these materials are transported in the substance we'll look at first within the cardiovascular system—blood.

The Composition of Blood: Formed Elements and Plasma

Collect a sample of blood from a human being, prevent it from clotting, and spin it in a centrifuge, and it will separate into two layers. The lower layer, accounting for some 45 percent of its volume, consists of blood *cells* and cell fragments, which collectively are known as the **formed elements** of blood (**see** FIGURE 28.2). The yellowish, upper layer, accounting for the rest of the volume, is called **plasma**, meaning the fluid portion of blood. Within an actual blood vessel, the formed elements are suspended in the flowing plasma, just as tiny bits of sediment might be suspended within a moving stream.

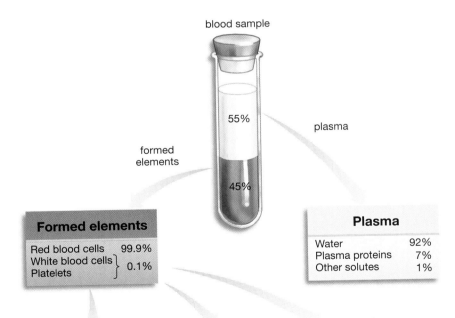
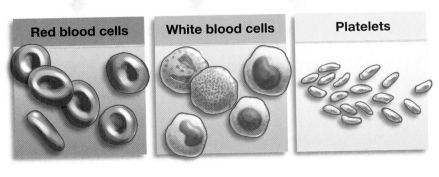

FIGURE 28.2 **The Composition of Blood** Blood has two primary constituent parts: formed elements and plasma. Formed elements are cells and cell fragments that help make up the blood—red blood cells, white blood cells, and platelets. Blood plasma, the liquid in which the formed elements are suspended, is composed mostly of water, but also contains proteins and other materials.

Formed Elements

There are three kinds of formed elements in the plasma. The most numerous by far are the **red blood cells**, which transport oxygen to, and carbon dioxide from, every part of the body. Then there are **white blood cells**, which are critical players in the immune system reviewed last chapter. Finally, there are **platelets**—small fragments of cells that are important in the blood-clotting process. Let's now look at each of these elements in more detail.

Red blood cells, or *erythrocytes*, are red because within them is the iron-containing protein **hemoglobin**. Molecules of hemoglobin that lie within a red blood cell (RBC) bind with oxygen as it comes in from the lungs and then hold onto it, releasing it only when the cell reaches a part of the body whose oxygen concentration is relatively low. RBCs then work with blood plasma in picking up the carbon dioxide that cells have produced through their metabolism; they carry this CO_2 back to the lungs, where it will be exhaled. Blood is deep red because RBCs make up 99.9 percent of all the formed elements in it. Indeed, the *number* of RBCs in an average person staggers the imagination. One cubic millimeter of blood—an amount that might scarcely be visible—contains 4.8 to 5.4 million RBCs. So numerous are these cells that they make up about a third of all the cells in the human body.

Apart from their numbers, RBCs are also notable for another quality, which is their odd structure. As they mature, RBCs lose the standard equipment that almost all cells have—a nucleus and other organelles. When fully mature, they essentially consist of a cell membrane that surrounds a mass of hemoglobin. Their lack of cellular machinery means, however, that RBCs can't maintain themselves very well; hence their average life span is only 120 days. This in turn means that about 180 million new RBCs have to enter the circulation *each minute* for us to have enough of them in our system.

The second formed element in blood, the white blood cells or *leukocytes*, have a completely different function from red blood cells. They have standard cellular equipment, but they are not in the oxygen or CO_2 transport business. Instead, in their many varieties, they are central to the body's immune-system operation. A typical cubic millimeter of blood contains about 6,000 white blood cells—a tiny number relative to the millions of red blood cells in such a volume.

The third formed element, platelets, can be thought of as enzyme-bearing packets. They are not cells, but fragments of cells that have broken away from a set of unusually large cells that exist in the bone marrow. The enzymes in them aid in blood clotting at sites of injury.

Blood's Other Major Component: Plasma

Formed elements are suspended in blood's other major component, plasma, which is 92 percent water. Proteins and a mixture of other materials are dissolved in this fluid as well. Why aren't these proteins and other materials "formed elements"? Well, note that they are not cells, nor even cell fragments, but are instead much smaller molecules. Plasma proteins come in a variety of forms. We saw one of these forms last chapter, with the antibodies that attack foreign invaders. Then there are the "transport" proteins, which do what their name implies: They transport substances from one part of the body to another. Special mention should be made of one variety of transport proteins, the *lipoproteins*, which essentially are capsules of protein that surround globules of fatty material—usually cholesterol. Lipoproteins are best known by the names of their two varieties: **low-density lipoproteins** (or **LDLs**), which carry cholesterol *to* outlying tissues from the liver and small intestine; and **high-density lipoproteins** (or **HDLs**), which carry lipids *from* these tissues to the liver. Keep both LDLs and HDLs in mind, because you'll be reading about them later in connection with heart attacks.

Taken together, the plasma proteins plus water make up 99 percent of blood plasma. The other 1 percent is a mixture of hormones, nutrients, wastes, and ions (which are better known as electrolytes). As you can imagine, wastes that are dissolved in the plasma are on their way out of the body altogether; eventually they will end up in the bladder as urine.

28.3 Blood Vessels

Having looked at what moves through the cardiovascular system, let's now look at the plumbing that carries these materials—our network of blood vessels.

Blood vessels are tubes, in a sense. To get a real idea of them, however, imagine a regular tube, such as a straw, that could change its internal diameter from second to second, that could take in some materials along its length while keeping others out, and that could repair itself when damaged. Because blood vessels are composed mostly of cells, they are dynamic entities that change their activities in accordance with the body's needs. Blood vessels carrying blood *away* from the heart are the **arteries**, while the vessels carrying blood back *to* the heart are the **veins**. The farther from the heart these vessels are, the smaller they tend to be. Connecting the arteries with the veins are the smallest blood vessels of all, the **capillaries**.

Arteries and veins are always made up of three distinct layers of tissue (**see FIGURE 28.3**). The innermost layer is composed of a group of flat cells, which you may remember from Chapter 25 are *epithelial cells*—cells that cover an exposed surface. The middle layer of arteries and veins contains smooth muscle. These are muscles that allow the blood vessel to shrink or expand in diameter. This very ability has health consequences, because the dangerous condition called hypertension or high blood pressure is brought about by blood vessels that are *persistently* constricted by muscle contraction. (As the vessels' diameter decreases, the pressure within them increases.) The outer layer of arteries and veins is a stabilizing sheath of connective tissue. Taken as a whole, these three layers give arteries and veins a great dual capability: strength combined with flexibility. The largest arteries have diameters of about an inch, but farther out from the heart, the smallest arteries have diameters that average about a thousandth of an inch.

Tiny as this diameter is, things get a lot smaller yet with the body's other main variety of blood vessel, the capillaries. Their diameters average about three *ten*-thousandths of an inch. Why so minuscule? Their central function is to have substances pass into them and out of them along their length. As such, their walls need to be very thin. Accordingly, each capillary is made of a *single* layer of cells. So small are capillaries that tiny red blood cells must move through them in single file.

A BROADER VIEW

Circulation and Gangrene

The blood vessels of the cardiovascular system could be thought of as a series of lifelines to the various tissues of our body. To grasp this, all we need do is consider what happens when the lifelines get blocked. The condition called gangrene is a death of tissue caused by a lack of blood supply to it. We usually associate gangrene with a traumatic injury, but one of its forms can be caused when blood vessels are blocked by conditions such as diabetes. When this happens neither oxygen nor nutrients can get to tissue, and it simply starts to dry up and die. If circulation cannot be restored to a gangrenous extremity, such as a finger or toe, it may be lost.

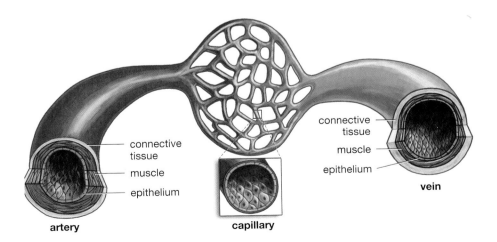

FIGURE 28.3 The Structure of Blood Vessels
Both arteries and veins have three layers to them: an inner epithelium, a layer of smooth muscle, and a layer of connective tissue. By contracting or relaxing, the smooth muscle can change the diameter of the vessels. Arteries and veins are linked by intervening capillary beds. Each capillary consists of a single layer of cells.

(a) The pulmonary and systemic circulation networks

(b) The circulation of blood through the heart

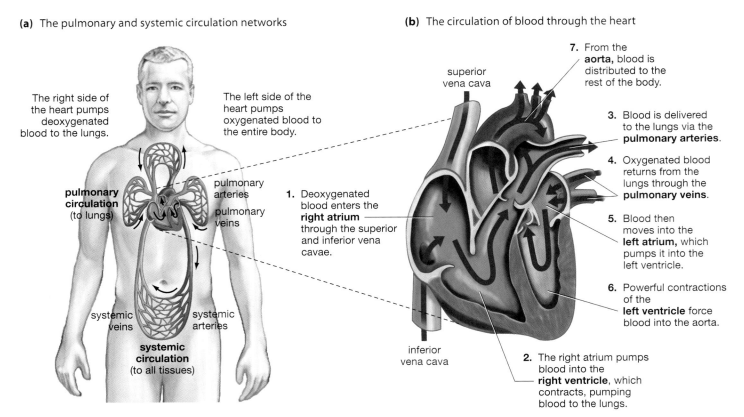

The right side of the heart pumps deoxygenated blood to the lungs.

The left side of the heart pumps oxygenated blood to the entire body.

pulmonary circulation (to lungs)

pulmonary arteries

pulmonary veins

systemic veins

systemic arteries

systemic circulation (to all tissues)

superior vena cava

inferior vena cava

7. From the **aorta,** blood is distributed to the rest of the body.

3. Blood is delivered to the lungs via the **pulmonary arteries**.

4. Oxygenated blood returns from the lungs through the **pulmonary veins**.

5. Blood then moves into the **left atrium,** which pumps it into the left ventricle.

6. Powerful contractions of the **left ventricle** force blood into the aorta.

1. Deoxygenated blood enters the **right atrium** through the superior and inferior vena cavae.

2. The right atrium pumps blood into the **right ventricle**, which contracts, pumping blood to the lungs.

FIGURE 28.4   Two Circulatory Networks and How the Heart Supports Them

(a) Veins in the systemic circulation bring blood into the right side of the heart; it is pumped out via the pulmonary circulation to the lungs, where it is oxygenated, and then returns to the left side of the heart. It is then pumped out from the left side of the heart, through the systemic circulation, to the rest of the body.

(b) The path of the blood's circulation through the four chambers of the heart.

28.4 The Heart and Blood Circulation

What makes blood flow through these vessels? The answer is the muscular pump we call the heart. A small organ, roughly the size of a clenched fist, the heart lies near the back of the chest wall, directly behind the bone known as the sternum.

The heart sends blood out, and then gets it back, through two circulation loops. As noted earlier, one major task of the blood-transport system is to get oxygen to all the various bodily tissues. For blood to carry oxygen to tissues, it first has to *take up* the oxygen that is brought into the body through the lungs. So imposing is this task of oxygenating blood that one of the heart's circulation loops is devoted solely to getting blood to the lungs and then back to the heart. The oxygenated blood that comes to the heart in this loop is then pumped out to the rest of the body in the heart's second major loop (**see** FIGURE 28.4a).

To put these things more formally, there are two general networks of blood vessels in the cardiovascular system. One of these is the **pulmonary circulation**, in which the blood flows between the heart and the lungs. Then there is the **systemic circulation**, in which the blood is moved between the heart and the rest of the body. Let's see how this two-network system works.

Following the Path of Circulation

The human heart contains four muscular chambers, two associated with the pulmonary circulation and two associated with the systemic circulation. These are the right atrium and right ventricle (pulmonary circulation) and the left atrium and left ventricle (systemic circulation). Let's take as the starting point the blood entering the right atrium from the veins called the superior and inferior venae cavae (on the *left* side of FIGURE 28.4b). These veins are returning blood to the heart from the upper and lower portions of the body, respectively. This is blood that is coming back after distributing oxygen and other blood-borne materials throughout the body—to our legs, our hands, our trunk, and so forth. As such it is deoxygenated blood, which actually is dark red in color; but it is indicated by the blue color of the veins it flows through. After the right atrium receives this blood, it pumps it the short distance to the right ventricle, which contracts, pushing the blood into the pulmonary arteries. This blood is now bound for the lungs, where it will pick up oxygen. Once it does this, it returns to the heart through the two pulmonary veins, which empty into the left atrium. The blood then is pumped down into the left ventricle, which contracts and sends the blood coursing up into the enormous artery called the **aorta**. This vessel has branches stemming from it that will carry the blood to all the tissues of the body.

Taking a step back from this, you can see that the right side of the heart pumps blood only to the lungs, while the left side of the heart pumps blood to all the body's tissues, as shown in Figure 28.4a. All four chambers of the heart pump blood, but the right and left *atria* pump blood only into another portion of the heart—their adjacent ventricles. The right ventricle is pumping blood to the lungs, so there is some greater contraction power needed there. But the *left* ventricle is pumping to the whole body, and its contractions are by far the most powerful. When we "take our pulse," what we are feeling is the surge of blood produced by the contraction of the left ventricle.

Valves Control the Flow of Blood

Because both ventricles contract with such force, you may wonder why blood doesn't go back up into their respective atria with each beat. The answer is that there are valves between the atria and ventricles that allow the blood to flow only one way—from atrium to ventricle. Occasionally, because of disease or genetic predisposition, people do have this kind of backflow into the atria. The fluid turbulence that results can be heard as a kind of gurgling sound, known as a "heart murmur" (which is not necessarily a sign of a serious heart problem). Valves also exist to prevent backflow between the ventricles and the arteries they pump blood into. The heart's two sets of valves are responsible for the familiar "lub-dub" sound of the heartbeat. The "lub" is the sound of the valves closing between the atria and ventricles, while the "dub" is the sound of the valves closing between the ventricles and the arteries.

So Far...

1. Blood is composed of formed elements, which are _____, and plasma, which is the _____. There are three kinds of formed elements: _____, _____, and _____. Plasma, meanwhile, is overwhelmingly composed of _____.

2. Arteries carry blood _____ the heart, while veins carry blood _____ the heart. The tiny blood vessels called _____ connect arteries to veins.

3. There are two networks of blood vessels in the body. In one of these, the pulmonary circulation, blood flows between the _____. In the other, the systemic circulation, blood flows between the _____.

What Is a Heart Attack?

The heart is a unique organ, but like any other organ, it needs a supply of blood and the oxygen and nutrients that come with it. Indeed, the heart is a voracious user of these materials because it essentially is a set of muscles that never stops working. Blood does not come to the muscles of the heart through any of the chambers you've been reviewing. (Think about the two ventricles; one is sending blood to the lungs, the other to the rest of the body, but not directly to the heart itself.) The way blood does come to the heart is through two arteries that branch off from the aorta just after it emerges from the left ventricle. Because these arteries

A BROADER VIEW

Taking Our Blood Pressure

What are physicians listening for when they take our blood pressure? The tightening of the cuff around our upper arm collapses a major artery—the brachial artery—whose blood pressure is being measured. This ensures that no blood is getting through the artery, which means that there is no sound of a pulse coming through the stethoscope. After a doctor starts to release pressure on the cuff and blood starts to flow, the point at which the doctor can first hear a pulse marks our systolic blood pressure—the highest pressure we have, brought about by the contraction of the left ventricle. The point at which the sound of the pulse disappears marks our diastolic blood pressure—our lowest blood pressure. Doctors watch the gauge on the blood-pressure device and note the blood-pressure levels at the two sound points.

So Far... Answers:

1. *blood cells and cell fragments; fluid portion of blood; red blood cells; white blood cells; platelets; water*

2. *away from; toward; capillaries*

3. *heart and lungs; heart and the rest of the body*

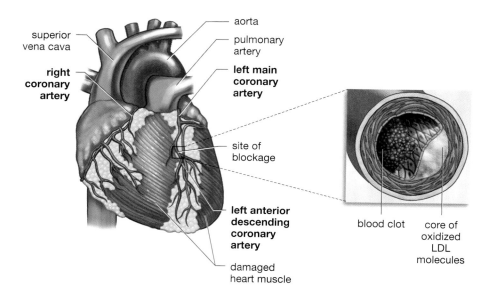

(a)

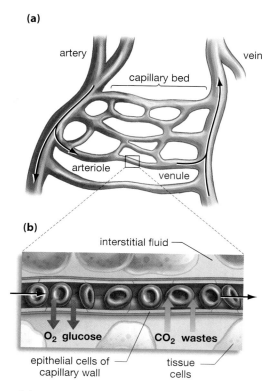

(b)

(c)

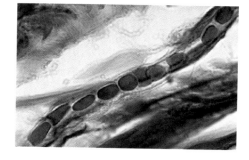

FIGURE 28.5 **Critical Vessels** The coronary arteries begin with a right and left coronary artery that branch off from the aorta. These arteries then undergo branching themselves, thus supplying blood to the whole heart. Blockage of the left anterior descending coronary artery figures in almost half of all heart attacks. The magnification of this artery shows the factors that lead to most heart attacks. A buildup of oxidized LDL molecules within the inner artery tissue layer and a resulting blood clot together bring about a complete blockage of the artery.

encircle or "crown" the heart before they start branching, they are known as the **coronary arteries** (**see** FIGURE 28.5).

It is difficult to overstate the importance of the coronary arteries, because about half of all deaths in the United States today are caused by a blockage of one or more of them. Such blockages generally have, as their starting point, the LDLs or low-density lipoproteins, noted earlier, that carry cholesterol molecules away from the liver and small intestine to varying locations in the body. One place these LDLs go is the coronary arteries. There they can lodge within one of the three layers of tissue that make up these arteries—the innermost layer that comes into contact with the flowing blood. Once present in large numbers, these LDLs can become damaged through oxidation (the same process that causes metal to rust), and this is where the real trouble starts. The immune system now regards these damaged LDLs as *foreign invaders* and thus initiates a complex attack against them. What follows is the "inflammatory response" that readers of Chapter 27 are familiar with. Inflammation that would take place with, say, a puncture wound causes tissue to become flushed and swollen. The same thing is happening here, only this tissue is swelling into the space through which blood flows in a coronary artery, thus forming a "cap" and beginning a blockage of it.

Even so, this swelling is rarely extensive enough to completely block an artery. Eighty-five percent of all heart attacks are set off by what happens next. The cap of swollen tissue ruptures, letting circulating blood mingle with the tissue within the cap, and the cap's immune cells react with this blood to make it clot. Now there is a blood clot combined with a swollen cap, and the result can be a *complete* blockage of a coronary artery—a **heart attack**. The effect of this blockage is that groups of heart-muscle cells no longer have a blood supply, which kills them if it goes on for long.

The good news about all this is that we can do some things to lessen the likelihood of this sequence of events getting started. A key concept is to lower the number of LDLs we have circulating within us while raising the number of the HDLs or high-density lipoproteins—the "good" lipoproteins that carry cholesterol away from the heart. Exercise does both things, and losing excess weight does both things. Meanwhile, eating saturated (animal) fat raises LDL levels, and smoking may damage the LDLs in ways that set off the immune response. It's worth noting

FIGURE 28.6 **Cardiovascular System Delivery Vehicle** Capillary beds are the sites of the exchange of materials between blood vessels and the body's tissues.

(a) An idealized view of how blood flows from arteries to arterioles, through capillary beds, and then into the venules and veins that return it to the heart.

(b) Oxygen and glucose are able to move out of the capillaries through simple diffusion. Meanwhile, carbon dioxide and wastes move into the capillaries, to be carried away and disposed of.

(c) Micrograph of red blood cells moving in single file through a capillary.

that fatty LDL deposits can begin forming in coronary arteries from childhood forward. Heart attacks may be an affliction of the elderly and middle-aged, but heart-disease prevention can be practiced at any age.

So Far... Answers:

1. *blockage; coronary artery*
2. *away from the heart*
3. *Capillaries*

28.5 Distributing the Goods: The Capillary Beds

Having looked at the players involved, you're now ready to see how the cardiovascular system actually carries out its central functions of bringing materials to tissues and taking materials away from them. Propelled by the heart through the aorta, blood flows through the tube-like arteries. These large-diameter arteries fork repeatedly, gradually decreasing in size until they become smaller-diameter arterioles. These then branch into the delivery vehicles of the cardiovascular system, the capillary beds. If you look at FIGURE 28.6a, you can see a representation of one of these beds. Note that at one end, the bed extends from the body's arterial system, but at the other end it feeds into the body's system of veins—the venous system that brings blood *back* to the heart. Looking at the venous system as it begins with the capillaries, you can see that it consists of its smallest vessels, the tiny venules, which then merge into veins. Smaller veins eventually merge into larger ones that finally come together into the largest veins of all, the two venae cavae that feed directly into the heart.

Thus we can see that blood comes "out" through the arteries and "back" through the veins, but the real action in the circulatory system takes place in the capillaries that lie *between* the arteries and the veins. As you've seen, capillaries are very thin. Think now about a line of oxygenated red blood cells, suspended in plasma and moving in single file through a capillary along with hormones, glucose, ions, and other materials needed by the cells *outside* the capillary (**see** FIGURE 28.6c). The oxygen, glucose, and all the rest move out through the capillary wall and into the interstitial fluid—the liquid in which both the capillary and the cells around it are immersed. (The movement of most of these materials is brought about by the simple diffusion covered in Chapter 5.) Once within this fluid, these needed materials will make their way to nearby cells. The distance they must travel to the cells is not far, however. So extensive is the body's capillary network that no cell is farther than two or three cells away from a capillary. As this is going on, carbon dioxide and wastes, produced by nearby cells, are diffusing *into* the capillaries from the interstitial fluid (**see** FIGURE 28.6b).

Muscles and Valves Work to Return Blood to the Heart

Blood pressure is relatively strong at the arterial end of a capillary bed. But the narrowness of the capillaries means that blood pressure is largely spent by the time blood approaches the venous end of a bed. So, with little pressure to propel it, how does blood get back to the heart? The answer is that our skeletal muscles do double duty. In contracting, they squeeze the veins in a way that moves the venous blood along. Blood can move only *toward* the heart in this system, because the veins have a series of valves in them that block movement of blood away from the heart (**see** FIGURE 28.7).

So Far...

1. A heart attack can be defined as a complete _____ of a _____.

2. High-density lipoproteins are regarded as healthy because they carry cholesterol _____.

3. _____ are the delivery vessels of the cardiovascular system.

A BROADER VIEW

What Is a Varicose Vein?

A varicose vein is one that has expanded beyond its normal size and then has become distorted into an odd shape. (Think of a water balloon that's been bent in a couple places.) Such a vein gets overfilled, in a sense, when one of the one-way circulation valves in it fails. This causes blood to pool in it, putting a pressure on it that distorts its shape. Genetics plays a part in the condition, as does smoking, lack of exercise, and standing for long hours. Sitting cross-legged, however, has nothing to do with getting varicose veins.

FIGURE 28.7 **One-Way Flow to the Heart** Skeletal muscle contraction is the primary force that drives blood back to the heart through the venous circulation. A system of valves guards against the backflow of this blood.

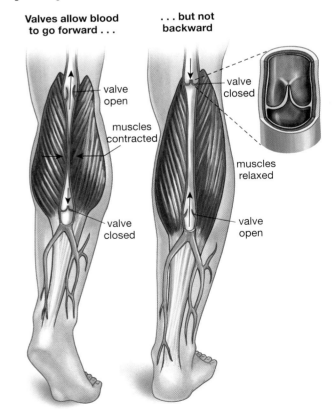

Valves allow blood to go forward . . .

. . . but not backward

valve open

valve closed

muscles contracted

valve closed

muscles relaxed

valve open

28.6 The Respiratory System

As you've seen, one of the things the cardiovascular system transports is oxygen and another is carbon dioxide. Now it's time to look at the bigger picture of how these substances are exchanged between our cells and the air around us.

People can live perhaps a couple of months without food, and generally a few days without water. But if they go 5 or 6 minutes without breathing, death is usually the result. This is because oxygen is in the energy transfer business; without oxygen, our cells simply don't have enough energy to function. When oxygen is present, however, cells carry out their work and *generate* carbon dioxide as a result, and this CO_2 has to be disposed of. Thus do we see the two central functions of breathing: The first is capturing and distributing oxygen; the second is disposing of carbon dioxide. Apart from these tasks, breathing also helps balance the pH level in our bloodstream. All three tasks are technically part of **respiration**: the exchange of gases between the atmosphere outside the body and the cells within it.

Structure of the Respiratory System

The lungs are but one part of the respiratory system, which you can see in FIGURE 28.8a. This system also includes the nose, nasal cavity, and sinuses, along with the pharynx (throat), the larynx (voice box), and the trachea (windpipe). In addition, there are air-conducting passageways: the left bronchus and right bronchus, and the many small air passageways they branch into, the bronchioles, which deliver air to the lungs.

FIGURE 28.8   Anatomy of the Respiratory System

(a) The large-scale components of the human respiratory system.

(b) In this magnification of the alveoli, the capillary network that all alveoli are surrounded by has been cut away, on the right, to reveal the structure of the alveolar sacs.

(c) A micrograph of a single bronchiole surrounded by many alveoli—essentially the "cutaway" view of this tissue as seen in part (b).

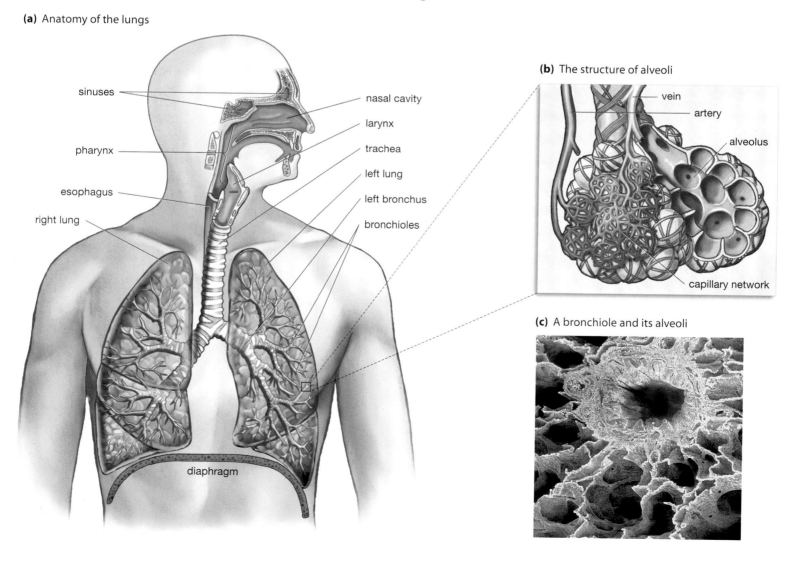

(a) Anatomy of the lungs

sinuses
pharynx
esophagus
right lung

nasal cavity
larynx
trachea
left lung
left bronchus
bronchioles

diaphragm

(b) The structure of alveoli

vein
artery
alveolus
capillary network

(c) A bronchiole and its alveoli

If we ask what the lungs themselves are, the answer is that they are made up mostly of tiny, hollow sacs sprouting in grape-like clusters from the end of each bronchiole. These sacs are the **alveoli**, the air-exchange chambers of the lungs. There are about 300 million of them in a typical set of lungs—a number that gives you an idea of just how tiny these sacs are. If you look at Figure 28.8b, you can see both what alveoli look like and how tightly linked they are with their partners in gas exchange, the capillary networks.

Steps in Respiration

The first step in respiration is **ventilation**, meaning the physical movement of air into and out of the lungs. To understand ventilation, look at the drawing, in FIGURE 28.9, of the old-fashioned "accordion bellows" that can be used to get a fire going. Pull the bellows handles apart, and the volume inside the bellows increases. This lowers the air pressure inside the bellows. Because gases always flow from areas of higher to lower pressure, the air flows into the bellows from the outside. Reverse this process—squeeze the bellows handles together—and the air pressure inside becomes higher than the pressure outside, meaning the air flows out of the bellows. Similarly, we have a space inside our chests (called the thoracic cavity) whose volume can increase or decrease. The contraction of a muscular sheet called the diaphragm causes the ribs to rise and, with this, the thoracic cavity expands like the inside of the bellows. Air then rushes in. Exhaling in this kind of "quiet" ventilation is a more passive process. When the diaphragm relaxes, the ribs drop due to the combined force of gravity and natural elasticity of the lungs. As a result, the thoracic cavity shrinks, and air is expelled. During the heavy breathing that takes place in exercise, exhalation is aided by a second set of chest muscles whose contractions help shrink the thoracic cavity.

Next Steps: Exchange of Gases

We've now seen how oxygen gets into the lungs, but how does it then get into the bloodstream? An oxygen molecule that is inhaled into one of the alveoli then diffuses across the thin alveolar tissue into an adjacent capillary. Upon entering the capillary, this oxygen molecule encounters a passing red blood cell, diffuses into

A BROADER VIEW
Surface Area of Our Lungs

Alveoli can do their job because they are small and numerous enough to have a tremendous surface area. What does being small and numerous have to do with surface area? Imagine a cube of Jell-O, 1 inch on each side. Its surface area is 6 square inches. Divide this cube in half, however, and the Jell-O's surface area goes to 8 square inches, even though its volume—the amount of space it takes up—stays the same. Repeat this kind of dividing millions of times, as with the alveoli, and the results are startling. The alveoli in a typical set of lungs have a surface area about the size of a tennis court.

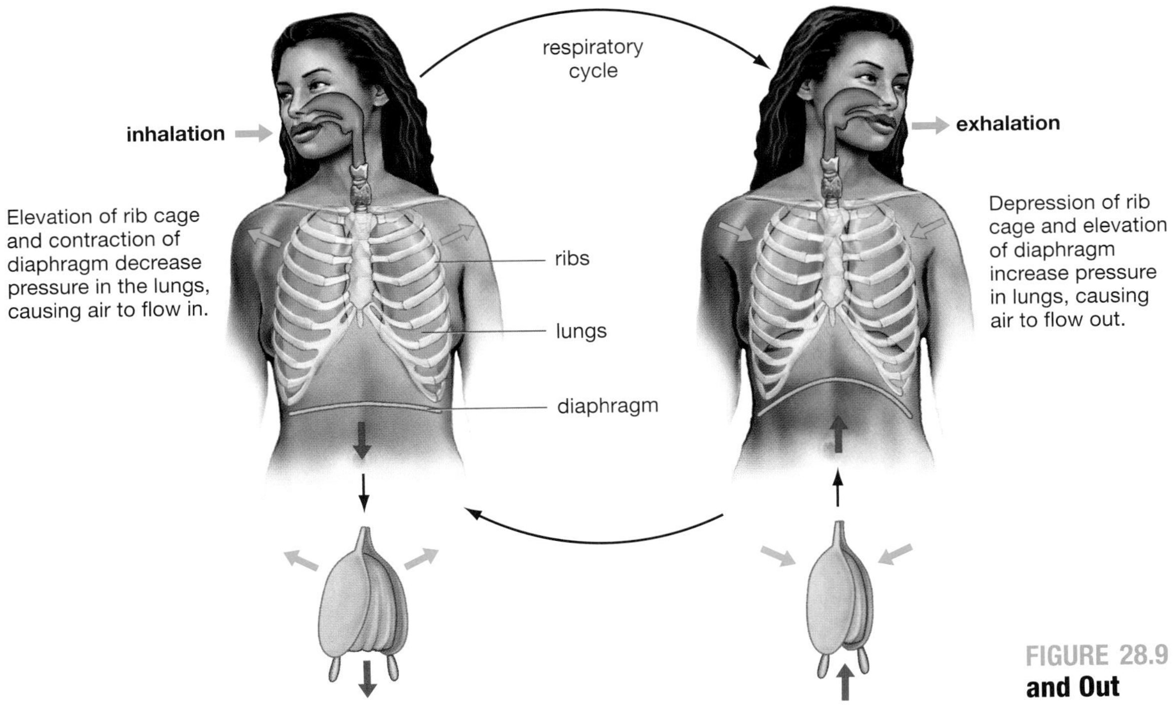

inhalation — Elevation of rib cage and contraction of diaphragm decrease pressure in the lungs, causing air to flow in.

exhalation — Depression of rib cage and elevation of diaphragm increase pressure in lungs, causing air to flow out.

respiratory cycle

ribs
lungs
diaphragm

FIGURE 28.9 **Ventilation—How Air Is Moved In and Out**

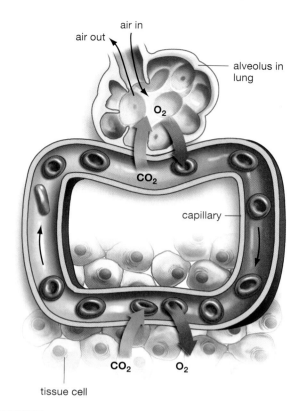

FIGURE 28.10 **Gas Exchange in the Body**

it, and binds with one of the hemoglobin molecules inside it. Locked up by this binding, the oxygen molecule moves with the red blood cell to the heart. In its ceaseless pumping, the heart is taking oxygenated blood cells like this one and sending them up through the aorta and from there out to the rest of the body. Moving through this route to a tissue somewhere in the body, the red blood cell drops off the oxygen molecule to be used.

This drop-off occurs once again by way of diffusion: The oxygen molecules have a natural tendency to move from an area of their higher concentration (inside the capillary) to an area of their lower concentration (outside the capillary, in the interstitial fluid). Other body cells are immersed nearby in this fluid, and they can now receive the oxygen. These same cells are, of course, producing carbon dioxide, which now diffuses from them into the interstitial fluid, and then diffuses again into a capillary. This CO_2 is returned to the lungs via the venous system. Then, at the same capillary-alveoli interface noted earlier, the CO_2 moves into the alveoli, from which it is expelled into the environment outside the body (**see FIGURE 28.10**).

As noted, all the oxygen that is loaded into red blood cells binds initially with hemoglobin molecules that exist in these cells. In contrast to this simplicity, the carbon dioxide that is being transferred to the lungs is transported by the blood in three different ways. We need only note here that, in a kind of chemical packing and unpacking, some CO_2 ends up traveling within the hemoglobin in red blood cells, while most of it travels outside of red blood cells, in the blood plasma.

So Far...

1. The two central functions of respiration are capturing and distributing _____ and disposing of _____.

2. The air-exchange chambers of the lungs are tiny sacs called _____, which are surrounded by _____.

3. Oxygen binds with the molecule called _____ inside _____.

28.7 The Digestive System

Now we shift gears again, moving this time from respiration to digestion. The digestive system is in one sense very simple. Its central structure is a muscular tube called the **digestive tract** that passes through the body from the mouth to the anus. Along its length, this tube receives input from various accessory organs, such as the salivary glands, gallbladder, liver, and pancreas. The central function of the digestive system is to get the foods the body ingests into a form the body can use. The system must also rid the body of the waste that remains once the useful material has been removed from the food.

You've seen already that nutrients are delivered to tissues throughout the body by the circulatory system. So the real question for the digestive system is: What does it need to do to get food into a form that can *move into* the circulatory system? For carbohydrates and proteins, the pathway is straightforward. Partway through a portion of the digestive tract called the small intestine, carbohydrates and proteins have been broken down enough that they move out from the inner lining of the intestine into capillaries that lie just outside the intestine. This network of capillaries then feeds into a common vein that goes to the liver. Most nutrients, then, are carried straight from the small intestine to the liver. In a controlled release, the liver then sends these nutrients out (through another vein) to the heart, which pumps them out to all the tissues of the body.

The digestion of fats takes a somewhat more complicated route. But the principle remains the same. Fats must be broken down to such a point that they can

So Far... Answers:

1   carbon dioxide

         ·pillaries

         · red blood cells

leave the digestive tract for the rest of the body. For all foods, *breakdown* means just what it sounds like: a transition from large to small. The transition goes from big, visible bites of food to smaller bits, and eventually from big *molecules* of food to the building blocks that make them up. It is only these smaller molecules— simple sugars, amino acids, and fatty acids—that can leave the digestive tract for the rest of the body. With this in mind, we can think of **digestion** as the breakdown of food into a chemical form that can be passed into circulation. Now let's take a look at the components of the digestive system, after which you'll follow the path of digestion.

Structure of the Digestive System

FIGURE 28.11 shows the locations and functions of the accessory glands and different subdivisions of the digestive tract. The digestive tract begins with the mouth (or oral cavity) and continues through the pharynx, esophagus, stomach, small intestine, and large intestine before ending at the rectum and anus. The digestive tract is sometimes called the alimentary canal, or the gastrointestinal tract.

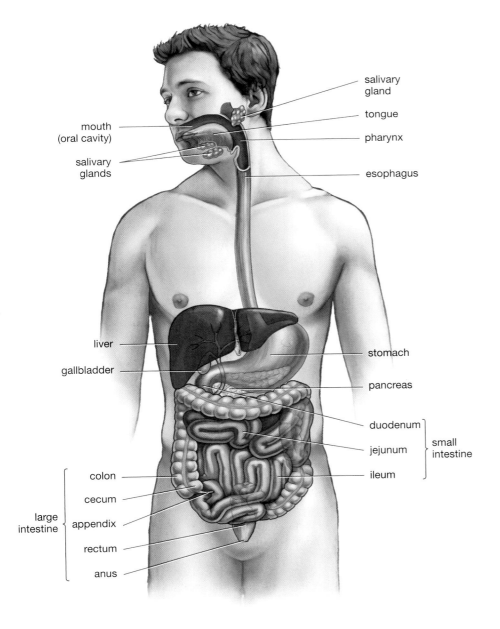

FIGURE 28.11 **Components of the Digestive System**

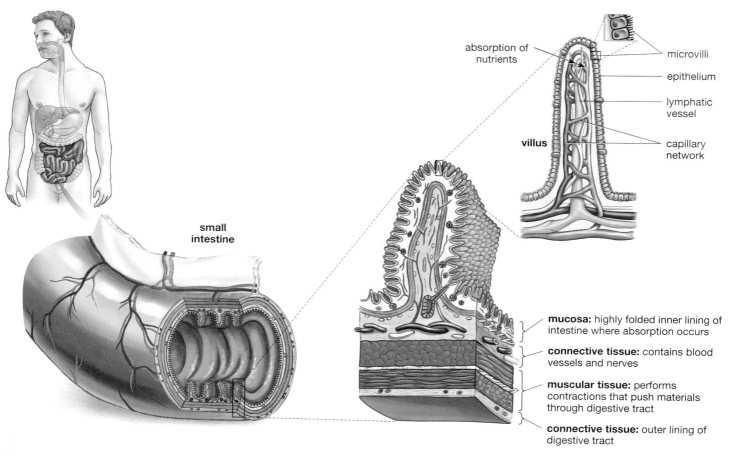

absorption of nutrients

microvilli

epithelium

lymphatic vessel

villus

capillary network

small intestine

mucosa: highly folded inner lining of intestine where absorption occurs

connective tissue: contains blood vessels and nerves

muscular tissue: performs contractions that push materials through digestive tract

connective tissue: outer lining of digestive tract

FIGURE 28.12 **Structure of the Digestive Tract** The digestive tract is a tube formed of several layers of tissue, each performing a different function. Note that the absorption of nutrients takes place when food that has been sufficiently digested moves across the epithelium and is taken up by either a capillary (in the case of proteins and carbohydrates) or a lymphatic vessel (in the case of fats).

The Digestive Tract in Cross Section

Thinking again of the digestive tract as a muscular tube, let's look at a cross section of it as it would appear within the small intestine. As you can see in FIGURE 28.12, the digestive tract is a tube composed of several layers. Taking things from the outside, there is a layer of connective tissue that then has, inside it, a layer composed of two sets of smooth muscles. In a process called **peristalsis**, these muscles take turns contracting in waves that push materials along the length of the digestive tract. Inside the muscles, there is yet another layer of connective tissue that contains large blood vessels and a network of nerve tissue that both coordinates the contractions of the smooth muscle layers and regulates the secretions of digestive juices. Finally, there is an epithelial layer, called the mucosa, that makes up the internal lining of the tract.

Along most of its length, the mucosa has a large number of folds that increase the surface area available for absorption of digested food; these folds also work like accordion folds, in that they permit expansion, which is helpful after a large meal. The mucosa of the small intestines also has on it finger-like projections, called villi, that further increase the area for absorption. These villi are the places where most of our digested food actually makes the transition out of the digestive system. If you look at the blow-up of the villus in Figure 28.12, you can see where nutrients leave the system. As you can see, they move from a villus either to an adjacent capillary or, in the case of fats, to a lymphatic vessel whose connecting vessels will put the nutrients into the bloodstream.

28.8 Steps in Digestion

Let's now look at the process of digestion from start to finish, which means taking it from the top, so to speak. Our mouths are the entry place for food, of course, and we immediately start breaking food down in them by the mechanical means of chewing. Food starts to get digested in the mouth by chemical means as well, because enzymes in the saliva have the ability to break down carbohydrates.

The Pharynx and Esophagus

Food begins to move out of the mouth when the tongue pushes it back toward the pharynx, which we might think of as the upper throat (see Figure 28.11). The pharynx has a trio of muscles wrapping partway around it that get activated in a 1-2-3 sequence to push the food into the esophagus. The esophagus in turn has its own muscles, which work with gravity to send food on a 6-second trip to the next stop in the digestive tract.

The Stomach

The **stomach** is an organ that functions in both the digestion and temporary storage of food. It carries out its digestion work partly by the mechanical means of churning food, thus breaking it into small bits. It also performs some chemical breakdown of foods, with proteins being the primary target of its digestive juices. When empty, the stomach resembles a tube with a narrow cavity, but it can expand to contain almost half a gallon of material. This expansion is possible because the stomach contains numerous folds, called *rugae* (**see** FIGURE 28.13).

Glands feed into the stomach by way of millions of small depressions, called gastric pits, that open onto its interior. Each day, about 45 ounces of gastric juice pour into the stomach through these pits. This gastric juice is *caustic*; it contains hydrochloric acid and, as a result, the stomach has a pH level that lies between that of lemon juice and battery acid. Why is such an extreme environment needed? It serves not only to break down food but also to kill bacteria and other invaders that ride in on the food. You may wonder why the stomach itself wouldn't be eaten away in this kind of environment. The answer is that the cells lining it produce a protective mucus.

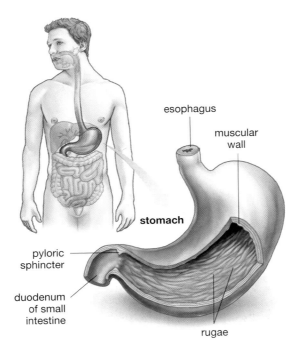

esophagus

muscular wall

stomach

pyloric sphincter

duodenum of small intestine

rugae

FIGURE 28.13 **External and Internal Views of the Stomach**

A few substances, such as alcohol and some drugs, can pass directly from the stomach into circulation, but in general the stomach is only preparing food for transfer within the digestive system. After several hours in the stomach, food becomes a soupy mixture of food and gastric juices known as **chyme**. A circular or "sphincter" muscle—the pyloric sphincter—regulates the flow of chyme between the stomach and the next structure in the digestive system.

The Small Intestine

The only thing small about the small intestine is its diameter, which is about 1.6 inches where it starts (at the stomach) and about an inch where it ends (at the large intestine). In between these two points, the "small" intestine winds on for about 20 feet. Eighty percent of our nutrients make the transition out of the digestive system along the length of the **small intestine**, which can be defined as the portion of the digestive tract that runs between the stomach and the large intestine. (About 10 percent of our nutrients are transferred from the stomach, and the remaining 10 percent are transferred from the large intestine.)

As you can see in **FIGURE 28.14**, the small intestine is made up of three regions. The first of these, the *duodenum*, is the 10 inches that are closest to the stomach. Think of the duodenum as that portion of the small intestine that receives not only chyme from the stomach, but digestive secretions from three of the specialized digestive organs noted earlier: the pancreas, the liver, and the gallbladder. Next there is the *jejunum*, which goes on for about eight feet. This is the site within the small intestine where most of our nutrients move into circulation. Such "nutrient absorption" continues to some extent in the third segment, the *ileum*, which is the longest, averaging about 12 feet in length. The ileum ends at a sphincter muscle that controls the flow of chyme into the large intestine. Now let's look at what happens at the start of the small intestine, where juices from the digestive glands come into the picture.

The Pancreas

Feeding into the small intestine is the pancreas, which lies behind the stomach and is an elongated, pinkish-gray organ about 6 inches long. The pancreas is best known as the organ that produces insulin, which it secretes directly into the bloodstream. But our concern is with materials the pancreas secretes

A BROADER VIEW
The Nature of Ulcers

What is an ulcer? It is a small area in the stomach or small intestine where the tissue that normally lines these organs (the epithelium) has been destroyed, such that underlying tissue is exposed. You could think of it as an internal sore—one that can become particularly inflamed a few hours after a meal. Though we generally associate ulcers with the stomach, most actually occur in the first part of the small intestine (the duodenum). The vast majority are caused by a bacterium (*Helicobacter pylori*) whose activity brings about an excessive secretion of acid that erodes the epithelial tissue. Thankfully, a seven-day regimen of antibiotics usually is sufficient to get rid of both the bacterium and the ulcer.

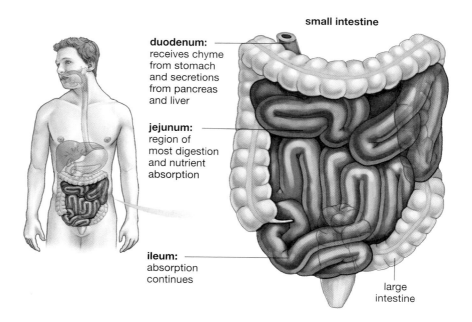

small intestine

duodenum: receives chyme from stomach and secretions from pancreas and liver

jejunum: region of most digestion and nutrient absorption

ileum: absorption continues

large intestine

FIGURE 28.14 **Structure of the Small Intestine**

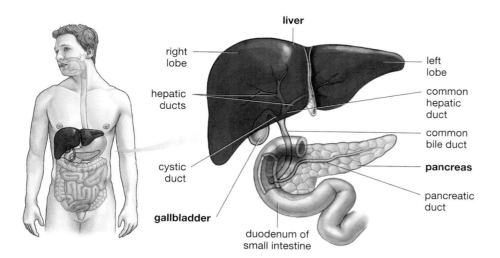

through a tube or duct. The duct in question, as you can see in FIGURE 28.15, is the pancreatic duct, which empties into the small intestine's duodenum alongside the common bile duct that stems from the liver and gallbladder. In digestion, the **pancreas** can be defined as a gland that secretes, into the small intestine, digestive enzymes, along with chemical buffers that help raise the pH of the chyme coming from the stomach. (Remember, this chyme is very acidic; when its pH is being "raised," that means it is becoming less acidic.) Pancreatic digestive enzymes are classified according to the kind of food they work on. *Lipases* break down lipids, *proteases* break proteins apart, and *carbohydrases* digest sugars and starches.

The Gallbladder and the Liver

As Figure 28.15 shows, lodged within a recess under the right lobe of the liver is a muscular sac called the **gallbladder**: an organ that stores and concentrates a digestive material called bile. Note the operative terms about the gallbladder's role: It stores and concentrates bile. But bile is *produced* by the liver. This is why people can have their gallbladders removed and still function perfectly well. **Bile** is a substance produced by the liver that facilitates the digestion of fats. It does two things that allow the digestive lipase enzymes to go to work on fats: It breaks up clusters of fat molecules, and it coats these molecules with a material that gives the lipases a greater ability to bind with them.

The **liver** itself can be defined as a reddish-brown organ that is central to the body's metabolism of nutrients. This large organ actually does hundreds of things for us, but here we will concentrate on a few of its digestive functions.

The liver clearly plays a central role in digestion because, as you may recall, all blood carrying nutrients from the digestive tract is channeled into a single blood vessel. This is the hepatic portal vein, which flows into the liver. We could thus think of a series of blood-vessel "streams"—stemming from the stomach, the small intestine, and so forth—all of them carrying digested nutrients, and all of them flowing like tributaries into the common river of the hepatic portal vein that goes to the liver. This arrangement makes the liver a first stop for most nutrients. From this position, the liver *controls* which nutrients it will send to the rest of the body and which nutrients it will store. The liver is, however, not only the first stop for nutrients, but the first stop for *toxins* such as alcohol. This is why people who have abused their bodies by drinking too much can end up with damaged livers. In addition to transferring and storing nutrients, the liver produces the bile noted earlier, and it packages waste products for removal by the kidneys.

A BROADER VIEW

Digestive Gas

What causes digestive gas? Carbohydrates that we're incapable of digesting arrive in the colon intact and serve as a nutrient source for bacteria who live there. It is these bacteria who actually produce the gas, in their metabolism of the food. Beans often trigger this series of events because they contain a high concentration of carbohydrates that we cannot digest, but that the bacteria can.

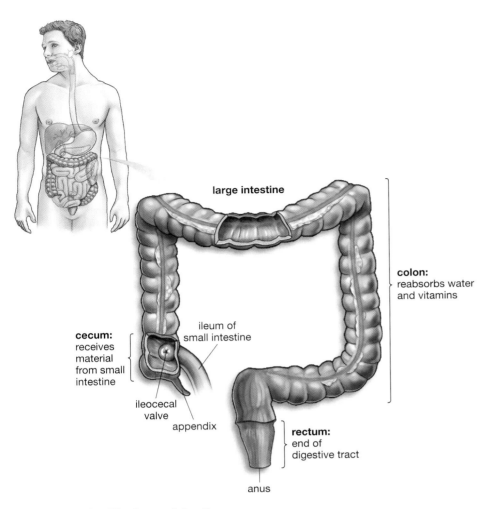

large intestine

colon:
reabsorbs water
and vitamins

cecum:
receives
material
from small
intestine

ileum of
small intestine

ileocecal
valve

appendix

rectum:
end of
digestive tract

anus

FIGURE 28.16   **The Large Intestine**

The Large Intestine

If most nutrients have been transferred out of the digestive system by the time the small intestine ends, what is there left for the large intestine to do? Its main task is to serve as a kind of trash compactor: It holds and compacts material that by now is largely refuse, turning it into the solid waste we know as feces. It also returns water to general circulation and absorbs vitamins produced by resident bacteria.

If you look at the upper drawing in FIGURE 28.16, you can see that the large intestine neatly frames the small intestine. The small intestine's last section, the ileum, feeds into the large intestine's first section by means of a sphincter (called the ileocecal valve) that you can see at about 7 o'clock in the lower drawing. The large intestine continues for about 5 feet from this point before ending at the anus. Compared to the 20-foot small intestine, the large intestine is large only in diameter.

The **large intestine** can be defined as that part of the digestive tract that begins at the small intestine and ends at the anus. It is divided into three major regions: the cecum, the colon, and the rectum. Note that the cecum has a pouch called the *appendix* attached to it; the walls of this pouch can become infected, resulting in appendicitis. (When healthy, the appendix is an arm of the immune system, though obviously not an essential one, as it can be removed.) The colon is the large intestine's second section, and by far its longest. The individual pouches that you can see in it allow it to expand and contract, depending on how full it is. The colon then empties into the expandable rectum, which serves as a storage site for feces.

So Far...

1. The central structure of the digestive system is a muscular tube called the _____ that passes through the body from the _____ to the _____.

2. Arrange the following parts of the digestive system in the order in which food passes through them: stomach, pharynx, large intestine, esophagus, small intestine.

3. Bile is a substance secreted by the _____ and stored and concentrated by the _____, which facilitates the digestion of _____. The pancreas secretes _____ into the_____.

28.9 The Urinary System in Overview

Here our subject changes once again (for the last time in this chapter), as our tour of the body's systems moves from digestion to the elimination of liquid wastes. At this point, you may be thinking: Didn't we just look at waste removal in the digestive system? We did, but the urinary waste removal coming up is fundamentally different. Think of it this way. The digestive system simply had to *retain* the waste from food, with this waste never leaving the confines of the digestive tract until it was eliminated from the body. As you'll see, the urinary system has to get rid of waste that is produced by cellular activity throughout the body. As such, it works by *filtering* waste from the blood and then passing this waste on to the bladder for elimination.

So Far... Answers:

1. *digestive tract; mouth; anus*

2. *pharynx, esophagus, stomach, small intestine, large intestine*

3. *liver; gallbladder; fats; digestive enzymes; small intestine*

Such waste removal obviously is important, but it turns out to be only one of several critical functions the urinary system undertakes. Equally important is the system's regulation of blood volume—how much blood we have within us—a task whose importance becomes obvious when we think about consuming lots of liquids. Suppose the body simply incorporated all the water in these liquids into circulating blood. We would eventually swell and burst like a water balloon. Instead, the urinary system *regulates* how much water the body retains and how much it excretes, with the kidneys being a critical player in this process. The urinary system also controls ion concentration and maintains pH balance in the body. In short, it is important in preserving the stable state known as homeostasis. We almost never think of the urinary system in this regard, essentially because it works so well. But a person who suffers kidney failure and must rely on a kidney "dialysis" machine to filter wastes knows the value of the urinary system in a personal way.

In carrying out these activities, the urinary system is very much involved in the *conservation* of bodily resources. Indeed, the kidneys are as much in the conservation business as the waste removal business. They conserve water, amino acids, ions, sugars—all kinds of things that the body needs. Consider that the kidneys process almost 48 gallons of fluid every day. Yet our urination is only about *half* a gallon daily. More than 47 gallons cycled back into the body each day! Now let's see how the kidneys and the larger urination system manage this feat.

Structure of the Urinary System

The pivotal structures in the urinary system, shown in FIGURE 28.17, are its two filtering organs, the **kidneys**, which produce urine while conserving useful blood-borne materials. Urine leaving the kidneys travels along two tubes, the left and right **ureters**, to the **urinary bladder,** the hollow, muscular organ that acts as a temporary, expandable storage site for urine. When urination occurs, contraction of the muscular bladder forces the urine through the conducting tube called the **urethra** and out of the body.

The kidneys are located on either side of the vertebral column, at about the level of the eleventh and twelfth ribs. They are shaped like kidney beans, but since they are each about 4 inches long, their size is more like that of a small pear.

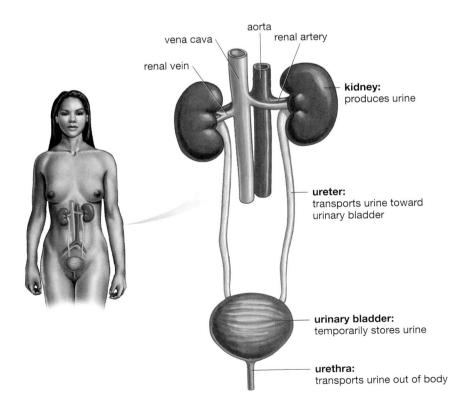

aorta

vena cava

renal artery

renal vein

kidney:
produces urine

ureter:
transports urine toward
urinary bladder

urinary bladder:
temporarily stores urine

urethra:
transports urine out of body

FIGURE 28.17   **The Urinary System in Overview**

FIGURE 28.18 **Structure of the Kidney and Its Working Unit, the Nephron**

(a) A kidney in cross section.

(b) A nephron in the kidney, composed of a nephron tubule and its associated blood vessels, which are immersed in fluid.

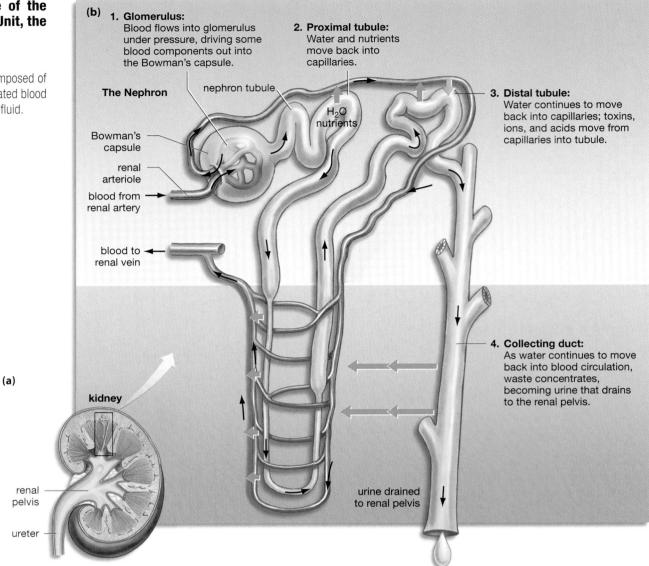

(b)

1. Glomerulus: Blood flows into glomerulus under pressure, driving some blood components out into the Bowman's capsule.

The Nephron

nephron tubule

Bowman's capsule

renal arteriole

blood from renal artery

blood to renal vein

2. Proximal tubule: Water and nutrients move back into capillaries.

H₂O nutrients

3. Distal tubule: Water continues to move back into capillaries; toxins, ions, and acids move from capillaries into tubule.

4. Collecting duct: As water continues to move back into blood circulation, waste concentrates, becoming urine that drains to the renal pelvis.

urine drained to renal pelvis

(a)

kidney

renal pelvis

ureter

In considering how the kidneys work, it's helpful to think first in terms of input and output. What comes into the kidney in the way of fluids? Blood from the circulatory system arrives by way of the renal arteries. As you can see in Figure 28.17, these branch off from the aorta, one renal artery going to the left kidney, one going to the right. Now, what comes *out* of the kidney? Two things: Blood flows out of the kidneys through the renal *veins* that are pictured, and urine exits each kidney through its ureter. Thus, each kidney has one input (the renal artery) but *two* outputs. This is a clue to what's happening inside. Some of the material coming into the kidney through the renal arteries is being channeled out of the blood vessels and into a separate system of vessels whose output tubes are the ureters. Now let's look more closely at the kidneys.

Seen in cross section in FIGURE 28.18a, each kidney has within it a chamber called the renal pelvis, whose numerous branches collect the urine produced by the kidneys and channel it into the ureter. If you look at FIGURE 28.18b, you can see a blow-up of the basic working unit of the kidney, a structure called a *nephron*. Each kidney has within it about 1.25 million nephrons, each serviced by a single arteriole that has branched off from the entering renal artery.

In the figure you can see how a nephron begins: An arteriole branching off the renal artery enters a hollow chamber, called the *Bowman's capsule*, and then promptly expands into a knotted network of capillaries, called the glomerulus, that is enclosed by the capsule. *Emerging* from this network, there is again an arteriole that goes on to surround something called a nephron tubule. (In reality, this arteriole branches into a network of capillaries that surrounds the tubule, but for simplicity's sake these capillaries are shown here as a single blood vessel.) Thus, we have two sets of tubes—the nephron tubule and the blood vessel—and these will become the two *outputs* just mentioned: The nephron tubule will transmit what the body doesn't need (waste sent to the bladder), while the blood vessel will first lose, and then take back, what the body does need (nutrients, etc.). All this put together adds up to a **nephron**: the functional unit of the kidneys, composed of a nephron tubule, its associated blood vessels, and the fluid in which both are immersed. At the "output" end of the nephron, you can see that its tubule is only one of several that are emptying into a common *collecting duct*. Eventually, many of these ducts will themselves merge to feed into the renal pelvis, which feeds into the ureter. Now let's look at what takes place within a nephron.

28.10 How the Kidneys Work

Imagine holding a washcloth, bag-like, underneath a running kitchen faucet. What happens? The water slowly flows out of the washcloth. What if this is a faucet that hasn't been turned on in a while, however, and the water thus contains some sediment? Now the washcloth acts to retain these larger particles while letting smaller substances (the water molecules) flow through. In short, the washcloth acts as a filter. The knotted network of nephron capillaries called the **glomerulus** takes on this same filtering role. Receiving blood that is under pressure, it is porous enough to let some smaller blood-borne materials pass out of it—and into the surrounding Bowman's capsule (see Figure 28.18b). What materials pass out? Water, first of all, which makes up such a large portion of the blood, and then many of the materials suspended in the water, such as vitamins, nutrients, and waste materials. All these substances now move from the glomerulus into the Bowman's capsule. What stays in the glomerulus (and hence in the bloodstream) are plasma proteins and blood cells, which are too big to pass out.

We should note at this point the nature of the waste products that are moving out of the glomerulus. Every time you use a muscle, you produce a waste product called creatinine that has to be eliminated from the body. Likewise, proteins are broken down mostly in the liver, and a by-product of this breakdown is a substance called urea. Both urea and creatinine are picked up by the bloodstream and move to a glomerulus, where they flow out, along with the water and other substances.

All the materials that have entered the Bowman's capsule constitute a fluid, called *filtrate*, that flows from the capsule into the nephron's next structure, its proximal tubule. With this tubule, we get to the essence of how the kidneys work as a filtering mechanism—getting rid of waste, but retaining what is valuable. The retention process begins immediately after the filtrate moves into the proximal tubule. Some active pumping of materials out of the tubule is coupled to the process of osmosis (reviewed in Chapter 5). The result of both actions is that about two-thirds of the water lost in filtration, as well as almost all the nutrients lost there, move *back* into blood circulation at the proximal tubule—back into the network of capillaries represented in Figure 28.18b by the single blood vessel. While this "reabsorption" is going on, however, the urea and other waste products have remained in the nephron tubule because, thanks to their chemical makeup, they cannot easily pass out of it.

If you just follow the nephron tubule on around, past the proximal tubule, you can see what comes next: Water continues to flow out of the tubule and back into circulation (blue arrows), while up at the distal tubule, toxins and other waste

A BROADER VIEW

Beer and Urination

Ever wonder why drinking beer makes you have to urinate more than drinking water does? The alcohol in the beer suppresses production of antidiuretic hormone. With this, the kidneys move less water back into circulation; or, to look at it another way, more water to the bladder.

A BROADER VIEW

Bladder Infections

You might expect that urine would be filled with bacteria, as is the case with feces, but urine is almost always germ free. Indeed, the movement of urine down the urinary tract is one of the reasons the kidneys and bladder tend to stay free from infection—the flow of urine washes away any encroaching bacteria. Bladder infections do occur, of course, particularly in women. Why are women more prone to such infections than men are? In part because their urethras are shorter—1 to 2 inches, compared to the 7–8 inches the male urethra extends from the bladder to the tip of the penis. In females, bacteria don't have as far to go to reach the bladder.

FIGURE 28.19   **The Urinary Bladder in a Male**

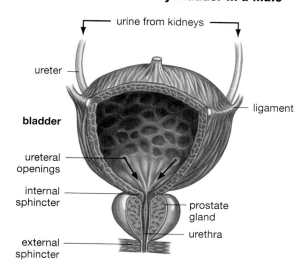

move the other way: from circulation into the tubule (green arrow). The end result is that, with movement through the tubule, the filtrate becomes an ever-more concentrated waste product. Water and nutrients are removed from it and sent back to circulation, while urea, creatinine, and other wastes remain in it. By the time the filtrate reaches the collecting duct, it contains such a high concentration of waste that it goes by another name: urine.

Hormonal Control of Water Retention

We noted earlier that the kidneys are key players in the body's control of the volume of fluids we have within us. The key to this control is a substance, produced by the brain's hypothalamus, called **antidiuretic hormone (ADH)**. If we get dehydrated, the hypothalamus releases ADH. When this hormone reaches the kidneys, it has the effect of allowing water to flow more freely out of the distal nephron tubule and the collecting duct. What happens next, of course, is that this water then moves back into blood circulation. Thus, ADH prompts the kidneys to recycle more water back into circulation, rather than allowing them to send this water to the bladder as urine. Of course, there are times when we have too much water in the body. In this event, the hypothalamus shuts down its production of ADH, and this prompts the distal tubule and collecting duct to hold on to more water, which means more of it is sent to the bladder. The end result of all this is that the body can tightly regulate the amount of water it retains.

28.11   Urine Storage and Excretion

Having seen how the kidneys produce urine, let's now look at the steps by which urine is then excreted from the body. Recall that urine is transported to the urinary bladder by the pair of tubes called the ureters—one stemming from each kidney. About every 30 seconds, a wave of muscular contraction moves down the ureters, thus squeezing the urine out of the renal pelvis and into the bladder. (This is a long journey, as each ureter is about 12 inches long.) Occasionally, solids develop along the route from a nephron through the ureters. These solids are kidney stones, which, in addition to being painful, can obstruct the flow of urine and reduce or eliminate filtration in the affected kidney.

Thanks to the muscles of the ureters, urine ends up in the bladder, which is shown in FIGURE 28.19. This hollow, muscular organ can hold up to 27 ounces of urine, but the urge to urinate usually occurs when the bladder contains about 7 ounces of urine. Note that the bladder empties into a tube called the urethra, within which the urine is transported to the external environment. Also note that two sphincters govern the release of urine—an external sphincter, which is under voluntary control, and an internal sphincter, which will not relax until the external sphincter does. We become aware of the need to urinate through nerve impulses sent to the brain from stretch receptors in the wall of the bladder. Stimulation of these receptors results in an involuntary contraction of the bladder that increases the fluid pressure within it. This is the sometimes uncomfortable signal telling us we "have to go." We then can choose to relax the external sphincter, the internal sphincter follows suit, and urination occurs. How much urination? On average, about half a gallon per day.

On to Reproduction

Human beings breathe and digest and move in ways that we have reviewed over the last four chapters. Each generation of people gives rise to another, of course, through the workings of a specialized system, the reproductive system. But how is this system able to create the specialized cells—eggs and sperm—that are critical to reproduction, how do these cells come together, and then how does a baby develop from the two cells that have become one? These are the subjects we'll turn to in the next chapter.

So Far...

1. The "input" into each kidney is _____ that flows through one of the body's _____. The "output" from each kidney is _____ that flows into one of the body's _____ and _____ that flows into the _____.

2. The working unit of the kidneys is the nephron, composed of a nephron tubule, its associated _____, and the _____ in which both are immersed.

3. Nephrons work by means of retaining _____ within the nephron tubules while returning _____ and _____ to the circulatory system.

So Far... Answers:

1. blood; renal arteries; blood; renal veins; urine; bladder

2. blood vessels; fluid

3. waste; water; nutrients

Chapter Review

Summary

28.1 Transport and Exchange in the Body

- **Integrated transport and exchange**—Four systems are involved in the transport and exchange of substances within the body and between the body and the outside world. These are the respiratory, digestive, urinary, and cardiovascular systems.   (page 514)

28.2 The Cardiovascular System

- **Overview of the system**—The human cardiovascular system is a fluid transport system that consists of the heart, all the body's blood vessels, the blood, and the bone-marrow tissue in which red blood cells are formed. This system transports substances both to and from the body's cells. Such substances include oxygen, carbon dioxide, nutrients, vitamins, hormones, waste products, and immune-system cells and proteins.   (page 515)

- **Blood components**—Blood has two primary components: formed elements and blood plasma. Formed elements are blood cells and cell fragments; blood plasma is the fluid portion of blood in which the formed elements are suspended. There are three kinds of formed elements: red blood cells, white blood cells, and platelets. Red blood cells carry oxygen to, and carbon dioxide from, every part of the body; white blood cells are central to the immune system; and platelets are small fragments of cells that are important in the blood-clotting process.   (page 515)

- **Plasma**—Blood plasma is 92 percent water, but it also contains dissolved materials, including proteins. These proteins include antibodies, which aid in the body's defenses, and transport proteins, which transport substances from one part of the body to another. Other plasma compounds include nutrients, wastes, hormones, and electrolytes. Two transport proteins are important in the health of the heart. Low-density lipoproteins (LDLs) carry lipids to bodily tissues from the liver and small intestines; high-density lipoproteins (HDLs) carry lipids from these tissues to the liver.   (page 516)

28.3 Blood Vessels

- **Types of vessels**—Blood vessels carrying blood away from the heart are arteries; blood vessels returning blood to the heart are veins. The smallest blood vessels, the capillaries, connect the arteries with the veins.   (page 517)

- **Makeup of vessels**—Arteries and veins are always made of three distinct layers of tissue, the middle layer of which is muscle that allows arteries and veins to widen or constrict in diameter. Capillaries, conversely, are composed of only a single layer of cells; the thinness that results allows the movement of blood-borne materials into and out of them, along their length.   (page 517)

28.4 The Heart and Blood Circulation

- **Pulmonary and systemic circulation**—The heart's contractions propel blood out to the various tissues of the body. Two blood-circulation loops exist in the body. The first loop is the pulmonary circulation, in which blood circulates between the heart and the lungs (with the result that blood is oxygenated). The second loop is the systemic circulation, in which blood circulates between the heart and the rest of the body (with the result that needed materials are transported to and from all parts of the body).   (page 518)

- **Chambers of the heart**—The human heart contains four muscular chambers—two for pulmonary circulation (the right atrium and right ventricle) and two for systemic circulation (the left atrium and left ventricle).   (page 518)

- **Factors in a heart attack**—About half of all deaths in the United States today are caused by the blockage of one or more of the coronary arteries that supply heart tissue with blood. Such blockages generally are caused by a buildup of low-density lipoprotein (LDL) molecules in a coronary artery, followed by an immune-system inflammation response to these LDLs and formation of a blood clot in the artery. A heart attack occurs when this process results in the complete blockage of a coronary artery, which cuts off the blood supply to groups of cells within the heart, thus killing them.   (page 519)

Web Tutorial 28.1 The Cardiovascular System

28.5 Distributing the Goods: The Capillary Beds

- **The circulatory loop**—Arteries near the heart branch into smaller arterioles, which feed into the delivery vehicles of the cardiovascular system, the capillary beds. The capillary beds then feed back into small venules, which come together to form the larger veins that return blood to the heart.   (page 521)

- **Two-way movement of materials**—Materials needed by the body's tissues move out of the capillaries and into the interstitial fluid that surrounds both the capillaries and nearby cells. Meanwhile, carbon dioxide and wastes from these cells flow into capillaries from the interstitial fluid.   (page 521)

- **Return of venous blood**—Blood pressure is at low levels by the time blood has moved through the capillaries. Blood returns to the heart by the contraction of skeletal muscles, which squeeze the veins in a way that moves the venous blood toward the heart. A system of valves in the veins ensures that this movement is one way—toward the heart.   (page 521)

28.6 The Respiratory System

- **The system in overview**—The central function of the respiratory system is to capture oxygen and to dispose of carbon dioxide. It also aids in controlling pH balance. Respiration can be defined as the exchange of gases between the atmosphere outside the body and the cells within it.   (page 522)

- **Respiratory structures**—The respiratory system includes the lungs; the nose, nasal cavity, and sinuses; the pharynx (upper throat); the larynx (voice box); the trachea (windpipe); and the conducting passageways, called bronchi and bronchioles, that lead to the lungs. The lungs themselves are largely composed of the tiny, hollow sacs called alveoli that lie at the end of each bronchiole and that are the air-exchange chambers of the body. The interface between the alveoli and their associated capillaries is used for the exchange of oxygen and carbon dioxide.   (page 522)

- **Ventilation**—The first step in respiration is breathing or ventilation, meaning the physical movement of air into and out of the lungs. (page 523)

- **Loading and transport**—Once in the lungs, oxygen diffuses across the thin wall of an alveolus into an adjacent capillary and binds with hemoglobin protein in red blood cells. Oxygen then moves with the blood cells to the heart, which pumps the blood to body tissues, where the oxygen diffuses into the interstitial fluid and then into nearby cells. The carbon dioxide produced in the body cells moves into nearby capillaries, to be carried to the lungs. All the oxygen loaded into red blood cells binds initially with the hemoglobin in them. Carbon dioxide, however, is transported both within red blood cells and in blood plasma.   (page 523)

Web Tutorial 28.2 Gas Exchange in the Lung

28.7 The Digestive System

- **The system in overview**—The central functions of the digestive system are (1) to get the foods the body ingests into a form the body can use and (2) to rid the body of the waste that remains after nutrients from the food have been absorbed. The central structure of the digestive system is a muscular tube called the digestive tract that passes through the body from the mouth to the anus and that receives input along its length from accessory digestive organs such as the gallbladder, liver, and pancreas.   (page 524)

- **Digestion**—Digestion is a process of breaking down foods—first into small pieces, and then into small molecules. These small molecules leave the digestive system for transport through the rest of the body. (page 524)

- **Structure of the digestive tract**—The digestive tract begins with the mouth (or oral cavity) and continues through the pharynx, esophagus, stomach, small intestine, and large intestine before ending at the rectum and anus. Through most of its length, the digestive tract is a tube composed of four layers. Digested food moves from the digestive tract to an adjacent capillary (in the case of carbohydrates and proteins) or a lymphatic vessel (in the case of fats).   (page 525)

Web Tutorial 28.3 The Digestion and Absorption of Food

28.8 Steps in Digestion

- **Early digestion**—Digestion begins in the mouth through both mechanical means (chewing) and chemical means (enzymes breaking down carbohydrates). Food is pushed by the tongue into the pharynx (or throat) and moves by muscle contraction from there to the esophagus, whose own muscle contractions work with gravity to push food into the stomach.   (page 527)

- **The stomach**—The stomach digests food partly by the mechanical means of churning it and partly by chemical means, with proteins being a primary target of the stomach's digestive juices. The stomach's contents are very acidic—a quality that is valuable both for breaking food down and for killing microorganisms that come in with the food. The material that leaves the stomach is a mixture of food and digestive juices called chyme.   (page 527)

- **The small intestine**—Eighty percent of the digestive tract's absorption of nutrients takes place within the small intestine—the portion of the digestive tract that begins at the stomach and ends at the large intestine.   (page 528)

- **Digestive glands and organs**—In digestion, the pancreas is an organ that secretes, into the small intestine, three classes of digestive enzymes that help break down fats, proteins, and carbohydrates. The gallbladder stores and concentrates bile, a substance produced by the liver that aids in the digestion of fats. The liver is a large organ that plays a central role in digestion. All blood carrying nutrients from the digestive tract is channeled through the hepatic portal vein to the liver, making the liver a first stop for digested material. The liver controls which nutrients it will store and which nutrients it will send to the rest of the body.   (page 528)

- **The large intestine**—The large intestine holds and compacts material left over from digestion, turning it into feces. It also returns water to general circulation and absorbs vitamins produced by resident bacteria. One of the large intestine's regions, the rectum, is usually empty except when peristaltic contractions force fecal materials into it.   (page 530)

28.9 The Urinary System in Overview

- **Structure and function**—The urinary system filters waste materials from the blood, regulates fluid levels (blood volume), and conserves useful materials, such as water, nutrients, and ions. The system consists of two kidneys that produce urine; the left and right ureters that the urine travels through upon leaving the kidneys; the muscular urinary bladder, which receives the urine from the ureters and temporarily stores it; and the tube called the urethra, through which urine passes from the bladder out of the body.   (page 530)

- **The nephron**—The working unit of the kidneys is the nephron, composed of a nephron tubule, its associated blood vessels, and the fluid in which both are immersed. Several nephron tubules empty into a common collecting duct, which will merge with other such ducts to feed into the renal pelvis, which feeds into the ureter.   (page 532)

Web Tutorial 28.4 The Mammalian Kidney

28.10 How the Kidneys Work

- **Filtration and reabsorption**—A knotted network of capillaries within a nephron, the glomerulus, receives arterial blood, but is porous enough to allow much of the fluid portion of the blood to flow out of it, along with smaller molecules such as vitamins, nutrients, and waste products. These materials enter the surrounding Bowman's capsule, thus moving into the nephron's tubule as a fluid called filtrate. At the nephron's next structure, called the proximal tubule, much of the original water and almost all the original nutrients are moved back into blood circulation. Waste products remain in the nephron tubule, however, because of their chemical composition. This general process continues over the length of the nephron tubule: Water and nutrients move back into circulation, while waste products become ever-more concentrated within the tubule. By the time the filtrate has reached the collecting duct, it has become urine.   (page 533)

- **Control of fluid levels**—The body is able to control how much water the kidneys send to the bladder (in urine) or retain in circulation. Retention of water by the kidneys is regulated by the brain structure called the hypothalamus, which controls the secretion of antidiuretic hormone (ADH). An increased secretion of ADH means that more water will move out of the kidney's tubules and collecting ducts and back into circulation.   (page 534)

28.11 Urine Storage and Excretion

- **Stages in urination**—Waves of muscle contraction squeeze urine out of the renal pelvis, thus moving it through the ureters and into temporary storage in the urinary bladder. The tube called the urethra then carries the urine from the urinary bladder to the exterior of the body. An internal sphincter muscle provides involuntary control over the discharge of urine. The urethra also contains an external sphincter that is under voluntary control. We become aware of the need to urinate when the bladder is stretched beyond a certain threshold. We then relax the voluntary, external sphincter, which relaxes the involuntary internal sphincter, and the urine moves out of the bladder and body.   (page 534)

Key Terms

alveoli   p. 523

antidiuretic hormone (ADH)   p. 534

aorta   p. 519

artery   p. 517

bile   p. 529

capillary   p. 517

cardiovascular system   p. 515

chyme   p. 528

coronary artery   p. 520

digestive tract   p. 524

digestion   p. 525

formed elements   p. 515

gallbladder   p. 529

glomerulus   p. 533

heart attack   p. 520

hemoglobin   p. 516

HDL (high-density lipoprotein)   p. 516

kidneys   p. 531

large intestine   p. 530

liver   p. 529

LDL (low-density lipoprotein)   p. 516

nephron   p. 533

pancreas   p. 529

peristalsis   p. 526

plasma   p. 515

platelet   p. 516

pulmonary circulation   p. 518

red blood cell (RBC)   p. 516

respiration   p. 522

small intestine   p. 528

stomach   p. 527

systemic circulation   p. 518

ureters   p. 531

urethra   p. 531

urinary bladder   p. 531

vein   p. 517

ventilation   p. 523

white blood cell (WBC)   p. 516

Testing Your Understanding

In Your Own Words   *(answers in the back of the book)*

1. Describe the essential difference between the heart's atrial chambers and its ventricular chambers.

2. Where, specifically, does oxygen enter the bloodstream?

3. Why don't blood cells and protein molecules get passed into the filtrate that moves through the kidneys' nephron tubules?

Thinking about What You've Learned

1. Heart attacks have sometimes been called a disease of modern civilization, because their incidence is much higher in affluent, developed societies than in poorer, less developed ones. Why should this be so?

2. Why does the term *circulation* so aptly describe the flow of blood in the body?

3. Suppose that instead of returning nearly all nutrients and 98 percent of fluids back into circulation, the kidneys could return only half these proportions. Could human beings function? How would it change the way they live?

Human Reproduction

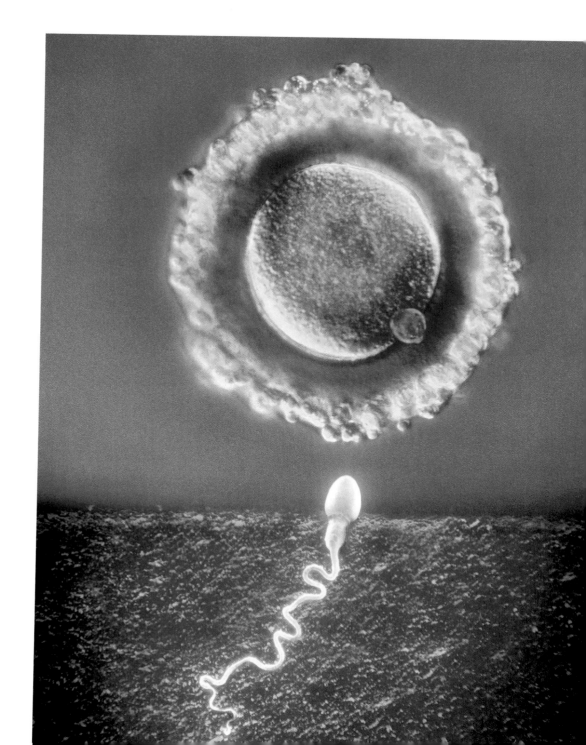

A human sperm near a human egg that is surrounded by its accessory cells.

ST. PAUL PIONEER PRESS

WOMAN, 66, GIVES BIRTH TO DAUGHTER—BOTH REPORTED IN GOOD CONDITION; ROMANIAN PROFESSOR HAD 9 YEARS OF FERTILITY TREATMENTS

By Alison Mutler, Associated Press
January 18, 2005

BUCHAREST, Romania—A 66-year-old professor who writes children's books claims to have become the world's oldest woman to give birth, and doctors said Monday she and her day-old baby daughter were in good condition with both in intensive care.

Doctors at the Giulesti Maternity Hospital in Bucharest said Adriana Iliescu became pregnant through in vitro fertilization using sperm and egg from anonymous donors.

29.1 Human Reproduction in Outline

When is a person too old to become a parent? Adriana Iliescu sparked a worldwide round of controversy about this question when, at age 66, she gave birth to a daughter in Bucharest, Romania. Arthur Caplan, director of the Center for Bioethics at the University of Pennsylvania, said Iliescu's advanced age made her conduct "completely unethical and immoral." Others noted that Iliescu will be 80 by the time her daughter is in high school. Conversely, one British newspaper columnist wrote that Iliescu was simply doing what men have been doing "from time immemorial": using affluence to have children late in life. Men do so by using wealth to attract younger wives, the argument went; Iliescu used her relative affluence to undergo years of fertility treatments. The response to this argument was that, in the case of older men, there is at least one parent (the wife) who is young, while Iliescu had no partner in parenting.

Seemingly lost in the controversy was an examination of what it was that Iliescu wanted. Her desire was not just to become a parent—she presumably could have done that years earlier by adopting a child. And her nine years of fertility treatments were not aimed at making her a biological mother in the sense of producing an egg that

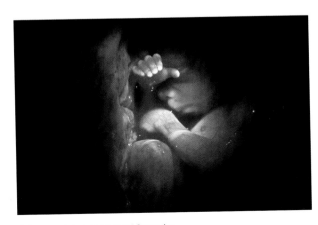

A human fetus at about 19 weeks.

could be fertilized by a sperm. That would have been a virtual impossibility at the age she *started* the fertility treatments, which was 57. Instead, what Iliescu got for her years of treatments was the ability to bring to term a child conceived with eggs donated by another woman. What she got was the ability to have a child *develop* within her. Since she had undergone menopause years earlier, her fertility treatments were, in a sense, aimed at reviving the ability of her uterus to hold onto and nurture an egg that implanted in it.

For our purposes, the Iliescu case is valuable because it separates out what might be called the component parts of reproduction. The first of these is producing either a viable egg (in the case of women) or viable sperm (in the case of men). Then there is the matter of bringing egg and sperm together, an act that normally takes place in the woman's uterine (or Fallopian) tube but that, in this age of reproductive technologies, can take place in a petri dish, as in Iliescu's case. Finally, there is the need for the fertilized egg to implant in the wall of the uterus and begin developing there into a child. Implantation normally happens when a fertilized egg moves from the uterine tube into the uterus, but in today's world it can happen when a physician uses a medical instrument to position a fertilized egg in a uterus.

The purpose of this chapter is to explain how all these steps of reproduction take place: egg and sperm production, egg and sperm union, implantation of the fertilized egg in the uterus, and development of the fertilized egg into a baby. You might say that what we'll be looking at is how a baby comes to be when the means are natural, rather than assisted by technology.

Steps in Reproduction

The first part of reproduction, the formation and delivery of egg and sperm, is a straightforward story in outline. The eggs a woman possesses exist in her in a precursor form, called oocytes; they lie within two walnut-shaped structures called ovaries. These exist just right and left of center within the female pelvic area (**see** FIGURE 29.1). On average, once every 28 days an oocyte is expelled from an ovary and thus begins a slow journey down a structure called the **uterine tube** (or *Fallopian tube*). The release of the oocyte is called **ovulation**. If, during this trip, the oocyte encounters a male sperm, the two may fuse. This is the act of fertilization, which also is known as conception. (For information on some of the methods humans use to *prevent* conception, see Essay 29.1, "Methods of Contraception," on page 547.)

Sperm in men are produced, in unfinished form, in sets of tubules in the two male testes (**see** FIGURE 29.2). They then are transported to a structure called the epididymis (one for each testis) for further development. Then, with sexual excitation, sperm are transported in a loop: up and over the urinary bladder through two ducts, each called a vas deferens. The contents of these ducts then empty into a single duct, called the urethra, from which they are ejaculated into the female vagina. Along the way, materials secreted by several glands join the sperm in a process that can be likened to tributaries feeding into a common stream. The resulting mixture of sperm and glandular materials is called **semen**.

Upon entering the vagina, the sperm move up into the uterine tubes. Of the millions of sperm that begin the journey, a few dozen may encounter the female oocyte as

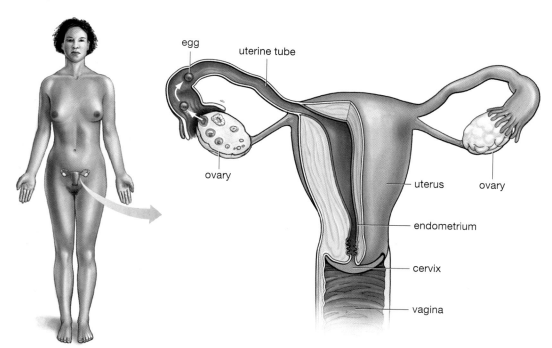

FIGURE 29.1 Reproductive Anatomy of the Human Female Eggs develop in the ovaries. Once every 28 days, on average, an egg is expelled from one ovary or the other and journeys through the uterine tube to the uterus. If the egg encounters sperm on the way, pregnancy will result when egg and sperm fuse. The fertilized egg then continues traveling through the uterine tube and implants in the lining of the uterus, the endometrium.

egg

uterine tube

ovary

uterus

ovary

endometrium

cervix

vagina

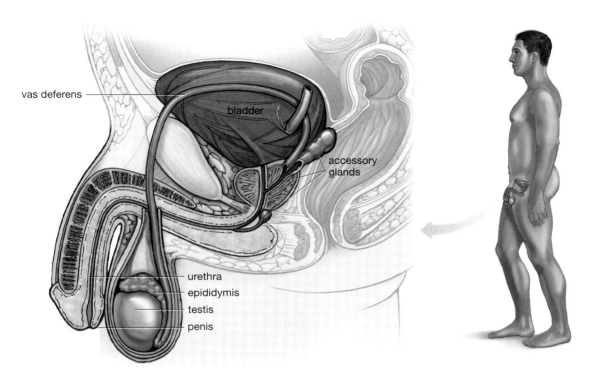

vas deferens
bladder
accessory glands
urethra
epididymis
testis
penis

FIGURE 29.2 **Reproductive Anatomy of the Human Male** Sperm are produced in the testes and transported to the epididymis, where they mature and are stored. With orgasm, the sperm are transported through the vas deferens and then the urethra, from which they are ejaculated. Along the way, they are joined by materials secreted by several accessory glands. The mixture of sperm and glandular secretions is called semen.

it continues its journey (**see** FIGURE 29.3). The first sperm to breach the coating that surrounds the oocyte fuses with it, while the rest are shut out through a kind of gate-slamming mechanism. The now-fertilized egg still floats free in the uterine tube, where it begins to develop: One cell becomes two, two become four, and so on, with cells becoming more specialized over time. Four days after fertilization, the developing embryo enters the uterus; three days more and it implants itself in the uterine wall and continues to develop, for an average of 38 weeks in total, until the time of birth.

Both the male and female reproductive systems are greatly influenced by the actions of hormones—chemical signaling molecules that move through the bloodstream. Indeed, the reproductive organs are primary sites of production for some of the most important of these hormones, testosterone in men and estrogen and progesterone in women.

29.2 The Female Reproductive System

The Female Reproductive Cycle

Although hormones operate in both sexes, in females they bring about a reproductive *cycle* that repeats itself about once every 28 days (though it can range from 21 to 42 days). In males, conversely, there is no cycle, but instead a steady production of sperm and a destruction or removal of the sperm that are not ejaculated. Why a cycle for females and not for males? The answer lies in the two roles the female reproductive system must fulfill. It must form an egg *and* create an environment in the uterus in which a fertilized egg can develop. An important part of this uterine environment is essentially built up and then dismantled once every 28 days—except in those rare instances in which an oocyte is fertilized. The dismantling of the uterine environment results in **menstruation**: a cyclical release of blood and a uterine tissue, the **endometrium**, from the female reproductive tract. The endometrium is tissue that lines the interior of the uterus; it is tissue that *would have* housed an embryo, had one been implanted there (see Figure 29.1).

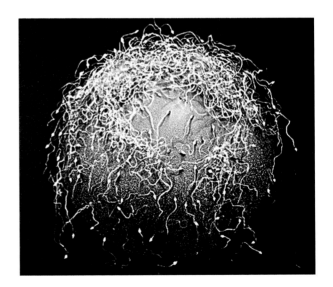

FIGURE 29.3 **Oocyte and Sperm** Numerous sperm surround a single oocyte. Note the tremendous difference in size between the two kinds of cells. The mature egg is the largest cell in the human body.

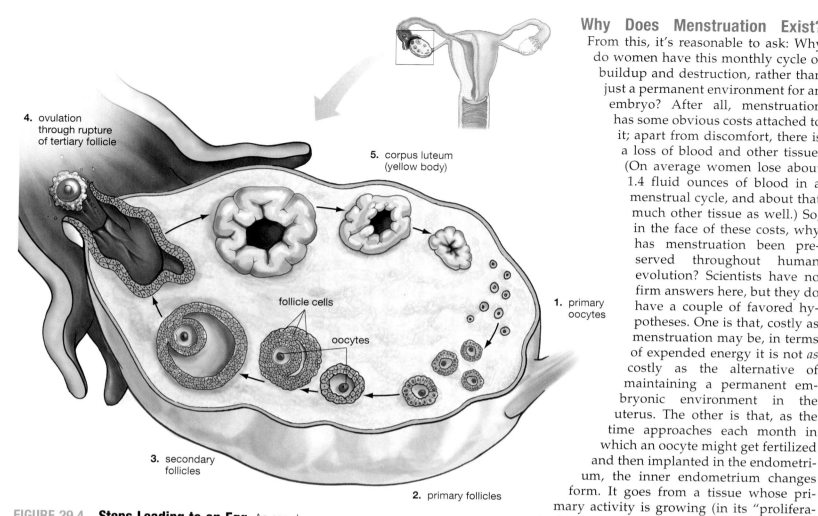

4. ovulation through rupture of tertiary follicle

5. corpus luteum (yellow body)

follicle cells

oocytes

1. primary oocytes

3. secondary follicles

2. primary follicles

FIGURE 29.4 **Steps Leading to an Egg** An egg develops from precursor cells called oocytes that exist near the surface of the ovary. Oocytes join with nurturing complexes of cells to form ovarian follicles. On average, only one of these follicles matures—once every 28 days—through the stages of primary, secondary, and tertiary follicle. The oocyte contained in the tertiary follicle is expelled at the end of the process and then begins a journey through the uterine tube. With the oocyte's expulsion, the tertiary follicle is transformed into the corpus luteum, which for a time produces hormones that facilitate pregnancy.

Why Does Menstruation Exist?

From this, it's reasonable to ask: Why do women have this monthly cycle of buildup and destruction, rather than just a permanent environment for an embryo? After all, menstruation has some obvious costs attached to it; apart from discomfort, there is a loss of blood and other tissue. (On average women lose about 1.4 fluid ounces of blood in a menstrual cycle, and about that much other tissue as well.) So, in the face of these costs, why has menstruation been preserved throughout human evolution? Scientists have no firm answers here, but they do have a couple of favored hypotheses. One is that, costly as menstruation may be, in terms of expended energy it is not *as* costly as the alternative of maintaining a permanent embryonic environment in the uterus. The other is that, as the time approaches each month in which an oocyte might get fertilized and then implanted in the endometrium, the inner endometrium changes form. It goes from a tissue whose primary activity is growing (in its "proliferative" phase) to one whose primary activity is secreting a nutrient-rich mucus (in its "secretory" phase). The assumption is that a fertilized egg can implant itself only in a secretory-phase endometrium, but a secretory-phase endometrium cannot perpetuate itself; it has a short life span and must develop from a proliferative-phase endometrium. Given this, it makes sense that the endometrium washes a part of itself out every 28 days and starts its growth anew.

How Does an Egg Develop?

As noted earlier, eggs develop from precursor cells called **oocytes**. If you look at FIGURE 29.4, you can see where oocytes develop: in the outer portion of each ovary. Each oocyte is surrounded by a collection of accessory cells, called *follicle cells*, that eventually provide the oocyte with nutrients. The complex of an oocyte and its follicle cells is called an **ovarian follicle**; it is the basic unit of oocyte development.

Follicle Development and Ovulation Every oocyte that matures into an egg must go through the developmental steps you see in Figure 29.4. This is a monthly process of selection, running from a large starting set of follicles, through fewer *primary follicles* that develop from them, to a greatly reduced number of *secondary follicles* that continue development, and finally to a single *tertiary follicle* that nurtures an oocyte through its time in the ovary. Responding to hormonal influence midway through the menstrual cycle, the tertiary follicle is squeezed until it ruptures, thus expelling the oocyte and some surrounding accessory cells not only from the follicle but from the ovary as well. With this event, ovulation has occurred. It takes place in this way in one ovary each month.

(a)

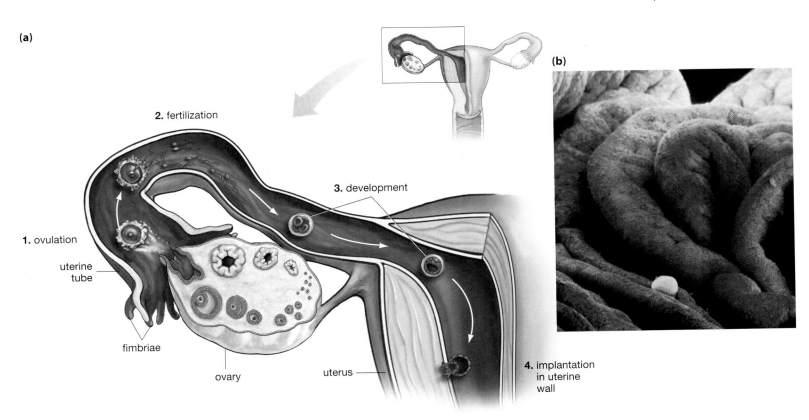

(b)

2. fertilization

3. development

1. ovulation

uterine tube

fimbriae

ovary

uterus

4. implantation in uterine wall

FIGURE 29.5 **Journey to Pregnancy**

(a) Surrounded by accessory cells, an oocyte is expelled from the ovary into the uterine tube, where it encounters sperm and is fertilized by one of them. Now a fertilized egg, it begins the process of development and completes its journey with implantation in the endometrium of the uterine wall.

(b) A micrograph of an unfertilized oocyte moving through folds of tissue in a uterine tube.

Movement through the Uterine Tube Freed from its ovarian housing, but still surrounded by accessory cells, the oocyte—technically not yet an egg—is now swept into the uterine tube through the actions of the finger-like *fimbriae* lying at the tube's funnel-shaped end (**see** FIGURE 29.5). Propelled by liquid currents and the uterine tube's hair-like cilia, the oocyte makes a journey of about 4 inches through the uterine tube over the next four days. While within the tube, the oocyte may encounter sperm and become fertilized prior to entering the uterus.

Meanwhile the ruptured tertiary follicle now takes on a new role: It develops into a body called the corpus luteum ("yellow body"), a structure that secretes hormones that help prepare the female reproductive tract for pregnancy and then maintain it during pregnancy. If pregnancy does not occur, the corpus luteum begins degenerating after about 12 days.

Changes through the Female Life Span

In between the creation of oocytes and the journey of one oocyte down the uterine tube lies a multistep process that could be characterized by the phrase: "Many are called, but few are chosen." A 6-month-old female fetus has in its ovaries about 7 million oocytes that have become part of follicle complexes. At birth, however, this number has been reduced to about 800,000, primarily because of the process of *atresia*, or the natural degeneration of follicles. By puberty, the number has declined to about 250,000. A small number of these follicles are "recruited" monthly for further development, but, as you've seen, only *one* of these will go on to develop into the oocyte that bursts forth from the ovary each month. The only exception to this rule comes in the 1 percent or so of ovarian cycles that result in *multiple* ovulations, in which case fraternal twins, triplets, and other multiple births are possible. In sum, the oocyte that begins the journey down the uterine tube each month has been selected from more than 7 million initially created, and it is one of only 500 or so that will make this trip in a woman's lifetime (about 13 a year for 35 or 40 years).

A BROADER VIEW

Big Eggs, Small Sperm

The eggs of females develop in a pattern opposite that of the sperm of males. Whereas eggs start large and get larger, sperm progressively lose most of their cellular material as they develop. By the time an individual sperm is ejaculated, it consists of little more than a set of DNA-containing chromosomes in front, a collection of energy-supplying "engines" (mitochondria) in the middle, and a propeller (its tail) in the back. Thus, each sperm is a kind of stripped-down DNA-delivery vehicle.

Follicle Loss and Female Fertility Given that a woman begins puberty with about 250,000 follicles, you might think that her supply would never run out. This is not the case, because the process of atresia continues throughout a woman's lifetime. By the time a woman reaches her early fifties, perhaps a thousand follicles remain; this scarcity is the primary factor that brings about **menopause**, or the cessation of the monthly ovarian cycle.

Taking a step back from this story of follicle loss, you can see that its underlying assumption is that, while a woman will lose follicles over the course of her lifetime, she will not gain any. Indeed, the assumption is that all the follicles a woman will ever have are produced in her prior to her birth. It's clear that men remain fertile throughout their lives because their sperm are produced each day by "stem cells," which is to say cells that never lose their ability to produce more sperm. But for decades a bedrock of reproductive science has been that women have no reproductive stem cells. In 2004, however, a research team from Harvard stunned the medical world when they announced that female mice *do* possess stem cells that are capable of giving rise to eggs and that these stem cells remain active throughout life. The hunt is now on, therefore, to see whether human females likewise possess such stem cells. If the answer is yes, then our ideas about female fertility may change fundamentally. If female reproductive stem cells exist—and if their power can be harnessed—fertility might be prolonged for older females or enhanced for females of any age.

The Mystery of Menopause

Irrespective of whether female reproductive stem cells can be found, the massive loss of follicles over the female life span is real and presents science with a mystery: Why should women steadily lose follicles and, as a result, become infertile in midlife? Put another way, why should menopause exist? The living world is shaped by success in reproduction. Traits that help an organism produce relatively *more* offspring will be retained over evolutionary time. Yet the trait of menopause developed in the course of evolution, and it seemingly leads to *fewer* offspring for women. Through it, contemporary women are taken out of reproduction altogether for the last 20 or 30 years of their lives, during which time they suffer from a number of menopause-related physical maladies, such as osteoporosis (which is a lessening of bone density).

As was the case with menstruation, we have no firm answers on why menopause exists, but we do have several candidate hypotheses. One is that menopause is simply a product of women living longer today than their evolutionary ancestors did. Proponents of this view point out that a number of female mammals undergo what amounts to menopause, but then live only long enough to raise one additional generation of offspring after this loss of reproductive capacity sets in. Human females may once have followed this pattern and lived to, say, 60 years old—just long enough to give birth to a final child in their late 40s and then raise that child to a self-sustaining age of 10 or so. Today, however, women don't live until 60; they live until 75 or 85 and thus have decades of life following menopause.

A very different hypothesis holds that menopause is "adaptive," as biologists say—it arose through evolution because it actively helps females perpetuate their genes. The starting point for this "grandmother hypothesis" is the observation that what counts in evolution is not just whether you have children; it is whether you have children who themselves go on to have healthy children. Now, how is a given woman most likely to achieve this? Under this hypothesis, she best achieves it not by continuing to give birth throughout life but by leaving direct reproduction behind at a certain point and becoming a grandmother who provides for her children as well as her grandchildren. We now have evidence, from eighteenth- and nineteenth-century groups in Finland and Canada, that things actually can work this way—that an increased duration of "post-menopausal" life in women can positively affect the reproductive success of their children.

A BROADER VIEW

Grandmothers and Provisioning

One factor consistent with the "grandmother hypothesis" is the degree to which human offspring are dependent on previous generations. Unlike other primates, human children cannot obtain their own food even after they have been weaned from breast-feeding; they must be "provisioned" with food by others. As researcher Kristen Hawkes has noted, this factor stands to increase the importance of a figure, such as a grandmother, who can help provide food for children.

So Far...

1. While they are in development, the reproductive cells a woman produces are known as _____ and are produced in the organs called _____.

2. Every 28 days, on average, one of these reproductive cells is released in the process called _____ and thereupon begins a journey down the woman's _____.

3. Menstruation, the release of blood and a portion of the uterine tissue called the _____, occurs when there has been no _____ during the reproductive cycle.

So Far... Answers:

1. *oocytes; ovaries*
2. *ovulation; uterine (or Fallopian) tube*
3. *endometrium; fertilization of an egg*

29.3 The Male Reproductive System

Switching now to males, if you look at FIGURE 29.6, you can see another view of the male reproductive system. Here you see the large-scale structures of the system.

There is the **testis**, in which sperm begin development; the **epididymis**, the tubule in which sperm mature and are stored; and the **vas deferens**, the tube through which sperm move in the process of ejaculation. (All three of these structures exist in

FIGURE 29.6 Development and Delivery of Sperm

(a) After the sperm develop in the testes and mature in the epididymis, orgasm prompts them to be transported through the vas deferens and then the urethra. Secretions from accessory glands join with the sperm to form semen before ejaculation. These glands are the seminal vesicles, the prostate gland, and the bulbourethral glands.

(b) How sperm develop. Sperm development begins within seminiferous tubules, located in the testes. Sperm start out as spermatogonia at the periphery of the tubules and end up as immature sperm in the interior cavity of the tubules. From there, they are transported to the epididymis for further development and storage. Note the stem cell capability of the spermatogonia: Each of them gives rise not only to a primary spermatocyte, but to another spermatogonium.

(a) Delivery of sperm

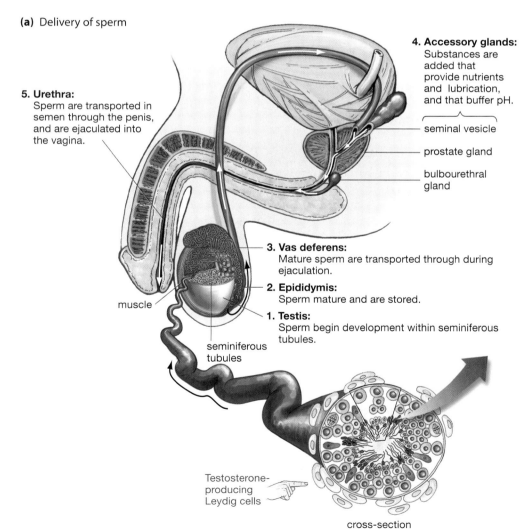

5. Urethra:
Sperm are transported in semen through the penis, and are ejaculated into the vagina.

4. Accessory glands:
Substances are added that provide nutrients and lubrication, and that buffer pH.

seminal vesicle

prostate gland

bulbourethral gland

muscle

3. Vas deferens:
Mature sperm are transported through during ejaculation.

2. Epididymis:
Sperm mature and are stored.

1. Testis:
Sperm begin development within seminiferous tubules.

seminiferous tubules

Testosterone-producing Leydig cells

cross-section of seminiferous tubule

(b) Development of sperm

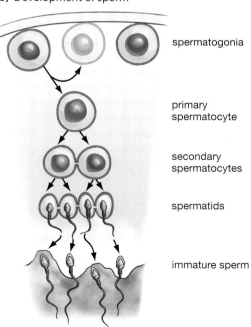

spermatogonia

primary spermatocyte

secondary spermatocytes

spermatids

immature sperm

pairs, one on each side of the body, though the second member of each pair is not visible in the figure.) Then there is the single **urethra**, the tube each vas deferens empties into and through which sperm leave the male body. There are also several accessory glands whose contributions are added to sperm as it proceeds. These are the seminal vesicles, the prostate gland, and the bulbourethral glands. (The seminal vesicles and bulbourethral glands also exist in pairs.)

Structure of the Testes

The testes are surrounded by a muscle whose contraction helps regulate the temperature of the sperm inside. Sperm development requires temperatures somewhat cooler than those found in the rest of the body, which is why the testes are *external* to the body's torso. Sperm cannot develop in temperatures that are too cold, however. Thus, when a male steps into, say, a cold swimming pool, the muscle surrounding the testes will contract, drawing the testes up closer to the torso—and warming the sperm.

Figure 29.6a shows that the testes are divided into a small series of lobes, and that within each lobe there are a number of highly convoluted tubes, the **seminiferous tubules**: structures that are the sites of initial sperm development. If you now look at the blow-up of the single seminiferous tubule in Figure 29.6b, you'll see how this development proceeds: from the outside of each tubule—where sperm precursors are in their *least* developed state—toward the central cavity of each tubule, where sperm are in a more developed state. Eventually, the immature sperm are released into the central cavity and are transported up it, moving then to the epididymis for further processing and storage. As you can see, there is an assembly-line quality to sperm production, starting with initial development at the periphery of the tubules, continuing in the interior of the tubules, and ending with storage in the epididymis. Older sperm that are not ejaculated are destroyed (by other cells) or eliminated in the urine. Inside the testes, surrounding the seminiferous tubules, there exist numerous *Leydig cells*, which are the cells that produce testosterone.

Male and Female Gamete Production Compared

The production of sperm bears an interesting comparison with the development of eggs. The cells from which sperm develop are called **spermatocytes**. In going back to the precursors of *these* cells, however, we see a significant difference between males and females. The cells that give rise to spermatocytes are called **spermatogonia**; they exist at the periphery of the seminiferous tubules and, through normal cell division, they actually lead to two kinds of cells. One is the primary spermatocyte, which goes on to develop into a sperm; the other, however, is another spermatogonium (see Figure 29.6b). Thus, spermatogonia are "sperm factories," if you will, that give rise not only to sperm, but to more sperm factories as well. Put another way, spermatogonia are reproductive stem cells. This is the reason, noted earlier, that males *keep* producing sperm throughout their lives. Although male sex hormone production lessens with age, and sperm decline in quality, it is possible for male reproductive capability to remain intact until death.

Further Development of Sperm

As you can see in Figure 29.6a, the seminiferous tubules eventually feed into the epididymis, a single convoluted tubule (one for each testis) that is about 23 feet long. This structure serves as the site of storage and final sperm development and as a kind of materials recycling site for damaged sperm. By the time sperm arrive at the bottom portion of the epididymis, they are fairly mature, but not yet fully mature. They won't become mobile until fluid from a type of supporting gland noted earlier—the seminal vesicles—mixes with them upon ejaculation, thus supplying them with nutrients. And, as you'll see, substances in the female reproductive tract will interact with them to prompt their final maturation.

Essay 29.1   Methods of Contraception

Selected methods of contraception are presented here. The rate of effectiveness for each method refers to effectiveness with correct use during every instance of sexual intercourse. In practice, effectiveness for each method tends to be lower, because use of the method may be incorrect or inconsistent.

Table 29.1 Methods of Contraception

	**Birth Control Pill**	A set of pills containing a combination of hormones that suppress the female ovulatory cycle. Effectiveness: 99.9 percent.
	**Condom**	Made of latex, plastic, or animal tissue, condoms are put on the erect penis prior to intercourse and catch sperm after it is ejaculated. Condoms are the only method of contraception that helps protect both partners from some sexually transmitted diseases. Effectiveness: 97 percent.
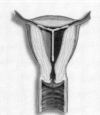	**Diaphragm**	A soft, circular rubber barrier inserted into the woman's vagina before sex. Covering the cervix, it prevents sperm from reaching the egg. Along with the related cervical caps, diaphragms offer females some protection against some sexually transmitted diseases, such as gonorrhea and chlamydia. Effectiveness: about 95 percent, when used with a spermicide (a chemical compound that immobilizes sperm).
	Intrauterine device (IUD)	A small device, made in some instances of plastic and copper, that is inserted into a woman's uterus by a physician. These IUDs can be left in place for several years. Other IUDs are made of plastic and contain a supply of a progestin hormone that is continuously released. These IUDs must be replaced once a year. Both types of IUDs help prevent fertilization of egg by sperm, in a process that may involve altering the movement of each type of gamete. Both types of IUDs also work to prevent implantation in the uterus of any embryo that might be produced. Effectiveness: copper IUD, 99.4 percent; hormone-releasing IUD, 98.5 percent.
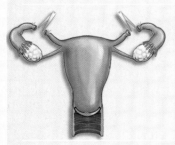	**Tubal ligation**	A surgical procedure that generally results in the sterilization of a woman, meaning permanent infertility. In it, the uterine tubes are first cut and then tied, thus blocking eggs from moving to a position in which they could be fertilized by sperm. Does not affect feminine characteristics, hormonal production, or sexual performance. Very rarely, tubes will reconnect themselves, which accounts for the (low) failure rate of the procedure. It is possible to reverse the operation in some instances. Effectiveness: 99.6 percent.
	Vasectomy	A surgical procedure that can result in the sterilization of a man, though vasectomies are sometimes reversible. In the procedure, the vas deferens are cut and tied, thus blocking the movement of sperm during ejaculation. Does not affect male hormonal production, physical capacity for erection, or ability to ejaculate. (Sperm make up a small portion of the semen that is ejaculated.) In rare instances, vas deferens will grow back together again, which accounts for the (low) failure rate of the procedure. Effectiveness: 99.9 percent.

Time Sequence and Sperm Competition How long does this whole process of sperm development take? About 2.5 months from the start of development to storage in the epididymis. About 200 million sperm are produced in this assembly line each day, which you may recall is about the number of sperm that are ejaculated in an average orgasm.

As noted, the watchword for oocyte selection could be "many are called but few are chosen." How much more extreme is the selection that goes on with sperm! The number expelled with *each ejaculation* vastly exceeds the number of oocytes a woman ever has stored in her ovaries. However, only a single sperm per ejaculation will be able to fertilize an egg—assuming an egg is present in the uterine tube at the right time.

Why is there such an enormous difference between the number of sperm produced and the number that will fertilize eggs? The essential answer here seems pretty clear: Just as males often are in competition with other males to mate with females, so the sperm of males is in competition with the sperm of other males to *fertilize* female eggs. This general phenomenon of *sperm competition* goes on throughout the animal kingdom—in insects, birds, and reptiles as well as in mammals. Its root cause is that there are very few animal species in which females or males are truly monogamous. As a result, mating by itself does not guarantee offspring. If a female who is fertile over a period of several days mates with a number of males during these days, then each male's chance of fatherhood is enhanced to the degree that his sperm can outcompete the sperm of the other males. This is the basis for a sperm "arms race" that has taken place over evolutionary time. Looking at various species, we find that their sperm has gotten longer, more active, or has been produced in greater numbers—all as a means of gaining a reproductive edge. This is why men produce so many sperm. Their sperm production has followed the lottery principle: the more tickets you have, the more likely you are to win.

From Vas Deferens to Ejaculation Maturing sperm are stored in both the epididymis and the vas deferens. With sufficient sexual stimulation, ejaculation takes place through the contractions of muscles that surround the penis at its base. This contraction pushes the sperm through the urethra and out the urethral opening at the tip of the penis.

Supporting Glands

As noted, several other kinds of materials join with sperm before ejaculation to form the substance known as semen. Indeed, given the number of sperm ejaculated, it may be surprising to learn that sperm don't account for much of the volume of semen; some 95 percent of the ejaculated material comes from the accessory glands already mentioned: two seminal vesicles, the prostate gland, and two bulbourethral glands. It's helpful to think of sperm as getting "outfitted" for their travels by these glands just as they are exiting from the body. In addition to the nutrients the seminal vesicles supply, accessory glands provide alkaline and lubricating substances (which neutralize the acidic environment of the vagina and facilitate transportation).

One of these supporting glands may sound familiar. The **prostate gland** is a structure that surrounds the male urethra near the bladder and contributes a substantial amount of material to semen. Later in life, a significant percentage of men develop either enlarged prostate glands, or worse, cancerous cells in the prostate, which is to say prostate cancer. Given that the prostate encircles the urethra—and that the urethra transmits not only semen but urine as well—you can see why a warning sign of these conditions is reduced urine flow. The sexual activity that is so important to human reproduction also carries with it the risk of sexually transmitted disease, which you can read about in Essay 29.2, "Sexually Transmitted Disease," on page 550.

A BROADER VIEW

Sperm Competition and Testicle Size

Sperm competition drives not only the nature of sperm, but the nature of the testicles that produce it. Across primate species, testicle size increases in accordance with the competition males face for fertilizing eggs. Female chimpanzees are very promiscuous, and chimp males consequently produce huge amounts of sperm, which requires large testicles. Indeed, as researcher Tim Birkhead has observed, if human males had testicles the same relative size as a chimp's, the human's testicles would be the size of grapefruits. Conversely, male gorillas with harems face no sexual competition from other males. If human males had the gorilla's testicles, they would be about the size of small beans. The actual size of human testicles—intermediate between the chimp's and the gorilla's—indicates that human males have faced an intermediate level of sperm competition.

So Far...

1. Sperm begin development in the _____ and continue development and are stored in the convoluted tubules called the _____.

2. The material that is ejaculated from the penis, _____, is composed of both _____ and fluids secreted by three supporting _____.

3. Males are able to remain fertile throughout their lives because the cells called spermatogonia are reproductive _____ that give rise to more of themselves and to the _____ that develop into sperm.

So Far... Answers:

1. *testes; epididymis*
2. *semen; sperm; glands*
3. *stem cells, spermatocytes*

29.4 The Union of Sperm and Egg

To this point, you have reviewed the process by which male sperm and female egg are first created and then *positioned* so that they can come together in the process called fertilization. Now let's look at this moment of fusion in somewhat greater detail. When the oocyte is expelled from the ovaries, there still is no solitary oocyte in transit, but rather an oocyte surrounded by a group of accessory cells and connective substances.

Now consider the sperm. As you can see in FIGURE 29.7a, each sperm has a "head" that contains not only the nucleus and its vital complement of chromosomes, but an outer compartment, called an acrosome, which contains enzymes. These enzymes are released *externally*, so that they break down the cells and materials that surround the oocyte. To prepare themselves for this release, sperm have to undergo a final process of maturation while swimming up the female reproductive tract. Substances within the tract cause the acrosome to lose its "cap," thus allowing it to discharge its enzymes. Many sperm mature fully through this process, but only one of them completes the next step, which is to latch onto the

A BROADER VIEW

Tight Timing for Fertilization

While only one sperm fuses with an oocyte, a single sperm cannot undertake this task by itself, because it takes the digestive enzymes of dozens of sperm to break down the cellular layers that surround the oocyte. Thus the timing on fertilization is tight: Many sperm must be close to the oocyte, but the binding of the first one of these sperm with the oocyte triggers the mechanism that shuts out the rest.

(a) Structure of a sperm cell

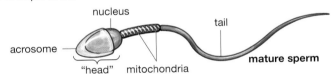

(b) How sperm fertilizes egg

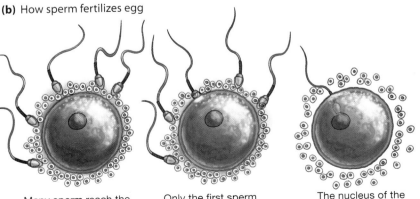

Many sperm reach the cells that surround the oocyte.

Only the first sperm that binds with receptors on the oocyte can fertilize it.

The nucleus of the sperm unites with the nucleus of the oocyte.

The two nuclei fuse together.

fertilized egg

FIGURE 29.7 Anatomy of a Sperm and Steps in Fertilization

(a) The "head" of each sperm includes not only its complement of chromosomes but also an acrosome, which contains enzymes that, when released externally, break down material surrounding the oocyte. Behind the sperm head are the mitochondria that supply the energy for the motion of the tail that propels the sperm.

(b) The process by which a sperm fertilizes an egg.

Essay 29.2 Sexually Transmitted Disease

Given the necessity of sex for the continuation of human life, it may seem surprising that so much risk can be associated with it, but such is the case. The root of the problem is that microorganisms that literally couldn't stand the light of day—that is, that couldn't exist for long in an exterior environment—are quite well adapted to life inside human beings. (The bacterium that causes gonorrhea lives nowhere in nature *except* inside human beings.) More important, these microbes can be transmitted from one human being to another during the act of sex.

The risk of sexually transmitted diseases (STDs) varies in accordance with the disease itself, the type of sexual activity a person engages in, and the precautions that are taken to avoid trouble. But one rule holds true for all STDs: As the number of partners goes up, so does the risk. Here are the most common STDs, their prevalence in the United States, and the methods of treatment for them.

Table 29.2 Sexually Transmitted Disease

Disease	Caused by	Possible consequences	New cases in the United States	Treatment
Chlamydia	*Chlamydia trachomatis* bacterium	Spread to uterine tubes in women, causing pelvic inflammatory disease (PID), which can lead to sterility	2.8 million (2003)	Curable with timely use of antibiotics
Gonorrhea	*Neisseria gonorrhoeae* bacterium	Arthritis, pelvic inflammatory disease	718,000 (2003)	Curable with timely use of antibiotics, though antibiotic-resistant strains now exist
Syphilis	*Treponema pallidum* bacterium	Heart, nervous system, bone damage if allowed to progress	34,000 (2003)	Curable with timely use of antibiotics, though antibiotic-resistant strains now exist
Genital herpes	Herpes simplex virus	Recurrent skin lesions, eye damage, pregnancy complications	1 million (2000)	Occurence and severity of outbreaks can be lessened with treatment, but no cure exists
Genital warts	Human papilloma virus	Cervical infection in women, association with increased risk of cervical and penile cancer	5.5 million (2000)	Several treatments exist for removing warts, but virus may persist
AIDS	Human immunodeficiency virus (HIV)	Death, dementia, injury from a variety of opportunistic infections	40,000 HIV infections; 1 million persons living with HIV in the U.S. (2003)	No cure; treatments to reduce symptoms exist

oocyte (**see** FIGURE 29.8). This happens by means of a sperm cell binding with receptors on the surface of the oocyte (FIGURE 29.7b). Once this binding has occurred, the plasma membranes of sperm and oocyte fuse; the sperm is then inside the oocyte, and sperm and egg nuclei combine. Fertilization is complete.

How Latecomers Are Kept Out

But what of the other dozens of sperm surrounding the oocyte? None of them can be allowed entry, because an embryo could not survive having two sets of male chromosomes, but only one set of female chromosomes. To guard against this possibility, a kind of gate-slamming process is put into motion. The fusion of the first

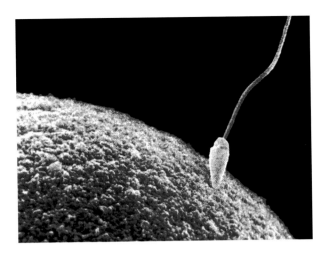

FIGURE 29.8 **Near the Moment of Conception** A human sperm and oocyte at the moment the sperm binds the oocyte. Very soon after this, the sperm will enter the oocyte and its nucleus will fuse with the oocyte nucleus, producing a human zygote.

sperm with the oocyte causes the oocyte to release thousands of tiny granules into the space just outside itself. These granules release substances that harden the material immediately outside the oocyte while inactivating sperm receptors on the oocyte. The result is that, with rare exceptions, only one sperm gets in.

29.5 Human Development Prior to Birth

The coming together of sperm and egg initiates the process of development: the process by which a fertilized egg is transformed, step by step, into a functioning organism.

Early Development

About four days after fertilization, the egg arrives in the uterus as a tightly packed ball of cells. Over the next two days, it will transform itself into a hollow, fluid-filled ball of cells—a *blastocyst*, which will implant itself in the endometrial lining of the mother's uterus. Implantation normally occurs in the wall of the uterus (**see** FIGURE 29.9). Occasionally, however, a blastocyst attaches itself inside the uterine

FIGURE 29.9 **Implantation Accomplished** Twelve days after conception, the embryo, now in its blastocyst stage, has implanted in the uterine wall.

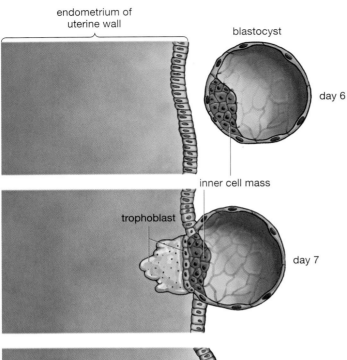

endometrium of
uterine wall

blastocyst

day 6

inner cell mass

trophoblast

day 7

day 8

future
extensions
of placenta

FIGURE 29.10 Establishing Links with the Uterus Six days after conception, the embryo, now in the blastocyst stage of development, arrives at the uterine wall. It is composed of both an inner cell mass, which develops into the baby, and a group of cells called a trophoblast, which carves out a cavity for the embryo in the uterine wall and then begins establishing physical links with the maternal tissue. The outgrowth of this linkage is the placenta.

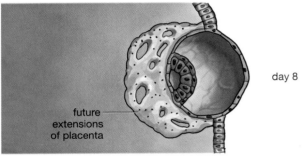

A BROADER VIEW

Blastocysts and Embryonic Stem Cells

The cells that make up the inner cell mass of the blastocyst go by another name that may sound familiar: embryonic stem cells. It is these cells whose power scientists are trying to harness to cure such conditions as Parkinson's disease and spinal cord injuries. The source of embryonic stem cells is fertility clinics, which each year discard thousands of excess embryos that have been produced for in vitro fertilizations. The power of these cells comes from their ability to differentiate into any type of cell in the adult body.

tube, or sometimes to the **cervix**, which is the lower part of the uterus that opens onto the vagina. This is an **ectopic pregnancy**: a pregnancy in which the embryo has implanted in a location other than the uterine wall. In general, embryos do not survive such an implantation.

Two Parts to the Blastocyst FIGURE 29.10 shows you the arrangement of cells in the blastocyst. On one side, you can see a group of cells called the **inner cell mass**: the portion of the blastocyst that develops into the baby. The periphery of the blastocyst, however, consists of a group of cells called the **trophoblast**. These are the cells that bring about implantation in the uterine wall—they release enzymes that actually carve out a cavity in the wall, thus allowing implantation to proceed. They then begin extending farther into the endometrium; in time, these trophoblast cells grow into the fetal portion of a well-known structure in pregnancy, the **placenta**: a network of maternal and embryonic blood vessels and membranes that allows nutrients and oxygen to diffuse *to* the embryo (from the mother), while allowing carbon dioxide and waste to diffuse *from* the embryo (to the maternal blood system). See FIGURE 29.11 for a look at how the placenta functions about two months after conception. The tissue that links the fetus with the placenta is the **umbilical cord**, a tube that houses fetal blood vessels that move blood between the fetus and the placenta.

The Origins of Twins Having learned a little about conception and the time period immediately following it, you're now in a position to understand how twins come about. Identical twins can develop at any time in the very early stages of pregnancy—from the time the fertilized egg undergoes its first division through the seventh day after fertilization. In all cases, identical twins result from two *parts* of the embryo developing into separate human beings. This may be a matter of the two cells produced by the fertilized egg's first division each developing as separate embryos; or it may entail the inner cell mass of the blastocyst separating into *two* masses, with each then giving rise to a different person. In contrast, **fraternal twins** (or triplets, and so on) are twins who are produced first through multiple ovulations in the mother, followed by multiple fertilizations from separate sperm of the father, and then multiple implantations of the resulting embryos in the uterus. This means that fraternal twins are like any other pair of siblings, except that they happen to develop at the *same time*. Meanwhile, **identical twins** are twins who develop from a single fertilized egg. Strictly speaking, they are a single organism at one point in their development and as such have exactly the same genetic makeup.

Embryonic Development over Time

During its first few weeks of existence, the fertilized egg is doing more than implanting in the uterus. It is starting the slow process of developing into a fully formed human being. All the cells in the blastocyst's inner cell mass are the same—there is no cell that, at this point, is a muscle cell or a brain cell. But this will soon change. The unspecialized cells of the blastocyst soon give rise to three different types of cells that are organized into tissue layers in the developing embryo. These tissues then start developing into the organs and organ systems that make up the human body.

What is perhaps surprising is how *early* many key developmental steps occur. Organs begin developing in an embryo in the fourth week following conception. At the end of the first month, the primitive heart has actually begun to beat, even though it is still only partially formed. By the end of 12 weeks, immature versions of all the major organ systems have formed. At about week 13, sex organs start to differentiate into male or female versions. To place some of these events in the context of what the mother experiences, organ formation is occurring about the

(a) Nutrient and gas exchange in the placenta

(b) The placenta in context

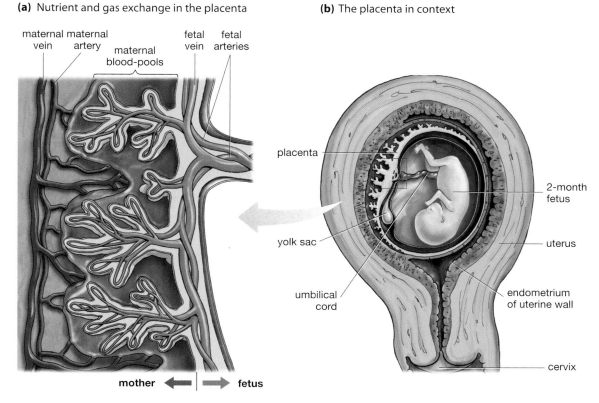

FIGURE 29.11 **Nutrient and Gas Exchange in the Placenta**

(a) Oxygen- and nutrient-rich blood comes from the mother through the maternal arteries and moves, through smaller arterial vessels, into the maternal blood pools in the placenta. Oxygen-poor blood from the embryo, transported through the umbilical cord's fetal arteries, does not flow into the maternal blood pools, but instead stays within the fetal blood vessels, which project into the maternal blood pools. Oxygen and nutrients in the maternal blood pools move into the fetal blood vessels, however, through such processes as diffusion. The resulting oxygen- and nutrient-rich blood is carried back to the embryo through the umbilical cord's fetal vein. Meanwhile, carbon dioxide from the embryo moves the other way—into the maternal blood pools—to be carried away through maternal veins.

(b) Larger structure of the placenta and uterine environment at 2 months.

time the first menstrual period is missed, and formation of the immature organ systems is being completed at about the same 12-week point at which morning sickness generally ends.

The embryo's development over time can be contrasted with its *growth* over time. By the end of the third week, a human embryo may be less than a tenth of an inch in diameter. Skip to the twelfth week after conception, when major portions of organ formation are concluding, and the fetus may still be only about 5 inches in length.

The message here is that development is concentrated early in pregnancy, while the simple growth of what *has* developed comes later. So pronounced are the changes during development that biologists use different terms for the growing organism over time. The original cell—the fertilized egg—is a **zygote**. In humans, about 30 hours after the zygote is formed, it undergoes its first cell division; at this point, the developing organism is referred to as an **embryo**, a name that will be applied to it through the eighth week of development. From the ninth week of development to birth, the organism is referred to as a **fetus**. The typical 9-month pregnancy is divided into periods of roughly three months each, called **trimesters**. The first trimester ends at about the time organ formation does. If you

A BROADER VIEW

Premature Infants

The lungs are among the last organs to develop in the unborn. Given this, one of the main risks to premature babies is a condition called respiratory distress syndrome (RDS), in which newborns labor to breathe. Sacs in their lungs that fill with air upon inhalation tend to collapse upon exhalation. As a result, these babies have to struggle to reopen the sacs and are quickly exhausted. Developmental timing is at the root of RDS: It is only at the 26th week of pregnancy that a fetus starts producing a fatty liquid that keeps its air sacs from collapsing.

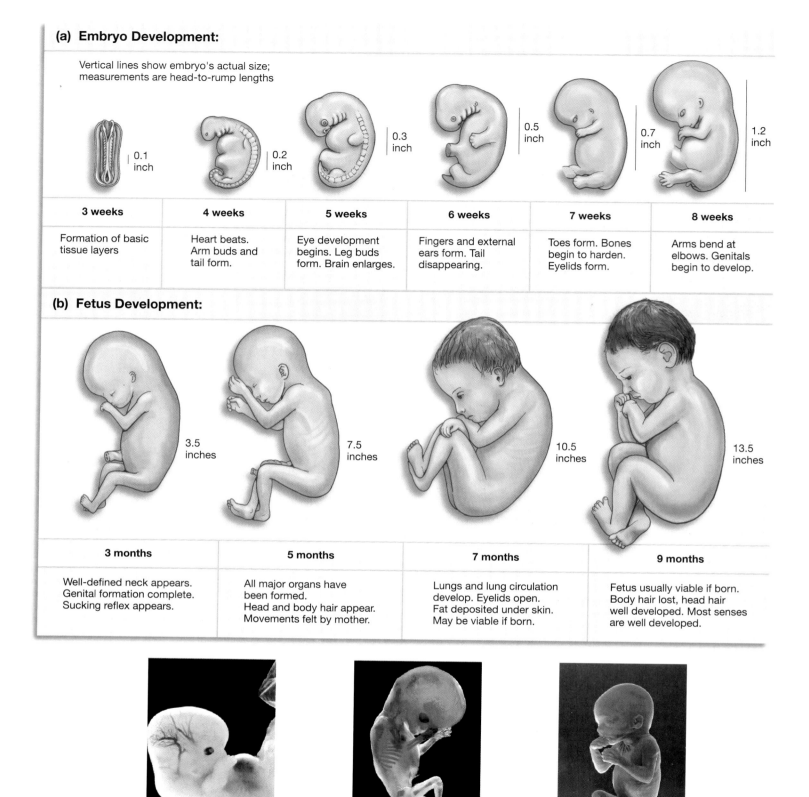

(a) Embryo Development:

Vertical lines show embryo's actual size; measurements are head-to-rump lengths

3 weeks	4 weeks	5 weeks	6 weeks	7 weeks	8 weeks
Formation of basic tissue layers	Heart beats. Arm buds and tail form.	Eye development begins. Leg buds form. Brain enlarges.	Fingers and external ears form. Tail disappearing.	Toes form. Bones begin to harden. Eyelids form.	Arms bend at elbows. Genitals begin to develop.

(b) Fetus Development:

3 months	5 months	7 months	9 months
Well-defined neck appears. Genital formation complete. Sucking reflex appears.	All major organs have been formed. Head and body hair appear. Movements felt by mother.	Lungs and lung circulation develop. Eyelids open. Fat deposited under skin. May be viable if born.	Fetus usually viable if born. Body hair lost, head hair well developed. Most senses are well developed.

6 weeks   11 weeks   5 months

FIGURE 29.12   Development through the Trimesters   (a) Development of the human embryo (3 weeks to 8 weeks). **(b)** Development of the human fetus (3 months to 9 months). Photos of a human embryo at 6 weeks (left), 11 weeks (middle), and 5 months (right).

look at the series of pictures and diagrams that make up FIGURE 29.12, you can see how an embryo and then fetus develops at selected points in the 38 weeks of pregnancy.

29.6 The Birth of the Baby

As the fetus grows, the uterus becomes an increasingly cramped place, and fetal movement becomes more restricted. In late pregnancy, the fetus' head normally will lodge near the base of the mother's spine (see FIGURE 29.13). In this "upside-down" position, the fetus is ready to make its entrance into the world. What triggers this journey? The immediate cause is **labor**: the regular contractions of the uterine muscles that sweep over the fetus from its legs to its head. The pressure this generates opens or "dilates" the cervix (to approximately 10 centimeters or about 4 inches). This creates an opening large enough for the baby to pass through. These things must occur in stages, however. The first phase of the uterine contractions is given over to cervical dilation; then comes the expulsion of the baby and finally expulsion of the placenta (also called the *afterbirth*, for obvious reasons). Once all of this has been completed, the baby takes its place, for the first time, in the world outside the mother's body.

On to Plant Physiology

In this chapter and the four that preceded it, you followed the workings of the cells and tissues that allow human beings to function. But it goes without saying that human beings are not the only organisms that have an anatomy and a physiology—a physical makeup and an interrelated set of working physical processes. Plants have both things as well: They have different kinds of cells that make up various kinds of tissues; they have hormones and reproductive cycles and ways of sensing time and means of responding to threats from the outside world. In short, plants have a wide range of what might be called components and capabilities. In the chapter coming up, we'll be looking at these things as they exist in the world's most widespread plants, the angiosperms.

So Far...

1. The number of sperm that can fuse with an oocyte is limited to _____. This is vital in order to make sure that a fertilized egg has a balanced number of _____.

2. With the implantation of the fertilized egg in the wall of the _____, cells in the egg grow into the fetal portion of a network of blood vessels and membranes that support the fetus. This is the _____.

3. Labor can be defined as regular _____ of uterine _____ that help bring about the birth of the baby.

So Far... Answers:

1. *one; chromosomes from the mother and father*

2. *uterus; placenta*

3. *contractions; muscles*

(a) Baby's position prior to birth

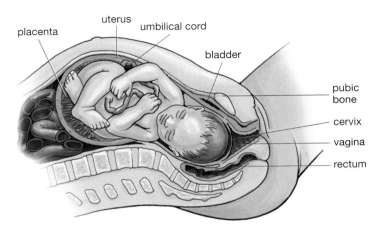

(b) Movement through birth canal

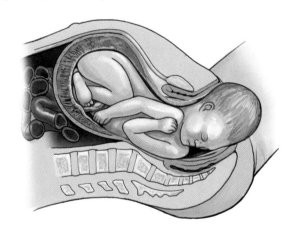

(c) Structure of "afterbirth"

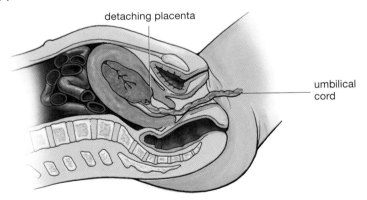

FIGURE 29.13   Birth of the Baby With uterine contractions, the cervix dilates, allowing room for the baby to pass through the birth canal and to the outside world. The placenta and its associated fluids and tissues are expelled shortly afterward as the "afterbirth."

Chapter Review

Summary

29.1 Human Reproduction in Outline

- **In females**—The precursors to eggs, called oocytes, exist in the organs called ovaries. On average, once every 28 days one oocyte is released from an ovary and then begins a journey through the uterine tube to the uterus. This release of the oocyte is called ovulation. Conception can occur if the oocyte encounters sperm on the way to the uterus.   (page 540)

- **In males**—Sperm are produced in the male testes and then pass through two types of ducts, the vas deferens and the urethra, after which the sperm are ejaculated from the penis into the female vagina. Materials secreted by three separate sets of glands will join with the sperm before ejaculation. The resulting mixture of sperm and glandular materials is called semen. The first of the sperm to reach the oocyte moving through the uterine tube may fertilize it. Over the succeeding 38 weeks, the fertilized egg develops into a child.   (page 540)

29.2 The Female Reproductive System

- **The reproductive cycle**—A portion of the female reproductive system is built up and then dismantled in a reproductive cycle that takes approximately 28 days to complete. The result of the dismantling is menstruation, which is a release of blood and the specialized uterine tissue it infuses, the endometrium. The endometrium, which lines the uterus, is tissue that an embryo implants itself in when pregnancy occurs. Given the costs of menstruation to the female, it is not clear why it has been preserved in evolution, though several hypotheses have been put forward to explain it.   (page 541)

- **Follicles**—The basic unit of oocyte development is the ovarian follicle, which includes the oocyte that develops into an egg and some surrounding accessory cells.   (page 542)

- **From ovary to uterus**—On average, a single follicle each month will develop into a tertiary follicle that ruptures, expelling the egg and its accessory cells from the ovary. The ruptured tertiary follicle develops into the corpus luteum, which secretes hormones that prepare the reproductive tract for pregnancy and maintain it during pregnancy. (page 543)

- **Follicle degeneration**—The number of follicles a woman possesses steadily decreases throughout her lifetime, primarily through the process of atresia or follicle degeneration. The scarcity of follicles a woman has in middle age is the primary factor that brings about menopause, the cessation of the monthly ovarian cycle.   (page 544)

- **Menopause**—Recent research has called into question the long-standing assumption that all the follicles a woman possesses develop in her prior to birth. It may be that, like men, women possess reproductive stem cells that are capable of generating new follicles. It is not clear why menopause exists, since it seems to carry a number of costs to females, but several hypotheses have been put forth to account for it.   (page 544)

Web Tutorial 29.1 The Female Reproductive System

29.3 The Male Reproductive System

- **Component parts**—Sperm begin development in the testes; development of sperm continues while they are stored in the adjacent epididymis. Sperm move from the epididymis through the vas deferens and then the urethra in ejaculation. The accessory glands that contribute material to the sperm before ejaculation are the two seminal vesicles, the prostate gland, and the two bulbourethral glands. (page 545)

- **Sperm development**—Within the testes, sperm development takes place in the seminiferous tubules. Development begins at the periphery of a given tubule, with sperm becoming more developed toward the interior of the tubule. Eventually, developing sperm are transported from the interior of a tubule to the epididymis for further development and storage.   (page 546)

- **Sperm precursors**—Sperm development in the testes begins with cells called spermatogonia, which develop into spermatocytes. These in turn give rise to spermatids and then to immature sperm. Spermatogonia give rise not only to spermatocytes, but to more spermatogonia. This self-generation of sperm precursors is the reason men keep producing sperm throughout their lifetimes.   (page 546)

- **Sperm production**—About 200 million sperm are produced each day—about the number of sperm ejaculated with each orgasm. Males produce such a huge number of sperm because of sperm competition: in species in which females mate with multiple partners, sperm has evolved various characteristics that serve to provide an advantage over other sperm in reaching and fertilizing eggs.   (page 548)

- **Semen**—Sperm make up only 5 percent of the semen that is ejaculated. Materials secreted by the seminal vesicles, the prostate gland, and the bulbourethral glands make up the other 95 percent of semen. These materials provide nutrients, pH neutralizers, and lubricating fluids for the sperm.   (page 548)

Web Tutorial 29.2 The Male Reproductive System

29.4 The Union of Sperm and Egg

- **Movement toward union**—Sperm release enzymes externally that break down layers of cells and materials that surround the oocyte. To release these enzymes, the sperm must undergo a final process of maturation brought about by substances in the female reproductive tract.   (page 549)

- **Sperm and egg union**—The first sperm cell to reach the oocyte binds with receptors on its surface, after which the plasma membranes of the two cells fuse. Then the sperm and egg nuclei combine. Fusion of sperm and oocyte cells brings about release of substances from the oocyte that block the entry of any other sperm.   (page 549)

29.5 Human Development Prior to Birth

- **Linkage with the uterus**—Shortly after arriving in the uterus, the fertilized egg is transformed into a hollow, fluid-filled ball of cells called the blastocyst—the form the egg has assumed when it implants in the mother's uterus. The blastocyst initially consists of an inner cell mass, which develops into the baby, and a peripheral group of cells called a trophoblast. The trophoblast is active in uterine implantation, and trophoblast cells grow into the fetal portion of the placenta. The placenta allows for the movement of nutrients, wastes, and gases between the developing embryo and the mother.   (page 551)

- **Stages of development**—Many of the most important processes in development are completed early in a pregnancy. Organs begin developing in the fourth week following conception and by the end of 12 weeks, immature versions of all the major organ systems have formed. In contrast, the growth of the fetus largely takes place in the later stages of pregnancy. The developing organism is termed a zygote at conception, an embryo from the zygote's first division through the eighth week of pregnancy, and a fetus from the ninth week of pregnancy to the birth of the baby.   (page 552)

29.6 The Birth of the Baby

- **Labor**—The immediate cause of birth is labor: the regular contractions of uterine muscles that sweep over the fetus from its legs to its head. The pressure that results from these contractions opens or dilates the cervix, thus creating an opening large enough for the baby to fit through.   (page 555)

Key Terms

cervix p. 552	inner cell mass p. 552	prostate gland p. 548	umbilical cord p. 552
ectopic pregnancy p. 552	labor p. 555	semen p. 540	urethra p. 546
embryo p. 553	menopause p. 544	seminiferous tubule p. 546	uterine tube p. 540
endometrium p. 541	menstruation p. 541	spermatocyte p. 546	vas deferens p. 545
epididymis p. 545	oocyte p. 542	spermatogonia p. 546	zygote p. 553
fetus p. 553	ovarian follicle p. 542	testis p. 545	
fraternal twins p. 552	ovulation p. 540	trimester p. 553	
identical twins p. 552	placenta p. 552	trophoblast p. 552	

Testing Your Understanding

In Your Own Words   *(answers in the back of the book)*

1. Discuss how the testes act to protect the developing sperm.

2. Menstruation is costly to females in some ways. Why do scientists believe it occurs?

3. What mechanism ensures that only one sperm will be able to fuse with a given oocyte?

Thinking about What You've Learned

1. It is widely accepted among anthropologists that our female evolutionary ancestors needed help from male partners in raising offspring. (Human children are so needy for so long that women could not have raised offspring on their own.) It is also generally assumed that there is a connection between this need and the fact that human females not only can mate at any time but also have hidden ovulation—the time when they are ovulating each month is unknown to both them and their partners. Can you think of what the connection might be between hidden ovulation and the need for a parenting partner? (For those who are stumped, see Jared Diamond's book, *Why Is Sex Fun?*)

2. Most mammals are "placental," which is to say their embryos get nourishment within the mother through the placental system of blood vessels (reviewed in the text). By contrast, amphibian and reptilian embryos are nourished by substances contained in eggs the mother lays outside her body. Can you think of any advantages mammals derive from the placental means of development? Can you think of any disadvantages?

3. Forty years after birth control pills for women came into widespread use, there is still no birth control pill for men. One reason for this is the basic difference between female release of an oocyte each month and the male production of sperm. Why should it be easier to suppress fertility in women than in men?

Flowering Plants

LOS ANGELES TIMES

STATES GETTING ANTIDOTES TO CHEMICAL WEAPONS

From Times Wire Reports
July 14, 2004

States will begin getting stocks of antidotes to chemical weapons under a long-awaited federal program to boost response to a potential terrorist attack, the Centers for Disease Control and Prevention said.

New York and Boston, sites of this summer's political conventions, are among the first areas that will get the "chem-packs."

The gurney-sized packs come with an assortment of antidotes to the many chemicals available to a terrorist; atropine to fight nerve agents, for instance, or amyl nitrite for cyanide.

COLUMBUS LEDGER-ENQUIRER

DROPS HELP KIDS WITH 'LAZY EYE'

By The Associated Press
March 13, 2002

CHICAGO—Eye drops are just as effective as eye patches for treating "lazy eye" and are less likely to be shunned by children, a study shows.

Lazy eye is a condition in which the brain favors one eye over the other. It is the most common cause of visual impairment in children, with symptoms including crossed eyes, farsightedness and nearsightedness. Standard treatment has been eye patches worn over the unaffected eye to stimulate better vision in the "lazy" eye. The same thing happens with atropine drops, which temporarily blur vision in the unaffected eye.

30.1 The Importance of Plants

How versatile can a drug be? As noted in the stories above, the drug called atropine can combat both terrorist gas attacks and children's "lazy eye"—a diverse set of uses, to be sure. But these two applications only begin to give a sense of what atropine has done for human beings over time. Cities and states are just now laying in stocks of this compound to

FIGURE 30.1 **A Plant That Has Served Many Purposes** The flowering plant *Atropa belladonna*.

guard against malicious poisonings, but for years armed forces units have been outfitted with atropine "autoinjectors" that can pump a fixed dose of the drug into the thigh of any soldier who has been gassed. Meanwhile, civilian paramedics are likely to carry a supply of atropine not because of its usefulness as a gas antidote, but because it can bring a pulse back to people who have suffered a heart attack. And the "lazy eye" use for atropine is an extension of the role it once played during routine eye exams—it was a drug that dilated patients' eyes.

All of these applications are relatively modern, though, as each of them came after 1831, when atropine was separated out or "isolated" from its natural source, a plant once known as sleepy nightshade. This name was apt, because extracts of sleepy nightshade can put people into the deepest of sleeps. In the eleventh century, as legend has it, the Scottish leader Earl Macbeth poisoned an invading army of Danish soldiers by lacing their drinks with a juice distilled from nightshade, thus killing them. Skip forward to the Renaissance, however, and nightshade elixirs were used for *cosmetic* purposes. Ladies of the Italian court would lift their eyes upward and apply to them a few drops of a substance they hoped would make them *bella donnas*—beautiful ladies. And what did this substance do? It dilated the ladies' eyes slightly, making their pupils large and entrancing. In time, the plant that aided the ladies was named partly for them. Today, sleepy nightshade is known as *Atropa belladonna* (**see** FIGURE 30.1).

Taking the long view, belladonna is something like a talented employee: It keeps getting called back to work by human beings. If you were keeping track, we humans have derived, from just this one plant, a poison, an antidote to poison, a cosmetic, an aid to visual diagnosis, an aid to an eye affliction, and a heart medicine. But that's the way with plants. The number of things they do for human beings is enormous. The best-known thing they do, of course, is provide us with food. But even in this role, their significance is not fully appreciated. In the world today, up to 90 percent of the calories that human beings consume come directly

(a)

(b)

(c)

FIGURE 30.2 **Uses Galore** Plants are used by human beings in innumerable ways.

(a) Food, here in the form of Concord grapes.

(b) Wood, being harvested from a forest in British Columbia.

(c) A foxglove plant, which yields the heart medicine digitalis.

from plants. Indeed, as some botanists have noted, there are about a dozen plants that stand between humanity and starvation. (The most important of these plants is rice, with wheat coming in second.) Plants are able to serve this function because they make their own food through photosynthesis. The bounty they produce through this process has made Earth a planet of *surplus*—a planet of grains and leaves and roots that are available for the taking. Along these same lines, plants also produce much of the oxygen that most living things require. Then there are the products that human beings have learned to derive from plants, among them lumber, medicines, fabrics, fragrances, and dyes (**see** FIGURE 30.2).

A Focus on Flowering Plants

Though the plant kingdom is vast and varied, it turns out that there are just four principal types of plants in it. These are the bryophytes, represented by mosses; the seedless vascular plants, represented by ferns; the gymnosperms, represented by coniferous trees; and the flowering plants, also known as angiosperms. In this chapter, we will focus strictly on flowering plants. Why pay so much attention to just one type of plant? Because of the overwhelming dominance of flowering plants. There are about 16,000 species of bryophytes, 13,000 species of seedless vascular plants, and 700 species of gymnosperms, but there are an estimated 260,000 species of flowering plant. The term *flowering plant* may bring to mind roses or orchids, and these plants are indeed angiosperms. But food crops such as rice and wheat are flowering plants, as are all cacti, almost all the leafy trees, innumerable bushes, pineapple plants, cotton plants, ice plant—the list is very long (**see** FIGURE 30.3). In this chapter, you'll get an overview of how flowering plants are structured, of how they function internally, and of how they respond to signals from the outside world. A formal definition of them will be provided once you've learned a little about them.

A BROADER VIEW

Corn in Fast Food

How important is just one plant, the corn plant, to your diet? You may think the answer is "not important at all," but in an interview with *California Monthly*, science writer Michael Pollan noted the part it plays in a typical fast-food meal. "A Chicken McNugget is corn upon corn upon corn, beginning with corn-fed chicken all the way through the obscure food additives and the corn starch that holds it together," he said. "All the meat at McDonald's is really corn. Chickens have become machines for converting two pounds of corn into one pound of chicken. The beef, too, is from cattle fed corn on feedlots. The main ingredient in the soda is corn—high-fructose corn syrup . . . . Even the dressing on the new salads at McDonald's is full of corn."

(a)

(b)

(c)

FIGURE 30.3   **Angiosperm Variety**

(a) A trumpet vine (*Clytostoma callistegioides*) spreading on a bush.

(b) A senita cactus, on the left, stands next to a saguaro cactus in the Arizona desert.

(c) The cereal grain rye (*Secale*) growing in a field.

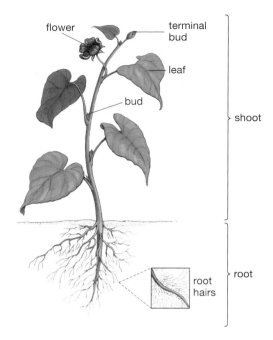

FIGURE 30.4 A Life-Form in Two Parts Flowering plants live in two worlds, with their shoots in the air and their roots in the soil. Although flowering plants differ enormously in size and shape, they generally possess the external features shown.

30.2 The Structure of Flowering Plants

Let's begin our tour of the flowering plants by looking at their component parts.

The Basic Division: Roots and Shoots

We'll first look at the larger-scale structures of these plants, starting with a simple, two-part division that rhymes: roots and shoots. Plants live in two worlds, air and soil, with their root system below ground and their shoot system above (**see FIGURE 30.4**).

The function of the root system is straightforward: Grow toward water and minerals, absorb them from the soil, and then begin transporting them up to the rest of the plant. (For an account of the importance of minerals to plants, see Essay 30.1, "What Are Plant Nutrients?" on page 564.) Most roots also serve as anchoring devices for plants, and some act as storage sites for food reserves. (This food-storage function is important not just for plants, but for people as well. Why? We eat some plant roots—sweet potatoes and carrots, for example.)

The shoot system of plants undertakes several tasks. Photosynthesis takes place in this system, primarily in leaves, which means leaves must be positioned to absorb sunlight. Plants are in *competition* with one another for sunlight, which is a primary reason so many plants are tall. The shoot system also must distribute the food that is produced in photosynthesis throughout the plant. It is not just food production and distribution that are centered in the shoot system, however; it is plant reproduction as well. Within the part of the shoot system called flowers, we find all the component parts of reproduction—seeds, pollen, and so forth. Now let's look in more detail at the root and shoot systems.

Roots: Absorbing the Vital Water

If you look at FIGURE 30.5, you can see pictures of the two basic types of plant root systems, one a *taproot system* consisting of a large central root and a number of smaller lateral roots, and the second a *fibrous root system* consisting of many roots that are all about the same size. FIGURE 30.6 then shows you not only the taproot of a young sweet-corn plant but also another feature common to root systems. These are **root hairs**: threadlike extensions of roots that greatly increase their absorptive surface area. Each root hair actually is an elongation of a single outer cell of the root. Between roots and root hairs, the root structure of a given plant can be extensive. One

FIGURE 30.5 Two Root Strategies

(a) Dandelions such as this one employ a taproot system—a large central root and a number of smaller lateral roots.

(b) The French marigold employs a fibrous root system—a collection of roots that are all about the same size.

(a) Taproot system

(b) Fibrous root system

famous analysis, conducted in the 1930s on a rye plant, concluded that its taproot and lateral roots alone, if laid end to end, would have stretched out for about 386 miles. Meanwhile, the root hairs collectively were almost 6,600 miles long. This from a plant whose *shoot* stood about 3 inches off the ground.

Why do plants lavish such resources on their roots? Essentially because plants have such a great need to take up water, and roots are the structures that allow them to do this. To perform photosynthesis, plants must have a constant supply of carbon dioxide, which enters the plants through microscopic pores, called *stomata*, that exist mostly on the underside of plant leaves. Stomata, however, don't just let carbon dioxide in; they let the plant's water *out*, as water vapor. To accommodate this loss while still keeping vital tissues moist, plants continually pass water through themselves, from roots, up through the stem, and into the leaves (at which point it exits as water vapor). The evaporation of water from a plant's shoot is known as **transpiration**; through it, more than 90 percent of the water that enters a plant evaporates into the atmosphere. The scale of transpiration can be immense, as the roots of a single tall maple tree can absorb nearly 60 gallons of water per *hour* on a hot summer day.

FIGURE 30.6 **Root Hairs** Root hairs enormously increase the surface area of a root. The greater the surface area, the more fluid absorption can occur. Shown here are root hairs on the taproot of a sweet-corn plant.

Shoots: Leaves, Stems, and Flowers

Now let's turn to the shoot system, looking first at the leaves within it.

Leaves: Sites of Food Production
The primary business of leaves is to absorb the sunlight that drives photosynthesis, which is why most leaves are thin and flat. This leaf shape maximizes the surface area that can be devoted to absorbing sunlight while minimizing the number of cells that are not used in photosynthesis. Beyond this, through their tiny stomata, leaves serve as the plant's primary entry and exit points for gases. As noted, the most important gas that's entering is carbon dioxide, which is one of the starting ingredients for photosynthesis. What's exiting is the by-product of photosynthesis, oxygen, and the water vapor.

The broad, flat leaves that are so common in nature have in essence a two-part structure—a *blade* (which we usually think of as the leaf itself) and a *petiole*, more commonly referred to as the leaf stalk (see FIGURE 30.7a). If you look at the idealized cross section of this leaf in FIGURE 30.7b, you can see that the blade can be likened to a kind of cellular sandwich, with layers of cuticle (a waxy outer covering) on the outside and a layer of epidermal cells just inside them. In the leaf's interior there are *vascular bundles*, which bear some relation to animal veins in that they are part of a fluid transport system. Then there are several layers of mesophyll cells. It is these cells that are the sites of most photosynthesis.

Now note, on the underside of the blade, the openings called **stomata** (singular, *stoma*). These are the pores, noted earlier, that let water vapor out and carbon dioxide in—but for most plants, only during the day. When the sun goes down photosynthesis can no longer be performed, and it is not cost effective for a plant to lose water without gaining carbon dioxide. As such, the stomata close up until photosynthesis begins again the following day. If you look at FIGURE 30.7c, you can see how this opening and

(a) Leaves tend to be broad and flat to increase the surface area exposed to sunlight.

(b) Photosynthesis occurs primarily within the mesophyll cells in the interior of the leaf. The vascular bundles carry water to the leaves and carry the product of photosynthesis, sugar, to other parts of the plant. Pores called stomata, mostly on the underside of leaves, allow for the passage of gases in and out of the leaf.

(c) Guard cells of the stomata control the opening and closing of the stomata. When sunlight shines on the leaf during the day, the guard cells engorge with water that makes them bow apart, thus opening the stomata. When sunlight is reduced, water flows out of the guard cells, causing the stomata to close.

FIGURE 30.7 **Site of Photosynthesis**

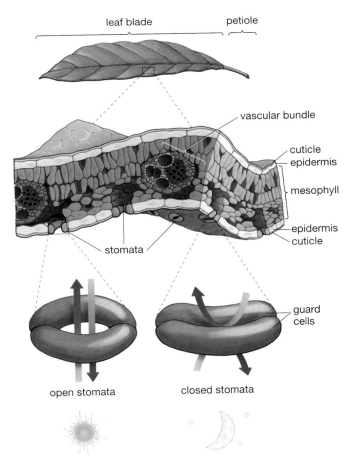

leaf blade   petiole

vascular bundle

cuticle
epidermis

mesophyll

epidermis
cuticle

stomata

guard cells

open stomata   closed stomata

Essay 30.1 What Are Plant Nutrients?

Plants are able to make their own food through photosynthesis if they have access to water, sunlight, and a few nutrients. Water and sunlight are no mystery, but what exactly are nutrients?

A **nutrient** is simply a chemical element that is used by living things to sustain life. Recall from Chapter 2 that "elements" are the most basic building blocks of the chemical world. Silver is an element, as is uranium; but these are not nutrients, because living things do not need them to live. In the broadest sense, carbon, oxygen, and hydrogen are nutrients. And it turns out that almost 96 percent of the weight of the average plant is accounted for by these three elements, which come to plants primarily from water and air.

Water and sunlight are no mystery, but what exactly are nutrients?

Plants need at least another 13 elements to live, however, and not all of these are supplied by water and air. When we think of common "plant food," we are generally referring to a group of nutrients known as minerals that plants can *use up* in their soil and that thus must be replenished to assure continued growth. Of these, the most important three are nitrogen, phosphorus, and potassium, whose chemical symbols are N, P, and K. When you look at a package of an average plant fertilizer, you will see three numbers in sequence on it (for example, 10-20-10). What these refer to are the percentages of the fertilizer's weight accounted for by these minerals (**see** FIGURE E30.1.1).

It is not just houseplants or lawns that benefit from fertilizer, but farm crops as well. Indeed, the planting of a crop on a parcel of land generally requires a large investment in fertilizer because a crop such as wheat removes a great deal of N, P, and K from the soil in a single season. Historically, fertilizers used on farms were *organic*, meaning they came from decayed living things, such as the fish that Native Americans taught the Pilgrims to use when planting corn. Nowadays, however, commercial fertilizers generally are *inorganic*, meaning they are mixtures of pure elements within binding materials, produced by chemical processes.

Nature is, of course, perfectly capable of producing a green bounty in such places as forests and marshes without the aid of any human-made fertilizer. Decaying plant and animal matter put minerals back into the soil, and bacteria are able to fix nitrogen, meaning to absorb it from the air and transform it into a form that plants can take up. Looked at one way, however, houseplants and farm crops are *isolated* plants whose participation in this web of life is very limited; thus they require a helping hand from humans in order to remain robust.

FIGURE E30.1.1 **Needed Nutrients** Plants make their own food from sunlight, water, carbon dioxide, and a few nutrients, some of which come from the soil. The most important of these nutrients are nitrogen (N), phosphorus (P), and potassium (K). Because garden plants and houseplants are isolated from natural ecosystems, they must often get these nutrients in the form of fertilizer.

closing of stomata is achieved: Two "guard cells" are arranged around each stoma. When sunlight strikes the leaf, these cells engorge with water, which makes them bow apart. Then, in the absence of light, water flows out and the door closes again. A given square centimeter of a plant leaf may contain from 1,000 to 100,000 stomata. FIGURE 30.8 gives you some idea of the varied forms of leaves.

Stems: Structure and Storage
We all generally understand what the stem of a plant is, though it is less generally appreciated that the trunk of a tree is simply one kind of stem. The main functions of stems are to give structure to the plant as a whole and to act as storage sites for food reserves. In addition, water, minerals,

compound leaves

simple leaves

leaves modified as spines

highly reduced leaves (needles)

leaves modified as tendrils

FIGURE 30.8 Leaves Come in Many Sizes, Shapes, and Colors Simple leaves have just one blade. In compound leaves, the blade is divided into little leaflets. Some leaves are so modified that the average person can hardly recognize them as leaves—for example, the spines of a cactus.

hormones, and food are constantly shuttling through (and to) the stem, with water on its way up from the roots and food on its way down from the leaves to the rest of the plant.

If you look at FIGURE 30.9, you can see a cross section of one type of plant stem, showing the vascular bundles or veins you first saw in the leaves and the outer, or *epidermal, tissue* just as the leaves had. You can also see so-called *ground tissue* of two types: an outer cortex and an inner pith, both of which can play a part in food storage and wound repair and provide structural strength to the plant.

Flowers: Many Parts in Service of Reproduction
Flowers are the reproductive structures of plants. A single flower generally has both male and female reproductive structures on it, which might make you think that a given plant would fertilize *itself*. This is indeed the case with some plants, such as Gregor Mendel's

FIGURE 30.9 The Stem and Its Parts Stems provide support to the rest of the plant, act as storage sites, and conduct fluids. Vascular plants have a "plumbing" system that transports food, water, minerals, and hormones. The vascular bundles in the figure are groups of tubes, running in parallel, that serve this transport function.

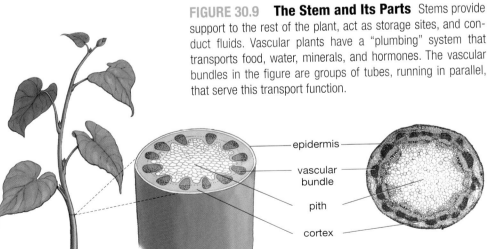

epidermis

vascular bundle

pith

cortex

FIGURE 30.10 **Parts of the Flower** Flowers are composed of four main parts: the sepals, petals, stamens, and a carpel. The sepals protect the young bud until it is ready to bloom. The petals attract pollinators. The stamens are the reproductive structures that produce pollen grains (which contain sperm cells). The carpel is the female reproductive structure; it includes an ovary that contains an egg.

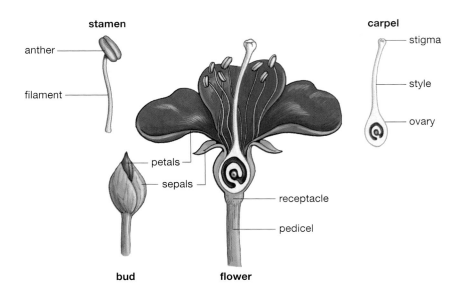

pea plants, reviewed in Chapter 11. In general, however, sperm from one plant fertilize eggs from another. Some of the illustrations you'll be seeing show a plant fertilizing itself, but this is done only for visual simplicity.

If you look at FIGURE 30.10, you can see the components of a typical flower. Taking things from the bottom, there is a modified stem, called a pedicel, which widens into a base called a receptacle, to which the other flower parts are attached. Flowers themselves can be thought of as consisting of four parts: sepals, petals, stamens, and a carpel. The sepals are the leaf-like structures that protect the flower before it opens. (Drying out is a problem, as are hungry animals.) The function of the colorful petals is to announce "food here" to pollinating animals.

The heart of the flower's reproductive structures consists of the stamens and the carpel. If you look at Figure 30.10, you can see that the **stamens** consist of a long, slender **filament** topped by an **anther**. These anthers contain cells that ultimately will yield sperm-bearing pollen grains. Thus, the anthers are the place in the flower where *male* reproductive cells are produced. Pollen grains ultimately will be released from the anther and then be carried—perhaps by a pollinating bee or bird—to the carpel of another plant. As Figure 30.10 shows, a **carpel** is a composite structure of a flower, composed of three main parts: the **stigma**, which is the tip end of the carpel, on which pollen grains are deposited; the **style**, a slender tube that raises the stigma to such a prominent height that it can easily catch the pollen; and the **ovary**, the area in which fertilization of the female egg and then early development of the plant embryo take place. Thus, the ovary is the structure in the flower that houses the *female* reproductive cells, and it is the structure in which the male and female reproductive cells come together.

So Far...

So Far... Answers:

1. absorb water and minerals and begin transporting them to the rest of the plant

2. serve as the sites of photosynthesis

3. reproductive

4. anthers; ovaries

1. The primary function of roots is to _____.

2. The primary function of leaves is to _____.

3. Flowers are the _____ structures of plants.

4. The male reproductive cells of plants are produced in their _____, while the female reproductive cells are produced in their _____.

30.3 Reproduction in Angiosperms

Having looked at the structure, or anatomy, of the angiosperms, let's now go over some of their activities. We'll think in terms of systems here: the reproductive system, the fluid transport system, and the hormonal system (which is linked to the ways plants grow). Because you've just reviewed the parts of a flower, we'll begin with reproduction.

A key event in angiosperm reproduction is that a plant sperm must fertilize a plant egg. As noted, the sperm in flowering plants is contained inside pollen grains. But how do sperm and pollen grains get produced? In the anthers of a flowering plant there are cells that undergo the type of cell division reviewed in Chapter 10, meiosis, thereby producing cells called microspores. Each pollen grain develops from one of these microspores—a process you can see pictured in FIGURE 30.11. In the type of plant shown there, the pollen grain will consist, by the time it leaves the anther, of a tough outer coat and three cells: one *tube cell* and two *sperm cells*.

You can also see in Figure 30.11 what happens next. When the pollen grain leaves the anther, it is bound for the stigma of another plant. But for a grain to merely *land* on a stigma doesn't mean that anything has been fertilized. The tube cell in the pollen grain must then begin to *germinate* on the stigma, sprouting a pollen tube that grows down through the style. Once this has taken place, one of the sperm cells in the grain travels through the tube, gets to the female egg, and fertilizes it. The other sperm moves down along with the first, but then spurs the production of *food* that will surround and nourish the fertilized egg. (What's this about? Stay tuned.)

So, how does the egg that gets fertilized come into being? Inside the plant's ovary, there is a type of cell that also goes through meiosis, thus producing a cell known as a megaspore (*mega*, because it's bigger than the male microspore). It in turn gives rise to a cluster of cells, one of which is the egg that the sperm from the pollen grain will fertilize. Once this happens—once sperm has fertilized egg—we're on our way to a new generation of plant. Many a step remains before arriving at something that *looks* like the original plant. The fertilized egg (now called an embryo) must first have a tough covering develop around it. The combination of embryo, its surrounding food supply, and the covering is called a **seed**. This seed must be released and then land on a suitable patch of earth, there to germinate and grow to a full flowering plant. But the fertilization of egg by sperm sets all this in motion.

The microspores and megaspores that were produced by our original flowering plant may appear to be nothing more than component parts of that plant, but it's important to recognize that these cells actually are the start of an alternate *generation* of plant. After all, it wasn't the original plant that produced the eggs and sperm that were so critical. It was the microspores (on the male side) and megaspores (on the female side) that produced these *gametes*, as they're called. Thus the microspores and megaspores represent the **gametophyte** or gamete-producing generation of plant, while our starting plant was a **sporophyte** or spore-producing generation of plant. (Remember how it produced microspores and a megaspore?) You can read more about the alternation of generations in plants on page 359 of Chapter 21.

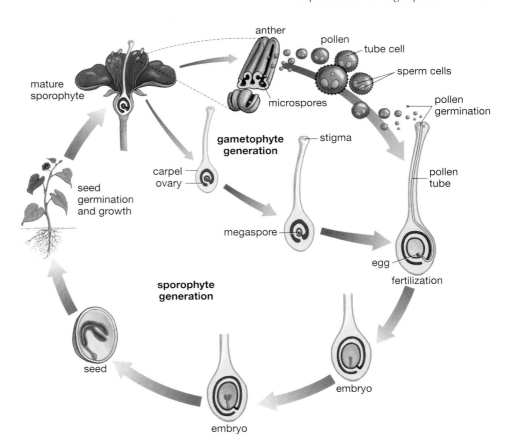

FIGURE 30.11 **The Angiosperm Life Cycle** The mature sporophyte flower produces many male microspores within the anther, and a female megaspore within the ovary. The male microspores then develop into pollen grains that contain the male gamete, sperm. Meanwhile the female megaspore produces the female gamete, the egg. Pollen grains, with their sperm inside, move to the stigma of a plant. There the tube cell of the pollen grain sprouts a pollen tube that grows down toward the egg. The sperm cells then move down through the pollen tube, and one of them fertilizes the egg. This results in an embryo that develops while protected inside a seed. This seed leaves the parent plant and sprouts or "germinates" in the earth, growing eventually into the type of sporophyte plant that began the cycle.

A BROADER VIEW

Seeds from Tiny to Large

The seeds of flowering plants come in a stunning array of shapes and sizes. Orchid seeds are scarcely larger than grains of dust and contain so little food that the embryos within them often are sustained only by nutrients provided by other plants (via underground fungal networks). Meanwhile, the world's largest seeds are produced by a palm tree found only on two islands in the Indian Ocean. Individual seeds of the coco-de-mer tree have weighed in at more than 36 pounds.

FIGURE 30.12 Double Fertilization in Angiosperms When the male pollen (shown greatly enlarged) lands on the stigma, the tube cell within it sprouts a pollen tube, through which the two sperm pass in moving to the ovule. One sperm cell fertilizes the egg, forming an embryo—the new sporophyte plant. The other sperm cell fuses with the two nuclei of the central cell, forming the endosperm that will later nourish the embryonic plant.

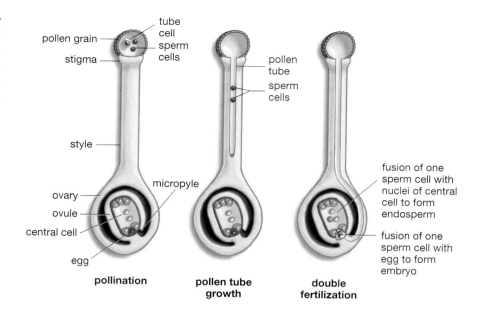

pollination

pollen tube growth

double fertilization

A BROADER VIEW

What Is "Whole Grain"?

What does it mean for bread to be "whole grain," rather than "white"? Well, a whole grain is essentially a whole seed, which is to say a seed complete with endosperm, plus the seed coat, plus the embryo. Food companies often talk about wheat "germ," which is the embryo, and wheat "bran," which is the seed coat. Both the germ and the bran are removed in the processing that produces white flour—and white bread.

FIGURE 30.13 Endosperm Is Food for the Embryo—and for People In angiosperms such as corn, double fertilization produces endosperm that serves as nutrition for a growing embryo. Endosperm like this provides human beings with much of their nutrition. The cotyledon that's noted in the figure is an embryonic leaf of the corn plant.

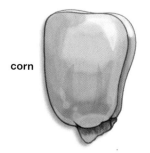

corn

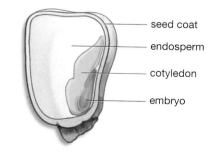

seed coat

endosperm

cotyledon

embryo

Now, what about that detail that was promised on the linkage between egg fertilization and food production within the plant's ovary? If you look at FIGURE 30.12, you can see a kind of close-up of a pollen grain that has landed on a stigma. Note that the pollen grain has in it not only the two sperm cells, but the tube cell that will sprout the pollen tube. If you now look at the bottom of the style, you can see some detail on the female reproductive structure. Note that the ovary has grown a structure internal to it called the ovule. Both ovary and ovule are part of the original, sporophyte generation of plant, but the ovule is enclosing the cells of the *gametophyte* generation—the cells that were produced from the original megaspore. It turns out there are eight of these cells in all. One of these is the egg, of course, but another is a large "central cell" that is unusual in that it has *two* nuclei.

With this setup in mind, here's the series of events that ensues. The pollen tube tunnels down through the style and through an opening in the ovule called a micropyle. Then the two sperm cells make the journey down the tube. One of these sperm fuses with the egg, yielding the fertilized egg that now becomes a developing embryo. Meanwhile, the other sperm enters the central cell (with its two nuclei), and fuses with it, thus producing a cell with three nuclei. This too is a fertilization, only what it sets in motion is the development of food, specifically *endosperm*, a tissue rich in nutrients that will sustain the growing embryo. Taking a step back from this, you can see that two fertilizations have taken place. This is known as **double fertilization**: a simultaneous fusion of gametes and nutritive cells in plants.

Double fertilization is an innovation that belongs almost solely to the angiosperms. It has, however, been a boon to *people*, in that it brings about much of the food we eat. The rice and wheat grains we're familiar with consist in large part of endosperm meant to sustain a plant embryo, as you can see in FIGURE 30.13.

Fruit Defines Angiosperms

If you look at Figure 30.12 again, you can see both the ovule that surrounds the gametophyte-generation plant and the ovary that in turn surrounds the ovule. The ovule, it turns out, will develop into the seed that will surround the growing embryo. But as this is happening, the surrounding ovary is also developing into something: a tissue called fruit. We are all familiar with fruit, of course, and sometimes the tissue surrounding the seed is fruit as we commonly understand that term—

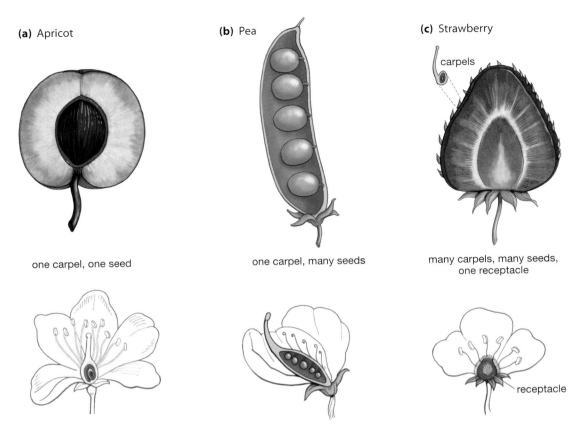

(a) Apricot

one carpel, one seed

(b) Pea

one carpel, many seeds

(c) Strawberry

carpels

many carpels, many seeds,
one receptacle

receptacle

FIGURE 30.14 Fruits Come in Many Forms A fruit, the mature ovary of a flowering plant, surrounds a seed or seeds. Thus **(a)** the flesh of an apricot fits this definition of fruit, but so does **(b)** the pod of a pea, which has several seeds inside. The structure we commonly think of as **(c)** the strawberry fruit actually is the strawberry flower receptacle, with each receptacle having many fruits on its surface. What are commonly thought of as strawberry "seeds" are tiny strawberry plant carpels, each complete with its own fruit and seed.

the flesh of an apricot, for example. But ovaries can mature into fruit in other forms. The pod that surrounds peas is fruit in the scientific sense, as is the outer covering of a kernel of corn. A **fruit** is simply the mature ovary of a flowering plant. All angiosperms have fruit in this sense, and it provides us with our definition of the flowering plants. An **angiosperm** is a flowering seed plant whose seeds are enclosed within the tissue called fruit. This "fruit" can take a lot of forms. Where is the fruit in a strawberry? The answer, shown in FIGURE 30.14, may surprise you. We'll now turn from how angiosperms reproduce to how they transport fluids within themselves.

So Far...

1. A pollen grain develops from a cell called a _____ and is composed, at maturity, of an outer coat, one _____, and two _____ cells.

2. The egg that is fertilized is one of a cluster of eight cells that develops from a cell called a _____ that is located in the plant's _____.

3. Once an egg is fertilized, the embryo that results will be sustained by the food called _____, whose development is sparked during the process called _____.

4. The plant embryo eventually will be surrounded by a tough protective covering called a _____, which in turn will be surrounded by the tissue called _____.

So Far... Answers:

1. *microspore; tube cell; sperm cells*

2. *megaspore; ovary*

3. *endosperm; double fertilization*

4. *seed; fruit*

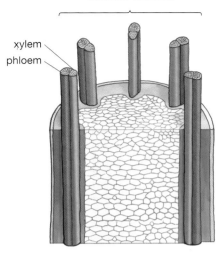

vascular bundles

xylem
phloem

stem section

Fluids move through plants in sets of tubes called vascular bundles. Two types of tissue, running in parallel, can be found in each vascular bundle: xylem, the tissue through which water and dissolved minerals flow; and phloem, the tissue through which the food produced in photosynthesis flows.

FIGURE 30.15 **Fluid-Transport Structure of Plants**

30.4 Fluid Transport, Growth, and Hormones

Plant Plumbing

In looking at leaves and stems earlier, you saw that they have something called *vascular bundles* running through them. This term refers to collections of tubes through which fluid materials move from one part of the plant to another. This transport system bears obvious similarities with the circulation systems of animals, but there are differences as well. Many animals employ blood as a transport medium, and animals such as ourselves have a transport *pumping* device, called a heart. Plants do not have blood, and they have no pumping system. (In fact, plants have almost no "moving parts" at all.) Yet think of the transport job plants have to do. Water that is transpired from the top of a redwood tree may have made a journey of more than 100 meters straight up—more length than a football field. What kind of power is at work here?

There are two essential components to the plant transport system. First there is **xylem**, which can be defined as the tissue through which water and dissolved minerals flow in vascular plants. (**Tissue** means a group of cells that perform a common function.) Second there is **phloem**, the tissue through which the *food* produced in photosynthesis—mostly sucrose—is conducted in plants, along with some hormones and other compounds. As noted before, water is making a one-way journey from root through leaf, but the plant must be able to transport food and hormones everywhere within itself.

If you look at FIGURE 30.15, you can see an idealized view of a plant's transport system as it would appear within a stem. You'll also see that the vascular bundles noted earlier are composed of bundles of linked xylem and phloem tubes running in parallel.

The movement of water straight up through a tall tree obviously requires a considerable expenditure of energy; but, surprisingly, it is not the *plant's* energy that is being expended. You may recall that water is constantly evaporating from the leaves of plants in the process called transpiration. As the water moves into the air, it creates a region of low pressure, and the resulting suction pulls a continuous column of

FIGURE 30.16 Suction Moves Water through Xylem When water evaporates from leaves, a low pressure is created that pulls more water up to fill the void. Water is so cohesive that it moves up in a continuous column. The sun's energy, rather than the plant's energy, fuels this process.

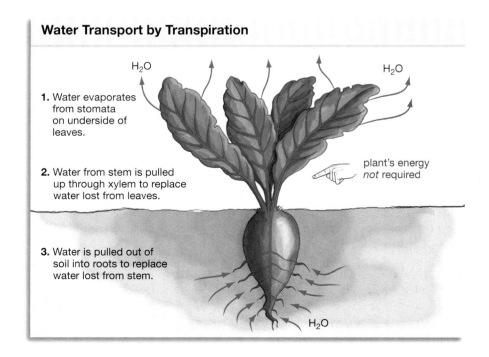

Water Transport by Transpiration

H_2O H_2O

1. Water evaporates from stomata on underside of leaves.

2. Water from stem is pulled up through xylem to replace water lost from leaves.

plant's energy *not* required

3. Water is pulled out of soil into roots to replace water lost from stem.

H_2O

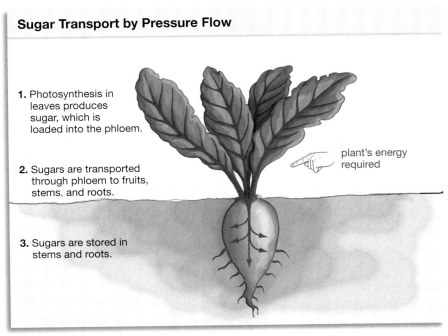

Sugar Transport by Pressure Flow

1. Photosynthesis in leaves produces sugar, which is loaded into the phloem.

2. Sugars are transported through phloem to fruits, stems, and roots.

3. Sugars are stored in stems and roots.

plant's energy required

FIGURE 30.17 **How Food Moves through the Plant** Sucrose is produced in the leaves by photosynthesis, after which the plant expends its own energy to load its sucrose (sugar) into the phloem. Water then moves into the phloem, through the process of osmosis, and the resulting pressure transports sucrose through the plant for use and storage.

A BROADER VIEW

Why Do Trees Stop Growing?

So, if plants can grow indefinitely, why does a tree like a redwood ever stop growing in height? Because water can only be pulled so far against gravity. In 2004, scientists published results of research they did at the top of five of the world's tallest trees (all redwoods). They found that water flow there was so poor that the leaves turned very thick and were inefficient at performing photosynthesis.

water upward through the xylem, from roots through leaves (**see** FIGURE 30.16). So, what is the energy source for this movement? The warm rays of the sun, which get the suction going by powering the evaporation of water at the leaf surface.

When we turn to the movement of nutrients through phloem, however, it is not the sun's energy but the plant's own energy that powers the process. The sugars that a plant produces, mostly in its leaves, must be actively "loaded" into phloem tissue, and the plant expends energy reserves to do this. Once this is accomplished, *water* will move into the phloem through the process of osmosis (reviewed in Chapter 5). With the addition of water and the sugars, the fluid in the phloem is under pressure; and it is this pressure that moves the sugars to locations where they can be used or stored (for example the roots, as you can see in FIGURE 30.17).

Plant Growth: Indeterminate and at the Tips

How do plants grow? Unlike animals, they do not grow globally, over their whole surface, but instead confine most of their vertical growth to special regions, called apical meristems, that exist at the *tips* of the roots and shoots. To visualize this aspect of plant growth, imagine a boy of about 8 years old who drives a long nail into the trunk of a secluded tree. He doesn't drive the nail in completely, but lets it stick out a bit. After 10 years pass, the boy happens to walk by the same tree; upon seeing it, a couple of things occur to him (**see** FIGURE 30.18). First, his height has increased, but the height of the nail has not; he would have to bend down now to pull the nail out, whereas when he drove it in he was standing up straight. Second, he would have a hard time actually pulling the nail out now because not much of it is visible; the tree has grown around it. Third, if the nail's height is any indication, the trunk of the tree may not have moved up, but the *top* certainly has; it's much taller than he remembered it.

The lessons here are that, in their vertical growth, plants do not grow the way people do—throughout their entire length—but instead grow only at their tips (of both shoots *and* roots, though the

FIGURE 30.18 **The Basic Characteristics of Plant Growth** If a young boy drives a long nail into a tree and returns 10 years later, two things will be apparent: The nail is at the same height it used to be despite the increased height of the tree, and it is almost completely buried in the tree. What does this say about tree growth? Trees do not elongate throughout their whole length, but only at their tips in "primary growth." Further, they widen laterally in "secondary growth."

When Billy was eight . . .   When Billy was eighteen . . .

Table 30.1 Plant Hormones

Hormone	Major functions	Where found or produced in plant
Auxins	Suppression of lateral buds; elongation of stems; growth and abscission (falling off) of leaves; differentiation of xylem and phloem tissue	Root and shoot tips; young leaves
Cytokinins	Stimulate cell division; active in the development of plant tissues from undifferentiated cells	Roots
Gibberellins	Stem elongation, growth of fruit, promotion of seed germination	Seeds, apical meristem tissue, young leaves
Ethylene	Ripening of fruit, retardation of lateral bud growth, promotion of leaf abscission	Nearly all plant tissue
Abscisic acid	Induces closing of leaf pores (stomata) in drought; promotes dormancy in seeds; counteracts growth hormones	Young fruit; leaves, roots

boy couldn't see the roots). In addition, as the half-buried nail attests, this tree has carried out the *lateral* growth, a growth in width that botanists refer to as *secondary growth*.

Beyond these things, it turns out that a plant's growth is indeterminate. In most plants it can go on indefinitely at the tips of roots and shoots. By contrast, animal growth is generally determinate: It comes to an end at a certain point in development. Imagine if, say, the tips of your fingers just kept growing long after your trunk and arms had reached their full extension.

Plant Communication: Hormones

We're so used to associating hormones with animal functioning that it may come as a surprise to learn that plants also have hormones. Taken as a whole, plant hormones do many of the same things that animal hormones do. Their most important roles are to regulate growth and development and to integrate the functioning of the various plant parts. In less abstract terms, hormones help buds grow and leaves fall and fruit ripen.

The Nature of Plant Hormones Though most people have an intuitive sense of what hormones are, a definition might be helpful. **Hormones** are chemical messengers; they are substances that, when released in one part of an organism, go on to prompt physiological activity in another part of that organism. In animals, hormones are generally synthesized in well-defined organs, called *glands*, whose main function is to produce the hormones (for example, the thyroid or adrenal glands). In plants, however, hormone production is a more decentralized process, taking place not in glands, but in collections of cells that carry out a range of functions.

If you look at **Table 30.1** you can see a list of five of the most important hormones that function in plants. Three of these actually are *classes* of hormones, and two are individual hormones. We'll now look at one of the roles that a hormone called IAA takes on. But a larger lesson about plant hormones will become clear as you progress through the chapter and see IAA popping up again and again, performing other functions. The message here is that one hormone can do many different things for a plant.

FIGURE 30.19 Uniform "A" Shapes The effects of apical dominance are clearly visible in this stand of blue spruce trees at a Michigan Christmas tree farm. The apical meristem tissue at the tip of each tree produces a hormone called IAA that inhibits the growth of lateral branches. Because the concentration of IAA is highest at the top near the apical meristem, and lowest at the bottom, the branches are very short near the top and longest at the bottom.

An Auxin Gives the "A" Shape to Trees Why do Christmas trees have their characteristic "A" shape? (See FIGURE 30.19.) They develop in this way thanks in large part to the effects of IAA, which is an auxin-class hormone. As noted earlier, plants confine most of their vertical growth to the apical meristem regions that exist at the tips of the roots and shoots. FIGURE 30.20 shows you the location of this meristem tissue in a typical plant shoot. As you can see, this tissue gives rise to a series of growth modules: more stem, one or more leaves, and a lateral bud that forms in tandem with each leaf. These lateral buds have meristem tissue in them and can grow into new branches, but they normally don't *when they are close to the apical meristem*. Why? The apical meristem is producing IAA, which works together with other hormones to suppress the growth of these buds. But, the farther away from the apical meristem, the *smaller* the concentration of IAA that is available to the lateral buds. The result is a greater budding of branches at the base of a tree, and a tapering of this growth going from the base up through the tree's apex. This phenomenon, called apical dominance, is most pronounced in certain conifers, such as the Douglas firs often used for Christmas trees. Meanwhile, trees that don't have such a strict IAA gradient don't get the strict "A" shape.

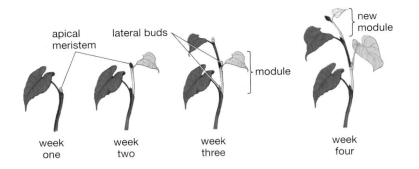

apical meristem

lateral buds

module

new module

week one
week two
week three
week four

FIGURE 30.20 **Modular Growth** Vertical growth in plants occurs in "modules" that develop from the apical meristem located at the top of the plant. Each module includes a new portion of stem, at least one leaf, and a lateral bud that can give rise to a branch or a flower. Here alternating dark-green and light-green segments indicate modules.

30.5 Responding to External Signals

Plants may lack a nervous system, but this doesn't mean they can't respond to their environment. They must do so, actually, because their life depends on it. They sense the march of the seasons by measuring available light and the surrounding temperature; they use internal clocks to synchronize their development and physical functioning; they respond to light, gravity, touch, heat, cold, and pH, among other things. We'll conclude the chapter by reviewing the plant's interactions with gravity, light, touch, and the seasons as examples of these environmental responses.

Responding to Gravity: Gravitropism

When a seed first starts to sprout, or "germinate," its roots and shoots must go in very specific directions: The root must go *down,* toward water and minerals, and the shoot must go *up* toward the sunlight. If you look at FIGURE 30.21, you can see dramatic evidence of a plant's ability to sense which way is up and which way down. The plant in Figure 30.21a was placed on its side and, within 16 hours, the shoot began curving upward, as you can see. Figure 30.21b shows the *root* of a germinating sweet-corn seed, likewise oriented horizontally (by researchers), but quickly beginning to bend downward. What's being exemplified here is **gravitropism**, a bending of a plant's root or shoot in response to gravity. Figure 30.21c again shows the root of a sweet-corn plant, only this one has had its very tip, or *root cap,* snipped off. Here instead of bending in accordance with gravity, the root continues its straight, horizontal growth. Conclusion? Cells or substances in the root cap are essential for root gravitropism.

But how do roots "know" which way is down, and how do they orchestrate a course correction when they need one? Logically, there must be at least two elements at work here: first, a gravitational *sensing* mechanism; and then a means of *responding* to gravitational cues, such that corrective bending points the root toward the center of Earth.

(a)

(b)

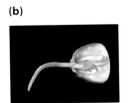

(c)

FIGURE 30.21 **Plants Respond to Gravity**

(a) Potted *Impatiens* plants demonstrate the effects of gravitropism. The plant on the right was laid on its side 16 hours before the photo was taken, by which time the shoot had curved, beginning to grow upward.

(b) Similarly, the root of a germinating sweet-corn seed begins to curve downward only hours after being placed on its side. How do these plants sense which way is up and which way is down?

(c) One hint is that if the very tip, or "root cap," of the root is removed, the root no longer bends downward. Researchers believe that organelle "sedimentation"—the movement of certain organelles in response to gravity—is the sensing mechanism in gravitropism.

The consensus among botanists today is that an important sensing mechanism in gravitropism is the "sedimentation" that various plant-cell organelles perform in response to gravity. Recall that organelles are small but highly organized structures inside cells. (Mitochondria and ribosomes are organelles, for example.) Like the bubble in a carpenter's level, some plant-cell organelles can be seen changing position in response to gravity. When these organelles move in response to gravity, it sets gravitropism in motion. The organelles "land" on other structures inside the cell, and the resulting impact triggers a redistribution of a substance you were recently introduced to: the hormone IAA. The result of this redistribution is differential growth within the plant—one *side* of the plant stem or root will grow more than the other side. With this growth, a root that is oriented horizontally will start bending toward the ground (because its top side is now growing more than its bottom side). Meanwhile, a stem that is oriented horizontally will start bending toward the sky (because its bottom side is now growing more than its top side).

Responding to Light and Contact: Phototropism and Thigmotropism

Many plants will bend toward a source of light, with the value of this perhaps being obvious: Plants produce their own food, and that production depends on light. Thus a plant needs to respond when its sunlight becomes *blocked* by another plant or some other physical object in its surroundings. As it turns out, it is once again IAA—this time produced in the shoot tips of growing plants—that controls this **phototropism**, defined as a curvature of shoots in response to light. When light strikes one side of the shoot, it causes IAA to migrate to the other side, where the IAA acts to promote the *elongation* of cells on this far side. The effect is to make the shoot curve toward the light (**see** FIGURE 30.22).

We've all seen plants that manage to climb upward by encircling the stem of another plant. Whole stems can undertake such encircling, but it is often a thin, modified stem or leaflet called a *tendril* that does so (**see** FIGURE 30.23). But how is a tendril able to wrap around another object? Once again through differential growth—more rapid growth on one side of a tendril than on the other. Contact with the object is perceived by outer, or epidermal, cells in the tendril. This sets into motion the differential growth, which probably is controlled by the hormones IAA and ethylene. This process is called **thigmotropism**, meaning growth of a plant in response to touch. Plants with this capability can piggyback on other plants in order to get more access to sunlight.

The Importance of Growth

Note that plants respond to gravity, light, and touch not by moving, as an animal might, but by growing in various ways. Likewise, the general "strategy" of plants is to grow as tall as possible, so as to catch more sunlight. Given all this, it's no wonder that plant growth is indeterminate; imagine the trouble a plant would be in if, like an animal, its growth came to a halt at a certain point in its life.

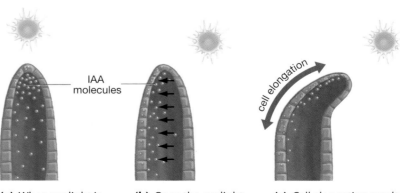

(a) When sunlight is overhead, the IAA molecules produced by the apical meristem are distributed evenly in the shoot.

(b) Once the sunlight shines on the shoot at an angle, the IAA molecules move to the far side and induce the elongation of cells on that side.

(c) Cell elongation results in the bending of the shoot toward the light.

FIGURE 30.22 Plants Respond to Light

Responding to the Passage of the Seasons

The profusion of brightly colored leaves in the fall is one of the great seasonal markers in temperate climates such as those in the American Midwest and most of the East Coast. As green leaves turn red, gold, and purple, we know that autumn has arrived (see FIGURE 30.24).

Dormancy in Winter
Trees that exhibit a coordinated, seasonal loss of leaves are called **deciduous** trees. But why should these broad-leafed trees lose their leaves, while the evergreen pines and firs keep the modified leaves known as needles?

First and foremost, there is relatively little water available in winter in cold climates (most of it being frozen in the ground), and flat-leafed trees transpire more than do evergreens. In addition, because of the way wind flows over flat leaves, they lose more *heat* than do pine needles. Thus, the deciduous trees have evolved a strategy that is a matter of straight economics. Any winter photosynthesis that they might perform would not be worth the water loss that would result. The strategy thus becomes to lose the leaves, grow new ones next spring, and in the meantime perform no photosynthesis but exist instead completely on food reserves. This state, in which growth is suspended and metabolic activity is low, is called **dormancy**. Evergreens exhibit dormancy too, in low temperatures, but deciduous trees are locked into it until the coming of spring.

Photoperiodism
So, how does a deciduous tree sense when it's time to begin preparing for cold weather? In general it is an interaction between two factors. First there is cold itself, which brings about changes in metabolism. In addition, there is a phenomenon known as **photoperiodism**, which is the ability of a plant to respond to changes it is experiencing in the daily duration of darkness, relative to light. Most plants in the Earth's temperate regions and many in the subtropics exhibit photoperiodism, which can affect such processes as the timing of flowering as well as the onset of dormancy.

To get a feel for how photoperiodism works, consider ragweed, a plant that is the bane of many hay-fever sufferers. After sprouting in the spring and reaching a

FIGURE 30.23 **Moving Up by Curling Around** This dodder plant is able to curl around its host plant because of thigmotropism—a plant's ability to grow in response to touch.

FIGURE 30.24 **Seasonal Marker** Brightly colored fall leaves in Michigan's Ottawa National Forest. Plants have a number of mechanisms that allow them to respond to the passage of the seasons.

A BROADER VIEW

Why Do Leaves Change Color?

Why do the leaves of deciduous trees turn colors in the fall? There are several hypotheses about this, but surprisingly few conclusions. Different leaf colors are not so much added in the fall as unmasked by breakdown of green chlorophyll. Given this, one hypothesis is that fall coloration does nothing for trees, but simply is a by-product of the chlorophyll loss. However, another hypothesis, supported by recent research, is that fall colors provide a signal to insects that effectively says: "Photosynthesis has stopped here; not much food is available; this is not a good place to eat or to lay eggs."

certain level of maturity, a ragweed flowers only when it is in darkness for 10 hours *or more* in a 24-hour day. When would this threshold be crossed? As the nights begin to get longer. Ragweed thus joins a group of plants, called long-night plants, whose flowering comes only with an *increased* amount of darkness. As you can imagine, most of these plants are like ragweed in that they flower only in late summer or early fall. Their counterparts are the so-called short-night plants, which flower only when the nights get short enough—that is to say, when night-time hours are *decreasing* as a proportion of the day. Predictably, these plants tend to flower in early to midsummer.

The mechanism by which photoperiodism works is now fairly well understood in one short-night plant, a member of the mustard family called *Arabidopsis*. Flowering in *Arabidopsis* is controlled by a protein, known as CO, that can switch genes on by binding to DNA. When enough CO accumulates in *Arabidopsis*, the genes that set flowering in motion are switched on. So, what can bring about CO levels that are high enough to get the flowering started? Shorter nights. Substances called photoreceptors, which are sensitive to light, act mostly to allow CO to accumulate during the day, but then predominantly act to destroy it at night. As the nights grow shorter, the balance is shifted toward CO accumulation. When this accumulation reaches a certain threshold, the relevant genes are switched on and flowering begins.

On to the Rest of Life

Every chapter of this book has ended with a little preview of what is to come in the next chapter. This time there is no such thing as a next chapter, because you have reached the end of the book. Hopefully, this chapter on plants has conveyed the same general message as the other chapters: Life has interesting rules, and the scientists called biologists are having some success in figuring out what those rules are. Readers who have been through large portions of this book know, however, that our ignorance of the living world at least matches our knowledge of it. The good news is that biologists are acquiring knowledge at a faster pace now than ever before. The average college student of today eventually will understand nature in ways that his or her grandparents could not have imagined. Stay tuned for what's coming up.

So Far...

1. Xylem is the plant tissue that conducts _____, while phloem is the plant tissue that conducts _____.

2. Vertical plant growth occurs only at the _____ of shoots and roots.

3. Plants can respond to gravity, touch, and the direction of sunlight through the mechanism of _____ of roots or stems.

4. Plants in temperate climates make seasonal adjustments based on the length of the _____ they are experiencing.

So Far... Answers:

1. water and minerals; the food produced in photosynthesis

2. tips

3. differential growth

4. nights

Chapter Review

Summary

30.1 The Importance of Plants

- **Roles of plants**—Plants are vital to almost all life on Earth. The photosynthesis they carry out indirectly feeds many other life-forms. The oxygen they produce as a by-product of photosynthesis is vital to many organisms. The lumber and paper that trees provide is important to human beings, as are the medicines, fabrics, fragrances, and dyes that other plants provide. (page 559)

- **Types of plants**—There are four principal varieties of plants: bryophytes, represented by mosses; seedless vascular plants, represented by ferns; gymnosperms, represented by coniferous trees; and the flowering plants or angiosperms. Of the four varieties, angiosperms are by far the most dominant on Earth. (page 561)

30.2 The Structure of Flowering Plants

- **Roots and shoots**—Plants live in two worlds, above the ground and below it. As such, their anatomy consists of the above-ground shoots and the below-ground roots. Roots absorb water and nutrients, anchor the plant, and often act as nutrient storage sites. Shoots include the plant's leaves, stems, and flowers. (page 562)

- **Leaves**—Leaves serve as the primary sites of photosynthesis in most plants. Leaves have a profusion of tiny pores, called stomata, that open and close in response to the presence or absence of light. In this way, the stomata control the flow of carbon dioxide into the plant and the flow of oxygen and water vapor out of the plant. (page 563)

- **Stems**—Stems give structure to plants and act as storage sites for food reserves. (page 563)

- **Flowers**—Flowers are the reproductive structures of plants, with most flowers containing both male and female reproductive parts. The male reproductive structure, called a stamen, consists of a slender filament topped by an anther. The anther's chambers contain the cells that will develop into sperm-containing pollen grains. The female reproductive structure, the carpel, is composed of a stigma, on which pollen grains are deposited; a tube called a style, which raises the stigma to such a height that it can catch pollen; and a structure called an ovary, where fertilization of the female egg and early development of the resulting embryo take place. (page 565)

Web Tutorial 30.1 Leaves: The Site of Photosynthesis

30.3 Reproduction in Angiosperms

- **Pollen and sperm**—Pollen grains develop from cells called microspores that reside inside the plant's anther. At maturity, each pollen grain consists of two sperm cells, one tube cell, and an outer coat. Pollen grains from one plant land on the stigma of a second plant and then germinate, developing a pollen tube that grows down through the second plant's style. The sperm cells from the grain then travel through the pollen tube, with one of the sperm cells reaching the egg in the ovary and fertilizing it. (page 567)

- **Egg and ovary**—A cell within the ovary called a megaspore develops into a small cluster of eight cells, one of which is the egg. Once the egg is fertilized by the sperm, it becomes an embryo that eventually will be surrounded by a tough outer covering. The combination of embryo, its food supply, and the outer covering is called a seed, which is capable of being implanted in the ground and growing into a new generation of plant. (page 567)

- **Double fertilization**—One of the eight cells that develop in the ovary from the megaspore is a large central cell. The second sperm from the pollen grain that enters into the ovary fuses with the central cell, fertilizing it and thus prompting the production of endosperm, which is food for the embryo that results from egg fertilization. Flowering plants thus undertake a double fertilization. (page 568)

- **Defining angiosperms**—The ovary of angiosperms develops into a tissue called fruit that surrounds the seed within which the embryo is growing. Angiosperms can be defined as plants whose seeds are surrounded by fruit. Fruit in turn is defined as the mature ovary of a flowering plant. The flesh of an apricot is fruit, as is the pod that surrounds peas. (page 568)

Web Tutorial 30.2 The Angiosperm Life Cycle

30.4 Fluid Transport, Growth, and Hormones

- **Plant plumbing**—Fluid transport in plants is handled through two kinds of tissue: xylem, through which water and dissolved minerals flow; and phloem, through which the food the plant produces flows, along with hormones and other compounds. The movement of water through xylem is powered by the sun's energy. But plants must expend their own energy to transport nutrients through phloem. (page 570)

- **Growth**—Plants do not grow vertically throughout their length, but instead grow almost entirely at the tips of both their roots and shoots. Some plants, such as trees, thicken through lateral or "secondary" growth. The growth of most plants is indeterminate, meaning it can go on indefinitely. (page 571)

- **Hormones**—Plant hormones regulate plant growth and development and integrate the functioning of various plant structures. Many fruits ripen under the influence of the plant hormone ethylene, while the hormone IAA is important in controlling plant growth. (page 572)

30.5 Responding to External Signals

- **Tropisms**—Plants are able to sense their orientation with respect to the Earth and direct the growth of their roots and shoots accordingly—roots into the Earth, shoots toward the sky. This is called gravitropism. Plants will also bend toward a source of light through the process of phototropism: a curvature of shoots in response to light. Some plants can climb upward on other objects by making contact with them and then encircling them in growth. This is thigmotropism, the growth of a plant in response to touch. Differential growth on one side of the root or stem makes all of these tropisms possible. (page 573)

- **Loss of leaves**—In temperate climates, deciduous trees undertake a coordinated, seasonal loss of leaves and enter into a state of dormancy, existing on stored nutrient reserves in colder months. (page 575)

- **Photoperiodism**—Plants can sense the passage of seasons and time their reproductive activities accordingly. One mechanism that assists in this process is photoperiodism, which is the ability of a plant to respond to changes it is experiencing in the daily duration of darkness, as opposed to light. Some plants that exhibit photoperiodism are long-night plants, meaning those whose flowering comes only with an increased amount of darkness—in late summer or early fall. Others are short-night plants, meaning those that flower only with a decreased amount of darkness—in early to midsummer.   (page 575)

Web Tutorial 30.3 Plant Hormones: Responding to External Signals

Key Terms

angiosperm p. 569	**fruit** p. 569	**photoperiodism** p. 575	**stomata** p. 563
anther p. 566	**gametophyte** p. 567	**phototropism** p. 574	**style** p. 566
carpel p. 566	**gravitropism** p. 573	**root hair** p. 562	**thigmotropism** p. 574
deciduous p. 575	**hormone** p. 572	**seed** p. 567	**tissue** p. 570
dormancy p. 575	**nutrient** p. 564	**sporophyte** p. 567	**transpiration** p. 563
double fertilization p. 568	**ovary** p. 566	**stamen** p. 566	**xylem** p. 570
filament p. 566	**phloem** p. 570	**stigma** p. 566	

Testing Your Understanding

In Your Own Words *(answers in the back of the book)*

1. The stomata on leaves allow important gas exchanges to take place in plants, but these exchanges also present plants with a problem they must solve. What gases are exchanged through the stomata, and in which direction? What problem must plants solve because of these exchanges, and how do they solve it?

2. List the main parts of a flower, and describe the functions performed by each part.

3. Distinguish between the gametophyte and sporophyte plant generations.

Thinking about What You've Learned

1. Gardeners often "pinch off" the apical meristems of young plants to make the plants bushier. Explain the physiological basis for this common horticultural practice.

2. The text makes clear that plants respond to their environments in sophisticated ways. Given this, can plants be said to be conscious beings, or are animals the only conscious beings?

3. The "race for sunlight" in plants led to the evolution of the tallest living things in existence, trees. But sunlight is not the only thing that stands to make a tree successful or not in reproducing. Think about how plants function, and then answer this question: What are the *costs* of being taller, as opposed to shorter?

Appendix

Metric-English System Conversions

Length

1 inch (in.) = 2.54 centimeters (cm)

1 centimeter (cm) = 0.3937 inch (in.)

1 foot (ft) = 0.3048 meter (m)

1 meter (m) = 3.2808 feet (ft) = 1.0936 yard (yd)

1 mile (mi) = 1.6904 kilometer

1 kilometer (km) = 0.6214 mile (mi)

Area

1 square inch (in.^{2}) = 6.45 square centimeters (cm^2)

1 square centimeter (cm^2) = 0.155 square inch (in.^{2})

1 square foot (ft^2) = 0.0929 square meter (m^2)

1 square meter (m^2) = 10.7639 square feet (ft^2) = 1.1960 square yards (yd^2)

1 square mile (mi^2) = 2.5900 square kilometers (km^2)

1 acre (a) = 0.4047 hectare (ha)

1 hectare (ha) = 2.4710 acres (a) = 10,000 square meters (m^2)

Volume

1 cubic inch (in.^{3}) = 16.39 cubic centimeters (cm^3 or cc)

1 cubic centimeter (cm^3 or cc) = 0.06 cubic inch (in.^{3})

1 cubic foot (ft^3) = 0.028 cubic meter (m^3)

1 cubic meter (m^3) = 35.30 cubic feet (ft^3) = 1.3079 cubic yards (yd^3)

1 fluid ounce (oz) = 29.6 milliliters (mL) = 0.03 liter (L)

1 milliliter (mL) = 0.03 fluid ounce (oz) = 1/4 teaspoon (approximate)

1 pint (pt) = 473 milliliters (mL) = 0.47 liter (L)

1 quart (qt) = 946 milliliters (mL) = 0.9463 liter (L)

1 gallon (gal) = 3.79 liters (L)

1 liter (L) = 1.0567 quarts (qt) = 0.26 gallon (gal)

Mass

1 ounce (oz) = 28.3496 grams (g)

1 gram (g) = 0.03527 ounce (oz)

1 pound (lb) = 0.4536 kilogram (kg)

1 kilogram (kg) = 2.2046 pounds (lb)

1 ton (tn), U.S. = 0.94 metric ton (t or tonne)

1 metric ton (t or tonne) = 1.10 tons (tn), U.S.

Metric Prefixes

Prefix	Abbreviation	Meaning
giga-	G	$10^9 = 1{,}000{,}000{,}000$
mega-	M	$10^6 = 1{,}000{,}000$
kilo-	k	$10^3 = 1{,}000$
hecto-	h	$10^2 = 100$
deka-	da	$10^1 = 10$
		$10^0 = 1$
deci-	d	$10^{-1} = 0.1$
centi-	c	$10^{-2} = 0.01$
milli-	m	$10^{-3} = 0.001$
micro-	μ	$10^{-6} = 0.000001$

Answers

CHAPTER 1
In Your Own Words
1. Science is a body of knowledge and a formal process for acquiring that knowledge. It is a unified collection of insights about the physical world, and also a way of learning about the physical world. Unlike religion, science does not acknowledge any absolute laws, nor does it acknowledge any unquestioned authority. Instead, all scientific principles are open to modification, based on the discovery of new evidence.
2. An experiment is said to be controlled when a scientist controls the factors in it that will change and those that will be held constant. All factors must be kept the same in a controlled experiment except for the variables that are being tested for. Should a new cancer drug be tested, for example, all patients getting the drug must have the same disease, and all must be getting the same level of general medical care, among other factors. The variable that is being tested for in this experiment is the effectiveness of the new drug. To judge its effectiveness, experiments would have to compare the health of the group of patients who got the drug with the health of a matched group of patients who did not get the drug. Having cancer but not getting the new cancer drug is the control condition in this experiment.
3. Pasteur investigated a widely held notion of the time that living things can arise spontaneously from mixtures of basic chemicals. He hypothesized that instances where life appeared to arise in a medium believed to lack it could be explained as the result of contamination from unseen airborne microscopic organisms. He sterilized meat broth in a flask with a long, S-shaped neck, by heating both the broth and the flask. Although the mouth of the long tube remained open, any incoming dust particles laden with microscopic organisms would be trapped in the bent neck and could not gain access to the meat broth. Under these conditions, the broth remained free of any visible signs of life. In other tests, the S-shaped neck was broken off or the flask tilted so that broth made contact with the neck of the flask; in these instances, the broth came "alive" with teeming bacteria and fungi. Pasteur inferred from these experiments that the appearance of life-forms depended on whether the broth was contaminated by airborne microscopic organisms. When the airborne organisms had no access to the broth, however, it remained sterile. There was no evidence, in other words, of life coming about through spontaneous generation.

CHAPTER 2
In Your Own Words
1. The forms are called isotopes. Isotopes of an element differ from one another in accordance with the number of neutrons they have. A regular carbon atom has 6 neutrons in its nucleus, while a carbon-14 atom has eight neutrons.
2.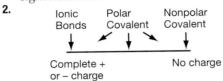
3. When the outer shell of an atom is filled, the atom is at a lower, more stable energy state.

CHAPTER 3
In Your Own Words
1. Water absorbs tremendous amounts of heat from the sun, and releases this heat slowly, with the onset of night's colder temperature. The perspiration human beings throw off carries with it a great deal of heat, which has been absorbed by the water.
2. LDL is associated with deposition of cholesterol in the arteries, so you should be concerned about high levels. HDL is associated with carrying cholesterol to the liver, away from the arteries where it does damage. A high HDL count relative to LDL count is in general a good sign.
3. Proteins function as enzymes, as hormones, and as structural molecules (keratin), transport molecules (hemoglobin), and contractile molecules (myosin and actin in the muscles).

CHAPTER 4
In Your Own Words
1. The cell would have difficulty obtaining materials from the external environment because its surface-to-volume ratio would be too low.
2. Atoms, amino acids, proteins, bacteria, animal cells.
3. The DNA code is copied onto an mRNA molecule, which leaves the nucleus through the pores in the nuclear envelope. The length of mRNA then arrives at a ribosome, which reads part of it before migrating to, and embedding in, the rough endoplasmic reticulum (RER). The ribosome then continues to read the RNA chain, and the resulting amino acid chain drops into the cisternal space of the RER. There, the amino acid chain folds up into a protein that undergoes editing. Once this editing has been completed in the RER, the protein is encased in a transport vesicle, which buds off from the RER and moves to the Golgi complex for further editing and for routing to its proper location. The finished protein then moves through the cytoplasm in a transport vesicle that fuses with the cell's outer or "plasma" membrane, thereby exporting the protein from the cell.

CHAPTER 5
In Your Own Words
1. Because the phospholipid's phosphate "heads" are charged, hydrophilic groups, they are attracted to the polar water molecules that lie on either side of the plasma membrane, and they orient themselves accordingly. Meanwhile, the phospholipid's "tails" are hydrophobic fatty-acid chains that do not bond with water and thus end up oriented away from it, toward the interior of the plasma membrane.
2. In receptor-mediated endocytosis (RME), a group of receptors extending from the surface of the plasma membrane bind with a given target substance (for example, cholesterol) and then migrate to a depression on the membrane surface, a coated pit, which folds inward, becoming a transport vesicle that carries the target substance into the cell.
3. You would not expect that communication and recognition would be as important inside the cell as outside. Thus the glycocalyx would be unnecessary, and the proteins in the membranes would serve different functions.

CHAPTER 6
In Your Own Words
1. Energy from the sun is captured by plants or other photosynthesizing organisms and transformed by the plants into chemical energy, which is stored in a plant structure such as fruit. Human beings eat the fruit and digest it into a simple form such as glucose, which now is storing some of the energy that came originally from the sun. The energy in glucose is transferred to ATP, which then provides the energy that powers any number of processes in the body. At each step in the energy capture, storage, and use process, some amount of the original solar energy is lost to heat.
2. In animals, the energy stored and released by ATP comes from food. This energy in food powers the "uphill" process by which a third phosphate group is attached to ADP, making it ATP. The attached phosphate group then powers chemical reactions in the body by splitting off from ATP.
3. A metabolic pathway is a series of interrelated chemical reactions in an organism that results in the synthesis of some product or the completion of some process. Each reaction in a metabolic pathway is likely to be catalyzed by a different enzyme, with the

product of one reaction becoming the substrate for the next reaction. Most large-scale processes in living things are carried out through metabolic pathways.

CHAPTER 7

In Your Own Words

1. The food that we eat contains energetic molecules, such as glucose. In cellular respiration, foods are oxidized—they have electrons removed from them—and the energy released by the "fall" of these electrons from a higher to a lower energetic state powers phosphate groups onto ADP molecules, making them ATP molecules.

2. We would expect people to have more mitochondria in their muscle cells. This is so because muscle cells use a great deal more energy than skin cells do, and the organelles that supply that energy are mitochondria.

3. No. In short bursts of activity, the energy we are using comes primarily from glycolysis that ends in lactate fermentation.

CHAPTER 8

In Your Own Words

1. The sunlight that makes it to Earth's surface is composed of a spectrum of energetic rays (measured by their wavelengths) that range from very short ultraviolet rays, through visible light rays, to the longer and less energetic infrared rays. Photosynthesis in plants is driven by part of the visible light spectrum—mainly by blue and red light of certain wavelengths.

2. A photosystem includes, first, several hundred antenna pigments, consisting of both chlorophyll *a* and accessory pigments. These molecules collect solar energy and pass it along to another part of the photosystem, the reaction center. The reaction center consists of a pair of special chlorophyll *a* molecules and an initial electron acceptor molecule. The reaction center receives both the solar energy from the antennae pigments and electrons derived from the splitting of water.

3. Yes, plants are as dependent on cellular respiration as we are. Plants make food—carbohydrates—in photosynthesis. But just as the food we eat supplies the energy for producing our ATP, so the food plants make supplies the energy for producing their ATP.

CHAPTER 9

In Your Own Words

1. The 23 pairs of human chromosomes are matched (or homologous) in the sense that the two members of each pair contain information about similar functions, such as hair color, metabolic processes, and so forth. The one exception to the matched-pair rule comes in human males, who have 22 pairs of matched chromosomes, but then one X chromosome and one Y chromosome, which are not matched. Human females have 23 matched pairs, including two X chromosomes.

2. Mitosis serves to equally divide a cell's genetic material, while cytokinesis is a physical separation of a parent cell into two daughter cells.

3. The four phases of mitosis are as follows:

Prophase—duplicated chromosomes condense and become visible, the duplicated centrosomes migrate to the poles, the nuclear membrane breaks down, and the mitotic spindle is formed.

Metaphase—chromosome pairs align along the cell midline or metaphase plate and attach to the microtubules extending from the cellular poles.

Anaphase—sister chromatid pairs are separated; one member of each pair migrates to one pole of the cell, while the other member migrates to the opposite pole.

Telophase—condensed chromosomes unwind at their respective poles; new nuclear membranes are formed. At this point, division of genetic material is complete. The cell is ready for cytokinesis.

4. The cell is in metaphase. There are eight chromatids present in this stage, and each daughter cell will have four chromosomes.

CHAPTER 10

In Your Own Words

1. In the process called crossing over, homologous chromosomes exchange reciprocal sections with one another. Thus a maternal chromosome—one inherited from an individual's mother—now has a portion of a paternal chromosome inserted into it, and vice versa. Then, in the process called independent assortment, homologous chromosome pairs line up in a random way, relative to one another, along the metaphase plate. This ensures that maternal and paternal chromosomes will randomly end up in separate gametes.

2. Somatic cells are diploid cells that undergo mitosis. They include every cell in the body except the reproductive cells. Gametes are haploid reproductive cells, resulting from the process of meiosis. Male gametes are sperm; female gametes are eggs.

3. The regular somatic cells of human females contain two X chromosomes. In meiosis, each female gamete gets one of these X chromosomes. Somatic cells in males contain one X chromosome and one Y chromosome. In meiosis, half of the male gametes get an X chromosome, while the other half of the gametes get a Y chromosome. When egg and sperm then fuse in fertilization, each egg that is fertilized by sperm bearing a Y chromosome will become a male, while each egg that is fertilized by a sperm bearing an X chromosome will become a female.

CHAPTER 11

In Your Own Words

1. (a) Phenotype is a behavior or a physical trait in an organism (that is, yellow vs. green peas), while genotype is the genetic makeup that influences phenotype (that is, *YY* or *Yy* = yellow peas, *yy* = green peas).

(b) Dominant refers to an allele that is expressed in the heterozygous genotype (i.e., *Yy* = yellow peas). Recessive refers to an allele that is expressed only in the homozygous genotype (i.e., *yy* = green peas). Recessive traits are masked in the heterozygous genotype (i.e., *Yy* = yellow peas).

2. Mendel's first law (the law of segregation) states that members of a gene pair (alleles) will segregate from each other during gamete formation, such that each gamete receives only one of the original two members of the gene pair. Mendel's second law (the law of independent assortment) states that, during gamete formation, gene pairs assort independently of one another.

3. In Mendelian genetics, two different genotypes can bring about a dominant phenotype. For example, *YY* and *Yy* both bring about the dominant phenotype of yellow peas.

Genetics Problems

1. a
2. b
3. b
4. b (purple and white)
5. c (*PP, Pp, pp*)
6. e (*pp* × *pp* yields only *pp*)
7. c
8. c
9. Since the purple allele is dominant, flowers will be purple whether the allele is present in two identical copies (homozygous state) or in one copy (heterozygous state). In contrast, expression of the recessive white allele is masked if a purple allele is present in the genotype. Therefore, it must be in homozygous form for the development of white flowers.
10. b
11. c

CHAPTER 12

In Your Own Words

1. Aneuploidy goes widely unrecognized because, in most instances, it is directly affecting embryos, rather than children or adults. The most common result of aneuploidy is a miscarriage of the embryo. In many instances, prospective mothers will not realize that a miscarriage has taken place.

2. Male. Sex-linked diseases are almost always associated with the X chromosome, which females have two copies of. A female can carry one faulty version (or allele) of an X-chromosome gene and still be protected against a recessive genetic condition by a functional allele on her second X chromosome. Males, however, have only a single copy of the X chromosome. If an allele on it is faulty, they have no second, protective copy of it.

3. An aneuploidy that results in cancer would take place during mitosis, or the division of a regular, non-sex cell in the body. In consequence, not every cell in the body would be affected by this aneuploidy; only the line of cells stemming from the original aneuploid cells would be affected. Conversely, an aneuploidy that gives rise to a Down syndrome

child takes place during meiosis—the formation of eggs or sperm—and hence affects every cell in the body.

Genetics Problems

1. None will be affected (although half will be carriers).
2. One-half of their sons are likely to be affected.
3. d
4. Recessive. Neither of the parents shows the autosomal trait, yet one of their offspring does.
5. His genotype is *aa*.
6. Her genotype must be *Aa*. (Reasoning: If she were *AA*, none of her children would be affected, whereas one is; if she were *aa*, she would manifest the trait herself, which she does not.)
7. One-half are likely to be affected.
8. d
9. Her genotype is X^+X^p (Reasoning: She is not affected, her mate is not affected, yet they have affected sons. This is possible only if she is a carrier.)
10. One-quarter of their children would be expected to show this trait—half of their sons and none of their daughters.
11.

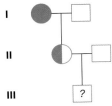

The chance that the son is color blind is $\frac{1}{2}$. This is because the woman in question must be a carrier.

12. One-half of sons and daughters will have hyperphosphatemia. This is in sharp contrast to what would be observed if the trait were X-linked recessive. In this case, half of the sons would likely be affected, and none of the daughters.

CHAPTER 13

In Your Own Words

1. "Something old, something new" is a shorthand means of conveying that, in DNA replication, each newly formed DNA double helix is a mixture of the old and new. Each one is made of an existing "template" strand of DNA and a new strand that is complementary to it.
2. Watson and Crick's discovery of the structure of the DNA molecule immediately suggested answers to two of the most important questions in biology. First, how can genetic information be passed on from one cell to the next or from one generation to the next? (Through base-pairing between template-strand nucleotides and complementary nucleotides.) Second, how can DNA be complex enough to contain information about the huge variety of proteins that living things produce? (Bases can be arranged in an extremely varied manner along the "rails" of the double helix.)

3. A mutation is a permanent alteration in a DNA base sequence. A change in a DNA base sequence becomes permanent only when it goes uncorrected by a cell's DNA error-correcting mechanisms, meaning the change will be copied over and over during DNA replication. Not all mutations are harmful. Most have no effect at all, and some are important in evolution.

CHAPTER 14

In Your Own Words

1. Transcription is the process by which the information encoded in DNA sequences is put into the form of mRNA sequences. Translation is the process by which the information encoded in mRNA sequences is put into the form of amino acid sequences; this results in the production of a protein.
2. A complex of proteins, the enzyme complex RNA polymerase, both unwinds the DNA that is to be transcribed and pairs DNA bases with their complementary RNA bases.
3. The three types of RNA reviewed in the chapter are (a) mRNA (messenger RNA), which takes DNA-encoded information to ribosomes; (b) rRNA (ribosomal RNA), which helps make up the ribosomes that act as the sites of protein synthesis; and (c) tRNA (transfer RNA), which brings amino acids to the ribosomes and binds with mRNA codons there.

CHAPTER 15

In Your Own Words

1. PCR is valuable because it allows scientists to quickly make a large quantity of a given segment of DNA from a very small starting sample.
2. The production of sticky ends is useful because this type of end allows different restriction fragments to be joined together easily to create recombinant DNA.

   #### Fragment sequences:

   ATCG   GATCCTCCG

   TAGCCTAG   GAGGC

3. Therapeutic cloning would be used to help individuals produce their own embryonic stem cells. Tissues derived from these cells would not be recognized as "foreign" by the body's immune system and hence would not set off an immune-system attack.

CHAPTER 16

In Your Own Words

1. Figure 16.12 shows that, in a gene shared by humans, pigs, and yeast, there are fewer DNA base differences between the forms of the gene found in humans and pigs than between the forms found in humans and yeast. This is what you would expect if humans and pigs shared a common ancestor more recently than did humans and yeast.
2. Congruent lines of evidence that are consistent with the theory of evolution: (a) The placement of fossils found in geological layers is consistent with the ages of those layers as determined by radiometric dating.

(b) Evidence from molecular clocks is consistent with fossil and other evidence regarding the degree of relatedness among species.
3. Darwin brought back many birds that he thought were the blackbirds, wrens, and warblers he was familiar with in Britain, but he found they were in fact all species of finches. Darwin perceived this as an example of descent with modification. Without taking the *Beagle* voyage, Darwin would not have been exposed to such extravagant examples of branching evolution, nor would he have made various other biological observations that spurred his thinking about evolution by natural selection.

CHAPTER 17

In Your Own Words

1. An individual may be selected for survival—and hence greater reproduction—because of the qualities that individual has, such as coloration or height. But this individual's characteristics do not change because it possesses these qualities. An individual does not evolve, in other words. An individual can pass on more of its genes relative to other individuals in a population, however; this causes the population's allele frequencies to change, meaning the population has evolved.
2. A gene pool is the sum total of alleles that exist in a population. The inheritable characteristics of living things are based on the alleles they possess. Thus, for a population to change in a way that is passed on from one generation to the next—for the population to evolve—the mixture of alleles in that population must change. Thus the evolution of a population is based at root on changes in that population's gene pool.
3. The five causes of microevolution are (a) Mutation—changes in the form of the genetic material that result in new alleles. (b) Gene flow—brings alleles from other populations into the target population. (c) Sexual selection—comes about when some members of a population mate (and hence reproduce) more than others. Can change the relative genetic contributions individuals make to a succeeding generation. (d) Genetic drift—the chance alteration of allele frequencies in a population. May occur if a new population is founded by only a few individuals (founder effect), or if the size of a population is decreased suddenly (the bottleneck effect). (e) Natural selection—differential survival and reproductive ability of individuals based on their fit with a particular environment. If the phenotypic variation is inherited, the population will evolve to be like the individuals that survive longest and reproduce most.

CHAPTER 18

In Your Own Words

1. Without the evolution of intrinsic isolating mechanisms, two populations that have become geographically separated would simply resume interbreeding, remaining

one species, if any physical barriers to interbreeding were removed. Thus, they can't be called separate species unless they possess intrinsic reproductive isolating mechanisms.

2. A first requirement of allopatric speciation is that two populations of the same species must become separated geographically, such that gene flow between them is restricted. This allows the forces operating in microevolution—natural selection, genetic drift, mutation, and so forth—to work separately on the two populations. Eventually, the populations may develop intrinsic reproductive isolating mechanisms, such that, should they be reunited, they would no longer interbreed. In sympatric speciation, intrinsic isolating mechanisms evolve in the absence of geographical separation. The quality that both sympatric and allopatric speciation have in common is the development of intrinsic reproductive isolating mechanisms.

3. Convergent evolution is the evolution of a similar character in two separate evolutionary lines. It comes about because unrelated organisms may live in similar environments and experience similar natural selection pressures. Litopterns and horses, for example, both needed to roam long distances over a grassy terrain. Long legs and undivided hooves worked well for herbivores in such terrain, and natural selection thus resulted in these characters evolving twice—in the separate evolutionary lines of horses and litopterns.

CHAPTER 19
In Your Own Words
1. The amniotic egg evolved in the Reptilia. It is an egg in which the embryo lives in a watery environment (albumin, the "white") while being enclosed in a protective leathery or hard shell. Special membranes facilitate gas exchange while protecting the embryo from drying out. This was a huge breakthrough for terrestrial animals, because it allowed them to lay eggs outside an aquatic environment without risking drying out.
2. Primates have forward-looking eyes and an opposable thumb. This facilitates depth perception and the ability to pick up and manipulate objects, paving the way for the use of tools.
3. No. Neanderthals and human beings may share a distant common ancestor in *Homo ergaster,* but since modern humans and Neanderthals appear not to have mated with each other, contemporary human beings are not descended from Neanderthals.

CHAPTER 20
In Your Own Words
1. Viruses cannot replicate themselves without being in a host cell. Indeed, without the aid of cells, they can carry on very few of life's central functions, such as acquiring energy and getting rid of wastes.
2. Animals, fungi, and plants all evolved from protists.

3. An amoeba is a protist that moves through use of pseudopodia or "false feet," meaning an extension of one part of the cell and a flowing of the rest of the cell into that extension. Many amoebae are parasites, and some of these are human parasites that cause the class of diseases known as amoebic dysentery.

CHAPTER 21
In Your Own Words
1. Only a tiny fraction of the spores that fungi release will be able to germinate, or grow into a new mycelium. The fungi's solution to this problem is to "buy many tickets"—to create a huge number of spores, which scatter to diverse environments strictly because of their numbers.
2. A spore is a reproductive cell that can give rise to a complete organism without fusing with another reproductive cell. A gamete (an egg or sperm) is a reproductive cell that must fuse with another reproductive cell for a complete organism to be produced—egg must fuse with sperm.
3. One fungal association takes the form of mycorrhizae, which are associations of fungal hyphae and plant roots. In this association, plants provide the fungi with food (made in photosynthesis) while the fungi supply plants with water and minerals (absorbed by the fungal hyphae). A second association is lichens, which are composite organisms made of upper and lower layers of fungi and a middle layer of algae (or sometimes bacteria) interspersed with fungi. The algae or bacteria supply the fungi with food (made in photosynthesis), and the fungi may keep the algae or bacteria from drying out, though it may be that the fungi are simply parasites.

CHAPTER 22
In Your Own Words
1. A coelom provides an animal with flexibility, allows for the expansion and contraction of organs such as the heart and stomach, and helps protect an animal's organs from being damaged by external blows.
2. Sponges have no brains or organs of any sort, no tissues, and each of their cells functions with a great deal of independence. Thus the cells in a sponge have a lower level of specialization and integration with one another than is the case in more advanced animals.
3. The amniotic egg first developed in reptiles. Its shell and some of its membranes provide protection against jarring or external blows; other membranes ensure a flow of oxygen to the embryo, a flow of carbon dioxide away from it, and storage of waste products. The amniotic egg freed reptiles from the need to lay their eggs in water; this allowed them to move inland.

CHAPTER 23
In Your Own Words
1. The natural world's four levels of group organization are population, community, ecosystem, and biosphere.

2. Environmental resistance is the term used to describe all the environmental forces that act to limit population growth. Organisms will run out of food or have their sunlight blocked; temperatures will plunge or rain will cease; greater numbers of predators will discover the population; the wastes produced by the organisms will begin to be toxic to the population.
3. This is an example of Batesian mimicry, in which one species (the red salamander) has evolved to acquire the appearance of a species with superior protective capabilities (the red newt). The red newt is the model, the red salamander is the mimic, and the dupe is any predator species that would avoid the red salamander, believing it to be a red newt.

CHAPTER 24
In Your Own Words
1. Within any ecosystem, the amount of available energy drops significantly at each trophic level. This is so because much of the energy locked up in organisms at each trophic level is used for the organisms' metabolic processes (building up tissue, staying warm), while additional energy is lost as heat. The drop in available energy at each trophic level means there can be fewer predator than prey animals.
2. Warm air (a) rises and (b) can carry more moisture than cold air. Near the equator, Earth gets warm due to the direct rays coming from the sun. As this warm air rises, it cools and drops moisture. This cooler air moves toward the poles in both directions, and descends at about 30° north and south, warming as it goes, thus absorbing moisture from the land, which is dry as a result.
3. It could be that eutrophication has occurred in the pond. If so, large algal blooms can result from the excess nutrients. As this dead organic matter falls to the lower depths, decomposing bacteria who consume the organic matter multiply in such numbers that they deplete the water of oxygen, thus killing fish.

CHAPTER 25
In Your Own Words
1. Lymphatic organs such as the spleen have high concentrations of immune cells within them that monitor passing lymphatic fluid for signs of invading microorganisms.
2. Our skin is water-resistant because the surface epidermal layer of the skin is formed of tightly interlocking cells made almost entirely of the protein keratin, which is water-resistant.
3. Ligaments hold bones together in stable arrangements. The anterior cruciate ligament (ACL) links the back of the femur (thigh bone) to the front of the tibia (shin bone). When the ACL is damaged, the two bones are not held in place properly.

CHAPTER 26
In Your Own Words
1. The sensory neuron receives input about conditions inside and outside the body and

conveys this received information to the central nervous system (CNS). The motor neuron exists in the peripheral nervous system (PNS) and sends instructions from the CNS to such structures as muscles or glands. The interneuron exists solely in the CNS and serves to interconnect neurons and integrate signals from them.

2. A simple reflex arc works by means of a sensory neuron sending an afferent message about some stimulation to the spinal cord. There, the sensory neuron will synapse with a motor neuron (or perhaps an interneuron), after which the motor neuron sends a signal that brings about some action at an effector such as a muscle. The brain is not involved in such an arc.

3. Amino-acid-based hormones are derived from a chemical modification of a single amino acid. Peptide hormones are built from chains of amino acids. The building block of all steroid hormones is the cholesterol molecule. Amino-acid-based and peptide hormones generally bind with receptors that are on the surface of target cells. Steroid hormones generally pass through the outer membrane of a target cell and bind with a receptor inside the cell.

CHAPTER 27

In Your Own Words

1. The immune system increases blood flow near the site of injury and increases the permeability of capillaries near the site (so that more substances can be transported out); it delivers cells and proteins to the site that kill the invaders; and it releases substances that help seal off the site of injury, thus limiting the area of the infection.

2. HIV has so many effects because the disease it causes, AIDS, weakens the immune system, which normally fights off so many kinds of invaders.

3. The immune system regards any transplanted organ as "non-self" and thus will initiate an attack against it.

CHAPTER 28

In Your Own Words

1. The left and right atrial chambers of the heart pump blood only into their respective ventricles, while the ventricles pump blood to locations outside the heart—the right

ventricle to the lungs, the left ventricle to all the tissues of the body.

2. Oxygen enters the bloodstream at a capillary that covers an alveolus in the lungs. Oxygen is inhaled into the sac-like alveolus and then diffuses across the alveolar tissue into an adjacent capillary.

3. Blood cells and most proteins are too large to pass out of the glomerulus capillaries and into the filtrate; they remain in the circulatory system.

CHAPTER 29

In Your Own Words

1. Sperm development is dependent on temperature. If temperatures get too hot, then sperm production is impaired. Similarly, cold temperatures also prevent sperm development. The structure of the testes acts to keep sperm production at the correct temperature. The sperm are protected from the high body temperatures because the testes are located outside the body's torso. In addition, a muscle that surrounds the testes contracts when temperatures drop, so that the testes are pulled in closer to the warm body.

2. Two hypotheses currently are favored to explain the existence of menstruation. One is that the cycle of tissue buildup and breakdown that underlies it is less costly, in terms of energy expenditure, than the alternative of keeping the endometrium in a permanent state of readiness for embryo implantation. The second hypothesis is that, when it is ready for embryo implantation, the endometrium is in a form that cannot be maintained permanently and that must arise from an earlier form. Thus it makes sense that the inner endometrium would wash away every 28 days and then enter a new cycle of growth and development.

3. Fusion of a sperm with an oocyte prompts the oocyte to release granules whose secretions harden the material surrounding the oocyte and inactivate sperm receptors on the surface of the oocyte.

CHAPTER 30

In Your Own Words

1. The stomata allow the movement of carbon dioxide into the plant and oxygen out.

These same openings also allow water to escape from the plant as water vapor. The plant solves the resulting problem of water loss by moving large quantities of water up through the roots and stem and out through the stomata.

2.

Flower Part	Function
pedicel	Flower stalk
sepals	Leaf-like structures that protect flower buds before they open
petals	Generally most conspicuous part of the flower; often brightly colored; sometimes fragrant; attract pollinators
stamens	Male reproductive structures; consist of a stalk called a filament, which holds aloft the pollen-producing sacs called anthers
pollen	The male gametophyte; the vehicle within which sperm are transported to the female reproductive structures
carpel	Female reproductive structure; consists of the ovary, style, and stigma—the ovary houses and protects the egg-containing female gametophyte, and after fertilization turns into the fruit; the stigma is the receptive surface that captures pollen; the slender style raises the stigma to make it accessible to pollinators

3. The gametophyte is the generation of plant that produces gametes—eggs and sperm—while the sporophyte is the generation that produces spores: megaspores (female reproductive cells) and microspores (male reproductive cells). In angiosperms, the sporophyte generation is the relatively large plant with which we are familiar. The gametophyte generation amounts, at maturity, to a pollen grain on the male side and to a cluster of eight cells (including the unfertilized egg) on the female side.

Glossary

1n Haploid; possessing a single set of chromosomes.

2n Diploid; possessing two sets of chromosomes.

A

acid Any substance that yields hydrogen ions when in solution. An acid has a number lower than 7 on the pH scale.

action potential A temporary reversal of neuronal cell membrane potential at one location that results in a conducted nerve impulse down an axon.

activation energy The energy required to initiate a chemical reaction. Enzymes lower the activation energy of a reaction, thereby greatly speeding up the rate of the reaction.

active site The area of an enzyme, usually a groove or pocket, that binds the substrate and changes its shape to affect its reactivity.

active transport Transport of materials across the plasma membrane in which energy must be expended. Through active transport, solutes can be moved against their concentration gradient. The sodium-potassium pump is an example of active transport.

actively acquired immunity Immunity developed as a result of accidental or deliberate exposure to an antigen. Accidental exposure: coming into contact with someone who has a transmissible disease. Deliberate exposure: being vaccinated.

adaptation An evolutionary modification in the structure or behavior of organisms over generations that makes them better suited to their environment.

adaptive radiation The rapid emergence of many species from a single species that has been introduced to a new environment. The different species specialize to fill available niches in the new environment.

adenosine triphosphate (ATP) The most important energy-transfer molecule in living things. ATP powers a broad range of chemical reactions by donating one of its three phosphate groups to these reactions. In the process, it becomes adenosine diphosphate (ADP), which reverts to being ATP when a third phosphate is added to it.

afferent division The division of the peripheral nervous system that carries sensory information toward the central nervous system, having gathered information about the body or environment.

alcoholic fermentation The process by which yeasts produce alcohol as a by-product of glycolysis they perform in oxygenless environments.

algae Protists that perform photosynthesis.

alkaline Basic, as in solutions. Alkaline (basic) solutions have numbers above 7 on the pH scale.

allele One of the alternative forms of a single gene. In pea plants, a single gene codes for seed color, and it comes in two alleles—one codes for yellow seeds, the other for green seeds.

allergen A foreign substance that triggers an allergic reaction. These substances are usually derived from living things, including pollen, dust mites, foods, and fur.

allergies Immune-system overreactions to a class of antigens, called allergens, resulting in the release of histamine.

allopatric speciation Speciation that involves the geographic separation of populations. Most speciation involves geographic separation, followed by the development of intrinsic isolating mechanisms in the separated populations.

alternation of generations A life-cycle practiced by plants, in which successive plant generations alternate between the diploid sporophyte condition and the haploid gametophyte condition.

alveoli (singular, *alveolus*) Tiny, hollow air-exchange sacs that exist in clusters at the end of each of the air-conducting passageways in the lungs, the bronchioles.

amino-acid-based hormones Hormones that are derived from a single amino acid. One of three principal classes of hormones, the other two being peptide and steroid hormones.

amniotic egg An egg with a hard outer casing and an inner series of membranes and fluids that form a padding around the growing embryo. The evolution of the amniotic egg, in reptiles, freed them from the constraint of having to reproduce near water.

analogy A structure found in different organisms that is similar in function and appearance, but is not the result of shared ancestry.

aneuploidy A condition in which an individual organism has either more or fewer chromosomes than is normal for its species, so that the individual has the wrong number of chromosomes in its set. Down syndrome is the result of aneuploidy—generally three copies of chromosome 21.

angiosperm A flowering seed plant whose seeds are enclosed within the tissue called fruit. Angiosperms are the most dominant and diverse of the four principal types of plants. Examples include roses, cacti, corn, and deciduous trees.

anther The part of a flower that produces pollen grains. The anther is on top of a filament, and together they are called the stamen.

antibiotics Chemical compounds produced by one microorganism that are toxic to another microorganism.

antibody A protein of the immune system, found on the surface of a B cell or circulating free in the plasma, that is formed in response to a particular antigen and reacts with it specifically, deactivating it.

antibody-mediated immunity An immune-system capability that works through the production of proteins called antibodies.

antidiuretic hormone (ADH) Substance that helps control how much water is either sent to the bladder (in urine) by the kidneys or retained in circulation. In a release controlled by the brain's hypothalamus, ADH increases the permeability of both the distal nephron tubule and the nephron's collecting duct to water, thus conserving it.

antigen Any foreign substance that elicits a response by the immune system. Certain proteins on the surface of an invading bacterial cell, for instance, act as antigens that trigger an immune response.

antigen-presenting cell (APC) Any immune-system cell that presents, on its surface, fragments of an antigen that it has ingested. Dendritic cells, macrophages, and B cells are the immune system's three classes of antigen-presenting cells.

aorta The enormous artery extending from the heart that receives all the blood pumped by the heart's left ventricle. Branches stemming from the aorta supply oxygenated blood to all the tissues in the body.

aquifer The porous underground rock in which groundwater is stored.

arithmetical increase An increase in numbers by an addition of a fixed number in each time period.

artery A blood vessel that carries blood away from the heart.

atmosphere The layer of gases that surrounds the Earth. The atmosphere is nonliving, but its presence enables life to exist on Earth.

autoimmune disorder An attack by the immune system on the body's own tissues.

autonomic nervous system That portion of the peripheral nervous system's efferent division that provides involuntary regulation of smooth muscle, cardiac muscle, and glands.

axon A single, large extension of the cell body of a neuron that carries signals away from the cell body toward other cells.

B

base Any substance that accepts hydrogen ions in solution. A base has a number higher than 7 on the pH scale.

Batesian mimicry A type of mimicry in which one species evolves to resemble a species that has superior protection against predators.

bell curve A distribution of values that is symmetrically largest around the average.

bilateral symmetry A bodily symmetry in which opposite sides of a sagittal plane are mirror images of one another. Animals generally are bilaterally symmetrical.

bile Substance produced by the liver that facilitates the digestion of fats. Bile can be released either directly by the liver or by the gallbladder, which stores and concentrates this substance.

binary fission   The form of reproduction carried out by prokaryotic cells in which the chromosome replicates and the cell pinches between the attachment points of the two resulting chromosomes to form two new cells. In this type of simple cell-splitting, each pair of daughter cells is an exact replica of the parental cell.

biodegradable   Capable of being broken down by living organisms.

biodiversity   The diversity of living things. Includes species diversity, diversity of distributions, and genetic diversity.

biogeochemical cycling   The movement of water and nutrients back and forth between biotic (living) and abiotic (nonliving) systems.

biological species concept   A definition of species that relies on the breeding behavior of populations in nature. Accordingly, populations that actually or potentially interbreed in nature and are reproductively isolated from other such groups constitute a species.

biology   The study of life.

biomass   Material produced by living things, generally measured by dry weight.

biome   An ecosystem dominated by a large vegetation formation, whose boundaries are largely determined by climate. The same biome type can occur on two different continents and have different species, but the two regions will bear striking similarities.

biosphere   The interactive collection of all the world's ecosystems. Also thought of as that portion of the Earth that supports life.

biotechnology   The use of technology to harness the power of biological processes as a means of meeting societal needs.

blastocyst   Hollow, fluid-filled ball of cells that is formed in the early stages of the embryonic development of humans and other mammals. In non-mammalian animals, the blastocyst is known as the blastula.

B-lymphocyte cells (B cells)   The central cells of antibody-mediated immune-system function. B cells produce antibodies, called antigen receptors, that bind with specific antigens while remaining embedded in the B cell. Conversely, these antibodies may be exported from B cells as free-standing entities to fight antigens.

bone   A connective tissue that provides support and storage capacity, and that is the site of blood-cell production.

bottleneck effect   A change in allele frequencies in a population due to chance, following a sharp reduction in the population's size. One of the factors that potentiates genetic drift.

bryophyte   A plant that lacks a vascular (fluid transport) structure. Bryophytes are the most primitive of the four principal varieties of plants. Mosses are the most familiar example.

budding   A form of asexual reproduction in which an offspring is produced as an outgrowth of an organism.

C

Calvin reactions   The second set of chemical reactions in photosynthesis, in which an energy-rich sugar is produced by means of two essential processes—the fixing of atmospheric carbon dioxide into a sugar, and the energizing of this sugar with the addition of electrons supplied by the light reactions of photosynthesis. Named for the biochemist Melvin Calvin, who elucidated its steps. Also known as the C_3 cycle.

Cambrian Explosion   A period of dramatic increase in the number of animal phyla that began about 540 million years ago, to judge by the fossil record.

capillary   The smallest type of blood vessel, connecting the arteries and veins in the body's tissues. Gases, nutrients, and wastes are exchanged between the blood and the tissues through the thin walls of capillaries.

carbohydrate   An organic molecule that always contains carbon, oxygen, and hydrogen and that, in many instances, contains nothing but carbon, oxygen, and hydrogen. Carbohydrates usually contain exactly twice as many hydrogen atoms as oxygen atoms. The building blocks or monomers of carbohydrates are monosaccharides, such as glucose, which combine to create the polymers of carbohydrates, the polysaccharides, such as starch and cellulose.

cardiac muscle   In the human body, striped or "striated" muscle, found only in the heart, whose contractions are governed by heart pacemaker cells.

cardiovascular system   A fluid transport system of the body, consisting of the heart, all the blood vessels in the body, the blood that flows through these vessels, and the bone marrow tissue in which red blood cells are formed.

carnivore   An animal that eats meat.

carpel   The female reproductive structure of a flower, consisting of an ovary, a style, and a stigma.

carrier   A person who does not suffer from a recessive genetic debilitation, but who carries an allele for the condition that can be passed along to offspring.

carrying capacity   The maximum population of a species that can be sustained in a given geographical area over time. In ecology, this is often denoted as K.

cartilage   A connective tissue that serves as padding in most joints, forms the human larynx (voice box) and trachea (windpipe), and links each rib to the breastbone.

catalyst   A substance that retains its original chemical composition while bringing about a change in a substrate. Enzymes are catalysts in chemical reactions; one enzyme can carry out hundreds or thousands of chemical transformations without itself being transformed.

cell   The fundamental unit of life; a generally microscopic entity, always enclosed within a plasma membrane, containing biological molecules that together can carry out life's basic processes. All life is either single-celled or multicelled, and all cells are produced by other cells.

cell cycle   The repeating pattern of growth, genetic duplication, and division seen in most cells.

cell wall   A relatively thick layer of material that forms the periphery of plant, bacterial, and fungal cells.

cell-mediated immunity   An immune-system capability that works through the production of cells that destroy other cells in the body—those that have been infected by a microbial invader.

cellular respiration   The three-stage, oxygen-dependent harvesting of energy that goes on in most cells. The three stages are glycolysis, the Krebs cycle, and the electron transport chain.

cellulose   A complex carbohydrate that is the largest single component of plant-cell walls. Cellulose is dense and rigid and provides structure for much of the natural world. Mammals cannot digest cellulose, so it serves as insoluble dietary fiber that helps move food through the digestive tract.

central nervous system (CNS)   The portion of the nervous system consisting of the brain and spinal cord.

centrosome   A cellular structure that acts as an organizing center for the assembly of microtubules. A cell's centrosome duplicates prior to mitosis and plays an important part in the development of the cell's mitotic spindle.

cerebral cortex   Thin outer covering of the region of the brain known as the cerebrum. Responsible for the highest human thinking and processing.

cerebrum   The largest region of the human brain, responsible for much of the human capacity for higher mental functioning.

cervix   The lower part of the uterus, a narrow neck that opens into the vagina.

chemical bonding   General term for a bond created when electrons of two atoms interact and rearrange into a new form that allows the atoms to become attached to each other. Ionic, covalent, and hydrogen bonds are all chemical bonds.

chitin   A complex carbohydrate that gives shape and strength to the external skeleton of arthropods, including insects, spiders, and crustaceans.

chlorophyll a   The primary pigment of chloroplasts, found embedded in its membranes. Together with the accessory pigments, chlorophyll a absorbs some wavelengths of sunlight in the first step of photosynthesis.

chloroplast   The organelle within plant and algae cells that is the site of photosynthesis.

cholesterol   A steroid molecule that forms part of the outer membrane of all animal cells, and that acts as a precursor for many other steroids, among them the hormones testosterone and estrogen.

chromatid   One of the two identical strands of chromatin (DNA plus associated proteins) that make up a chromosome in its duplicated state.

chromatin   A molecular complex of DNA and its associated proteins that makes up the chromosomes of eukaryotic organisms.

chromosome   Structural unit containing part or all of an organism's genome, consisting of DNA and its associated proteins (chromatin). The human genome is made up of 23 pairs of chromosomes, or 46 chromosomes in all.

chyme The soupy mixture of food and gastric juices that passes from the stomach to the small intestine.

cilia (singular, _cilium_) Hair-like extensions of a cell, composed of microtubules. Many cilia occur on the surface of a given cell, and they move rapidly back and forth to propel the cell or to move material around the cell.

citric acid cycle Another name for the Krebs cycle, one of the three main sets of steps in cellular respiration; named for the first product of the cycle, citric acid.

class A taxonomic grouping subordinate to phylum and superordinate to order. Humans are in the class Mammalia.

climate The average weather conditions, including temperature, precipitation, and wind, in a particular region.

climax community The relatively stable community that develops at the end of any process of ecological succession.

clone An exact genetic copy. Also used as a verb—to make one of these copies. A single gene or a whole, complex organism can be cloned.

closed circulation A type of circulation, found in all vertebrates and in some invertebrates, in which circulatory fluid stays within vessels.

coastal zone The region lying between the point on shore where the ocean's waves reach at high tide to the point offshore where the continental shelf drops off.

codon An mRNA triplet that codes for a single amino acid or a start or a stop command in the translation stage of protein synthesis.

coelom A central body cavity, found in most animals.

coenzyme A molecule other than an amino acid that facilitates the work of enzymes by binding with them.

coevolution The interdependent evolution of two or more species. Coevolution can benefit both species, as in flowering plants and their animal pollinators, or it can manifest as an arms race between species, as in a plant and its predators.

coexistence The condition in which two species can live in the same habitat, dividing up resources in a way that allows both to survive.

colonial multicellularity A form of life in which individual cells form stable associations with one another but do not take on specialized roles.

commensalism An interaction between two species in which one benefits while the other is neither harmed nor helped.

common descent with modification The process by which species of living things undergo modification in successive generations, with such modification sometimes resulting in the formation of new, separate species.

community All the populations of all species that inhabit a given area. A community is sometimes defined as all the populations in a given area that potentially interact with each other.

compact bone A relatively dense form of bone tissue, always found toward the periphery of bones. Compare to spongy bone, which is a less dense form of bone tissue, found in the interior of bones.

competitive exclusion principle When two populations compete for the same limited, vital resource, one will always outcompete the other and thus bring about the local extinction of the latter species.

concentration gradient A gradient within a given medium defined by the difference between the highest and lowest concentration of a solute. The solute will have a natural tendency to move from the areas of higher concentration to lower, thus diffusing.

connective tissue A tissue, active in the support and protection of other tissues, whose cells are surrounded by a material that they have secreted. In humans, one of the four principal types of tissue.

consumer Any organism that eats other organisms rather than producing its own food.

convergent evolution Evolution that occurs when similar environmental influences shape two separate evolutionary lines in similar ways.

copulation In animal sexual reproduction, the release of sperm within a female by a male reproductive organ.

coronary artery An artery that delivers oxygenated blood to the muscles of the heart. Blockage of coronary arteries causes heart attacks.

covalent bond A type of chemical bond in which two atoms are linked through a sharing of electrons.

Cretaceous Extinction A mass extinction event that occurred at the boundary between the Cretaceous and Tertiary periods. This event, which included an asteroid impact, resulted in the extinction of the dinosaurs along with many other organisms.

cross-pollinate In plant breeding, the transfer of pollen from the anther of one plant to the stigma of another.

crossing over (genetic recombination) A process, occurring during meiosis, in which homologous chromosomes exchange reciprocal portions of themselves.

cytokinesis The physical separation of one cell into two daughter cells.

cytoplasm The region of a cell inside the plasma membrane and outside the nucleus. Usually, this region is filled with the jelly-like cytosol containing the cell's extranuclear organelles.

cytoskeleton A network of protein filaments that functions in cell structure, cell movement, and the transport of materials within the cell. Microfilaments, intermediate filaments, and microtubules are all parts of the cytoskeleton.

cytosol The protein-rich, jelly-like fluid in which a cell's organelles that are outside the nucleus are immersed.

cytotoxic T cell (killer T cell) Type of T lymphocyte cell that binds to and kills the body's own cells when they have become infected.

D

deciduous Refers to plants that show a coordinated, seasonal loss of leaves. This strategy allows plants to conserve water during a time they could perform little photosynthesis anyway.

decomposer A type of detritivore that, in feeding on dead or cast-off organic material, breaks it down into its inorganic components. Most decomposers are fungi or bacteria.

dendrites Extensions of a neuron that carry signals toward the neuronal cell body.

density-dependent In ecology, effects on a population that increase or decrease in accordance with the size of that population. Density-dependent effects tend to involve biological factors.

density-independent In ecology, effects on a population that are not related to the size of that population. Density-independent effects tend to involve physical forces, such as temperature and rain.

dermis In certain animals, the thick layer of the skin—composed mostly of connective tissue—that underlies, nourishes, and supports the epidermis.

desert A biome in which rainfall is less than 10 inches per year and water evaporation rates are high, relative to precipitation.

detritivore An organism that feeds on the remains of dead organisms or the cast-off material from living organisms.

diffusion The movement of molecules or ions from areas of their higher concentration to areas of their lower concentration. Over time, the random movement of molecules will result in the even distribution of the material.

digestion The breakdown of food into a chemical form that can be passed into circulation.

digestive system The organ system that transports food into the body, secretes digestive enzymes that help break down food to allow it to be absorbed by the body, and excretes waste products. This system consists of the esophagus, stomach, and large and small intestines, plus the accessory glands that produce the enzymes along the way.

digestive tract A muscular tube that is the central structure of the digestive system. The digestive tract passes through the body from the mouth to the anus, along the way receiving input from accessory organs that aid in digestion, such as the gallbladder, the liver, and the pancreas.

dikaryotic phase Phase during a life cycle, common in fungi, in which single cells contain two haploid nuclei.

diploid Possessing two sets of chromosomes. All human cells are diploid with the exception of human gametes (eggs and sperm), which are haploid. Such haploid cells possess only a single set of chromosomes.

directional selection In evolution, the type of natural selection that moves a character toward one of its extremes. Compare to stabilizing and disruptive selection.

disruptive selection In evolution, the type of natural selection that moves a character toward both of its extremes, operating against individuals that are average for that character. This type of selection seems to be less common in nature than either stabilizing or directional selection.

DNA (deoxyribonucleic acid) The primary information-bearing molecule of life, composed of two chains of nucleotides, linked together in the form of a double helix. Proteins are put together in accordance with the information encoded in DNA.

DNA polymerase   An enzyme that is active in DNA replication, separating strands of DNA, bringing bases to the parental strands, and correcting errors by removing and replacing incorrect base pairs.

domain   The highest-level taxonomic grouping of organisms. There are only three domains: Archaea, Bacteria, and Eukarya.

dominant   Term used to designate an allele that is expressed in the heterozygous condition.

dominant disorder   Genetic conditions in which a single faulty allele can cause damage, even when a second, functional allele exists.

dormancy   A state in which growth is suspended and there is a prolonged low level of metabolic activity. Dormancy allows organisms to conserve energy during times of unfavorable environmental conditions.

double fertilization   In plants, the fusion of one sperm with the egg and another sperm with the central cell; the first of these fertilizations results in the embryo, the other in endosperm to provide food for the embryo. Double fertilization occurs almost exclusively in flowering plants.

Down syndrome   A disorder in humans in which affected individuals usually have three copies of chromosome 21. Individuals with this syndrome have short stature, shortened life span, and low IQ.

E

ecological dominant   A species that is abundant and obvious in a given community. In any community, a few species, usually plants, will dominate in numbers.

ecology   The study of the interactions that living things have with each other and with their environment.

ecosystem   A community of living things and the physical environment with which they interact.

ectopic pregnancy   An abnormal pregnancy in which the blastocyst attaches inside the uterine tube or to the cervix, rather than in the normal position in the dorsal uterine wall.

efferent division   The division of the peripheral nervous system that carries motor commands from the central nervous system (CNS) toward the effectors (muscles and glands).

electron   A basic constituent of an atom that has negative electrical charge. Electrons are distributed in an atom at a distance from the nucleus. Electrons interact to form chemical bonds between atoms.

electron carrier   A molecule that serves to transfer electrons from one molecule to another in ATP formation. The most important electron carrier in ATP formation is NAD^+ (NADH in its reduced form). The other major carrier is FAD ($FADH_2$).

electron transport chain (ETC)   The third stage of cellular respiration, occurring within the inner membrane of the mitochondria, in which most of the ATP are formed.

electronegativity   The measure of the strength of attraction an atom has for electrons. An atom with higher electronegativity will tend to pull electrons away from atoms with lower electronegativity.

element   A substance that cannot be reduced to any simpler set of components through chemical processes. An element is defined by the number of protons in its nucleus.

embryo   A developing organism. In humans, the developing organism from the time a zygote undergoes its first division through the end of the eighth week of development.

embryonic stem cell   A cell from the blastocyst stage of a human embryo that is capable of giving rise to almost all of the cells or tissues in the body.

endocrine gland   A gland that releases its materials directly into surrounding tissues or into the bloodstream, without using ducts. Many hormones are produced by endocrine glands.

endocrine system   The organ system that sends signals throughout the body using the chemical messengers called hormones.

endocytosis   The process by which the plasma membrane folds inward and pinches off, bringing relatively large materials into the cell enclosed inside a vesicle.

endomembrane system   An interactive group of membrane-lined organelles and transport vesicles that functions within eukaryotic cells.

endometrium   The tissue lining the interior of the uterus in mammals, which thickens in response to progesterone secretion during ovulation and is shed during menstruation. If pregnancy occurs, this tissue houses the embryo.

environmental resistance   All the forces in the environment that act to limit the size of a population. Limited food or sunshine, low temperature or rainfall, and predators are some components of environmental resistance.

enzyme   A chemically active type of protein that speeds up, or in practical terms enables, chemical reactions in living things.

epidermis   The outermost layer of skin in animals, or the outermost cell layer in plants.

epididymis   The mass of tubules near the testis in which sperm complete their development and are stored.

epithelial tissue   Tissue that covers an internal or external surface.

estuary   An aquatic area, partly enclosed by land, where a stream or river flows into the ocean.

eukaryote   An organism composed of one or more cells whose primary complement of DNA is contained within a nucleus. All plants, animals, fungi, and protists are eukaryotes. Contrast with prokaryote: a single-celled organism whose complement of DNA is not contained within a nucleus. Bacteria and archaea are prokaryotes.

eukaryotic cell   A cell whose primary complement of DNA is contained within a membrane-lined nucleus. Eukaryotic cells have several other organelles in addition to the nucleus. All organisms except bacteria and archaea either are single eukaryotic cells or are composed of eukaryotic cells.

evolution   Any genetically based phenotypic change in a population of organisms over successive generations. Evolution can also be thought of as the process by which species of living things can undergo modification over successive generations, with such modification sometimes resulting in the formation of new species.

exocytosis   The process in which a transport vesicle fuses with the plasma membrane of a cell and the contents of the vesicle are ejected outside the cell. Exocytosis can be used to expel waste products in single-celled organisms or to export protein products in multicelled organisms.

exoskeleton   An external material covering the animal body, providing support and protection.

exponential growth   A form of population growth in which the rate of growth increases with time. Exponential growth results in a J-shaped growth curve because, when plotted on a graph, the population's increase resembles the letter J.

exponential increase   An increase in numbers that is proportional to the number already in existence. This type of increase occurs in populations of living things, and it carries the potential for enormous growth of populations.

F

facilitated diffusion   The passage of materials through the cell's plasma membrane, aided by a concentration gradient and a transport protein.

family   A taxonomic grouping of related genera. This category is subordinate to order and superordinate to genus.

fatty acid   A molecule found in many lipids that is composed of a hydrocarbon chain bonded to a COOH group.

female fertility   The average number of children born to each woman in a population. The most important statistic used in predicting human population changes. The "replacement" level of female fertility in developed countries is 2.1.

fetus   An organism at a later stage of development. In humans, the developing organism from the start of the ninth week of development to the moment of birth.

filament   The part of a stamen (male reproductive part of a flower) that is shaped like and functions as a stalk, and has an anther at the top.

first law of thermodynamics   Energy cannot be created or destroyed, but only transformed from one form to another. The process of transformation of energy is never totally efficient, and some energy is always converted into heat.

fitness   The success of an organism, relative to other members of its population, in passing on its genes to offspring. Fitness is a relative concept only; some organisms are better than others at passing on their genes in a given environment at a given point in time.

fixation   The process of a gas being incorporated into an organic molecule.

flagella (singular, *flagellum*)   The relatively long, tail-like extensions of some cells, composed of microtubules, that function in cell movement.

Often cells will have but a single flagellum, as with mammalian sperm cells.

formed elements   Cells and cell fragments that form the non-fluid portion of blood. Formed elements include red blood cells, white blood cells, and platelets. Contrast with plasma, the fluid portion of blood, consisting mostly of water.

founder effect   The phenomenon by which an initial gene pool for a population is established by means of that population migrating to a new and isolated area. One of the conditions that potentiates genetic drift.

fraternal twins   Twins who are produced first through multiple ovulations in the mother, followed by multiple fertilizations from separate sperm of the father, and then by multiple implantations of the resulting embryos in the uterus.

fruit   The mature ovary of any flowering plant. Many fruits protect the underlying seeds, and many attract animals that will eat the fruit and disperse the seeds.

G

gallbladder   Organ of the body that stores and concentrates the digestive material bile, which is produced by the liver. Bile facilitates the breakdown of fats by digestive enzymes.

gamete   A haploid reproductive cell, either egg or sperm.

gametophyte generation   The generation in a plant's life cycle that produces gametes (sperm or egg). This generation is microscopic in angiosperms. Contrast with sporophyte generation.

gene flow   The movement of genes from one population to another.

gene pool   The entire collection of alleles in a population.

genetic code   The inventory that specifies which nucleotide triplets code for which particular amino acids, or for start or stop commands in protein synthesis. With few exceptions, the genetic code is universal in living things.

genetic drift   The chance alteration of allele frequencies in a population, with such alterations having greatest impact on small populations.

genetics   The study of physical inheritance among living things.

genome   The complete haploid set of an organism's chromosomes.

genotype   The genetic makeup of an organism, including all the genes that lie along its chromosomes.

genus   A taxonomic grouping of related species. This category is subordinate to family and superordinate to species. Humans are in the genus *Homo.*

germ-line cell   A cell that becomes an egg or a sperm cell.

gland   An organ or group of cells that secretes one or more substances.

glomerulus   Knotted network of capillaries in each of the kidneys' nephrons that receives blood and lets some smaller blood-borne materials pass out of it (and into the surrounding Bowman's capsule) while retaining larger materials.

glycocalyx   An outer layer of the plasma membrane, composed of short carbohydrate chains that attach to membrane proteins and phospholipid molecules. Such chains serve as the actual binding sites on many membrane proteins, act to lubricate the cell, and can form an adhesion layer that allows one cell to stick to another.

glycogen   A complex carbohydrate that serves as the primary form in which carbohydrates are stored in animals.

glycolysis   The first stage of cellular respiration, occurring in the cytosol. For some organisms, glycolysis is the sole means of extracting energy from food. In most organisms, it is a means of extracting some energy and a necessary precursor to the other two stages of cellular respiration, the Krebs cycle and the electron transport chain.

glycoprotein   A molecule that combines protein and carbohydrate. Glycoproteins play important roles as cell receptors and some types of hormones, among other functions.

Golgi complex   A network of membranes, found in the cytoplasm of eukaryotic cells, that processes and distributes proteins that come to it from the rough endoplasmic reticulum.

gravitropism   The bending of a plant's roots or shoots in response to gravity. This capability helps a plant orient roots and shoots properly—roots toward the center of the Earth, shoots away from it.

groundwater   Water that is found underground in porous rock saturated with water.

gymnosperm   A seed plant whose seeds do not develop within fruit. One of the four principal varieties of plants, gymnosperms reproduce through wind-aided pollination. Coniferous trees, such as pine and fir, are the most familiar examples.

H

habitat   The surroundings in which individuals of a species are normally found.

haploid   Possessing a single set of chromosomes. Human gametes (eggs and sperm) are haploid cells, because they have only a single set of chromosomes. All other cells in the human body are diploid, meaning they possess two sets of chromosomes. Some organisms are strictly haploid. Bacteria, for example, have only a single chromosome, making them haploid.

HDL (high-density lipoprotein)   A type of protein that transports fat or lipid molecules (usually cholesterol) from various tissues in the body to the liver. Sometimes known as the "good cholesterol," HDLs help remove, and possibly neutralize, the LDLs or low-density lipoproteins that can harmfully reside in coronary arteries.

heart attack   A complete blockage of one of the heart's coronary arteries, resulting in the death of groups of heart cells from lack of a blood supply.

helper T cell   Type of T lymphocyte cell that stimulates both T-cell and B-cell immunity. Referred to in AIDS therapy as a CD-4 cell.

hemoglobin   The iron-containing protein in red blood cells that binds to both oxygen and carbon dioxide, thus assisting in their transportation.

herbivore   An animal that eats only plants.

hermaphrodite   An animal that possesses both male and female sex organs.

heterozygous   Possessing two different alleles of a gene for a given character.

histamine   A compound that, in the immune system's inflammatory response, brings about blood-vessel dilation and increased blood-vessel permeability.

homeostasis   The maintenance of a relatively stable internal environment in living things.

homologous   In anatomy, having the same structure owing to inheritance from a common ancestor. Forelimb structures in whales, bats, cats, and gorillas are homologous.

homologous chromosomes   Chromosomes that are the same in size and function. Species that are diploid (have two sets of chromosomes) have matching pairs of homologous chromosomes; one member of each homologous pair is inherited from the male, the second member of each homologous pair is inherited from the female.

homology   A structure that is shared in different organisms owing to inheritance from a common ancestor. Homologies are used to help decipher evolutionary relationships.

homozygous   Having two identical alleles of a gene for a given character.

hormone   A substance that, when released in one part of an organism, goes on to prompt physiological activity in another part of the organism. Both plants and animals have hormones.

host   The prey in a parasitic relationship.

hydrocarbon   A compound made of hydrogen and carbon. Hydrocarbons are nonpolar covalent molecules and therefore are not easily dissolved in water.

hydrogen bond   A chemical bond that links an already covalently bonded hydrogen atom with a second, relatively electronegative atom.

hydrophilic   Property of a compound indicating that it will interact with water. Table salt ($NaCl$), which dissolves readily in water, is hydrophilic.

hydrophobic   Property of a compound indicating that it will not interact with water. Oil is hydrophobic and will not readily dissolve in water.

hydrostatic skeleton   A skeleton in which the force of muscular contraction is applied through a fluid.

hydroxide ion   The OH^- ion. Compounds that yield hydroxide ions are strongly basic, so they can be used to counteract acids and shift solutions toward neutral or basic on the pH scale.

hypertonic solution   A solution that has a higher concentration of solutes than another.

hyphae (singular, *hypha*)   The slender filaments that make up the bulk of most fungi.

hypothalamus   Portion of the brain important in drives, and in the maintenance of homeostasis, the latter capacity coming about because of the hypothalamus' control over a good deal of the body's hormonal release.

hypothesis   A tentative, testable explanation of an observed phenomenon.

hypotonic solution   A solution that has a lower concentration of solutes than another.

I

identical twins Twins who develop from a single fertilized egg.

immune system A collection of cells and proteins whose central function is to rid the body of invading microbes.

immunity A state of long-lasting protection that the immune system develops against specific microorganisms.

independent assortment The random distribution of homologous chromosome pairs during meiosis.

inner cell mass In mammalian development, the group of cells in the embryo that will develop into the baby, rather than into the placenta.

integumentary system The organ system that protects the body from the external environment and assists in regulation of body temperature. This system consists of the skin and associated structures, such as glands, hair, and nails.

interneuron A type of neuron, located only within the brain or spinal cord, that connects other neurons. These neurons are responsible for the analysis of sensory inputs and the coordination of motor commands.

interphase That portion of the cell cycle in which the cell simultaneously carries out its work and—in preparation for division—duplicates its chromosomes. The other primary phase of the cell cycle is mitotic phase (or M phase), which includes both mitosis (in somatic cells) and cytokinesis.

intertidal zone The region within the coastal zone of the ocean that extends from the ocean's low-tide mark to its high-tide mark.

intrinsic isolating mechanism A difference in anatomy, physiology, or behavior that prevents interbreeding between individuals of the same species or of closely related species. One or more intrinsic isolating mechanisms must exist for two populations of the same species to begin the process of evolving into separate species.

invertebrate An animal without a vertebral column.

ion An atom that has positive or negative charge because it has fewer or more electrons than protons. Ions will be attracted to other ions with the opposite charge, thus forming ionic compounds.

ionic bonding A linkage in which two or more ions are bonded to each other by virtue of their opposite charge.

isotonic solution A solution that has the same concentration of solutes as another.

isotope A form of an element as defined by the number of neutrons contained in its nucleus. Different isotopes of an element have varying numbers of neutrons but the same number of protons.

K

karyotype A pictorial arrangement of a full set of an organism's chromosomes.

keratin A flexible, water-resistant protein, abundant in the outer layers of skin, that also makes up hair and fingernails.

keystone species A species whose impact on the composition of a community is large, relative to its abundance within that community.

kidneys The filtering organs of the urinary system, which produce urine while conserving useful blood-borne materials.

kingdom A taxonomic grouping superordinate to every other grouping except domain. There are four kingdoms in domain Eukarya: protista, fungi, animalia, and plantae.

Krebs cycle The second stage of cellular respiration, occurring in the inner compartment of mitochondria. The Krebs cycle is the major source of electrons for the third stage of respiration, the electron transport chain.

K-selected species A species that tends to be relatively long-lived, that tends to have relatively few offspring for whom it provides a good deal of care, and whose population size tends to be relatively stable, remaining at or near its environment's carrying capacity (K). Also known as an equilibrium species.

L

labor The regular contractions of the uterine muscles that sweep over the fetus, creating pressure that opens the cervix and expels the baby and the placenta.

lactate fermentation The process in animal cells by which pyruvic acid, the product of glycolysis, accepts electrons from NADH to form lactic acid.

large intestine That portion of the digestive tract that begins at the small intestine and ends at the anus. The large intestine serves largely to compact and store material left over from the digestion of food, turning this material into the solid waste known as feces.

Law of Independent Assortment During gamete formation, gene pairs assort independently of one another. Also known as Mendel's Second Law, this is one of the principles of inheritance formulated by Gregor Mendel.

Law of Segregation Differing characters in organisms result from two genetic elements (alleles) that separate in gamete formation, such that each gamete gets only one of the two alleles. Also known as Mendel's First Law, this is one of the principles of inheritance formulated by Gregor Mendel.

LDL (low-density lipoprotein) A type of protein that transports fat or lipid molecules (usually cholesterol) from the liver and small intestine to various tissues throughout the body. Sometimes known as the "bad cholesterol," LDLs can initiate heart disease by coming to reside within coronary arteries.

lichen A composite organism composed of a fungus and either algae or photosynthesizing bacteria.

ligament A type of tissue in the human body that links one bone to another.

light reactions The first set of chemical reactions in photosynthesis, in which solar energy strips water of electrons and powers the process by which those electrons are transferred to the primary electron acceptors of photosystems II and I.

lipid A type of biological molecule whose defining characteristic is its relative insolubility in water. Examples include triglycerides, cholesterol, steroids, and phospholipids.

lipoprotein A molecule that combines lipid and protein. Lipoproteins transport fat molecules through the bloodstream to all parts of the body.

liver A large, reddish-brown organ that is central to the body's metabolism of nutrients.

logistic growth In ecology, growth of a population in which the rate of growth slows and finally ceases altogether, in response to environmental resistance.

lysosome An organelle found in animal cells that digests worn-out cellular materials and foreign materials that enter the cell.

M

macroevolution Evolution that results in the formation of new species or other groupings of living things.

meiosis A process in which a single diploid cell divides to produce four haploid reproductive cells. Meiosis produces the gametes for sexual reproduction.

membrane potential The electrical charge difference that exists from one side of a neuron's plasma membrane to the other.

memory B cell One of two types of cells that the B-lymphocyte or B cell develops into (the other type being a plasma cell). Memory B cells remain in the system long after a first infection by an organism has ended and serve to produce more plasma B cells quickly, should the organism ever invade again.

menopause The cessation of the monthly ovarian cycle that occurs when women reach about 50 years of age.

menstruation The release of blood and the specialized uterine lining, the endometrium; occurs about once every 28 days in human females, except when the oocyte is fertilized.

messenger RNA (mRNA) A type of RNA that encodes, and carries to ribosomes, information for the synthesis of proteins.

metabolic pathway A sequential set of enzymatically controlled reactions in which the product of one reaction serves as the substrate for the next.

metabolism The sum of all chemical reactions carried out by a cell or larger organism.

microevolution A change of allele frequencies in a population over a short period of time. The basis for all large-scale or macroevolution.

migration A regular movement of animals from one location to a distant location. Also, the movement of individuals from one population into the territory of another population. Migration is the basis of gene flow among populations.

mimicry A phenomenon in which one species evolves to resemble another species.

mitochondria (singular, mitochondrion) Organelles that are the primary sites of energy conversion within eukaryotic cells.

mitosis The separation of a somatic cell's duplicated chromosomes prior to cytokinesis.

mitotic phase (M phase) That portion of the cell cycle that includes both mitosis and cytokinesis. Mitosis is the separation of a somatic cell's duplicated chromosomes; cytokinesis is the physical separation of one cell into two daughter cells.

mitotic spindle   The microtubules active in cell division, including those that align and move the chromosomes.

molecular biology   The investigation of life at the level of its individual molecules.

molecule   A structure with a defined number of atoms in a particular spatial arrangement. The atoms and their arrangement determine how the molecule interacts with other molecules.

molting   A periodic shedding of an old skeleton.

monomer   A small molecule that can be combined with other similar or identical molecules to make a larger polymer.

monosaccharide   A building block or monomer of carbohydrates. Monosaccharides combine to form complex carbohydrates, or polysaccharides. Glucose is an example of a monosaccharide.

monounsaturated fatty acid   A fatty acid with one double bond between the carbon atoms of its hydrocarbon chain.

morphology   The physical form of an organism.

motor neuron   A neuron that carries instructions from the central nervous system to an effector (muscle or gland) in the body.

Müllerian mimicry   A type of mimicry in which several species that have protection against predators evolve to look alike.

multiple alleles   Three or more alleles—alternative forms of a gene—occurring in a population.

muscle tissue   A tissue that has the ability to contract. In humans, one of the four principal types of tissue.

muscular system   The organ system that produces movement and maintains posture of the body. This system consists of all the skeletal muscles of the body that are under voluntary control.

mutation   A permanent alteration in a DNA base sequence.

mutualism   A form of relationship between two organisms in which both organisms benefit.

mycelium   A web of fungal hyphae that makes up the major part of a fungus.

mycorrhizae   Associations of plant roots and fungal hyphae. The fungal hyphae absorb minerals, growth hormones, and water that are then available to the plant, and the fungus gets carbohydrates from the photosynthesizing plant.

myelin   Membranous covering of some neuronal axons, provided by glial cells, that allows faster nerve-signal transmission through these axons.

N

natural selection   A process in which the differential adaptation of organisms to their environment selects those traits that will be passed on with greater frequency from one generation to the next.

negative feedback   A system of control in which the product of a process reduces the activity that led to the product.

nephron   Functional unit of the kidneys, composed of a nephron tubule, its associated blood vessels, and the interstitial fluid in which both are immersed.

nerve   A bundle of axons in the peripheral nervous system that transmits information to or from the central nervous system.

nervous system   The organ system that monitors an animal's internal and external environment, integrates the sensory information received, and coordinates the animal's responses. This system consists of all the body's neurons, plus the supporting neuroglia cells, plus the sensory organs.

nervous tissue   A tissue specialized for the rapid conduction of electrical impulses. In humans, one of the four principal tissue types.

neurotransmitter   A chemical, secreted into a synaptic cleft by a neuron, that affects another neuron or an effector by binding with receptors on it.

neutron   A basic constituent of an atom, possessing no electrical charge and found in the atom's nucleus. Isotopes are defined by the number of neutrons in an atom.

niche   A characterization of an organism's way of making a living that includes its habitat, food, and behavior.

nicotinamide adenine dinucleotide (NAD)   The most important intermediate electron carrier in cellular respiration.

nitrogen fixation   The conversion of atmospheric nitrogen into a form that can be taken up by living things. Bacteria fix nitrogen, which is essential to life.

nondisjunction   The failure of homologous chromosomes or sister chromatids to separate during meiosis, resulting in unequal numbers of chromosomes in the daughter cells. Nondisjunction results in aneuploidy.

nonspecific defenses   Immune-system defenses that do not discriminate between one invader and the next.

nuclear envelope   The double membrane that lines the nucleus in eukaryotic cells.

nucleotide   The building block of nucleic acids, including DNA and RNA, consisting of a phosphate group, a sugar, and a nitrogen-containing base.

nucleus (atomic)   The core of an atom, containing its protons and neutrons. Nearly all of the mass of the atom resides in its nucleus.

nucleus (cell)   The membrane-lined compartment that encloses the primary complement of DNA in eukaryotic cells.

nutrient   A chemical element that is used by living things to sustain life.

O

oil   Fat in liquid form.

omnivore   An animal that eats both plants and animals.

oocyte   The precursor cell of eggs in the vertebrate ovary. Diploid primary oocytes develop from oogonia and in turn give rise to haploid secondary oocytes in the process of meiosis.

open circulation   A type of circulation, found in many invertebrate animals, in which circulatory fluid does not stay confined in vessels but flows into open spaces, called sinuses, where it bathes tissues before being returned to vessels.

order   A taxonomic grouping subordinate to class and superordinate to family. Humans are in the order Primates.

organ   A highly organized unit within an organism, performing one or more functions, that is formed of several kinds of tissue. Kidneys, heart, lungs, and liver are all familiar examples of organs in humans.

organelle   A highly organized structure within a cell that carries out specific cellular functions. Almost all organelles (meaning "tiny organs") are bound by membranes. The lone exception is the ribosome, which has no membrane and is the single organelle possessed by prokaryote cells (bacteria and archaea). Organelles in eukaryotic cells include the cell nucleus, mitochondria, lysosomes, chloroplasts, and ribosomes.

organic chemistry   A branch of chemistry concerned with compounds that have carbon as their central element.

organ system   A group of interrelated organs and tissues that serve a particular set of functions in the body. For example, the digestive system consists of mouth, stomach, and intestines, and functions in digesting food and eliminating waste.

osmosis   The net movement of water across a semipermeable membrane from an area of lower solute concentration to an area of higher solute concentration.

ovarian follicle   The complex of the oocyte (developing egg) and a set of accessory cells that surrounds it.

ovary   In flowering plants, the area, located at the base of the carpel, where fertilization of the egg and early development of the embryo occur. In animals, the female reproductive organ in which eggs develop.

oviparous   A condition, seen in all birds and many reptiles, in which fertilized eggs are laid outside the mother's body and then develop there.

ovulation   The release of an oocyte from the ovary in animals.

oxidation   Loss of one or more electrons by an atom or a molecule. Important in energy transfer in living things as part of redox reactions, in which one substance undergoes reduction (a gain in electrons) by oxidizing another.

ozone   A gas in the Earth's atmosphere consisting of three oxygen atoms bonded together that serves to protect living things from the sun's ultraviolet radiation.

P

pancreas   In digestion, a gland that secretes, into the small intestine, digestive enzymes, along with buffers that raise the pH of chyme. In nutrient metabolism, a gland that secretes the hormones insulin and glucagon.

parasitism   A type of predation in which the predator gets nutrients from the prey, but does not kill the prey immediately and may never kill it.

parasympathetic division   The division of the autonomic nervous system that generally has relaxing effects on the body.

parthenogenesis   A form of asexual reproduction in which an unfertilized egg develops into an adult organism.

passively acquired immunity   Immunity gained by the administration of antibodies produced by another individual (e.g., the administration of a gamma globulin shot).

passive transport   Transport of materials across the cell's plasma membrane that involves no expenditure of energy. Simple and facilitated diffusion are examples of passive transport.

pathogenic   Disease-causing. Viruses, bacteria, protists, and fungi are spoken of as being pathogenic or nonpathogenic.

pedigree   A familial history. Pedigrees can be created, generally in the form of diagrams, that trace the history of inheritable diseases through a family.

peptide hormones   Hormones composed of chains of amino acids, with these chains ranging from small polypeptides to proteins composed of hundreds of amino acids. One of three principal classes of hormones, the other two being amino-acid-based and steroid hormones.

peripheral nervous system (PNS)   The part of the nervous system that includes all of the neural tissue outside the central nervous system (brain and spinal cord). The PNS brings information to and carries it from the central nervous system; it also provides voluntary control of the skeletal muscles and involuntary control of the smooth muscles, cardiac muscles, and glands.

peristalsis   Waves of contraction carried out by two sets of muscles in the digestive tract that push material through the tract.

Permian Extinction   The greatest mass-extinction event in Earth's history. This event, in which up to 96 percent of all species on Earth were wiped out, occurred about 245 million years ago.

phagocyte   An immune-system cell capable of ingesting another cell, parts of cells, or other materials. Phagocytes ingest both invading microorganisms and tissue fragments or the body's own cells when they have become damaged.

phagocytosis   A process of bringing relatively large materials into a cell by means of wrapping extensions of the plasma membrane around the materials to be brought in and fusing the extensions together.

phenotype   A physical function, bodily characteristic, or behavior of an organism.

phloem   The tissue through which the food produced in photosynthesis, along with some hormones and other compounds, is conducted in vascular plants.

phosphate group   A phosphorus atom surrounded by four oxygen atoms.

phospholipid   A charged lipid molecule composed of two fatty acids, glycerol, and a phosphate group. The phospholipid's phosphate group is hydrophilic, while its fatty-acid chains are hydrophobic. Phospholipids are a major constituent of cell membranes.

phospholipid bilayer   A chief component of the plasma membrane, composed of two layers of phospholipids, arranged with their fatty-acid chains pointing toward each other.

photoperiodism   The ability of a plant to respond to changes it experiences in the daily duration of darkness, relative to light.

photosynthesis   The process by which certain groups of organisms capture energy from sunlight and convert this solar energy into chemical energy that is initially stored in a carbohydrate.

photosystem   An organized complex of molecules within a thylakoid membrane that, in photosynthesis, collects solar energy and transforms it into chemical energy.

phototropism   The bending of a plant's shoots in response to light. Generally, this capability helps a plant grow toward the sun to get the most available sunlight.

pH scale   A scale utilized in measuring the relative acidity or alkalinity of a solution. The scale, ranging from 0 to 14, quantifies the concentration of hydrogen ions in a solution. The lower the pH number, the more acidic the solution; the higher the number, the more basic the solution.

phylogeny   A hypothesis about the evolutionary relationships of a group of organisms.

phylum   A category of living things, directly subordinate to the category of kingdom, whose members share traits as a result of shared ancestry.

phytoplankton   Small photosynthesizing organisms that drift in the upper layers of oceans or bodies of freshwater, often forming the base of aquatic food webs.

pinocytosis   A form of endocytosis that brings into the cell a small volume of extracellular fluid and the solutes suspended in it.

placenta   A complex network of maternal and embryonic blood vessels and membranes that develops in mammals in pregnancy. The placenta allows nutrients and oxygen to flow to the embryo from the mother, while allowing carbon dioxide and waste to flow from the embryo to the mother.

plasma   The fluid portion of blood, consisting mostly of water but also containing proteins and other molecules. Contrast with formed elements—the cells and cell fragments that form the non-fluid portion of blood.

plasma cell   One of two types of cells the B-lymphocyte develops into (the other type being a memory B cell). Plasma cells are specialized to produce the free-standing antibodies that fight invaders.

plasma membrane   A membrane forming the outer boundary of many cells, composed of a phospholipid bilayer interspersed with proteins and cholesterol molecules and coated, on its exterior face, with short carbohydrate chains associated with proteins and lipids.

plasmid   A ring of DNA that lies outside the chromosome in bacteria. Plasmids can move into bacterial cells in the process called transformation, thus making them a valuable tool in biotechnology.

platelet   One of the three varieties of formed elements within blood, platelets are small fragments of cells that facilitate blood clotting by means of releasing clotting enzymes and clumping together at the site of an injury.

point mutation   A mutation of a single base-pair in a genome.

polar covalent bond   A type of covalent bond in which electrons are shared unequally between atoms, so that one end of the molecule has a slight negative charge and the other end a slight positive charge.

polarity   A difference in electrical charge at one end of a molecule, as compared to the other.

polygenic inheritance   Inheritance of a genetic character that is determined by the interaction of multiple genes, with each gene having a small additive effect on the character.

polymer   A large molecule made up of many similar or identical subunits, called monomers.

polymerase chain reaction (PCR)   A technique for generating many copies of a DNA sequence.

polypeptide   A series of amino acids linked in linear fashion. Polypeptide chains fold up to become proteins.

polyploidy   A process by which one or more sets of chromosomes are added to the genome of an organism. Human beings cannot survive in a polyploid state, but many plants flourish in it. Polyploidy is a means by which speciation can occur (most often in plants) in a single generation.

polysaccharides   The polymers of carbohydrates, composed of many monosaccharides. Examples are starch and cellulose.

polyunsaturated fatty acid   A fatty acid with two or more double bonds between the carbon atoms of its hydrocarbon chain.

population   All the members of a species that live in a defined geographic region at a given time.

predation   One organism feeding on parts or all of a second organism.

primary consumer   Any organism that eats producers (organisms that make their own food).

primary succession   In ecology, a succession in which the starting state is one of little or no life and no soil.

producer   Any organism that manufactures its own food. Plants, algae, and certain bacteria are producers. By converting the sun's energy into biomass, producers capture energy that is then passed along in food webs.

product   A substance formed in a chemical reaction. The products are written on the right side of a chemical equation.

prokaryote   A single-celled organism whose complement of DNA is not contained within a nucleus.

prokaryotic cell   A cell whose DNA is not located in the membrane-bound organelle known as a nucleus. Prokaryotes are microscopic forms of life, and all are either bacteria or archaea.

prostate gland   In human males, a gland surrounding the urethra near the urinary bladder that contributes fluids to the semen.

protein   A large polymer of amino acids, composed of one or more polypeptide chains. Proteins come in many forms, including enzymes, structural proteins, and hormones.

proton   A basic constituent of an atom, found in the nucleus of the atom and having positive electrical charge. Elements are defined by the number of protons in their nucleus.

pulmonary circulation The system that circulates blood between the heart and the lungs. This system brings oxygen into and takes carbon dioxide away from the body.

R

radial symmetry A type of animal symmetry in which body parts are distributed evenly about a central axis. Sea stars are radially symmetrical.

radiometric dating A technique for determining the age of objects by measuring the decay of the radioactive elements within them. The age of fossils can be determined with this technique.

reaction center A molecular complex in a chloroplast that, in photosynthesis, transforms solar energy into chemical energy.

receptor protein A plasma membrane protein that binds with a specific signaling molecule.

receptor-mediated endocytosis (RME) A form of endocytosis in which receptors on the surface of a cell bind to a substance and then move laterally across the plasma membrane, joining other similarly bound receptors of their type at a cellular depression that pinches off and moves into the cell.

recessive Term used to designate an allele that is not expressed in the heterozygous condition.

recessive disorder A condition that will not occur when an organism possesses a single functional allele for a given trait. Red-green color blindness is an example of a recessive disorder, because only one set of functional alleles need be present for normal color vision.

recombinant DNA Two or more segments of DNA that have been combined by humans into a sequence that does not exist in nature.

red blood cell (RBC) The blood cells, also known as erythrocytes, that transport oxygen to and carry carbon dioxide from every part of the body.

redox reaction A combination of a reduction and an oxidation reaction in which the electrons lost from one substance in oxidation are gained by another in reduction.

reduction Gain of one or more electrons by an atom or a molecule. Important in energy transfer in living things as part of redox reactions, in which one substance is reduced by oxidizing (or removing electrons from) another substance.

reflex Automatic nervous-system response that helps an organism avoid danger or preserve a stable physical state. The knee-jerk response is a well-known reflex.

regulatory T cell A type of immune-system cell that acts to limit the body's immune-system response, thus protecting the body's own tissues from attack.

reproductive cloning The cloning of whole, complex living things.

reproductive isolating mechanism Any factor that, in nature, prevents interbreeding between individuals of the same or closely related species. These factors keep species separate.

reproductive system The organ system that develops gametes and delivers them to a location where they can fuse with other gametes to produce a new individual.

resource partitioning The dividing of scarce resources among species that have similar requirements. Such partitioning allows species to coexist in the same habitat.

respiration The exchange of gases between the atmosphere outside the body and the cells within it.

respiratory system The organ system that brings oxygen into the body and expels carbon dioxide from the body. In humans, this system includes the lungs and passageways that carry air to the lungs.

restriction enzyme A type of enzyme, occurring naturally in bacteria, that recognizes a specific series of DNA bases and cuts the DNA strand at that site. Restriction enzymes are used in biotechnology to cut DNA in specific places.

retina An inner layer of tissue in the eye, containing cells that convert light signals into neural signals.

ribosomal RNA (rRNA) A type of RNA that, along with proteins, forms ribosomes.

ribosome An organelle, located in the cell's cytoplasm, that is the site of protein synthesis. The translation phase of protein synthesis takes place within ribosomes.

RNA (ribonucleic acid) A nucleic acid that is active in the synthesis of proteins and that forms part of the structure of ribosomes. Varieties include messenger RNA (mRNA), transfer RNA (tRNA), and ribosomal RNA (rRNA).

RNA polymerase The enzyme that unwinds the DNA double helix and puts together a chain of RNA nucleotides complementary to the exposed DNA nucleotides.

root hair In plants, a thread-like extension of a root cell. Root hairs greatly increase the surface area of roots, thus allowing greater absorption of water and nutrients.

rough endoplasmic reticulum (rough ER or RER) A network of membranes, found in the cytoplasm of eukaryotic cells, that aids in the processing of proteins.

r-selected species A species that tends to be relatively short-lived, that tends to produce relatively many offspring for whom it provides little or no care, and whose population size tends to fluctuate widely in reaction to an environment that it experiences as highly variable. Also known as an opportunist species.

S

sarcomere The functional unit of a striated muscle, which contracts when thin filaments slide past thick filaments. The sarcomeres shorten, thus contracting the whole muscle.

saturated fatty acid A fatty acid with no double bonds between the carbon atoms of its hydrocarbon chain.

science A process of learning about nature by observation and experiment, as well as a collection of knowledge and insights about nature.

scientific method The process by which scientists investigate the natural world. The scientific method involves the testing of hypotheses through observation and experiment, as aided by the tools of statistics.

second law of thermodynamics Energy transfer always results in a greater amount of disorder (entropy) in the universe.

secondary consumer Any organism that eats a primary consumer.

secondary succession In ecology, a succession that begins in a habitat that has been disturbed by some force, but in which life remains and soil exists.

seed A reproductive structure in plants that includes a plant embryo, its food supply, and a tough protective casing.

seedless vascular plant A plant that has a vascular (fluid transport) structure but does not reproduce through use of seeds. One of the four principal varieties of plants. Ferns are the most familiar example.

semen The mixture of sperm and glandular materials that is ejaculated from the human male through the urethra.

seminiferous tubule A convoluted tubule inside the testes where sperm development begins. Immature sperm are eventually released into the interior cavity and travel to the epididymis, where sperm maturation is completed.

sensory neuron A neuron that senses conditions both inside and outside the body and conveys that information to neurons in the central nervous system.

sessile Fixed in location; organisms such as mushrooms and sponges are sessile.

sex chromosome The chromosomes that determine the sex of an organism. The X or Y chromosomes in humans.

sexual selection In evolution, a form of natural selection that produces differential reproductive success based on differential success in obtaining mating partners.

simple diffusion Diffusion through the plasma membrane that requires only concentration gradients, as opposed to concentration gradients and special protein channels. Water, oxygen, carbon dioxide, and steroid hormones can all cross the plasma membrane through simple diffusion.

simple sugar The building block of carbohydrates. Also known as a monosaccharide; glucose and fructose are familiar examples.

skeletal muscle In the human body, muscle that is attached to bone and that is under conscious control.

skeletal system The animal organ system that forms an internal supporting framework for the body and protects delicate tissues and organs. This system consists of all the bones and cartilages in the body, and the connective tissues and ligaments that connect the bones at the joints.

skin An organ consisting of two layers, epidermis and dermis, and covering the outside of an animal. The skin protects the body and receives signals from the environment.

small intestine That portion of the digestive tract that runs between the stomach and large intestine.

smooth endoplasmic reticulum (smooth ER or SER) A network of membranes within the cell that is the site of the synthesis of various lipids, and the site at which potentially harmful substances are detoxified.

smooth muscle In the human body, muscle that is not under conscious control and whose fibers are not striped or "striated" in appearance. Smooth muscle lines, for example, blood vessels and the digestive tract.

solute The substance being dissolved in a solvent to form a solution. For example, sugar is the solute in the sugar-water nectar you put in your hummingbird feeder.

solution A homogeneous mixture of two or more substances in the same phase (gas, liquid, or solid). Frequently, solutions consist of a solute dissolved in water, and these are called aqueous solutions.

solvent The substance in which a solute is dissolved to form a solution. In an aqueous solution, the solvent is water.

somatic cell Any cell that is not and will not become an egg or sperm cell.

somatic nervous system That portion of the peripheral nervous system's efferent division that provides voluntary control over skeletal muscle.

speciation The development of new species through evolution.

species Groups of actually or potentially interbreeding natural populations which are reproductively isolated from other such groups.

specific defenses Immune-system defenses that provide protection against particular invaders.

specific heat The amount of energy needed to raise the temperature of 1 gram of a substance by 1 degree Celsius.

spermatocyte An immature sperm cell. Spermatocytes develop from spermatogonia; they develop into spermatids and eventually into mature sperm.

spermatogonia Diploid cells that are the starting cells in sperm production in males. Each division of a spermatogonium produces a spermatocyte—a precursor to sperm—along with another spermatogonium.

sperm cell In flowering plants, either of two cells in a pollen grain, one of which fertilizes an egg, the other of which fertilizes the central cell in an embryo sac. In animals, the male gamete, which fertilizes the female gamete (the egg).

spongy bone Type of bone that is porous and less dense than compact bone. Spongy bone fills the expanded ends of long bones.

spore A reproductive cell that can develop into a new organism without fusing with another reproductive cell. (The term *spore* also refers to a dormant, stress-resistant form of a bacterial or fungal cell.)

sporophyte generation The spore-producing plant generation. This generation is the dominant, visible generation in flowering plants. Contrast with *gametophyte generation*.

stabilizing selection In evolution, the type of natural selection in which intermediate forms of a given character are favored over either extreme.

This process tends to maintain the average for the character. Compare to *directional* and *disruptive selection*.

stamen The male reproductive structure in a flower; consists of a filament (stalk) with an anther, bearing pollen, at the top.

starch A complex carbohydrate that serves as the major form of carbohydrate storage in plants. Starches—found in such forms as potatoes, rice, carrots, and corn—are important sources of food for animals.

stem cells Cells with the capacity to produce more cells of their own type, along with at least one type of specialized daughter cell.

steroids A class of lipid molecules that have, as a central element in their structure, four carbon rings. One steroid differs from another in accordance with the varying side chains that can be attached to these rings. Examples include cholesterol, testosterone, and estrogen.

steroid hormones Hormones constructed around the chemical framework of the cholesterol molecule. One of three principal classes of hormones, the other two being peptide and amino-acid-based hormones.

stigma The tip end of the carpel of a flower, where pollen grains are deposited before fertilization occurs.

stomach An organ that performs digestion and that serves as a temporary, expandable storage site for food.

stomata Microscopic pores in the epidermal cells of plants, found mostly on the underside of leaves. Carbon dioxide enters and water vapor leaves through these openings.

stratosphere The layer of the Earth's atmosphere situated above the troposphere, at about 20 to 35 kilometers (13 to 21 miles) above sea level. The ozone layer lies in this level.

stroma The liquid material of chloroplasts.

style The slender tube structure in the carpel of a flower, connecting the stigma and the ovary.

substrate The substance that is worked on by an enzyme.

succession In ecology, a series of replacements of community members at a given location until a stable final state is reached.

symmetry An equivalence of size, shape, and relative position of parts across a dividing line or around a central point.

sympathetic division The division of the autonomic nervous system that generally has stimulatory effects on the body.

sympatric speciation A type of speciation that occurs without geographic separation of populations. Polyploidy is one form of sympatric speciation.

synapse An area in the nervous system in which one neuron communicates either with another neuron or with an effector, such as a muscle cell. The synapse consists of the sending neuron, the receiving cell, and a tiny gap between them called the synaptic cleft.

synaptic cleft Tiny gap that exists between a neuron sending a nervous-system signal and a neuron (or effector cell) that is receiving this signal.

systemic circulation One of the body's two general networks of blood vessels, the systemic circulation transports blood between the heart and the rest of the body, following oxygenation of this blood in the body's pulmonary circulation.

T

taiga A northern latitude biome characterized by low rainfall, cold winter temperatures, and the ecological dominance of a few types of coniferous trees.

target cells Cells that can be affected by a given hormone owing to their capacity to bind with that hormone.

temperate deciduous forest A biome characterized by seasonal growth patterns of both trees and an understory of woody and herbaceous plants, and relatively abundant rainfall—between 30 and 78 inches per year.

temperate grassland A biome whose ecological dominants tend to be seasonal grasses, rather than trees.

tendon The connective tissue that attaches a skeletal muscle to a bone.

tertiary consumer Any organism that eats secondary consumers.

testis The organ in the male reproductive system in which sperm begin development and testosterone is produced.

thalamus Portion of the brain through which most sensory perceptions are channeled before being relayed to the cerebral cortex.

theory A general set of principles, supported by evidence, that explains some aspect of nature.

therapeutic cloning The use of cloning to produce human embryonic stem cells that can be used to treat disease.

thigmotropism The growth of a plant in response to touch. This capability allows tendrils to wrap around other objects, thus helping a plant climb upward toward light.

thylakoids Membrane-bound sacs found in the interior of chloroplasts that are the sites of the light-dependent reactions in photosynthesis.

tissue An organized assemblage of similar cells that serves a common function. Nervous, epithelial, and muscle tissue are some familiar examples.

top predator A species that preys on other species but is not preyed upon by any species.

transcription The process in which DNA's information is copied onto mRNA. This process, the first stage in protein synthesis, occurs in the cell nucleus.

transfer RNA (tRNA) A form of RNA that, in protein synthesis, bonds with amino acids, transfers them to ribosomes, and then bonds with messenger RNA.

transgenic organism An organism whose genome has stably incorporated one or more genes from another species.

translation The process in which a polypeptide chain is produced within a ribosome, based on the information encoded in messenger RNA. This process, the second major stage in protein synthesis (after transcription), occurs in the cytoplasm.

transpiration The process by which plants lose water when water vapor leaves the plant through open stomata. More than 90 percent of the water that enters a plant evaporates into the atmosphere via transpiration.

transport protein Proteins that form hydrophilic channels through the hydrophobic interior of the cell's plasma membrane, allowing hydrophilic materials to pass through the membrane.

transport vesicle A membrane-lined sphere that moves within the cell's endomembrane system, carrying within it proteins or other molecules.

triglyceride A lipid molecule formed from three fatty acids bonded to glycerol.

trimester A period lasting about 3 months during a human pregnancy. There are three trimesters during the 9-month pregnancy.

trophic level A position in an ecosystem's food chain or web, with each level defined by a transfer of energy from one kind of organism to another. Plants and other photosynthesizers are producers of food and thus occupy the first trophic level. Organisms that consume producers are primary consumers, and occupy the second trophic level, and so on.

trophoblast The cells at the periphery of the developing mammalian embryo that establish physical links with the mother's uterine wall and eventually develop into the fetal portion of the placenta.

tropical rain forest A biome characterized by heavy rain and stable, warm temperatures.

troposphere The lowest layer of the Earth's atmosphere, extending from sea level to about 12 kilometers (7.4 miles) above sea level. This layer contains most of the gases in the atmosphere.

true multicellularity A form of life in which individual cells exist in stable groups, with different cells in a group specializing in different functions.

tundra The biome of Earth's far northern latitudes, characterized by little rainfall, a short summer growing season, extremely cold winter temperatures, and a treeless vegetation pattern.

U

umbilical cord The tissue linking the fetus with the placenta in mammals and containing the fetal blood vessels that carry blood between the fetus and the placenta.

ureters Two tubes within which urine is transported from the kidneys to the urinary bladder.

urethra Tube in which urine flows from the bladder to the outside of the body in both males and females. In males, the urethra also transmits semen.

urinary bladder Hollow, muscular organ that serves as a temporary, expandable storage site for the waste product urine.

urinary system The organ system that eliminates waste products from the blood through formation of urine. In mammals, this system consists of the kidneys where the urine is formed; and the ureters, urinary bladder, and urethra, which transport the urine outside the body.

uterine tube (Fallopian tube) The tube that transports an ovulated egg from the ovary to the uterus in mammals. Fertilization occurs inside this tube.

V

variable An element of an experiment that is changed compared to an initial condition.

vas deferens In human males, the tube that carries the sperm from the epididymis on top of the testis to the urethra for ejaculation.

vein A blood vessel that carries blood toward the heart.

ventilation The physical movement of air into and out of the lungs, brought about by contractions and relaxations of muscles in the chest.

vestigial character A structure in an organism whose original function has been lost during the course of evolution.

virus A noncellular replicating entity that must invade a living cell to replicate itself.

viviparous A condition in animals in which fertilized eggs develop inside a mother's body.

W

wetlands Lands that are wet for at least part of the year. Wetlands are sites of great biological productivity, and they provide vital habitat for migrating birds.

white blood cell (WBC) The central cells of the immune system. Types of white blood cells include T cells, B cells, neutrophils, eosinophils, and macrophages, each playing a specific role in immune responses.

X

xylem The tissue through which water and dissolved minerals flow in vascular plants.

Z

zygote A fertilized egg. In humans, the developing organism from the time of fertilization through the time of the first cell division (about 30 hours after fertilization).

Photo Credits

CHAPTER 1

CO1.1: Jeremy Walker/Photo Researchers, Inc.
CO1.2: Getty Images/Digital Vision
1.1: AP Wide World Photos
1.2a: James King-Holmes/Science Photo Library/Photo Researchers, Inc.
1.2b: palmOne
1.6a: Dr. Gopal Murti/Visuals Unlimited
1.6b: Alfred Pasieka/Visuals Unlimited
1.6c: Jeff Hunter/Getty Images Inc.-Image Bank
1.7a: Frans Lanting/Minden Pictures
1.7b: Mark Moffett/Minden Pictures
1.7c: George Holton/Photo Researchers, Inc.
1.8a: © Steve Kaufman/DRK PHOTO
1.8b: © Robert J. Erwin/DRK PHOTO

CHAPTER 2

CO2.1: Getty Images, Inc.
CO2.2: Mike Powell/Getty Images Inc.-Stone Allstock
2.3: Peter M. Fisher/Corbis/Stock Market
2.4: Courtesy of Project Exploration
2.6: National Cancer Institute/Photo Researchers, Inc.
2UN.1a,b: Aleksandra Trifunovic, Karolinska Institutet, Stockholm

CHAPTER 3

CO3.1: Time & Life Pictures/Getty Images
CO3.2: Ruaridh Stewart/ZUMA Press, Inc./Keystone Press Agency
3.2: Flip Nicklin/Minden Pictures
3.3: David Woodfall/Getty Images Inc.-Stone Allstock
3.5: Getty Images, Inc.-Photodisc
3.7: © Don & Pat Valenti/DRK Photo
3.9: Richard Megna/Fundamental Photographs, NYC
3.10a: Dr. Jeremy Burgess/Science Photo Library/Photo Researchers, Inc.
3.10b: Don W. Fawcett/Visuals Unlimited
3.10c: Biophoto Associates/Photo Researchers, Inc.
3.10d: U.S. National Tick Collection/Getty Images, Inc.-Photodisc
3.15a: ROB & SAS/Corbis/Bettmann
3.15b: Getty Images, Inc.
3.18a,b: Courtesy Peter M. Colman, The Walter and Eliza Hall Institute of Medical Research, Melbourne. (Tulip et al (1992) *J Mol Biol* 227,122)
3.19b: Kenneth Eward/BioGrafx/Science Source/Photo Researchers, Inc.
3UN.1b: Ben S. Kwiatkowski/Fundamental Photographs, NYC

CHAPTER 4

CO4.1: Photo Lennart Nilsson/Albert Bonniers Forlag AB
CO4.2: National Cancer Institute/Photo Researchers, Inc.
4.1a: Prof. P. Motta/Dept. of Anatomy/University La Sapienza, Rome/Science Photo Library/Photo Researchers, Inc.
4.1b: Photo Researchers, Inc.
4.1c: Andrew Syred/Science Photo Library/Photo Researchers, Inc.

4.6: Biophoto Associates/Photo Researchers, Inc.
4.7: K.G. Murti/Visuals Unlimited
4.8: P. Motta & T. Naguro/Science Photo Library/Photo Researchers, Inc.
4.10: P. Motta & T. Naguro/Science Photo Library/Photo Researchers, Inc.
4.11: K.G. Murti/Visuals Unlimited
4.12: Photo Lennart Nilsson/Albert Bonniers Forlag
4.13b: Karl Auderheide/Visuals Unlimited
4.13c: Leroy Francis/SPL/Photo Researchers, Inc.
4.17: Eldon H. Newcomb, University of Wisconsin, and William P. Wergin, Agricultural Research Service, U.S.D.A.
4UN.1a-c: Tony Brain & David Parker/Science Photo Library/Photo Researchers, Inc.

CHAPTER 5

CO5.1: Denis Kunkel/Phototake NYC
CO5.2: A. Menashe/humanistic-photography.com
5.4a-c: Mary Teresa Giancoli
5.9b: L.A. Hufnagel, Ultrastructural Aspects of Chemoreception in Ciliated Protists (Ciliophora), *Journal of Electron Microscopy Technique*, 1991. Photomicrograph by Jurgen Bohmer and Linda Hufnagel, University of Rhode Island.
5.10a: Dennis Kunkel/Phototake NYC
5.10ba-d: M.M. Perry
5.10c: Biology Media/Science Source/Photo Researchers, Inc.

CHAPTER 6

CO6.1: Liysa/PacificStock.com
CO6.2: Maria Stenzel/National Geographic Image Collection
6.8: Thomas A. Steitz/Yale University

CHAPTER 7

CO7.1: Scott Fisher/Woodfin Camp & Associates
CO7.2: Froomer, Brett/Getty Images Inc.-Image Bank
7.2: Victor De Schwanberg/Science Photo Library/Photo Researchers, Inc.
7UN.1: Weinberg Clark Photography

CHAPTER 8

CO8.1: Alamy Images
CO8.2: PHILIPPE BAYLE/Peter Arnold, Inc.
8.1a: Tom Mareschal/Creative Eye/MIRA.com
8.1b: Randy Morse/Tom Stack & Associates, Inc.
8.1c: Sherman Thomson/Visuals Unlimited
8.8: Nick Garbutt/Nature Picture Library

CHAPTER 9

CO9.1: Alfred Pasieka/Photo Researchers, Inc.
CO9.2: Biophoto Associates/Photo Researchers, Inc.
9.1: Yann Arthus-Bertrand/CORBIS-NY
9.7: Michael Speicher and David C. Ward
9.8a: Biophoto Associates/Photo Researchers, Inc.
9.8b: Biophoto Associates/Science Source/Photo Researchers, Inc.
9.10a-f: Ed Reschke/Peter Arnold, Inc.
9.11: David M. Phillips/Visuals Unlimited
9UN.1: Photo Researchers, Inc.

CHAPTER 10

CO10.1: Getty Images, Inc.
CO10.2: Photo Lennart Nilsson/Albert Bonniers Forlag AB
10.3: SuperStock, Inc.
10.4: Andrew Syred/Photo Researchers, Inc.

CHAPTER 11

CO11.1: AP Wide World Photos
CO11.2: Getty Images, Inc.
11.1: Getty Images Inc.-Hulton Archive Photos
11.4: Dr. Madan K. Bhattacharyya
11.11a: Peter Morenus, University of Connecticut
11.12: Horticultural Photography

CHAPTER 12

CO12.1: Darrell Walker/Icon Sports Media, Inc.
CO12.2: Library of Congress
12.1: Ishihara, Test for Color Deficiency. Courtesy Kanehara & Co., Ltd. Offered Exclusively in The USA by Graham-Field, Inc., Bay Shore, New York
12.3: Oliver Meckes & Nicole Ottawa/Photo Researchers, Inc.
12.7b: © Dr. Dennis Kunkel/CNRI/Phototake
12UN.1a: Pascal Goetgheluck/Photo Researchers, Inc.
12UN.1b: Dept. of Clinical Cytogenetics, Addenbrookes Hospital/Science Photo Library/Photo Researchers, Inc.

CHAPTER 13

CO13.1: © D. VoTrung/Phototake
CO13.2: A. Barrington Brown/Photo Researchers, Inc.
13.1: Omikron/Photo Researchers, Inc.
13.2a: By courtesy of the National Portrait Gallery, London.
13.2b: Science Source/Photo Researchers, Inc.

CHAPTER 14

CO14.1: Genentech, Inc.
CO14.2: AP Wide World Photos
14.9b: E. Kiselva & Don W. Fawcett/Visuals Unlimited
14.12a: Photo Researchers, Inc.
14.12b: Oliver Meckes/Photo Researchers, Inc.
14.12c: James King-Holmes/Science Photo Library/Photo Researchers, Inc.
14.12d: Fred Habegger/Grant Heilman Photography, Inc.

CHAPTER 15

CO15.1: AP Wide World Photos
CO15.2: Matthias Schrader/dpa/Landov LLC
15.1: AQUA Bounty Farms Inc.
15.3: Dr. Gopal Murti/Science Photo Library/Photo Researchers, Inc.
15.5: Peter Beyer, University of Freiburg, Germany
15.6: Photograph courtesy of The Roslin Institute.
15.8: AP Wide World Photos
15.10: Courtesy of Orchid Cellmark, Inc., Germantown, Maryland.
15.11: Dr. Yorgos Nikas/Phototake NYC
15.12: Monsanto UK Ltd.

CHAPTER 16

CO16.1: Getty Images, Inc.
CO16.2: Corbis Digital Stock
16.1: Painting by John Collier, 1883. London, National Portrait Gallery. © Archiv/Photo Researchers, Inc.
16.2: Christopher Ralling
16.4a: Ken Lucas/Visuals Unlimited
16.4b: Breck P. Kent/Animals Animals/Earth Scenes
16.5: Yoav Levy/Phototake NYC
16.6: Tui De Roy/Bruce Coleman Inc.
16.7a: Tui De Roy/Bruce Coleman Inc.
16.7b: Frans Lanting/Minden Pictures
16.7c: Tui Dee Roy/Minden Pictures
16.8: Mark Kauffman/Getty Images/Time Life Pictures
16.10: Sinclair Stammers/Science Photo Library/Photo Researchers, Inc.

CHAPTER 17

CO17.1: BERNARD CASTELEIN/Nature Picture Library
CO17.2: John W. Warden/The Stock Connection
17.4a: S.J. Krasemann/Peter Arnold, Inc.
17.4b: Bruce G. Baldwin
17.7: Ray Richardson/Animals Animals/Earth Scenes
17.8: Tui De Roy/Bruce Coleman Inc.
17.13: Hans Reinhard/Bruce Coleman Inc.

CHAPTER 18

CO18.1: Bob Abraham/PacificStock.com
CO18.2: Frans Lanting/Minden Pictures
18.1(inset): Dr. Gerald Carr/CBC/University of Hawaii-Manoa
18.1a: Tim Davis/Davis/Lynn Images
18.1b: Jeff Lepore/Photo Researchers, Inc.
18.2: David Welling
18.4a: © Tom & Pat Leeson/DRK PHOTO
18.4b: Pat & Tom Leeson/Photo Researchers, Inc.
18.5: Gerard Lacz/Animals Animals/Earth Scenes
18.6: Guy L. Bush, Michigan State University
18.7a: David Barron/Animals Animals/Earth Scenes
18.7b: John Cancalosi/Peter Arnold, Inc.
18.7c: Tui De Roy/Minden Pictures

CHAPTER 19

CO19.1: Dr. Robert F. Walters
CO19.2: Michael Collier/Stock Boston
19.1: Jack Dermid/Bruce Coleman Inc.
19.3: Frank T. Awbrey/Visuals Unlimited
19.4: B. Murton/Southampton Oceanography Centre/Science Photo Library/Photo Researchers, Inc.
19.5: Jim Brandenburg/Minden Pictures
19.7: James L. Amos/Corbis/Bettmann
19.8c: Phototake/Carolina Biological Supply Company
19.9: The Natural History Museum, London
19.11: Matt Meadows/Peter Arnold, Inc.
19.12: David L. Dilcher, University of Florida
19.15: Gregory S. Paul
19.16: M. Loup/Jacana/Photo Researchers, Inc.
19.20: From: *Nature* 418; pp 133-135. Bernard Wood, July 11, 202.
19.21a-c: Jay H. Matternes
19.23: Kenneth Garrett/Kenneth Garrett Photography
19.24: Courtesy of artist Peter Schouten and the National Geographic Society.

CHAPTER 20

CO20.1: Photo Researchers, Inc.
CO20.2: E. Nagele/Edmund Nagele F. R. P. S.
20.1a: Tim Davis/Davis/Lynn Images
20.1b: Norbert Wu/Peter Arnold, Inc.
20.1c: © Stanley Breeden/DRK PHOTO
20.1d: Roland Birke/Peter Arnold, Inc.
20.4b: Biozentrum, University of Basel/Science Photo Library/Photo Researchers, Inc.
20.5: Sercomi/Photo Researchers, Inc.
20.7a: David Scharf/Peter Arnold, Inc.
20.7b: © Gary Gaugler/Visuals Unlimited
20.7c: CNRI/Science Photo Library/Photo Researchers, Inc.
20.8a: Karl O. Stetter, University of Regensburg, Germany
20.8b: Norman R. Pace, University of Colorado
20.9: Dennis Kunkel Microscopy, Inc.
20.10a: John Walsh/Micrographia
20.10b: Manfred Kage/Peter Arnold, Inc.
20.10c: Dave King © Dorling Kindersley
20.11a: Karl Aufderheide/Visuals Unlimited
20.11b: P.M. Motta and F.M. Magliocca/Science Photo Library/Photo Researchers, Inc.
20.12a: Michael Abbey/Visuals Unlimited
20.12b: Grant Heilman Photography, Inc.
20.13a: Visuals Unlimited
20.13b: Carolina Biological Supply Company/Phototake NYC
20UN.1: Getty Images, Inc.

CHAPTER 21

CO21.1: © Dorling Kindersley
CO21.2: Landov LLC
21.1a: Fred Bruemmer/Peter Arnold, Inc.
21.1b: Science Photo Library/Photo Researchers, Inc.
21.1c: © Michael Fogden/DRK Photo
21.3a,b: Elm Research Institute
21.5: Breck P. Kent
21.6a: VU/S. Fleger/Visuals Unlimited
21.6b: Larry L. Miller/Photo Researchers, Inc.
21.7: John Eastcott & Yva Momatiuk/The Image Works
21.9: Dr. Gerald Van Dyke/Visuals Unlimited
21.12a: Paul Wakefield/Getty Images Inc.-Stone Allstock
21.12b: © John Gerlach/DRK PHOTO
21.12c: Werner H. Muller/Peter Arnold, Inc.
21.12d: H. Richard Johnston/Getty Images Inc.-Stone Allstock
21.13a: Susan G. Drinker/Corbis/Stock Market
21.13b: Stephen Dalton/Photo Researchers, Inc.
21.13c: Robert & Linda Mitchell Photography
21.14a: Carr Clifton/Minden Pictures
21.14b: Michael Gadomski/Animals Animals/Earth Scenes
21.14c: Dwight R. Kuhn Photography
21.15a: Geoff Bryant/Photo Researchers, Inc.
21.15b: Kathy Merrifield/Photo Researchers, Inc.
21.18a,b: Carr Clifton/Minden Pictures
21.18c: Charlie Waite/Getty Images Inc.-Stone Allstock
21.19a: Tim Fitzharris/Minden Pictures
21.19b: Copyright 1990 Robert A. Tyrrell
21.19c: Merlin D. Tuttle/Bat Conservation International/Photo Researchers, Inc.
21.20a: Erwin and Peggy Bauer/Bruce Coleman Inc.
21.20b: Scott Camazine/Photo Researchers, Inc.

CHAPTER 22

CO22.1: Andy Harmer/Photo Researchers, Inc.
CO22.2: © Jett Britnell/DRK Photo
22.1: Kjell B. Sandved/Visuals Unlimited
22.1a: Bob Cranston/Animals Animals/Earth Scenes
22.1b: Fred Bavendam/Minden Pictures
22.6: Dr. Clive Kocher/Photo Researchers, Inc.
22.7: Dr. John D. Cunningham/Visuals Unlimited
22.9: Bruce Watkins/Animals Animals/Earth Scenes
22.11: Ron Borland/Bruce Coleman Inc.
22.14: Michael & Patricia Fogden/Minden Pictures
22.17: Dr. Liu Jinyuan/Dalian Museum of Natural History
22.18a: Gunter Ziesler/Peter Arnold, Inc.
22.18b: John Cancalosi/Peter Arnold, Inc.
22.18c: Kennan Ward Photography

CHAPTER 23

CO23.1: SERGIO HANQUET/Peter Arnold, Inc.
CO23.2: © Paul A. Souders/CORBIS All Rights Reserved
23.1: Mark Moffett/Minden Pictures
23.5a: Frans Lanting/Minden Pictures
23.5b: Kim Taylor/Bruce Coleman Inc.
23.11a: Konrad Wothe/Minden Pictures
23.11b: Grant Heilman Photography, Inc.
23.12: Nancy Sefton/Photo Researchers, Inc.
23.15: Chuck Pratt/Bruce Coleman Inc.
23.17: B. Runk/S. Schoenberger/Grant Heilman Photography, Inc.
23.19: Fred Bavendam/Minden Pictures
23.20a: Milton Tierner, Jr./Visuals Unlimited
23.20b: Art Wolfe/Getty Images Inc.-Stone Allstock
23.21: Dr. James L. Castner
23.22: Rob Plowes/Beth Plowes-Proteapix
23.23a: Stephen J. Krasemann/Photo Researchers, Inc.
23.23b: Fred Bavendam/Minden Pictures
23.24a,b: Thomas Eisner, Cornell University
23.25a,b: John Marshall
23UN.1: Peter J. Bryant/Biological Photo Service

CHAPTER 24

CO24.1: Getty Images, Inc.
CO24.2: Alaska Stock
24.3: Holt Studios International (Inga Spence)/Photo Researchers, Inc.
24.9: Wesley Bocxe/Photo Researchers, Inc.
24.15: Provided by the SeaWIFS Project, NASA/Goddard Space Flight Center and ORBIMAGE.
24.20a,b: Lonnie Thompson/Byrd Polar Research Center
24.25: Ed Reschke/Peter Arnold, Inc.
24.27: Johnny Johnson/DRK Photo
24.28: Daniel J. Cox/Getty Images Inc.-Stone Allstock
24.29: Tom Till/DRK Photo
24.30: Frank Oberle/Getty Images Inc.-Stone Allstock
24.31: Tim Fitzharris/Minden Pictures
24.32: Frans Lanting/Minden Pictures
24.34a: Darryl Torckler/Getty Images Inc.-Stone Allstock
24.34b: Tui De Roy/Minden Pictures
24.34c: Fred Bavendam/Minden Pictures
24.36: Scott Camazine/Photo Researchers, Inc.
24.38a: Jim Brandenburg/Minden Pictures
24.38b: Jim Steinberg/Photo Researchers, Inc.
24UN.1: Masa Ushioda/Image Quest 3-D

CHAPTER 25

CO25.1: AGE Fotostock America, Inc.
CO25.2: AP Wide World Photos
25.5a: John D. Cunningham/Visuals Unlimited
25UN.1: Photo Researchers, Inc.

CHAPTER 26

CO26.1: Biophoto Associates/Photo Researchers, Inc.
CO26.2: Photo Researchers, Inc.
26.3: C. Raines/Visuals Unlimited
26.5: Dennis Kunkel/Phototake NYC
26UN.1: Interfoto USA/SIPA/SIPA Press

CHAPTER 27

CO27.1: AP Wide World Photos
CO27.2: Photo Lennart Nilsson/Albert Bonniers Forlag
27.1: © Manfred Kage/Peter Arnold, Inc.
27.9: Russ Kinne/Comstock Images

CHAPTER 28

CO28.1: Dave King © Dorling Kindersley
CO28.2: Science Photo Library

28.6c: Ed Reschke/Peter Arnold, Inc.
28.8c: Dr. Fred Hoessler/Visuals Unlimited

CHAPTER 29

CO29.1,2: Lennart Nilsson/Albert Bonniers Forlag AB/Photo Lennart Nilsson/Albert Bonniers Forlag AB
29.3: Francis LeRoy/Photo Researchers, Inc.
29.5b: Photo Lennart Nilsson/Albert Bonniers Forlag AB
29.8: Don W. Fawcett/Photo Researchers, Inc.
29.9: Photo Lennart Nilsson/Albert Bonniers Forlag
29.12L: Neil Harding/Getty Images Inc.-Stone Allstock
29.12M: John Watney/Photo Library/Photo Researchers, Inc.
29.12R: James Stevenson/Science Photo Library/Photo Researchers, Inc.

CHAPTER 30

CO30.1: Jerry Alexander/Getty Images Inc.-Stone Allstock

CO30.2: Mark Harvey/D. Donne Bryant Stock Photography
30.1: Photo Researchers, Inc.
30.2a: © David Cavagnaro/DRK PHOTO
30.2b: Al Harvey/Masterfile Corporation
30.2c: Jim Strawser/Grant Heilman Photography, Inc.
30.3a: © Dorling Kindersley
30.3b: Murray, Patti/Animals Animals/Earth Scenes
30.3c: Photo Researchers, Inc.
30.5a,b: Runk/Schoenberger/Grant Heilman Photography, Inc.
30.6: Malcolm B. Wilkins
30.9b: Ed Reschke/Peter Arnold, Inc.
30.19: Ed Ruschke/Peter Arnold, Inc.
30.21a,b: Malcolm B. Wilkins
30.21c: Runk/Schoenberger/Grant Heilman Photography, Inc.
30.23: Runk/Schoenberger/Grant Heilman Photography, Inc.
30.24: Carr Clifton/Minden Pictures
30UN.1: Diane Schiumo/Fundamental Photographs, NYC

Article Credits

Science in the News

CHAPTER 1
Juliet Eilperin, "Alarm Sounded on Global Warming; Researchers Say Dangers Must Be Addressed Immediately," *The Washington Post*, 16 June 2004.

Stephen Dinan, "GOP Disputes Global-Warming Cause," *The Washington Times*, 30 July 2003.

CHAPTER 2
Jere Longman and Joe Drape, "Decoding a Steroid: Hunches, Sweat, Vindication," *The New York Times*, 2 November 2003.

CHAPTER 3
Christy Oglesby, "What's Behind the Curb-Your-Carbs Craze?" CNN, 18 June 2004.

CHAPTER 4
Angela Stewart, "Brain Nerve Cells Trigger Addicts' Relapse: Rutgers Prof Links Environmental Stimuli to Memories That Revive Craving for Drugs," *The Star-Ledger* (Newark, N.J.), 14 August 2003.

Aaron Nathans, "UW Study: Stem Cells Grow Into Heart Cells," *The Capital Times* (Madison, Wisc.), 26 June 2003.

CHAPTER 5
Jewell Cardwell, "Ex-NFL Star on Mark in Cystic Fibrosis Fight: Boomer Esiason Makes Rounds to Champion Cause for Son, Gunnar," *Akron Beacon Journal* (Ohio), 10 December 2003.

CHAPTER 6
Associated Press, "Arctic People Evolved to Produce More Heat," *The Cincinnati Post*, 18 June 2004.

CHAPTER 7
David Fogarty, "2 Americans Among 8 Feared Dead on Mt. Everest; Eight Said to Succumb on Mt. Everest; Copter Saves American and Taiwanese," *The Washington Post*, 14 May 1996.

CHAPTER 8
David A. Fahrenthold *(Washington Post)*, "Chesapeake Bay Grasses Dying Off: Pollution Blotting Out Sunlight for Important Plant Life," *Houston Chronicle*, 30 May 2004.

CHAPTER 9
Associated Press, "Genetic Mutation Produces Toddler with Twice the Muscle," *The Maryland Gazette* (Glen Burnie, MD), 30 June 2004.

Keay Davidson, "Researchers Discover 'Jekyll and Hyde' Cancer Gene: Amount of a Specific Protein Determines Whether a Tumor Is Created or Suppressed," *San Francisco Chronicle*, 13 September 2004.

CHAPTER 10
Rosie Mestel (*Los Angeles Times* Service), "Women's Supplies of Eggs May Not Be Fixed Number," *The Miami Herald*, 11 March 2004.

CHAPTER 11
Susan Fitzgerald, "Babies' Deaths in Amish Area Tied to SIDS Gene," *The Advocate* (Baton Rouge, La.), 20 July 2004.

Laura Billings, "From Bottle Blonde to Honest Brunette," *St. Paul Pioneer Press* (Minn.), 6 October 2002.

CHAPTER 12
Tribune News Services, "FSU Death Linked to Genetic Disease," *Chicago Tribune*, 9 June 2004.

CHAPTER 13
Michelle Morgante (Associated Press) "Crick Model Cracked Code of Life; Obituary: 1950s Work with James Watson Led to DNA 'Double Helix'," *Long Beach Press-Telegram* (Calif.), 30 July 2004.

CHAPTER 14
Judy Silber, "Genentech Wins Cancer Drug OK: Avastin Approval by FDA Drives Up Stock, Is Seen as Big Profit-Driver," *Contra Costa Times* (Walnut Creek, Calif.), 27 February 2004.

CHAPTER 15
Michael B. Farrell, "Cloning Kitty: A California Company Is Selling Cloning Technology to Pet Owners. But Opponents Say This Technology Holds False Promises," *The Christian Science Monitor*, 24 November 2004.

Associated Press, "Dandruff, DNA Solve 1993 Robbery," *Orlando Sentinel*, 23 November 2004.

CHAPTER 16
Ellen Barry *(Los Angeles Times)*, "'Evolution' to Reappear in Georgia Curriculum," *The Seattle Times*, 6 February 2004.

Guy Gugliotta, "Fossil May Show Ape-Man Ancestor; Bones Found in Spain Are Called Landmark in Evolution," *The Washington Post*, 19 November 2004.

CHAPTER 17
Associated Press, "It's Survival of Fittest in Broiling Nasdaq-100 Open," *Houston Chronicle*, 21 March 2003.

Penni Crabtree, "Biotech Meets Darwin: It's Survival of the Fittest for Many Private San Diego Biotechnology Companies as They Wait—And Hope—For More Financing," *The San Diego Union-Tribune*, 25 February 2003.

CHAPTER 18
John Heilprin (Associated Press), "Jumping Mouse Loses Federal Protection," Yahoo! News, 29 January 2005.

CHAPTER 19
David Perlman, "Study Shows Dinosaur Could Fly: Winged Creature Had Birdlike Senses, Fossil X-Rays Reveal," *San Francisco Chronicle*, 5 August 2004.

CHAPTER 20
Linda A. Johnson (Associated Press), "Staph Getting Resistant: As Bacteria Defy Antibiotics, More Patients Are Hospitalized," *Long Beach Press-Telegram* (Calif.), 30 September 2004.

William Hathaway, "Bacteria Play Vital Role in Healthy Intestines, Yale Study Shows," *The Hartford Courant* (Conn.), 27 July 2004.

CHAPTER 21
Bill Hendrick (Cox News Service), "It's Hard to Beat Mold—But You Can Try to Contain It," *The Chicago Tribune*, 19 November 2004.

Associated Press, "London Restaurant Forks Over Record $52,000 for White Truffle," *The Record*, 26 November 2004.

CHAPTER 22
Associated Press, "A Whale of a Find: DNA Tests Show the Discovery of a New Species in the Indian Ocean and Near Japan," *The Grand Rapids Press* (Mich.), 19 November 2003.

Henry Fountain *(The New York Times)*, "Lurking Below: Myth, Mystery of the Giant Squid," *Milwaukee Journal Sentinel*, 17 May 2004.

CHAPTER 23
Associated Press, "Wildlife Group Backs Deer Kills—The N.J. Audubon Society Said Lethal Means Could Help Preserve Forests from the Growing Herds," *The Philadelphia Inquirer*, 15 March 2005.

CHAPTER 24
Alexandra Witze, "Studies: Global Warming to Be Felt for Centuries—Temperatures, Oceans to Rise Even If Gas Levels Steady, Research Shows," *The Dallas Morning News*, 18 March 2005.

Katy Human and Kim McGuire, "Ozone Decline Stuns Scientists—Thinning in Arctic—Solar Storms, Bitter Cold, and Manmade Chemicals Are Believed to Be Behind the Phenomenon," *The Denver Post*, 2 March 2005.

CHAPTER 25
Pamela LeBlanc (Cox News Service), "With ACL Tears, It's a Long Road to Recovery," *Austin American-Statesman* (Tex.), 21 February 2005.

Tim Vacek, "Torn ACL Ends Year for Spencer: Fourth Lady Vols Player to Hurt Knee This Season," *The Tennessean* (Nashville, Tenn.), 25 February 2005.

CHAPTER 26
Duff Wilson, "McGwire Offers No Denials at Steroid Hearings," *The New York Times*, 18 March 2005.

Jane Norman, "Young Athletes' Steroid Abuse Called Perilous," *The Des Moines Register* (Iowa), 14 July 2004.

CHAPTER 27
Lois M. Collins, "Utah 'Bubble Boy' a Girl," *The Deseret News* (Utah), 5 June 2004.

CHAPTER 28
Eagle News Services, "Beatle George Harrison Dies," *The Wichita Eagle* (Kans.), 1 December 2001.

Anthony Breznican (Associated Press), "Performer Warren Zevon Dies After Battling Cancer," *The Deseret News* (Utah), 9 September 2003.

CHAPTER 29
Alison Mutler (Associated Press), "Woman, 66, Gives Birth to Daughter: Both Reported in Good Condition; Romanian Professor Had 9 Years of Fertility Treatments," *St. Paul Pioneer Press* (Minn.), 18 January 2005.

CHAPTER 30
Times Wire Reports, "States Getting Antidotes to Chemical Weapons," *Los Angeles Times*, 14 July 2004.

Associated Press, "Drops Help Kids with 'Lazy Eye'," *The Columbus Ledger-Enquirer* (Ga.), 13 March 2002.

Index